Regulatory RNAs

Bibekanand Mallick • Zhumur Ghosh

Editors

Regulatory RNAs

Basics, Methods and Applications

Springer

Editors
Bibekanand Mallick
Wadsworth Center
New York State Department of Health
New Scotland Avenue
Albany, New York - 12208
USA
rnabiz@iscb.org
vivek.iitian@gmail.com

Zhumur Ghosh
Bose Institute
A.J.C. Bose Centenary Building
P-1/12, CIT Scheme - VII M
Kolkata - 700 054
India
zhumur@boseinst.ernet.in
ghosh.jhumur@gmail.com

ISBN 978-3-642-22516-1   e-ISBN 978-3-642-22517-8
DOI 10.1007/978-3-642-22517-8
Springer Heidelberg Dordrecht London New York

Library of Congress Control Number: 2012931506

Printed on acid-free paper

Springer is part of Springer Science+Business Media (www.springer.com)

Preface

RNA molecules participate in and regulate a vast array of cellular processes besides being the physical link between DNA and proteins. They play several other key roles, which include RNA catalysis and gene regulation mediated mainly by noncoding RNAs. This regulation occurs at some of the most important levels of genome function, such as chromatin structure, chromosome segregation, transcription, RNA processing, RNA stability, and translation. Further, harnessing the potential of RNA as a therapeutic or diagnostic tool, or as a central player in a fundamental biological process is becoming increasingly important to the modern day scientific community. Previously scientists imagined that there was an "RNA World," in which primitive RNA molecules assembled themselves randomly from building blocks in the primordial ooze and accomplished some very simple chemical chores. But these molecules were thought only to be carrying information from DNA to ribosomes. Discovery of catalytic RNAs changed this idea and opened up a wealth of opportunities to allow investigators to modulate gene expression post-transcriptionally using ribozymes and derivatives. In addition to ribozymes, a new RNA-based strategy for regulating gene expression in mammalian cells has recently been described. This strategy is known as RNA interference (RNAi). Although much is known about the mechanisms of RNAi, there lie a number of hurdles that need to be overcome along the applicative path of gene-silencing technology which includes the activation of innate immunity, off-target effects, and in vivo delivery.

Currently, high-throughput sequencing, bioinformatic and biochemical approaches are identifying an increasing number of regulatory RNAs. Unfortunately, our ability to characterize the detailed story of regulatory RNAs is significantly lacking. Extensive research of these RNAs is an emergent field that is unraveling the molecular underpinnings of how RNA fulfills its multitude of roles in sustaining cellular life. The resulting understanding of the physical and chemical processes at the molecular level is critical to our ability to harness RNA for use in biotechnology and human therapy, a prospect that has recently spawned a multibillion-dollar industry.

Nevertheless, RNA research can be daunting, and without a thorough understanding of the challenges and complexities inherent in handling this fragile nucleic acid, forays into the RNA world can be quite frustrating.

In this book, we have made an attempt to bring together the contributions of the leading noncoding RNA researchers to embellish the story of regulatory RNAs and provide a snapshot of the current status of this dynamic field.

The book consisting of 21 chapters offers a comprehensive overview of our current understanding of the regulatory noncoding RNAs, namely, small interfering RNAs (siRNAs), microRNAs (miRNAs), Piwi-interacting RNAs (piRNAs), small nucleolar RNAs (snoRNAs), long noncoding RNAs (lncRNAs), small RNAs (sRNAs), etc., and their applications in understanding biological systems and diseases, including therapeutics. This book is divided into three major sections as per its title. The first section "Basics" consists of eight chapters (Chaps. 1–8). The first chapter gives an overview of the entire landscape of noncoding RNAs, mainly highlighting their history and functions with a focus on the current status of research and future perspectives. This is followed by chapters on discovery, biogenesis, evolution, regulatory functions, and molecular mechanisms of different category of noncoding RNAs.

The "Methods" section provides state-of-the-art experimental and computational methodologies for noncoding RNA detection using different techniques and experimental analysis of noncoding RNA regulatory networks in different systems. This part includes Chaps. 9–15 and provides different bioinformatic, high-throughput RNA sequencing, ncRNA-specific microarrays, and biochemical approaches to identify these RNAs as well as protocols for transfection, gene knockout experiments, and regulatory RNA–based cellular reprogramming and pathways in different species. Further, some chapters are devoted to methods and protocols that have been developed by the authors themselves.

The "Applications" section includes Chaps. 16–21, which cover applicative areas of various noncoding RNAs within a biological system. These serve as biomarkers for different diseases like cancer, target cancer stem cells, act as regulators in cell lineage determination, etc. Further, RNAi therapeutics is applied against solid organ malignancies, cellular reprogramming, and stem cell–based regenerative therapy.

We are grateful to our friends and colleagues who have encouraged and supported us in many ways towards preparation of this book. We acknowledge them, with sincere thanks and appreciation. We take this opportunity to thank all the authors who have contributed excellent chapters to this book and the reviewers for their critical comments to improve the quality and integrity of the chapters. Their special effort has made this book a valuable resource for scientists and aspiring research students interested in the intersection of RNA biology and clinical research. We would like to express our sincere appreciation to Sabine Schwarz and Ursula Gramm of Springer Heidelberg for their invitation to initiate this book and their continuing support and commitments in making this book a reality and to other staff members involved in the production of the book.

Bibekanand Mallick and Zhumur Ghosh

Contents

Part II   Methods

Part III   Applications

Part I
Basics

Chapter 1
Renaissance of the Regulatory RNAs

Zhumur Ghosh and Bibekanand Mallick

Abstract "Non-coding RNAs (ncRNAs)" originate from various types of regulatory DNA, which lie deep in the wilderness of so-called junk DNA present within the genomes. Far from being humble messengers, a group of ncRNAs are powerful players in how genomes operate and are better termed as "regulatory RNAs". The new regulatory role of RNA began to emerge recently as researchers discovered different classes of regulatory RNA molecules, namely, small interfering RNAs (siRNAs), microRNAs (miRNAs), PIWI-interacting RNAs (piRNAs), small nucleolar RNAs (snoRNAs), long noncoding RNAs (lncRNAs), etc. These versatile RNA molecules appear to comprise a hidden layer of internal signals that control various levels of gene expression in physiology and development, including chromatin architecture/epigenetic memory, transcription, RNA splicing, editing, translation, and turnover. RNA regulatory networks may determine most of our complex characteristics, play a significant role in diseases, and constitute an unexplored world of genetic variation both within and between species. In this chapter, we have attempted to provide a snapshot of the entire landscape of these versatile molecules.

Keywords Gene expression • long noncoding RNA • microRNA • noncoding RNA • regenerative therapy • regulatory RNA • ribozymes • RNA world • siRNA

Z. Ghosh (✉)
Bose Institute, Kolkata, India
e-mail: zhumur@boseinst.ernet.in; ghosh.jhumur@gmail.com

B. Mallick (eds.), *Regulatory RNAs*, DOI 10.1007/978-3-642-22517-8_1,
© Springer-Verlag Berlin Heidelberg (outside the USA) 2012

1.1   Introduction

Beginning of life on earth is one of those *big events* where the prime role is played by the *tiniest ones*. To depict the exact scenario regarding what happened millions of years ago, scientists study how life works now and then trace back. Today's cells keep all-inclusive instruction manual – the DNA under tight wraps in the nucleus. Tiny pores in the wrapping are the only way in or out. When the cell needs directions, the DNA makes a copy of the particular pages required in the form of a short, single strand of ribonucleic acid – the messenger RNA that can leave the nucleus. Outside the nucleus lies the cell's framework – the cytoplasm. Messenger RNA (mRNA) wends its way through the maze looking for the nearest relay station: a ribosome. Ribosomes call in their interpreters: transfer RNA (tRNA). These recognize parts of the mRNA message and give it to the ribosome. Ribosomes get the instructions from DNA to make proteins, which carry out functions in the cell and in the body ranging from digesting the burger you had for lunch to determining your skin color. Ribosomes assemble proteins from building blocks, called amino acids that tRNAs line up in the correct order. Yet another kind of RNA in the ribosome (rRNA) helps move the assembly line along.

Researchers wondered regarding which came first, DNA, RNA, or protein? This classic "chicken-and-egg" problem made it immensely difficult to conceive of any plausible prebiotic chemical pathway to the molecular biological system. It is obvious that the first information molecule must have been able to reproduce itself and carry out tasks similar to those done by proteins today, and this limited the choice. Among the options, RNAs were found to perform numerous functions, which were once thought to be domains of proteins. Their unique properties bagged appreciation of the scientific community and obligated them to revise the tenets of "central dogma". Hence, they imagined an "RNA World," in which primitive RNA molecules assembled themselves randomly from building blocks in the primordial ooze and accomplished some very simple chemical chores. This concept originated in late 1960s and was supported by different groups (Woese 1967; Crick 1968). RNA molecules mainly garnered attention with the discovery of ribozymes – the catalytic RNAs in 1980s (Guerrier-Takada et al. 1983; Kruger et al. 1982). Tom Cech and his group discovered that an intron within a pre-rRNA from *Tetrahymena thermophila* catalyzes its own cleavage (called self-splicing) to form the mature rRNA product. This explained why some RNAs act as natural RNA enzymes with self-splicing activity, which is a favorable prerequisite factor for origin of life on earth (Kruger et al. 1982).

The breakthrough discovery of catalytic RNAs entailed a remarkable increase in knowledge about the folding of RNA molecules and their functional activities. Moving a bit further along the landscape of present day research, the explosion of high-throughput next generation sequencing methods (Mortazavi et al. 2008), large-scale genome sequencing, and genome-wide transcriptome studies (Lao et al. 2009) in various organisms has led to the discovery of the RNAi (RNA interference) phenomenon (A. Fire and G. Mello, Nobel Prize in Medicine or

Physiology, 2006) and the role of noncoding RNAs (ncRNAs) in it, that act as transcriptional and posttranscriptional regulators. Apart from regulating gene expression, these ncRNAs also play a dominating role in maintaining genome stability (Moazed 2009) and have led to novel insights into the biological systems. This "regulatory RNA" field is presently expanding at an unprecedented rate, and exciting new developments will undoubtedly emerge over the next years.

Recently, it has been revealed from deep sequencing data of Encyclopedia of DNA Elements Consortium (ENCODE) transcriptome projects that eukaryotes transcribe up to 90% of their genomes, whose large fraction includes large and short RNAs with no coding ability (Birney et al. 2007). Earlier, there was a belief that more complex organisms would have a greater number of protein-coding genes; however, it is now well established that human and mouse have approximately the same number of genes as that of the microscopic organism, *Caenorhabditis elegans* (Taft et al. 2007). The complexity of cellular functions in advanced organisms and their small percentage of coding genome (~2–3%) was always a tough question to explain their correlation, but not anymore because of the discovery of thousands of different types of ncRNAs in recent years. It is now clear that biological complexity probably correlates to non-protein coding genes, not protein coding genes as thought earlier (Taft et al. 2007).

1.2 The Serendipity

With the discovery of RNAi in 1998 by Andrew Fire, Craig Mello, and colleagues (Fire et al. 1998), the long-believed concept about RNA became complicated. They observed silencing of gene expression by double-stranded RNAs (dsRNAs) in nematodes. This serendipitous phenomenon, termed as RNAi, was discerned when they injected dsRNAs into the *Caenorhabditis elegans* and observed silencing of a gene whose sequence was complementary to that of the dsRNAs. Since then, RNA has become the heart and soul of a scientific study and created a new revolution in the field of biological sciences. This revolution started unnoticed in the late 1980s and early 1990s when plant biologists working with purple petunia were surprised to find that introducing numerous copies of a gene that codes for deep purple flowers led to plants with white or patchy flowers, which was not expected (Napoli et al. 1990; van der Krol et al. 1990). Somehow, the inserted transgenes silenced both themselves and the plants' own "purple-flower" genes. These observations mystified the biologists for a few years but were readily deciphered after the findings by Fire and Mello in 1998. This RNAi phenomenon was originally thought to be confined to exogenous dsRNAs; however, it gradually became clear that genomes of plants and animals encode various types of endogenous dsRNAs, namely, small interfering RNAs (siRNAs), microRNAs (miRNAs), etc. The canonical RNAi pathway in animals has been described in details in Chapter 5. New classes of ncRNAs and more members of existing classes continue to be elucidated in past years and are yet to be discovered in future.

1.3   Regulatory RNAs to Date

The world of ncRNAs keeps expanding with the advent of new molecular and genomic technologies in recent years. Figure 1.1 depicts the different types of regulatory ncRNAs identified till date.

There have been recent discovery of new ncRNAs sitting adjacent to transcription start sites, e.g., promoter-associated small RNAs (PASRs) (Kapranov et al. 2007), transcription initiation RNAs (tiRNAs) (Taft et al. 2009a), and termini-associated small RNAs (TASRs) located near 3′ end of the genes (Kapranov et al. 2007; Kapranov et al. 2010), aside from identification of other regulatory ncRNAs such as small nucleolar RNAs (snoRNAs) and processed snoRNAs (psnoRNAs) (see Chap. 3), small RNAs (sRNAs) in bacteria (see Chap. 4), long noncoding RNAs (lncRNAs) (see Chap. 6), siRNAs, miRNAs, piRNAs, small modulatory RNAs (smRNAs), tiny noncoding RNAs (tncRNAs), etc. While many of these ncRNAs remain undeciphered at an appreciable level, miRNAs, siRNAs, and

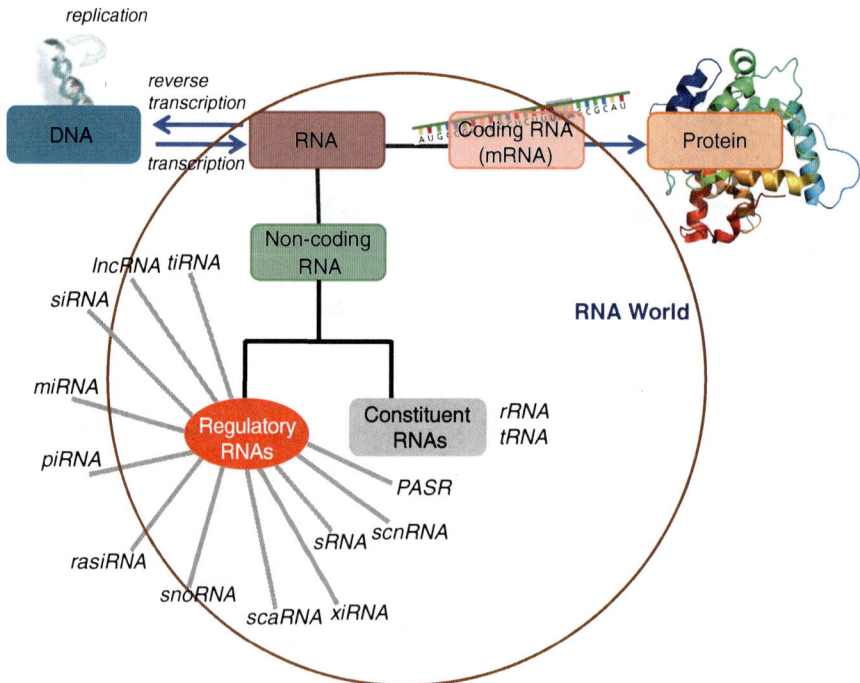

Fig. 1.1   The expanding noncoding RNA landscape

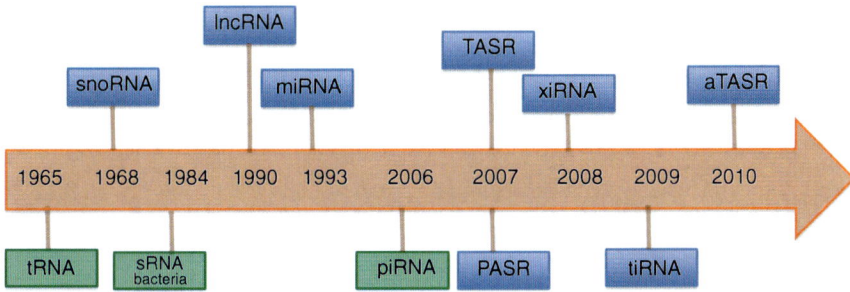

Fig. 1.2 Chronological trajectory of the major discoveries related to RNAs

piRNAs have been most thoroughly investigated to infer their evolution, function, and applications in myriad areas of biological systems. In humans, there are over 1,000 miRNAs, hundreds of siRNAs, and millions of piRNAs, complying with the observation of ncRNAs occupying a substantial portion of the genome. They contribute significantly to complex regulatory systems of a higher organism by coordinating important cellular functions at the transcriptional and/or posttranscriptional level. These regulatory RNAs have expanded the RNA world in due course of time (see Fig. 1.2 for chronological trajectories for major RNA discoveries). And, Table 1.1 provides an overview of various types of regulatory RNAs discovered till date.

1.4   Mini Silencers

Extensive research in the past few years on gene silencing has revealed that the argonaute protein family members are key players in these pathways (Hutvagner and Simard 2008; Peters and Meister 2007) guided by different types of small RNAs. Argonaute proteins are evolutionarily conserved and phylogenetically classified into the AGO subfamily and the PIWI subfamily. AGO proteins, ubiquitously expressed in cells bind to siRNAs and miRNAs and regulate posttranscriptional gene silencing either by destabilization of the mRNA or by translational repression. Although miRNAs and siRNAs were independently discovered, they share a common chemical composition, biogenesis-related events, RISC complex assembly Table 1.2, and mechanism of action (see details in Chaps. 5 and 2). They also differ in their evolutionary conservation process, and possibly they target different genes (Bartel and Bartel 2003). piRNAs are the most recent development in the RNAi field that were first reported in 2006 by four independent studies (Aravin et al. 2006; Girard et al. 2006; Grivna et al. 2006; Watanabe et al. 2006) through cloning of small RNAs associated with PIWI proteins. PIWI proteins are mostly expressed in the germlines (Seto et al. 2007) and bind to these novel class of ncRNAs and facilitate silencing of transposons, the mobile genetic elements, and serve diverse functions in germline development and gametogenesis (refer to Chap. 5).

Table 1.1 Overview of different types of regulatory RNAs

ncRNAs	Organisms	Length (nt)	Characteristics/functions	References
siRNAs	Plants, animals, protists, fungi	21–22	Dicer-dependent cleavages of complementary dsRNA duplexes produce siRNAs. siRNAs associate with argonaute proteins and are involved in posttranscriptional gene regulation, transposon control, and viral defense	Hamilton and Baulcombe (1999), Malone and Hannon (2009)
miRNAs	Plants, animals, algae, protists, viruses	18–25	Associate with argonaute proteins and are involved in posttranscriptional gene regulation	Carthew and Sontheimer (2009), Huntzinger and Izaurralde (2011), Czech and Hannon (2011)
tiRNAs (transcription initiation RNAs)	Human, chicken, *Drosophila*	~18–22	Regulation of chromatin modifications and protein recruitment involved in transcription initiation	Taft et al. (2009a)
piRNAs	Metazoans excluding *T. adhaerens*	24–30	Associate with PIWI-clade argonaute proteins and regulate transposon activity and chromatin state and regulate DNA methylation affecting gene expression	Seto et al. (2007), Lin (2007), Thomson and Lin (2009)
snoRNAs	Eukaryotes, archaea	80–200	Guides chemical modifications (methylation, pseudouridylation) of RNAs (such as rRNAs and tRNAs)	Bachellerie et al. (2002), Dieci et al. (2009)
psnoRNAs (processed snoRNAs)	Eukaryotes	20–100	Regulate alternative pre-mRNA splicing, function as miRNA-like regulators of translation	Khanna and Stamm (2010), Taft et al. (2009b)
sRNAs	Bacteria	50–300	Posttranscriptional gene regulation	Repoila and Darfeuille (2009), Waters and Storz (2009)

Name	Organism	Size	Function	References
moRNAs (miRNA-offset RNAs)	Chordate, human	~20	Derived from regions adjacent to pre-miRNAs; function is unknown	Langenberger et al. (2009), Shi et al. (2009)
tel-sRNAs (telomere small RNAs)	Mammals	~24	Derived from G-rich strand of telomeric repeats; probably involved in maintenance of telomeres	Cao et al. (2009)
natsiRNA (natural antisense transcript-derived siRNA)	Plants	21–24	Regulate stress-response genes	Vaucheret (2006); Yoshikawa et al. (2005)
crasiRNAs (centromere repeat-associated small interacting RNAs)	Vertebrates	34–42	Derived from centrosomes; involved in chromatin modifications, and formation and maintenance of centromeric heterochromatin	Carone et al. (2009)
rasiRNA (repeat-associated small interfering RNA)	*Drosophila*, unicellular eukaryote, plants	24–29	Associate with argonaute and PIWI argonaute protein subfamily; maintain heterochromatin structure, silencing of transposons, and retrotransposons in the germline	Gunawardane et al. (2007)
hc-siRNAs (heterochromatin siRNAs)	Plants	~24	Derived from repeat sequences of genomes (transposons, retroelements, centromeric repeats) and involved in DNA and histone methylation	Daxinger et al. (2009)
scnRNAs (small scanRNAs)	Protozoa	~28	Expressed during conjugation and participate in chromatin modifications	Kurth and Mochizuki (2009)
qiRNAs (quelling deficient, QDE2-interacting small RNAs)	Plants	20–21	Mostly originate from ribosomal RNA locus, contribute to DNA damage checkpoints by inhibiting protein synthesis	Lee et al. (2009)
casiRNAs (chromatin-associated siRNAs)	Plants	24	Guide de novo cytosine methylation at the homologous genomic DNA region and transcriptional events	Kasschau et al. (2007)

(continued)

Table 1.1 (continued)

ncRNAs	Organisms	Length (nt)	Characteristics/functions	References
tasiRNAs (trans-acting siRNAs)	Plants	21	Posttranscriptional gene regulation	Vazquez et al. (2004), Allen et al. (2005)
lncRNAs (long (regulatory) noncoding RNAs)	Animals	$> \sim 200$	Epigenetic regulation	Ponting et al. (2009), Mercer et al. (2009), De Lucia and Dean (2011)
xiRNAs (X-inactivation RNAs)	Mammals	25–42	Processed from duplexes of two lncRNAs, Xist and Tsix that are responsible for X chromosome inactivation in placental mammals; probably involved in translational silencing	Kanduri et al. (2009)
PASRs (promoter-associated small RNAs)	Animals	22–200	Located at 5′ end of the genes; may be involved in regulation of gene expression	Kapranov et al. (2007)
TASRs (termini-associated small sRNAs)	Animals	22–200	Located at 3′ boundaries of genes	Kapranov et al. (2007)
non-PASRs	Animals		Located at a site distant from PASRs	Fejes-Toth et al. (2009)
aTASRs (antisense TASRs)	Human	< 200	Found within 50 bp and antisense to 3′ UTR of the transcripts; are closely associated with 3′ end of known RNAs, pointing to the existence of RNA-copying mechanism	Kapranov et al. (2010)

The information provided in this table are up to date as per our knowledge (till the time of drafting of the book). These information are likely to change in due of time because of new research discoveries

Table 1.2 Comparison between three most popular small regulatory RNAs

Features	siRNAs	miRNAs	piRNAs
Organisms	Eukaryotes	Eukaryotes, viruses	Worm, zebra fish, mammals
Origin	Endogenous and Exogenous: transposons, viruses, DNA heterochromatin	Endogenous	Endogenous
Length	21–22	18–25	24–30
Nature of precursors	dsRNA	dsRNA	ssRNA
Genomic location	Dispersed throughout	Dispersed throughout	Clustered
Site of biogenesis	Cytoplasm/nucleus	Nucleus/cytoplasm	Not clearly known
Argonaute	AGO1–AGO4	AGO2	PIWI/aubergine, AGO3
Site of expression	All tissues	All tissues	Germlines
Type of transcripts	Polycistronic	Polycistronic/ monocistronic	Polycistronic
Phylogenetic conservation	Rarely conserved	Highly conserved	Not conserved

The information provided in this table are up to date as per our knowledge (till the time of drafting of the book). These information are likely to change in due of time because of new research discoveries

piRNAs are found in the testes and ovaries of zebra fish and *Drosophila* as well as in the testes of mammals. In the germline, these small RNAs ensure genomic stability by silencing endogenous selfish genetic elements (retrotransposons and repetitive sequences). piRNA biogenesis is driven by two distinct processes that have been revealed by deep sequencing and genetic studies (Siomi et al. 2010). Majority of unique piRNAs are derived from transposon-rich heterochromatic clusters (Brennecke et al. 2007; Yin and Lin 2007). There is a "ping-pong" amplification cycle which is needed to amplify siRNA triggers in plants, nematodes, and yeast (Verdel et al. 2009) (see Chap. 2). The ping-pong model was developed from observations in *Drosophila*, but a similar mechanism appears to function in other animal groups (Aravin et al. 2007; Houwing et al. 2007; Grimson et al. 2008; Palakodeti et al. 2008; Lau et al. 2009).

A considerable fraction of the piRNAs isolated to date map to transposon-encoding regions (although this is highly variable from species to species) (Girard and Hannon 2008), and piRNA mutations lead to massive transposon over-expression. piRNA–PIWI complexes are therefore assumed to directly control transposon activity. piRNAs bound to PIWI proteins direct homology-dependent target cleavage *in vitro*, suggesting that transposons are silenced through posttranscriptional transcript destruction (Gunawardane et al. 2007; Saito et al. 2006; Nishida et al. 2007). piRNAs bound to Aub and AGO3 direct homology dependent cleavage of mature transposon transcripts after export from the nucleus. Mutations in piRNA pathway genes disrupt germline development, often producing complex and poorly understood phenotypes that are difficult to directly associate with transposon targets of the pathway (see Chap. 5). There are also evidences, which show that piRNAs might also have a role in regulating translation (Grivna et al. 2006; Unhavaithaya et al. 2009). These ncRNAs are also assumed to play a significant role in regulating gene expression, which might be restricted to specific tissues or developmental stages. Majority of the piRNAs map to the unannotated regions of the genome in poriferans, cnidarians, worm, and mouse and only a limited set match transposons and other repeats (Aravin et al. 2006; Girard et al. 2006; Grimson et al. 2008; Batista et al. 2008; Ruby et al. 2006) which supports this hypothesis. All these hints toward the biological function for this novel class of small RNAs well beyond transposons and germline development.

1.5   Exploring the Genomic Dark Matter

In recent years, novel strategies – both computational and experimental – have been undertaken to identify a great number of novel ncRNA candidates in various model organisms from *Escherichia coli* to *Homo sapiens* (Storz 2002; Washietl et al. 2005; Huttenhofer et al. 2001; Wassarman et al. 2001). These findings demonstrated that the number of ncRNAs in genomes of model organisms is much higher than it had been anticipated.

Among the different experimental strategies for identifying novel ncRNAs, RNA sequencing is one of the most powerful and widely adopted approaches and relies on the generation of specialized cDNA libraries, e.g., RNP libraries (see details in Chap. 9). Other methods include microarrays for identifying ncRNAs expressed under a given experimental condition (see Chap. 12) and/or ncRNAs of various sizes in a single experiment employing hybrid LNA/DNA microarrays (see details in Chap. 9), "genomic SELEX" to select ncRNA candidates from the sequence space represented by the genome of an organism of interest, or targeted deep sequencing of classes of RNA with distinct $5'$ and $3'$ ends or affinity for specific proteins after extraction with immunoprecipitation (see Chap. 10). Apart from such biochemical methods, bioinformatics tools are also employed to identify various types of ncRNAs from different species and model organisms (Washietl et al. 2005; Vogel and Sharma 2005; Eddy 2002). These bioinformatic tools are often based on sequence, secondary structure, and thermodynamic identities, and/or conservation features of ncRNAs revealed through comparative genomics approaches. For comprehensive understanding of the principles and methods for prediction of small RNAs among bacteria and their targets refer to chapter 11 for biocomputational approaches, and chapter 14 for experimental approaches. The long ncRNAs represent another major unexplored component of genomes of great potential biological importance (see Chap. 8), but they are not properly acknowledged and explored unlike other small RNAs (Carninci and Hayashizaki 2007). Moreover, lncRNAs surprisingly have no significant homology identified across each lncRNA in their sequences and mechanisms of function, unlike other ncRNAs such as miRNAs. These raise questions regarding diversity in their functions and origins. Therefore, many methods, both computational and experimental have come up in these years to identify and characterize lncRNAs (refer to Chap. 13) and make comprehensive catalogs of these ncRNAs for better understanding of their functions in gene regulation and human diseases.

However, without a clue to their biological functions, the newly identified ncRNA molecules raise the burning questions: what are the functions of all of these RNA transcripts? Or, if they are not functional, why does the cell devote its resources to producing them? Hence, next to "novel approaches" for identifying ncRNAs in different organisms comes the novel methods preferably high-throughput methods (Willingham et al. 2005; Krutzfeldt et al. 2005) to understand their biological roles in those organisms.

1.6 ncRNAs: A Password to Future Personalized Therapy

Continual discoveries of ncRNAs have changed the landscape of human genetics and molecular biology. Over the past 10 years, it has become clear that ncRNAs are involved in all developmental processes (see Chap. 7), including stem cell and germline maintenance, development and differentiation, and when dysfunctional, underpin disease (Lee and Calin 2011; Qureshi and Mehler 2011). Several classes of ncRNAs, such as siRNAs, miRNAs, piRNAs, snoRNAs, etc., are implicated in different diseases, namely, cancer, heart diseases, immune disorders, and neurodegenerative disorders (see Chap. 18) and metabolic diseases, etc. (Galasso et al. 2010; Taft et al. 2010).

ncRNAs also play a dominant role towards shaping the epigenetic program in human embryonic stem cells and adult cells (Lunyak and Rosenfeld 2008). This has opened up the avenue to understand how cells remember their own fates and hence can improve regenerative medicine in several ways. Specific ncRNAs can be used as markers to track and predict when cells are acquiring or forgetting specific cell fates (see Chap. 17). For instance, it may be possible to learn from the pattern of ncRNAs that an embryonic stem cell is ready to become cardiac cells, which can be used to treat a patient with cardiac hypertrophy. Further, beyond tracking cell fate, ncRNAs may be used for direct manipulation of stem or adult cell fates. They can be used for reprogramming pluripotent stem cells into desired cell types (see Chap. 15). While these potential applications are far in the future, we believe that better knowledge of this new level of gene regulation will lead to more facile and efficient manipulation of cell fates for regenerative medicine in future.

Moreover, siRNAs have become not only an exciting new tool in molecular biology but also the next frontier in molecular therapeutic applications. In this volume, we have described the types of choices that must be made in the development of siRNA therapeutics, the features of the siRNA molecule that are important for maximizing silencing activity, how to design delivery vehicles to transport siRNAs to their intended location, and examples of ongoing clinical trials utilizing siRNA therapeutics to treat solid tumors, acute kidney failures, and some of the acute and dreadful viral infections (see Chap. 19). Furthermore, it has been observed that some cellular pathways are altered in cancer stem cells (CSC), and these preferentially offer targets for RNAi therapy against cancers (see Chap. 16). RNAi provides a unique opportunity to silence cancer-causing stem cell genes at the pretranslation level, which is otherwise not possible with conventional therapies such as cytotoxic chemotherapy, small molecule inhibitors, or monoclonal antibodies.

Since ncRNAs are linked to pathological conditions and, in particular, disease development and progression, ncRNAs might become useful biomarkers for diagnostic purposes. For example, miRNAs have been found to be associated with disease prognosis, survival, and mortality in biopsies (Schetter et al. 2009; Bloomston et al. 2007). Their expression levels can be determined by in situ hybridization and microarray, e.g., on a tumor section and its normal adjacent counterparts (see Chap. 21). Major challenge lies in translating the molecular signatures determined in the laboratory to the clinical setting.

The fundamental roles of ncRNAs in development, differentiation, and malignancy suggest that these classes of molecules are potential targets for novel therapeutics. Antisense oligonucleotide approaches used for inhibition, and siRNA-like technologies used for replacement are currently being explored for therapeutic modulation of miRNAs. Several approaches are currently adopted to silence ncRNAs. Table 1.3 enlists the different approaches for the purpose, which has been mostly applied to miRNAs till today.

Table 1.3 Approaches employing ncRNAs in therapeutic applications

Approaches	Name of the tools/methods	Applications	References
Inhibition of mature miRNAs	microRNA sponges	Silence oncomiR family	Ebert et al. (2007)
	2'-Ome AMOs	Silence oncomiR	Krutzfeldt et al. (2005)
	2'-MOE AMOs	Silence oncomiR	Weiler et al. (2006)
	LNA-antagomir	Silence oncomiR	Elmen et al. (2008a)
Manipulation of miRNA precursor	amiRNAs	Silencing of target genes involved in metastasis	Liang et al. (2007) Zhang et al. (2006)
Inhibition of pri-miRNA	AMOs (RNase H-based)	Silence polycistronic clusters of miRNAs	Wu et al. (2004)
Replacement of mature miRNAs	Pre-miRNA-like shRNAs	Restore tumor suppressor miRNAs	Brummelkamp et al. (2002)
	Double-stranded miRNA mimetics	Restore tumor suppressor miRNAs	Tsuda et al. (2006)
Designing small oligonucleotides with perfect complementary to the seed	Target protectors	Inhibit functions of oncomiR	Choi et al. (2007)

AMOs anti-miRNA oligonucleotides, *2'-Ome* 2'-O-methyl, *2'-MOE* 2'-O methoxyethyl, *LNA* locked nucleic acid, *oncomiR* oncogenic miRNAs, *amiRNAs* artificial miRNAs

Specific knockdown of miRNAs by anti-miRNA oligonucleotides (AMOs), double-strand miRNA mimetics, and overexpression of miRNA duplexes have been conducted *in vitro* and *in vivo*. Inhibition of specific miRNAs in mouse model has been performed by antagomirs. Also RNase H-based AMOs have been found useful for targeting polycistronic pri-miRNAs, like the miR-17–92 cluster (Wu et al. 2004). Specific dose-dependent silencing of miR-122 has been performed by systemic administration of 16-nucleotide unconjugated locked nucleic acid (LNA)-AMO which is complementary to the 5′ end of miR-122 (Elmen et al. 2008b). Another de novo engineered ncRNA inhibitors are "miRNA sponges" which inhibit miRNAs with a complementary heptameric seed, such that a single sponge can be used to block an entire miRNA family with the same seed. Inhibition of Drosha, Dicer, or any other components in the maturation pathway is another method for therapeutic targeting of ncRNAs. This method however might be difficult to be made specific in its therapeutic effect. Moreover, artificial miRNAs (amiRNAs) are recently developed miRNA-based tools to silence endogenous genes. These are created from an endogenous miRNA precursor by exchanging the miRNA/miRNA sequence of it with a sequence designed to match the target gene of interest (see Chap. 20).

An alternative therapeutic strategy of replacement of defective/absent RNA effectors is needed if there is a loss in the activity of ncRNAs in the diseased/affected cells. Lentiviral delivery of short hairpin RNAs is one of the systems for the delivery of shRNA constructs designed to mimic the pri-miRNA by including the miRNA flanking sequence into the shRNA stem (Chang et al. 2006; Zeng et al. 2005). Further, there has been activation of tumor suppressor miRNAs, such as miR-127, by chromatin-modifying drugs which can inhibit tumor growth through downregulation of their target oncogenes (Grunweller and Hartmann 2007).

All these highlight the clinical potential of ncRNAs as biomarkers for diagnosis, prognosis, and prediction of therapeutic outcome.

1.7 Future Perspectives

The possibility of self-replicating ribozymes emerging from pools of random polynucleotides and surviving in a prebiotic soup has put forth these RNAs to be a challenging molecule, which leads us to an "RNA world." The logical order of events begins with prebiotic chemistry and ending with DNA/protein-based life. The present challenge lies in decoding the genomic dark matter. Further, the absolute number of protein-coding genes encoded by a genome is essentially static across all animals from simple nematodes to humans (Taft et al. 2007), which hints for additional genetic elements that must be involved in the development of the increasingly complex cellular, physiological, and neurological systems. Noncoding RNAs are the likely candidates, which can resolve such discrepancy within the genomic content and illuminate on the genomic dark matter, as they are adaptively plastic, capable of regulating processes both broadly and sequence-specifically, and are now known to be components of nearly all cellular and developmental systems.

It is becoming clear that a comprehensive understanding of human biology must include both small and large noncoding RNAs. With new systems biology approaches, and in-depth investigation of other important players and their interactions, we may see an emerging integration of RNA-dependent regulatory networks into normal cell physiology. It is perhaps only through inclusion of these elements in the biomedical research agenda along with studies to determine the mechanistic basis of the causative variations (identified by genome-wide association studies), that complex human diseases will be completely deciphered.

References

Allen E, Xie Z, Gustafson AM, Carrington JC (2005) MicroRNA-directed phasing during trans-acting siRNA biogenesis in plants. Cell 121(2):207–221. doi:S0092-8674(05), 00345-4 [pii] 10.1016/j.cell.2005.04.004

Aravin A, Gaidatzis D, Pfeffer S, Lagos-Quintana M, Landgraf P, Iovino N, Morris P, Brownstein MJ, Kuramochi-Miyagawa S, Nakano T, Chien M, Russo JJ, Ju J, Sheridan R, Sander C, Zavolan M, Tuschl T (2006) A novel class of small RNAs bind to MILI protein in mouse testes. Nature 442(7099):203–207. doi:nature04916 [pii] 10.1038/nature04916

Aravin AA, Sachidanandam R, Girard A, Fejes-Toth K, Hannon GJ (2007) Developmentally regulated piRNA clusters implicate MILI in transposon control. Science 316(5825):744–747. doi:1142612 [pii] 10.1126/science.1142612

Bachellerie JP, Cavaille J, Huttenhofer A (2002) The expanding snoRNA world. Biochimie 84 (8):775–790. doi:S0300908402014025 [pii]

Bartel B, Bartel DP (2003) MicroRNAs: at the root of plant development? Plant Physiol 132 (2):709–717. doi:10.1104/pp. 103.023630132/2/709 [pii]

Batista PJ, Ruby JG, Claycomb JM, Chiang R, Fahlgren N, Kasschau KD, Chaves DA, Gu W, Vasale JJ, Duan S, Conte D Jr, Luo S, Schroth GP, Carrington JC, Bartel DP, Mello CC (2008) PRG-1 and 21U-RNAs interact to form the piRNA complex required for fertility in C. elegans. Mol Cell 31(1):67–78. doi:S1097-2765(08), 00391-2 [pii] 10.1016/j.molcel.2008.06.002

Birney E, Stamatoyannopoulos JA, Dutta A, Guigo R, Gingeras TR, Margulies EH, Weng Z, Snyder M, Dermitzakis ET, Thurman RE, Kuehn MS, Taylor CM, Neph S, Koch CM, Asthana S, Malhotra A, Adzhubei I, Greenbaum JA, Andrews RM, Flicek P, Boyle PJ, Cao H, Carter NP, Clelland GK, Davis S, Day N, Dhami P, Dillon SC, Dorschner MO, Fiegler H, Giresi PG, Goldy J, Hawrylycz M, Haydock A, Humbert R, James KD, Johnson BE, Johnson EM, Frum TT, Rosenzweig ER, Karnani N, Lee K, Lefebvre GC, Navas PA, Neri F, Parker SC, Sabo PJ, Sandstrom R, Shafer A, Vetrie D, Weaver M, Wilcox S, Yu M, Collins FS, Dekker J, Lieb JD, Tullius TD, Crawford GE, Sunyaev S, Noble WS, Dunham I, Denoeud F, Reymond A, Kapranov P, Rozowsky J, Zheng D, Castelo R, Frankish A, Harrow J, Ghosh S, Sandelin A, Hofacker IL, Baertsch R, Keefe D, Dike S, Cheng J, Hirsch HA, Sekinger EA, Lagarde J, Abril JF, Shahab A, Flamm C, Fried C, Hackermuller J, Hertel J, Lindemeyer M, Missal K, Tanzer A, Washietl S, Korbel J, Emanuelsson O, Pedersen JS, Holroyd N, Taylor R, Swarbreck D, Matthews N, Dickson MC, Thomas DJ, Weirauch MT, Gilbert J, Drenkow J, Bell I, Zhao X, Srinivasan KG, Sung WK, Ooi HS, Chiu KP, Foissac S, Alioto T, Brent M, Pachter L, Tress ML, Valencia A, Choo SW, Choo CY, Ucla C, Manzano C, Wyss C, Cheung E, Clark TG, Brown JB, Ganesh M, Patel S, Tammana H, Chrast J, Henrichsen CN, Kai C, Kawai J, Nagalakshmi U, Wu J, Lian Z, Lian J, Newburger P, Zhang X, Bickel P, Mattick JS, Carninci P, Hayashizaki Y, Weissman S, Hubbard T, Myers RM, Rogers J, Stadler PF, Lowe TM, Wei

CL, Ruan Y, Struhl K, Gerstein M, Antonarakis SE, Fu Y, Green ED, Karaoz U, Siepel A, Taylor J, Liefer LA, Wetterstrand KA, Good PJ, Feingold EA, Guyer MS, Cooper GM, Asimenos G, Dewey CN, Hou M, Nikolaev S, Montoya-Burgos JI, Loytynoja A, Whelan S, Pardi F, Massingham T, Huang H, Zhang NR, Holmes I, Mullikin JC, Ureta-Vidal A, Paten B, Seringhaus M, Church D, Rosenbloom K, Kent WJ, Stone EA, Batzoglou S, Goldman N, Hardison RC, Haussler D, Miller W, Sidow A, Trinklein ND, Zhang ZD, Barrera L, Stuart R, King DC, Ameur A, Enroth S, Bieda MC, Kim J, Bhinge AA, Jiang N, Liu J, Yao F, Vega VB, Lee CW, Ng P, Yang A, Moqtaderi Z, Zhu Z, Xu X, Squazzo S, Oberley MJ, Inman D, Singer MA, Richmond TA, Munn KJ, Rada-Iglesias A, Wallerman O, Komorowski J, Fowler JC, Couttet P, Bruce AW, Dovey OM, Ellis PD, Langford CF, Nix DA, Euskirchen G, Hartman S, Urban AE, Kraus P, Van Calcar S, Heintzman N, Kim TH, Wang K, Qu C, Hon G, Luna R, Glass CK, Rosenfeld MG, Aldred SF, Cooper SJ, Halees A, Lin JM, Shulha HP, Xu M, Haidar JN, Yu Y, Iyer VR, Green RD, Wadelius C, Farnham PJ, Ren B, Harte RA, Hinrichs AS, Trumbower H, Clawson H, Hillman-Jackson J, Zweig AS, Smith K, Thakkapallayil A, Barber G, Kuhn RM, Karolchik D, Armengol L, Bird CP, de Bakker PI, Kern AD, Lopez-Bigas N, Martin JD, Stranger BE, Woodroffe A, Davydov E, Dimas A, Eyras E, Hallgrimsdottir IB, Huppert J, Zody MC, Abecasis GR, Estivill X, Bouffard GG, Guan X, Hansen NF, Idol JR, Maduro VV, Maskeri B, McDowell JC, Park M, Thomas PJ, Young AC, Blakesley RW, Muzny DM, Sodergren E, Wheeler DA, Worley KC, Jiang H, Weinstock GM, Gibbs RA, Graves T, Fulton R, Mardis ER, Wilson RK, Clamp M, Cuff J, Gnerre S, Jaffe DB, Chang JL, Lindblad-Toh K, Lander ES, Koriabine M, Nefedov M, Osoegawa K, Yoshinaga Y, Zhu B, de Jong PJ (2007) Identification and analysis of functional elements in 1% of the human genome by the ENCODE pilot project. Nature 447 (7146):799–816. doi:10.1038/nature05874

Bloomston M, Frankel WL, Petrocca F, Volinia S, Alder H, Hagan JP, Liu CG, Bhatt D, Taccioli C, Croce CM (2007) MicroRNA expression patterns to differentiate pancreatic adenocarcinoma from normal pancreas and chronic pancreatitis. JAMA 297(17):1901–1908. doi:297/17/1901 [pii] 10.1001/jama.297.17.1901

Brennecke J, Aravin AA, Stark A, Dus M, Kellis M, Sachidanandam R, Hannon GJ (2007) Discrete small RNA-generating loci as master regulators of transposon activity in Drosophila. Cell 128(6):1089–1103. doi:S0092-8674(07), 00257-7 [pii] 10.1016/j.cell.2007.01.043

Brummelkamp TR, Bernards R, Agami R (2002) A system for stable expression of short interfering RNAs in mammalian cells. Science 296(5567):550–553. doi:10.1126/science.10689991068999 [pii]

Cao F, Li X, Hiew S, Brady H, Liu Y, Dou Y (2009) Dicer independent small RNAs associate with telomeric heterochromatin. RNA 15(7):1274–1281. doi:rna.1423309 [pii] 10.1261/rna.1423309

Carninci P, Hayashizaki Y (2007) Noncoding RNA transcription beyond annotated genes. Curr Opin Genet Dev 17 (2):139-144. doi:S0959-437X(07)00034-2 [pii] 10.1016/j.gde.2007.02.008

Carone DM, Longo MS, Ferreri GC, Hall L, Harris M, Shook N, Bulazel KV, Carone BR, Obergfell C, O'Neill MJ, O'Neill RJ (2009) A new class of retroviral and satellite encoded small RNAs emanates from mammalian centromeres. Chromosoma 118(1):113–125. doi:10.1007/s00412-008-0181-5

Carthew RW, Sontheimer EJ (2009) Origins and Mechanisms of miRNAs and siRNAs. Cell 136 (4):642–655. doi:S0092-8674(09)00083-X [pii] 10.1016/j.cell.2009.01.035

Chang K, Elledge SJ, Hannon GJ (2006) Lessons from Nature: microRNA-based shRNA libraries. Nat Methods 3(9):707–714. doi:nmeth923 [pii] 10.1038/nmeth923

Choi WY, Giraldez AJ, Schier AF (2007) Target protectors reveal dampening and balancing of nodal agonist and antagonist by miR-430. Science 318(5848):271–274. doi:1147535 [pii] 10.1126/science.1147535

Crick FH (1968) The origin of the genetic code. J Mol Biol 38(3):367–379. doi:0022-2836(68) 90392-6 [pii]

Czech B, Hannon GJ (2011) Small RNA sorting: matchmaking for Argonautes. Nat Rev Genet 12 (1):19–31. doi:nrg2916 [pii] 10.1038/nrg2916

Daxinger L, Kanno T, Bucher E, van der Winden J, Naumann U, Matzke AJ, Matzke M (2009) A stepwise pathway for biogenesis of 24-nt secondary siRNAs and spreading of DNA methylation. EMBO J 28(1):48–57. doi:emboj2008260 [pii] 10.1038/emboj.2008 260

De Lucia F, Dean C (2011) Long non-coding RNAs and chromatin regulation. Curr Opin Plant Biol 14(2):168–173. doi:S1369-5266(10), 00177-9 [pii] 10.1016/j.pbi.2010.11.006

Dieci G, Preti M, Montanini B (2009) Eukaryotic snoRNAs: a paradigm for gene expression flexibility. Genomics 94(2):83–88. doi:S0888-7543(09), 00106-2 [pii] 10.1016/j.ygeno.2009.05.002

Ebert MS, Neilson JR, Sharp PA (2007) MicroRNA sponges: competitive inhibitors of small RNAs in mammalian cells. Nat Methods 4(9):721–726. doi:nmeth1079 [pii] 10.1038/nmeth1079

Eddy SR (2002) Computational genomics of noncoding RNA genes. Cell 109(2):137–140. doi: S0092867402007274 [pii]

Elmen J, Lindow M, Schutz S, Lawrence M, Petri A, Obad S, Lindholm M, Hedtjarn M, Hansen HF, Berger U, Gullans S, Kearney P, Sarnow P, Straarup EM, Kauppinen S (2008a) LNA-mediated microRNA silencing in non-human primates. Nature 452(7189):896–899. doi: nature06783 [pii] 10.1038/nature06783

Elmen J, Lindow M, Silahtaroglu A, Bak M, Christensen M, Lind-Thomsen A, Hedtjarn M, Hansen JB, Hansen HF, Straarup EM, McCullagh K, Kearney P, Kauppinen S (2008b) Antagonism of microRNA-122 in mice by systemically administered LNA-antimiR leads to up-regulation of a large set of predicted target mRNAs in the liver. Nucleic Acids Res 36 (4):1153–1162. doi:gkm1113 [pii] 10.1093/nar/gkm1113

Fejes-Toth K, Sotirova V, Sachidanandam R, Assaf G, Hannon G, Kapranov P, Foissac S, Willingham A, Duttagupta R, Dumais E, Gingeras T (2009) Post-transcriptional processing generates a diversity of 5′-modified long and short RNAs. Nature 457(7232):1028–1032. doi: nature07759 [pii] 10.1038/nature07759

Fire A, Xu S, Montgomery MK, Kostas SA, Driver SE, Mello CC (1998) Potent and specific genetic interference by double-stranded RNA in Caenorhabditis elegans. Nature 391 (6669):806–811. doi:10.1038/35888

Galasso M, Elena Sana M, Volinia S (2010) Non-coding RNAs: a key to future personalized molecular therapy? Genome Med 2(2):12. doi:gm133 [pii] 10.1186/gm133

Girard A, Sachidanandam R, Hannon GJ, Carmell MA (2006) A germline-specific class of small RNAs binds mammalian Piwi proteins. Nature 442(7099):199–202. doi:nature04917 [pii] 10.1038/nature04917

Girard A, Hannon GJ (2008) Conserved themes in small-RNA-mediated transposon control. Trends Cell Biol 18 (3):136-148. doi:S0962-8924(08)00042-1 [pii] 10.1016/j.tcb.2008.01.004

Grimson A, Srivastava M, Fahey B, Woodcroft BJ, Chiang HR, King N, Degnan BM, Rokhsar DS, Bartel DP (2008) Early origins and evolution of microRNAs and Piwi-interacting RNAs in animals. Nature 455(7217):1193–1197. doi:nature07415 [pii] 10.1038/nature07415

Grivna ST, Beyret E, Wang Z, Lin H (2006) A novel class of small RNAs in mouse spermatogenic cells. Genes Dev 20(13):1709–1714. doi:gad.1434406 [pii] 10.1101/gad.1434406

Grunweller A, Hartmann RK (2007) Locked nucleic acid oligonucleotides: the next generation of antisense agents? BioDrugs 21(4):235–243. doi:2144 [pii]

Guerrier-Takada C, Gardiner K, Marsh T, Pace N, Altman S (1983) The RNA moiety of ribonuclease P is the catalytic subunit of the enzyme. Cell 35(3 Pt 2):849–857. doi:0092-8674(83), 90117-4 [pii]

Gunawardane LS, Saito K, Nishida KM, Miyoshi K, Kawamura Y, Nagami T, Siomi H, Siomi MC (2007) A slicer-mediated mechanism for repeat-associated siRNA 5′ end formation in Drosophila. Science 315(5818):1587–1590. doi:1140494 [pii] 10.1126/science.1140494

Hamilton AJ, Baulcombe DC (1999) A species of small antisense RNA in posttranscriptional gene silencing in plants. Science 286(5441):950–952. doi:7953 [pii]

Houwing S, Kamminga LM, Berezikov E, Cronembold D, Girard A, van den Elst H, Filippov DV, Blaser H, Raz E, Moens CB, Plasterk RH, Hannon GJ, Draper BW, Ketting RF (2007) A role for Piwi and piRNAs in germ cell maintenance and transposon silencing in Zebrafish. Cell 129 (1):69–82. doi:S0092-8674(07), 00392-3 [pii] 10.1016/j.cell.2007.03.026

Huntzinger E, Izaurralde E (2011) Gene silencing by microRNAs: contributions of translational repression and mRNA decay. Nat Rev Genet 12(2):99–110. doi:nrg2936 [pii] 10.1038/nrg2936

Huttenhofer A, Kiefmann M, Meier-Ewert S, O'Brien J, Lehrach H, Bachellerie JP, Brosius J (2001) RNomics: an experimental approach that identifies 201 candidates for novel, small, non-messenger RNAs in mouse. EMBO J 20(11):2943–2953. doi:10.1093/emboj/20.11.2943

Hutvagner G, Simard MJ (2008) Argonaute proteins: key players in RNA silencing. Nat Rev Mol Cell Biol 9(1):22–32. doi:nrm2321 [pii] 10.1038/nrm2321

Kanduri C, Whitehead J, Mohammad F (2009) The long and the short of it: RNA-directed chromatin asymmetry in mammalian X-chromosome inactivation. FEBS Lett 583 (5):857–864. doi:S0014-5793(09), 00102-1 [pii] 10.1016/j.febslet.2009.02.004

Kapranov P, Cheng J, Dike S, Nix DA, Duttagupta R, Willingham AT, Stadler PF, Hertel J, Hackermuller J, Hofacker IL, Bell I, Cheung E, Drenkow J, Dumais E, Patel S, Helt G, Ganesh M, Ghosh S, Piccolboni A, Sementchenko V, Tammana H, Gingeras TR (2007) RNA maps reveal new RNA classes and a possible function for pervasive transcription. Science 316 (5830):1484–1488. doi:1138341 [pii] 10.1126/science.1138341

Kapranov P, Ozsolak F, Kim SW, Foissac S, Lipson D, Hart C, Roels S, Borel C, Antonarakis SE, Monaghan AP, John B, Milos PM (2010) New class of gene-termini-associated human RNAs suggests a novel RNA copying mechanism. Nature 466(7306):642–646. doi:nature09190 [pii] 10.1038/nature09190

Kasschau KD, Fahlgren N, Chapman EJ, Sullivan CM, Cumbie JS, Givan SA, Carrington JC (2007) Genome-wide profiling and analysis of Arabidopsis siRNAs. PLoS Biol 5(3):e57. doi:1544-9173-5-3-e57 [pii] 10.1371/journal.pbio.0050057

Khanna A, Stamm S (2010) Regulation of alternative splicing by short non-coding nuclear RNAs. RNA Biol 7(4):480–485. doi:12746 [pii]

Kruger K, Grabowski PJ, Zaug AJ, Sands J, Gottschling DE, Cech TR (1982) Self-splicing RNA: autoexcision and autocyclization of the ribosomal RNA intervening sequence of Tetrahymena. Cell 31(1):147–157. doi:0092-8674(82), 90414-7 [pii]

Krutzfeldt J, Rajewsky N, Braich R, Rajeev KG, Tuschl T, Manoharan M, Stoffel M (2005) Silencing of microRNAs in vivo with 'antagomirs'. Nature 438(7068):685–689. doi: nature04303 [pii] 10.1038/nature04303

Kurth HM, Mochizuki K (2009) 2'-O-methylation stabilizes Piwi-associated small RNAs and ensures DNA elimination in Tetrahymena. RNA 15(4):675–685. doi:rna.1455509 [pii] 10.1261/rna.1455509

Langenberger D, Bermudez-Santana C, Hertel J, Hoffmann S, Khaitovich P, Stadler PF (2009) Evidence for human microRNA-offset RNAs in small RNA sequencing data. Bioinformatics 25(18):2298–2301. doi:btp419 [pii] 10.1093/bioinformatics/btp419

Lao KQ, Tang F, Barbacioru C, Wang Y, Nordman E, Lee C, Xu N, Wang X, Tuch B, Bodeau J, Siddiqui A, Surani MA (2009) mRNA-sequencing whole transcriptome analysis of a single cell on the SOLiD system. J Biomol Tech 20(5):266–271

Lau NC, Ohsumi T, Borowsky M, Kingston RE, Blower MD (2009) Systematic and single cell analysis of Xenopus Piwi-interacting RNAs and Xiwi. EMBO J 28(19):2945–2958. doi: emboj2009237 [pii] 10.1038/emboj.2009.237

Lee HC, Chang SS, Choudhary S, Aalto AP, Maiti M, Bamford DH, Liu Y (2009) qiRNA is a new type of small interfering RNA induced by DNA damage. Nature 459(7244):274–277. doi: nature08041 [pii] 10.1038/nature08041

Lee SK, Calin GA (2011) Non-coding RNAs and cancer: new paradigms in oncology. Discov Med 11(58):245–254

Liang Z, Wu H, Reddy S, Zhu A, Wang S, Blevins D, Yoon Y, Zhang Y, Shim H (2007) Blockade of invasion and metastasis of breast cancer cells via targeting CXCR4 with an artificial microRNA. Biochem Biophys Res Commun 363(3):542–546. doi:S0006-291X(07), 01935-3 [pii] 10.1016/j.bbrc.2007.09.007

Lin H (2007) piRNAs in the germ line. Science 316(5823):397. doi:316/5823/397 [pii] 10.1126/science.1137543

Lunyak VV, Rosenfeld MG (2008) Epigenetic regulation of stem cell fate. Hum Mol Genet 17 (R1):R28–R36. doi:ddn149 [pii] 10.1093/hmg/ddn149

Malone CD, Hannon GJ (2009) Small RNAs as guardians of the genome. Cell 136(4):656–668. doi:S0092-8674(09), 00127-5 [pii] 10.1016/j.cell.2009.01.045

Mercer TR, Dinger ME, Mattick JS (2009) Long non-coding RNAs: insights into functions. Nat Rev Genet 10(3):155–159. doi:nrg2521 [pii] 10.1038/nrg2521

Moazed D (2009) Small RNAs in transcriptional gene silencing and genome defence. Nature 457 (7228):413–420. doi:nature07756 [pii] 10.1038/nature07756

Mortazavi A, Williams BA, McCue K, Schaeffer L, Wold B (2008) Mapping and quantifying mammalian transcriptomes by RNA-Seq. Nat Methods 5(7):621–628. doi:nmeth.1226 [pii] 10.1038/nmeth.1226

Napoli C, Lemieux C, Jorgensen R (1990) Introduction of a chimeric chalcone synthase gene into petunia results in reversible co-suppression of homologous genes in trans. Plant Cell 2 (4):279–289. doi:10.1105/tpc.2.4.279 2/4/279 [pii]

Nishida KM, Saito K, Mori T, Kawamura Y, Nagami-Okada T, Inagaki S, Siomi H, Siomi MC (2007) Gene silencing mechanisms mediated by Aubergine piRNA complexes in Drosophila male gonad. RNA 13(11):1911–1922. doi:rna.744307 [pii] 10.1261/rna.744307

Palakodeti D, Smielewska M, Lu YC, Yeo GW, Graveley BR (2008) The PIWI proteins SMEDWI-2 and SMEDWI-3 are required for stem cell function and piRNA expression in planarians. RNA 14(6):1174–1186. doi:rna.1085008 [pii] 10.1261/rna.1085008

Peters L, Meister G (2007) Argonaute proteins: mediators of RNA silencing. Mol Cell 26 (5):611–623. doi:S1097-2765(07), 00257-2 [pii] 10.1016/j.molcel.2007.05.001

Ponting CP, Oliver PL, Reik W (2009) Evolution and functions of long noncoding RNAs. Cell 136 (4):629–641. doi:S0092-8674(09), 00142-1 [pii] 10.1016/j.cell.2009.02.006

Qureshi IA, Mehler MF (2011) Non-coding RNA networks underlying cognitive disorders across the lifespan. Trends Mol Med 17(6):337–346. doi:S1471-4914(11), 00029-3 [pii] 10.1016/j. molmed.2011.02.002

Repoila F, Darfeuille F (2009) Small regulatory non-coding RNAs in bacteria: physiology and mechanistic aspects. Biol Cell 101(2):117–131. doi:BC20070137 [pii] 10.1042/BC20070137

Ruby JG, Jan C, Player C, Axtell MJ, Lee W, Nusbaum C, Ge H, Bartel DP (2006) Large-scale sequencing reveals 21U-RNAs and additional microRNAs and endogenous siRNAs in C. elegans. Cell 127(6):1193–1207. doi:S0092-8674(06), 01468-1 [pii] 10.1016/j.cell.2006.10.040

Saito K, Nishida KM, Mori T, Kawamura Y, Miyoshi K, Nagami T, Siomi H, Siomi MC (2006) Specific association of Piwi with rasiRNAs derived from retrotransposon and heterochromatic regions in the Drosophila genome. Genes Dev 20(16):2214–2222. doi:gad.1454806 [pii] 10.1101/gad.1454806

Schetter AJ, Nguyen GH, Bowman ED, Mathe EA, Yuen ST, Hawkes JE, Croce CM, Leung SY, Harris CC (2009) Association of inflammation-related and microRNA gene expression with cancer-specific mortality of colon adenocarcinoma. Clin Cancer Res 15(18):5878–5887. doi:1078-0432.CCR-09-0627 [pii] 10.1158/1078-0432.CCR-09-0627

Seto AG, Kingston RE, Lau NC (2007) The coming of age for Piwi proteins. Mol Cell 26 (5):603–609. doi:S1097-2765(07)00322-X [pii] 10.1016/j.molcel.2007.05.021

Shi W, Hendrix D, Levine M, Haley B (2009) A distinct class of small RNAs arises from pre-miRNA-proximal regions in a simple chordate. Nat Struct Mol Biol 16(2):183–189. doi: nsmb.1536 [pii] 10.1038/nsmb.1536

Siomi M, Miyoshi T, Siomi H (2010) piRNA-mediated silencing in Drosophila germlines. Semin Cell Dev Biol 21(7):754–759

Storz G (2002) An expanding universe of noncoding RNAs. Science 296(5571):1260–1263. doi:10.1126/science.1072249 296/5571/1260 [pii]

Taft RJ, Pheasant M, Mattick JS (2007) The relationship between non-protein-coding DNA and eukaryotic complexity. Bioessays 29(3):288–299. doi:10.1002/bies.20544

Taft RJ, Glazov EA, Cloonan N, Simons C, Stephen S, Faulkner GJ, Lassmann T, Forrest AR, Grimmond SM, Schroder K, Irvine K, Arakawa T, Nakamura M, Kubosaki A, Hayashida K, Kawazu C, Murata M, Nishiyori H, Fukuda S, Kawai J, Daub CO, Hume DA, Suzuki H,

Orlando V, Carninci P, Hayashizaki Y, Mattick JS (2009a) Tiny RNAs associated with transcription start sites in animals. Nat Genet 41(5):572–578. doi:ng.312 [pii] 10.1038/ng.312

Taft RJ, Glazov EA, Lassmann T, Hayashizaki Y, Carninci P, Mattick JS (2009b) Small RNAs derived from snoRNAs. RNA 15(7):1233–1240. doi:rna.1528909 [pii] 10.1261/rna.1528909

Taft RJ, Pang KC, Mercer TR, Dinger M, Mattick JS (2010) Non-coding RNAs: regulators of disease. J Pathol 220(2):126–139. doi:10.1002/path.2638

Thomson T, Lin H (2009) The biogenesis and function of PIWI proteins and piRNAs: progress and prospect. Annu Rev Cell Dev Biol 25:355–376. doi:10.1146/annurev. cellbio.24.110707.175327

Tsuda N, Ishiyama S, Li Y, Ioannides CG, Abbruzzese JL, Chang DZ (2006) Synthetic microRNA designed to target glioma-associated antigen 1 transcription factor inhibits division and induces late apoptosis in pancreatic tumor cells. Clin Cancer Res 12(21):6557–6564. doi:12/21/6557 [pii] 10.1158/1078-0432.CCR-06-0588

Unhavaithaya Y, Hao Y, Beyret E, Yin H, Kuramochi-Miyagawa S, Nakano T, Lin H (2009) MILI, a PIWI-interacting RNA-binding protein, is required for germ line stem cell self-renewal and appears to positively regulate translation. J Biol Chem 284(10):6507–6519. doi: M809104200 [pii] 10.1074/jbc.M809104200

van der Krol AR, Mur LA, Beld M, Mol JN, Stuitje AR (1990) Flavonoid genes in petunia: addition of a limited number of gene copies may lead to a suppression of gene expression. Plant Cell 2(4):291–299

Vaucheret H (2006) Post-transcriptional small RNA pathways in plants: mechanisms and regulations. Genes Dev 20(7):759–771. doi:20/7/759 [pii] 10.1101/gad.1410506

Vazquez F, Vaucheret H, Rajagopalan R, Lepers C, Gasciolli V, Mallory AC, Hilbert JL, Bartel DP, Crete P (2004) Endogenous trans-acting siRNAs regulate the accumulation of Arabidopsis mRNAs. Mol Cell 16(1):69–79. doi:S1097276504005817 [pii] 10.1016/j.molcel.2004.09.028

Verdel A, Vavasseur A, Le Gorrec M, Touat-Todeschini L (2009) Common themes in siRNA-mediated epigenetic silencing pathways. Int J Dev Biol 53(2–3):245–257. doi:082691av [pii] 10.1387/ijdb.082691av

Vogel J, Sharma CM (2005) How to find small non-coding RNAs in bacteria. Biol Chem 386 (12):1219–1238. doi:10.1515/BC.2005.140

Washietl S, Hofacker IL, Lukasser M, Huttenhofer A, Stadler PF (2005) Mapping of conserved RNA secondary structures predicts thousands of functional noncoding RNAs in the human genome. Nat Biotechnol 23(11):1383–1390. doi:nbt1144 [pii] 10.1038/nbt1144

Wassarman KM, Repoila F, Rosenow C, Storz G, Gottesman S (2001) Identification of novel small RNAs using comparative genomics and microarrays. Genes Dev 15(13):1637–1651. doi:10.1101/gad.901001

Watanabe T, Takeda A, Tsukiyama T, Mise K, Okuno T, Sasaki H, Minami N, Imai H (2006) Identification and characterization of two novel classes of small RNAs in the mouse germline: retrotransposon-derived siRNAs in oocytes and germline small RNAs in testes. Genes Dev 20 (13):1732–1743. doi:gad.1425706 [pii] 10.1101/gad.1425706

Waters LS, Storz G (2009) Regulatory RNAs in bacteria. Cell 136(4):615–628. doi:S0092-8674 (09), 00125-1 [pii] 10.1016/j.cell.2009.01.043

Weiler J, Hunziker J, Hall J (2006) Anti-miRNA oligonucleotides (AMOs): ammunition to target miRNAs implicated in human disease? Gene Ther 13(6):496–502. doi:3302654 [pii] 10.1038/ sj.gt.3302654

Willingham AT, Orth AP, Batalov S, Peters EC, Wen BG, Aza-Blanc P, Hogenesch JB, Schultz PG (2005) A strategy for probing the function of noncoding RNAs finds a repressor of NFAT. Science 309(5740):1570–1573. doi:309/5740/1570 [pii] 10.1126/science.1115901

Woese C (1967) The evolution of the genetic code. In: The genetic code. Harper and Row, New York, pp. 179–195

Wu H, Lima WF, Zhang H, Fan A, Sun H, Crooke ST (2004) Determination of the role of the human RNase H1 in the pharmacology of DNA-like antisense drugs. J Biol Chem 279 (17):17181–17189. doi:10.1074/jbc.M311683200 M311683200 [pii]

Yin H, Lin H (2007) An epigenetic activation role of Piwi and a Piwi-associated piRNA in Drosophila melanogaster. Nature 450(7167):304–308. doi:nature06263 [pii] 10.1038/nature06263

Yoshikawa M, Peragine A, Park MY, Poethig RS (2005) A pathway for the biogenesis of trans-acting siRNAs in Arabidopsis. Genes Dev 19(18):2164–2175. doi:gad.1352605 [pii] 10.1101/gad.1352605

Zeng Y, Yi R, Cullen BR (2005) Recognition and cleavage of primary microRNA precursors by the nuclear processing enzyme Drosha. EMBO J 24(1):138–148. doi:7600491 [pii] 10.1038/sj.emboj.7600491

Zhang B, Pan X, Cobb GP, Anderson TA (2006) Plant microRNA: a small regulatory molecule with big impact. Dev Biol 289(1):3–16. doi:S0012-1606(05), 00764-5 [pii] 10.1016/j.ydbio.2005.10.036

Chapter 2
Diversity, Overlap, and Relationships in the Small RNA Landscape

Michelle S. Scott

Abstract Rapidly evolving, of high abundance and great diversity, small RNAs are increasingly found to play central cellular regulatory roles, the extent of which we are only now starting to comprehend. The evolutionary association of diverse classes of small RNAs and transposable elements is offering clues about the origin, abundance, biogenesis pathways, and target acquisition mechanisms of small RNAs. And as well as a similar relationship with transposable elements, different types of small RNAs show commonalities in their processing pathways while displaying a wide degree of diversity and variation within their biogenesis pathways and amongst their precursors, likely allowing for flexible regulation. This book chapter examines the evolutionary relationship between small RNAs and transposable elements through the role transposable elements play in the expansion of small RNA classes as well as the acquisition of novel targets. The great diversity but also overlap in both the small RNA biogenesis pathways and functional entities are also explored.

Keywords Biogenesis pathways • microRNAs • transposable elements

2.1 Introduction and Overview

During the past decade, numerous members of diverse classes of small RNAs have been associated with transposable elements, offering clues about the origin, abundance, biogenesis pathways, target acquisition mechanisms, and rapid evolution of groups of small RNAs. Large numbers of small RNAs, and in particular miRNAs,

M.S. Scott (✉)
Division of Biological Chemistry and Drug Discovery, College of Life Sciences, University of Dundee, Dundee, UK

Current address: Department of Biochemistry, University of Sherbrooke, Sherbrooke, Québec, Canada
e-mail: michelle.scott@usherbrooke.ca

B. Mallick (eds.), *Regulatory RNAs*, DOI 10.1007/978-3-642-22517-8_2,
© Springer-Verlag Berlin Heidelberg (outside the USA) 2012

have been recently found to originate from noncanonical precursors, display variation in their biogenesis pathways, or exert novel regulatory functionality, making clear-cut classification of these molecules increasingly difficult.

The great diversity of small RNA precursors, biogenesis pathways, and targeting mechanisms allows for combinatorial complexity and high flexibility in the regulation of multiple aspects of cellular function. Increased understanding of the evolutionary and regulatory relationships between these different noncoding RNAs will be central to unlocking the multiple layers of regulation underlying the cell's complexity. This understanding will also be important to determine how we can harness this knowledge to treat the numerous diseases likely to result from defects in the regulation of these pathways. For example, numerous miRNAs display deregulated expression and a functional involvement in cancers (reviewed in Garzon et al. 2006).

The first half of this chapter explores the relationships between transposable elements and small RNAs, examining in particular how they have served in the expansion of small RNA classes as well as the acquisition of novel targets, leading to both overlap and diversity. The rapid expansion of these elements has led to an RNA landscape displaying overlapping but also varied and diverse biogenesis pathways and functional entities which are explored, from a microRNA perspective, in the second half.

2.2 Evolutionary Relationship Between Small Noncoding RNAs and Transposable Elements

Members of diverse classes of small RNAs have strong ties with transposable elements (TEs), generating both small RNAs employed by the cell to suppress TE expression and transposition but also small RNAs which have acquired new functionality and serve other and diverse cellular roles, as described in Sect. 2.2.2. Also known as repeat elements and "jumping genes," TEs have provided some small noncoding RNAs with a mechanism for expansion and acquisition of novel targets and functions, as explored in Sect. 2.2.3.

2.2.1 Transposable Elements

TEs are highly abundant genetic sequences which have the capacity of both moving and proliferating within and between genomes. It is estimated that between 30% and 50% of the sequence in mammalian genomes (45% in human), and even higher proportions in some plants, is derived from TEs, although most are currently inactive (reviewed in Cordaux and Batzer 2009; Mourier and Willerslev 2009; Tenaillon et al. 2010). As illustrated in Fig. 2.1, two main types of TEs have been

Class I (retrotransposons)	Class II (DNA transposons)
a. LTR LTR | gag | pol | LTR	d. Autonomous element TIR | transposase | TIR
b. SINE	e. Non-autonomous elements TIR | TIR
c. LINE 5'UTR | ORF1 | ORF2 | 3'UTR	TIR | TIR MITE

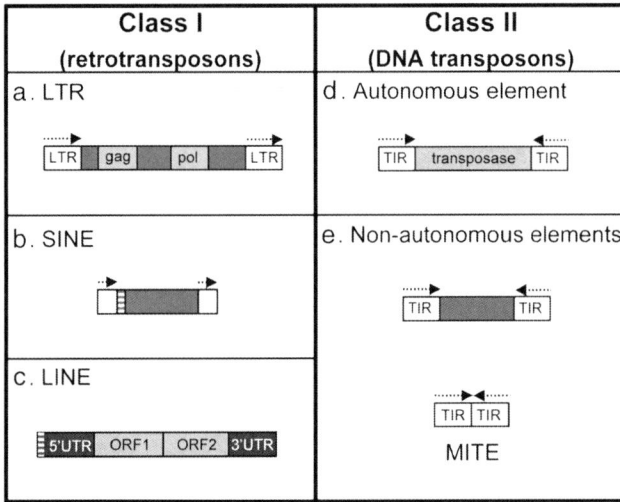

Fig. 2.1 *Classes and structures of TEs.* TEs can be classified into two groups based on their requirement of reverse transcription for transposition. LTRs (**a**) are flanked by direct repeats at their ends and encode proteins (gag and pol) which closely resemble retroviral proteins. SINEs (**b**) are flanked by direct repeats and encode an RNA polymerase III promoter, whereas LINEs (**c**) contain two open reading frames necessary for transposition (ORF1 and ORF2) which are flanked by untranslated regions (UTRs). Unlike LINEs, SINEs do not encode proteins and are believed to use the LINE retrotranscription machinery for reverse transcription. Class II elements do not utilize an RNA step and instead employ a transposase, which recognizes terminal inverted repeats (TIRs) for excision from the donor site and integration into an acceptor site. Autonomous DNA transposons (**d**) encode their own transposase, whereas nonautonomous DNA transposons (**e**) encode either a mutated version of the autonomous transposase gene, an unrelated portion of the host genome, or even a deleted version which consists simply of TIRs in a tail-to-tail orientation (MITE). *Arrows* represent direct or inverted repeats. Small striped boxes represent RNA polymerase promoters. *Light gray* boxes represent genes important for the transposition. Classes and characteristics of TEs are reviewed in Cordaux and Batzer (2009), Deininger and Batzer (2002), Richard et al. (2008), Slotkin and Martienssen (2007)

described, the retrotransposons and the DNA transposons, also referred to as class I and class II transposons respectively (reviewed in Cordaux and Batzer 2009; Slotkin and Martienssen 2007). DNA transposons can move to new genomic locations, either autonomously or nonautonomously, by excising themselves from their current location as a DNA molecule and inserting themselves elsewhere. In contrast, retrotransposons copy themselves using an RNA intermediate which is reversed transcribed and inserted back into the genome in a different location.

Three main subclasses of retrotransposons have been described (reviewed in Deininger and Batzer 2002; Richard et al. 2008; Cordaux and Batzer 2009):

– Long terminal repeats (LTRs)
– Long interspersed nuclear elements (LINEs)
– Nonautonomous retrotransposons including the abundant short interspersed nuclear elements (SINEs).

LTRs are very abundant in plant genomes. However, they have low activity in organisms like humans. In contrast, members of the LINEs and SINEs are believed to be currently active in humans (Lander et al. 2001; Mills et al. 2007). TEs range in length from less than one hundred to a few thousand nucleotides, and some TEs have been identified in very large copy numbers (e.g., over 1,000,000 copies of Alu elements, TEs of the SINE subclass of ~300 nucleotides in length, have been found in the human genome, representing approximately 10% of the genome) (Deininger and Batzer 2002; Richard et al. 2008; Cordaux and Batzer 2009). The very large copy number of different TEs in diverse organisms testifies to the great impact they had in shaping their host genomes.

While TEs can lead to genomic instability if inserted for example in functional genomic sequences such as protein coding or regulatory regions, they have also been found to cause the emergence of new regulatory features and genes, likely playing an important role in evolution and defining organism-specific characteristics (reviewed in Cordaux and Batzer (2009)). In addition to their role in the modification and duplication of protein-coding genes, TEs are emerging as drivers in the creation of novel noncoding genes in numerous organisms.

2.2.2   Association Between Small Noncoding RNAs and Transposable Elements

Various classes of small RNAs have been described as displaying an association with one or several types of TEs, as summarized in Table 2.1. Some small RNAs, such as germ line piRNAs (PIWI-interacting RNAs) and somatic endo-siRNAs (endogenous small interfering RNAs), derive from TEs and functionally interact with them, serving the purpose of suppressing TE expression and duplication (Saito and Siomi 2010; Ghildiyal et al. 2008; Siomi et al. 2008). In contrast, other small RNAs probably either evolved from TEs that subsequently acquired new functionality (a subset of miRNAs for example) or used TE transposition mechanisms for duplication from existing parental small RNA molecules such as has been described for some small nucleolar RNA (snoRNA) copies.

2.2.2.1   piRNAs and TEs

Present both in vertebrates and invertebrates, piRNAs were identified as small RNA interactors of PIWI proteins, a family originally characterized through genetic studies as playing a role in the maintenance of germ line integrity (Cox et al. 1998; Brennecke et al. 2007). piRNAs show a strong bias for uridine residues at their 5' end but no clear secondary structure features in flanking regions in the genome (O'Donnell and Boeke 2007). Most piRNAs map to a small number of position-conserved TE-rich clusters which express up to several thousand piRNAs

Table 2.1 Relationships between TEs and small RNAs

Small RNA class	Position in genome	Associated TEs	Relationship with TEs	References
piRNAs	TEs, repeats and piRNA clusters	Class I and class II	Transcribed from TEs, piRNAs serve in TE silencing	Kim et al. (2009), Saito and Siomi (2010), Siomi et al. (2008)
Endogenous siRNAs	TEs, repeats and endo-siRNA clusters	Class I and class II	Transcribed from TEs, endo-siRNAs serve in TE silencing	Ghildiyal et al. (2008), Kim et al. (2009), Saito and Siomi (2010), Siomi et al. (2008)
miRNAs	In introns of protein-coding and non-protein coding host genes. Others are encoded in intergenic transcription units	Class I and class II	TE expression and secondary structure appropriate for evolution into miRNA and generation of targets	Baskerville and Bartel (2005), Borchert et al. (2006), Kim et al. (2009), Lee et al. (2004), Rodriguez et al. (2004)
snoRNAs	Intergenic units or in introns of host genes. Can be clustered (frequent in plants) or individual (frequent in animals)	Class I (non-LTR retro-transposons)	Transposition machinery used for duplication	Dieci et al. (2009), Filipowicz and Pogacic (2002), Luo and Li (2007), Weber (2006)

displaying low sequence conservation (Malone and Hannon 2009; Brennecke et al. 2007). In both vertebrates and invertebrates, piRNAs have been detected almost uniquely in germ line cells, where they are believed to function in the silencing of TEs (Aravin et al. 2006; Brennecke et al. 2008; Das et al. 2008; Houwing et al. 2007). piRNAs are processed from single-stranded precursors derived from both sense and antisense TE transcripts (Brennecke et al. 2007; Saito et al. 2006). The biogenesis of piRNAs in flies and mammals has been proposed to involve primary and secondary processing in a mechanism referred to as the ping-pong cycle (illustrated in Fig. 2.2). This cycle involves primary piRNAs binding to their targets and recruiting PIWI family proteins, leading to target cleavage and TE transcript destruction and resulting in the production of secondary piRNAs, which can perpetuate the cycle (Brennecke et al. 2007; Gunawardane et al. 2007). In addition, piRNAs have been found to regulate the DNA methylation of TEs, thus also exerting epigenetic control over these elements (Brennecke et al. 2008; Kuramochi-Miyagawa et al. 2008).

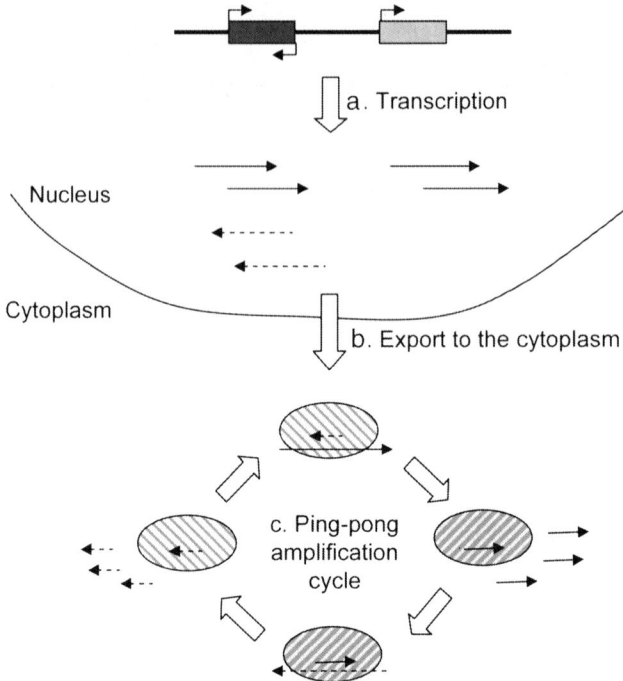

Fig. 2.2 *piRNA biogenesis and ping-pong amplification cycle.* (**a**) piRNA clusters (*dark gray boxes*) and functional TEs (*light gray boxes*) are transcribed in the nucleus. (**b**) Sense transcripts (*solid arrows*) and antisense transcripts (*broken arrows*) are exported to the cytoplasm. (**c**) Primary piRNAs complexed with PIWI family proteins (represented by lined ovals) bind to complementary TE and piRNA cluster transcripts leading to cleavage and amplification. These resulting short RNAs are bound by different PIWI family proteins and base pair to complementary antisense piRNA cluster transcripts resulting in antisense piRNAs, thus completing the proposed cycle. Block arrows represent progression to subsequent steps in the pathway. piRNA biogenesis is reviewed in Khurana and Theurkauf (2010), Kim et al. (2009)

2.2.2.2 Endogenous siRNAs and TEs

Originally observed during virus- and transgene-induced silencing in plants, canonical small interfering RNAs (siRNAs) originate from double-stranded RNAs (Carthew and Sontheimer 2009). More recently, endogenous siRNAs, including many derived from TEs, have been uncovered in a wide range of organisms including animals, plants, and fungi. Endogenous TE-derived siRNA precursors are believed to originate from diverse TEs including read-through transcription of DNA transposons as well as bidirectional transcription of LINE1 $5'$ UTR and are found located in both intronic and intergenic regions (Slotkin and Martienssen 2007; Sunkar et al. 2005). As for piRNAs, endogenous siRNAs are believed to function in TE silencing, but they have been predominantly identified in somatic tissues (Ghildiyal et al. 2008; Kawamura et al. 2008; Chung et al. 2008). However, some loci-producing piRNAs in flies have also been found to generate endogenous siRNAs (Kawamura et al. 2008). A cross talk between the piRNA and endogenous siRNA pathways has been described in worm (Das et al. 2008). The relationships between these two types of small RNAs and TEs are further described in Chap. 5.

2.2.2.3   MicroRNAs and TEs

Widely expressed in animals and plants, microRNAs (miRNAs) are ~22 nucleotide-long single-stranded RNAs that are processed out of hairpin precursors of variable length through an extensively characterized biogenesis pathway (see Sect. 2.3.1). Encoded in introns of protein-coding genes or independent transcription units, miRNAs are involved in gene silencing through the regulation of the stability and translation of target messenger RNAs (mRNAs), usually by base pairing with their 3′ UTR (untranslated region) or their coding region (Bartel 2009; Lai 2005).

Numerous studies have reported TE-derived miRNAs in several different organisms. As early as 2002, miRNA-like molecules were described encoded in TEs in *Arabidopsis* (Llave et al. 2002). Following that report, through sequence analyses and computational searches, hundreds of previously identified as well as novel mammalian miRNAs were described as derived from TEs (Borchert et al. 2006; Piriyapongsa et al. 2007; Smalheiser and Torvik 2005; Yuan et al. 2011). In mammals, a large proportion of miRNAs is found clustered in the genome, in regions highly enriched in TEs (Yuan et al. 2011). It has been proposed that the close proximity of TEs of similar sequence inserted in reverse orientations would result in structures resembling miRNA precursors which if expressed, might be recognized as substrates by the miRNA biogenesis pathway (Mourier and Willerslev 2009). Such a clustered TE region has been described on human chromosome 19 (referred to as C19MC) which encodes interspersed miRNAs and Alu elements. It was found that many of the Alu elements contain intact RNA polymerase III promoters which could ensure the expression of the miRNAs (Borchert et al. 2006). This suggests that TEs might represent not only a template from which small noncoding RNAs can evolve but also a mechanism to ensure their expression (Fig. 2.3).

The analysis of TE-derived miRNAs has revealed that miRNAs have evolved from all types of TEs described in the previous section. In human, while most frequently found associated with the L2 (from the LINE subclass) and MIR (from

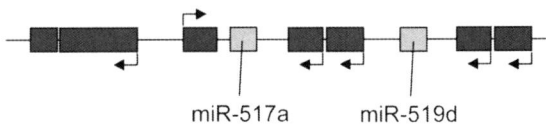

miR-517a   miR-519d

Fig. 2.3 *Three kilobase portion of the human C19MC cluster* (Borchert et al. 2006; Kent et al. 2002). *Light* and *dark gray boxes* respectively represent miRNAs and Alu elements. The direction of transcription is indicated with arrows

the SINE subclass) elements, miRNAs have also been described as derived from LTRs and DNA transposons such as DNA mariners as well as other SINE and LINE elements including the B2, Alu, and L1 elements (Mourier and Willerslev 2009; Piriyapongsa et al. 2007; Smalheiser and Torvik 2005). The proportion of miRNAs derived from different TE types varies depending on the organism (Yuan et al. 2011). A family of human miRNAs has also been described as originating from MITEs (see Fig. 2.1), leading to the formation of stable hairpins resembling miRNA precursors (Piriyapongsa and Jordan 2007; Slotkin and Martienssen 2007). TE-derived miRNAs have recently been found to be significantly less conserved within mammals than miRNAs not derived from TEs, likely due to relatively recent acquisition (Yuan et al. 2011). The evolution of miRNAs from TEs is depicted in Fig. 2.5.

2.2.2.4 snoRNAs and TEs

snoRNAs are an ancient family of highly conserved and abundant small noncoding RNAs that predominantly function as guides for the chemical modification of ribosomal RNA (rRNA) (reviewed in Matera et al. 2007). Like miRNAs, snoRNAs are either encoded in introns of protein-coding genes or in independent transcription units. Two main types of snoRNAs have been described, the box C/D snoRNAs and the box H/ACA snoRNAs, which differ in terms of the chemical modification they catalyze. While several hundred snoRNAs, most of them highly conserved, have been described in mammalian organisms, computational searches have identified hundreds and even thousands of additional snoRNAs displaying TE characteristics (Weber 2006; Luo and Li 2007; Schmitz et al. 2008), described in Fig. 2.4. These TE-derived snoRNAs, referred to as snoRTs (snoRNA retroposons), result from the retroposition of existing (parental) snoRNA transcripts which employed LINE machinery to duplicate and transpose themselves to new genomic locations (Weber 2006).

While many computationally identified snoRTs have not been found previously and have not been experimentally validated, others have previously been described as functional snoRNAs (Luo and Li 2007).

2.2.3   A Driving Force of Evolution

TEs have been described as the most nonconserved regions and represent the most lineage-specific elements in eukaryotic genomes (Lander et al. 2001). Their recent contribution to animal and plant genomes and their high abundance suggest that transposition is a common occurrence that is likely a strong driving force in evolution, providing a mechanism for the emergence of organism-specific regulatory elements. Recent studies have revealed that transposition likely leads to both the creation of small RNAs with new functionality as well as new targets for small RNAs as depicted in Fig. 2.5.

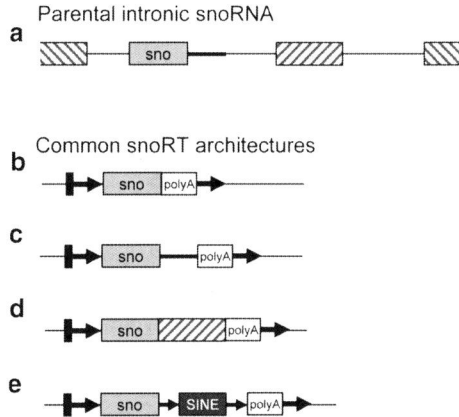

Fig. 2.4 *Common snoRT genomic architectures.* In all panels, snoRNAs and exons are represented respectively by *light gray boxes* and diagonally lined boxes, while the intronic region immediately flanking the parental snoRNA is depicted by a thick black line. snoRTs often display LINE characteristics including upstream L1 consensus recognition site (shown with a thin black box), poly A tails at their 3′ end, and flanking direct repeats referred to as target-site duplications (TSDs, depicted by black arrows). (**a**) In mammals, parental snoRNAs are typically encoded in intronic regions of host genes. (**b**) The simplest form of snoRTs consists of a snoRNA followed by a poly A tail at its 3′ end and flanked by TSDs. (**c**) Part of the intronic region downstream of the parental snoRNA can also be retroposed (represented by the thick black line). The intronic region flanking the 5′ end of the parental snoRNA can also be retroposed (not shown here). (**d**) Exonic regions from the parental snoRNA host gene have been identified in some snoRTs. (**e**) Repeat elements of the SINE class, flanked by TSDs, are found downstream of the snoRNA sequence in some snoRTs. Examples of snoRTs are described in Luo and Li (2007), Weber (2006)

2.2.3.1 Creation of New Small RNAs

The integration of TEs in new genomic locations can lead to the creation of new small RNAs provided the TE is expressed and its transcript displays an appropriate structure that is recognized as a substrate in a small RNA biogenesis pathway. As described in the previous subsection, diverse small RNAs have been found to be TE-derived, generally following insertion into intergenic or intronic regions (illustrated in Fig. 2.5a, b). Some such TE-derived miRNAs display a high level of conservation, but many are lineage specific (Piriyapongsa et al. 2007; Smalheiser and Torvik 2005). Studies of the pattern of occurrence of TE-derived miRNA members of specific families throughout multiple organisms have led to the hypothesis that these families follow the birth-and-death model of evolution. As opposed to concerted evolution in which members of a family evolve in a similar, concerted way, in the birth-and-death model, new members of the family created by duplication either remain in the genome over long periods or are deleted or inactivated (Nei and Rooney 2005). Several miRNA families display gains and losses of members when closely related organisms are considered, suggesting a birth-and-death evolutionary process, functional diversification of the family, and a role in evolution and lineage-specific traits (Zhang et al. 2008; Yuan et al. 2010).

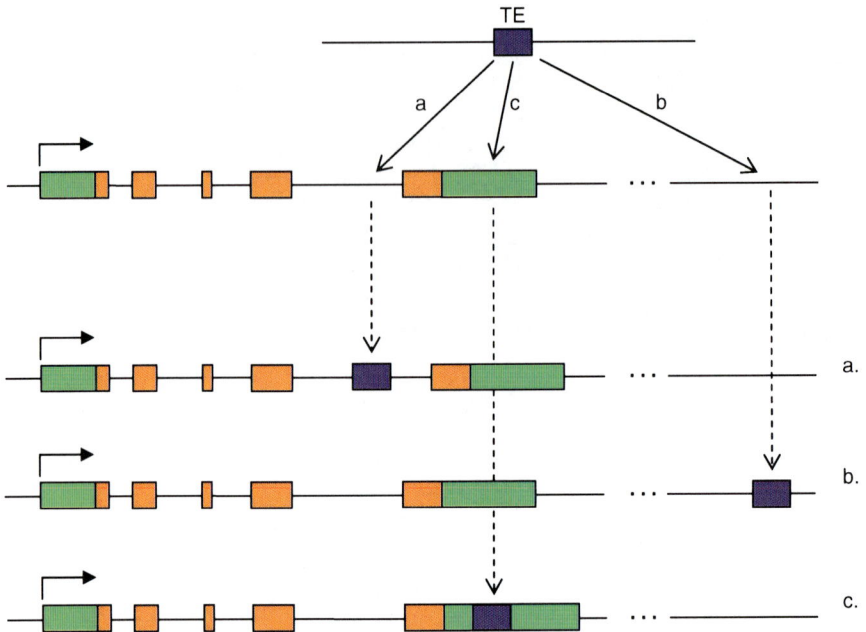

Fig. 2.5 *Transposition events generating new small RNAs or new small RNA targets.* The insertion of TEs can result in the emergence of novel features including intronic TE-derived small RNAs (**a**), intergenic TE-derived small RNAs (**b**), and TE-derived targets of small RNAs if inserted in reverse orientation into 3′ untranslated regions (UTRs) (**c**), or coding regions (not shown), for example. TEs are depicted as purple boxes, UTRs as green boxes, and coding exons as orange boxes. Straight solid arrows represent transposition events and broken arrows show the genomic region depicted after the transposition

In addition to providing templates for the creation of new small RNAs, TEs have also been suggested to have allowed the expansion of specific small RNA families. The C19MC region of human chromosome 19 described in Sect. 2.2.2.3 and Fig. 2.3 consists of a cluster of alternating Alu elements and miRNAs of a primate-specific family. Alu elements from this cluster are proposed to have not only led to the creation of novel miRNAs through the reverse orientation of adjacent elements but are also believed to have facilitated the amplification of the region through a recombination event, leading to the generation of many additional copies of these miRNAs and the rapid expansion of this family. This large group of primate-specific miRNAs was found to be evolving rapidly, resulting in some nonfunctional miRNAs (pseudo-miRNAs) but also novel lineage-specific miRNAs (Zhang et al. 2008). Other families of miRNAs are also believed to have been amplified by Alu-mediated recombination events including placental-specific miRNAs derived from the MER53 DNA transposon (Yuan et al. 2010).

The duplication of functional small RNAs likely results in molecules that are under less evolutionary pressure to avoid mutation than a single-copy small RNA. In the case of snoRNAs, the numerous snoRTs originating from snoRNAs have been proposed to serve two roles: safeguarding against mutation in parental copies and possibly also allowing for the rapid evolution of snoRNAs with novel targets (Weber 2006). In support of this hypothesis, numerous "orphan" snoRNAs and snoRTs have been identified, displaying typical snoRNA characteristics and features but with no known targets, possibly due to sequence diversification. Some such orphan snoRNAs are ubiquitously expressed in mammals suggesting they might have alternate functions and/or noncanonical targets (Bachellerie et al. 2002; Luo and Li 2007). And indeed, one large mammalian family of snoRNAs, the HBII-52 in human, displays a conserved region of complementarity to several transcripts including the serotonin 5-HT2C receptor and has been found to regulate their alternative splicing through base pairing (Bachellerie et al. 2002; Kishore and Stamm 2006).

2.2.3.2 Creation of New Targets

Small RNAs generally exert their function by base pairing with their targets and bringing them in close proximity to effector proteins. TEs have been found to play an important role in generating not only novel small regulator RNAs but also novel targets for these molecules (Fig. 2.5c).

The large abundance of TEs and transposition events in genomes can likely rapidly generate regulatory networks if TEs with high sequence similarity to a TE-derived small RNA are inserted into protein-coding genes in reverse orientation, readily creating target sites for the small RNA. In human, many miRNAs display complementarity to conserved Alu elements in 3′ UTRs of mRNAs (Smalheiser and Torvik 2006). Amongst all 3′ UTR targets of human TE-derived miRNAs, approximately 10% were estimated to be TE-derived (Piriyapongsa et al. 2007). As discussed in the study, this is likely to be an underestimate because miRNA target-site prediction methods consider conservation in their prediction, and TEs are among the most lineage-specific elements of genomes. However, some TE-derived miRNAs were found to have up to 80% of their targets derived from TEs (Piriyapongsa et al. 2007). Other types of TEs have also been found to generate both miRNAs and their targets including LINE elements for which an example of conserved mammalian miRNA and its human-specific target mRNAs were described (Smalheiser and Torvik 2005).

2.2.3.3 A Prevalent Evolutionary Mechanism

Although often described as harmful and selfish elements in genomes, TEs have also been found to be beneficial to genomes, and mutualistic relationships between genomes and their colonizing TEs have been described (Faulkner and Carninci 2009; Cordaux and Batzer 2009; Malone and Hannon 2009). Genomes have evolved mechanisms to control and limit TE expansion, in great part through the use of small RNAs such as members of piRNAs and siRNAs (Malone and Hannon

2009) as described in Sect. 2.2.2. However, a diverse and growing body of evidence is now suggesting that transposition and TE activity are also a prevalent evolutionary mechanism in the de novo creation of small RNAs as well as in their expansion and generation of targets, providing additional regulatory layers in the control of cellular networks. As such, TE-derived miRNAs and snoRNAs represent a beneficial consequence of TE activity and expansion. As our understanding of the contribution of TEs to genome function and evolution increases, so too will our understanding of their contribution to small RNA evolution and organism-specific regulation.

2.3 Diversity and Overlap in the Small RNA Landscape, a miRNA Perspective

The best characterized small RNA biogenesis pathway is the miRNA biogenesis pathway which has been extensively investigated. In recent years however, numerous unrelated reports have identified miRNAs that deviate from the canonical pathway in terms of their precursors, biogenesis pathway, characteristics of the mature molecule, or in their functionality. Together with the contribution of TEs to miRNA and target diversity, variations from the canonical biogenesis pathway and diverse characteristics of the mature molecules add flexibility to miRNA regulatory networks.

2.3.1 Variation and Overlap in miRNA Biogenesis Pathways

In animals, most miRNAs are encoded in introns or within independent transcription units and are transcribed by the RNA polymerase II (RNA pol II) (Baskerville and Bartel 2005; Lee et al. 2004; Rodriguez et al. 2004). The resulting primary miRNA transcripts are then cleaved by the microprocessor complex which contains the nuclear RNase type III enzyme Drosha and the double-stranded RNA binding domain protein DGCR8 (Kim and Kim 2007; Lee et al. 2003). Processing of the primary miRNA transcript by the microprocessor complex generates the miRNA precursor hairpin, which is exported to the cytoplasm by exportin-5 where it is further processed by the RNase type III enzyme Dicer (Bernstein et al. 2001; Hutvagner et al. 2001; Ketting et al. 2001). Dicer releases the miRNA duplex, one strand of which is loaded onto specific argonaute (AGO) proteins forming the RNA-induced silencing complex (RISC) (Hutvagner and Zamore 2002; Mourelatos et al. 2002). The canonical miRNA biogenesis is reviewed in Bartel (2004) and Kim et al. (2009) and depicted in Fig. 2.6a. Although the majority of miRNAs in mammals have been found to be both DGCR8 and Dicer dependent (Babiarz et al. 2008), numerous examples of deviation from this canonical pathway have

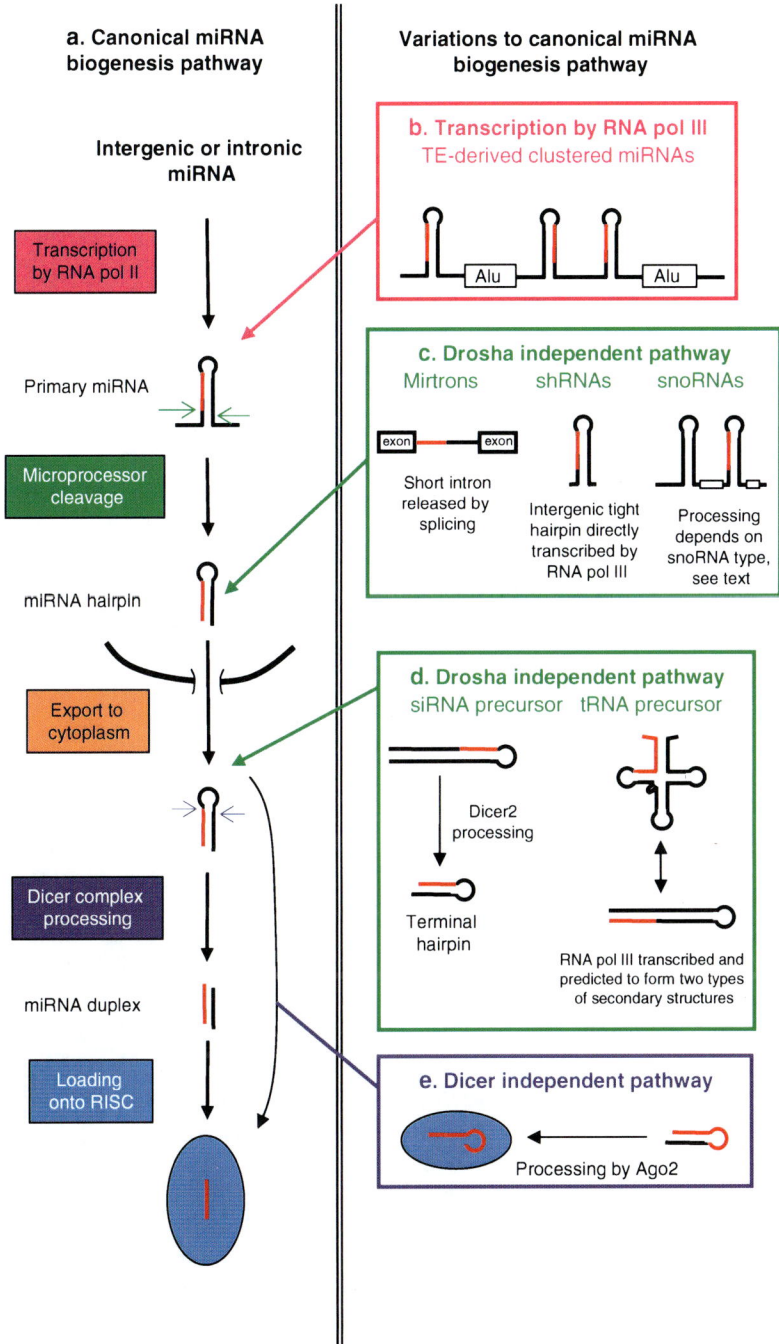

Fig. 2.6 *Diversity and variation in the miRNA biogenesis pathway.* The canonical miRNA biogenesis pathway (**a**) has been extensively investigated. Variations from the canonical pathway have been identified at most steps including at the level of transcription (**b**), Drosha processing (**c** and **d**), and Dicer processing (**e**)

been described, and numerous miRNAs have been shown to originate from diverse noncanonical precursors (illustrated in Fig. 2.6b–e).

2.3.1.1 Transcription of miRNA Genes by the RNA pol III

As described above, most animal miRNAs, both intronic and intergenic, are transcribed by the RNA pol II. However, the human chromosome 19 cluster (C19MC) of interspersed miRNAs and Alu elements described in Sect. 2.2.2.3 contains RNA pol III promoters, and miRNAs encoded within this region were found to be transcribed by the RNA pol III (Borchert et al. 2006), see Fig. 2.6b. Other TE-derived human miRNAs in similar genomic contexts are also likely to be transcribed by the RNA pol III (Borchert et al. 2006). In addition, viral miRNAs adjacent to a transfer RNA were also recently found to be transcribed by the RNA pol III as described further below (Bogerd et al. 2010). And shRNA-derived miRNAs were also found to be transcribed by the RNA pol III as described below (Babiarz et al. 2008). Thus, although a large majority of animal miRNAs are transcribed by the RNA pol II, other pathways are used for the generation of primary miRNA transcripts and are likely under the control of alternate regulatory networks.

2.3.1.2 Microprocessor-Independent miRNA Biogenesis

Subsets of noncanonical miRNA precursors have been found to bypass the microprocessor cleavage step. Included among them are mirtrons, small hairpin RNAs (shRNAs), and long hairpin siRNA precursors.

Mirtrons

Canonical intronic miRNAs are processed by Drosha prior to the host transcript intron splicing (Kim and Kim 2007). However, a group of short introns of size <150 nucleotides which fold into hairpins has been found to serve as precursors for miRNAs. Referred to as mirtrons, these precursors result from splicing and debranching of the intron lariat by the spliceosome, thus bypassing the Drosha cleavage step (reviewed in Winter et al. 2009). The resulting hairpin precursor is then exported to the cytoplasm and further processed following the canonical miRNA biogenesis pathway (Fig. 2.6c). Although not found in large numbers, mirtrons have been identified throughout the animal kingdom (Babiarz et al. 2008; Carthew and Sontheimer 2009; Winter et al. 2009).

Endogenous shRNA Precursors

A second type of miRNA precursors that was found to exhibit independence from
Drosha is endogenous shRNAs, which form short tight stem loop structures
(Fig. 2.6c). A subset of mammalian miRNAs was annotated as derived from
shRNAs for three main reasons: they were found to be Dicer dependent, they are
not encoded in introns and do not display splicing signals, and they show characteristic read position patterns within the full-length molecule. Such shRNA-derived
miRNA precursors seem to be generated directly as short hairpins by RNA pol
III transcription (Babiarz et al. 2008). In mammals, shRNA-derived miRNAs
make up a much larger number of reads than mirtron-derived miRNAs (Babiarz
et al. 2008).

Long Hairpin siRNA Precursors

In *Drosophila*, miRNAs are typically processed by Dicer1, while endogenous
siRNAs are processed by Dicer2 (reviewed in Miyoshi et al. 2010). However, it
is believed that the terminal hairpins resulting from processing by Dicer2 of a
subset of endogenous siRNAs can serve as miRNA precursors (Fig. 2.6d) which are
recognized as substrates by Dicer1, producing a mature miRNA. Thus these siRNA
precursors would generate both siRNAs and miRNAs as a result of sequential
Dicer2 and Dicer1 processing (reviewed in Miyoshi et al. 2010).

tRNase Z-Derived Precursors

Several animal viruses encode viral miRNAs which display miRNA biogenesis
features characteristic of canonical cellular miRNAs including transcription by the
RNA polymerase II and processing by Drosha and Dicer. However, the murine
gamma-herpesvirus 68 (MHV68) has recently been shown to encode miRNAs
which employ noncanonical biogenesis pathways, including transcription by the
RNA polymerase III of the miRNA precursors linked to an adjacent transfer RNA
(tRNA). The resulting primary transcripts are cleaved by the tRNase Z, releasing
the tRNA from the miRNA precursors and bypassing the Drosha cleavage step. The
miRNA precursors are then further processed by Dicer to generate the mature
miRNA (Bogerd et al. 2010).

2.3.1.3 Dicer-Independent miRNA Biogenesis

A small group of miRNAs processed in a Dicer-independent manner has
been recently described in both mouse and zebra fish (reviewed in Suzuki and
Miyazono 2011). Drosha has been shown to process the primary transcript of the
well-conserved miR-451 miRNA resulting in a short hairpin precursor with a

17 nucleotide stem, which is shorter than the length required by Dicer for efficient processing. In the absence of Drosha, the mature form of the miRNA was highly reduced while its levels were not affected by the absence of Dicer (Cheloufi et al. 2010). As illustrated in Fig. 2.6e, following cleavage by Drosha, this noncanonical miRNA precursor is believed to be loaded directly onto the RISC complex where it is sliced by Ago2, thus bypassing Dicer processing (Suzuki and Miyazono 2011; Cheloufi et al. 2010). This suggests Dicer processing and RISC loading might not always be coupled (Babiarz et al. 2008; Miyoshi et al. 2010).

Thus over the past 3 years, diverse examples of alternate miRNA biogenesis pathways, typically with partial overlap with the canonical pathway, have been described, suggesting both that deviations from the canonical pathway are relatively common and that more such deviations will be found over the next few years. These examples demonstrate that diverse variations of the miRNA processing pathway exist and that no component of the canonical miRNA biogenesis pathway is essential for the biogenesis of all miRNAs. It should be noted however that most miRNAs are transcribed by the RNA pol II and require both the microprocessor and Dicer complexes for proper biogenesis (Babiarz et al. 2008).

2.3.1.4   Additional Noncanonical miRNA Precursors

The examples described above of variations to the canonical miRNA biogenesis pathway generally relate to miRNA precursors whose main function is the generation of miRNAs. However, subsets of two types of abundant cellular RNAs, small nucleolar RNAs (snoRNAs), and transfer (tRNAs), which play seemingly unrelated primary functions in the cell, have also been found to act as noncanonical miRNA precursors. Some long noncoding RNAs have also been found to serve as precursors to miRNAs.

Small Nucleolar RNAs (snoRNAs)

Several independent studies and lines of evidence suggest that snoRNAs can serve as precursors for miRNAs. Small RNAs derived from snoRNAs were identified in a deep sequencing dataset of small human RNAs associated with argonaute proteins (Ender et al. 2008). In particular, small RNAs of miRNA size were found derived from the box H/ACA snoRNA ACA45 and shown to display functional miRNA characteristics including gene silencing capabilities and endogenous targets. The processing of ACA45 was shown to depend on Dicer but not Drosha (Ender et al. 2008). Box C/D snoRNA–derived small RNAs were also identified and found to display miRNA characteristics including incorporation into RISC complexes and gene silencing capabilities in human and *Giardia lamblia* (Saraiya and Wang 2008; Brameier et al. 2011).

The processing of snoRNAs into small RNAs of size generally less than 30 nucleotides (referred to as sdRNAs for snoRNA-derived small RNAs) has recently been shown to be widespread and conserved from most snoRNA loci in animal, *Arabidopsis*, and yeast genomes (Taft et al. 2009). The processing pathways responsible for the generation of sdRNAs were investigated, revealing that a subset of sdRNAs (those derived from box C/D snoRNAs) were only mildly downregulated in the absence of either DGCR8 or Dicer1, while some box H/ACA sdRNAs showed a pronounced response, displaying a downregulation in the absence of Dicer1 but an upregulation in the absence of DGCR8 (Taft et al. 2009). Thus, multiple processing pathways might be used for the generation of miRNAs and other small RNAs from snoRNAs (Fig. 2.6c).

While numerous examples of snoRNA-derived small RNAs and snoRNA-derived miRNAs have now been described, several reports have also identified known miRNA precursors with snoRNA-like features. Numerous reported, and in several cases extensively validated, miRNA precursors have been identified displaying sequence, structure, and functional snoRNA characteristics, suggesting a possible evolutionary relationship between subsets of miRNAs and snoRNAs (Ono et al. 2011; Scott et al. 2009). In addition, evidence of cross talk between the miRNA and snoRNA pathways has also been observed including core snoRNA-binding proteins in argonaute complexes (Hock et al. 2007).

Transfer RNAs

Several independent studies have recently reported small RNAs derived from transfer RNAs (tRNAs) and displaying miRNA processing characteristics (reviewed in (Pederson 2010; Suzuki and Miyazono 2011)). In mouse, miRNAs were identified originating from a tRNA-Ile gene which encodes a primary transcript with the capacity to form not only a mature tRNA cloverleaf secondary structure but also, alternatively, a long hairpin, as illustrated in Fig. 2.6d. The miRNAs originating from this precursor were found to be Drosha independent and Dicer dependent (Babiarz et al. 2008). In human, tRNA-derived small RNAs appear to be generated with a clear preference for 5′ ends of the full-length molecule, indicating directed processing and accumulation as opposed to nonspecific degradation (Cole et al. 2009). Several studies found evidence of miRNA processing and binding characteristics for the tRNA-derived small RNAs such as processing by Dicer and binding to argonaute proteins (reviewed and discussed in Pederson 2010).

Long Noncoding RNAs

Long noncoding RNAs are generally defined as transcripts of size greater than 200 nucleotides that do not encode proteins. The imprinted and maternally expressed H19 long noncoding RNA was recently found to encode a previously reported miRNA in both mouse and human, providing another example of a noncanonical miRNA precursor. The primary transcript was found to be processed by Drosha

(Cai and Cullen 2007), but further investigation will be required to fully characterize its processing. In addition to this example, a computational analysis of tens of
thousands of long messenger-like noncoding RNAs in mouse predicted they encode
dozens of likely miRNA candidates, including 20 previously reported miRNAs
(He et al. 2008). miRNA-encoding long noncoding RNAs were also described in
Arabidopsis (Hirsch et al. 2006).

Thus, several types of noncanonical miRNA biogenesis pathways and precursor types exist and can lead to functional mature miRNAs. These diverse
sources of miRNAs likely offer the possibility of variety and flexibility in their
regulation.

2.3.2 Diversity of Targets and Targeting Mechanisms

Processing of miRNA precursor hairpins by Dicer generates short miRNA duplexes
of approximately 22 base pairs. One strand preferentially associates with an
argonaute protein, forming the core of the RISC complex (illustrated in Figs. 2.6a
and 2.7a). Other small RNAs such as endogenous siRNAs are also loaded onto
argonaute proteins (reviewed in Kim et al. 2009). The argonaute family member to
which small RNAs are complexed depends on several factors including the cytoplasmic Dicer that processed the small RNA, as well as the extent of complementarity of the double-stranded precursor. The argonaute interactor plays an important
role in determining the functionality of the complex. Argonaute loading and sorting
is reviewed in Czech and Hannon (2011).

The RISC complex carries out gene silencing by base pairing to its messenger
RNA (mRNA) targets. High complementarity of a miRNA to its target, which is
typically seen in plants, results in mRNA cleavage, while lower complementarity,
more characteristic of animals, promotes deadenylation of the mRNA followed by
either maintenance in a translation-repressed state or decapping and 5'–3' decay
(reviewed in Bartel 2004; Kim et al. 2009; Huntzinger and Izaurralde 2011). The
target sequences of animal miRNAs typically lie in the 3' untranslated regions (3'
UTRs) of mRNAs while in plants, they are predominantly found in coding regions
(Carthew and Sontheimer 2009). However, mature miRNAs displaying atypical
characteristics and functionality have been reported recently.

2.3.2.1 5' UTR Binding and Translation Activation

An affinity-based target-identification method revealed that a highly conserved
murine miRNA, miR-10a, preferentially binds to ribosomal protein transcripts.
Further investigations showed that regions in the 5' UTRs of these transcripts,
and not in their 3' UTRs, are directly bound by the miRNA and that rather than
translation repression, this binding leads to translation activation, thus resulting in
stimulation of global protein synthesis (Orom et al. 2008). Similarly, a human

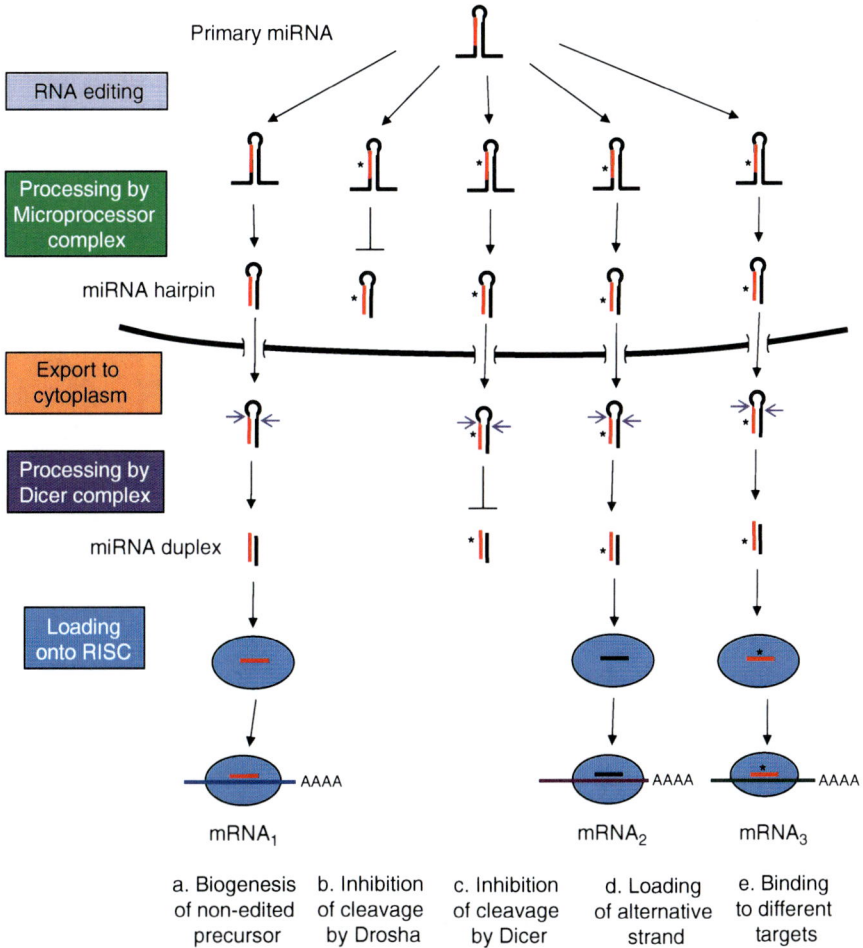

Fig. 2.7 *Diversity generated by RNA editing.* Primary miRNA transcripts can be modified by ADARs resulting in variation in the miRNA sequence (marked with a star). Such modifications can lead to diversity in the biogenesis pathway and in miRNA functionality

miRNA was found to bind to 5′ UTRs in the hepatitis C virus genome resulting in translation activation of viral proteins (Henke et al. 2008). Several miRNAs have also been shown to bind targets in coding sequences in both mouse and human (Chi et al. 2009; Duursma et al. 2008; Forman et al. 2008; Tay et al. 2008).

In addition to variation in the position of targets within transcripts, a small number of human miRNAs have been also found to activate translation of their targets in a cell cycle-dependent manner (Vasudevan et al. 2007). Thus, a small number of miRNAs can cause the upregulation of translation rather than its repression.

2.3.2.2 Sequence Editing

RNA editing is the directed modification of specific positions in RNA transcripts resulting in molecules that differ from the template DNA. RNA editing is thus an important mechanism driving diversity in the small RNA landscape. Estimates of the proportion of human miRNA transcripts altered by RNA editing range from 6% to 16% (Blow et al. 2006; Kawahara et al. 2008). RNA editing of primary miRNA transcripts is carried out by ADARs (adenosine deaminases acting on RNA) and leads to the modification of adenosine (A) residues into inosines (I) (reviewed in Cai et al. 2009; Wulff and Nishikura 2010). RNA editing of miRNA transcripts can affect their biogenesis by blocking processing by Drosha or Dicer and thus plays a regulatory role in the biogenesis of miRNAs (Fig. 2.7b, c). RNA editing has also been proposed to cause a change in the miRNA duplex strand most predominantly chosen for incorporation in the RISC complex (Wulff and Nishikura 2010), see Fig. 2.7d. RNA editing can also lead to the binding of alternative targets when compared to the nonedited miRNA, thus extending the repertoire of targets that are regulated by a single miRNA locus (reviewed in Cai et al. 2009 and illustrated in Fig. 2.7e).

2.3.2.3 Nonprototypical miRNA Characteristics and Functions

Nuclear miRNAs

As described previously, the final steps in processing of miRNA precursors and release of mature miRNAs take place in the cytoplasm. However, a large number of human mature miRNAs have been detected in both the cytoplasm and nucleus, and some accumulate predominantly in the nucleus (Liao et al. 2010; Winter et al. 2009). In addition, many mammalian miRNA precursors as well as mature miRNAs have been detected in the nucleolus, some accumulating strongly in this compartment (Politz et al. 2009; Ono et al. 2011; Scott et al. 2009). The nuclear/nucleolar localization mechanism of miRNAs and their function in these compartments are currently not known. A subset of miRNAs might be processed by a noncanonical pathway and/or from noncanonical precursors (including snoRNAs, see Sect. 2.3.1.4) in the nucleus. Alternatively, they might localize in this compartment after processing in the cytoplasm for further modification including RNA editing or proper packaging into ribonucleoprotein complexes (RNPs) as discussed in Politz et al. (2009). However, evidence is also mounting that small RNAs are involved in a number of functions not directly related to translation regulation.

Functions Outside Posttranscriptional Regulation

Small RNAs identify their targets by base pairing to complementary sequences and wield their functionality by bringing effector proteins in close proximity to the targets. Small RNAs thus have access to the whole spectrum of RNA molecules present in the cell to exert their regulatory roles and are not limited to the small range of specific targets that have been thus far extensively characterized. In recent years, small RNAs have been found to regulate numerous other cellular processes from chromatin structure to transcription and RNA processing (reviewed in Costa 2010; Carthew and Sontheimer 2009; Taft et al. 2010). Newly identified classes of small noncoding RNAs include promoter-associated short RNAs, transcription initiation RNAs (tiRNAs), and transcription-start-site-associated RNAs which regulate diverse aspects of the transcription of protein-coding genes or help to maintain these genes in an active state (Costa 2010). Different classes of small noncoding RNAs have also been proposed to play a role in the regulation of alternative splicing, X chromosome inactivation, and possibly the maintenance of telomeres (reviewed in Khanna and Stamm 2010; Taft et al. 2010). Some of these small RNAs with newly described functions derive from other well-characterized RNA molecules while others represent novel types. For example, small RNAs derived from specific snoRNAs have been found to regulate splicing of several transcripts (Kishore and Stamm 2006) while tiRNAs, derived from regions adjacent to transcription start sites in metazoans, have not been shown to produce other types of previously characterized small RNAs (Taft et al. 2009a).

2.4 Conclusions and Outlook

Rapidly evolving, abundant and highly diverse, small RNAs play fundamental cellular regulatory roles, the extent of which we are only now starting to grasp. Though only a small number of types have been extensively characterized, it is becoming clear that they are dynamic molecules, often displaying a strong association with transposable elements. Different types of small RNAs show commonalities in their origin and biogenesis pathways, making it often difficult to classify them in a clear-cut manner. In addition, a wide diversity and variation exist in small RNA biogenesis pathways, even when considering only one type of small RNAs. And though the canonical miRNA pathway has been extensively described, no component has been found essential for the biogenesis of all miRNAs. The diversity that exists in small RNA biogenesis pathways likely allows for differential regulation for molecules within the same class and in different cell types.

Small RNAs are central regulators in a steadily increasing number of fundamental cellular processes. They regulate a diverse and growing number of types of targets and are responsible for multiple layers in cellular regulatory networks. Understanding their functions and how they are themselves regulated will thus be central to a global comprehension of cellular networks and the numerous diseases caused by their deregulation.

Acknowledgments MSS is the recipient of a Caledonian Research Foundation Fellowship.

References

Aravin A, Gaidatzis D, Pfeffer S, Lagos-Quintana M, Landgraf P, Iovino N, Morris P, Brownstein MJ, Kuramochi-Miyagawa S, Nakano T, Chien M, Russo JJ, Ju J, Sheridan R, Sander C, Zavolan M, Tuschl T (2006) A novel class of small RNAs bind to MILI protein in mouse testes. Nature 442(7099):203–207. doi:nature04916 [pii] 10.1038/nature04916

Babiarz JE, Ruby JG, Wang Y, Bartel DP, Blelloch R (2008) Mouse ES cells express endogenous shRNAs, siRNAs, and other microprocessor-independent, dicer-dependent small RNAs. Genes Dev 22(20):2773–2785. doi:22/20/2773 [pii] 10.1101/gad.1705308

Bachellerie JP, Cavaille J, Huttenhofer A (2002) The expanding snoRNA world. Biochimie 84 (8):775–790

Bartel DP (2004) MicroRNAs: genomics, biogenesis, mechanism, and function. Cell 116 (2):281–297. doi:S0092867404000455 [pii]

Bartel DP (2009) MicroRNAs: target recognition and regulatory functions. Cell 136(2):215–233. doi:S0092-8674(09), 00008-7 [pii] 10.1016/j.cell.2009.01.002

Baskerville S, Bartel DP (2005) Microarray profiling of microRNAs reveals frequent coexpression with neighboring miRNAs and host genes. RNA 11(3):241–247

Bernstein E, Caudy AA, Hammond SM, Hannon GJ (2001) Role for a bidentate ribonuclease in the initiation step of RNA interference. Nature 409(6818):363–366. doi:10.1038/35053110

Blow MJ, Grocock RJ, van Dongen S, Enright AJ, Dicks E, Futreal PA, Wooster R, Stratton MR (2006) RNA editing of human microRNAs. Genom Biol 7(4):R27. doi:gb-2006-7-4-r27 [pii] 10.1186/gb-2006-7-4-r27

Bogerd HP, Karnowski HW, Cai X, Shin J, Pohlers M, Cullen BR (2010) A mammalian herpesvirus uses noncanonical expression and processing mechanisms to generate viral microRNAs. Mol Cell 37(1):135–142. doi:S1097-2765(09), 00920-4 [pii] 10.1016/j.molcel.2009.12.016

Borchert GM, Lanier W, Davidson BL (2006) RNA polymerase III transcribes human microRNAs. Nat Struct Mol Biol 13(12):1097–1101

Brameier M, Herwig A, Reinhardt R, Walter L, Gruber J (2011) Human box C/D snoRNAs with miRNA like functions: expanding the range of regulatory RNAs. Nucleic Acids Res 39 (2):675–686. doi:gkq776 [pii] 10.1093/nar/gkq776

Brennecke J, Aravin AA, Stark A, Dus M, Kellis M, Sachidanandam R, Hannon GJ (2007) Discrete small RNA-generating loci as master regulators of transposon activity in Drosophila. Cell 128(6):1089–1103. doi:S0092-8674(07), 00257-7 [pii] 10.1016/j.cell.2007.01.043

Brennecke J, Malone CD, Aravin AA, Sachidanandam R, Stark A, Hannon GJ (2008) An epigenetic role for maternally inherited piRNAs in transposon silencing. Science 322 (5906):1387–1392. doi:322/5906/1387 [pii] 10.1126/science.1165171

Cai X, Cullen BR (2007) The imprinted H19 noncoding RNA is a primary microRNA precursor. RNA 13(3):313–316. doi:rna.351707 [pii] 10.1261/rna.351707

Cai Y, Yu X, Hu S, Yu J (2009) A brief review on the mechanisms of miRNA regulation. Genom Proteom Bioinf 7(4):147–154. doi:S1672-0229(08), 60044-3 [pii] 10.1016/S1672-0229(08) 60044-3

Carthew RW, Sontheimer EJ (2009) Origins and mechanisms of miRNAs and siRNAs. Cell 136 (4):642–655. doi:S0092-8674(09)00083-X [pii] 10.1016/j.cell.2009.01.035

Cheloufi S, Dos Santos CO, Chong MM, Hannon GJ (2010) A Dicer-independent miRNA biogenesis pathway that requires Ago catalysis. Nature 465(7298):584–589. doi:nature09092 [pii] 10.1038/nature09092

Chi SW, Zang JB, Mele A, Darnell RB (2009) Argonaute HITS-CLIP decodes microRNA-mRNA interaction maps. Nature 460(7254):479–486. doi:nature08170 [pii] 10.1038/nature08170

Chung WJ, Okamura K, Martin R, Lai EC (2008) Endogenous RNA interference provides a somatic defense against Drosophila transposons. Curr Biol 18(11):795–802. doi:S0960-9822 (08), 00603-9 [pii] 10.1016/j.cub.2008.05.006

Cole C, Sobala A, Lu C, Thatcher SR, Bowman A, Brown JW, Green PJ, Barton GJ, Hutvagner G (2009) Filtering of deep sequencing data reveals the existence of abundant Dicer-dependent small RNAs derived from tRNAs. RNA 15(12):2147–2160. doi:rna.1738409 [pii] 10.1261/rna.1738409

Cordaux R, Batzer MA (2009) The impact of retrotransposons on human genome evolution. Nat Rev Genet 10(10):691–703. doi:nrg2640 [pii] 10.1038/nrg2640

Costa FF (2010) Non-coding RNAs: meet thy masters. Bioessays 32(7):599–608. doi:10.1002/bies.200900112

Cox DN, Chao A, Baker J, Chang L, Qiao D, Lin H (1998) A novel class of evolutionarily conserved genes defined by piwi are essential for stem cell self-renewal. Genes Dev 12 (23):3715–3727

Czech B, Hannon GJ (2011) Small RNA sorting: matchmaking for Argonautes. Nat Rev Genet 12 (1):19–31. doi:nrg2916 [pii] 10.1038/nrg2916

Das PP, Bagijn MP, Goldstein LD, Woolford JR, Lehrbach NJ, Sapetschnig A, Buhecha HR, Gilchrist MJ, Howe KL, Stark R, Matthews N, Berezikov E, Ketting RF, Tavare S, Miska EA (2008) Piwi and piRNAs act upstream of an endogenous siRNA pathway to suppress Tc3 transposon mobility in the *Caenorhabditis elegans* germline. Mol Cell 31(1):79–90. doi:S1097-2765(08), 00392-4 [pii] 10.1016/j.molcel.2008.06.003

Deininger PL, Batzer MA (2002) Mammalian retroelements. Genome Res 12(10):1455–1465. doi:10.1101/gr.282402

Dieci G, Preti M, Montanini B (2009) Eukaryotic snoRNAs: a paradigm for gene expression flexibility. Genomics 94 (2):83–88. doi:S0888-7543(09)00106-2 [pii]

Duursma AM, Kedde M, Schrier M, le Sage C, Agami R (2008) miR-148 targets human DNMT3b protein coding region. RNA 14(5):872–877. doi:rna.972008 [pii] 10.1261/rna.972008

Ender C, Krek A, Friedlander MR, Beitzinger M, Weinmann L, Chen W, Pfeffer S, Rajewsky N, Meister G (2008) A Human snoRNA with microRNA-like functions. Mol Cell 32(4):519–528

Faulkner GJ, Carninci P (2009) Altruistic functions for selfish DNA. Cell Cycle 8(18):2895–2900. doi:9536 [pii]

Filipowicz W, Pogacic V (2002) Biogenesis of small nucleolar ribonucleoproteins. Curr Opin Cell Biol 14 (3):319–327

Forman JJ, Legesse-Miller A, Coller HA (2008) A search for conserved sequences in coding regions reveals that the let-7 microRNA targets Dicer within its coding sequence. Proc Natl Acad Sci USA 105(39):14879–14884. doi:0803230105 [pii] 10.1073/pnas.0803230105

Garzon R, Fabbri M, Cimmino A, Calin GA, Croce CM (2006) MicroRNA expression and function in cancer. Trends Mol Med 12(12):580–587. doi:S1471-4914(06), 00242-5 [pii] 10.1016/j.molmed.2006.10.006

Ghildiyal M, Seitz H, Horwich MD, Li C, Du T, Lee S, Xu J, Kittler EL, Zapp ML, Weng Z, Zamore PD (2008) Endogenous siRNAs derived from transposons and mRNAs in Drosophila somatic cells. Science 320(5879):1077–1081. doi:1157396 [pii] 10.1126/science.1157396

Gunawardane LS, Saito K, Nishida KM, Miyoshi K, Kawamura Y, Nagami T, Siomi H, Siomi MC (2007) A slicer-mediated mechanism for repeat-associated siRNA 5′ end formation in Drosophila. Science 315(5818):1587–1590. doi:1140494 [pii] 10.1126/science.1140494

He S, Su H, Liu C, Skogerbo G, He H, He D, Zhu X, Liu T, Zhao Y, Chen R (2008) MicroRNA-encoding long non-coding RNAs. BMC Genom 9:236. doi:1471-2164-9-236 [pii] 10.1186/1471-2164-9-236

Henke JI, Goergen D, Zheng J, Song Y, Schuttler CG, Fehr C, Junemann C, Niepmann M (2008) MicroRNA-122 stimulates translation of hepatitis C virus RNA. EMBO J 27(24):3300–3310. doi:emboj2008244 [pii] 10.1038/emboj.2008.244

Hirsch J, Lefort V, Vankerssschaver M, Boualem A, Lucas A, Thermes C, d'Aubenton-Carafa Y, Crespi M (2006) Characterization of 43 non-protein-coding mRNA genes in Arabidopsis, including the MIR162a-derived transcripts. Plant Physiol 140(4):1192–1204. doi:pp.105.073817 [pii] 10.1104/pp.105.073817

Hock J, Weinmann L, Ender C, Rudel S, Kremmer E, Raabe M, Urlaub H, Meister G (2007) Proteomic and functional analysis of Argonaute-containing mRNA-protein complexes in human cells. EMBO Rep 8(11):1052–1060. doi:7401088 [pii] 10.1038/sj.embor.7401088

Houwing S, Kamminga LM, Berezikov E, Cronembold D, Girard A, van den Elst H, Filippov DV, Blaser H, Raz E, Moens CB, Plasterk RH, Hannon GJ, Draper BW, Ketting RF (2007) A role for Piwi and piRNAs in germ cell maintenance and transposon silencing in Zebrafish. Cell 129 (1):69–82. doi:S0092-8674(07), 00392-3 [pii] 10.1016/j.cell.2007.03.026

Huntzinger E, Izaurralde E (2011) Gene silencing by microRNAs: contributions of translational repression and mRNA decay. Nat Rev Genet 12(2):99–110. doi:nrg2936 [pii] 10.1038/nrg2936

Hutvagner G, Zamore PD (2002) A microRNA in a multiple-turnover RNAi enzyme complex. Science 297(5589):2056–2060. doi:10.1126/science.10738271073827 [pii]

Hutvagner G, McLachlan J, Pasquinelli AE, Balint E, Tuschl T, Zamore PD (2001) A cellular function for the RNA-interference enzyme Dicer in the maturation of the let-7 small temporal RNA. Science 293(5531):834–838. doi:10.1126/science.10629611062961 [pii]

Kawahara Y, Megraw M, Kreider E, Iizasa H, Valente L, Hatzigeorgiou AG, Nishikura K (2008) Frequency and fate of microRNA editing in human brain. Nucleic Acids Res 36 (16):5270–5280. doi:gkn479 [pii] 10.1093/nar/gkn479

Kawamura Y, Saito K, Kin T, Ono Y, Asai K, Sunohara T, Okada TN, Siomi MC, Siomi H (2008) Drosophila endogenous small RNAs bind to Argonaute 2 in somatic cells. Nature 453 (7196):793–797. doi:nature06938 [pii] 10.1038/nature06938

Kent WJ, Sugnet CW, Furey TS, Roskin KM, Pringle TH, Zahler AM, Haussler D (2002) The human genome browser at UCSC. Genome Res 12(6):996–1006

Ketting RF, Fischer SE, Bernstein E, Sijen T, Hannon GJ, Plasterk RH (2001) Dicer functions in RNA interference and in synthesis of small RNA involved in developmental timing in C. elegans. Genes Dev 15(20):2654–2659. doi:10.1101/gad.927801

Khanna A, Stamm S (2010) Regulation of alternative splicing by short non-coding nuclear RNAs. RNA Biol 7(4):480–485. doi:12746 [pii]

Khurana JS, Theurkauf W (2010) piRNAs, transposon silencing, and Drosophila germline development. J Cell Biol 191(5):905–913. doi:jcb.201006034 [pii] 10.1083/jcb.201006034

Kim YK, Kim VN (2007) Processing of intronic microRNAs. EMBO J 26(3):775–783

Kim VN, Han J, Siomi MC (2009) Biogenesis of small RNAs in animals. Nat Rev Mol Cell Biol 10(2):126–139. doi:nrm2632 [pii] 10.1038/nrm2632

Kishore S, Stamm S (2006) The snoRNA HBII-52 regulates alternative splicing of the serotonin receptor 2 C. Science 311(5758):230–232

Kuramochi-Miyagawa S, Watanabe T, Gotoh K, Totoki Y, Toyoda A, Ikawa M, Asada N, Kojima K, Yamaguchi Y, Ijiri TW, Hata K, Li E, Matsuda Y, Kimura T, Okabe M, Sakaki Y, Sasaki H, Nakano T (2008) DNA methylation of retrotransposon genes is regulated by Piwi family members MILI and MIWI2 in murine fetal testes. Genes Dev 22(7):908–917. doi:22/7/908 [pii] 10.1101/gad.1640708

Lai EC (2005) miRNAs: whys and wherefores of miRNA-mediated regulation. Curr Biol 15(12): R458–R460

Lander ES, Linton LM, Birren B, Nusbaum C, Zody MC, Baldwin J, Devon K, Dewar K, Doyle M, FitzHugh W, Funke R, Gage D, Harris K, Heaford A, Howland J, Kann L, Lehoczky J, LeVine R, McEwan P, McKernan K, Meldrim J, Mesirov JP, Miranda C, Morris W, Naylor J et al (2001) Initial sequencing and analysis of the human genome. Nature 409(6822):860–921

Lee Y, Ahn C, Han J, Choi H, Kim J, Yim J, Lee J, Provost P, Radmark O, Kim S, Kim VN (2003) The nuclear RNase III Drosha initiates microRNA processing. Nature 425(6956):415–419. doi:10.1038/nature01957 nature01957 [pii]

Lee Y, Kim M, Han J, Yeom KH, Lee S, Baek SH, Kim VN (2004) MicroRNA genes are transcribed by RNA polymerase II. EMBO J 23(20):4051–4060

Liao JY, Ma LM, Guo YH, Zhang YC, Zhou H, Shao P, Chen YQ, Qu LH (2010) Deep sequencing of human nuclear and cytoplasmic small RNAs reveals an unexpectedly complex subcellular distribution of miRNAs and tRNA 3′ trailers. PLoS One 5(5):e10563. doi:10.1371/journal. pone.0010563

Llave C, Kasschau KD, Rector MA, Carrington JC (2002) Endogenous and silencing-associated small RNAs in plants. Plant Cell 14(7):1605–1619

Luo Y, Li S (2007) Genome-wide analyses of retrogenes derived from the human box H/ACA snoRNAs. Nucleic Acids Res 35(2):559–571

Malone CD, Hannon GJ (2009) Small RNAs as guardians of the genome. Cell 136(4):656–668. doi:S0092-8674(09), 00127-5 [pii] 10.1016/j.cell.2009.01.045

Matera AG, Terns RM, Terns MP (2007) Non-coding RNAs: lessons from the small nuclear and small nucleolar RNAs. Nat Rev Mol Cell Biol 8(3):209–220. doi:nrm2124 [pii] 10.1038/nrm2124

Mills RE, Bennett EA, Iskow RC, Devine SE (2007) Which transposable elements are active in the human genome? Trends Genet 23(4):183–191. doi:S0168-9525(07), 00059-5 [pii] 10.1016/j.tig.2007.02.006

Miyoshi K, Miyoshi T, Siomi H (2010) Many ways to generate microRNA-like small RNAs: non-canonical pathways for microRNA production. Mol Genet Genom 284(2):95–103. doi:10.1007/s00438-010-0556-1

Mourelatos Z, Dostie J, Paushkin S, Sharma A, Charroux B, Abel L, Rappsilber J, Mann M, Dreyfuss G (2002) miRNPs: a novel class of ribonucleoproteins containing numerous microRNAs. Genes Dev 16(6):720–728. doi:10.1101/gad.974702

Mourier T, Willerslev E (2009) Retrotransposons and non-protein coding RNAs. Brief Funct Genom Proteomic 8(6):493–501. doi:elp036 [pii] 10.1093/bfgp/elp036

Nei M, Rooney AP (2005) Concerted and birth-and-death evolution of multigene families. Annu Rev Genet 39:121–152. doi:10.1146/annurev.genet.39.073003.112240

O'Donnell KA, Boeke JD (2007) Mighty Piwis defend the germline against genome intruders. Cell 129(1):37–44. doi:S0092-8674(07), 00396-0 [pii] 10.1016/j.cell.2007.03.028

Ono M, Scott MS, Yamada K, Avolio F, Barton GJ, Lamond AI (2011) Identification of human miRNA precursors that resemble box C/D snoRNAs. Nucleic Acids Res. doi:gkq1355 [pii] 10.1093/nar/gkq1355

Orom UA, Nielsen FC, Lund AH (2008) MicroRNA-10a binds the 5′UTR of ribosomal protein mRNAs and enhances their translation. Mol Cell 30(4):460–471. doi:S1097-2765(08), 00328-6 [pii] 10.1016/j.molcel.2008.05.001

Pederson T (2010) Regulatory RNAs derived from transfer RNA? RNA 16(10):1865–1869. doi:rna.2266510 [pii] 10.1261/rna.2266510

Piriyapongsa J, Jordan IK (2007) A family of human microRNA genes from miniature inverted-repeat transposable elements. PLoS One 2(2):e203

Piriyapongsa J, Marino-Ramirez L, Jordan IK (2007) Origin and evolution of human microRNAs from transposable elements. Genetics 176(2):1323–1337

Politz JC, Hogan EM, Pederson T (2009) MicroRNAs with a nucleolar location. RNA 15 (9):1705–1715. doi:rna.1470409 [pii] 10.1261/rna.1470409

Richard GF, Kerrest A, Dujon B (2008) Comparative genomics and molecular dynamics of DNA repeats in eukaryotes. Microbiol Mol Biol Rev 72(4):686–727. doi:72/4/686 [pii] 10.1128/MMBR.00011-08

Rodriguez A, Griffiths-Jones S, Ashurst JL, Bradley A (2004) Identification of mammalian microRNA host genes and transcription units. Genome Res 14(10A):1902–1910

Saito K, Siomi MC (2010) Small RNA-mediated quiescence of transposable elements in animals. Dev Cell 19(5):687–697. doi:S1534-5807(10), 00464-8 [pii] 10.1016/j.devcel.2010.10.011

Saito K, Nishida KM, Mori T, Kawamura Y, Miyoshi K, Nagami T, Siomi H, Siomi MC (2006) Specific association of Piwi with rasiRNAs derived from retrotransposon and heterochromatic regions in the Drosophila genome. Genes Dev 20(16):2214–2222. doi:gad.1454806 [pii] 10.1101/gad.1454806

Saraiya AA, Wang CC (2008) snoRNA, a novel precursor of microRNA in Giardia lamblia. PLoS Pathog 4(11):e1000224. doi:10.1371/journal.ppat.1000224

Schmitz J, Zemann A, Churakov G, Kuhl H, Grutzner F, Reinhardt R, Brosius J (2008) Retroposed SNOfall – a mammalian-wide comparison of platypus snoRNAs. Genome Res 18 (6):1005–1010. doi:gr.7177908 [pii] 10.1101/gr.7177908

Scott MS, Avolio F, Ono M, Lamond AI, Barton GJ (2009) Human miRNA precursors with box H/ ACA snoRNA features. PLoS Comput Biol 5(9):e1000507. doi:10.1371/journal.pcbi.1000507

Siomi MC, Saito K, Siomi H (2008) How selfish retrotransposons are silenced in Drosophila germline and somatic cells. FEBS Lett 582(17):2473–2478. doi:S0014-5793(08), 00508-5 [pii] 10.1016/j.febslet.2008.06.018

Slotkin RK, Martienssen R (2007) Transposable elements and the epigenetic regulation of the genome. Nat Rev Genet 8(4):272–285. doi:nrg2072 [pii] 10.1038/nrg2072

Smalheiser NR, Torvik VI (2005) Mammalian microRNAs derived from genomic repeats. Trends Genet 21(6):322–326

Smalheiser NR, Torvik VI (2006) Alu elements within human mRNAs are probable microRNA targets. Trends Genet 22(10):532–536. doi:S0168-9525(06), 00260-5 [pii] 10.1016/j. tig.2006.08.007

Sunkar R, Girke T, Zhu JK (2005) Identification and characterization of endogenous small interfering RNAs from rice. Nucleic Acids Res 33(14):4443–4454. doi:33/14/4443 [pii] 10.1093/nar/gki758

Suzuki HI, Miyazono K (2011) Emerging complexity of microRNA generation cascades. J Biochem 149(1):15–25. doi:mvq113 [pii] 10.1093/jb/mvq113

Taft RJ, Glazov EA, Cloonan N, Simons C, Stephen S, Faulkner GJ, Lassmann T, Forrest AR, Grimmond SM, Schroder K, Irvine K, Arakawa T, Nakamura M, Kubosaki A, Hayashida K, Kawazu C, Murata M, Nishiyori H, Fukuda S, Kawai J, Daub CO, Hume DA, Suzuki H, Orlando V, Carninci P, Hayashizaki Y, Mattick JS (2009a) Tiny RNAs associated with transcription start sites in animals. Nat Genet 41 (5):572–578. doi:ng.312 [pii]

Taft RJ, Glazov EA, Lassmann T, Hayashizaki Y, Carninci P, Mattick JS (2009) Small RNAs derived from snoRNAs. RNA 15(7):1233–1240. doi:rna.1528909 [pii] 10.1261/rna.1528909

Taft RJ, Pang KC, Mercer TR, Dinger M, Mattick JS (2010) Non-coding RNAs: regulators of disease. J Pathol 220(2):126–139. doi:10.1002/path.2638

Tay Y, Zhang J, Thomson AM, Lim B, Rigoutsos I (2008) MicroRNAs to nanog, Oct4 and Sox2 coding regions modulate embryonic stem cell differentiation. Nature 455(7216):1124–1128. doi:nature07299 [pii] 10.1038/nature07299

Tenaillon MI, Hollister JD, Gaut BS (2010) A triptych of the evolution of plant transposable elements. Trends Plant Sci 15(8):471–478. doi:S1360-1385(10), 00096-8 [pii] 10.1016/j. tplants.2010.05.003

Vasudevan S, Tong Y, Steitz JA (2007) Switching from repression to activation: microRNAs can up-regulate translation. Science 318(5858):1931–1934

Weber MJ (2006) Mammalian small nucleolar RNAs are mobile genetic elements. PLoS Genet 2(12):e205

Winter J, Jung S, Keller S, Gregory RI, Diederichs S (2009) Many roads to maturity: microRNA biogenesis pathways and their regulation. Nat Cell Biol 11(3):228–234. doi:ncb0309-228 [pii] 10.1038/ncb0309-228

Wulff BE, Nishikura K (2010) Substitutional A-to-I RNA editing. WIREs RNA 1(1):90–101. doi:10.1002/wrna.10

Yuan Z, Sun X, Jiang D, Ding Y, Lu Z, Gong L, Liu H, Xie J (2010) Origin and evolution of a placental-specific microRNA family in the human genome. BMC Evol Biol 10:346. doi:1471-2148-10-346 [pii] 10.1186/1471-2148-10-346

Yuan Z, Sun X, Liu H, Xie J (2011) MicroRNA genes derived from repetitive elements and expanded by segmental duplication events in mammalian genomes. PLoS One 6(3):e17666. doi:10.1371/journal.pone.0017666

Zhang R, Wang YQ, Su B (2008) Molecular evolution of a primate-specific microRNA family. Mol Biol Evol 25(7):1493–1502. doi:msn094 [pii] 10.1093/molbev/msn094

Chapter 3
Fragments of Small Nucleolar RNAs as a New Source for Noncoding RNAs

Marina Falaleeva and Stefan Stamm

Abstract Small nucleolar RNAs (snoRNAs) are small, nonprotein-coding RNAs that accumulate in the nucleolus. So far, these RNAs have been implicated in modification of rRNAs, tRNAs, and snRNAs. snoRNAs can be grouped into two classes: C/D box and H/ACA box snoRNAs that direct 2′-O-methylation and pseudouridylation, respectively. However, for numerous snoRNAs, no target RNAs have been identified. High-throughput sequencing and detailed analysis of RNase protection experiments have demonstrated that some snoRNAs are processed into smaller RNAs. These processed snoRNAs are 20–100 nt in length, are mostly nuclear and do not form canonical snoRNPs, that is, they do not associate with methylase or pseudouridylation activity. They can act by binding to pre-mRNAs in the nucleus where they regulate alternative pre-mRNA splicing. Thus, processed snoRNAs (psnoRNAs) represent a novel class of regulatory RNAs.

Abbreviations

snRNA   Small nuclear RNA
snoRNA   Small nucleolar RNA
psnoRNAs   Processed snoRNAs

S. Stamm (✉)
Department of Molecular and Cellular Biochemistry, University of Kentucky, Lexington, KY, USA
e-mail: Stefan@stamms-lab.net

B. Mallick (eds.), *Regulatory RNAs*, DOI 10.1007/978-3-642-22517-8_3,
© Springer-Verlag Berlin Heidelberg (outside the USA) 2012

3.1 Overview of Full-Length snoRNAs

3.1.1 Introduction

Small nucleolar RNAs are an abundant class of nonprotein-coding RNAs found in archaea and all eukaryotic lineages. They range in size from 80 to 200 nt. Based on their sequence elements and the function of representative members, snoRNAs are subdivided into C/D box and H/ACA box snoRNAs. A generalized function of snoRNAs is that they form a ribonuclear protein complex that acts on other RNAs. C/D box snoRNAs assemble proteins that perform 2′-O-methylation of their target RNAs, and the H/ACA snoRNA protein complex causes pseudouridylation of targeted RNAs. C, D, H, and ACA boxes are characteristic sequence elements that define snoRNAs. Most published examples show that snoRNAs target ribosomal RNAs, but they have also been found to target snRNAs (small nuclear RNAs) (Tycowski et al. 1998; Kiss et al. 2004) and tRNAs in archaea (Singh et al. 2004). Within these ribonuclear protein complexes, the snoRNAs serve two functions: They act as a scaffold allowing the formation of a protein complex and serve as a guide that direct this complex to its RNA targets by hybridization.

Whereas the general function of directing RNA modifying enzymes to their targets is conserved in snoRNAs from all species, their genomic localization, mode of transcription, and processing are highly variable. In humans, there are at least 257 C/D box snoRNA genes and 181 H/ACA snoRNA genes. More than 90% of human snoRNAs are located with in the introns of hosting genes (Dieci et al. 2009). In contrast, most of the 76 yeast snoRNAs are driven by their own promoter in a monocistronic gene expression unit. The majority (about 70%) of plant snoRNAs are transcribed by their own promoters (Fig. 3.1a). Thus, the flexibility in genomic organization and mode of transcription appears to be an evolutionary hallmark of snoRNAs (Dieci et al. 2009).

snoRNAs have been previously covered in excellent reviews (Filipowicz and Pogacic 2002; Matera et al. 2007; Dieci et al. 2009). This chapter will review the common genesis and function of snoRNAs and summarize our current knowledge of the emerging class of processed snoRNAs, psnoRNAs.

3.1.2 Generation of snoRNAs from Pre-snoRNAs

Like all mature RNA forms, snoRNAs are generated from precursor RNAs, the pre-snoRNAs. In general, snoRNAs are excised from their hosting RNAs by nucleases that remove unprotected RNA. The known processing enzymes are summarized in Table 3.1.

Despite their different genomic organization, snoRNAs share common features in this processing pathway. snoRNAs can be located between introns, or expressed as individual units under their own promoter that generate pre-snoRNAs (Fig. 3.1a).

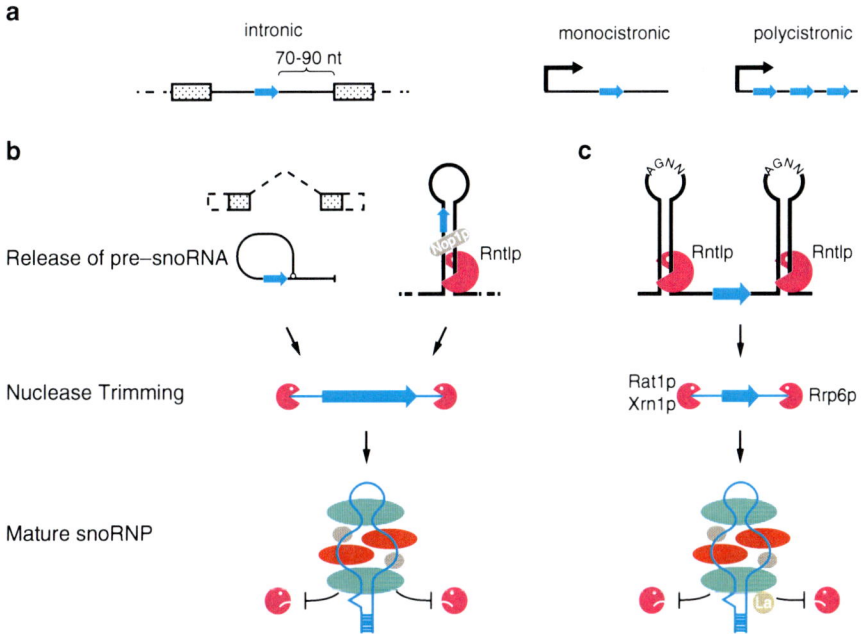

Fig. 3.1 *Processing of snoRNAs.* (**a**) Genomic arrangements of snoRNAs. The snoRNA is shown as a blue line. Exons are shown as boxes and promoters as thick arrows. The intronic localization is found in mammals, monocistronic and polycistronic arrangements in yeast and plants. RNases are shown as red "faces". (**b**) Processing of intron-dependent C/D box snoRNAs. On the left side, the most common pathway of intron-located snoRNAs is shown. During the splicing reaction, the intron is released as a lariat which contains the snoRNA. In some cases, the snoRNA is located in a stem structure that is cleaved by an RNase III activity (Rnt1p) which is stabilized by Nop1p/fibrillarin. Both pathways generate an RNA with unprotected ends that is further degraded by exonucleases. The snoRNA assembles a protein complex (see Fig. 3.2 for details), which prevents further RNase action. (**c**) Processing of cistronic snoRNAs. Cistronic snoRNAs are often located between RNA stems that contain an AGNN tetraloop. The yeast RNase Rnt1p cleaves these stems, which generates unprotected ends that are attached by exonucleases. Similar to intron-encoded snoRNAs, a protein complex assembles on the RNA and prevents further nuclease action. In addition to constitutive snoRNA proteins, auxiliary proteins such as La help protecting the snoRNA during the snoRNP assembly. These proteins dissociate from the final RNP

Individual classes of pre-snoRNAs then undergo different types of processing that leads to a linear RNA with free, unprotected 5' and 3' ends. These RNAs are then subject to nuclease cleavage by 5' and 3' exonucleases. Proteins assembling on the future mature snoRNA eventually prevent further exonuclease activity, which finally generates the snoRNA particle (snoRNP).

Almost all human snoRNAs are located in introns, which are transcribed by RNA polymerase II acting on the hosting gene. Some hosting genes are noncoding, and the production of the snoRNA is their only currently known function. For example, the H/ACA snoRNA U17a (Pelczar and Filipowicz 1998) and 10 C/D box

Table 3.1 Proteins associated with snoRNAs

	Vertebrate	Yeast	Archaea	Function
Core proteins of C/D box snoRNP	15.5 kD	Snu13p	L7Ae	Binding to C/D motif, snoRNP assembly nucleation
	Fibrillarin	Nop1p	Fibrillarin	Methyltransferase
	Nop56	Nop56p	Nop5 (single Nop56p/Nop58p-like protein)	Fibrillarin recruitment
	Nop58	Nop58p/Nop5p		
Core proteins of H/ACA snoRNP	NHP2	Nhp2p	L7Ae	Kink-turn motif binding, snoRNP assembly nucleation
	NAP57/dyskerin	Cbf5p	Cbf5	Pseudouridine synthase
	GAR1	Gar1p	Gar1	Binding and/or release of target RNA
	NOP10	Nop10p	Nop10	Interaction with L7Ae and core RNA
Proteins participating in snoRNP biogenesis	Naf1/Shq1-complex	Naf1p/Shq1p-complex		Coupling of H/ACA snoRNP assembly to transcription and intranuclear trafficking
	Bcd1	Bcd1p		Coupling of C/D snoRNP assembly to transcription
	TIP48/TIP49-complex	TIP48p/TIP49p-complex	Single TIP48/TIP49-like protein	Interactions between 15.5 kD and NOP56/58 during snoRNP biogenesis
	Xrn1	Xrn1p		$5' \rightarrow 3'$ trimming
	Xrn2	Rat1p/Xrn2p		$5' \rightarrow 3'$ trimming
	Rrp6	Rrp6p		$3' \rightarrow 5'$ trimming
	Rrp46	Rrp46p		$3' \rightarrow 5'$ trimming
	–	Rnt1p		Excision of snoRNA from hairpin
	Lsm	Lsmp		Stabilization of mature 3' end of snoRNA
	La	Lhp1p		
	Nopp140	Srp40p		Nucleocytoplasmic transport factor
	IBP160			Coupling between pre-mRNA splicing and snoRNA processing

The names of the orthologous proteins in vertebrate, yeast, and archaea are indicated

snoRNAs hosted by growth arrest-specific transcript 5 (gas 5) (Smith and Steitz 1998) are hosted by genes that do not encode a protein.

For the majority of intronic snoRNAs, processing starts with the splicing of the introns which generates a lariat structure (Fig. 3.1b), that is than opened by the debranching enzyme (Dbr1). The debranching enzyme hydrolyzes the $2' \rightarrow 5''$-branched phosphodiester bond that forms at the branch point and thus converts the lariat structure into a linear RNA molecule (Chapman and Boeke 1991).

By opening the lariat, the debranching enzyme generates the linear pre-snoRNA that undergoes further $5'-3'$ and $3'-5'$ exonuclease trimming. Excised introns that are debranched are typically subject to fast degradation. To protect the intron-encoded future snoRNA from these exonucleases, proteins assemble on the pre-snoRNA. In mammalian systems, where most snoRNAs are encoded by introns, the snoRNA generation is, therefore, coupled with the pre-mRNA-splicing process. Most intron-encoded C/D box snoRNAs are located 70–90 nucleotides upstream of the $3'$ splice site. This position defines the optimal distance between the snoRNA position and the branch point of the hosting intron, which is 50 nucleotides upstream of the $3'$ splice site (Fig. 3.1a). snoRNP assembly occurs at an advanced stage of spliceosomal assembly when the pre-mRNA binds to U2, U5, and U6 snRNPs. Thus, snoRNP components could interact directly or indirectly with U2 snRNA or another splicing component associated with the branch point (Hirose et al. 2003) (Fig. 3.1b).

An alternative pathway was described in *Xenopus* for the intron-encoded U16 snoRNA. Here, proteins binding to the C and D boxes promote the entry of an endonuclease that cleaves the pre-snoRNA (Caffarelli et al. 1996). Experiments in intron-containing snoRNAs in yeast suggest that this protein is Nop1p/fibrillarin that stimulates the endoribonuclease Rnt1p (Giorgi et al. 2001). The cleavage of the intron prevents proper splicing of the adjacent exons, indicating that snoRNA formation can prevent expression of the proper hosting mRNA. Therefore, as this example shows, snoRNA formation can be in competition with host mRNA formation.

The majority of snoRNAs in yeast and plants are controlled by their own promoter. Often, these snoRNAs are in a polycistronic arrangement. They therefore undergo a slightly different processing pathway than their intron-embedded relatives (Fig. 3.1c). The future mature snoRNA is typically flanked by a short hairpin structure that forms around an AGNN tetraloop. Similar to *Xenopus* U16 snoRNA, the double-stranded RNA stem serves as the entry point for the endoribonuclease Rnt1p, which is the orthologue of bacterial RNase III that cleaves double-stranded RNAs. Rnt1p cleavage generates unprotected $5'$ and $3'$ ends, which are further trimmed by the $5'$ endonucleases Xrn1p and Rat1p, as well as the $3'$ endonuclease Rrp6p (Chanfreau et al. 1998). As with other snoRNAs, proteins assemble on the final snoRNA and prevent further nuclease cleavage. These proteins can remain associated with the snoRNP, as in the case of Nop1p, Nop58p, and Nop56p (Lafontaine and Tollervey 1999), or protect the RNA only during the processing steps, as exemplified by the association of the Lhp1p/La protein in U3 snoRNA processing (Kufel et al. 2000) (Fig. 3.1c, Table 3.1).

3.1.3   Structure of C/D and H/ACA Box snoRNAs

Despite different genomic organization and assembly pathways, the two classes of snoRNAs form ribonuclear protein complexes that are similar across species. Both snoRNA types have RNA "boxes," which refer to short sequence elements that bind proteins or target RNAs.

C/D box snoRNAs. These RNAs contain C (UGAUGA) and D boxes (CUGA), and a highly conserved kink-turn element prior to the C box (Fig. 3.2a). C and D boxes bind to the protein 15.5 kD (Snu13p in yeast, L7e in archaea). Adjacent to the D boxes is the RNA-interacting sequence, called antisense box or antisense element. The arrangement of a C box, antisense box, and D box is the functional unit of the snoRNA. Most C/D box snoRNAs contain additional units, which are labeled C′ and D′. The protein 15.5 kD/Snu13p binds to both the C and D boxes, which brings these two RNA elements in close proximity. This newly formed RNA: protein complex then forms a new binding site for Nop56/58. The formation of this complex allows the entry of NOP1p/fibrillarin, which possesses the catalytic activity that performs the 2′-O-methylation of the target RNA (Omer et al. 2002). The elaborate assembly of the complex allows the precise alignment of the target RNA with the fibrillarin catalytic activity. As a result, the ribose on the target RNA that base pairs with the nucleotide located five nucleotides upstream of the D box will be methylated (Fig. 3.2a).

H/ACA box snoRNAs contain the H (ANANNA) box and the ACA (ACA) box that are located in single-stranded regions of a "hairpin-hinge-hairpin-tail" configuration. One or both hairpins contain a loop structure that allows interaction with the target RNA (Fig. 3.2b). Dyskerin (centromere binding factor 5, Cbf5 in archaea) contains a PUA (pseudouridine and archaeosine transglycosylase) domain that binds directly to the ACA and lower stem of the H/ACA snoRNA (Charpentier et al. 2005). Dyskerin also contains the catalytic activity, pseudouridine synthase, that converts a uridine to a pseudouridine in the target RNA. Two other proteins, GAR1 and NOP10, bind subsequently to dyskerin and increase enzyme activity. Similar to C/D box snoRNAs, the 15.5 kD (Snu13p) protein binds to a kink-turn motif of the H/ACA RNA, which bends the RNA (Li and Ye 2006) (Fig. 3.2b).

The idealized structure of C/D box and H/ACA box snoRNAs are shown in Fig. 3.2c.

3.1.4   Canonical Functions of snoRNAs

The structures of mature snoRNAs allow the identification of their target mRNAs by identifying RNAs that exhibit sequence complementarity toward their antisense boxes. The majority of the highly expressed snoRNAs cause modification of ribosomal RNAs (Steitz and Tycowski 1995). In addition, snoRNAs play a role in snRNA modification and telomere maintenance, which are summarized in Table 3.2. Recently, the number of known snoRNAs has been expanded by

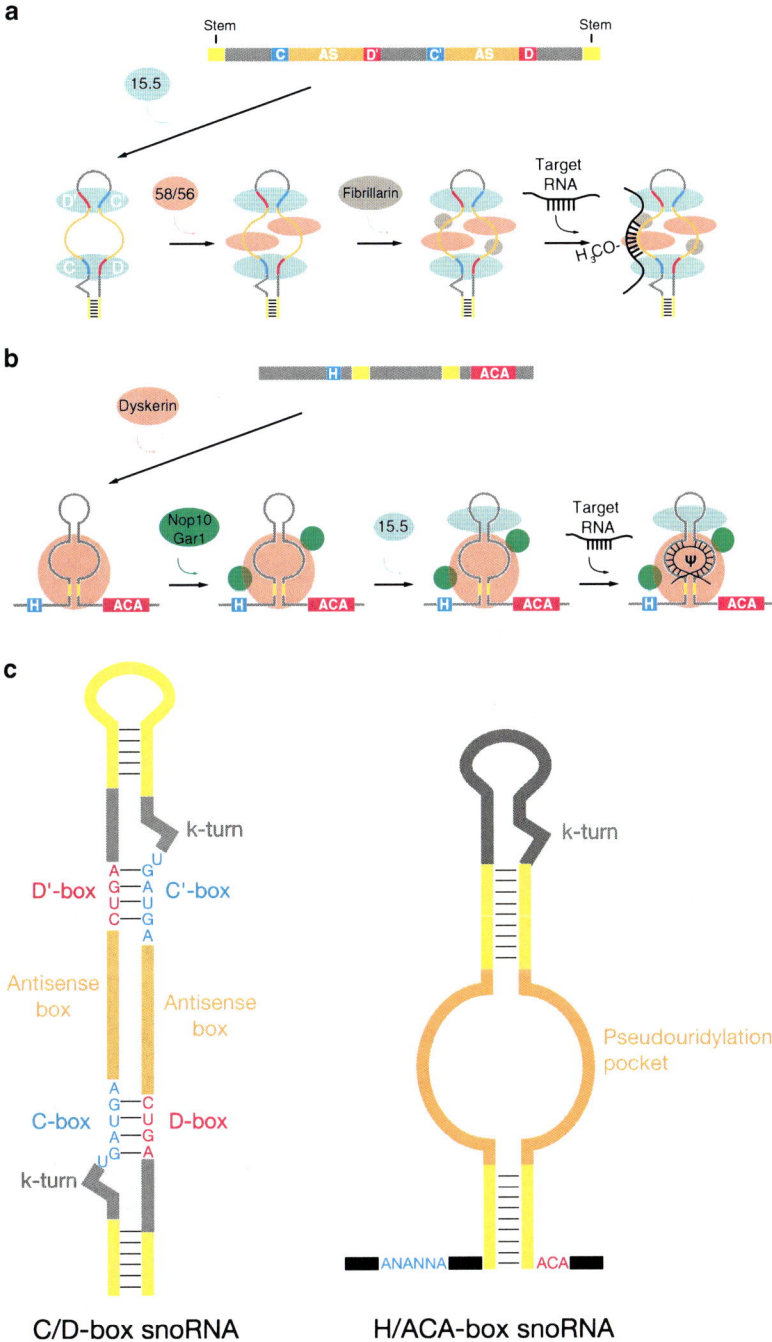

Fig. 3.2 *Formation and function of snoRNPs.* The structure of the snoRNAs is schematically shown on the left. Colors indicate the various sequence elements. C, C′: C box (UGAUGA); D, D′:

Table 3.2 Examples of functions of eukaryotic snoRNAs

snoRNA group	Functions	Members	Cellular localization	Conservative motives
C/D box snoRNA	rRNA, snRNA methylation Pre-rRNA processing	Many snoRNA species U3, U8, U14, U22	Nucleolar	C box (UGAUGA), D box (CUGA)
H/ACA box snoRNA	rRNA, snRNA pseudouridylation Pre-rRNA processing	Many snoRNA species snR10, snR30	Nucleolar	H box (ANANNA), ACA box (ACA)
H/ACA box scaRNA	snRNA pseudouridylation	Many scaRNA species	Cajal body	H box (ANANNA), ACA box (ACA), CAB box (ugAG)
C/D-H/ACA box scaRNA	snRNA methylation and pseudouridylation	U85, U87, U88, U89	Cajal body	H box (ANANNA), ACA box (ACA), CAB box (ugAG)

more sensitive experimental cloning efforts that use high-throughput sequencing (reviewed in Gardner et al. (2010)). A bioinformatic analysis of these snoRNAs revealed that a large proportion of them have no known target sites on ribosomal RNAs, snRNAs, or tRNAs (Bazeley et al. 2008; Hertel et al. 2008). Since they lack known sequence complementarity, these snoRNAs are considered orphan and their function is elusive.

rRNA processing and modification is the best-studied function of snoRNAs. In eukaryotes, snoRNPs bind to ribosomal RNA using Watson-Crick base pairing between their antisense boxes and targets. About half of the estimated 300 rRNA processing factors are snoRNAs (Fatica and Tollervey 2002).

Both C/D and H/ACA box snoRNPs help in identifying cleavage sites of rRNA precursors. They do not contain any ribonucleolytic activity on their own and likely function as RNA chaperones, by promoting the correct folding of rRNA that is subsequently subjected to cleavage (Beltrame and Tollervey 1995). Examples are the U3, U14, U8, U22 C/D box snoRNPs, and U17 H/ACA snoRNP. Inhibition of mature 18S rRNA accumulation occurs in U3, U14, and U17 snoRNPs depleted *Saccharomyces cerevisiae* cells and U14 and U22 depleted *Xenopus* oocytes (Atzorn et al. 2004).The U8 snoRNA was shown to function in 5.8S and 28S rRNA processing in *Xenopus* (Peculis and Steitz 1993).

Other snoRNPs functions are chemical modifications of rRNA, which are believed to be essential for the correct folding and function of the ribosome (Decatur and Fournier 2003). C/D box snoRNPs are responsible for 2′-O-methylation, and

Fig. 3.2 (continued) D box (CUGA); AS: antisense box, stem: terminal sequences that form a stem-loop end structure. H: H box (ANANNA), ACA: ACA box (ACA). Proteins are indicated by oval, colored shapes. The target RNA is a *thick black line*; hybridization to the snoRNA is indicated by *shorter lines*. (**a**) Assembly of C/D box snoRNAs. (**b**) Assembly of H/ACA box snoRNAs. (**c**) Schematic structure of C/D box and H/ACA box snoRNAs

H/ACA snoRNPs convert uridine into pseudouridine in the mature rRNAs. There are at least 106 methylated and 95 pseudouridinated residues found in the human rRNA (Tollervey and Kiss 1997) that are likely targeted by snoRNA-rRNA interaction. Interestingly, the modifications are concentrated in functionally important regions of rRNA. For example, the V domain of 25S rRNA, which is thought to associate with peptidyl transferase activity, is the subject to massive modification in yeast cells (Decatur and Fournier 2002).

snRNA modifications. There are five major spliceosomal RNAs (U1, U2, U4, U5, and U6) which play a critical role in pre-mRNA splicing. U1, U2, U3, and U4 are RNA polymerase II transcripts while U6 is transcribed by RNA polymerase III. These RNAs are also modified to achieve optimal function. The modifications include 2′-O-methylation and pseudouridylation which are mostly located in the functionally important regions involved in interactions with other snRNAs, pre-mRNA, or spliceosomal proteins (Yu et al. 1998).

Telomere maintenance. Human telomerase RNA serves as a template for the replication of chromosome ends. A characteristic H/ACA snoRNA "hairpin-hinge-hairpin-tail" motif was found in the 3′ end of human telomerase RNA (Mitchell et al. 1999). It spans 240 out of 451 nucleotides of telomerase RNA. Despite of the fact that the H/ACA domain of telomerase RNA binds to all H/ACA snoRNP proteins, there is no experimental data that indicate a participation of telomerase RNA in pseudouridylation (Meier 2005). It is possible that the H/ACA motif is responsible for the nuclear localization of telomerase RNA since mutation of key elements of the box H/ACA motif results in localization of the RNA in the cytoplasm (Lukowiak et al. 2001).

These examples illustrate that snoRNAs can acquire multiple functions in RNA metabolism. The underlying theme is that snoRNAs identify target RNAs using short RNA-RNA interaction. The target RNA can be chemically modified, cleaved, or translocated in the cell.

3.1.5   Noncanonical Functions of snoRNAs

Cloning efforts for small, non-polyadenylated RNAs discovered new, brain-specific C/D box snoRNAs. One of these snoRNAs, HBII-52 (SNORD 115 in humans), exhibited an 18 nt sequence complementarity toward the serotonin receptor 2 C pre-mRNA. The complementary sequences were located in the antisense box of the C/D box snoRNA and in the alternative exon Vb of the serotonin receptor 2 C (Cavaille et al. 2000). Transfection experiments using reporter genes showed that HBII-52 promotes the inclusion of this alternative exon, indicating that snoRNAs can regulate alternative splicing by interacting with pre-mRNAs (Kishore and Stamm 2006). The idea that snoRNAs can act on mRNAs was further supported by proof-of-principle experiments, where the antisense box of a C/D box snoRNA was engineered to cause a 2′-O-methylation of the adenosine branch point of an alternative exon. As expected, transfection of this snoRNA caused a change in alternative splicing, by preventing the phosphodiester bond formation to the modified 2' hydroxyl group of the branch point (Semenov et al. 2008). These experiments indicate that snoRNAs can act on pre-mRNAs, but their mechanism of action is hard to understand.

3.2   snoRNA Expressing Units Give Rise to Shorter RNAs, psnoRNAs

3.2.1   Discovery of snoRNA fragments

The analysis of high-throughput sequencing data from human, mouse, chicken, *Drosophila, Arabidopsis,* and *Schizosaccharomyces pombe* revealed the existence RNA fragments that were derived from known H/ACA snoRNAs (Taft et al. 2009; Cole et al. 2009) and C/D box snoRNAs. Collectively, these RNAs were between 17 and 27 nt in length (Taft et al. 2009). These studies were supported by bioinformatic analyses that showed that numerous H/ACA snoRNAs could be precursors for experimentally confirmed miRNAs (Scott et al. 2009). The snoRNA precursors of these miRNAs could bind dyskerin and were localized in the nucleolus indicating that they have functional properties of snoRNAs (Scott et al. 2009). Deep sequencing of RNAs associated with argonaute proteins identified H/ACA snoRNA fragments, indicating that these RNA fragments are functionally important as they are associated with components of the miRNA machinery (Ender et al. 2008).

snoRNA fragments have been found in other species, including the ancient eukaryote *Giradia lamblia* that expresses four RNA fragments derived from C/D box snoRNAs (Saraiya and Wang 2008) and the Epstein Barr virus that expressed an RNA fragment of a C/D box snoRNA (Hutzinger et al. 2009).

The analysis of whole libraries of noncoding RNAs showed a slight bias toward short and abundant RNAs, as these RNAs can be more efficiently cloned. In addition, some of the libraries used to identify fragments of snoRNAs were specifically size-selected to enrich for miRNAs (Brameier et al. 2011). Therefore, the currently known psnoRNAs might represent only shorter members of this class of RNAs.

RNase protection experiments allow for the identification of the RNA fragments generated from a precursor RNA in an unbiased way. A major problem in analyzing RNase protection experiments was the difficulty in cloning the generated dsRNA fragments. Using a new cloning technique that overcomes these problems, the processing pattern of the brain-specific C/D box snoRNA HBII-52 (SNORD115) and HBII-85 (SNORD 116) could be determined (Kishore et al. 2010; Shen et al. 2011). This direct analysis showed that HBII-52 is processed into at least six shorter RNAs ranging from 37 to 73 nt. All these RNAs remained soluble in the biochemically defined nuclear fraction, indicating that they can have function outside the nucleolus (Kishore et al. 2010; Soeno et al. 2010). For the HBII-52 class of RNAs, it was found that the processed form and not the previously reported full-length snoRNA are the predominant product of the gene expression unit (Kishore et al. 2010). These RNAs were termed psnoRNAs for processed snoRNAs.

The psnoRNAs known to date are summarized in Table 3.3.

Currently, most psnoRNAs are derived from C/D box snoRNAs. A schematic alignment of the known psnoRNAs with a hypothetical, "generic" C/D box snoRNA precursor is shown in Fig. 3.3. It is striking that most of the psnoRNAs contain C and D boxes, which could indicate that the 15.5 kD protein that binds these sequence elements plays a role in their biogenesis.

This data suggest that psnoRNAs are generated from the two major classes of snoRNAs. Shorter, processed psnoRNAs appear in different phyla: human, mouse, *Giardia lamblia, Drosophila, Arabidopsis* and *Schizosaccharomyces pombe*, indicating that they represent a ubiquitous form of RNA. In contrast to canonical miRNAs, psnoRNAs appear to be nuclear. psnoRNA processing appears to stop at defined sites, indicating a regulated biogenesis.

3.2.2   Mechanism of psnoRNA Formation

The mechanism of psnoRNA formation is only beginning to emerge, mainly for C/D box snoRNAs. Previously, it was shown that shortening of the terminal stem-loop structures destabilizes snoRNAs (Darzacq and Kiss 2000), and one discriminating factor between snoRNAs and psnoRNAs could be the length and stability of the stem. The mutation of C and D boxes completely abolished snoRNA production from expression clones (Kishore et al. 2010), indicating that psnoRNAs share processing pathways with traditional snoRNAs. Several scenarios for psnoRNA generation are possible. The first model suggests that psnoRNAs are derived through further processing of mature snoRNAs in the nucleus. Related to this option is that psnoRNAs represent "recycled" degradation products of canonical snoRNAs. It is not known how snoRNAs are degraded, but, similar to other RNAs, they will be cleaved by nucleases to allow for an RNA turnover.

psnoRNAs could thus represent degradation products of canonical snoRNAs that associate with different hnRNPs.

A second scenario is that psnoRNA formation diverges from canonical snoRNA formation during the processing step. Since most psnoRNAs contain C and D boxes, it is possible that they still interact with the 15.5/Snu13 protein (Fig. 3.2). However, for unknown reasons, these RNAs do not form a canonical snoRNP by further association of NOP58/56, but instead, associate with hnRNPs that protect the psnoRNA from further degradation. Finally, it can be imagined that a mixture of these scenarios is at work in the cell.

Table 3.3 Summary of currently known processed snoRNAs (psnoRNAs)

	snoRNA name, organism	sequence	Reference
C/D-box snoRNA	MBII-52, Mus musculus	ACUGGGUCAAUGAUGACAACCCAAUGUCAUGAGAGAAAGGUGAUGACAUAAAAUUCAUGCUCAUAGGAUUACGCUGAGGCCCAACCA AAUGAUGACAACCCAAUGUCAUGAGAGAAAGGUGAUGACAUAAAAUUCAUGCUCAUAGGAUUACGCUGAGGCC GGUCAAUGAUGACAACCCAAUGUCAUGAGAGAAAGGUGAUGACAUAAAAUUCAUGCUCAUU UGGGUCAAUGAUGACAACCCAAUGUCAUGAGAGAAAGGUGAUGAC GGUCAAUGAUGACAACCCAAUGUCAUGAGAGAAAGGUGA AUGAUGACAACCCAAUGUCAUGAGAGAAAGGUGAUGA	(Kishore et al. 2010)
	MBII-85 cluster 2, Mus musculus	GGAUCUAUGAUGAUCCCAGUCAAACAUUCCUUGGAAAAGCUGAACAAAAUGAGUGAAAACUCUGACCGCCACUCUCAUCGGAACUGAGUCCA AUCUAUGAUGAUCCCAGUCAAACAUUCCUUGGAAAAGCUGAACAAAAUGAGUGAAAACUCUGACCGCCACUCUCAUCGGAACUGAGG AUCUAUGAUGAUCCCAGUCAAACAUUCCUUGGAAAAGCUGAACAAAAUGAGUGAAAACUCUGUA CCUUGGAAAAGCUGAACAAAAUGAGUGAAAACUCUGUA	(Shen et al. 2011)
	MBII-85 cluster 10, Mus musculus	GGAUCUAUGAUGAUUCCCAGUCAAACAUUCCUUGGAAAAGCUGAACAAAAUGAGUGAAAACUCUGACCGCCACUCUCAUCGGAACUGAGUCCA AUCUAUGAUGAUUCCCAGUCAAACAUUCCUUGGAAAAGCUGAACAAAAUGAGUGAAAACUCUGUA GGAUCUAUGAUGAUUCCCAGUCAAACAUUCCUUGGAAAAGCUGAACAAAAUGAGUGAAAACUCUGUA AAAACUGAACAAAAUGAGUGAAAACUCUGUA CUCUCAUCGGAACUGAGUCC	(Shen et al. 2011)
	MBII-85 cluster 23, Mus musculus	GGAUCUAUGAUGAUUCCAGUCAAACAUUCCUUGGAAAAGCUGAACAAAAUGAGUGAAAACUCUGACCGCCACUCUCAUCGGAACUGAGUCCA CUAUGAUGAUUUCCAGUCAAACAUUCCUUGGAAAAGCUGAACAAAAUGAGUGAAAACUCUGUA CAGUCAAACAUUCCUUGGAAAAGCUGAACAAAAUGAGUGAAAACUCUGUA AACUGAACAAAAUGAGUGAAAACUCUGUA GGAAAAGCUGAACAAAAUGAGUGAAAACUCUGUACCGCCACUCUCAUCGGAACU	(Shen et al. 2011)
	Snord2 Homo sapiens	AAGUGAAAUGAUGGCAAUCAUCUUUCGGACUGACUGAAAUGAGAGAAUAUCAUUGCUGAUCACUUG AGUGAAAUGAUGGCAAUCAUCUUUCG AAUGAAGAGAAUAUCACUAUG	(Brameier et al. 2011)
	X14945 Homo sapiens	. . GAAGUUUCUCUGACGUGUAGAGC. . UCUCCUGAUC. . AUUGAUGAUCGUUC. . GUAUUGGGGAGUGAGUGAGGAGGAGAGAGAAGACGCGGUCUGAGUGGU AGUUUCUCUGACGUGUAGAGC GGGAGUGAGGGAGGAGGAGAGAGAAGACGCGGUCUGAGUG	(Brameier et al. 2011)
	SNORD78 Homo sapiens	GGUAAUGAUGUUGAUCAAAUGUCUGACAUGAGCAUGUAGACAAAGGUAACACUGAGAA GGUAAUGAUGUUGAUCAAAUGU AAUGACAUGUAGACAAAGGUAACACUGAAG	(Brameier et al. 2011)
	SNORD93 Homo sapiens	UGGCCAAGGAUGAGAACUCUAAUCUGAUUUUAUGUGCUUCUGCUGUGAUGGAUUAAAGGAUUUACUGAGGCCA GGCCAAGGAUGAGAACUCUAAU	(Brameier et al. 2011)
	SNORD 17 Homo sapiens	GUGAAAUGAUGAUUCAGUUAUUCAUUCGCUGUAGUGCCUGCACUGAGC. . . CUCUGAGACCCAUUCUACAAAGAUGAGUGUGUGAAAAUCUGAUCAC GAAAUGAUGAUUCAGUUUAUUCCAUUCG ACAAAGAUGAGUGUGUGAAAAUCU	(Brameier et al. 2011)
	SNORD 66 Homo sapiens	UUCCUCUGAUGACUUCCGUUAGUGCCACGUGUCUGGCCCACUGAGACCACCAUUGAGGAACUGAGGAUCUGAGG GACACCAUUGAUGGAACU	(Brameier et al. 2011)
	SNORD 48 Homo sapiens	AGUGAUGAUGACCCCAGGUAACUCUGAGUGUGUGUGAUGGCCAAUCACCUGAGCGCGUCUGACC AGUGAUGAUGACCCCAGGUAACUCU UGCCAUCACCGCAGCGCU	(Brameier et al. 2011)

	Name	Sequence	Reference
	SNORD 51 Homo sapiens	GUUGCAUGAUGAAUAAAAUCAAAUCACCAUCUUUCGCCUGAGUUCGUGAUGGAUUUGCUUUUUUCUGAUU UUGCAUGAUGAAUAAAAUCAAAUCACCA AGUUCGUGAUGGAUUUGCUUUUUU	(Brameier et al. 2011)
	SNORD 21 Homo sapiens	GCUGAAUGAUGAUAUCCCACUAACUGAGCAGUCAGUAGUGUGGUUCCUUUGGUUGCAUAUGAUGCGAUAAAUUGGUUCAAGACGGACGACUGAUGGCAGC CUGAAUGAUGAUAUCCCACUAACUGA UGCGAUAAAUUGGUUCAAGACGGGA	(Brameier et al. 2011)
	SNORD 27 Homo sapiens	ACUCCAUGAUGAACACAAAAUGACAAGCAUAUGGCCUGAACUUUCAAGUGAUGUCAUCUACUACUGAGAAGU CUCCAUGAUGAACACAAAAUGAC AACUUUCAAGUGAUGUCAUCUACUAC	(Brameier et al. 2011)
	SNORD 44 Homo sapiens	CCUGGAUGAUGAUAAGCAAAUGCUGACUAAACAUGAAGGUCUAAUUAGCUCUAACUGACUAA CUGGAUGAUGAUAAGCAAAUGCUGA ACAUGAAGGUCUAAUUAGCUCUAA	(Brameier et al. 2011)
	SNORD100 Homo sapiens	GCUGUACAUGAUGACAACUGGCUCCCUACUGAACUGCAUGGAGAAACUGCCAUGUCACCCUCUGAUUACAGC GUACAUGAUGACAACUGGCUCCCUCUACUG ACUGCCAUGAGAGAAACUGCCAUGUCACCCUU	(Brameier et al. 2011)
	SNORD59b Homo sapiens	UAUUCCUCACUGAUGAGUACGUUCUGACUUUCGUCUCUGACUUCUGAAGCCAGAUGCAAUUCUGAGAAGG CUCACUGAUGAGUACGUU UUGCUGAAGCCAGAUGCAAUU	(Brameier et al. 2011)
	SNORD83a Homo sapiens	GCUGUUCGUGAUGAGGCUCAGAGUGAGCGCCGGGUUACAGCGCCGAAUCGACAGUGAAACCAUUCUACUCGUCCUUCCUUCUGAGAACAGC GUUCGUGAUGAGGCUCAGAGU	(Brameier et al. 2011)
	SNORD15a Homo sapiens	CUUCGAUGAAGAUGAAUGACAGGAUGAUCUUGGGGGAUGUGUCU...CGGUUGAAGUAGUAAUUUCCUAAAGAUGACUAGAGGCAUUGUGUCUGAGAAGG UUCGAUGAAGAUGAAUGACAGGAUGUCU UAAAGAUGACUUAGAGGCAUUUGU	(Brameier et al. 2011)
	SNORD74 Homo sapiens	CUGCCUGAUGAUGAACCGUGUGUUGGAUUAUAUCUGAGAGUAAUGAUGAAUGCCAACCGCUCUGAUGGUGG GCCUCUGAUGAAGCCGUGUGGUUGGAGGGACA UAAUGAUGAAUGCCAACCGCUCUGA	(Brameier et al. 2011)
	SNORD45a Homo sapiens	GGUCAAUGAUGUUGGGCAUGUUAUAUCUGAUAUGUGGUAAUAACACUUUAGCUCUAGAAUUACUCUGAGACCU GGUCAAUGAUGUUGGGCAUGUAUAU UUAGCUCUAGAAUUACU	(Brameier et al. 2011)
	GlsR17 Giardia lamblia	GUUUCUAGA...AUCCGGGCACUGAGCAUCCCCAGGACACAGGCGAGGCACGGCCUGCGCCACCGCCUAAUCACCGCCCCUAUAGUC CAGCCUAAUCACCGCCCCUAUAGUC	(Saraiya and Wang 2008)
	GlsR16 Giardia lamblia	UAAAACUAUGAUGAGGUUAGCGAUCCCAAGCGGGCCUGGCCCUGCGUUGCAACGCAUCACCGCUCGACCUU GCAGCAACGCAUCACCGCUCGACC CAACGCAUGACCGCUCGACC	(Saraiya and Wang 2008)
	v-snoRNA1 Epstein–Barr virus	CCCGAUGAUGAUGACAACCGCGGCUGUCUGAAGCGGCUGACGAAAUCGGUUGAGAUUCUGAUG ACGAAAUCGGUUGAGAUUCUGAUG	(Hutzinger et al. 2009)
	HBII-180A Homo sapiens	..GGGCCUCCAUGAUGUCCAGCACUGGGCUCCGACUCGCCACUGAGGACACGUGCCCCCGGGACCUUUGACACCGGGGCUCGGUCUGAGGGGCCCUGG UGCCACUGAGGACACGGUGCC	(Ono et al. 2011)
	HBII-180B Homo sapiens	..GGACCCCGUGAUGUCCAGCACUGGGCUCGCCACCCGACCCGCACCCGGACCUUUGACACUGCCGGGGUUCUGAGGGGCCCCAC UGCCCCUGAGGACACGGUGCA	(Ono et al. 2011)
	HBII-180C Homo sapiens	..GGGCUCCCAUGAUGUCCAGCACUGGGCUCGACACCCGGGGGGUCUGAGGACACCCCAGGACCUUUGACACCUGGGGGGUCUGAGGGGCCCCAG CACCCCUGAGGACACAGUGCA	(Ono et al. 2011)

(continued)

Table 3.3 (continued)

	Name	Sequence	Reference
H/ACA snoRNA	HBII-239 / Homo sapiens	UGUGUGUUGGAGGAUGAAAGUACGGGAUGAUCCAUCGGCUAAGUGUCUUGUCACAAGUGCUGACACUCAAACUGCUGACAGCACACG / GUUGGAGAGGAUGAAAGUACGGAGUGAU / UCACAAUGCUGACACUCAAACUGCUGAC	(Ono et al. 2011)
	mgU6-77 / Homo sapiens	GCUCUGUGAUGGAGCCCAUGCGUGUCAUCUGAGCCUCUGGCUUC…CGGGAAGGUCCUGGGCAAAGGAUCUUUGUACUCUGAGAGCAGACUA / GCUCUGUGAUGGAGCCCAUGCGUGUCAUCUGAGC	(Kawaji et al. 2008)
	ACA 45 / Homo sapiens	..UCUGACUCUGCCUUUAGCCUCCUAAAUGAAAAGGUAGAAGAACAGGUCUUUGUCAAAAUAAAAUUCAAGACCUACUAUCUACCAACACG / AAGGUAGAUAGAACAGGUCU / AGACCUACUAUCUACCAACAC	(Ender et al. 2008)
	ACA36B / Homo sapiens	..GCUCUGUUAAUUAAAACUUUGGAACAUUGAAACUGGCUAGGGAAAAUGAUUGGAUAGAAACAUAUUAUCUAUUCAUUUAUCCCCAGCCUACAAAA / CUGGCUAGGGAAAAUGAUUGGA / AAUCAUUUAUCCCCAGCCUACA	(Ender et al. 2008)
	U92 / Homo sapiens	…CUACUUGUCACCAUGCCUCCCUAGAAUAAAACUGCCUUUUGACUGGGACGAAUUGAGUGAAAAUCGUAACGGGGGAGACAGAU / CUGCCCUUUGAUGACCGGGACGA / AACGACAGAUACGGGGCAGACAG	(Ender et al. 2008)
	HBI-100 / Homo sapiens	AAUGGAAGCUGGAACCAGCCAUG…AAAUAUAUGGAGGGUCUCGUCUGGCUGCUUAGGACAGCAGGUGGCUAAGUCUGGUCGUCCCCUCCGUACAGCC / UGGAAGCGCUGGAACCAGC / AUGGAGGGUCUCGUCUGGCUUA	(Ender et al. 2008)
	ACA56 / Homo sapiens	CGGC…GGGAGUCAGUCGCAACAGUAAGUGGUGAGUCUCGUCCAGCGUCACGAUAUUUGAUGGUGGCUUUAGACUUGCCAGAUAACA / UAAGUGGUGAGUCUCGUCCAGC / UGGUGGCUUUAGACUUGCCAGA	(Ender et al. 2008)
	ACA3 / Homo sapiens	AUCGAGGCUAGAGUCACGCUUGGGUAUCGCGUAUUGCCUGA…AUUCACUGCCUGUGGGCUUAUGGCACAGUCAGUCUCACCAGGUUAGAGACAUGC / AUCGAGGCUAGAGUCACGCUUGG / CACCAGGUUAGAGACAUGC	(Ender et al. 2008)
	ACA50 / Homo sapiens	AAGCACUGCCUUUGAACCUGAUGUGUCUUGUUUGUAGCUUCACGGGCCAAGCAACAGUCUAGACUAUAAC…GCUUUAAAAACAAGGUACAUUU / AGCACUGCCUUUGAACCUGAUGUGU / CGGGCCAAGCAACAGUCUAGA	(Ender et al. 2008)
	ACA47 / Homo sapiens	ACGGUCUGGGGAAAGGCUCCUGUGUUGUUGAGCCU…GAAUUUGCAGUAACAGGUGUAGCAUUCUA…UGAUCAUGUAGUAUACUGCAAACUGGA / ACGGUCUGGGGAAAGGCUCCUG / UUUGCAGUAACAGGUGUGAGC	(Ender et al. 2008)
	ACA34 / Homo sapiens	GUGGCCCUGACGAAGACCAGCAGUUGUACUGGUGGCUGGUUCUCAGGCUGUGGGUUUCAAGC…GGGACCUCAGUAGGAAUGGCUAUUUCAUUUGGAAGAAACAACC / GGCCCGACUGAAAGACCAGCAGU	(Scott et al. 2009)
	HBI-61 / Homo sapiens	ACCUUCUUGUAUAAGCACUGUGCUAAAAUUGCGAGACACUAGGACCAUGUGCUU…UAAUGUCCAAUGUCCUUUUGUGUCCUUUAAGAGAUUUGGUGCAAUAUCU / CUCUUGUAUAAAGCACUGUGCUAAA	(Scott et al. 2009)

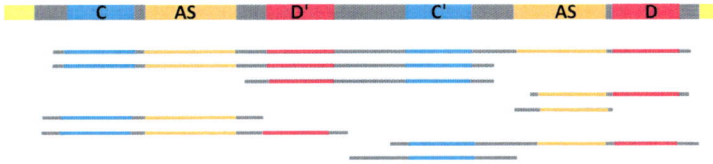

Fig. 3.3 *Schematic alignment of psnoRNAs to a generic C/D box snoRNA.* A Hypothetical, generic C/D box snoRNAs is shown on the top. The RNA elements are colored as in Figs. 3.1 and 3.2. C: C-box, D: D-box, D′ and C′: D′ and C′ boxes, AS: antisense box. The yellow shading at the end reflects the terminal stem structures. The psnoRNAs shown in Table 3.3 are schematically indicated by showing the RNA elements that they contain. Note that most of them include C and D boxes and lack the stems

3.2.3 Methods to Study psnoRNA Processing and Function

Most of the current knowledge of psnoRNAs comes from the analysis of high-throughput sequencing data, which is the method of choice to identify psnoRNAs. In order to avoid a size bias, RNAs in the range of 20–200 nt should be analyzed, as the focus on miRNAs currently favors RNAs in the 20–30 nt range. To analyze the processing of psnoRNAs, they can be overexpressed in cells, and cis-acting RNA elements can be determined by mutagenesis (Kishore et al. 2010; Cavaille et al. 2000). RNA targets can be determined by overexpressing the psnoRNAs in cell lines followed by an analysis of RNA expression. psnoRNA-dependent targets can be identified by a candidate approach that analyzes RNAs that exhibit sequence complementarity to the psnoRNA. Here, it should be taken into account that potentially the entire psnoRNA, not just the antisense boxes, can interact with the target, as psnoRNAs do not form a conventional snoRNP. Finally, a psnoRNA-dependent change in expression can be determined genome-wide using expression arrays.

3.3 Function of Processed snoRNAs

The identification of psnoRNAs raised the question whether they are functional or represent degradation products of canonical snoRNAs. The first indications that psnoRNAs acquire functions came from the analysis of the MBII-52-derived psnoRNA. It was shown that this snoRNA promotes inclusion of the alternative exon of the serotonin receptor 2 C (Kishore and Stamm 2006), as well as six other alternative exons (Kishore et al. 2010). Since all the snoRNAs from the MBII-52 expression units are processed into smaller psnoRNAs, it is likely that psnoRNAs, not the snoRNAs, influence alternative splicing. To further gain insight into the mechanism of action, the protein complexes associated with MBII-52 psnoRNAs were investigated. Surprisingly, none of the canonical C/D box-associated proteins

Table 3.4 Proteins associated with psnoRNAs C/D box snoRNAs

Protein name	HEK 293T nuclear extract from an expression construct (Kishore et al. 2010)	Total mouse brain (Soeno et al. 2010)	Protein function
Pur-alpha	+	+	Transcription activation
Pur-beta	–	+	
TDP-43	+	+	Transcriptional repression, nuclear pre-mRNA splicing, mRNA export/import, translational regulation
hnRNPs A1/A2/A3/ B1/D0	+	+	Transcriptional regulation, nuclear pre-mRNA splicing via spliceosome, mRNA localization, translation and turnover
ATP-dependent RNA helicase A	–	+	Transcriptional regulation and RNA processing
Matrin-3	–	+	Nuclear pre-mRNA splicing via spliceosome, rRNA editing
Nucleolin	–	+	Chromatin silencing, nuclear pre-mRNA splicing via spliceosome, nucleosome mobilization, rRNA processing, transcription from Pol I promoter
ELAV-like protein I	–	+	mRNA stabilization, nuclear pre-mRNA splicing via spliceosome
Centromere protein V	–	+	Regulation of chromosome organization

The table lists proteins associated with the psnoRNAs derived from the MBII-52 cluster. PsnoRNAs derived from H/ACA snoRNA ACA45 are associated with Ago1 and Ago2 (Ender et al. 2008)

(15.5 kD, fibrillarin, NOP 56/58) were found. Instead the psnoRNAs were associated with hnRNPs and other nuclear proteins, which are listed in Table 3.4.

PsnoRNAs that resembled miRNAs have been tested experimentally, and CDC2L6 (CDK11) was identified as a target for an H/ACA-derived psnoRNA. The psnoRNA was argonaute-associated and blocked translation of CDC2L6 (CDK11) via an element in the 3′ UTR, similar to a miRNA (Ender et al. 2008). Other psnoRNAs were found to block translation in luciferase assays, using synthetic binding sites (Saraiya and Wang 2008; Brameier et al. 2011). This data indicates that snoRNA fragments can act similar to canonical miRNAs.

The association with proteins and the metabolic stability of the psnoRNAs argue that they are functional noncoding RNAs. Given the variability of snoRNA expression, it is not surprising that psnoRNAs can be recruited to several functions, such as a change in nuclear pre-mRNA processing and in cytosolic translational control. Due to its structure, the double-stranded RNase dicer generates fragments that are typically 22 nt in length. The larger size (>30 nt) of most psnoRNAs indicates that they are formed by a dicer-independent pathway. This suggests the existence of a novel pathway to generate noncoding regulatory RNAs. This idea is supported by the comparison of psnoRNAs derived from dicer and DGCR8 knockout embryonic stem cells with wild-type cells. It was found that C/D box psnoRNAs have similar length distributions, indicating that their formation is independent from Dicer: DGCR8 complex. However, H/ACA-derived psnoRNAs length was different between these cells, suggesting that dicer plays a role in their processing (Taft et al. 2009). The dicer dependency was experimentally confirmed for one H/ACA psnoRNA (Ender et al. 2008).

Finally, orphan snoRNAs have been defined as snoRNAs without targets for their antisense boxes. Since psnoRNAs often do not contain antisense boxes, a targeting of other RNAs that relies on imperfect RNA-RNA interactions will be missed by bioinformatic prediction programs that focus on the antisense-target RNA interaction.

3.4   Role of psnoRNAs in Diseases

The cluster of HBII-52 and HBII-85 psnoRNA expression units is subject to intense analysis, as their loss of expression is linked to Prader-Willi syndrome. Prader-Willi syndrome (PWS) is a human congenital disease with an incidence of about 1 in 8,000–20,000 live births. A characteristic of the disease in older children is the inability to gain satiety after a meal, and people with PWS are subsequently hyperphagic. The hyperphagia causes weight gain and makes PWS the most common genetic cause of marked obesity in humans. PWS is caused by the loss of gene expression from a maternally imprinted region on chromosome 15q11–q13 (reviewed in Butler et al. (2006)). Recently, three PWS patients with microdeletions have been identified that only lack expression of HBII-52 and HBII-85 (Sahoo et al. 2008; de Smith et al. 2009; Duker et al. 2010). The comparison of the microdeletions suggests that the loss of HBII-85 expression is the strongest contributor to the phenotype (de Smith et al. 2009). HBII-85 and HBII-52 are human C/D box snoRNAs that are expressed in 27 and 48 expression units, respectively. Each expression unit consists of two exons flanking a hosting intron that contains the snoRNA. The snoRNAs are evolutionary highly conserved, whereas the hosting noncoding exons are poorly conserved. In mice, both the MBII-85 and MBII-52 expression units are processed into smaller psnoRNAs (Shen et al. 2011; Brameier et al. 2011), and the full-length snoRNA from the MBII-85 cluster appears as only a minor intermediate form (Kishore et al. 2010). MBII refers to the mouse orthologous of the human HB-II snoRNAs. HB-II simply indicates human brain library number II, and the added number represents the clone number sequenced.

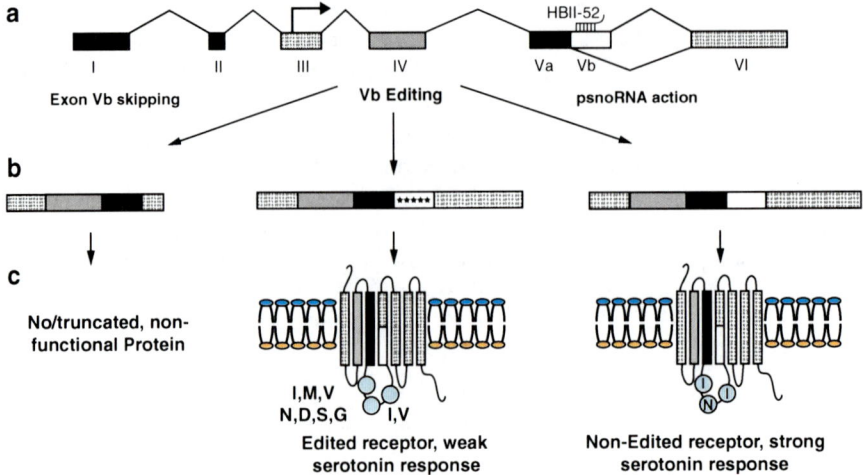

Fig. 3.4 *Regulation of serotonin receptor 5-HT2C by HBII-52-derived psnoRNAs.* (**a**) The genomic structure of the 5-HT2C receptor. The arrow in exon III indicated the translational start point. HBII-52-derived psnoRNAs interact with an 18 nucleotide complementarity region in exon Vb. (**b**) Protein-coding parts of the mRNAs derived from different pre-mRNA processing events. Exon Vb skipping results in a shortened mRNA that encodes a truncated protein but is most likely subject to nonsense-mediated mRNA decay. Exon Vb can be edited at five positions (indicated as arrows). The editing event promotes inclusion of the exon but changes the amino acid sequence at three points. The psnoRNAs cause inclusion of exon Vb without editing, which generates a receptor with the highest agonist efficacy. (**c**) Structure of the encoded proteins. Editing of exon Vb leads to a change a potentially three amino acids, which are located in the second intracellular loop that couples to the effector G protein. The editing events weaken the receptor-G protein interaction and lead to a weak serotonin response. The non-edited receptor features the amino acids I, N, and I at the positions that could be edited and shows the strongest coupling to the G protein and response to serotonin.

As the serotonin-receptor 2 C-target RNA for MBII-52 could be predicted from its sequence, its possible contribution to the Prader-Willi phenotype is best understood. The perfect 18 nt-long sequence complementarity between the MBII-52 antisense box and the pre-mRNA is located in the alternative exon Vb of the receptor. Failing to include exon Vb into the pre-mRNA generates a nonfunctional receptor, due to a frameshift (Fig. 3.4). Transfection experiments show that the HBII-52 expression unit with a 5-HT2C reporter gene promotes exon Vb inclusion. Mutagenesis studies show that exon Vb contains splicing silencers that normally prevent the inclusion of the exon. Expression of the snoRNA blocks the action of the silencers and promotes exon inclusion. The silencers located on the pre-mRNA can also be modified by RNA editing that changes adenosine to inosine residues (Kishore and Stamm 2006). As a result, there are two ways of generating a full-length serotonin 5-HT2C receptor: blocking the silencers through expression of the snoRNA and weakening the silencers by editing some of its bases. The nature of the silencing element is not clear and it could be a protein, RNA, or an RNA structure.

The RNA editing events change the amino acid composition of the receptor at three sites. These sites are located in an area critical for protein function, namely, in a loop that couples to the G protein. The editing of the receptor pre-mRNA decreases the coupling of the G protein to the receptor and thus reduces its efficacy. The mRNA containing the non-edited version of the 5-HT2C receptor encodes a receptor that couples optimal to its effector G protein and shows the highest response to serotonin stimulation. Analysis of limited brain samples from PWS patients showed a reduction of the non-edited isoform (Kishore and Stamm 2006), which has also been observed in mouse models lacking HBII-52 snoRNA expression (Doe et al. 2009). A molecular link between a defect in the 5-HT2C production and PWS is an attractive hypothesis, as the 5-HT2C receptor plays a crucial role in hunger control and satiety, which is the major problem in PWS. Since HBII-52 snoRNA promotes the generation of the most active receptor, it acts like a "genetic agonist" of the serotonin receptor. The administration of selective 5-HT2CR agonists, such as d-fenfluramine, has a strong appetite-suppressing effect (Vickers et al. 2001). Underlining the importance of the 5-HT2C receptor for hunger control, the mouse knockout of 5-HT2CR is hyperphagic and develops obesity. Expression of the 5-HT2CR in the arcuate nucleus, a major hunger control center, reverses the hyperphagic phenotype (Xu et al. 2008). Conversely, when a mutant of the receptor that represents the fully edited 5-HT2CR is expressed in knockout mice, the resulting mice remain hyperphagic (Kawahara et al. 2008). Collectively, the data strongly supports a model where the loss of HBII-52 causes a loss of the mRNA isoform that encodes the most active form of the receptor, which is necessary for proper hunger control. Finally, overexpression of MBII-52 in mouse brain causes an autistic-like phenotype, which further underlines the importance of MBII-52 in normal brain function (Nakatani et al. 2009).

Prader-Willi syndrome represents a loss of function of psnoRNAs. The findings clearly show that psnoRNA expression is physiologically important. It remains to be seen whether point mutations in psnoRNA can have detrimental effects for human health, similar to the mutations of the H/ACA snoRNA domain of human telomerase RNA that result in dyskeratosis congenital (Vulliamy et al. 2001; Vulliamy et al. 2008).

3.5 Summary and Outlook

The identification of psnoRNAs indicates the existence of a new class of noncoding regulatory RNAs. Several studies aimed at finding new miRNAs identified snoRNAs as possible miRNA precursors. These snoRNA-derived miRNAs are dependent on dicer and are loaded on argonaute proteins. This suggests that some snoRNAs could be recruited to form miRNAs.

However, the majority of the currently characterized psnoRNAs is larger than 22 nt in length and is in contrast to miRNAs predominantly nuclear. It is therefore expected that their processing is independent of dicer, and their function is independent of binding to argonaute proteins. Therefore, they likely represent a new class of nuclear regulatory RNAs. The expression of MBII-52 and MBII-85 snoRNAs was found to be regulated in response to neuronal activity in a fear-conditioning mouse model. It is therefore possible that psnoRNA expression is plastic in a physiological system (Rogelj et al. 2003), similar to some neuronal miRNAs (Krol et al. 2010).

The function of psnoRNAs is only beginning to emerge: Similar to other RNAs, they appear to be coated with hnRNPs, suggesting that they can form hnRNPs. Such hnRNPs could influence numerous pre-mRNAs, as has been shown for MBII-85. Some of the associated hnRNPs have been shown to be involved in chromatin reorganization, which together with their nuclear localization could indicate that psnoRNAs influence chromatin structure.

Since it is now clear that every part of a snoRNA, not just the antisense boxes, can be recruited to target other nucleic acids, improved prediction programs can be devised to bring targets to so far orphan snoRNAs.

Acknowledgements This work was supported by the NIH RO1 GM083187 and the Binational US-Israel Science Foundation.

References

Atzorn V, Fragapane P, Kiss T (2004) U17/snR30 is a ubiquitous snoRNA with two conserved sequence motifs essential for 18S rRNA production. Mol Cell Biol 24(4):1769–1778

Bazeley PS, Shepelev V, Talebizadeh Z, Butler MG, Fedorova L, Filatov V, Fedorov A (2008) snoTARGET shows that human orphan snoRNA targets locate close to alternative splice junctions. Gene 408(1–2):172–179

Beltrame M, Tollervey D (1995) Base pairing between U3 and the pre-ribosomal RNA is required for 18S rRNA synthesis. EMBO J 14(17):4350–4356

Brameier M, Herwig A, Reinhardt R, Walter L, Gruber J (2011) Human box C/D snoRNAs with miRNA like functions: expanding the range of regulatory RNAs. Nucleic Acids Res 39(2): 675–686

Butler MG, Hanchett JM, Thompson TE (2006) Clinical findings and natural history of Prader-Willi syndrome. In: Butler MG, Lee PDK, Whitman BY (eds) Managment of Prader-Willi syndrome. Springer, New York, pp 3–48

Caffarelli E, Fatica A, Prislei S, De Gregorio E, Fragapane P, Bozzoni I (1996) Processing of the intron-encoded U16 and U18 snoRNAs: the conserved C and D boxes control both the processing reaction and the stability of the mature snoRNA. EMBO J 15(5):1121–1131

Cavaille J, Buiting K, Kiefmann M, Lalande M, Brannan CI, Horsthemke B, Bachellerie JP, Brosius J, Huttenhofer A (2000) Identification of brain-specific and imprinted small nucleolar RNA genes exhibiting an unusual genomic organization. Proc Natl Acad Sci USA 97: 14311–14316

Chanfreau G, Legrain P, Jacquier A (1998) Yeast RNase III as a key processing enzyme in small nucleolar RNAs metabolism. J Mol Biol 284(4):975–988. doi:10.1006/jmbi.1998.2237

Chapman KB, Boeke JD (1991) Isolation and characterization of the gene encoding yeast debranching enzyme. Cell 65(3):483–492

Charpentier B, Muller S, Branlant C (2005) Reconstitution of archaeal H/ACA small ribonucleoprotein complexes active in pseudouridylation. Nucleic Acids Res 33(10):3133–3144

Cole C, Sobala A, Lu C, Thatcher SR, Bowman A, Brown JW, Green PJ, Barton GJ, Hutvagner G (2009) Filtering of deep sequencing data reveals the existence of abundant Dicer-dependent small RNAs derived from tRNAs. RNA 15(12):2147–2160

Darzacq X, Kiss T (2000) Processing of intron-encoded box C/D small nucleolar RNAs lacking a 5′,3′-terminal stem structure. Mol Cell Biol 20(13):4522–4531

de Smith AJ, Purmann C, Walters RG, Ellis RJ, Holder SE, Van Haelst MM, Brady AF, Fairbrother UL, Dattani M, Keogh JM, Henning E, Yeo GS, O'Rahilly S, Froguel P, Farooqi IS, Blakemore AI (2009) A deletion of the HBII-85 class of small nucleolar RNAs (snoRNAs) is associated with hyperphagia, obesity and hypogonadism. Hum Mol Genet 18(17): 3257–3265

Decatur WA, Fournier MJ (2002) rRNA modifications and ribosome function. Trends Biochem Sci 27(7):344–351

Decatur WA, Fournier MJ (2003) RNA-guided nucleotide modification of ribosomal and other RNAs. J Biol Chem 278(2):695–698. doi:10.1074/jbc.R200023200

Dieci G, Preti M, Montanini B (2009) Eukaryotic snoRNAs: a paradigm for gene expression flexibility. Genomics 94(2):83–88

Doe CM, Relkovic D, Garfield AS, Dalley JW, Theobald DE, Humby T, Wilkinson LS, Isles AR (2009) Loss of the imprinted snoRNA mbii-52 leads to increased 5htr2c pre-RNA editing and altered 5HT2CR-mediated behaviour. Hum Mol Genet 18(12):2140–2148

Duker AL, Ballif BC, Bawle EV, Person RE, Mahadevan S, Alliman S, Thompson R, Traylor R, Bejjani BA, Shaffer LG, Rosenfeld JA, Lamb AN, Sahoo T (2010) Paternally inherited microdeletion at 15q11.2 confirms a significant role for the SNORD116 C/D box snoRNA cluster in Prader-Willi syndrome. Eur J Hum Genet 18:1196–1201

Ender C, Krek A, Friedlander MR, Beitzinger M, Weinmann L, Chen W, Pfeffer S, Rajewsky N, Meister G (2008) A human snoRNA with microRNA-like functions. Mol Cell 32(4):519–528

Fatica A, Tollervey D (2002) Making ribosomes. Curr Opin Cell Biol 14(3):313–318

Filipowicz W, Pogacic V (2002) Biogenesis of small nucleolar ribonucleoproteins. Curr Opin Cell Biol 14(3):319–327

Gardner PP, Bateman A, Poole AM (2010) SnoPatrol: how many snoRNA genes are there? J Biol 9(1):4

Giorgi C, Fatica A, Nagel R, Bozzoni I (2001) Release of U18 snoRNA from its host intron requires interaction of Nop1p with the Rnt1p endonuclease. EMBO J 20(23):6856–6865

Hertel J, Hofacker IL, Stadler PF (2008) SnoReport: computational identification of snoRNAs with unknown targets. Bioinformatics 24(2):158–164

Hirose T, Shu MD, Steitz JA (2003) Splicing-dependent and -independent modes of assembly for intron-encoded box C/D snoRNPs in mammalian cells. Mol Cell 12(1):113–123

Hutzinger R, Feederle R, Mrazek J, Schiefermeier N, Balwierz PJ, Zavolan M, Polacek N, Delecluse HJ, Huttenhofer A (2009) Expression and processing of a small nucleolar RNA from the Epstein-Barr virus genome. PLoS Pathog 5(8):e1000547

Kawahara Y, Grimberg A, Teegarden S, Mombereau C, Liu S, Bale TL, Blendy JA, Nishikura K (2008) Dysregulated editing of serotonin 2 C receptor mRNAs results in energy dissipation and loss of fat mass. J Neurosci 28(48):12834–12844

Kawaji H, Nakamura M, Takahashi Y, Sandelin A, Katayama S, Fukuda S, Daub CO, Kai C, Kawai J, Yasuda J, Carninci P, Hayashizaki Y (2008) Hidden layers of human small RNAs. BMC Genom 9:157

Kishore S, Stamm S (2006) The snoRNA HBII-52 regulates alternative splicing of the serotonin receptor 2 C. Science 311(5758):230–232

Kishore S, Khanna A, Zhang Z, Hui J, Balwierz P, Stefan M, Beach C, Nicholls RD, Zavolan M, Stamm S (2010) The snoRNA MBII-52 (SNORD 115) is processed into smaller RNAs and regulates alternative splicing. Hum Mol Genet 19:1153–1164

Kiss AM, Jady BE, Bertrand E, Kiss T (2004) Human box H/ACA pseudouridylation guide RNA machinery. Mol Cell Biol 24(13):5797–5807

Krol J, Busskamp V, Markiewicz I, Stadler MB, Ribi S, Richter J, Duebel J, Bicker S, Fehling HJ, Schubeler D, Oertner TG, Schratt G, Bibel M, Roska B, Filipowicz W (2010) Characterizing light-regulated retinal microRNAs reveals rapid turnover as a common property of neuronal microRNAs. Cell 141(4):618–631

Kufel J, Allmang C, Chanfreau G, Petfalski E, Lafontaine DL, Tollervey D (2000) Precursors to the U3 small nucleolar RNA lack small nucleolar RNP proteins but are stabilized by La binding. Mol Cell Biol 20(15):5415–5424

Lafontaine DL, Tollervey D (1999) Nop58p is a common component of the box C + D snoRNPs that is required for snoRNA stability. RNA 5(3):455–467

Li L, Ye K (2006) Crystal structure of an H/ACA box ribonucleoprotein particle. Nature 443(7109):302–307

Lukowiak AA, Narayanan A, Li ZH, Terns RM, Terns MP (2001) The snoRNA domain of vertebrate telomerase RNA functions to localize the RNA within the nucleus. RNA 7(12): 1833–1844

Matera AG, Terns RM, Terns MP (2007) Non-coding RNAs: lessons from the small nuclear and small nucleolar RNAs. Nat Rev Mol Cell Biol 8(3):209–220

Meier UT (2005) The many facets of H/ACA ribonucleoproteins. Chromosoma 114(1):1–14. doi:10.1007/s00412-005-0333-9

Mitchell JR, Cheng J, Collins K (1999) A box H/ACA small nucleolar RNA-like domain at the human telomerase RNA 3′ end. Mol Cell Biol 19(1):567–576

Nakatani J, Tamada K, Hatanaka F, Ise S, Ohta H, Inoue K, Tomonaga S, Watanabe Y, Chung YJ, Banerjee R, Iwamoto K, Kato T, Okazawa M, Yamauchi K, Tanda K, Takao K, Miyakawa T, Bradley A, Takumi T (2009) Abnormal behavior in a chromosome-engineered mouse model for human 15q11–13 duplication seen in autism. Cell 137(7):1235–1246

Omer AD, Ziesche S, Ebhardt H, Dennis PP (2002) In vitro reconstitution and activity of a C/D box methylation guide ribonucleoprotein complex. Proc Natl Acad Sci USA 99(8):5289–5294

Ono M, Scott MS, Yamada K, Avolio F, Barton GJ, Lamond AI (2011) Identification of human miRNA precursors that resemble box C/D snoRNAs. Nucleic Acids Res 39(9):3879–3891

Peculis BA, Steitz JA (1993) Disruption of U8 nucleolar snRNA inhibits 5.8S and 28S rRNA processing in the Xenopus oocyte. Cell 73(6):1233–1245

Pelczar P, Filipowicz W (1998) The host gene for intronic U17 small nucleolar RNAs in mammals has no protein-coding potential and is a member of the 5′-terminal oligopyrimidine gene family. Mol Cell Biol 18(8):4509–4518

Rogelj B, Hartmann CE, Yeo CH, Hunt SP, Giese KP (2003) Contextual fear conditioning regulates the expression of brain-specific small nucleolar RNAs in hippocampus. Eur J Neurosci 18(11):3089–3096

Sahoo T, del Gaudio D, German JR, Shinawi M, Peters SU, Person RE, Garnica A, Cheung SW, Beaudet AL (2008) Prader-Willi phenotype caused by paternal deficiency for the HBII-85 C/D box small nucleolar RNA cluster. Nat Genet 40(6):719–721

Saraiya AA, Wang CC (2008) snoRNA, a novel precursor of microRNA in Giardia lamblia. PLoS Pathog 4(11):e1000224

Scott MS, Avolio F, Ono M, Lamond AI, Barton GJ (2009) Human miRNA precursors with box H/ACA snoRNA features. PLoS Comput Biol 5(9):e1000507

Semenov DV, Vratskih OV, Kuligina EV, Richter VA (2008) Splicing by exon exclusion impaired by artificial box c/d RNA targeted to branch-point adenosine. Ann N Y Acad Sci 1137:119–124

Shen M, Eyras E, Wu J, Khanna A, Josiah S, Rederstorff M, Zhang MQ, Stamm S (2011) Direct cloning of double-stranded RNAs from RNAse protection analysis reveals processing patterns of C/D box snoRNAs and provides evidence for widespread antisense transcript expression. Nucleic Acids Res. in press

Singh SK, Gurha P, Tran EJ, Maxwell ES, Gupta R (2004) Sequential 2′-O-methylation of archaeal pre-tRNATrp nucleotides is guided by the intron-encoded but trans-acting box C/D ribonucleoprotein of pre-tRNA. J Biol Chem 279(46):47661–47671

Smith CM, Steitz JA (1998) Classification of gas5 as a multi-small-nucleolar-RNA (snoRNA) host gene and a member of the 5′-terminal oligopyrimidine gene family reveals common features of snoRNA host genes. Mol Cell Biol 18(12):6897–6909

Soeno Y, Taya Y, Stasyk T, Huber LA, Aoba T, Huttenhofer A (2010) Identification of novel ribonucleo-protein complexes from the brain-specific snoRNA MBII-52. RNA 16(7): 1293–1300

Steitz JA, Tycowski KT (1995) Small RNA chaperones for ribosome biogenesis. Science 270(5242): 1626–1627

Taft RJ, Glazov EA, Lassmann T, Hayashizaki Y, Carninci P, Mattick JS (2009) Small RNAs derived from snoRNAs. RNA 15(7):1233–1240

Tollervey D, Kiss T (1997) Function and synthesis of small nucleolar RNAs. Curr Opin Cell Biol 9(3):337–342

Tycowski KT, You ZH, Graham PJ, Steitz JA (1998) Modification of U6 spliceosomal RNA is guided by other small RNAs. Mol Cell 2:629–638

Vickers SP, Dourish CT, Kennett GA (2001) Evidence that hypophagia induced by d-fenfluramine and d-norfenfluramine in the rat is mediated by 5-HT2C receptors. Neuropharmacology 41(2): 200–209

Vulliamy T, Marrone A, Goldman F, Dearlove A, Bessler M, Mason PJ, Dokal I (2001) The RNA component of telomerase is mutated in autosomal dominant dyskeratosis congenita. Nature 413(6854):432–435

Vulliamy T, Beswick R, Kirwan M, Marrone A, Digweed M, Walne A, Dokal I (2008) Mutations in the telomerase component NHP2 cause the premature ageing syndrome dyskeratosis congenita. Proc Natl Acad Sci USA 105(23):8073–8078

Xu Y, Jones JE, Kohno D, Williams KW, Lee CE, Choi MJ, Anderson JG, Heisler LK, Zigman JM, Lowell BB, Elmquist JK (2008) 5-HT2CRs expressed by pro-opiomelanocortin neurons regulate energy homeostasis. Neuron 60(4):582–589

Yu YT, Shu MD, Steitz JA (1998) Modifications of U2 snRNA are required for snRNP assembly and pre-mRNA splicing. EMBO J 17(19):5783–5795. doi:10.1093/emboj/17.19.5783

Chapter 4
Small Regulatory RNAs (sRNAs): Key Players in Prokaryotic Metabolism, Stress Response, and Virulence

Sabine Brantl

Abstract Small RNAs (sRNAs) gained worldwide attention in the late 2002, when the journal Science published a special issue entitled "Small RNAs – Breakthrough of the Year." However, small antisense RNAs in bacteria involved in the regulation of plasmid replication and maintenance, phage life cycles, and transposition had been investigated in great depth for more than 20 years. Whereas these sRNAs were discovered only fortuitously, systematic computer-based searches have only been used since 2001. Currently, it is estimated that a bacterial genome encodes ~200–300 sRNAs with diverse functions. To date (2011), about 140 sRNAs are known in *Escherichia coli* and *Salmonella*. However, only about 25 of these have been assigned a biological function, indicating that defining their functions continues to be a challenging issue. Systematic searches have also been performed for a few Gram-positive bacterial species.

sRNAs in bacteria can be divided into two major groups: The first group regulates gene expression by a base-pairing mechanism with target mRNA, whereas the second group acts by binding of small proteins. This chapter covers mechanisms of action, biological functions, integration in regulatory circuits, and evolutionary aspects of base-pairing and protein-binding sRNAs.

Keywords Antisense RNA • cis-encoded sRNA • Hfq • mRNA stability • riboregulator • RNA thermometer • small regulatory RNA • sRNA • trans-encoded sRNA • translation activation • translation inhibition

S. Brantl (✉)
AG Bakteriengenetik, Friedrich-Schiller-Universität Jena, Jena, Germany
e-mail: Sabine.Brantl@uni-jena.de

B. Mallick (eds.), *Regulatory RNAs*, DOI 10.1007/978-3-642-22517-8_4,
© Springer-Verlag Berlin Heidelberg (outside the USA) 2012

4.1 Introduction

In bacteria, small noncoding RNAs are the most abundant class of post-transcriptional regulators. Whereas sRNAs have been studied in plasmids, phages, and transposons, systematic computer-based searches of bacterial chromosomes began in earnest in 2001. These studies have identified ~140 sRNAs in *E. coli* and *Salmonella,* and it is estimated that a bacterial genome encodes ~200–300 sRNAs with diverse functions (Hershberg et al. 2003). Small RNAs can be classified into several major groups. They encompass sRNAs that act by a base-pairing mechanism, by protein binding, and by sensing environmental factors such as RNA thermometers (Narberhaus et al. 2006) and riboswitches (Roth and Breaker 2009). Additionally, sRNAs with specific functions like tmRNA, RNase P, and 4.5S RNA have been identified.

sRNAs that act by a base-pairing mechanism can be cis- or trans-encoded (Fig. 4.1a). Cis-encoded antisense RNAs are completely complementary to their target (sense) RNAs and can, therefore, form complete duplexes with them. By contrast, trans-encoded antisense RNAs are only partially complementary to their – often multiple – target RNAs yielding only partial duplexes between both molecules. In both cases, the interaction between antisense RNA and target mRNA results in inhibition or activation of target RNA function. The second group of sRNAs comprises RNAs that regulate gene expression by binding to proteins (Fig. 4.1b). Here, mechanisms of action, biological functions, integration in regulatory circuits, and evolutionary aspects of base-pairing and protein-binding sRNAs are discussed.

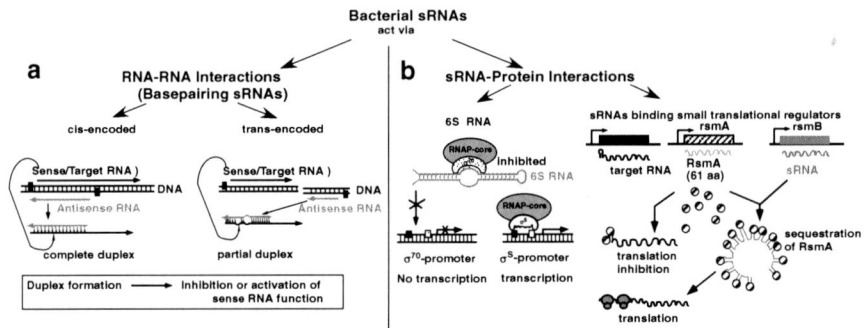

Fig. 4.1 *Overview of base-pairing and protein-binding sRNAs.* (**a**) Cis-encoded antisense RNAs form complete duplexes with their target RNAs, whereas trans-encoded sRNAs can only form partial duplexes with their target(s). Antisense RNAs are drawn in *grey*, sense RNAs in *black*. Black rectangles, promoters. (**b**) Protein-binding sRNAs. *Left*: 6S RNA interacts with vegetative RNAP containing σ^{70}, thus inhibiting its promoter binding. Stationary phase RNAP with σ^S is not bound by 6S RNA facilitating recognition of σ^S promoters. *Black and white* rectangles, -35 and -10 boxes, respectively. *Right*: The RsmA-RsmB regulatory system of *Erwinia carotovora*. Free RsmA protein inhibits target mRNA translation. Binding of RsmA protein to RsmB depletes the pool of free RsmA, thereby allowing target mRNA translation

4.2 cis-Encoded Antisense RNAs

The first cis-encoded antisense RNAs were discovered in 1981 in the *E. coli* plasmids ColE1 (Tomizawa et al. 1981) and R1 (Stougaard et al. 1981) where they regulate replication, and hence, control plasmid copy numbers. Subsequently, in a wide variety of plasmids, phages, and transposons, cis-encoded antisense RNAs were found and intensively characterized over the past 30 years (reviewed in Wagner et al. 2004). In plasmids, antisense RNAs regulate replication, conjugation frequency, maintenance, and segregational stability. In phages, they have a fine-tuning function in the decision between lysis and lysogeny. In transposons, they control transposition frequency. In addition, a growing number of chromosomally cis-encoded sRNAs have been found. Among them are RatA from the *B. subtilis* genome that regulates the toxin TxpA (Silvaggi et al. 2005, Fig. 4.2b) and IsrR from the cyanobacterium *Synechococcus* that controls the amount of the photosynthesis component IsiA (Dühring et al. 2006). Other examples from *E. coli* include GadY that is involved in the acid stress response (Opdyke et al. 2004, Fig. 4.2c) and SymR that controls the expression of a toxin, the SOS-induced endonuclease SymE (Kawano et al. 2007, Fig. 4.2a). In *Salmonella*, the 1,200-nt long sRNA AmgR

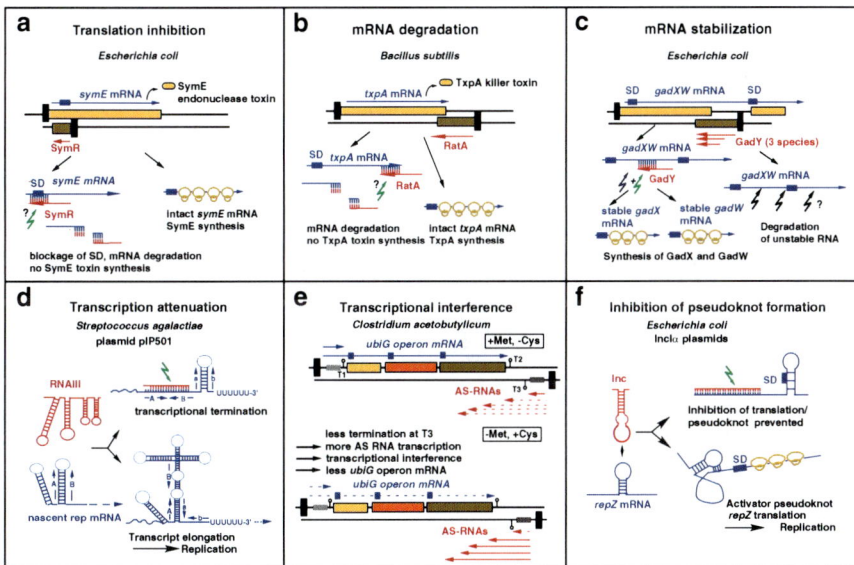

Fig. 4.2 *Overview of regulatory mechanisms employed by cis-encoded antisense RNAs.* Antisense RNAs are drawn in *red*, sense RNAs in *blue*. *Black* rectangles denote promoters, *yellow* and *brown* boxes sense and antisense RNA genes, respectively. Open *yellow* symbols indicate ribosomes. *Green arrows* denote RNase III cleavage, *black arrows* the putative action of other RNases. *Violet arrow* in Fig. 4.2c is an unidentified RNase. Details are described in the text. (**a–c**) are based on Brantl (2009)

controls *mgtC* (Lee and Groisman 2010, see 8) and in *Clostridium acetobutylicum*, four antisense RNAs of different length regulate the sulfur-metabolic *ubiG* operon (André et al. 2008, Fig. 4.2d). In *Salmonella typhimurium* (Padalon-Brauch et al. 2008), 19 novel sRNAs associated with pathogenicity islands were identified, many of which are cis-encoded. Recently, in the cyanobacterium *Synechocystis* sp. PCC6803, 73 cis-encoded sRNAs have been found, among them SyR7 which possibly modulates murein biosynthesis (Voss et al. 2009). In *Staphylococcus aureus*, pyrosequencing identified 30 small RNAs (Bohn et al. 2010), among them one cis-encoded sRNA, RsaOW, which is perfectly complementary to the IS1181 transposase 5' UTR. By screening cDNA libraries prepared from low-molecular-weight RNA of *Mycobacterium tuberculosis*, nine putative sRNAs, among them at least one cis-encoded (ASdes), were identified (Arnvig and Young 2009). ASdes might also act as trans-encoded sRNA. Table 4.1 summarizes all currently known cis-encoded antisense RNAs from bacterial chromosomes for which targets have been identified.

4.2.1   Mechanisms of Action

Inhibition of target RNA function prevails, but, in a few cases, activating mechanisms have been found, too. All currently known regulatory mechanisms employed by cis-encoded antisense RNAs are discussed below and – except for inhibition of primer formation – summarized in Fig. 4.2.

4.2.1.1   Translation Inhibition

The conceptually simplest mechanism, inhibition of translation of the sense mRNA by direct blocking of the ribosome binding site, has been found in control of plasmid replication and maintenance (Wagner et al. 2004). Examples include replication control by RNAII in streptococcal plasmid pLS1, control of conjugative transfer in plasmids R1 and F by FinP, maintenance control of plasmid R1 by Sok, and transposition control of IS10 by RNA-OUT. In some cases, the antisense RNAs inhibit translation of a leader peptide that itself is required for efficient Rep translation (rev. in Brantl 2007).

The SymR/*symE* antitoxin/toxin system from the *E. coli* chromosome shows that translational inhibition is not restricted to plasmids (Kawano et al. 2007, Fig. 4.2a).

4.2.1.2   Promotion of mRNA Degradation

One of the few antisense RNAs known to influence mRNA stability without effects on translation are λ OOP RNA that facilitates RNase III–dependent decay of the *cIIO* mRNA and RNAα expressed from plasmid pJM1 that affects the stability of

Table 4.1 Overview of cis-encoded antisense RNAs from bacterial chromosomes

Antisense RNA/ target RNA	Length of sRNA (nt)	Species	Biological function	Mechanism of action	Peculiarity
P3 RNA/glnA	43	C. acetobutylicum	Glutamine synthetase	Translation inhibition[a]	
Sof/gef	350	E. coli	Toxin/antitoxin system	Translation inhibition[a]	
Istf/sulA		E. coli	SOS response?	?	
RdlD/ldrD	66	E. coli/relatives	Toxin/antitoxin system	mRNA stability[a]	
GadY/gadXW	105/90/59	E. coli	Acid response	mRNA stabilization	3 GadY species of different length
RatA/txpA	222	B. subtilis	Antitoxin/toxin	mRNA degradation	75 nt overlap of ratA and txpA
IsrR/isiA	176	Synechocystis spec.	Photosynthesis component	mRNA degradation[a]	Regulation by iron stress
SymR/symE	77	E. coli	Antitoxin/toxin	Translation inhibition mRNA degradation	SOS induced
ASdes/desAl	75/110	M. tuberculosis	Lipid metabolism		Induced upon bacterial uptake
AmgR/mgtC	1,200	S. typhimurium	Mg^{2+} homeostasis virulence	RNA degradation	AmgR and mgtC under PhoP control
RyjB/sgcA	90	E. coli	Phosphotransferase II component	?	
RyeA, B/pphA	100, 275	E. coli	Phosphatase I	?	

?, No mechanism proposed

[a] Mechanism proposed but not experimentally substantiated

fatA/B-mRNA in *Vibrio anguillarum* (rev. in Brantl 2007). For the IsrR/*isiA* system involved in photosynthesis in *Synechocystis* sp., *isiA*-mRNA degradation has been suggested but not yet confirmed (Dühring et al. 2006). By contrast, antisense RNA–mediated mRNA degradation is supported experimentally for the RatA/txpA antitoxin/toxin system from *B. subtilis*, although an involvement of RNase III is still elusive. A deletion of the *ratA* promoter and 5' region led to a dramatic increase in *txpA*-mRNA levels, whereas a truncated 177-nt *txpA* RNA detected in the presence of RatA might result from RNase III cleavage of the RatA/*txpA* duplex (Silvaggi et al. 2005, Fig. 4.2b). For SymR/*symE*, a decrease in target RNA stability accompanies translation inhibition by the unusually long-lived SymR (half-life 60 min). RNase III was not required for inhibition (Kawano et al. 2007).

4.2.1.3 mRNA Stabilization Due to a Processing Event

For the *E. coli* GadY/*gadXW* system, a stabilizing effect of GadY on the *gadX* mRNA has been proposed. Both RNAs overlap at their 3' regions. A *gadY*-overexpressing strain displayed 20-fold higher levels of *gadX*-mRNA, whereas a strain with *gadY* promoter mutation showed 4.5-fold reduced *gadX*-mRNA levels. The *gadX* 3' UTR was required for this effect (Opdyke et al. 2004). Recently, base pairing between GadY and the *gadX* 3' UTR was found to stimulate RNase III–dependent cleavage of the unstable *gadXW* mRNA resulting in two short stable products. Another, still unknown RNase is required additionally to RNase III (Opdyke et al. 2011, Fig. 4.2c). How can GadY-dependent cleavage stabilize the *gadX* and *gadW* transcripts? (a) GadY could remain base-paired to the processed 3' end of *gadX* and block recognition of instability determinants in *gadXW* mRNA, (b) GadY-directed cleavage could lead to removal of instability determinants, and (c) full-length *gadXW* mRNA and transcripts produced by GadY-dependent cleavage could fold into secondary structures with different susceptibility to degradation (Opdyke et al. 2011).

4.2.1.4 Transcription Attenuation

This mechanism seemed to be confined to Gram-positive bacteria comprising replication control of staphylococcal plasmid pT181 (Novick et al. 1989; Brantl and Wagner 2000) and streptococcal plasmids pIP501 (Brantl et al. 1993, Fig. 4.2d) and pAMβ1 (rev. in Brantl 2007). The nascent *rep* mRNA can adopt two alternative conformations: Upon binding of the antisense RNA, a terminator stem-loop is induced in the nascent *rep* mRNA, and, consequently, transcription is terminated prematurely upstream of the *rep* SD sequence preventing Rep protein synthesis, and, hence, replication. If the nascent *rep* RNA does not encounter an antisense RNA, it can refold by complementary base pairing between two alternative sequences preventing terminator formation and allowing transcriptional read-through. Subsequently, Rep protein can be synthesized and replication occurs.

The antisense RNA binds and exerts its inhibitory effect only during a short time window (Brantl and Wagner 1994). Recently, the first case for Gram-negative bacteria was reported in *Shigella* (RnaG/*icsA*; Giangrossi et al. 2010, see 8).

4.2.1.5 Transcriptional Interference

To date, this unique mechanism where the antisense RNA acts exclusively *in cis* has been found in only one case (Fig. 4.2e): Four antisense RNAs with different 3′ ends are transcribed convergently to the *ubiGmccBA* operon mRNA in *Clostridium acetobutylicum* (André et al. 2008) from a promoter downstream of the operon terminator T2. This antisense promoter is located downstream of an S-box and is ≈3-fold stronger than the sense promoter. In the presence of methionine, premature transcription termination at T3 of the S-box riboswitch 3′ of the *ubiG* operon leads to a low level of antisense RNA and, due to lack of transcriptional interference, a concomitant increase of sense RNA transcription. By contrast, the absence of methionine results in refolding in the S-box allowing transcription from the anti-sense promoter to proceed through T3 and yields – due to colliding polymerases or accumulation of positive supercoils ahead of the transcribing RNAP – a reduction in *ubiGmccBA* transcription. In some transcription attenuation systems, transcrip-tional interference plays an additional, secondary role (RNAIII/II of pIP501, Brantl and Wagner 1997; RnaG/*icsA*, see 8).

4.2.1.6 Prevention of Formation of an Activator RNA Pseudoknot

In some cases, such as IncB and IncIα plasmids, the efficient translation of the replication initiator protein Rep requires a long-distance activator RNA pseudoknot (Fig. 4.2f). A leader peptide ORF, *repY*, must be translated to allow RepZ synthesis to disrupt an inhibitory stem-loop at the *rep* RBS (rev. in Brantl 2004). This permits formation of a short helix between the target loop and disrupted stem, located 100 nt apart. This long-distance pseudoknot activates *repZ* translation. The corresponding antisense RNAs block both leader peptide translation and pseudoknot formation.

4.2.1.7 Inhibition of Primer Maturation

This mechanism is limited to ColE1 and its related plasmids (Tomizawa et al. 1981; rev. in Brantl 2004) that require a plasmid encoded replication primer that is synthesized as a 550-nt pre-primer (RNAII). For the formation of a persistent RNAII/DNA hybrid within the origin, RNAII must fold into specific secondary and tertiary structures which form during RNAII synthesis in a well-characterized series of events. Afterward, the mature primer, which will be extended by

DNA polymerase I, is generated by RNase H cleavage of the RNA strand of the RNAII/DNA hybrid. Binding of the antisense RNA (RNAI) induces a change in the nascent primer, thereby preventing primer maturation. The kissing complex between RNAI and RNAII is stabilized by the plasmid-encoded Rom protein.

4.2.2   Binding Kinetics, Binding Pathway, and Requirement of RNase III

Antisense/sense RNA binding pathways have been studied in a variety of systems and binding kinetics have been measured (rev. in Brantl 2007). Usually, pairing rate constants were determined to be $\sim 10^6$ M^{-1} s^{-1}. The initial contact between antisense and sense RNA that often form complementary structures, can either occur between two complementary loops (many replication control systems) or between a loop and a single-stranded region (e.g., RNA-IN/RNA-OUT of IS10). In the first case, simple helix progression in both directions is topologically impossible due to accumulating torsional stress. Therefore, loop-loop initiating systems require a subsequent interaction at a distal site to circumvent this limitation. Independent of a one-step or multistep pathway, the final result of the interaction is a complete duplex that is often degraded by the double-strand-specific RNase III (shown for CopA/CopT, RNA-OUT/RNA-IN, and *hok*/Sok). However, formation of complete duplexes is too slow to account for the observed biological effects. Instead, many antisense RNAs mediate inhibition by forming complexes that involve limited numbers of base pairs with their targets. As has been shown for R1, ColE1, pIP501, pT181, and the IncB/Inc1α type plasmids, full duplex formation is not required for control (Wagner and Brantl 1998; Brantl 2007).

For the replication control system of plasmid R1, the binding pathway between antisense RNA CopA and sense RNA CopT has been elucidated in detail (rev. in Brantl 2007). Binding initiates with an unstable loop-loop interaction (kissing complex) that is transformed into an extended kissing complex. Later, a single-stranded region is required to overcome the torsional stress created upon the unidirectional progression of this loop-loop interaction. Afterward, a binding intermediate is formed which contains a four-helical junction. This intermediate is converted into a stable inhibitory complex which is only a partial duplex and is slowly transformed into a stable duplex, which is cleaved by RNAse III. This stepwise binding pathway is, apparently, conserved, and the four-helix junction, although comprising different sequences, is also found as a binding intermediate in Inc1α and related plasmids.

In contrast to R1 and IS10, λ OOP RNA depends on RNase III cleavage to exert control (Krinke and Wulff 1987, 1990), most probably, because the mechanism exerted by OOP is mRNA degradation, whereas in the other systems, steps preceding degradation, i.e., translation initiation, are inhibited.

4.3 Trans-Encoded sRNAs

The first trans-encoded RNA from the bacterial chromosome, MicF, was discovered in 1984 (Mizuno et al. 1984). It forms a 20-bp imperfect RNA duplex with the translation-initiation region of *E. coli ompF* mRNA encoding an outer membrane porin, thereby inhibiting *ompF* translation (Andersen et al. 1990). Until 1999, only a handful of trans-encoded sRNAs were known. Systematic genome searches using various methodologies have revealed that bacteria encode a plethora of trans-encoded sRNAs (e.g., *E. coli*, Argamann et al. 2001; Wassarman et al. 2001; Vogel et al. 2003; *Salmonella*, Sittka et al. 2008; Sittka et al. 2009). Many of the *E. coli* and *Salmonella* sRNA genes are conserved in closely related pathogens. Interestingly, the majority of sRNAs with known functions regulate outer membrane porins (Vogel and Papenfort 2006). However, many sRNA still await the identification of their targets. Two systematic searches in *Pseudomonas aeruginosa* (Livny et al. 2006; Sonnleitner et al. 2008) detected 17 sRNAs. In Gram-positives, five searches have been carried out in *B. subtilis* (Lee et al. 2001; Licht et al. 2005; Saito et al., 2009; Rasmussen et al. 2009; Irnov et al. 2010). Rasmussen et al. found 84 putative noncoding trans-encoded sRNAs in the *B. subtilis* genome, and Irnov et al. increased the total number to 100. Three searches were performed in *S. aureus* (Pichon and Felden 2005; Geissmann et al. 2009; Bohn et al. 2010), three in *Listeria monocytogenes* (Christiansen et al. 2006; Mandin et al. 2007; Toledo-Arana et al. 2009), and two in *Streptococcus* (Halfmann et al. 2007; Perez et al. 2009). Pyrosequencing approaches have allowed the detection of sRNAs in a variety of pathogenic bacteria (e.g., *Chlamydia trachomatis*, Albrecht et al. 2010, *Helicobacter pylori*, Sharma et al. 2010). In addition, in aerobically grown *Rhodobacter sphaeroides*, 20 sRNAs have been detected, four of which are involved in the response to $^1O_2$ (Berghoff et al. 2009). Table 4.2 provides an overview of all currently known trans-encoded sRNAs for which targets were identified.

4.3.1 Biological Functions

Trans-encoded sRNAs have been implicated in a variety of biological functions. Examples include iron transport and storage (RyhB, Massé et al. 2007), phosphosugar stress (SgrS, Rice and Vanderpool 2011), oxidative stress (FnrS, Boysen et al. 2010; Durand and Storz 2010), quorum sensing (Qrr1 to 4, Tu et al. 2007), SOS response (IstR-1, Darfeuille et al. 2007), curli synthesis (OmrA/B, Holmqvist et al. 2010), and plant/*Agrobacterium* interaction (AbcR1, Wilms et al. 2011). Frequently, sRNAs are involved in fine-tuning of metabolic processes which is reflected by the lack of severe phenotypes upon their deletion or overexpression.

One sRNA often regulates a set of mRNAs implicated in the same metabolic pathway: Spot42, initially found to control galactose degradation (Møller et al.

Table 4.2 Overview of trans-encoded antisense RNAs

sRNA (nt)	target RNA(s)	Biological function	Mechanism of action	Control of expression
Escherichia coli				
MicA (72)	*ompA, phoPQ*	Membrane, TCS	TI[a] and RD[b]	σ^E, stationary phase
MicC (109)	*ompC*	Membrane composition	TI and RD	Low temperature
MicF (93)	*ompF*	Membrane composition	TI	High temperature, salt, HU, H-NS, Lrp, OmpR
				SoxS, MarA, Rob
RseX (≈90)	*ompA, ompC*	Membrane composition		?
OmrA/B (88/82)	*cirA, fecA, fepA, ompR, ompT, csgD*	Iron transport, membrane protease, curli synthesis		High osmolarity, OmpR
IpeX (167)	*ompC, ompF*	Membrane composition	RD?	phage-encoded porin
RybB (78)	*ompC, ompW*	Membrane composition		σ^E
RybC=MicM (80)	*dpiBA*	TCS (SOS response)	TI?	*chBCD*
	chiP (ybfM)	Chitoporin	TI, RD	
RyhB (90)	*sodB, sdhD, ftn, acnB, fumA, bfr, fur*	Iron metabolism	TI and RD	Fur, iron
	cysE	Serin acetyl transferase	TI	
	shiA	Shikimate permease	TA[c]	
	entCEBAH	Enterobactin biosynthesis	TA?	
Spot42 (109)	*galK, fucI/K, gltA, maeA gltA, nanC/T, srlA, sthA*	Central and secondary metabolism, redox balance	TI	cAMP-CRP, glucose
SgrS (227)	*ptsG, manXYZ*	Glucose/mannose transp.	TI and (RD)	SgrR, phosphosugar stress
DsrA (87)	*rpoS*	Stationary phase	TA	Low temperature, osmotic shock, LeuO?
	hn-S	Stationary phase	TI	Osmotic shock, RcsB
RprA (106)	*rpoS*	Stationary phase	TA	
ArcZ (55/88/120)	*rpoS*	Stationary phase	TA	ArcA/B
OxyS (109)	*rpoS*	Stationary phase	Hfq binding	Oxidative stress
	fhlA	Formate metabolism	TI	

sRNA	Target genes	Function/role		Regulator/condition
IstR-1 (75)	*tisB*	Antitoxin	TI	Constitutive expression
GlmZ (210/155)	*glmS*	Amino-sugar-metabolism	TA	GlmY prevents processing
CyaR (86)	*ompX, luxS, nadE,yqaE*	Membrane, QS, NAD-syn	TI	CRP
FnrS (122)	*sodA/B, maeA, gpmA, folE/X, cydDC, metE*	Enzymes linked to anaerobic conditions / Oxidative stress		FNR
Salmonella typhimurium/enterica				
MgrR (98)	*eptB, ygdQ*	LPS modification		PhoPQ, Mg2+, Ca2+, Mn^{2+}
SibA-E (110/150)	*ibs*	Toxic peptide regulation	?	Minimal medium
OhS (not published)	*shoB*	Toxic peptide regulation	?	
RydC (62-64)	*yejABEF*	ABC transporter?	RD	
MicA (72)	*ompA*	Membrane composition	TI?	σ^{E}
RybB (78)	*ompC, D, F, N, S, chiP*	Membrane, chitobiose	TI? and RD?	σ^{E}
GcvB (130)	*oppA, dppA,gltI livK, livJ, argT*	Peptide transport ABC transporter	TI	GcvA, GcvR, *STM4351*
InvR (80)	*ompD*	Membrane composition	TI	HilD
CyaR (86)	*ompX*	Membrane composition	TI	CRP
ChiX (80)	*chiP*	Chitoporin	TI	*chBCD*
ArcZ (110/50)	*sdaCB, tpX, STM3216*	Serin uptake, chemotaxis	TI	Constitutive expression
IsrJ	74	Virulence-associat. effector	?	Low O_2 and Mg++
Pseudomonas aeruginosa				
PrrF1/PrrF2	*sodB, sdhD, bfr*	Iron metabolism	TI and RD	
PhrS (213)	*pqsR*	Quorum sensing, virulence	TA	ANR
Agrobacterium tumefaciens				
AbcR1 (117)	*atu2422/1879, frcCAK*	Plant/bacterial interaction	TI and RD	Stationary phase
Chlamydia trachomatis				
IhtA (120)	*hctA*	Histon homologue	TI	
Borrelia burgdorferi				
DsrA$_{BB}$(213-352)	*rpoS*	σ-factor, virulence factor	TA	Temperature

(continued)

Table 4.2 (continued)

sRNA (nt)	target RNA(s)	Biological function	Mechanism of action	Control of expression
Bacillus subtilis				
SR1 (205)	*ahrC*	Arginine catabolism	TI	CcpN, CcpA
FsrA (≈84)	*sdhCAB, citB, yvfW, leuCD*	Iron metabolism and storage	TI	Fur, iron; some targets need FbpA, B or C
Staphylococcus aureus				
RNAIII (514)	*hla*; *spa, rot, sa1000/2353, coa*	Hemolysine synthesis; Host–pathogen interaction	TA; TI and RD	AgrC, AgrA, stationary phase
SprA (202)	*SA2216-ORF*	ABC transporter?	Posttranslational?	Strain-specific
SprD (142)	*sbi*	Immune response	TI, RD	Growth phase
Streptococcus pyogenes				
Pel RNA (459)	*speB, emm, sic, nga*	Cysteine protease M- and -related proteins	Post-transcriptional transcriptional control	Stationary phase, conditioned media
FasX (205)	*ska*	Streptokinase	RNA stabilization	FasA
RivX (289,237,189)	*mga*	Virulence	TA?	CovR
Vibrio cholerae/ harveyi				
Qrr1-4 (96-108)	*hapR*	QS	RD	LuxO-P, σ^{54}
VrtA (140)	*ompA, ompT, tcpA*	Membrane, pili	TI	σ^{E}
Clostridium perfringens				
VR (386)	*colA, plc*	Collagenase, phospholipase	RNA processing	VirR/S
Neisseria mengitidis				
NrF	*sdhC, sdhA*	Iron metabolism	TI	

[*] Only sRNAs for which target genes have been identified allowing to conclude that they are trans-encoded. *QS* quorum sensing, *TI*[a] translation inhibition, *RD*[b] mRNA degradation, *TA*[c] translation activation. Control of expression: All proteins and growth conditions known to regulate sRNA expression are listed. It is not indicated whether these factors promote or inhibit sRNA expression. TCS, two-component system

2002), eventually turned out to be a regulator of central and secondary metabolism and redox balance (Beisel and Storz 2011). In the envelope stress response in *E. coli* and *Salmonella*, two σ^E-controlled sRNAs, MicA and RybB, are involved. Whereas MicA facilitates selectively the decay of *ompA* mRNA (Udekwu et al. 2005; Rasmussen et al. 2005), RybB accelerates the decay of at least eight *omp* mRNAs encoding outer membrane porins (Papenfort et al. 2006). Another example is *E. coli* GcvB that regulates seven ABC transporter mRNAs by targeting C/A-rich elements inside and upstream of RBS using its G/U-rich single-stranded central region (Sharma et al. 2007). In *Bacillus subtilis*, FsrA – similar to *E. coli* RyhB – is repressed by Fur and regulates at least four target mRNAs involved in iron metabolism and storage (Gaballa et al. 2008). In contrast to RyhB, FsrA cooperates with one or more Fur-regulated small basic proteins FbpA, FbpB, and FbpC. On the other hand, one mRNA may be targeted by several sRNAs under different environmental conditions. *E. coli rpoS* mRNA encoding sigma factor σ^S is translationally activated by DsrA at low temperatures (Sledjeski et al. 1996), by RprA at osmotic shock and cell surface stress (Majdalani et al. 2001), and by ArcZ under aerobic conditions (Mandin and Gottesman 2010). Under oxidative stress, *rpoS* is downregulated by OxyS (Altuvia et al. 1997).

Interestingly, a few trans-encoded sRNAs have dual functions: They act both as base-pairing sRNAs and as peptide-encoding mRNAs. The first reported example was the *S. aureus* RNAIII encoding δ-hemolysin (26 aa) (Morfeldt et al. 1995). Later, the streptolysin SLS-ORF of *Streptococcus* Pel RNA (Mangold et al. 2004) and the 43 codon-SgrT ORF on *E. coli* SgrS (Wadler et al. 2007) were identified. SgrS and SgrT downregulate PtsG glucose transporter activity and have a physiologically redundant, but mechanistically distinct function in inhibition (Wadler and Vanderpool 2007). Recently, *B. subtilis* SR1-ORF was found to encode a 39 aa peptide (SR1P) which interacts with GapA (Gimpel et al. 2010). Both SgrT and SR1P are evolutionarily conserved in Gram-negatives (Horler and Vanderpool 2009) and in *Bacillus/Geobacillus* species (Gimpel et al. 2010), respectively. The functions of *hyp7* ORF on *Clostridium perfringens* VR (Shimizu et al. 2002), the 37 codon PhrS-ORF of *Pseudomonas aeruginosa* (Sonnleitner et al. 2008), the 32 codon RivX-ORF (Roberts and Scott 2007), and the RSs0019-ORF of *Rhodobacter sphaeroides* (Berghoff et al. 2009) are still unknown.

Computer approaches to identify bacterial sRNAs and their target binding sites are summarized in the chapter 11 of this book. Furthermore, chapter 14 summarizes and explains in detail methods for the experimental analysis of RNA-based regulations.

4.3.2 Mechanisms Employed by trans-Encoded Antisense RNAs

The most prevalent mechanisms used by trans-encoded sRNAs are inhibition of translation and promotion of RNA degradation. Furthermore, a few trans-encoded sRNAs activate translation of their target mRNAs or stabilize them.

More indirect mechanisms comprise RNA trapping and inhibition of leader peptide translation. Currently known mechanisms are discussed below and summarized in Fig. 4.3.

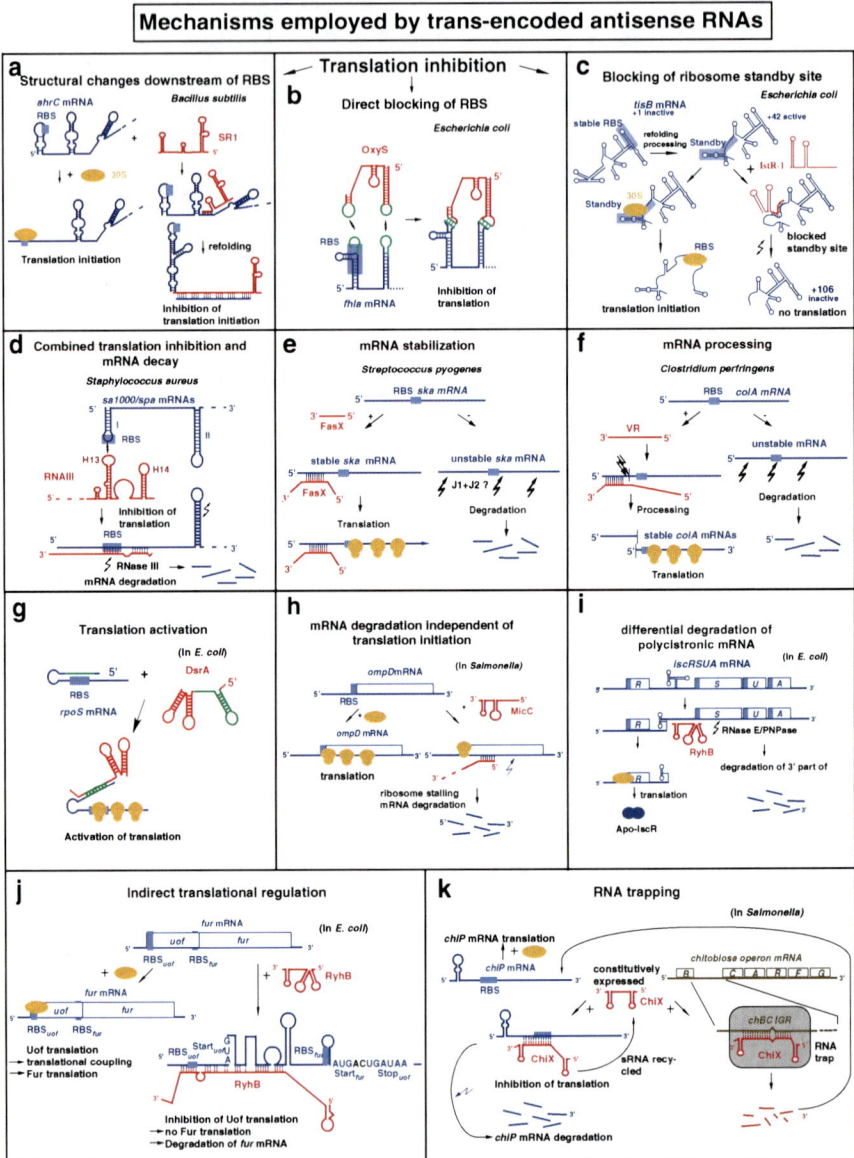

Fig. 4.3 *Overview of regulatory mechanisms employed by trans-encoded antisense RNAs.* Antisense RNAs are drawn in *red*, sense RNAs in *blue*. *Green* denotes regions complementary between sense and antisense RNA, *light blue*, ribosome binding sites. *Violet arrows* indicate RNase E cleavage; *black arrows* indicate unknown RNase action. *Yellow* symbols denote ribosomes. For details see text. (**a–d**) are based on Brantl (2009)

4.3.2.1 Translational Regulation

Translation Inhibition by Direct Blocking of the RBS

Translation inhibition by direct blocking of the RBS is the most widespread mechanism. Examples include OxyS, Spot42, MicA, MicC, MicF, RyhB, RybB, OmrA/B, and SgrS from *E. coli*. The complementary regions of the sRNAs and their target mRNAs overlap the RBS or the RBS and/or the adjacent 5′ or 3′ regions. For MicA and MicF, base pairing includes the *ompA* and *ompF* RBS, respectively, whereas MicC uses two complementary regions (6 and 16 continuous bp) immediately upstream of the *ompC* RBS (Chen et al. 2004). For OxyS/*fhlA*, two interacting regions of 7 and 9 bp overlapping the RBS and about 25 nt downstream from the *fhlA* start codon, were found (Altuvia et al. 1998, Fig. 4.3b). For SgrS, out of 23 complementary bp, only 6 bp around the RBS are crucial for inhibition (Kawamoto et al. 2006). This aspect is very similar to the cis-encoded antisense RNAs, where a nucleation step is essential for fast recognition and efficient regulation. Reminiscent of miRNA regulation in eukaryotes, *E. coli,* and *Salmonella* RybB use a 5–7-nt seed sequence at their 5′ end for multitarget recognition (Papenfort et al. 2010; Balbontín et al. 2010). A similar pattern was found for OmrA/B (Guillier and Gottesman 2008).

Translation Inhibition by Induction of Structural Changes Downstream from the RBS

To date, the only known example is SR1 from *B. subtilis*, which interacts with *ahrC* mRNA encoding the transcriptional activator of the arginine catabolic operons (Heidrich et al. 2006). Both RNAs share seven complementary regions A to G in the 3′ half of SR1 and the central part of *ahrC*. Region G is located ≈100 nt downstream from the *ahrC* RBS. Binding of SR1 induces structural alterations not only in all complementary regions but also immediately downstream from the *ahrC* RBS and upstream of region G, resulting in inhibition of translation initiation (Heidrich et al. 2007, Fig. 4.3a). Apparently, even base pairing far downstream from the RBS can prevent binding of the 30S ribosomal subunit.

Translation Inhibition by Blocking of a Ribosome Standby Site

So far, the only example is IstR-1/*tisAB* of *E. coli* (Darfeuille et al. 2007, Fig. 4.3c). Here, ≈100 nt upstream of the *tisB* RBS a standby site for ribosomes has been found, which is required for efficient synthesis of the TisB toxin from this highly structured RBS. The antisense RNA, IstR-1, is complementary to this site and competes with standby ribosomes for binding. The IstR-1/*tisB* interaction generates a cleavage site for RNase III which in turn results in a 5' truncated *tisB* mRNA which cannot be translated anymore.

Combined Translation Inhibition and mRNA Degradation

Frequently, translation inhibition is accompanied by target RNA degradation by RNase E or RNAse III. Degradation by RNase III is necessary for inhibition by *S. aureus* RNAIII (Huntzinger et al. 2005). As a base-pairing sRNA, RNAIII does not only activate translation of the α-hemolysin mRNA (Morfeldt et al. 1995) but also inhibits translation of several targets as the main surface adhesion protein Spa (see Fig. 4.3d), fibrinogen-binding protein SA1000, pleiotropic transcriptional factor Rot (Boisset et al. 2007), and staphylocoagulase *Coa* (Chevalier et al. 2010). Whereas the 5' domain of RNAIII encodes δ-hemolysin, the 3' domain carrying two redundant hairpin loop motifs (Benito et al. 2000) is decisive for base-pairing interactions. For all targets, the formation of RNAIII/mRNA duplexes results in inhibition of ribosome binding and favors recognition by RNase III (see Fig. 4.3). Specificity for RNAIII is obtained by either propagating the first loop-loop contact at the RBS into the stem regions (*sa1000* and *sa2353* mRNAs) or by addition of a second loop-loop interaction (*rot* and *coa* mRNAs). The *coa, spa,* and *SA1000* mRNAs carry a 5' hairpin structure which stabilizes these RNAs in the absence of RNAIII. Therefore, the coordinated action of RNAIII and RNase III is needed to irreversibly repress virulence factor synthesis. RNAIII guides RNase III to the repressed mRNAs in vivo (Huntzinger et al. 2005).

In many Gram-negative bacteria, degradation of translationally repressed mRNAs is assumed to be a consequence of ribosome exclusion rather than the primary event, because translation inhibition can occur in the absence of mRNA degradation. This was demonstrated for SgrS/*ptsG*, RyhB/*sodB* (Morita et al. 2006), and IstR-1/*tisAB* (Darfeuille et al. 2007). However, recently it was shown that RyhB/*sodB* base pairing initiates RNase E cleavage of *sodB* mRNA independent of translation and, surprisingly, >350 nt downstream from the RBS (Prévost et al. 2011) supporting an active cleavage model. A distal cleavage site may prevent RNase E action before ribosomes have cleared the upstream mRNA sequence.

Indirect Regulation by Inhibition or Activation of Leader Peptide Translation

An indirect manner of regulation can be used when translation of a target RNA with a suboptimal RBS is coupled to translation of a leader peptide. This is the case for *E. coli fur* mRNA encoding the negative regulator of iron uptake (Fig. 4.3j). Binding of RyhB to the RBS and first codons of *uof* RNA inhibits synthesis of the 28 aa leader peptide and, thereby, prevents *fur* translation (Vecerek et al. 2007). A similar but activating mechanism was found for *P. aeruginosa* PhrS which promotes translation of a 40-codon ORF *uof* which is translationally coupled to *pqsR*. PqsR is the key regulator of quorum sensing and virulence (Sonnleitner et al. 2011).

Translation Activation

In some cases, the ribosome binding site is located in double-stranded structures which prevent the access of ribosomes. Melting of these regions is promoted by binding of the antisense RNA to one strand, thus liberating the complementary strand containing the RBS and, hence, activating translation. To date, seven examples are known: *S. aureus* RNAIII activates translation of *hla* mRNA encoding α-hemolysin (Morfeldt et al. 1995). In *E. coli*, DsrA (Fig. 4.3g), RprA, and ArcZ promote translation of σ^S (see above), RyhB supports translation of the shikimate permease *shiA* (Prévost et al. 2007), and GlmZ activates translation of *glmS* encoding glucosamine-6-phosphate synthase (Urban and Vogel 2008; Reichenbach et al. 2008). Streptococcal RivX probably enhances *mgA* translation (Roberts and Scott 2007, see 8).

4.3.2.2 Effects on mRNA Stability

mRNA Stabilization by sRNA Binding

FasX RNA from *Streptococcus pyogenes* was shown to bind upstream of the RBS to *ska* mRNA encoding streptokinase. The resulting partial duplex stabilizes *ska* mRNA and allows translation (Ramirez-Peña et al. 2010; Fig. 4.3e). In the absence of FasX binding, *ska* mRNA is rapidly degraded, presumably by RNases J1 and J2.

Specific mRNA Processing Generating Two Stable RNAs

Binding of VR RNA from *Clostridium perfringens* to the 5′ UTR of *colA* mRNA encoding collagenase induces a specific processing event upstream of the *colA* RBS (Obana et al. 2010), which generates two stable RNAs and allows *colA* translation (Fig. 4.3f). In the absence of VR, *colA* mRNA is rapidly degraded. Interestingly, base pairing between VR and *colA* occurs within the VR *hyp7* ORF.

Differential Degradation of a Polycistronic mRNA

RyhB binds between *iscR* and *iscS* to the intergenic region of *iscRSUA* mRNA encoding the enzymes for the biosynthesis of Fe-S clusters. However, in contrast to VR action on *colA*, RyhB-induced processing entails differential mRNA degradation, since only the 5′ processing product is stable, whereas the 3′ product is degraded by RNase E and PNPase (Fig. 4.3i, Desnoyers et al. 2009).

mRNA Degradation Independent of Translation Initiation

Whereas *E. coli* MicC inhibits translation initiation of *ompC*, *Salmonella typhimurium* MicC targets *ompD* by binding within the coding sequence (codons 23–26) resulting in accelerated mRNA decay (Pfeiffer et al. 2009). MicC/*ompD* base pairing does not inhibit translation initiation but leads to transient ribosome stalling and reveals an RNase E site 4–5 nt downstream of the complementary region (Fig. 4.3h). Since the elongating 70S ribosome has a strong helicase activity, it is unlikely that the MicC/*ompD* duplex can permanently stall elongating ribosomes explaining while MicC fails to inhibit translation.

4.3.2.3 RNA Trapping

The observation that an sRNA was not destabilized upon overexpression of its target RNA but acts catalytically resulted in discovery of RNA trapping. This unusual mechanism in which an sRNA is converted from regulator to target has been found for *E. coli* MicM and its *Salmonella* homologue ChiX (Overgaard et al. 2009; Figuera-Bossi et al. 2009, Fig. 4.3k). In the absence of inducer, ChiX inhibits translation of *chiP* mRNA by binding to its RBS. Whereas *chiP*-mRNA is degraded, ChiX is recycled. Its amount is not controlled at transcriptional level. Instead, in the presence of chitooligosaccharides, a trap mRNA – *chBCARFG* mRNA – is transcribed that pairs with and promotes ChiX degradation, thereby abolishing silencing of the cognate ChiX target *chiP*.

4.3.3 Role of Hfq

One important hallmark of many trans-encoded antisense RNAs from *E. coli* is their ability to bind the RNA chaperone Hfq. An excellent review summarizes the important properties of Hfq (Brennan and Link 2007): Hfq was identified in *E. coli* as host factor for bacteriophage Qβ replication and is present in half of all sequenced bacterial species. *Bacillus anthracis*, *Ralstonia metallidurans,* and *Burkholderia cenocepacia* encode even two Hfq proteins, the latter ones having two distinct biological roles (Ramos et al. 2011). Hfq comprises between 70 and 110 amino acids. In stationary phase, up to 60.000 Hfq monomers/*E. coli* cell have been measured. The major fraction is associated with the ribosomes, whereas a minor Hfq fraction appears to be associated with the nucleoid. Hfq binds to AU-rich sequences in single-stranded regions generally flanked by one or two stem-loops. The Hfq homohexamer is very similar to the eukaryotic Sm and Sm-like proteins involved in splicing. It forms a toroid with an outer diameter of ≈70 Å and a thickness of 25 Å. The central pore is 8–12 Å wide. The N-terminal α-helix is followed by five β-strands that form a tightly bent sheet. The repetition of identical binding pockets on the Hfq hexamer suggests that the binding surface can accommodate more than just a single RNA target. This would allow simultaneous binding of two RNA strands and could promote the interaction between these strands, which is important in sRNA/target mRNA interactions. Hfq is involved in mRNA stability and polyadenylation, translation, virulence, bacteriocin production, and nitrogen fixation. It interacts with the 30S ribosomal subunit, the ribosomal protein S1, PNP, PAPI, and the C-terminal scaffold domain of RNase E.

Many trans-encoded sRNAs, for example DsrA, Spot42, and RyhB, require Hfq for their stability (Valentin-Hansen et al. 2004). In other cases, for example OxyS/*fhlA*, Spot42/*galK*, RyhB/*sodB*, MicA/*ompA,* and SgrS/*ptsG*, Hfq was shown to promote the interaction between sRNAs and their targets (Zhang et al. 2003; Kawamoto et al. 2006). A FRET study on DsrA/*rpoS* demonstrated that Hfq accelerates strand exchange and subsequent annealing between sRNA and *rpoS* mRNA, which results in exposure of the *rpoS* RBS (Arluison et al. 2007). New data revealed that C-terminally truncated *E. coli* Hfq variants are fully capable of promoting post-transcriptional control indicating that the C-terminal tail of *E. coli* Hfq plays a small or no role in riboregulation (Olsen et al. 2010). Recently, it was shown that RNAs displace each other on Hfq on a short time scale by RNA concentration–driven (active) cycling (Fender et al. 2010). This explains the paradox of an Hfq-RNA Kd value in nM range, long half-lives of Hfq-RNA complexes but the necessity for a 1–2-min response time for regulation in vivo. Lately, a DEAD box helicase required at low temperatures for *rpoS*-DsrA annealing additionally to Hfq has been found (Resch et al. 2011).

Among sRNAs from Gram-positive bacteria, *Listeria monocytogenes* LhrA is the only example for Hfq-dependent antisense regulation (Nielsen et al. 2010). Whereas some of the identified sRNAs bind Hfq, in at least two cases, *B. subtilis* SR1/*ahrC* and *S. aureus* RNAIII/*spa*, no influence of Hfq has been found on sRNA/ target interaction (Heidrich et al. 2007; Bohn et al. 2007). Interestingly, *Streptococci* do not encode Hfq, and *S. aureus* Hfq is not highly expressed. It is conceivable that other RNA binding proteins fulfill the role of Hfq in Gram-positive bacteria.

4.4   Differences Between cis- and trans-Encoded Antisense RNAs

Cis-encoded RNAs are complementary to their targets over a large nucleotide stretch and can, therefore, form stable duplexes with their target RNAs. Although in two plasmid cases, ColE1 and R1/F, plasmid-encoded RNA binding proteins (Rom, and FinO, respectively) were shown to have an effect (rev. in Brantl 2007), cis-encoded RNAs usually do not require an additional protein to facilitate complex formation with their targets. ColE1 Rom promotes sRNA/target RNA pairing only fivefold, since the inhibition rate is primarily determined by the binding rate constant and not the binding affinity between the loop-loop complexes. The FinO protein of F plasmid acts by promoting strand exchange between FinP and *traJ* mRNA (Arthur et al. 2003, 2011), but its key function is to protect the antisense RNA FinP against RNase E degradation (Jerome et al. 1999), so that the repression effect of FinO is 5–20-fold. Plasmid R1 and ColE1 replication control was found to be functional in a Δ*hfq E. coli* strain, although both antisense RNAs bound Hfq.

In contrast, many trans-encoded antisense RNAs from Gram-negative bacteria need Hfq either for stabilization or for complex formation, most probably to facilitate the interaction with their only partially complementary target RNA(s) (see above). It is not clear, whether at least in some Gram-positive bacteria another protein fulfills the role of Hfq.

The existence of U-turn motifs (5′ YUNR) has, so far, only been studied for cis-encoded antisense RNAs and their targets in plasmids: Here, one loop in either the antisense or sense RNA forms a U-turn structure (Franch et al. 1999) that is characterized by a sharp bend in the phosphosugar backbone 3′ of the YUNR motif that presents the following three or four bases in a solvent-exposed stacked conformation providing a scaffold for the rapid interaction with the complementary RNA. Both for *hok*/sok of plasmid R1 (Franch et al. 1999) and RNAIII/RNAII of plasmid pIP501 (Heidrich and Brantl 2003), it has been demonstrated that a U-turn structure in one sense RNA loop is important for efficient interaction with the antisense RNA.

Each cis-encoded antisense RNA uses one defined regulatory mechanism on its single target. By contrast, trans-encoded sRNAs can employ different mechanisms on different target mRNAs. For instance, RNAIII, DsrA, and RyhB inhibit translation of one/several targets, but activate translation of others (see above).

Whereas control of translation and RNA stability have been found as regulatory mechanisms for both cis- and trans-encoded sRNAs, transcriptional interference (André et al. 2008, Fig. 4.2d) can be exclusively used by cis-encoded sRNAs, because it requires a cis-acting sRNA.

Notably, a trans-encoded antisense RNA may be under control of another small RNA: The GlmY/GlmZ sRNA pair acts hierarchically to regulate GlmS synthesis in *E. coli*. Thereby, GlmY inhibits processing of GlmZ from a 210-nt into an active 155-nt species that in turn activates translation of *glmS* mRNA (Urban and Vogel 2008; Reichenbach et al. 2008). It remains to be seen whether RNA-controlled antisense RNAs will be also found for cis-encoded sRNAs.

4.5 Regulatory Circuits Involving Base-Pairing sRNAs

Regulatory sRNAs acting via base pairing may be integrated in global regulatory networks. Typically, the regulator is a protein that responds to environmental stimuli. Alternatively, sRNAs levels can be regulated by competition with other RNAs, as in the case of GlmZ or ChiX (see 4). Four types of regulatory circuits incorporate the action of base-pairing sRNAs: single-input module (SIM),

Regulatory circuits for basepairing sRNAs

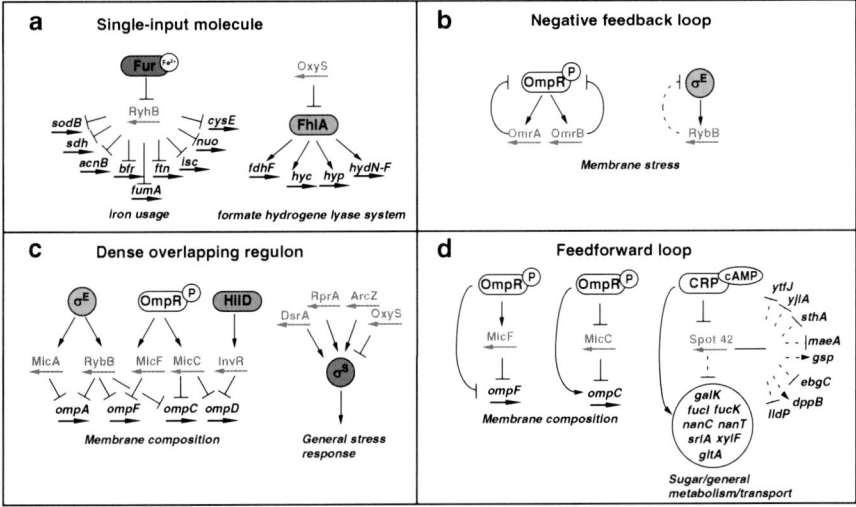

Fig. 4.4 *Regulatory circuits for base-pairing sRNAs.* Examples are from *E. coli* and *Salmonella enterica. Circles and ellipses* denote sigma factors and transcription regulators, respectively. The regulatory protein is sensitive to an environmental stimulus, e.g., intracellular iron (Fur) or osmotic shock (OmpR phosphorylation). *Thick grey arrows* indicate sRNAs; *thick black arrows,* target RNAs. Thin *arrows* indicate activation; thin bars, repression. Dashed, indirect regulation. Based on Beisel and Storz (2010). For details see text (5)

dense-overlapping regulon (DOR), negative feedback loop (NFL), and feedforward loop (FFL) (Beisel and Storz 2010; Fig. 4.4).

In SIMs, a single regulator activates or represses the expression of multiple genes to produce a coordinated response to environmental changes. None of the target genes regulate each other. The sRNAs often reverse the relationship between the environmental sensor and the sRNA targets: the repressor becomes an indirect activator as in the case of Fur/RyhB (Fig. 4.4a) and vice versa. RyhB is repressed by Fur under high intracellular iron concentrations. When iron is scarce, derepressed RyhB downregulates at least 18 operons encoding iron-using proteins involved in iron storage, TCA, dismutation of superoxide radicals, synthesis of siderophores, etc. (Massé et al. 2007; Salvail et al. 2010). sRNAs targeting multiple genes may establish a hierarchical order of regulation. First, sRNAs could delay the regulation of particular target genes to provide a temporal response. Second, under a low or transient stimulus, only the expression of genes with extensive base pairing to the sRNA would be affected. Eventually, sRNA-based SIMs can indirectly influence global expression by controlling other regulators: *E. coli* OxyS represses the synthesis of FhlA, an activator involved in formate metabolism (Fig. 4.4a).

DORs coordinate the response to multiple biological signals by combining various overlapping SIMs. The best characterized example is the regulation of outer membrane proteins in *E. coli* and *Salmonella enterica* (Fig. 4.4c). Omps are prevalent sRNA targets as they play pivotal roles in cell survival acting as gatekeepers for small molecules entering and leaving the cell, as recognition elements for the host immune system and phage infection (Valentin-Hansen et al. 2007). Another DOR involves σ^S, the master transcription regulator of the general stress response. With the help of sRNAs, it integrates distinct stress signals (see 4, Fig. 4.4c).

Hfq-binding sRNAs can only participate in mixed feedback loops with transcription factors, because they cannot directly influence their own transcription. Such mixed loops are exclusively NFLs. In direct feedback loops, the sRNA targets its own regulator, as in the OmrA/B case, while in indirect feedback loops, the sRNA affects the activity or expression of its regulator by targeting other genes (Fig. 4.4b): Cell envelope stress frees sequestered σ^E, which activates the transcription of RybB. RybB, in turn, downregulates the expression of major Omps, thereby reducing the buildup of Omps that contributes to envelope stress (Vogel and Papenfort 2006). One of the most complex NFLs is provided by quorum-sensing networks in *Vibrio* species (Ng and Bassler 2009).

Three FFLs including sRNAs are known so far (Fig. 4.4d): OmpR participates in two FFLs; it downregulates the expression of *ompF* and upregulates the expression of MicF, which translationally represses *ompF*. On the other hand, OmpR upregulates *ompC* expression and downregulates MicC expression which translationally represses *ompC*. OmpR is activated through phosphorylation in response to increased osmolarity, for example (Vogel and Papenfort 2006). Recently, the involvement of Spot 42 in a multi-output FFL was shown to help enact catabolite repression (Beisel and Storz 2011). Spot 42 represses multiple genes, many of which are activated by CRP, and, in turn, is transcriptionally repressed by CRP. These genes are involved in central and secondary metabolism, redox balancing, and consumption of nonpreferred carbon sources. By reducing the leaky expression of genes unnecessary for glucose catabolism, Spot 42 helps divert metabolic resources toward cell growth and glucose consumption. OmpR-containing FFLs introduce a lag in repression after removal of the detected signal which depends on the sRNA half-life. The lag was relatively shorter for sRNA-based FFLs comprising a short-lived sRNA (Shimoni et al. 2007).

Why are sRNAs used instead of regulatory proteins (transcription factors = TFs)? The following explanations (for details see Beisel and Storz 2010) are conceivable: (a) chance incorporation without regulatory advantage, (b) reduced metabolic costs, (c) need for additional layers of regulation, (d) faster regulation, and (e) unique regulatory properties of sRNAs. (a) The first hypothesis would predict that TFs and regulatory sRNAs can occupy the same niche. This question cannot be answered at the moment. (b) The reduced metabolic costs, due to the small size of an sRNA and its encoding gene, the limited energy to transcribe the ≈100–200-nt sRNA, and no energy required to translate the sRNA would confer a selective advantage on sRNAs over TFs. However, the total metabolic cost also includes transcription of the target gene and relative levels of either sRNA or the TF mRNA. The relative metabolic cost may be sRNA-specific and even target gene–specific. (c) Layered regulation is especially important for genes that must be tightly controlled or are critical in multiple cellular responses. By targeting an entirely separate part of the gene compared to TFs sRNAs expand the number of sites at which regulation can occur. The average number of TF binding sites is independent of genome size, and, therefore, transcriptional regulatory architecture in larger genomes is more complex than in smaller genomes (Molina and van Nimwegen 2008). It is still unclear whether sRNAs play a larger role in larger or in smaller genomes. (d) Faster regulation may be beneficial in coordinated regulatory processes or under sudden environmental changes. Since sRNAs act at the post-transcriptional level, gene expression is modulated at a point closer to protein production and, consequently, less time is required to affect target protein levels compared to TFs. Bioinformatic studies are needed to compare the regulatory dynamics of genes controlled by TFs or sRNAs. (e) The response curves and noise profiles of protein-based and sRNA-based regulation are significantly different. While protein-based regulation shows a graded response to repressor levels, sRNA-based regulation for a fast rate of sRNA action shows a two-regime response to the rate of sRNA and mRNA production with an ultrasensitive transition (sRNA/mRNA production ≈ 1) between regimes. However, for a slow-rate sRNA action, there is less distinction between both systems. When sRNA production dominates mRNA production, target protein levels are low and noise (cell-to-cell-variability) is dampened. At sRNA/mRNA production ≈ 1, noise builds and maximizes at the transition point. Consequently, when sRNA production dominates, sRNA-based regulation is advantageous due to tight repression and low noise. This is particularly important when target gene products are harmful to viability or important to establish a certain phenotype.

4.6 sRNAs that Act Via Protein Binding

Three groups of protein-binding sRNAs are known; 6S RNA that binds RNA polymerase, sRNAs that sequester small proteins which regulate target mRNA translation, and sRNAs that modulate enzyme activity. The latter group is small and represented only by Rcd of *E. coli* plasmid ColE1 (Chant and Summers 2007), and will not be discussed here.

4.6.1 6S RNA

In 2000, it was discovered that *E. coli* 6S RNA might interact with RNA polymerase (Wassarman and Storz 2000). To date, more than 100 putative 6S RNA homologues have been identified in diverse bacteria (rev. in Wassarman 2007). 6S RNA has a characteristic secondary structure consisting of a central region with a largely single-stranded internal loop, which is flanked by two long irregular double-stranded stem regions that are interrupted by small bulges. This secondary structure resembles a DNA transcription bubble leading to a hypothesis for the function of 6S RNA (Fig. 4.1). 6S RNA forms a stable complex with RNA polymerase, thereby acting as an open promoter DNA mimic that interferes with the formation of transcription initiation complexes. Its levels increase $\approx$10-fold to about 10.000 molecules during stationary phase. 6S RNA interacts preferentially with RNA polymerase containing the exponential phase sigma factor σ^{70}. Therefore, it was proposed that it participates in shifting global gene expression from exponential to stationary phase. However, a recent demonstration of weak contacts between 6S RNA and stationary phase sigma factor σ^{38} and the fact that 6S can inhibit both RNA polymerase holoenzymes (Gildehaus et al. 2007) show that the molecular details for the action of 6S RNA have still to be elucidated. Surprisingly, 6S RNA can act as a template for the production of small de novo transcripts: 14–22-nt transcripts (pRNA) were observed at high 6S concentration (Wassarman and Saecker 2006) whereas 170-nt transcripts of unknown function were observed at high RNA polymerase (RNAP) concentrations (Gildehaus et al. 2007). Synthesis of the pRNA during outgrowth from stationary phase destabilizes the 6S-RNAP complexes and leads to release of the pRNA-6S RNA hybrid, thus liberating RNAP from 6S RNA in response to nutrient availability. 6S RNA sensitive promoters were suggested to have weak -35 boxes or extended -10 boxes or both, whereas transcription from promoters with strong -35 boxes was not inhibited by 6S RNA. A model of competition between 6S RNA and promoters for σ^{70} binding was proposed (Cavanagh et al. 2008), and binding of 6S RNA was shown to require a positively charged surface of σ^{70} region 4.2 (Klocko and Wassarman 2009). However, a recent transcriptome analysis with an *ssrS* strain revealed 245 genes during exponential phase and 273 genes during early stationary phase to be $\geq$1.5-fold differentially expressed. This suggested that there are additional functions of 6S RNA different from downregulation of certain σ^{70}-dependent genes during stationary phase. Interestingly, 6S depletion causes a decrease in the expression of the translation machinery, in particular rRNA transcription (Neusser et al. 2010). In *E. coli*, 6S RNA transcription from its two promoters P1 and P2 depends on four global regulators and is inhibited by H-NS and LRP and, to a lesser extent, by StpA, while FIS seems to act as a dual regulator (Neusser et al. 2008).

4.6.2 Small RNAs that Act by Sequestration of Translational Regulators

In 1997, it was shown that a small untranslated RNA, CsrB from *E. coli*, does not act via base pairing but exerts its function by binding of the small protein CsrA, a repressor of gluconeogenesis and biofilm production and activator of glycolysis, motility, and acetate metabolism (Liu et al. 1997). CsrA binds to the untranslated leader of the *glgCAP* transcript, where it blocks translation and causes rapid mRNA degradation. Positive control of gene expression by CsrA involves binding to the leader and stabilization of the corresponding mRNAs, for example of *flhDC*, but the detailed mechanism is still elusive. Highly conserved CsrA homologues with monomer sizes of about 7 kD are found in diverse eubacteria and regulate virulence factors in animal and plant pathogens (rev. in Babitzke et al. 2007). CsrA homologues act as dimers and encompass five β-strands and a C-terminal α-helix, of which β1 and β5 are important for RNA binding. The dimer contains two symmetrical surfaces that function in RNA recognition.

The CsrA/CsrB ribonucleoprotein complex is comprised of 18 CsrA subunits and a single CsrB molecule. In 2003, a second RNA, CsrC, has been found that acts by binding 9 CsrA molecules (Weilbacher et al. 2003). Null mutations of either CsrB or CsrC cause a modest increase of each other's levels. The minimal binding motif of CsrA is 5′ GGA present on both sRNAs (CsrB and CsrC) and the corresponding mRNAs. However, the SELEX-derived consensus sequence is RU*AC*ARGGAU*GU*, with the underlined AC and GU residues being always base-paired to one another (Babitzke et al. 2007). Recently, it was discovered that CsrD mediates CsrB/C turnover together with RNase E (Suzuki et al. 2006). CsrD is not an endoribonuclease but requires GGDEF and EAL domains for its action. However, unlike other GGDEF/EAL proteins, it does not act via c-di-GMP signalling. Two CsrA homologues (RsmA and RsmE) and three redundant sRNAs (RsmX, RsmY, and RsmZ) that function as antagonists of RsmA and RsmE have been identified in *Pseudomonas fluorescens*, whereas only RsmY and RsmZ were found in *Pseudomonas aeruginosa* (rev. in Lapouge et al. 2008). The size of CsrB-type sRNAs in different bacteria varies between 100 and 479 nt, all share multiple unpaired GGA motifs, and are under control of the GacS/GacA TCS. Furthermore, the sRNAs feedback inhibit the transcription of their own genes by interfering with the function of the GacS/GacA system in an unknown manner. It is not clear, whether the sRNA families are functional homologues or have common ancestors. The three sRNAs from *Pseudomonas fluorescens* regulate secondary metabolism and biocontrol traits (e.g., antifungal metabolites and extracellular enzymes protecting plant roots from pathogenic fungi). An *rsmY/rsmZ* double mutant is strongly impaired in the synthesis of extracellular enzymes, whereas single mutants do not show significant effects. A similar pathway is present in *Erwinia carotovora* with RsmB sRNA (Fig. 4.1b). Homologous elements in *Salmonella typhimurium* control the expression of genes related to cellular invasion. In *Yersinia pseudotuberculosis*, a Csr-type system regulates virulence (Heroven et al. 2008). In *Pseudomonas aeruginosa*, the sRNA CrcZ transcribed under control of σ^{54} and CbrA/B contains five CA motifs (AAC/XAACAA) bound by the Crc protein, the mediator of catabolite repression of degradative genes. Similar CA motifs in their 5′ UTRs of these genes, for example *amiE*, are recognized by Crc (Sonnleitner et al. 2009). The CbrA-CbrB-CrcZ-Crc system is comparable to GacS-GacA-CsrB-CsrA in *E. coli* and allows the differential adaptation to various carbon sources.

4.7   Role of sRNAs in Pathogenesis

Two excellent reviews have covered the role of sRNAs in pathogenic bacteria
(Toledo-Arana et al. 2007; Papenfort and Vogel 2010). Therefore, only a few
examples will be described here. Among the recently detected cis-encoded antisense RNAs are several sRNAs in *Mycobacterium tuberculosis* (e.g., AsDes,
Arnvig and Young 2009) and *S. typhimurium* (e.g., IsrC, Padalon-Brauch et al.
2008), whose expression negatively correlates with convergent virulence genes (see
Table 4.1). The long 1,200-nt AmgR antisense RNA of *S. typhimurium* promotes
degradation of *mgtC* RNA required for Mg^{2+} homeostasis and virulence and
prevents bacterial hypervirulence in mice. Interestingly, both sRNA and its target
are transcriptionally activated by PhoP (Lee and Groisman 2010). RnaG from a
virulence plasmid in *Shigella flexneri* regulates *icsA* encoding an invasion protein
(Giangrossi et al. 2010).

Phenotypic alterations in *hfq* mutants range from loss of effector secretion in
Salmonella and *Yersinia pseudotuberculosis* to effector overproduction in pathogenic *E. coli*, *Yersinia enterocolitica*, *Pseudomonas aeruginosa,* and *Vibrio* species
(Papenfort and Vogel 2010) indicating that Hfq-dependent sRNAs directly regulate
bacterial virulence factors. *Salmonella* InvR RNA from the pathogenicity island
regulates the core genome–encoded *ompD* encoding an outer membrane porin
(Pfeiffer et al. 2007). In *Vibrio cholerae,* the virulence regulatory cascade controls
glucose uptake through ToxT-dependent activation of TarA RNA, a functional
homologue of *E. coli* SgrS that also regulates *ptsG*. In contrast to SgrS, TarA
does not contain an ORF (Richard et al. 2010). A Δ*tarA V. cholerae* mutant is
compromised for infant mouse colonization. *Borrelia burgdorferi* RpoS is required
for virulence gene expression. It is regulated by the functional homologue of *E. coli*
DsrA, $DsrA_{Bb}$, a molecular thermometer that responds to temperature (Lybecker
and Samuels 2007). Its secondary structure is closed at $23^{\circ}C$ but opens at $37^{\circ}C$ to
base pair with the *rpoS* 5′ UTR to activate translation. In *Neisseria meningitidis*, a
functional homologue of *E. coli* RyhB, NrrF, controls Fur-mediated regulation of
sdhA and *sdhC* (Mellin et al. 2007). In the intracelluar pathogen *Chlamydia
trachomatis*, IhtA specifically represses *hctA* encoding histone-like protein Hc1 in
replicating reticulate body (RB), thereby avoiding chromatin condensation during
the replicative stage (Grieshaber et al. 2006). The Qrr sRNAs in *Vibrio* species
govern quorum-sensing control (Ng and Bassler 2009). Trans-encoded sRNAs were
found in two species lacking Hfq: *H. pylori* and *Campylobacter jejuni*. HPnc5490
targets the 5′ UTR of a *H. pylori* chemotaxis receptor mRNA through a 13-bp GC-
rich RNA duplex. In pathogenic Gram-positives, the most prominent example is
RNAIII from *S. aureus (*see 4*)*. Several *S. aureus* sRNAs recognize their target
mRNAs at the RBS via C-rich loops (e.g., Geissmann et al. 2009). Lately, in *S.
aureus*, SprD RNA regulating translation initiation of the immune-evasion molecule *sbi* (Chabelskaya et al. 2010) and RsaE, a riboregulator of central metabolism
(Bohn et al. 2010) have been identified. In *Streptococcus pyogenes*, Pel regulates M
and M-related proteins (Mangold et al. 2004), FasX stabilizes the *ska* mRNA

encoding the secreted virulence factor streptokinase (Ramirez-Peña et al. 2010), and RivX activates the *mga* regulon by enhancing translation of the virulence gene regulator MgA (Roberts and Scott 2007). *Clostridium perfringens* VR RNA (see 4) is part of the VirR/S regulon that controls toxin production and acts as an inducer of collagenase (K-toxin) and β2-toxin synthesis (Okumura et al. 2008). In the foodborne pathogen *Listeria monocytogenes*, a trans-acting riboswitch (SreA) that can function like a trans-encoded sRNA was reported (Loh et al. 2009). It upregulates *argD* and represses translation of the virulence regulator PrfA by base pairing 85–24 nt upstream of the RBS. Furthermore, 15 of the 29 recently discovered novel sRNAs in *L. monocytogenes* (Toledo-Arana et al. 2009) were absent in the nonpathogenic *L. innocua*. Among them, Rli38 showed a 25-fold increased expression in blood and in the presence of H_2O_2 and might base pair with three mRNAs associated with virulence, among them *fur*. Seven of the 29 sRNAs were cis-encoded, and three of them covered more than one ORF.

Protein-binding sRNAs are also involved in pathogenesis. Lack of 6S RNA alters in *E. coli* expression of 5% of all genes and in *Legionella pneumophila* expression of type IV secretion effectors and replication in human macrophages or amoeba (Faucher et al. 2010). Furthermore, deletion of *csrA/rsmA* in *Pseudomonas aeruginosa* modifies invasion of human airway epithelial cells (Burrowes et al. 2006). CsrB/C of *Salmonella enterica* regulates expression of the genes of pathogenicity island 1 (SPI1) required for the invasion of epithelial cells (Fortune et al. 2006).

4.8   Evolutionary Considerations

Based on a recent review (Gottesman and Storz 2010), three possible scenarios are conceivable for the origin of sRNAs. First, if some low-level promoter activity – as detected antisense to genes and within spacer regions (Kawano et al. 2005) – was sufficient for advantageous regulation under some condition, the transcribed RNA might evolve into a highly expressed regulatory RNA. Secondly, transcripts with different primary functions such as tRNAs might be transformed into regulatory molecules. tRNAs are folded, stable RNAs of a similar size to many regulatory sRNAs and can specifically interact with Hfq (Lee and Feig 2008). Results from eukaryotes support that tRNA fragments might also have regulatory roles in bacteria (e.g., Cole et al. 2009). Recently, tmRNA – related to tRNA – was shown to act as a regulatory sRNA in *S. aureus* (Liu et al. 2010). Thirdly, mRNAs, many of which have Hfq binding sites in their 5′ UTRs, might be sources of sRNAs. Separation of the UTR from the mRNA or loss of the downstream ORF might convert an mRNA into an sRNA.

An interesting question is whether bacterial sRNAs evolved from the RNA world or more recently as new regulators (as discussed above). Protein-binding sRNAs such as 6S or CsrB, are more broadly conserved than base-pairing sRNAs. However, although 6S function is well conserved, the sequence is not (Barrick et al.

2005). The evolution of Hfq-binding regulatory sRNAs appears to be rapid. As a result, neither sequence nor structural similarities are sufficient to provide a clear picture of their evolution. However, since the Hfq-binding sRNAs pair with specific target mRNAs, the evolution of both interaction partners must be linked. Apparently, pairing helps constrain evolution as sRNAs expressed in different bacterial species contain a highly conserved core region required for pairing with the target (Sharma et al. 2007). In cases where multiple sRNAs control the same mRNA (e.g., *rpoS* mRNA) they may have a common ancestor.

Conserved regulation can be used to trace the evolution of sRNAs. Well-studied examples are Fur-regulated sRNAs involved in iron metabolism which have been found in Gram-negative bacteria like *E. coli* (RyhB), *Pseudomonas* (PrrF), or *Neisseria* (NrrF, Mellin et al. 2007) and Gram-positive bacteria like *B. subtilis* (FsrA). PrrF, NrrF, or FsrA bear no sequence similarity to RyhB, suggesting either independent evolution or rapid divergence.

The identification of base-pairing sRNAs in related species has been facilitated by using conserved gene neighborhood (e.g., SgrS, Horler and Vanderpool 2009). Enterobacterial homologues of Spot42 are all located in an intergenic region between two highly conserved protein-coding genes. In *Pseudomonas*, in the same region, a functional Spot42 homologue with no sequence similarity has been found, which might be evolutionarily related but have diverged rapidly. Alternatively, this location is prone to insertion of sRNA genes from different sources.

The discovery of sRNAs on cryptic prophages (*E. coli* IpeX) or pathogenicity islands (*S. aureus* SprD) are indicative for horizontal transfer. Furthermore, several sRNA genes are located in the neighborhood of potential prophage or transposon integration sites (De Lay and Gottesman 2009) and might be picked up accidentally upon excision of these elements.

4.9 Future Perspective

It can be anticipated that hundreds to thousands of novel sRNAs will be found in a multitude of bacterial genomes in the near future. The detection of new mechanisms of action for both cis- and trans-encoded base-pairing sRNAs can be expected. It is conceivable that some RNAs might act *in cis* on one target and *in trans* on several other targets, thereby using two or more different mechanisms of action on different targets.

Furthermore, it cannot be ruled out that – similar to siRNAs or miRNAs in eukaryotes – novel classes of very short or very long sRNAs with up to now unknown functions might be discovered. Moreover, sRNAs might be found that function directly on the genome like the siRNAs involved in chromatin silencing in *Schizosaccharomyces pombe* (Grewal 2011).

Interestingly, many concepts established lately for sRNA action in Gram-negative bacteria cannot be applied to Gram-positive hosts. Important differences include the role of RNA chaperones, discrepancies in the RNA degradation machineries, and the preferential use of transcriptional riboswitches in Gram-positives and translational ones in Gram-negatives. It can be expected that new RNA chaperones will be detected in both Gram-positive and Gram-negative bacteria, which may be equivalents to Hfq. Most probably, differences in the set of endoribonucleases will entail differences in the mechanisms of action of sRNAs found in enterobacteria vs. those in Gram-positive bacteria.

Additionally, it can be anticipated that more dual-function sRNAs will be identified and new unprecedented functions for small proteins or peptides encoded by these sRNAs or acting on them or their targets might be discovered, adding a new layer to the interplay between peptides and RNA. New representatives of the class of protein-binding sRNAs that use hitherto unknown mechanisms can be imagined, for example sRNAs that directly modulate the activity of enzymes or transcriptional regulators.

Since in prokaryotes, transcription, translation, and RNA degradation are coupled, new results on the role of mRNA structure can be expected. For instance, sRNAs might exert activating functions during the transcription process, such as controlling a transcriptional riboswitch, or interacting with RNA polymerase. The recent observation that RNA polymerase binds several sRNAs and reacts with them, mapping the interaction site of both molecules to the active center cleft of the enzyme (Windbichler et al. 2008), would be in support of the latter hypothesis.

References

Albrecht M, Sharma CM, Reinhardt R, Vogel J, Rudel T (2010) Deep sequencing-based discovery of the *Chlamydia trachomatis* transcriptome. Nucleic Acids Res 38:868–877

Altuvia S, Weinstein-Fischer D, Zhang A, Postow L, Storz G (1997) A small, stable RNA induced by oxidative stress: role as a pleiotropic regulator and antimutator. Cell 90:43–53

Altuvia S, Zhang A, Argaman L, Tiwari A, Storz G (1998) The *Escherichia coli* OxyS regulatory RNA represses *fhlA* translation by blocking ribosome binding. EMBO J 17:6069–6075

Andersen J, Delihas N (1990) micF RNA binds to the 5′ end of *ompF* mRNA and to a protein from *Escherichia coli*. Biochemistry 29:9249–9256

André G, Even S, Putzer H, Burguière P, Croux C, Danchin A, Martin-Verstraete I, Soutourina O (2008) S-box and T-box riboswitches and antisense RNA control a sulfur metabolic operon of *Clostridium acetobutylicum*. Nucleic Acids Res 36:5955–5969

Antal M, Bordeau V, Couchin V, Felden B (2005) A small bacterial RNA regulates a putative ABC transporter. J Biol Chem 280:7901–7908

Argamann L, Hershberg R, Vogel J, Bejerano G, Wagner EGH, Margalit H, Altuvia S (2001) Novel small RNA-encoding genes in the intergenic regions of *Escherichia coli*. Curr Biol 11:941–950

Arluison V, Hohng S, Roy R, Pellegrini O, Régnier P, Ha T (2007) Spectroscopic observation of RNA chaperone activities of Hfq in post-transcriptional regulation by a small non-coding RNA. Nucleic Acids Res 35:999–1006

Arnvig KB, Young DB (2009) Identification of small RNAs in *Mycobacterium tuberculosis*. Mol Microbiol 73:397–408

Arthur DC, Ghetu AF, Gubbins MJ, Edwards RA, Frost LS, Glover JN (2003) FinO is an RNA chaperone that facilitates sense-antisense RNA interactions. EMBO J 22:6346–6355

Arthur CD, Edwards RA, Tsutakawa S, Tainer JA, Frost LS, Glover JNM (2011) Mapping interactions between the RNA chaperone FinO and its RNA targets. Nucleic Acids Res 39:4450–4463

Babitzke P, Romeo T (2007) CsrB sRNA family: sequestration of RNA-binding regulatory proteins. Curr Opin Microbiol 10:156–163

Balbontín R, Fiorini F, Figueroa-Bossi N, Casadesús J, Bossi L (2010) Recognition of heptameric seed sequence underlies multi-target regulation by RybB small RNA in *Salmonella enterica*. Mol Microbiol 78:380–394

Barrick JE, Sudarsan N, Weinberg Z, Ruzzo WL, Breaker RR (2005) 6S RNA is a widespread regulator of eubacterial RNA polymerase that resembles an open promoter. RNA 11:774–784

Beisel CL, Storz G (2010) Base pairing small RNAs and their roles in global regulatory networks. FEMS Microbiol Rev 34:866–882

Beisel CL, Storz G (2011) The base-pairing RNA Spot 42 participates in a multioutput feedforward loop to help enact catabolite repression in *Escherichia coli*. Mol Cell 41:286–297

Benito Y, Kolb FA, Romby P, Lina G, Etienne J, Vandenesch F (2000) Probing the structure of RNAIII, the *Staphylococcus aureus agr* regulatory RNA, and identification of the RNA domain involved in repression of protein A expression. RNA 5:6668–6679

Berghoff B, Glaeser J, Sharma CM, Vogel J, Klug G (2009) Photooxidative stress-induced and abundant small RNAs *in Rhodobacter sphaeroides*. Mol Microbiol 74:1497–1512

Bohn C, Rigoulay C, Bouloc P (2007) No detectable effect of RNA-binding protein Hfq absence in *Staphylococcus aureus*. BMC Microbiol 7:10

Boisset S, Geissmann T, Huntzinger E et al (2007) *Staphylococcus aureus* RNAIII coordinately represses synthesis of virulence factors and the transcription regulator Rot by an antisense mechanism. Genes Dev 21:1353–1366

Bohn C, Rigoulay C, Chabelskaya S, Sharma CM, Marchais A, Skorski P et al (2010) Experimental discovery of small RNAs in *Staphylococcus aureus* reveals a riboregulator of central metabolism. Nucleic Acids Res 38:6620–6636

Boysen A, Moller-Jensen J, Kallipolitis B, Valentin-Hansen P, Overgaard M (2010) Translational regulation of gene expression by an anaerobically induced small non-coding RNA in *Escherichia coli*. J Biol Chem 285:10690–10702

Brantl S, Wagner EGH (1994) Antisense RNA-mediated transcriptional attenuation occurs faster than stable antisense/target RNA pairing: an *in vitro* study of plasmid pIP501. EMBO J 13:3599–3607

Brantl S (2004) Plasmid replication controlled by antisense RNAs. In: Funnel B, Phillips G (eds) The Biology of Plasmids. ASM Press, Herndon, pp 47–62

Brantl S (2007) Regulatory mechanisms employed by cis-encoded antisense RNAs. Curr Opin Microbiol 10:102–109

Brantl S (2009) Bacterial chromosome-encoded small regulatory RNAs. Future Microbiol 4:85–103

Brantl S, Birch-Hirschfeld E, Behnke D (1993) RepR protein expression on plasmid pIP501 is controlled by an antisense RNA-mediated transcription attenuation mechanism. J Bacteriol 175:4052–4061

Brantl S, Wagner EGH (1994) Antisense RNA-mediated transcriptional attenuation occurs faster than stable antisense/target RNA pairing: an *in vitro* study of plasmid pIP501. EMBO J 13: 3599–3607

Brantl S, Wagner EGH (1997) Dual function of the *copR* gene product of plasmid pIP501. J Bacteriol 179:7016–7024

Brantl S, Wagner EGH (2000) Antisense-RNA-mediated transcriptional attenuation: an *in vitro* study of plasmid pT181. Mol Microbiol 35:1469–1482

Brennan RG, Link TM (2007) Hfq structure, function and ligand binding. Curr Opin Microbiol 10:12–133

Burrowes E, Baysse C, Adams C, O'Gara F (2006) Influence of the regulatory protein RsmA on cellular functions in *Pseudomonas aeruginosa* PAO1, as revealed by transcriptome analysis. Microbiology 152:405–418

Cavanagh AT, Klocko AD, Liui X, Wassarman KM (2008) Promoter specificity for 6 S RNA regulation of transcription is determined by core promoter sequences and competition for region 4.2 of σ^{70}. Mol Microbiol 67:242–256

Chabelskaya S, Gaillot O, Felden B (2010) A *Staphylococcus aureus* small RNA is required for bacterial virulence and regulates the expression of an immun-evasion molecule. PLoS Pathog 6:e1000927

Chant EL, Summers DK (2007) Indole signalling contributes to the stable maintenance of *Escherichia coli* multicopy plasmids. Mol Microbiol 63:35–43

Chen S, Zhang A, Blyn LB, Storz G (2004) MicC, a second small-RNA regulator of Omp protein expression in *Escherichia coli*. J Bacteriol 186:6689–6697

Chevalier C, Boisset S, Romilly C, Masquida B, Fechter P, Geissmann T, Vandenesch F, Romby P (2010) *Staphylococcus aureus* RNAIII binds to two distant regions of *coa* RNA to arrest translation and promote RNA degradation. PLoS Pathog 6:e1000809

Christiansen JK, Nielsen JS, Ebersbach T, Valentin-Hansen P, Søgaard-Andersen B, Kallipolitis BH (2006) Identification of small Hfq-binding RNAs in *Listeria monocytogenes*. RNA 12:1–14

Cole C, Sobala A, Lu C, Thatcher SR, Bowman A, Brown JWS, Green PJ, Barton GJ, Hutvagner G (2009) Filtering of deep sequencing data reveals the existence of abundant Dicer-dependent small RNAs derived from tRNAs. RNA 15:2147–2160

Darfeuille F, Unoson C, Vogel J, Wagner EGH (2007) An antisense RNA inhibits translation by competing with standby ribosomes. Mol Cell 26:381–392

De Lay N, Gottesman S (2009) The Crp-activated small noncoding regulatory RNA CyaR (RyeE) links nutritional status to group behavior. J Bacteriol 191:461–476

Desnoyers G, Morissette A, Prévost K, Massé E (2009) Small RNA-induced differential degradation of the poylcistronic mRNA *iscRSUA*. EMBO J 28:1551–1561

Dühring U, Axmann IM, Hess WR, Wilde A (2006) An internal antisense RNA regulates expression of the photosynthesis gene *isiA*. Proc Natl Acad Sci USA 103:7054–7058

Durand S, Storz G (2010) Reprogramming of anaerobic metabolism by the FnrS small RNA. Mol Microbiol 75:1215–1231

Faucher SP, Friedlander G, Livny J, Margalit H, Shuman HA (2010) *Legionella pneumophila* 6S RNA optimizes intracellular multiplication. Proc Natl Acad Sci USA 107:7533–7538

Fender A, Elf J, Hampel K, Zimmermann B, Wagner EGH (2010) RNAs actively cycle on the Sm-like protein Hfq. Genes Dev 24:2621–2626

Figuera-Bossi N, Valentini M, Malleret L, Bossi L (2009) Caught at its own game: regulatory small RNA inactivated by an inducible transcript mimicking its target. Genes Dev 23:2004–2015

Fortune DR, Suyemoto M, Altier C (2006) Identification of CsrC and characterization of its role in epithelial cell invasion in *Salmonella enterica* serovar *Typhimurium*. Infect Immun 74:331–339

Franch T, Petersen M, Wagner EGH, Jacobsen JP, Gerdes K (1999) Antisense RNA regulation in prokaryotes: rapid RNA/RNA interaction facilitated by a general U-turn-loop structure. J Mol Biol 294:1115–1125

Gaballa A, Antelmann H, Aguilar C, Khakh S-K, Song K-B, Smaldone GT, Helmann JD (2008) The *Bacillus subtilis* iron-sparing response is mediated by a Fur-regulated small RNA and three small, basic proteins. Proc Natl Acad Sci USA 105:11927–11932

Geissmann T, Chevalier C, Cros M-J, Boisset S, Fechter P, Noirot C et al (2009) A search for small noncoding RNAs in *Staphylococcus aureus* reveals a conserved sequence motif for regulation. Nucleic Acids Res 37:7239–7257

Giangrossi M, Prosseda G, Tran CN, Brani A, Colonna B, Falconi M (2010) A novel antisense RNA regulates at transcriptional level the virulence gene *icsA* of *Shigella flexneri*. Nucleic Acids Res 38:3362–3375

Gildehaus N, Neusser T, Wurm R, Wagner R (2007) Studies on the function of the riboregulator 6S RNA from *E. coli*: RNA polymerase binding, inhibition of *in vitro* transcription and synthesis of RNA-directed *de novo* transcripts. Nucleic Acids Res 35:1885–1896

Gimpel M, Heidrich H, Mäder U, Krügel H, Brantl S (2010) A dual-function sRNA from *B. subtilis*: SR1 acts as a peptide encoding mRNA on the *gapA* operon. Mol Microbiol 76:990–1009

Gottesman S, Storz G (2010) Bacterial small RNA regulators: versatile roles and rapidly evolving variations. Cold Spring Harb. Perspect Biol [Epub ahead of print]

Grewal SI (2011) RNAi-dependent formation of heterochromatin and its diverse functions. Curr Opin Genet Dev 20:134–141

Grieshaber NA, Grieshaber SS, Fischer ER, Hackstadt T (2006) A small RNA inhibits translation of the histone-like protein Hc1 in *Chlamydia trachomatis*. Mol Microbiol 59:541–550

Guillier M, Gottesman S (2008) The 5′ end of two redundant sRNAs is involved in the regulation of multiple targets, including their own regulator. Nucleic Acids Res 36:6781–6794

Halfmann A, Kovacs M, Hakenbeck R, Brückner R (2007) Identification of the genes directly controlled by the response regulator CiaR in *Streptococcus pneumoniae*: five out of 15 promoters drive expression of small non-coding RNAs. Mol Microbiol 66:110–126

Heidrich N, Brantl S (2003) Antisense-RNA mediated transcriptional attenuation: importance of a U-turn loop structure in the target RNA of plasmid pIP501 for efficient inhibition by the antisense-RNA. J Mol Biol 333:917–929

Heidrich N, Chinali A, Gerth U, Brantl S (2006) The small untranslated RNA SR1 from the *B. subtilis* genome is involved in the regulation of arginine catabolism. Mol Microbiol 62:520–536

Heidrich N, Moll I, Brantl S (2007) *In vitro* analysis of the interaction between the small RNA SR1 and its primary target *ahrC* mRNA. Nucleic Acids Res 35:4331–4346

Heroven AK, Böhme K, Rohde M, Dersch P (2008) A Csr-type regulatory system, including small non-coding RNAs, regulates the global virulence regulator RovA of *Yersinia pseudotuberculosis* through RovM. Mol Microbiol 68:1179–1195

Hershberg R, Altuvia S, Margalit H (2003) A survey of small RNA-encoding genes in *Escherichia coli*. Nucleic Acids Res 31:1813–1820

Holmqvist E, Reimegård J, Sterk M, Grantcharova N, Römling U, Wagner EGH (2010) Two antisense RNAs target the transcriptional regulator CsgD to inhibit curli synthesis. EMBO J 29:1840–1850

Horler RS, Vanderpool CK (2009) Homologs of the small RNA SgrS are broadly distributed in enteric bacteria but have diverged in size and sequence. Nucleic Acids Res 37:5465–5476

Huntzinger E, Boisset S, Saveanu C et al (2005) *Staphylococcus aures* RNA III and endoribonuclease III coordinately regulate *spa* gene expression. EMBO J 24:824–835

Irnov K, Sharma CM, Vogel J, Winkler WC (2010) Identification of regulatory RNAs in *Bacillus subtilis*. Nucleic Acids Res 38:6637–6651

Jerome LJ, van Biesen T, Frost LS (1999) Degradation of FinP antisense RNA from F-like plasmids: the RNA binding protein, FinO, protects FinP from ribonuclease E. J Mol Biol 285:1457–1473

Kawamoto H, Koide Y, Morita T, Aiba H (2006) Base-pairing requirement for RNA silencing by a bacterial small RNA and acceleration of duplex formation by Hfq. Mol Microbiol 61:1013–1022

Kawano M, Storz G, Rao BS, Rosner JL, Martin RG (2005) Detection of low-level promoter activity within open reading frame sequences of *Escherichia coli*. Nucleic Acids Res 33:6268–6276

Kawano M, Aravind L, Storz G (2007) An antisense RNA controls synthesis of an SOS-induced toxin evolved from an antitoxin. Mol Microbiol 64:738–754

Klocko AD, Wassarman KM (2009) 6S RNA binding to $E\sigma^{70}$ requires a positively charged surface of σ^{70} region 4.2. Mol Microbiol 73:152–164

Krinke L, Wulff DL (1987) OOP RNA, produced from multicopy plasmids, inhibits *cII* gene expression through an RNase III-dependent mechanism. Genes Dev 1:1005–1012

Krinke L, Wulff DL (1990) RNase III-dependent hydrolysis of *cII-O* gene mRNA by OOP antisense RNA. Genes Dev 4:2223–2233

Lapouge K, Schubert M, Allain FH-T, Haas D (2008) Gac/Rsm signal transduction pathway of γ-proteobacteria: from RNA recognition to regulation of social behaviour. Mol Microbiol 67:241–253

Lee T, Feig AL (2008) The RNA binding protein Hfq interacts specifically with tRNAs. RNA 14:514–523

Lee EJ, Groisman EA (2010) An antisense RNA that governs the expression kinetics of a multifunctional virulence gene. Mol Microbiol 76:1020–1033

Lee M, Zhang S, Saha S, Santa Anna S, Jiang C, Perkins J (2001) RNA expression analysis using an antisense *Bacillus subtilis* genome array. J Bacteriol 183:7371–7380

Licht A, Preis S, Brantl S (2005) Implication of CcpN in the regulation of a novel untranslated RNA (SR1) in *B. subtilis*. Mol Microbiol 58:189–206

Liu MY, Gui G, Wei B et al (1997) The RNA molecule CsrB binds to the global regulatory protein CsrA and antagonizes its activity in *Escherichia coli*. J Biol Chem 272:17502–17510

Liu Y, Wu N, Dong J, Gao Y, Zhang X, Shao N, Yang G (2010) SsrA (tmRNA) acts as an antisense RNA to regulate *Staphylococcus aureus* pigment synthesis by base pairing with *crtMN* mRNA. FEBS Lett 584:4325–4329

Livny J, Brencic A, Lory S, Waldor MK (2006) Identification of 17 *Pseudomonas aeruginosa* sRNAs and prediction of sRNA-encoding genes in 10 diverse pathogens using the bioinformatic tool sRNAPredict2. Nucleic Acids Res 34:3484–3493

Loh E, Dussurget O, Gripenland J, Vaitkevicius K, Tiensuu T, Mandin P, Repoila F et al (2009) A trans-acting riboswitch controls expression of the virulence regulator PrfA in *Listeria monocytogenes*. Cell 139:770–779

Lybecker MC, Samuels DS (2007) Temperature-induced regulation of RpoS by a small RNA in *Borrelia burgdorferi*. Mol Microbiol 64:1075–1089

Majdalani N, Chen S, Murrow J, St John K, Gottesman S (2001) Regulation of RpoS by a novel small RNA: the characterization of RprA. Mol Microbiol 39:1382–1394

Mandin P, Repoila F, Vergassola M, Geissmann T, Cossart P (2007) Identification of new noncoding RNAs in *Listeria monocytogenes* and prediction of mRNA targets. Nucleic Acids Res 35:962–974

Mandin P, Gottesman S (2009) A genetic approach for finding small RNAs regulators of genes of interest identifies RybC as regulating the DpiA/DpiB two component system. Mol Microbiol 72:551–565

Mandin P, Gottesman S (2010) Integrating anaerobic/aerobic sensing and the general stress response through the ArcZ small RNA. EMBO J 29:3094–3107

Mangold M, Siller M, Roppenser B et al (2004) Synthesis of group A streptococcal virulence factors is controlled by a regulatory RNA molecule. Mol Microbiol 53:1515–1527

Massé E, Salvail H, Desnoyers G, Arguin M (2007) Small RNAs controlling iron metabolism. Curr Opin Microbiol 10:140–145

Mellin JR, Goswami S, Grogan S, Tjaden B, Genco CA (2007) A novel Fur- and iron-regulated small RNA, NrrF, is required for indirect Fur-mediated regulation of the *sdhA* and *sdhC* genes in *Neisseria meningitidis*. J Bacteriol 189:3686–3694

Mizuno T, Chou MY, Inouye M (1984) A unique mechanism regulating gene expression: translational inhibition by a complementary RNA transcript (micRNA). Proc Natl Acad Sci USA 81:1966–1970

Molina N, van Nimwegen E (2008) Universal patterns of purifying selection at noncoding positions in bacteria. Genome Res 18:148–160

Møller T, Franch T, Udesen C, Gerdes K, Valentin-Hansen P (2002) Spot 42 RNA mediates discoordinate expression of the *E. coli* galactose operon. Genes Dev 16:1696–1706

Morfeldt E, Taylor D, von Gabain A, Arvidson S (1995) Activation of alpha-toxin translation in *Staphylococcus aureus* by the trans-encoded antisense RNA, RNAIII. EMBO J 14:4569–4577

Morita T, Mochizuki Y, Aiba H (2006) Translational repression is sufficient for gene silencing by bacterial small noncoding RNAs in the absence of mRNA destruction. Proc Natl Acad Sci USA 103:4858–4863

Narberhaus F, Waldminghaus T, Chowdhury S (2006) RNA thermometers. FEMS Microbiol Rev 30:3–16

Neusser T, Gildehaus N, Wurm R, Wagner R (2008) Studies on the expression of 6S RNA from *E. coli*: involvement of regulators important for stress and growth adaptation. Biol Chem 389:285–297

Neusser T, Polen T, Geissen R, Wagner R (2010) Depletion of the non-coding regulatory 6S RNA in *E. coli* causes a surprising reduction in the expression of the translation machinery. BMC Genomics 11:165

Ng WL, Bassler BL (2009) Bacterial quorum-sensing network architectures. Annu Rev Genet 43:197–222

Nielsen JS, Lei LK, Ebersbach T, Olsen AS, Klitgaard JK, Valentin-Hansen P, Kallipolitis BH (2010) Defining a role for Hfq in Gram-positive bacteria: evidence for Hfq-dependent antisense regulation in *Listeria monocytogenes*. Nucleic Acids Res 38:907–919

Novick RP, Iordanescu S, Projan SJ, Kornblum J, Edelman I (1989) pT181 plasmid replication is regulated by a countertranscript-driven transcriptional attenuator. Cell 59:395–404

Obana N, Shirahama Y, Aboe K, Nakamura K (2010) Stabilization of *Clostridium perfringens* collagenase mRNA by VR-RNA-dependent cleavage in 5′ leader sequence. Mol Microbiol 77:1416–1428

Okumura K, Ohtani K, Hayashi H, Shimizu T (2008) Characterization of genes regulated directly by the VirR/VirS system in Clostridium perfringens. J Bacteriol 190:7719–7727

Olsen AS, Møller-Jensen J, Brennan RG, Valentin-Hansen P (2010) C-terminally truncated derivatives of *Escherichia coli* Hfq are proficient in riboregulation. J Mol Biol 404:173–182

Opdyke JA, Kang JG, Storz G (2004) GadY, a small-RNA regulator of acid response genes in *Escherichia coli*. J Bacteriol 186:6698–6705

Opdyke JA, Fozo EM, Hemm MR, Storz G (2011) RNase III participates in GadY-dependent cleavage of the *gadX-gadW* mRNA. J Mol Biol 406:29–43

Overgaard M, Johansen J, Møller-Jensen J, Valentin-Hansen P (2009) Switching off small RNA regulation with trap-mRNA. Mol Microbiol 73:790–800

Padalon-Brauch G, Hershberg R, Elgrably-Weiss M et al (2008) Small RNAs encoded within genetic islands of *Salmonella typhimurium* show host-induced expression and role in virulence. Nucleic Acids Res 36:1913–1927

Papenfort K, Pfeiffer V, Mika F, Lucchini S, Hinton JC, Vogel J (2006) SigmaE-dependent small RNAs of *Salmonella* respond to membrane stress by accelerating global *omp* mRNA decay. Mol Microbiol 62:1674–1688

Papenfort K, Said N, Welsink T, Lucchini S, Hinton JC, Vogel J (2009) Specific and pleiotropic patterns of mRNA regulation by ArcZ, a conserved, Hfq-dependent small RNA. Mol Microbiol 74:139–158

Papenfort K, Bouvier M, Mika F, Sharma CM, Vogel J (2010) Evidence for an autonomous 5′ target recognition domain in an Hfq associated small RNA. Proc Natl Acad Sci USA 107:20435–20440

Papenfort K, Vogel J (2010) Regulatory RNA in bacterial pathogens. Cell Host Microbe 8:116–127

Perez N, Trevino J, Liu Z, Ho SC, Babitzke P, Sumby P (2009) A genome-wide analysis of small regulatory RNAs in the human pathogen group A *Streptococcus*. PloS One 4:e7668

Pichon C, Felden B (2005) Small RNA genes expressed from *Staphylococcus aureus* genomic and pathogenicity islands with specific expression among pathogenic strains. Proc Natl Acad Sci USA 102:14249–14254

Pfeiffer V, Sittka A, Tomer R, Tedin K, Brinkmann V, Vogel J (2007) A small non-coding RNA of the invasion gene island (SPI-1) represses outer membrane protein synthesis from the *Salmonella* core genome. Mol Microbiol 66:1174–1191

Pfeiffer V, Papenfort K, Lucchini S, Hinton JCD, Vogel J (2009) Coding sequence targeting by MicC RNA reveals bacterial mRNA silencing downstream of translational initiation. Nat Struct Mol Biol 16:840–846

Prévost K, Salvail H, Desnoyers G, Jacques JF, Phaneuf E, Massé E (2007) The small RNA RyhB activates the translation of *shiA* mRNA encoding a permease of shikimate, a compound involved in siderophore synthesis. Mol Microbiol 64:1260–1273

Prévost K, Desnoyers G, Jacques J-F, Lavoie F, Massé E (2011) Small RNA-induced mRNA degradation achieved through both translation block and activated cleavage. Genes Dev 25:385–396

Ramirez-Peña E, Treviño J, Liu Z, Perez N, Sumby P (2010) The group A *Streptococcus* small regulatory RNA FasX enhances streptokinase activity by increasing the stability of the *ska* mRNA transcript. Mol Microbiol 78:1332–1347

Ramos CG, Sousa SA, Grilo AM, Feliciano JR, Leitão JH (2011) The second RNA chaperone, Hfq2, is also required for survival under stress and full virulence of *Burkholderia cenocepacia* J2315. J Bacteriol 193:1515–1526

Rasmussen AA, Erikssen M, Gilany K, Udesen C, Franch T, Petersen C, Valentin-Hansen P (2005) Regulation of *ompA* mRNA stability: the role of a small regulatory RNA in growth phase dependent control. Mol Microbiol 58:1421–1429

Rasmussen S, Nielsen HB, Jarmer H (2009) Transcriptionally active regions in the genome of *Bacillus subtilis*. Mol Microbiol 73:1043–1057

Reichenbach B, Maes A, Kalamorz F, Hajnsdorf E, Görke B (2008) The small RNA GlmY acts upstream of the sRNA GlmZ in the activation of *glmS* expression and is subject to regulation by polyadenylation in *Escherichia coli*. Nucleic Acids Res 36:2570–2580

Resch A, Vecerek B, Palavra K, Bläsi U (2011) Requirement of the CsdA DEAD-box helicase for low temperature riboregulation of *rpoS* mRNA. RNA Biol 7:96–102

Rice JB, Vanderpool CK (2011) The small RNA SgrS controls sugar-phosphate accumulation by regulating multiple PTS genes. Nucleic Acids Res 39:3806–3819

Richard AL, Withey JH, Beyhan S, Yildiz F, DiRita VJ (2010) The *Vibrio cholerae* virulence regulatory cascade controls glucose uptake through activation of TarA, a small regulatory RNA. Mol Microbiol 78:1171–1181

Roberts SA, Scott JR (2007) RivR and the small RNA RivX: the missing links between the CovR regulatory cascade and the Mga regulon. Mol Microbiol 66:1506–1522

Roth A, Breaker RR (2009) The structural and functional diversity of metabolite-binding riboswitches. Annu Rev Biochem 78:305–334

Saito S, Kakeshita H, Nakamura K (2009) Novel small RNA-encoding genes in the intergenic regions of *Bacillus subtilis*. Gene 428:2–8

Salvail H, Lanthier-Bourbonnais P, Sobota JM, Caza M, Benjamin J-AM, Sequeira Mendieta ME et al (2010) A small RNA promotes siderophore production through transcriptional and metabolic remodeling. Proc Natl Acad Sci USA 107:15223–15228

Sharma CM, Darfeuille F, Plantinga TH, Vogel J (2007) A small RNA regulates multiple ABC transporter mRNAs by targeting C/A-rich elements inside and upstream of ribosome-binding sites. Genes Dev 21:2804–2817

Sharma CM, Hoffmann S, Darfeuille F, Reignier J, Findeiss S, Sittka A et al (2010) The primary transcriptome of the major human pathogen *Helicobacter pylori*. Nature 464:250–255

Shimizu T, Yaguchi H, Ohtani K, Banu S, Hayasi H (2002) Clostridial VirR/VirS regulon involves a regulatory RNA molecule for expression of toxins. Mol Microbiol 43:257–265

Shimoni Y, Friedlander G, Hetzroni G, Niv G, Altuvia S, Biham O, Margalit H (2007) Regulation of gene expression by small non-coding RNAs: a quantitative view. Mol Syst Biol 3:138

Silvaggi JM, Perkins JB, Losick R (2005) Small untranslated RNA antitoxin in *Bacillus subtilis*. J Bacteriol 187:6641–6650

Sittka A, Lucchini S, Papenfort K, Sharma CM, Rolle K, Binnewies TT, Hinton JC, Vogel J (2008) Deep sequencing analysis of small noncoding RNA and mRNA targets of the global post-transcriptional regulator Hfq. PLoS Genet 4:e1000163

Sittka A, Sharma CM, Rolle K, Vogel J (2009) Deep sequencing of *Salmonella* RNA associated with heterologous Hfq proteins *in vivo* reveals small RNAs as a major target class and identifies RNA processing phenotypes. RNA Biol 6:266–275

Sledjeski D, Gupta A, Gottesman S (1996) The small RNA, DsrA, is essential for the low temperature expression of RpoS during exponential growth in *Escherichia coli*. EMBO J 15:3993–4000

Sonnleitner E, Sorger-Domenigg T, Madej MJ et al (2008) Detection of Hfq-binding small RNAs by RNomics and by structure-based bioinformatic tools in *Pseudomonas aeruginosa*. Microbiology 154:3175–3187

Sonnleitner E, Abdou L, Haas D (2009) Small RNA as global regulator of carbon catabolite repression in *Pseudomonas aeruginosa*. Proc Natl Acad Sci USA 106:21866–21871

Sonnleitner E, Gonzalez N, Sorger-Domenigg T, Heeb S, Richter AS, Backofen R et al (2011) The small RNA PhrS stimulates synthesis of the *Pseudomonas aeruginosa* quinolone signal. Mol Microbiol 80:868–885

Stougaard P, Molin S, Nordström K (1981) RNAs involved in copy-number control and incompatibility of plasmid R1. Proc Natl Acad Sci USA 78:6008–6012

Suzuki K, Babitzke P, Kushner SR, Romeo T (2006) Identification of a novel regulatory protein (CsrD) that targets the global regulatory RNAs CsrB and CsrC for degradation by RNase E. Genes Dev 20:2605–25617

Toledo-Arana A, Repoila F, Cossart P (2007) Small noncoding RNAs controlling pathogenesis. Curr Opin Microbiol 10:182–188

Toledo-Arana A, Dussurget A, Nikitas G, Sesto N, Guet-Revillet H, Balestrino D, Loho E et al (2009) The *Listeria* transcriptional landscape from saprophytism to virulence. Nature 459:950–956

Tomizawa J, Itoh T, Selzer G, Som T (1981) Inhibition of ColE1 RNA primer formation by a plasmid-specified small RNA. Proc Natl Acad Sci USA 78:1421–1425

Udekwu KI, Darfeuille F, Vogel J, Reimegard J, Homqvist E, Wagner EGH (2005) Hfq-dependent regulation of OmpA synthesis is mediated by an antisense RNA. Genes Dev 19:2355–2366

Urban JH, Vogel J (2008) Two seemingly homologous noncoding RNAs act hierarchically to activate *glmS* mRNA translation. PLoS Biol 6:e64

Valentin-Hansen P, Eriksen M, Udesen C (2004) The bacterial Sm-like protein Hfq: a key player in RNA transactions. Mol Microbiol 51:1525–1533

Valentin-Hansen P, Johansen J, Rasmussen AA (2007) Small RNAs control outer membrane porins. Curr Opin Microbiol 10:152–155

Vecerek B, Moll I, Bläsi U (2007) Control of Fur synthesis by the non-coding RNA RyhB and iron-responsive decoding. EMBO J 26:965–975

Vogel J, Bartels V, Tang TH, Churakov G, Slagter-Jäger JG, Hüttenhofer A, Wagner EGH (2003) RNomics in *Escherichia coli* detects new sRNA species and indicates parallel transcriptional output in bacteria. Nucleic Acids Res 31:6435–6443

Vogel J, Papenfort K (2006) Small noncoding RNAs and the bacterial outer membrane. Curr Opin Microbiol 9:605–611

Voss GJ, Scholz I, Mitschke J, Wilde A, Hess WR (2009) Evidence for a major role of antisense RNAs in cyanobacterial gene regulation. Mol Syst Biol 5:305

Wadler CS, Vanderpool CK (2007) A dual function for a bacterial small RNA: SgrS performs base-pairing dependent regulation and encodes a functional polypeptide. Proc Natl Acad Sci USA 104:20454–20459

Wagner EGH, Brantl S (1998) Kissing and RNA stability in antisense control of plasmid replication. Trends Biochem Sci 23:451–454

Wagner EGH, Altuvia S, Romby P (2004) Antisense RNAs in bacteria and their genetic elements. Adv Genet 46:361–398

Wassarman KM, Repoila F, Rosenow C, Storz G, Gottesman S (2001) Identification of novel small RNAs using comparative genomics and microarrays. Genes Dev 15:1637–1651

Wassarman KM, Storz G (2000) 6S RNA regulates *E. coli* RNA polymerase activity. Cell 101:613–623

Wassarman KM, Saecker RM (2006) Synthesis-mediated release of a small RNA inhibitor of RNA polymerase. Science 314:1601–1603

Wassarman KM (2007) 6S RNA, a small RNA regulator of transcription. Curr Opin Microbiol 10:164–168

Weilbacher T, Suzuki K, Dubey AK et al (2003) A novel sRNA component of the carbon storage regulatory system of *Escherichia coli*. Mol Microbiol 48:657–670

Wilms I, Voss B, Hess WR, Leichert LI, Narberhaus F (2011) Small RNA-mediated control of the *Agrobacterium tumefaciens* GABA binding protein. Mol Microbiol 80:492–506

Windbichler N, von Pelchrzim F, Mayer O, Csaszar E, Schroeder R (2008) Isolation of small RNA-binding proteins from *E. coli*: evidence for frequent interaction of RNAs with RNA polymerase. RNA Biol 5:30–40

Zhang A, Wassarman KM, Rosenow C, Tjaden BC, Storz G, Gottesman S (2003) Global analysis of small RNA and mRNA targets of Hfq. Mol Microbiol 50:1111–1124

Chapter 5
The Canonical RNA Interference Pathway in Animals

Jana Nejepinska, Matyas Flemr, and Petr Svoboda

Abstract The canonical RNA interference (RNAi) pathway is defined as a sequence-specific mRNA degradation mediated by short RNA molecules which are generated from long double-stranded RNA. Since its discovery in 1998, RNAi has become a popular tool for experimental silencing of gene expression. On the other hand, its natural role received less attention. Recent studies in animal systems, particularly the use of the next generation sequencing and analysis of animals defective in some aspect of small RNA biogenesis, revealed novel functions of RNAi and cross talks between RNAi and other pathways employing small RNAs. This chapter provides a comprehensive view of the natural canonical RNAi pathway in animals including its molecular mechanism and different biological roles.

Keywords Argonaute • Dicer • dsRNA • RNAi • siRNA

5.1 Introduction

RNA silencing is a common term for repression guided by small RNA molecules (20–30 nucleotides (nt) long). Suppressive effects of RNA silencing include mRNA degradation, translational repression, formation of repressive chromatin, and DNA deletion (reviewed in Czech and Hannon 2011; Ketting 2011). Some forms of RNA silencing exist in almost all eukaryotes. RNA interference (RNAi) is one of the best characterized RNA silencing pathways. The term RNAi has been coined for sequence-specific mRNA degradation mediated by small RNAs produced from a long double-stranded RNA (dsRNA). Although the term RNAi is sometimes used

P. Svoboda (✉)
Institute of Molecular Genetics AS CR, Prague, Czech Republic
e-mail: svobodap@img.cas.cz

B. Mallick (eds.), *Regulatory RNAs*, DOI 10.1007/978-3-642-22517-8_5,   111
© Springer-Verlag Berlin Heidelberg (outside the USA) 2012

for a broad range of RNA silencing pathways (Ketting 2011), we will use it in its original connotation here. Since our aim is to provide a detailed overview of RNAi in animals, we will not discuss data from plants, fungi, and protists. However, it must be acknowledged that research in these models provided a fundamental contribution to understanding RNAi and related pathways in animals. Likewise, other animal RNA silencing pathways, such as the microRNA (miRNA, reviewed in Kim 2005) pathway and piwi-associated RNA (piRNA, reviewed in Aravin et al. 2007) pathway, will be discussed only in the context of the canonical RNAi pathway.

RNAi was first described in the nematode *Caenorhabditis elegans* (*C. elegans*) by Andrew Fire and Craig Mello, who observed that injected long dsRNA induced sequence-specific mRNA degradation in the whole animal (Fire et al. 1998). Initial studies of effects of dsRNA in animals showed RNAi effects in a wide range of animal taxa, including mammals (Svoboda et al. 2000; Wianny and Zernicka-Goetz 2000; Lohmann et al. 1999; Sanchez Alvarado and Newmark 1999; Kennerdell and Carthew 1998). The molecular mechanism of RNAi was deciphered using a combination of genetic and biochemical approaches. An important step was the establishment of the biochemical model in *Drosophila* embryo lysates, which revealed that long dsRNA is processed by an RNase III Dicer into short interfering RNAs (siRNA) that guide specific cleavage of cognate mRNAs in positions corresponding to the center of the siRNA:miRNA duplex (Bernstein et al. 2001; Tuschl et al. 1999; Zamore et al. 2000). Genetic studies in *C. elegans* discovered numerous components of RNAi, including the Argonaute protein family and RNA-dependent RNA polymerases (RdRPs) (Ketting et al. 1999; Smardon et al. 2000; Vastenhouw et al. 2003). The final gap in understanding the RNAi mechanism was bridged by structural analyses of Argonaute 2, which revealed that this protein carries the endonucleolytic activity, which cleaves cognate mRNAs (Liu et al. 2004; Meister et al. 2004; Song et al. 2004). The following text summarizes the current knowledge of the RNAi mechanism and its function in the three most studied animal model systems: *C. elegans*, *Drosophila*, and mammals.

5.2 The Mechanism of RNAi

In this section, we summarize current understanding of ribonucleoprotein complexes involved in RNAi. The RNAi pathway (Fig. 5.1) can be divided into three steps (1) cleavage of long dsRNA by Dicer into siRNAs, (2) loading of small RNAs on the effector complex known as the RNA-induced silencing complex (RISC), and (3) recognition and cleavage of cognate RNAs by the RISC. In addition to the core pathway, two extensions of the pathway, which are restricted to some animal species, should be mentioned (1) an amplification step, in which RdRPs generate secondary siRNAs (transitive RNAi) and (2) the systemic RNAi, where an RNAi response can spread across cellular boundaries.

Fig. 5.1 A schematic view of the RNAi pathway. The "core" RNAi pathway involves a processing of long dsRNA into siRNAs by Dicer and loading of siRNAs on the RISC complex, which cleaves cognate transcripts. The *gray* area represents an amplification loop where RdRPs generate secondary siRNAs or dsRNA templates for Dicer processing

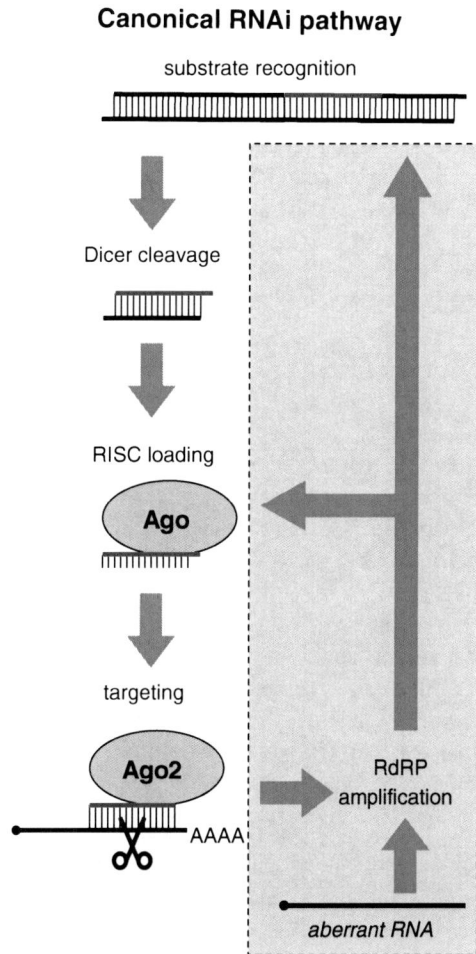

Canonical RNAi pathway

substrate recognition

Dicer cleavage

RISC loading

Ago

targeting

Ago2

AAAA

RdRP
amplification

aberrant RNA

5.2.1   dsRNA Recognition and Cleavage

RNAi is triggered by dsRNA, a unique helical structure formed by two antiparallel RNA strands. dsRNA is frequently formed by repetitive sequences or during viral infections. dsRNA can form by base pairing two single-stranded RNAs as an intramolecular duplex (a hairpin, reviewed in Svoboda and Cara 2006), or the second strand can be synthesized on a single-stranded RNA template by an RdRP (reviewed in Ng et al. 2008).

Recognition of dsRNA is mediated by the dsRNA-binding domain (dsRBD), which is found in a diverse group of proteins (dsRNA-binding proteins, dsRBPs) involved in various responses to dsRNA. Some proteins contain a single dsRBD while others carry multiple copies of the domain. Specific structural features of dsRNA (such as dsRNA termini or mismatches) and individual dsRBDs present in specific proteins contribute to the routing of dsRNA molecules into specific pathways (reviewed in Doyle and Jantsch 2002; Gantier and Williams 2007; Tian et al. 2004).

The key protein for dsRNA processing into siRNAs is Dicer, which typically carries a dsRBD at the C-terminus and can directly recognize and cleave dsRNA in vitro (Ketting et al. 2001; Zhang et al. 2002; Provost et al. 2002). However, the production of small RNAs in the RNAi pathway is also assisted by proteins carrying tandemly arrayed dsRBDs (for more details, see Sect. 5.2.1.2).

5.2.1.1 Dicer Structure and Function

Dicer generates small RNAs in RNAi and many other RNA silencing pathways (reviewed for example in Jaskiewicz and Filipowicz (2008)). Dicer is a large (~200 kDa), multi-domain RNase III endonuclease, which cleaves both strands of a duplex dsRNA (Fig. 5.2). This cleavage produces small (21–27 nt long) RNA duplexes with two nucleotide 3′ overhangs and 5′-monophosphate and 3′-hydroxyl groups at RNA termini. Dicer is a member of a class of RNase III enzymes, which carry two RNase III domains. Several other domains are typically found in Dicer-like proteins in multicellular eukaryotes. These include N-terminal DEAD-like (DExD) and helicase superfamily C domains, piwi/argonaute/zwille (PAZ) domain, domain of unknown function DUF283, and C-terminal dsRBD.

The ribonuclease activity of Dicer requires magnesium ions. Dicer can efficiently cleave dsRNA longer than 30 base pairs (bp), yielding siRNA of approximately 20 bp (Provost et al. 2002; Zhang et al. 2002). Dicer preferentially cleaves dsRNA at the termini but it can also cleave internally with low efficiency (Zhang et al. 2002). Dicer functions as a molecular ruler, measuring the length of the substrate from the PAZ domain to RNase III domains where it is cleaved. The PAZ

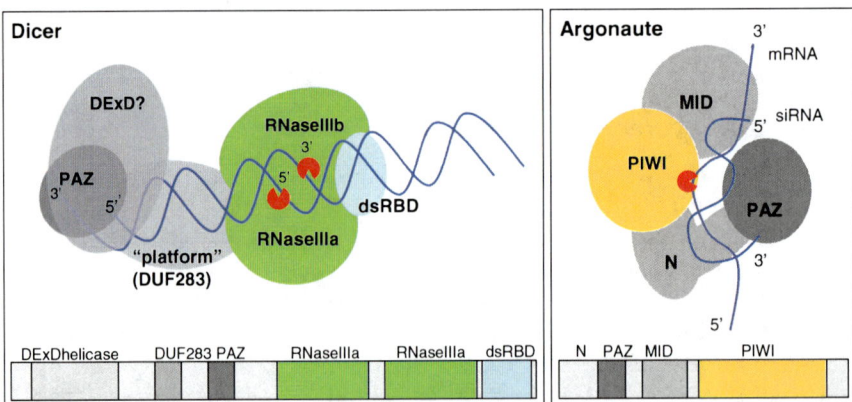

Fig. 5.2 Key components of RNA silencing. A schematic domain organization of Dicer and Argonaute. The schematic structural organization based on crystalography and single-particle electron microscopy structural analysis of Dicer (*left*) and Argonaute (*right*) is shown (Du et al. 2008; MacRae et al. 2007; Macrae et al. 2006; Wang et al. 2009a)

domain binds the end of dsRNA with high affinity to 3′ protruding overhangs (Ma et al. 2004; Lingel et al. 2003; Song et al. 2003; Yan et al. 2003). RNase IIIa and IIIb domains form a single processing center containing two catalytic "half-sites," each cleaving one strand of the duplex and producing short dsRNA with two nt 3′ overhang. The RNase IIIa domain processes the protruding 3′-OH-bearing strand, and the RNase IIIb cuts the opposite 5′-phosphate-containing strand (Zhang et al. 2004). This model was validated by the crystal structure of the full length Dicer from *Giardia intestinalis* (Macrae et al. 2006; MacRae et al. 2007), which showed that the RNase III domains form a catalytic center connected with the PAZ domain by a long α-helix ("connector" helix), which is implicated in determining the product length. The connector helix is supported by a platform-like structure containing the DUF283 domain, which has a dsRBD-like fold (Dlakic 2006) and perhaps mediates protein–protein interaction (Qin et al. 2010).

Most Dicer enzymes contain a DExD helicase domain suggesting that it might be involved in ATP-dependent binding and remodeling of nucleic acids. Although *Drosophila* DCR-2 seems to require ATP for processive cleavage of dsRNA, mammalian Dicer proteins do not require ATP (Provost et al. 2002; Zhang et al. 2002; Nykanen et al. 2001). A kinetic analysis of Dicer mutants showed that the DExD domain could have an autoinhibitory role because a deletion or a mutation of this domain increased catalytic efficiency of Dicer in both single- and multiple-turnover assays (Ma et al. 2008). A modest stimulation of Dicer catalysis was observed in the presence of TRBP, which interacts with the DExD domain (Ma et al. 2008). These and other data (Soifer et al. 2008) suggest that the DExD domain may be a part of the system selecting dsRNA substrates that efficiently enter the RNAi pathway. Further studies will be necessary to establish if and how the DExD domain participates in dsRNA processing and loading of small RNAs on the RISC.

Some organisms, like mammals, *C.elegans* or *Trypanosoma*, utilize a single Dicer protein to produce both siRNAs and miRNAs. In contrast, *Drosophila* utilizes two Dicer paralogs, DCR-1 to produce miRNAs and DCR-2 to produce siRNAs (Lee et al. 2004). DCR-2 contains an N-terminal helicase motif and hydrolyzes ATP. ATP hydrolysis is needed for the processing of dsRNA but not miRNA precursors, and a model has been proposed that the helicase domain is important for DCR-2 processivity (Cenik et al. 2011). Some species utilize even more Dicer paralogs with distinct functions and different cleavage product lengths (e.g., four Dicer paralogs in *Arabidopsis thaliana* (reviewed in Meins et al. 2005)).

Animal Dicer usually localizes to the cytoplasm but it can be also found in the nucleus (Billy et al. 2001; Sinkkonen et al. 2010). However, the significance of the nuclear localization of mammalian Dicer remains unknown at the moment, and it is commonly accepted that siRNA production in animals takes place in the cytoplasm.

5.2.1.2 Proteins with Multiple dsRBDs

RDE-4, the first RNAi protein with multiple dsRBDs (RNAi dsRBP), was identified by a systematic screen for *C. elegans* RNAi-deficient mutants (Tabara et al. 1999).

The *rde-4* mutant was completely deficient in RNAi but failed to show any discernible phenotype, including the absence of transposon activation, which was observed in some other *rde* mutants (Tabara et al. 1999). *Rde-4* encodes a 385-amino acid protein carrying two N-terminal dsRBDs and a third degenerate dsRBD at the C-terminus. A similar organization is found in other proteins implicated in RNA silencing, including *Drosophila* R2D2 and Loquacious and mammalian TRBP and PACT. However, while the phylogenetic analysis suggests that these proteins probably evolved from a common ancestral protein, they significantly diverged (Murphy et al. 2008) and play distinct roles in dsRNA processing and RISC loading.

Biochemical characterization of recombinant RDE-4 showed that it preferentially binds long dsRNA, and its dimerization is necessary for the cleavage of dsRNA into siRNAs (Parker et al. 2006). According to the model supported by mutants and biochemical analyses, RDE-4 dimers bind cooperatively to dsRNA and, together with Dicer, RDE-1 (an Argonaute protein), and DRH-1/2 (Dicer-related *h*elicase from the DExH helicase superfamily), forms a complex initiating the RNAi (Tabara et al. 2002; Parker et al. 2006, 2008). The presence of an Argonaute protein in the complex suggests that dsRNA recognition, processing into siRNA, and loading of the Argonaute-containing effector complex could be integrated in one complex. RDE-4 is involved in siRNA production from dsRNA but is not essential for later steps of RNAi because RDE-4 immunoprecipitates with trigger dsRNA but not siRNA (Tabara et al. 2002) and *rde-4* loss can be rescued with injection of synthetic siRNA (Parrish and Fire 2001).

It should be mentioned that this model of RDE-4 function is based on the analysis of the RNAi responding to exogenous dsRNA (exo-RNAi), which is one of the several RNAi-related pathways in *C. elegans*. RNA silencing in *C. elegans* evolved into an extremely complex system, which utilizes 27 Argonaute proteins and numerous classes of small RNAs (Ketting 2011). The endogenous RNAi pathway (endo-RNAi), which targets endogenous genes (Ambros et al. 2003), employs a distinct mechanism of siRNA production involving DCR-1 and RDE-4 but not RDE-1 and DRH-1/2 (Gent et al. 2010; Lee et al. 2006).

Drosophila employs two RNAi-RBPs: R2D2 and Loquacious. R2D2 was co-purified with Dicer-2 during purifying siRNA-generating activity from *Drosophila* S2 cell lysates (Liu et al. 2003). R2D2 bears 33% identity to RDE-4 but its role is different. DCR-2/R2D2 association does not affect DCR-2 processing. Instead, DCR-2/R2D2 bind siRNAs to promote AGO2-containing RISC loading (Liu et al. 2006) (for more details, see the Sect. 5.2.2.2). The second RNAi dsRBP in *Drosophila* is Loquacious, which was found to associate with DCR-1, suggesting that the miRNA pathway in *Drosophila* employs a distinct dsRBP in substrate routing (Saito et al. 2005; Forstemann et al. 2005). Surprisingly, deep sequencing of small RNAs in *Drosophila* revealed that siRNAs in the endogenous RNAi pathway are produced by DCR-2 but depend preferentially on Loquacious and not on R2D2, the canonical DCR-2 partner (Czech et al. 2008). It was further shown that endo-siRNAs in *Drosophila* predominantly bind AGO2 and can arise from perfect duplexes formed from overlapping sense and antisense transcripts as well as from long hairpins containing bulges and mismatches (Czech et al. 2008). Detailed analysis of Loquacious gene expression revealed specific protein isoforms, which associate with miRNA (LOQS-PB) and RNAi (LOQS-PD) pathways (Hartig et al. 2009; Zhou et al. 2009). Two studies proposed a model where Loquacious and R2D2 function sequentially and non-redundantly in the endogenous RNAi pathway (Marques et al. 2010; Hartig and Forstemann 2011). LOQS-PD functions upstream, stimulating DCR-2-mediated processing of dsRNA whereas R2D2 acts during RISC loading.

TRBP, one of the two known mammalian RNAi dsRBPs, was identified as a Dicer-interacting partner involved in miRNA processing and RISC loading (Chendrimada et al. 2005; Haase et al. 2005). Notably, the role of TRBP in RNA silencing has been studied in cells where the physiological substrate for Dicer processing and RISC loading are miRNA precursors and where long dsRNA readily activates the protein kinase R (PKR) and interferons (IFN). Thus, while the RISC loading role of TRBP may be common for miRNA and RNAi pathways, it is not clear if an isoform of TRBP plays any specific role in recognition and processing of long dsRNA in the canonical mammalian RNAi pathway. Since TRBP also interacts with and inhibits PKR (Cosentino et al. 1995; Park et al. 1994), it was speculated that TRBP could be a component of a network of protein–protein interactions underlying a reciprocal regulation of RNAi/miRNA and IFN-PKR pathway (Haase et al. 2005). This notion is further supported by PACT, a paralog of TRBP, which exerts a positive effect on PKR. PACT was shown to interact with TRBP and Dicer and to facilitate siRNA production (Kok et al. 2007).

5.2.1.3 Helicases and Other Auxiliary Factors

There are several other protein factors which interact with Dicer and facilitate dsRNA recognition and cleavage. In *C. elegans*, numerous DCR-1-interacting proteins were identified by proteomic analyses (Tabara et al. 2002; Duchaine et al. 2006). Among these proteins are the aforementioned Dicer-related helicases (DRH-1, DRH-2, and DRH-3), which interact with DCR-1. DRH-1/2 interact with RDE-4 and DCR-1, and they are essential for exo-RNAi (Tabara et al. 2002). DRH-3 is essential for viability and it is involved in the endo-RNAi pathway (Duchaine et al. 2006). Thus, distinct DRH/DCR-1 complexes recognize different dsRNA triggers and mediate processing and loading on distinct primary Argonaute proteins.

The closest mammalian homologues of *Drh* genes are helicases *Ddx58, Dhx58*, and *Ifih1*, which are expressed in the immune system. DDX58, which is most similar to DRH1, is a helicase also known as RIG-I, which recognizes blunt-ended dsRNA and induces the interferon response (Marques et al. 2006). In fact, the presence of 3′ overhangs in Dicer products impairs RIG-I ability to unwind the dsRNA substrate and activates downstream signaling to the transcription factor IRF-3. The porcine ortholog of *Ddx58, Rhiv-1*, was initially identified as a locus responding to porcine reproductive and respiratory syndrome virus (PPRSV) infection (Zhang et al. 2000). Thus, while DRH proteins facilitate Dicer processing in *C. elegans*, their closest mammalian homologues also respond to dsRNA but acquired novel roles in the immune system, which include an ability to tolerate Dicer products but respond to blunt-ended nonself dsRNAs, such as by-products of viral replication.

While other Dicer-associated proteins were found in *C. elegans*, *Drosophila*, and mammals, they seem to function downstream of Dicer cleavage and will be discussed further below.

5.2.2   RISC Complex Formation

The next step after siRNA production is selection and loading of one of its strand onto the RISC. The key component of RISC is an Argonaute family protein (AGO), which binds the selected siRNA strand and uses it as a sequence-specific guide for recognizing mRNAs that will be degraded. Upon formation of a perfect duplex, AGO protein endonucleolytically cleaves the cognate RNA in the middle of the duplex.

5.2.2.1   Argonaute Proteins

Argonaute proteins have a molecular weight of ~100 kDa and carry two distinct domains: the central PAZ domain and the PIWI (P-element induced wimpy testis) domain at the carboxy-terminus. Two additional domains are recognized, the N-terminal domain and the MID domain between PAZ and PIWI domains (Fig. 5.2). The PAZ domain binds the 3′ end of a short RNA in a sequence-independent manner (Lingel et al. 2003, 2004; Ma et al. 2004; Song et al. 2003). Structural studies of archeal Argonaute homologues showed that the PIWI domain has an RNase H-like fold (Song et al. 2004; Ma et al. 2005; Yuan et al. 2005; Parker et al. 2004). siRNA is anchored with its 3′ end in the PAZ domain. The 5′-phosphate of the siRNA is buried in a pocket at the interface between the MID domain and the PIWI domain (reviewed in Jinek and Doudna 2009). The 5′ end of the base pairing cognate mRNA enters between the N-terminal and PAZ domains, and its 3′ end exits between the PAZ and MID domains. Argonaute was identified to be a "slicer" (Liu et al. 2004; Meister et al. 2004; Song et al. 2004), i.e., the enzyme catalyzing the cleavage of the cognate mRNA in the canonical RNAi pathway. The active site in the PIWI domain is positioned to cleave the mRNA opposite the middle of the siRNA guide (Song et al. 2004).

Argonaute proteins can be divided into three distinct groups (reviewed in Faehnle and Joshua-Tor 2007) (1) AGO proteins, found in all kingdoms, (2) PIWI proteins, found in animals, and (3) WAGO proteins, found only in worms. The WAGO subfamily was described only recently (Yigit et al. 2006), so it is not recognized in the older literature, which typically divides Argonaute proteins into

AGO and PIWI subgroups (Carmell et al. 2002). Specific Argonaute proteins functioning in the RNAi pathway include RDE-1 (exo-RNAi) and ERGO-1 (endo-RNAi) in *C. elegans*, AGO-2 in *Drosophila*, and AGO2 in mammals. Other Argonautes act in the miRNA and other pathways employing small RNAs.

5.2.2.2   RISC Assembly

The key step in RISC formation is the loading of a short RNA produced by Dicer into the complex. In vitro experiments with mammalian proteins suggest that Dicer, TRBP, and AGO2 are critical for RISC loading, and the minimal RISC is composed of the AGO2 protein loaded with an siRNA (Martinez et al. 2002; Gregory et al. 2005; MacRae et al. 2008). A complex of Dicer, an RNAi dsRBP, and an Argonaute protein participates in RISC formation in all studied animal models (Tabara et al. 2002; Tomari et al. 2004b; Gregory et al. 2005).

RISC assembly is best understood in *Drosophila* and human models (Tomari et al. 2004a, b; Pham et al. 2004; Gregory et al. 2005; MacRae et al. 2008). The model of RISC loading in *Drosophila* suggests that RISC assembly occurs in several steps which involve a number of described complexes (Tomari and Zamore 2005). The first complex, formed by siRNA, R2D2, and DCR-2, is also known as R1 or R2/D2/DCR-2 initiator (RDI) complex (Pham et al. 2004; Kim et al. 2007) and develops into a mature form of the RISC loading complex RLC (Tomari and Zamore 2005). The RLC determines strand selection and recruits AGO2 (and other proteins) to form a pre-RISC (Kim et al. 2007), which contains duplex siRNA. Finally, the release of the passenger strand from the duplex produces holo-RISC, which can base pair with complementary mRNA substrates.

The coupling of dsRNA cleavage and RISC assembly is a matter of debate. It was suggested that, after cleavage, small-RNA duplexes need to dissociate from Dicer and then rebind to a sensor of the thermodynamic asymmetry of the duplex, because the siRNA guide strand will be at random orientation (Tomari et al. 2004a). Indeed, small RNA sorting in *Drosophila* suggests that dicing and RISC assembly are uncoupled (Tomari et al. 2007). In contrast, the immunopurified or reconstituted human AGO2 complex can use pre-miRNAs but not siRNA duplexes for a target cleavage suggesting that Dicer cleavage and RISC assembly are functionally coupled in humans (Gregory et al. 2005). However, newer data indicate that, just as in flies, human RISC assembly is uncoupled from dicing and ATP facilitates RISC loading of small-RNA duplexes (Yoda et al. 2010). Interestingly, all four human AGO proteins showed similar structural preferences for small-RNA duplexes, which were highly reminiscent of *Drosophila* AGO1 but not of AGO2 (Yoda et al. 2010).

Fly AGO1 and AGO2 require ATP for the RISC loading (Nykanen et al. 2001; Pham et al. 2004; Tomari et al. 2004a; Kawamata et al. 2009). ATP is presumably used to trigger the dynamic conformational opening of AGO proteins so that they can accept small RNA duplexes (Kawamata et al. 2009). Earlier studies suggested a difference between fly and human systems because human RISC assembly using immunopurified or reconstituted human RLC containing AGO2, Dicer, and TRBP did not require ATP hydrolysis (Gregory et al. 2005; MacRae et al. 2008; Maniataki and Mourelatos 2005). Recent data suggest that ATP facilitates also human RISC loading while it is dispensable for unwinding (Yoda et al. 2010).

Strand selection in the fly RLC is controlled by R2D2. Analysis of the interaction of DCR-2/R2D2 complex with siRNA duplexes showed that R2D2 orients the complex according to thermodynamic stabilities of siRNA strands and binds the 5′-phosphate of the passenger strand at the thermodynamically more stable end (Tomari et al. 2004b). Thus, R2D2 functions as a licensing factor for routing siRNAs into the RNAi pathway. Interestingly, a thorough analysis of AGO2 complexes revealed that, unlike mature miRNAs which are loaded on AGO1, complementary strands of mature miRNAs (miRNA*) are efficiently loaded on AGO2 in DCR2/R2D2-dependent manner (Ghildiyal et al. 2010; Okamura et al. 2011). Thus, the role of R2D2 in sorting small RNAs is wider and extends into the miRNA pathway.

Mammals differ from *Drosophila* because they do not separate Dicer and Argonaute proteins dedicated to RNAi and miRNA pathways, and it is assumed that both pathways use a similar if not the same RLC. However, it should be kept in mind that our knowledge of the mammalian RLC comes either from cells where RLC normally loads miRNAs or from in vitro reconstitution of the RLC with purified proteins. The mammalian RLC is functionally similar to that of *Drosophila*. It is composed of Dicer, TRBP, and AGO2 (Gregory et al. 2005; MacRae et al. 2008). In vitro reconstituted mammalian RLC contains one copy of each protein and has dicing, guide-strand selection, AGO2-loading, and slicing activities.

Biochemical and structural analysis suggests that TRBP is flexibly bound to the Dicer DExH/D domain (Wang et al. 2009a; Daniels et al. 2009). TRBP seems to bridge the release of the siRNA by Dicer and the loading of the duplex onto AGO2. Binding by TRBP may allow the siRNA intermediate to stay associated with RLC after being released from Dicer and may also help in orientation of the siRNA for AGO2 loading. It was also predicted that TRBP acts as a sensor of the thermodynamic stability of 5′ siRNA in strand selection during RISC loading, similarly to DCR-2 and R2D2 (a TRBP homologue) in *Drosophila* (Wang et al. 2009a). However, the supporting evidence is inconclusive (Haase et al. 2005) although some argue that TRBP can indeed act as a sensor (Gredell et al. 2010). Furthermore, while TRBP function is similar to that of R2D2, TRBP sequence is more closely related to Loquacious than R2D2 (Murphy et al. 2008). It is tempting to speculate that the closer evolutionary distance between TRBP and Loquacious and the lack of an R2D2 ortholog reflect the fact that the mammalian endogenous RNAi is restricted to a few cell types and employs proteins normally functioning in the miRNA pathway.

The final step in the assembly of an active RISC is the release of the passenger strand from the siRNA duplex. A helicase activity was proposed to separate the two siRNA strands while the guide remains bound to AGO2 (Sontheimer 2005; Tomari and Zamore 2005; Meister and Tuschl 2004). A candidate for such a helicase in *Drosophila* is Armitage helicase (Tomari et al. 2004a). However, experimental data support a simple solution where passenger strand cleavage by AGO2 slicer activity liberates the single-stranded guide siRNA strand from the pre-RISC complex (Matranga et al. 2005; Miyoshi et al. 2005; Kim et al. 2007). Removal of siRNA passenger strand cleavage products is assisted by C3PO endoribonuclease, which was identified as a RISC-enhancing factor that promotes RISC activation (Liu et al. 2009). The cleavage-assisted mechanism is typical for AGO2-loaded fly and human siRNAs in the RNAi pathway while passenger strand cleavage is not important for loading miRNAs (Matranga et al. 2005).

5.2.2.3 RISC Composition: AGO2-Interacting Proteins

While the minimal active RISC contains only the "slicing" Argonaute protein and the guide siRNA strand (Martinez et al. 2002; MacRae et al. 2008, Rand, 2004 #424), RISC activity was found in different models and cell types to reside in ~200 kDa, ~500 kDa, or 80S complexes (Martinez et al. 2002; Nykanen et al. 2001; Mourelatos et al. 2002; Pham et al. 2004). Various protein components of RISC complexes either contribute to RISC formation or might regulate RISC activity, stability, target selection, mode of repression, or otherwise contribute to RISC function.

In *C. elegans*, the RISC complex was not biochemically purified, so protein complexes containing Argonaute proteins RDE-1 (exo-RNAi) and ERGO-1 (endo-RNAi) remain uncharacterized.

Analysis of RISC in *Drosophila* embryo lysate identified the several additional components of AGO2-containing complexes:

dFMR1: *Drosophila* ortholog of human fragile X mental retardation protein (FMRP) (Caudy et al. 2002; Ishizuka et al. 2002; Pham et al. 2004). dFMR1 is associated with ribosomes through interaction with ribosomal proteins L5 and L1 and with complexes containing miRNAs (Ishizuka et al. 2002). dFMR1 is not a conserved RISC component involved in RNAi. While depletion of dFMR1 reduces RNAi efficiency in *Drosophila* S2 cells (Caudy et al. 2002), the loss of mammalian FMRP has no apparent direct impact on RISC function (Didiot et al. 2009).

VIG: Vasa Intronic Gene (Caudy et al. 2002; Pham et al. 2004). VIG is a conserved protein, which encodes for a putative RNA-binding protein, whose depletion reduces RNAi efficiency (Caudy et al. 2002). *Vig* mutants are more susceptible to viral infections in *Drosophila* (Zambon et al. 2006). Whether this role of VIG is coupled with its presence in the RISC complex is not known. There is no evidence that SERBP1, the closest mammalian VIG homologue, would be associated with RISC.

Armitage: RNA helicase, which was identified as a maternal effect gene required for RNAi (Tomari et al. 2004a). Armitage is probably not required for the RISC activity. Instead, it was proposed to facilitate the removal of the passenger strand during RISC formation (Tomari et al. 2004a). The mammalian homologue of Armitage is an Argonaute-associated protein MOV10 (Meister et al. 2005).

TSN: Tudor Staphylococcal Nuclease is a protein containing five staphylococcal/microccocal nuclease domains and a tudor domain. It is a component of the RISC in *C. elegans, Drosophila*, and mammals (Caudy et al. 2003; Pham et al. 2004). The role of TSN in RISC RNAi remains enigmatic. TSN is not the "slicer" (Schwarz et al. 2004), and it has been implicated in promoting cleavage of hyper-edited dsRNA (Scadden 2005). This observation is surprising considering that RNA editing and RNAi pathways appear mutually antagonistic (Tonkin and Bass 2003; Yang et al. 2005). TSN is also connected with the miRNA pathway, where it mediates degradation of edited miRNA precursors (Yang et al. 2006; Kawahara et al. 2007a).

DMP68 (RM62): This conserved helicase was co-purified with AGO1 and dFMR1 (Ishizuka et al. 2002). It seems to be required for RNAi in S2 cells where depletion of DMP68 results in inhibition of RNAi (Ishizuka et al. 2002). Whether DMP68 is needed for RISC formation or for RISC activity/stability is not known.

Many AGO2-associated proteins were identified in mammalian cells (reviewed in detail in Peters and Meister 2007). These include MOV10, DDX6 (Rck/p54), DDX20 (Gemin3), TNRC6A (GW182), and many others. However, experiments concerning mammalian AGO2-associated proteins were performed in somatic cells where these proteins are loaded with miRNAs. Accordingly, some of the AGO-associated proteins clearly associate with miRNA-mediated repression, while it is not known if any AGO2-associated protein is required for RNAi.

5.2.3   Target Recognition and Cleavage

The RISC complex uses the loaded siRNA as a guide for recognizing its target for cleavage. As mentioned above, the first cleavage actually targets the passenger strand of a loaded siRNA duplex to free the guiding strand, so it can base pair to cognate mRNAs (Matranga et al. 2005; Kim et al. 2007). Whether the first cleavage has any further effects on the RISC structure is not known.

Target recognition requires that the loaded RISC finds and hybridizes to the complementary targets. Target recognition by siRNAs exhibits a distinct 5' bias. Analysis of miRNA-targeted mRNAs revealed that miRNA bases 2–8 form a distinct "seed," which base pairs perfectly to the target transcript (Lewis et al. 2003; Enright et al. 2003). It was found that the 5' half of a small RNA provides most of the binding energy that tethers RISC to a target RNA (Doench et al. 2003; Haley and Zamore 2004). Biochemical analysis of target recognition by mammalian RISC showed that the RISC is apparently not systematically scanning transcripts and it is unable to unfold structured RNA. Thus, RISC randomly transiently contacts single-stranded RNA and promotes siRNA-target base pairing (Ameres et al. 2007). The 5' end of the loaded siRNA creates a thermodynamic threshold for a stable association of RISC with its target (Ameres et al. 2007).

The fact that 5' and 3' ends of an siRNA are bound by distinct binding pockets and that both ends contribute differently to binding to the target lead to a "two-state model of Argonaute function" (Tomari and Zamore 2005). In this model, the 3' end is bound in the PAZ domain and the 5' end in a pocket at the interface between the MID and the PIWI domains. The 5' end is pre-organized to interact with the cognate mRNA and, upon binding, the 3' end is dislodged from the binding pocket to allow for base pairing of the 3' end. Base pairing in the middle of siRNA results in a correct orientation and cleavage of the cognate strand in the active site. This model is supported by recent structural data (Wang et al. 2009b).

Notably, siRNAs can mediate other silencing effects than the cleavage. An imperfect complementarity in the middle to the base pairing may result in translational repression, which is a typical effect of miRNAs (Doench et al. 2003). In addition, out of four mammalian AGO proteins, which can associate with siRNAs, only AGO2 has the "slicer" activity (Liu et al. 2004; Meister et al. 2004; Song et al. 2004).

5.2.4 RdRP Amplifier and Its Loss During Animal Evolution

RdRP is an ancestral component of RNAi because RdRP orthologs were identified in RNA silencing pathways in plants, fungi, and some animals: QDE-1 in *Neurospora crassa* (Cogoni and Macino 1999), EGO-1 and RRF-1 in *C. elegans* (Grishok et al. 2001; Smardon et al. 2000), SDE1/SGS2 in *Arabidopsis* (Dalmay et al. 2001; Mourrain et al. 2000), and Rdp1 in *Schizosaccharomyces pombe* (Hall et al. 2002; Volpe et al. 2002).

C. elegans is the main animal model for studying RdRPs. While an earlier study of RdRP in *C. elegans* suggested that dsRNA synthesis can be primed by primary siRNAs (a model of "degradative PCR") (Sijen et al. 2001), later studies demonstrated that RdRPs do not require the priming by primary siRNAs and produce short RNAs using RISC-targeted mRNAs as templates (Sijen et al. 2007; Pak and Fire 2007). Surprisingly, a sequencing of small RNAs associated with ongoing RNAi in *C. elegans* showed that Dicer-independent secondary siRNAs constitute the majority of cloned siRNAs (Pak and Fire 2007). These secondary siRNAs are only antisense, carry 5'-di- or triphosphates, and are not bound by RDE-1 but by other Argonaute proteins (Sijen et al. 2007; Pak and Fire 2007). *C. elegans* genome encodes for 27 Argonaute proteins (Yigit et al. 2006) and four putative RdRPs, three of which were implicated in RNA silencing (Duchaine et al. 2006; Lee et al. 2006; Sijen et al. 2001; Smardon et al. 2000). A systematic analysis of small RNAs combining different cloning strategies with the next generation sequencing and analysis of Argonaute proteins revealed an amazing complexity of RNA silencing pathways in *C. elegans*. Different RNA substrates are processed in Dicer-dependent and Dicer-independent manner to produce numerous classes of small RNAs, which are loaded on different AGO proteins (Gent et al. 2010; Yigit et al. 2006; Correa et al. 2010; Vasale et al. 2010). Thus, the core RNAi pathway (Sects. 5.2.1–5.2.3) in *C. elegans* can be seen as a starter followed by the main course made by RdRPs.

Homologues of these RdRPs exist in numerous eumetazoan phyla, including *Nematoda* (e.g., *C. elegans*), *Cnidaria* (*Hydra*), *Chelicerata* (tick), *Hemichordata* (acorn worm), *Urochordata* (sea squirt), and *Cephalochordata* (lancelet), but appear absent in others, including *Platyhelminthes* (*Planaria*), *Mandibulata* (*Drosophila*), and *Craniata* (vertebrates). Phylogenetic data suggest that RdRPs in RNA silencing pathways have a monophyletic origin, i.e., evolved from a single ancestral RdRP (Murphy et al. 2008; Cerutti and Casas-Mollano 2006). The fact that RdRP orthologs are found in other protostomes and deuterostomes but not in *Drosophila* or mammals suggests a repeated loss of the ancestral RdRP component of RNA silencing. Whether RdRP activity completely disappeared from RNAi in *Drosophila* and mammals is unclear. The missing RdRP orthologs in RNA silencing in *Drosophila* or vertebrates could be replaced by another RdRP, for example by horizontal transfer of some viral RdRP. In fact, there is evidence for RdRP analogs in *Drosophila* and vertebrates (Lipardi and Paterson 2009; Sam et al. 1998; Maida et al. 2009; Pelczar et al. 2010). Whether and how these activities participate in RNAi remains unresolved. An earlier report of RdRP activity in *Drosophila* (Lipardi et al. 2001) was contradicted by experiments demonstrating the absence of transitive RNAi generating secondary sequences upstream of the region targeted by siRNAs (Roignant et al. 2003; Schwarz et al. 2002). A similar lack of transitive RNAi was observed in mammals (Stein et al. 2003).

Two recent reports propose that two different RdRP activities in *Drosophila* and human cells can generate dsRNA that can be processed by Dicer. One of the RdRPs is ELP1, a noncanonical RdRP conserved in all eukaryotes, which associates with DCR-2, and its loss results in reduction of endo-siRNAs and upregulation of transposon transcripts (Lipardi and Paterson 2009). The second RdRP is a ribonucleoprotein complex of the human telomerase reverse transcriptase (TERT) and the RNA component of mitochondrial RNA processing endoribonuclease (RMRP). RMRP shows a strong preference for substrates that have $3'$ fold-back structures and produces dsRNA that can be processed by Dicer (Maida et al. 2009). In both cases, additional experiments are needed to confirm that ELP1 or TERT-RMRP participates in RNAi and what their exact role is. These include analysis of small RNAs in mutants lacking *Elp1* or TERT and further characterization of complexes containing ELP1 or TERT and RNAi components. Since the transitive RNAi was not detected in *Drosophila* and mammals, it is possible that analogous RdRP activities are not involved in the production of secondary siRNAs. Instead, they could play a role in siRNA-independent production of dsRNA substrates for RNAi. That could explain why ELP1 was not found in any of the previous biochemical studies of RNAi in *Drosophila* and mammals.

5.2.5   Systemic and Environmental RNAi

RNAi can either act in a cell autonomous manner, i.e., affecting only cells directly exposed to dsRNA, or can propagate across cell boundaries (Fig. 5.3). The non-cell autonomous RNAi was observed already during the first RNAi experiments in

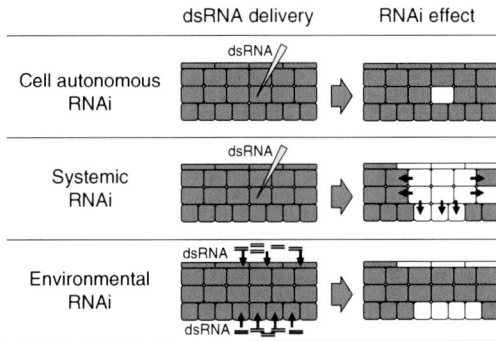

Fig. 5.3 A schematic overview of the different types of RNAi explained on a model of silencing a "*green*" gene in cells. The *first row* shows the cell-autonomous RNAi where the RNAi agent targeting the "*green*" gene is delivered into a cell (e.g., by injection or viral transduction). The silencing effect remains only in the cells where the RNAi was induced initially. Systemic RNAi (the *middle row*) includes processes where the silencing is transported from the cell where RNAi was induced to other cells or even to different tissues where the silencing also takes place. In case of environmental RNAi (the *bottom row*), the dsRNA is taken up from the environment of the cell (external or internal environment in relation to the animal). The silencing effect is observed in all cells which can take up the dsRNA

C. elegans (Fire et al. 1998). It worked like a magic trick: when animals were microinjected with dsRNA into head, tail, intestine, or gonad arm, or if they were even just soaked in dsRNA solution or fed by bacteria expressing dsRNA, a specific null phenotype was induced in the whole animal and even in its progeny, demonstrating a surprising ability of dsRNA to cross cellular boundaries (Fire et al. 1998; Tabara et al. 1998; Timmons and Fire 1998). Two modes of non-cell autonomous RNAi are recognized (1) environmental RNAi involves processes in which dsRNA is taken up by a cell from the environment; (2) systemic RNAi includes processes where a silencing signal spreads from a cell across cellular boundaries into other cells. The studies of RNAi in *C. elegans* show that both modes can be combined and environmental RNAi can be followed by systemic RNAi.

At least two pathways for dsRNA uptake were described (1) a specific transmembrane channel-mediated uptake and (2) an alternative endocytosis-mediated uptake (reviewed in Huvenne and Smagghe 2010; Whangbo and Hunter 2008).

The best understood systemic RNA mechanism is that of *C. elegans* where the transport of the silencing signal to neighboring cells is controlled by dsRNA-transporting channels. *Sid-1* and *sid-2* genes were identified in forward genetic screen to be responsible for systemic RNAi in *C. elegans* (Winston et al. 2002).

SID-1 (systemic RNAi deficient-1) is a conserved transmembrane protein that has homologues in a wide range of animals, including mammals. *Sid-1* mutants have intact cell autonomous RNAi, but are unable to perform either systemic RNAi or environmental RNAi in response to feeding, soaking, or injection of dsRNA (Winston et al. 2002). SID-1 sensitizes *Drosophila* cells to RNAi induced by soaking, enabling concentration-dependent cellular uptake of dsRNA suggesting that SID-1 forms a dsRNA channel (Feinberg and Hunter 2003; Shih et al. 2009). SID-2 is a transmembrane protein localized to an apical membrane of intestinal cells. It is necessary for the initial import of dsRNA from gut lumen, but not for the systemic spread of silencing signals among cells. *Sid-2* homologues have been identified only in two other *Caenorhabditis* species (Winston et al. 2007).

To date, non-cell autonomous RNAi has also been discovered in parasitic nematodes (Geldhof et al. 2007), *Hydra* (Chera et al. 2006), *Planaria* (Newmark et al. 2003; Orii et al. 2003), or insects (Tomoyasu et al. 2008; Xu and Han 2008). However, non-cell autonomous RNAi is not present uniformly. For example, only one of eight tested *Caenorhabditis* species showed efficient environmental RNAi (Winston et al. 2007). Diverse non-cell autonomous RNAi also exists in insects where different taxa have up to three expressed or silent *sid-1* orthologs (Huvenne and Smagghe 2010; Whangbo and Hunter 2008). Some insects, such as red flour beetle *Tribolium*, have efficient systemic RNAi where injection of adults causes RNAi effects in the progeny (Bucher et al. 2002). In *Drosophila*, the natural role of non-autonomous RNAi seems to be coupled with antiviral role of RNAi in adult flies (Saleh et al. 2009). Experimentally induced non-cell autonomous RNAi in *Drosophila* was achieved in some (Dzitoyeva et al. 2003; Eaton et al. 2002) but not all cases (Roignant et al. 2003).

Non-cell autonomous RNA with an extent similar to that of *C. elegans* or of some insects is highly unlikely in vertebrates. However, a limited environmental or systemic RNAi may exist therein as the homologues of *sid-1* have been found in all sequenced vertebrate genomes (Jose and Hunter 2007). Two *sid-1* homologues (*SidT1* and *SidT2*) are present in mice and humans with a documented role for *SidT1* in dsRNA uptake in humans (Duxbury et al. 2005; Wolfrum et al. 2007). Furthermore, experimental overexpression of human *SidT1* significantly facilitated cellular uptake of siRNAs and resulted in increased RNAi efficacy (Duxbury et al. 2005). At the same time, it should be kept in mind that the mammalian immune system employs a number of proteins responding to dsRNA independently of RNAi (Gantier and Williams 2007), the canonical RNAi pathway efficiently operates in a limited number of cell types, and mammalian RNAi does not seem to participate in the innate immunity (Cullen 2006). Thus, the primary role of a dsRNA uptake mechanism in mammals is likely not involving RNAi even though it could have served such a role in an ancestral organism.

5.3   Roles of RNAi in Animals

Because dsRNA often originates from harmful sources, such as viruses and mobile elements, the role of RNAi is often viewed as a form of innate immunity. While this role is experimentally supported, analysis of RNAi mutants suggests additional roles. This part summarizes the current knowledge of the role of RNAi in combating viruses, maintaining genome integrity, and control of gene expression (see also chapter 2 of this volume).

5.3.1 Antiviral Role of RNAi

RNA viruses generate dsRNA during their replication cycle in host cells; DNA viruses often produce complementary sense and antisense transcripts, which can form dsRNA upon annealing. Thus, dsRNA is a common marker of viral infection and it is recognized by various mechanisms mediating an innate immune response. A role of RNA silencing in the innate immunity is supported by several lines of evidence, which were first found in plants and later also in invertebrates (reviewed in Xie and Guo 2006; Marques and Carthew 2007) (1) siRNAs derived from viral sequences were found in infected organisms (Hamilton and Baulcombe 1999), (2) inhibition of RNA silencing results in increased viral replication (Mourrain et al. 2000), and (3) some viruses produce suppressors of RNA silencing (SRS) (Voinnet et al. 1999).

Several studies addressed the role of RNAi in viral suppression in *C. elegans*. As endogenous viral pathogens of *C. elegans* were unknown, this problem was bypassed by using an "artificial" infection with viruses, which had a broad host range and could infect *C. elegans* under laboratory conditions. Model viral infections were based on the (+)ssRNA flock house virus (FHV) (Lu et al. 2005) or the (−)ssRNA vesicular stomatitis virus (VSV) (Schott et al. 2005; Wilkins et al. 2005). Infection with the recombinant VSV was augmented in strong RNAi mutant animals (*rde-1* and *rde-4*), and mutants produced higher viral titers. Furthermore, VSV infection was attenuated in *rrf-3* and *eri-1* mutants that are hypersensitive to RNAi (Wilkins et al. 2005). Similar results were obtained from infected cultured cells (Schott et al. 2005) and from an FHV infection of *rde-1* mutants (Lu et al. 2005). The antiviral role of exo-RNAi in nematodes was recently demonstrated also for a newly discovered natural viral infection of *C. elegans* and *C. briggsae* (Felix et al. 2011).

An antiviral role of RNAi has also been demonstrated in insects. It was shown that FHV is an initiator and a target of RNA silencing in *Drosophila* host cells (Li et al. 2002). Infection of 14 different *Drosophila* RNA silencing mutants with a dsRNA X virus (DXV) showed that all but three lines were significantly more susceptible to viral infection (reduced survival and elevated viral titers) than normal flies. Moreover, replication of DXV was sequence-specifically inhibited (but not absolutely blocked) by "immunizing" *Drosophila* S2 cells with dsRNA from the coding region of DXV before infection (Zambon et al. 2006). Interestingly, increased susceptibility was observed not only for mutants of the RNAi pathway, such as *r2d2*, *armi*, or *ago2*, but also for mutants of the piRNA pathway (*aubergine* and *piwi*), suggesting that RNAi is not the only RNA silencing pathway in *Drosophila* dedicated to the antiviral response. A number of studies provides ample evidence that RNAi plays an essential role in antiviral response in insects (Galiana-Arnoux et al. 2006; Keene et al. 2004; Nayak et al. 2010; Sanchez-Vargas et al. 2009; Wang et al. 2006).

In contrast to nematodes and insects, data supporting involvement of mammalian RNAi in antiviral defense is weak (reviewed in detail in Cullen 2006). It is unlikely that RNAi substantially acts as an antiviral mechanism in mammals where long dsRNA induces a complex sequence-independent antiviral response, commonly known as the interferon response (reviewed in Gantier and Williams 2007). Consistent with this, no siRNAs of viral origin have been found in human cells infected with a wide range of viruses (Pfeffer et al. 2005). Occasional observations, such as detection of a single siRNA in HIV-1 infected cells (Bennasser et al. 2005) does not provide any conclusive evidence that RNAi is processing viral dsRNA and suppresses viruses under physiological conditions in vivo.

It must be stressed that any circumstantial evidence suggesting the role of RNAi in viral suppression must be carefully examined and interpreted. Since viruses coevolve with different hosts and explore all possible strategies to maintain and increase their fitness, it is not surprising that viral reproductive strategies come into contact with mammalian RNA silencing pathways, particularly the miRNA pathway, which shares components with the RNAi pathway. For example, Epstein-Barr virus (EBV) and several other viruses encode their own miRNAs (Pfeffer et al. 2004, 2005; Sullivan et al. 2005) or take advantage of host cell miRNAs to enhance their replication (Jopling et al. 2005; Pfeffer et al. 2005).

Another evidence for an interaction between viruses and RNA silencing is the presence of putative SRS in various viruses. As viral genomes rapidly evolve, SRS should be functionally relevant. For example, B2 protein in Nodaviruses (e.g., FHV) is essential for replication and inhibits Dicer function, and B2-deficient FHV can be rescued by artificial inhibition of RNAi response (Li et al. 2002). B2 protein also enhances the accumulation of Nodaviral RNA in infected mammalian cells (Fenner et al. 2006; Johnson et al. 2004). Other potential SRS molecules have been identified in viruses infecting vertebrates, such as Adenovirus VA1 noncoding RNA (Lu and Cullen 2004), Influenza NS1 protein (Li et al. 2004), Vaccinia virus E3L protein (Li et al. 2004), Ebola virus VP35 protein (Haasnoot et al. 2007), Tas protein in primate foamy virus (Lecellier et al. 2005), or HIV-1 Tat protein (Bennasser et al. 2005).

The existence of SRS in viruses infecting mammals does not prove that these viruses are targeted by RNAi in mammalian cells. First, viruses may have a broader range of hosts (or vectors), e.g., blood sucking insects. Thus, a virus can be targeted by RNAi in one host and by a different defense mechanism in another one. For example, the Dengue virus, whose life cycle takes place in humans and mosquitoes, is targeted by RNAi in mosquitoes and it likely evolved some adaptation to circumvent the response (Sanchez-Vargas et al. 2009). Second, viral SRS in mammalian cells may have other purpose than counteracting viral suppression by RNAi. Since biogenesis and mechanism of action of mammalian miRNAs overlaps with RNAi, it is possible that the role of such SRS is to modify cellular gene expression by suppressing the activity of miRNAs. Third, the primary effect of SRS may be aimed at other defense mechanisms recognizing and responding to dsRNA and, as a consequence, SRS effects on RNA silencing are observed.

5.3.2   RNAi and Suppression of Mobile Elements

In *C. elegans*, mutant screens for components of the RNAi pathway and for the repressors of Tc1 transposon revealed that RNAi controls the activity of transposable elements (TEs) (Ketting et al. 1999; Tabara et al. 1999). Consistent with these results, primary endo-siRNAs derived from Tc1 transposon were observed in the germline (Sijen and Plasterk 2003). High-throughput sequencing identified additional small RNA species involved in TE silencing. These include Dicer-independent piRNAs expressed in all animal phyla (reviewed in Malone and Hannon 2009a) and secondary siRNAs generated by RdRPs in *C. elegans* (Pak and Fire 2007; Sijen et al. 2007). An unexpected connection linking piRNAs to endo-siRNA biogenesis has been observed in *C. elegans* germline (Das et al. 2008), where endo-siRNAs targeting Tc3 element were dependent on piRNAs while Tc1-derived endo-siRNAs were still present in piwi mutants. These findings suggest a cross talk between small RNA pathways in worms to ensure efficient TE silencing.

In *Drosophila*, the small RNA pathways are believed to play distinct roles in TE silencing in the germline and in somatic tissues. While piRNAs are responsible for genome surveillance predominantly in the germline, TE-derived endo-siRNAs have been identified in somatic tissues and cultured cell lines (Czech et al. 2008; Okamura et al. 2008; Kawamura et al. 2008; Ghildiyal et al. 2008; Chung et al. 2008). However, piRNAs from ovarian somatic cells have also been described, arising specifically from the flamenco locus (Malone et al. 2009; Li et al. 2009). As endo-siRNAs from the flamenco locus and several other piRNA loci were also detected, it is likely that the *Drosophila* piRNA and endo-siRNA pathways might be interdependent, similarly to *C. elegans*, in their task to efficiently silence TEs (Ghildiyal et al. 2008).

Abundant piRNAs and potential endo-siRNAs derived from transposons and repetitive elements were also found in *Xenopus tropicalis* (Armisen et al. 2009). Here, the piRNAs are restricted solely to the germline while endo-siRNAs originating from similar genomic loci were found in both oocytes and somatic tissues, with most endo-siRNAs mapping to the palindromic sequences of Polinton DNA transposons.

RNAi-mediated TE silencing has also been documented in the mouse germline (Watanabe et al. 2006, 2008; Tam et al. 2008). Mutations in the piRNA pathway components are detrimental to sperm development, suggesting that piRNAs are a dominant class of small RNAs controlling TE activity in the male germline (Malone and Hannon 2009b). In contrast, female mice lacking functional piRNA pathway are fertile with no obvious defects in oocytes (Carmell et al. 2007). Endo-siRNAs suppress TE silencing in mammalian oocytes as documented by derepression of some retrotransposons in oocytes depleted of Dicer or AGO2 (Murchison et al. 2007; Watanabe et al. 2008). As already proposed for invertebrates, the piRNA and endo-siRNA pathways likely cooperate in creating a complex silencing network against TEs in the mammalian germline. Long terminal repeat MT elements and SINE elements are strongly upregulated in *Dicer-/-* oocytes, while the levels of IAP retrotransposon are elevated in the absence of Mili protein but not in *Dicer-/-* oocytes (Watanabe et al. 2008; Murchison et al. 2007). Still many loci composed of other types of TEs, e.g., LINE retrotransposons, give rise to both piRNAs and endo-siRNAs, again suggesting that the biogenesis of these small RNAs is interdependent. The role of endogenous RNAi in TE silencing extends from germ cells to preimplantation embryo stages. Apart from maternally derived piRNAs and endo-siRNAs, which persist in the embryos for a large part of preimplantation development, zygotic endo-siRNAs are generated de novo, mainly to control the activity of zygotically activated MuERV-L retrotransposon (Ohnishi et al. 2010; Svoboda et al. 2004). SINE-derived endo-siRNAs also increase in abundance in early embryo stages, which is consistent with the observation that B1/Alu SINE endo-siRNAs account for a vast majority of endo-siRNAs sequenced from mouse embryonic stem cells (mESCs) (Babiarz et al. 2008). Whether these SINE endo-siRNAs play an active role in TE silencing in mESCs similarly to other TE-derived endo-siRNAs in oocytes remains to be determined. RNAi-dependent silencing of LINE transposons has also been described in cultured HeLa cells, where endo-siRNAs derived from bidirectional transcripts of sense and antisense L1 promoter were proposed to control L1 activity (Yang and Kazazian 2006). Although some evidence for retrotransposon-derived endo-siRNAs from mammalian somatic cells was obtained from deep sequencing data (Kawaji et al. 2008), a convincing support for the function of endo-siRNAs in TE silencing in mammalian somatic tissues has yet to be provided. Further reading on small RNAs and TEs is provided in chapter 2 of this volume.

5.3.3   *RNAi and Regulation of Endogenous Genes and Development*

A role of endogenous RNAi in shaping the protein-coding transcriptomes during development has been challenged by mutant worms and flies lacking essential components of the RNAi pathway, which were viable and produced healthy offspring (Tabara et al. 1999; Lee et al. 2004; Okamura et al. 2004). In view of those findings, RNAi had been viewed solely as a defense mechanism against invasive nucleic acids. However, deep sequencing analyses revealed that endo-siRNAs with sequence complementarity to hundreds of protein-coding mRNAs are present in *C. elegans* (Ambros et al. 2003; Ruby et al. 2006). Gene regulating endo-siRNAs in *C. elegans* differ in biogenesis and in requirements for functional components of the RNAi exerting machinery (Pak and Fire 2007; Sijen et al. 2007). This is consistent with microarray analysis of mutant worms lacking various RNAi-related factors, which identified non-overlapping sets of differentially expressed genes, supporting the idea of functionally distinct RNAi pathways in nematodes (Lee et al. 2006). In search for the cellular and developmental processes, which might be controlled by endo-siRNAs in *C. elegans*, spermatogenesis-associated genes were found enriched in the group of transcripts matching endo-siRNAs (Ruby et al. 2006). Further studies revealed that mutations in RNAi-related genes result in defects in meiotic chromosome disjunction, spindle formation, or microtubule organization during sperm development and ultimately lead to male sterility or embryonic lethality of the offspring (Han et al. 2008, 2009; Pavelec et al. 2009; Gent et al. 2009).

A fraction of *Drosophila* endo-siRNAs maps to protein-coding regions (Ghildiyal et al. 2008; Kawamura et al. 2008; Okamura et al. 2008; Czech et al. 2008). However, only endo-siRNAs derived from a small number of loci are produced in sufficient amount to reduce target mRNA levels, as exemplified by the *esi-2* locus-derived endo-siRNAs targeting DNA damage-response gene *Mus-308* (Czech et al. 2008; Okamura et al. 2008). The second type of *Drosophila* endo-siRNAs arise from overlapping antisense transcripts observed in hundreds of protein-coding loci. Abundance of these endo-siRNAs is generally low, likely reflecting the fact that RdRP does not significantly contribute to endo-siRNA biogenesis in flies. In addition, their potential mRNA targets are not upregulated in *Ago2*-deficient flies, suggesting that these endo-siRNAs are not involved in posttranscriptional control of mRNA levels under physiological conditions (Czech et al. 2008). Interestingly, a dsRNA/endo-siRNA-binding protein Blanks, which associates with DCR-2 and forms an alternative Argonaute-independent functional RISC complex, has a role in spermatogenesis (Gerbasi et al. 2011). As Blanks deletion does not affect transposon activity, this finding would imply a role for endo-siRNAs in regulation of protein-coding mRNAs in *Drosophila* sperm development.

Although *Drosophila* mutants lacking *Dcr-2* or *Ago-2* develop to normal adults with no specific phenotype under standard laboratory conditions, severe defects in embryonic development have been noted in these mutants upon exposure to temperature perturbations (Lucchetta et al. 2009). When two halves of a living embryo were maintained at different temperatures, the mutant embryos were not able to compensate for faster development in the anterior part exposed to elevated temperature, which lead to segmentation abnormalities. This indicates that endo-siRNA pathway is needed to stabilize embryonic development under environmental stress (Lucchetta et al. 2009). Strikingly, the endo-siRNA-linked defects in *C. elegans* sperm development were also observed upon elevated temperature (Duchaine et al. 2006; Kennedy et al. 2004; Han et al. 2008; Pavelec et al. 2009; Gent et al. 2009). It is currently unknown whether the temperature-sensitive endo-siRNA pathway mobilization might be linked to increased endo-siRNA production or to enhanced activity of the RNAi machinery. However, it is tempting to speculate that support of development during unfavorable conditions is another role for endo-siRNA pathway in animals.

In mice, perturbation of the endo-siRNA pathway in oocytes is responsible for severe meiotic defects and resulting female infertility. Oocyte-specific knockout of either *Dicer* or *Ago2* leads to similar phenotypes, including chromosome misalignment and defective spindle (Murchison et al. 2007; Kaneda et al. 2009; Tang et al. 2007). These effects were originally attributed to the loss of maternal miRNAs.

However, miRNA pathway is suppressed in mouse oocytes and nonessential as oocytes lacking canonical miRNAs are fertile (Ma et al. 2010; Suh et al. 2010). Interestingly, transcriptomes of oocytes lacking either *Dicer* or *Ago2* are similarly affected and genes matching pseudogene-derived endo-siRNAs (Tam et al. 2008; Watanabe et al. 2008) are enriched in the group of upregulated genes in both knockouts (Kaneda et al. 2009; Ma et al. 2010; Tang et al. 2009). In addition, putative endo-siRNA targets are enriched in cell cycle regulators and genes involved in microtubule organization and dynamics (Tam et al. 2008). These findings suggest that regulation of protein-coding genes by endo-siRNAs controls the equilibrium of protein factors required for proper spindle formation, chromosome segregation, and meiosis progression in mouse oocytes. As pseudogenes are a rapidly evolving source of dsRNA for endo-siRNA production, it will be interesting to investigate whether the role of RNAi in spindle formation during meiotic maturation of oocytes is conserved in mammals.

Endo-siRNAs have also been proposed to contribute to the self-renewal and proliferation of mESCs, since the proliferation and differentiation defects observed in *Dicer-/-* mESCs are stronger than in *Dgcr8-/-* mESCs (Kanellopoulou et al. 2005; Murchison et al. 2005; Wang et al. 2007). A population of endo-siRNAs derived mostly from hairpin-forming B1/Alu subclass of SINE elements was identified in mESCs (Babiarz et al. 2008). Fragments of SINE elements are commonly present in untranslated regions of protein-coding transcripts; therefore, it is possible that SINE-derived endo-siRNAs participate in posttranscriptional gene silencing in mESCs. However, this hypothesis has not been tested experimentally.

Scarce evidence is available for potential role of endo-siRNAs in the regulation of protein-coding mRNAs in mammalian somatic tissues. Endo-siRNAs derived from natural antisense transcripts of *Slc34a* gene were identified in mouse kidney, where this sodium/phosphate cotransporter exerts its physiological function (Carlile et al. 2009). However, changes in expression levels of *Slc34a* upon suppression of the endo-siRNA pathway have not been addressed. Deep sequencing revealed a set of potential endo-siRNAs generated from overlapping sense/antisense transcripts and from hairpin structures within introns of protein-coding genes in the mouse hippocampus (Smalheiser et al. 2011). The most abundant endo-siRNAs from *SynGAP1* gene locus were also found in complexes with AGO proteins and FMRP in vivo. Interestingly, a large part of potential hippocampal endo-siRNA targets encode for proteins involved in the control of synaptic plasticity, and the number of endo-siRNAs derived from these gene loci increased significantly during olfactory discrimination training (Smalheiser et al. 2011). Given the fact that the vast majority of identified endo-siRNA sequences mapped to intronic regions, the endo-siRNAs could act co-transcriptionally on nuclear pre-mRNAs, perhaps similarly to the mechanism of RNAi-mediated inhibition of RNA polymerase II elongation recently described in *C. elegans* (Guang et al. 2010). Alternatively, endo-siRNAs could control a correct distribution of target mRNAs as unspliced pre-mRNA can be exported from the neuronal nucleus and transported to dendrites for processing (Glanzer et al. 2005). In any case, these findings open an attractive hypothesis that endo-siRNAs participate in synaptic plasticity during learning process, and the neuronal endo-siRNA pathway might be also linked to various neurodegenerative disorders (Smalheiser et al. 2011).

5.3.4 Interaction of RNAi with Other dsRNA-Induced Pathways

Diversity of dsRBPs shows that dsRNA plays other roles apart from serving as a trigger in RNAi. In this section, we will discuss two dsRNA responding pathways: A-to-I editing and interferon response, which coexist and interact with RNAi.

5.3.4.1 A-to-I Editing

A-to-I editing is mediated by adenosine deaminases acting on RNA (ADARs), enzymes that carry dsRBD and recognize both inter- and intramolecular dsRNAs longer than 20–30 bp (Nishikura et al. 1991). ADARs convert adenosines to inosines, which are interpreted as guanosines during translation. It was predicted that more than 85% of pre-mRNAs may be edited, predominantly in the noncoding regions (Athanasiadis et al. 2004). Many long perfect dsRNAs (>100 bp) undergo extensive editing with a conversion of approximately 50% of adenosines to inosines (Nishikura et al. 1991; Polson and Bass 1994). On the other hand, short RNAs (~20–30 bp) or imperfect long dsRNAs are edited selectively; usually only a few adenines at specific sites are deaminated (Lehmann and Bass 1999).

RNA editing can negatively influence RNAi in several ways. First, ADARs can compete with RNAi for dsRNA substrates including siRNAs. The ADAR1 isoform (ADAR1p150) strongly binds siRNA and reduces the availability of dsRNA for RNAi, resulting in less efficient RNAi in normal cells compared to *Adar1*-/- cells (Yang et al. 2005). Interestingly, injection of high doses of siRNAs enhances ADAR1 expression, suggesting a role of ADAR1 in a cellular feedback mechanism in response to siRNA (Hong et al. 2005).

A change of a single base in a sequence may result either in destabilization of dsRNA structure (inosine–uridine pair) or in its stabilization (inosine–cytidine pair) (Nishikura 2010). This transition in the local and global stability of dsRNA structure can influence further processing of dsRNA, such as the selection of the effective siRNA strand (Bartel 2004; Du and Zamore 2005; Meister and Tuschl 2004). While moderate deamination (one I–U pair per siRNA) does not prevent Dicer processing to siRNAs (Zamore et al. 2000), hyperditing (~50% of deaminated adenosines) can make dsRNA resistant to Dicer processing (Scadden and Smith 2001). Hyperedited dsRNA is also degraded by TSN (Scadden 2005).

Moreover, editing affects target recognition as a single nucleotide mismatch between siRNA and target mRNA can reduce RNAi efficacy (Scadden and Smith 2001) or modify target specificity (Kawahara et al. 2007b). This has been well documented for miRNAs. Several pri-miRNAs (e.g., miR-142) are known to undergo editing, which inhibits Drosha cleavage or even causes degradation of pri-miRNA by

TSN (Nishikura 2010; Scadden 2005; Yang et al. 2006). In other cases, pri-miRNA editing does not influence Drosha activity but inhibits processing of pre-miRNA by Dicer (e.g., miR-151) (Kawahara et al. 2007a). Last but not least, RNA editing might also inhibit export of miRNAs from the nucleus (Nishikura 2010).

An important connection between A-to-I editing and RNAi has been documented in *C. elegans*. In contrast to mice, where *Adar1* or *Adar2* deletion is lethal, in *Drosophila* and *C. elegans Adr* null phenotype causes only weak phenotypic alterations (Palladino et al. 2000; Tonkin et al. 2002). *Adr-1* or *adr-2* mutant worms exhibit a defective chemotaxis, but the phenotype is reverted when worms lacking *Adar* are crossed with RNAi-defective strains (Tonkin and Bass 2003). This indicates a necessity of a balance between RNAi and ADAR-mediated editing.

5.3.4.2 Interferon Response

Mammalian somatic cells can respond to dsRNA in a sequence-independent manner. A pioneering work by Hunter et al. showed that different types of dsRNA can block translation in reticulocyte lysates (Hunter et al. 1975). Analysis of the phenomenon identified PKR that is activated upon binding to dsRNA and blocks translation by phosphorylating the alpha subunit of eukaryotic initiation factor 2 (eIF2α) (Meurs et al. 1990). Activation of PKR represents a part of a complex response to foreign molecules known as the interferon response (reviewed in Sadler and Williams 2007), which includes an activation of the NFκB transcription factor and a large number of interferon-stimulated genes (ISGs) (Geiss et al. 2001). In addition to PKR, several other proteins recognizing dsRNA are integrated to the interferon response, including helicases RIG-I and MDA5, which sense cytoplasmic dsRNA and activate interferon expression, and the 2′,5′-oligoadenylate synthetase (OAS), which produces 2′,5′-linked oligoadenylates that induce general degradation of RNAs by activating latent RNase L (reviewed in Gantier and Williams 2007; Sadler and Williams 2007).

Interactions between RNAi and interferon response are poorly understood. There are two clear mechanistical connections between these two pathways. First, TRBP and PACT, two dsRNA-binding proteins, which were mentioned earlier as Dicer-interacting proteins, interact also with PKR. Notably, while TRBP inhibits PKR (Cosentino et al. 1995; Park et al. 1994), PACT activates it (Patel and Sen 1998). While cytoplasmic long dsRNA in somatic cells apparently triggers the interferon response, it is not clear if the same dsRNA is also routed into the RNAi pathways. Experiments in oocytes and undifferentiated mESCs (Stein et al. 2005; Yang et al. 2001) suggest that RNAi dominates in response to cytoplasmic long dsRNA in the absence of a strong interferon response and that the interferon pathway dominates when its relevant components are present. Nevertheless, this view may be too simplistic because there are several reports showing induction of RNAi with experimental intracellular expression of long dsRNA in transformed and primary somatic cells (Elbashir et al. 2001; Diallo et al. 2003; Gan et al. 2002; Shinagawa and Ishii 2003; Tran et al. 2004; Yi et al. 2003). A recent study of effects of ubiquitous long dsRNA expression in transgenic mice shows that an expressed long dsRNA is well tolerated in mammalian somatic cells and a robust RNAi response is observed only in oocytes (Nejepinska et al. in press). Further research is needed to understand mechanisms routing long dsRNA into RNAi and interferon pathways.

The second connection between RNAi and interferon response is evolutionary. As mentioned earlier, mammalian RNA helicases *Ddx58*, *Dhx58*, and *Ifih1*, which are involved in immune response, are the closest homologues of helicases involved in the processing of long dsRNA during RNAi in *C. elegans*. Notably, DDX58, also known as RIG-I, is an established component of the interferon response to long dsRNA (Yoneyama et al. 2004). This suggests that the interferon response, which has a common trigger and evolved after the RNAi pathway, adopted several components from the latter pathway. It remains to be determined whether these and other components of RNAi lost their function in RNAi entirely or mediate a certain form of a cross talk between RNAi and the interferon response.

5.4   Closing Remarks

Most biologists likely consider RNAi as an excellent tool to study gene function. At the same time, RNAi is a complex natural phenomenon with numerous physiological functions. We have summarized the current knowledge of the molecular mechanism of RNAi and its biological roles. RNAi in animals likely originated from an ancient innate immune response allowing dealing with viral infections and parasitic sequences in the genome, which have a capacity to generate dsRNA. The original role of RNAi still persists in invertebrates. In mammals, the defensive role of RNAi was largely replaced by a more recent form of immune system and RNAi retained its traditional function in suppression of mobile elements in the female germline. While mammals do not combat viruses by RNAi anymore, they adopted RNAi for a regulation of endogenous genes in the female germline. The story of RNAi provides another example of ingenuity of nature in evolving novel functions for old mechanisms.

Acknowledgments We thank Radek Malik for help with manuscript preparation. Our research is supported by the following grants: EMBO SDIG project 1483, GACR 204/09/0085, GACR P305/10/2215, and Kontakt ME09039. P.S. is a holder of the J.E. Purkyne Fellowship.

References

Ambros V, Lee RC, Lavanway A, Williams PT, Jewell D (2003) MicroRNAs and other tiny endogenous RNAs in *C. elegans*. Curr Biol 13(10):807–818. doi:S0960982203002872 [pii]
Ameres SL, Martinez J, Schroeder R (2007) Molecular basis for target RNA recognition and cleavage by human RISC. Cell 130(1):101–112. doi:S0092–8674(07)00583–1 [pii], 10.1016/j.cell.2007.04.037

Aravin AA, Hannon GJ, Brennecke J (2007) The Piwi-piRNA pathway provides an adaptive defense in the transposon arms race. Science 318(5851):761–764. doi:318/5851/761 [pii], 10.1126/science.1146484

Armisen J, Gilchrist MJ, Wilczynska A, Standart N, Miska EA (2009) Abundant and dynamically expressed miRNAs, piRNAs, and other small RNAs in the vertebrate Xenopus tropicalis. Genome Res 19(10):1766–1775. doi:gr.093054.109 [pii]. 10.1101/gr.093054.109

Athanasiadis A, Rich A, Maas S (2004) Widespread A-to-I RNA editing of Alu-containing mRNAs in the human transcriptome. PLoS Biol 2(12):e391

Babiarz JE, Ruby JG, Wang Y, Bartel DP, Blelloch R (2008) Mouse ES cells express endogenous shRNAs, siRNAs, and other Microprocessor-independent, Dicer-dependent small RNAs. Genes Dev 22(20):2773–2785. doi:22/20/2773 [pii]. 10.1101/gad.1705308

Bartel DP (2004) MicroRNAs: genomics, biogenesis, mechanism, and function. Cell 116 (2):281–297. doi:S0092867404000455 [pii]

Bennasser Y, Le SY, Benkirane M, Jeang KT (2005) Evidence that HIV-1 encodes an siRNA and a suppressor of RNA silencing. Immunity 22(5):607–619. doi:S1074–7613(05)00105–6 [pii]. 10.1016/j.immuni.2005.03.010

Bernstein E, Caudy AA, Hammond SM, Hannon GJ (2001) Role for a bidentate ribonuclease in the initiation step of RNA interference. Nature 409(6818):363–366

Billy E, Brondani V, Zhang H, Muller U, Filipowicz W (2001) Specific interference with gene expression induced by long, double-stranded RNA in mouse embryonal teratocarcinoma cell lines. Proc Natl Acad Sci USA 98(25):14428–14433

Bucher G, Scholten J, Klingler M (2002) Parental RNAi in Tribolium (Coleoptera). Curr Biol 12 (3):R85–86. doi:S0960982202006668 [pii]

Carlile M, Swan D, Jackson K, Preston-Fayers K, Ballester B, Flicek P, Werner A (2009) Strand selective generation of endo-siRNAs from the Na/phosphate transporter gene Slc34a1 in murine tissues. Nucleic Acids Res 37(7):2274–2282. doi:gkp088 [pii]. 10.1093/nar/gkp088

Carmell MA, Xuan Z, Zhang MQ, Hannon GJ (2002) The Argonaute family: tentacles that reach into RNAi, developmental control, stem cell maintenance, and tumorigenesis. Genes Dev 16 (21):2733–2742

Carmell MA, Girard A, van de Kant HJ, Bourc'his D, Bestor TH, de Rooij DG, Hannon GJ (2007) MIWI2 is essential for spermatogenesis and repression of transposons in the mouse male germline. Dev Cell 12(4):503–514. doi:S1534–5807(07)00100–1 [pii]. 10.1016/j.devcel.2007.03.001

Caudy AA, Myers M, Hannon GJ, Hammond SM (2002) Fragile X-related protein and VIG associate with the RNA interference machinery. Genes Dev 16(19):2491–2496

Caudy AA, Ketting RF, Hammond SM, Denli AM, Bathoorn AM, Tops BB, Silva JM, Myers MM, Hannon GJ, Plasterk RH (2003) A micrococcal nuclease homologue in RNAi effector complexes. Nature 425(6956):411–414

Cenik ES, Fukunaga R, Lu G, Dutcher R, Wang Y, Hall TM, Zamore PD (2011) Phosphate and R2D2 Restrict the Substrate Specificity of Dicer-2, an ATP-Driven Ribonuclease. Mol Cell. doi:S1097–2765(11)00178-X [pii]. 10.1016/j.molcel.2011.03.002

Cerutti H, Casas-Mollano JA (2006) On the origin and functions of RNA-mediated silencing: from protists to man. Curr Genet 50(2):81–99. doi:10.1007/s00294–006–0078-x

Chendrimada TP, Gregory RI, Kumaraswamy E, Norman J, Cooch N, Nishikura K, Shiekhattar R (2005) TRBP recruits the Dicer complex to Ago2 for microRNA processing and gene silencing. Nature 436(7051):740–744

Chera S, de Rosa R, Miljkovic-Licina M, Dobretz K, Ghila L, Kaloulis K, Galliot B (2006) Silencing of the hydra serine protease inhibitor Kazal1 gene mimics the human SPINK1 pancreatic phenotype. J Cell Sci 119(Pt 5):846–857. doi:jcs.02807 [pii]. 10.1242/jcs.02807

Chung WJ, Okamura K, Martin R, Lai EC (2008) Endogenous RNA interference provides a somatic defense against Drosophila transposons. Curr Biol 18(11):795–802. doi:S0960–9822 (08)00603–9 [pii]. 10.1016/j.cub.2008.05.006

Cogoni C, Macino G (1999) Gene silencing in Neurospora crassa requires a protein homologous to RNA-dependent RNA polymerase. Nature 399(6732):166–169. doi:10.1038/20215

Correa RL, Steiner FA, Berezikov E, Ketting RF (2010) MicroRNA-directed siRNA biogenesis in *Caenorhabditis elegans*. PLoS Genet 6(4):e1000903. doi:10.1371/journal.pgen.1000903

Cosentino GP, Venkatesan S, Serluca FC, Green SR, Mathews MB, Sonenberg N (1995) Double-stranded-RNA-dependent protein kinase and TAR RNA-binding protein form homo- and heterodimers in vivo. Proc Natl Acad Sci USA 92(21):9445–9449

Cullen BR (2006) Is RNA interference involved in intrinsic antiviral immunity in mammals? Nat Immunol 7(6):563–567. doi:ni1352 [pii]. 10.1038/ni1352

Czech B, Hannon GJ (2011) Small RNA sorting: matchmaking for Argonautes. Nat Rev Genet 12 (1):19–31. doi:nrg2916 [pii]. 10.1038/nrg2916

Czech B, Malone CD, Zhou R, Stark A, Schlingeheyde C, Dus M, Perrimon N, Kellis M, Wohlschlegel JA, Sachidanandam R, Hannon GJ, Brennecke J (2008) An endogenous small interfering RNA pathway in Drosophila. Nature 453(7196):798–802. doi:nature07007 [pii]. 10.1038/nature07007

Dalmay T, Horsefield R, Braunstein TH, Baulcombe DC (2001) SDE3 encodes an RNA helicase required for post-transcriptional gene silencing in Arabidopsis. EMBO J 20(8):2069–2078. doi:10.1093/emboj/20.8.2069

Daniels SM, Melendez-Pena CE, Scarborough RJ, Daher A, Christensen HS, El Far M, Purcell DF, Laine S, Gatignol A (2009) Characterization of the TRBP domain required for dicer interaction and function in RNA interference. BMC Mol Biol 10:38. doi:1471–2199–10–38 [pii]. 10.1186/1471–2199–10–38

Das PP, Bagijn MP, Goldstein LD, Woolford JR, Lehrbach NJ, Sapetschnig A, Buhecha HR, Gilchrist MJ, Howe KL, Stark R, Matthews N, Berezikov E, Ketting RF, Tavare S, Miska EA (2008) Piwi and piRNAs act upstream of an endogenous siRNA pathway to suppress Tc3 transposon mobility in the Caenorhabditis elegans germline. Mol Cell 31(1):79–90. doi: S1097–2765(08)00392–4 [pii]. 10.1016/j.molcel.2008.06.003

Diallo M, Arenz C, Schmitz K, Sandhoff K, Schepers U (2003) Long endogenous dsRNAs can induce complete gene silencing in mammalian cells and primary cultures. Oligonucleotides 13 (5):381–392. doi:10.1089/154545703322617069

Didiot MC, Subramanian M, Flatter E, Mandel JL, Moine H (2009) Cells lacking the fragile X mental retardation protein (FMRP) have normal RISC activity but exhibit altered stress granule assembly. Mol Biol Cell 20(1):428–437. doi:E08–07–0737 [pii]. 10.1091/mbc.E08–07–0737

Dlakic M (2006) DUF283 domain of Dicer proteins has a double-stranded RNA-binding fold. Bioinformatics 22(22):2711–2714. doi:btl468 [pii]. 10.1093/bioinformatics/btl468

Doench JG, Petersen CP, Sharp PA (2003) siRNAs can function as miRNAs. Genes Dev 17 (4):438–442. doi:10.1101/gad.1064703

Doyle M, Jantsch MF (2002) New and old roles of the double-stranded RNA-binding domain. J Struct Biol 140(1–3):147–153. doi:S1047847702005440 [pii]

Du T, Zamore PD (2005) microPrimer: the biogenesis and function of microRNA. Development 132(21):4645–4652. doi:132/21/4645 [pii]. 10.1242/dev.02070

Du Z, Lee JK, Tjhen R, Stroud RM, James TL (2008) Structural and biochemical insights into the dicing mechanism of mouse Dicer: a conserved lysine is critical for dsRNA cleavage. Proc Natl Acad Sci USA 105(7):2391–2396. doi:0711506105 [pii]. 10.1073/pnas.0711506105

Duchaine TF, Wohlschlegel JA, Kennedy S, Bei Y, Conte D Jr, Pang K, Brownell DR, Harding S, Mitani S, Ruvkun G, 3rd Yates JR, Mello CC (2006) Functional proteomics reveals the biochemical niche of C. elegans DCR-1 in multiple small-RNA-mediated pathways. Cell 124(2):343–354, S0092–8674(05)01394–2 [pii]. 10.1016/j.cell.2005.11.036

Duxbury MS, Ashley SW, Whang EE (2005) RNA interference: a mammalian SID-1 homologue enhances siRNA uptake and gene silencing efficacy in human cells. Biochem Biophys Res Commun 331(2):459–463. doi:S0006–291X(05)00672–8 [pii]. 10.1016/j.bbrc.2005.03.199

Dzitoyeva S, Dimitrijevic N, Manev H (2003) Gamma-aminobutyric acid B receptor 1 mediates behavior-impairing actions of alcohol in Drosophila: adult RNA interference and pharmacological evidence. Proc Natl Acad Sci USA 100(9):5485–5490. doi:10.1073/pnas.0830111100. 0830111100 [pii]

Eaton BA, Fetter RD, Davis GW (2002) Dynactin is necessary for synapse stabilization. Neuron 34(5):729–741. doi:S0896627302007213 [pii]

Elbashir SM, Harborth J, Lendeckel W, Yalcin A, Weber K, Tuschl T (2001) Duplexes of 21-nucleotide RNAs mediate RNA interference in cultured mammalian cells. Nature 411 (6836):494–498

Enright AJ, John B, Gaul U, Tuschl T, Sander C, Marks DS (2003) MicroRNA targets in Drosophila. Genome Biol 5(1):R1. doi:10.1186/gb-2003–5–1-r1. gb-2003–5–1-rl [pii]

Faehnle CR, Joshua-Tor L (2007) Argonautes confront new small RNAs. Curr Opin Chem Biol 11 (5):569–577. doi:S1367–5931(07)00116–0 [pii]. 10.1016/j.cbpa.2007.08.032

Feinberg EH, Hunter CP (2003) Transport of dsRNA into cells by the transmembrane protein SID-1. Science 301(5639):1545–1547. doi:10.1126/science.1087117. 301/5639/1545 [pii]

Felix MA, Ashe A, Piffaretti J, Wu G, Nuez I, Belicard T, Jiang Y, Zhao G, Franz CJ, Goldstein LD, Sanroman M, Miska EA, Wang D (2011) Natural and experimental infection of Caenorhabditis nematodes by novel viruses related to nodaviruses. PLoS Biol 9(1): e1000586. doi:10.1371/journal.pbio.1000586

Fenner BJ, Thiagarajan R, Chua HK, Kwang J (2006) Betanodavirus B2 is an RNA interference antagonist that facilitates intracellular viral RNA accumulation. J Virol 80(1):85–94. doi:80/1/85 [pii]. 10.1128/JVI.80.1.85–94.2006

Fire A, Xu S, Montgomery MK, Kostas SA, Driver SE, Mello CC (1998) Potent and specific genetic interference by double-stranded RNA in Caenorhabditis elegans. Nature 391 (6669):806–811

Forstemann K, Tomari Y, Du T, Vagin VV, Denli AM, Bratu DP, Klattenhoff C, Theurkauf WE, Zamore PD (2005) Normal microRNA maturation and germ-line stem cell maintenance requires Loquacious, a double-stranded RNA-binding domain protein. PLoS Biol 3(7):e236. doi:05-PLBI-RA-0205R1 [pii]. 10.1371/journal.pbio.0030236

Galiana-Arnoux D, Dostert C, Schneemann A, Hoffmann JA, Imler JL (2006) Essential function in vivo for Dicer-2 in host defense against RNA viruses in Drosophila. Nat Immunol 7 (6):590–597. doi:ni1335 [pii]. 10.1038/ni1335

Gan L, Anton KE, Masterson BA, Vincent VA, Ye S, Gonzalez-Zulueta M (2002) Specific interference with gene expression and gene function mediated by long dsRNA in neural cells. J Neurosci Methods 121(2):151–157. doi:S0165027002002303 [pii]

Gantier MP, Williams BR (2007) The response of mammalian cells to double-stranded RNA. Cytokine Growth Factor Rev 18(5–6):363–371. doi:S1359–6101(07)00085–8 [pii]. 10.1016/j.cytogfr.2007.06.016

Geiss G, Jin G, Guo J, Bumgarner R, Katze MG, Sen GC (2001) A comprehensive view of regulation of gene expression by double-stranded RNA-mediated cell signaling. J Biol Chem 276(32):30178–30182. doi:276/32/30178 [pii]

Geldhof P, Visser A, Clark D, Saunders G, Britton C, Gilleard J, Berriman M, Knox D (2007) RNA interference in parasitic helminths: current situation, potential pitfalls and future prospects. Parasitology 134(Pt 5):609–619. doi:S0031182006002071 [pii]. 10.1017/S0031182006002071

Gent JI, Schvarzstein M, Villeneuve AM, Gu SG, Jantsch V, Fire AZ, Baudrimont A (2009) A Caenorhabditis elegans RNA-directed RNA polymerase in sperm development and endogenous RNA interference. Genetics 183(4):1297–1314. doi:genetics.109.109686 [pii]. 10.1534/genetics.109.109686

Gent JI, Lamm AT, Pavelec DM, Maniar JM, Parameswaran P, Tao L, Kennedy S, Fire AZ (2010) Distinct phases of siRNA synthesis in an endogenous RNAi pathway in C. elegans soma. Mol Cell 37(5):679–689. doi:S1097--2765(10)00041--9 [pii]. 10.1016/j.molcel.2010.01.012

Gerbasi VR, Preall JB, Golden DE, Powell DW, Cummins TD, Sontheimer EJ (2011) Blanks, a nuclear siRNA/dsRNA-binding complex component, is required for Drosophila spermiogenesis. Proc Natl Acad Sci USA 108(8):3204–3209. doi:1009781108 [pii]. 10.1073/pnas.1009781108

Ghildiyal M, Seitz H, Horwich MD, Li C, Du T, Lee S, Xu J, Kittler EL, Zapp ML, Weng Z, Zamore PD (2008) Endogenous siRNAs derived from transposons and mRNAs in Drosophila somatic cells. Science 320(5879):1077–1081. doi:1157396 [pii]. 10.1126/science.1157396

Ghildiyal M, Xu J, Seitz H, Weng Z, Zamore PD (2010) Sorting of Drosophila small silencing RNAs partitions microRNA* strands into the RNA interference pathway. RNA 16(1):43–56. doi:rna.1972910 [pii]. 10.1261/rna.1972910

Glanzer J, Miyashiro KY, Sul JY, Barrett L, Belt B, Haydon P, Eberwine J (2005) RNA splicing capability of live neuronal dendrites. Proc Natl Acad Sci USA 102(46):16859–16864. doi:0503783102 [pii]. 10.1073/pnas.0503783102

Gredell JA, Dittmer MJ, Wu M, Chan C, Walton SP (2010) Recognition of siRNA asymmetry by TAR RNA binding protein. Biochemistry 49(14):3148–3155. doi:10.1021/bi902189s

Gregory RI, Chendrimada TP, Cooch N, Shiekhattar R (2005) Human RISC couples microRNA biogenesis and posttranscriptional gene silencing. Cell 123(4):631–640. doi:S0092–8674(05)01109–8 [pii]. 10.1016/j.cell.2005.10.022

Grishok A, Pasquinelli AE, Conte D, Li N, Parrish S, Ha I, Baillie DL, Fire A, Ruvkun G, Mello CC (2001) Genes and mechanisms related to RNA interference regulate expression of the small temporal RNAs that control C. elegans developmental timing. Cell 106(1):23–34

Guang S, Bochner AF, Burkhart KB, Burton N, Pavelec DM, Kennedy S (2010) Small regulatory RNAs inhibit RNA polymerase II during the elongation phase of transcription. Nature 465 (7301):1097–1101. doi:nature09095 [pii]. 10.1038/nature09095

Haase AD, Jaskiewicz L, Zhang H, Laine S, Sack R, Gatignol A, Filipowicz W (2005) TRBP, a regulator of cellular PKR and HIV-1 virus expression, interacts with Dicer and functions in RNA silencing. EMBO Rep 6(10):961–967

Haasnoot J, de Vries W, Geutjes EJ, Prins M, de Haan P, Berkhout B (2007) The Ebola virus VP35 protein is a suppressor of RNA silencing. PLoS Pathog 3(6):e86. doi:06-PLPA-RA-0347 [pii]. 10.1371/journal.ppat.0030086

Haley B, Zamore PD (2004) Kinetic analysis of the RNAi enzyme complex. Nat Struct Mol Biol 11(7):599–606. doi:10.1038/nsmb780. nsmb780 [pii]

Hall IM, Shankaranarayana GD, Noma K, Ayoub N, Cohen A, Grewal SI (2002) Establishment and maintenance of a heterochromatin domain. Science 297(5590):2232–2237

Hamilton AJ, Baulcombe DC (1999) A species of small antisense RNA in posttranscriptional gene silencing in plants. Science 286(5441):950–952. doi:7953 [pii]

Han W, Sundaram P, Kenjale H, Grantham J, Timmons L (2008) The Caenorhabditis elegans rsd-2 and rsd-6 genes are required for chromosome functions during exposure to unfavorable environments. Genetics 178(4):1875–1893. doi:178/4/1875 [pii]. 10.1534/genetics.107.085472

Han T, Manoharan AP, Harkins TT, Bouffard P, Fitzpatrick C, Chu DS, Thierry-Mieg D, Thierry-Mieg J, Kim JK (2009) 26G endo-siRNAs regulate spermatogenic and zygotic gene expression in Caenorhabditis elegans. Proc Natl Acad Sci USA 106(44):18674–18679. doi:0906378106 [pii]. 10.1073/pnas.0906378106

Hartig JV, Forstemann K (2011) Loqs-PD and R2D2 define independent pathways for RISC generation in Drosophila. Nucleic Acids Res. doi:gkq1324 [pii]. 10.1093/nar/gkq1324

Hartig JV, Esslinger S, Bottcher R, Saito K, Forstemann K (2009) Endo-siRNAs depend on a new isoform of loquacious and target artificially introduced, high-copy sequences. EMBO J 28 (19):2932–2944. doi:emboj2009220 [pii]. 10.1038/emboj.2009.220

Hong J, Qian Z, Shen S, Min T, Tan C, Xu J, Zhao Y, Huang W (2005) High doses of siRNAs induce eri-1 and adar-1 gene expression and reduce the efficiency of RNA interference in the mouse. Biochem J 390(Pt 3):675–679. doi:BJ20050647 [pii]. 10.1042/BJ20050647

Hunter T, Hunt T, Jackson RJ, Robertson HD (1975) The characteristics of inhibition of protein synthesis by double-stranded ribonucleic acid in reticulocyte lysates. J Biol Chem 250 (2):409–417

Huvenne H, Smagghe G (2010) Mechanisms of dsRNA uptake in insects and potential of RNAi for pest control: a review. J Insect Physiol 56(3):227–235. doi:S0022–1910(09)00342–4 [pii]. 10.1016/j.jinsphys.2009.10.004

Ishizuka A, Siomi MC, Siomi H (2002) A Drosophila fragile X protein interacts with components of RNAi and ribosomal proteins. Genes Dev 16(19):2497–2508

Jaskiewicz L, Filipowicz W (2008) Role of Dicer in posttranscriptional RNA silencing. Curr Top Microbiol Immunol 320:77–97

Jinek M, Doudna JA (2009) A three-dimensional view of the molecular machinery of RNA interference. Nature 457(7228):405–412. doi:nature07755 [pii]. 10.1038/nature07755

Johnson KL, Price BD, Eckerle LD, Ball LA (2004) Nodamura virus nonstructural protein B2 can enhance viral RNA accumulation in both mammalian and insect cells. J Virol 78 (12):6698–6704. doi:10.1128/JVI.78.12.6698–6704.2004. 78/12/6698 [pii]

Jopling CL, Yi M, Lancaster AM, Lemon SM, Sarnow P (2005) Modulation of hepatitis C virus RNA abundance by a liver-specific MicroRNA. Science 309(5740):1577–1581. doi:309/5740/ 1577 [pii]. 10.1126/science.1113329

Jose AM, Hunter CP (2007) Transport of sequence-specific RNA interference information between cells. Annu Rev Genet 41:305–330. doi:10.1146/annurev.genet.41.110306.130216

Kaneda M, Tang F, O'Carroll D, Lao K, Surani MA (2009) Essential role for Argonaute2 protein in mouse oogenesis. Epigenetics Chromatin 2(1):9. doi:1756–8935–2–9 [pii]. 10.1186/1756– 8935–2–9

Kanellopoulou C, Muljo SA, Kung AL, Ganesan S, Drapkin R, Jenuwein T, Livingston DM, Rajewsky K (2005) Dicer-deficient mouse embryonic stem cells are defective in differentiation and centromeric silencing. Genes Dev 19(4):489–501

Kawahara Y, Zinshteyn B, Chendrimada TP, Shiekhattar R, Nishikura K (2007a) RNA editing of the microRNA-151 precursor blocks cleavage by the Dicer–TRBP complex. EMBO Rep 8 (8):763–769. doi:7401011 [pii]. 10.1038/sj.embor.7401011

Kawahara Y, Zinshteyn B, Sethupathy P, Iizasa H, Hatzigeorgiou AG, Nishikura K (2007b) Redirection of silencing targets by adenosine-to-inosine editing of miRNAs. Science 315 (5815):1137–1140. doi:315/5815/1137 [pii]. 10.1126/science.1138050

Kawaji H, Nakamura M, Takahashi Y, Sandelin A, Katayama S, Fukuda S, Daub CO, Kai C, Kawai J, Yasuda J, Carninci P, Hayashizaki Y (2008) Hidden layers of human small RNAs. BMC Genomics 9:157. doi:1471–2164–9–157 [pii]. 10.1186/1471–2164–9–157

Kawamata T, Seitz H, Tomari Y (2009) Structural determinants of miRNAs for RISC loading and slicer-independent unwinding. Nat Struct Mol Biol 16(9):953–960. doi:nsmb.1630 [pii]. 10.1038/nsmb.1630

Kawamura Y, Saito K, Kin T, Ono Y, Asai K, Sunohara T, Okada TN, Siomi MC, Siomi H (2008) Drosophila endogenous small RNAs bind to Argonaute 2 in somatic cells. Nature 453 (7196):793–797. doi:nature06938 [pii]. 10.1038/nature06938

Keene KM, Foy BD, Sanchez-Vargas I, Beaty BJ, Blair CD, Olson KE (2004) RNA interference acts as a natural antiviral response to O'nyong-nyong virus (Alphavirus; Togaviridae) infection of Anopheles gambiae. Proc Natl Acad Sci USA 101(49):17240–17245. doi:0406983101 [pii]. 10.1073/pnas.0406983101

Kennedy S, Wang D, Ruvkun G (2004) A conserved siRNA-degrading RNase negatively regulates RNA interference in C. elegans. Nature 427(6975):645–649. doi:10.1038/nature02302. nature02302 [pii]

Kennerdell JR, Carthew RW (1998) Use of dsRNA-mediated genetic interference to demonstrate that frizzled and frizzled 2 act in the wingless pathway. Cell 95(7):1017–1026. doi:S0092– 8674(00)81725–0 [pii]

Ketting RF (2011) The many faces of RNAi. Dev Cell 20(2):148–161. doi:S1534–5807(11) 00040–2 [pii]. 10.1016/j.devcel.2011.01.012

Ketting RF, Haverkamp TH, van Luenen HG, Plasterk RH (1999) Mut-7 of C. elegans, required for transposon silencing and RNA interference, is a homolog of Werner syndrome helicase and RNaseD. Cell 99(2):133–141

Ketting RF, Fischer SE, Bernstein E, Sijen T, Hannon GJ, Plasterk RH (2001) Dicer functions in RNA interference and in synthesis of small RNA involved in developmental timing in C. elegans. Genes Dev 15(20):2654–2659. doi:10.1101/gad.927801

Kim VN (2005) MicroRNA biogenesis: coordinated cropping and dicing. Nat Rev Mol Cell Biol 6 (5):376–385

Kim K, Lee YS, Carthew RW (2007) Conversion of pre-RISC to holo-RISC by Ago2 during assembly of RNAi complexes. RNA 13(1):22–29. doi:rna.283207 [pii]. 10.1261/rna.283207

Kok KH, Ng MH, Ching YP, Jin DY (2007) Human TRBP and PACT directly interact with each other and associate with dicer to facilitate the production of small interfering RNA. J Biol Chem 282(24):17649–17657. doi:M611768200 [pii]. 10.1074/jbc.M611768200

Lecellier CH, Dunoyer P, Arar K, Lehmann-Che J, Eyquem S, Himber C, Saib A, Voinnet O (2005) A cellular microRNA mediates antiviral defense in human cells. Science 308 (5721):557–560. doi:308/5721/557 [pii]. 10.1126/science.1108784

Lee YS, Nakahara K, Pham JW, Kim K, He Z, Sontheimer EJ, Carthew RW (2004) Distinct roles for Drosophila Dicer-1 and Dicer-2 in the siRNA/miRNA silencing pathways. Cell 117 (1):69–81

Lee RC, Hammell CM, Ambros V (2006) Interacting endogenous and exogenous RNAi pathways in Caenorhabditis elegans. RNA 12(4):589–597. doi:rna.2231506 [pii]. 10.1261/rna.2231506

Lehmann KA, Bass BL (1999) The importance of internal loops within RNA substrates of ADAR1. J Mol Biol 291(1):1–13. doi:10.1006/jmbi.1999.2914. S0022–2836(99)92914–5 [pii]

Lewis BP, Shih IH, Jones-Rhoades MW, Bartel DP, Burge CB (2003) Prediction of mammalian microRNA targets. Cell 115(7):787–798

Li H, Li WX, Ding SW (2002) Induction and suppression of RNA silencing by an animal virus. Science 296(5571):1319–1321. doi:10.1126/science.1070948. 296/5571/1319 [pii]

Li WX, Li H, Lu R, Li F, Dus M, Atkinson P, Brydon EW, Johnson KL, Garcia-Sastre A, Ball LA, Palese P, Ding SW (2004) Interferon antagonist proteins of influenza and vaccinia viruses are suppressors of RNA silencing. Proc Natl Acad Sci USA 101(5):1350–1355. doi:10.1073/pnas.0308308100. 0308308100 [pii]

Li C, Vagin VV, Lee S, Xu J, Ma S, Xi H, Seitz H, Horwich MD, Syrzycka M, Honda BM, Kittler EL, Zapp ML, Klattenhoff C, Schulz N, Theurkauf WE, Weng Z, Zamore PD (2009) Collapse of germline piRNAs in the absence of Argonaute3 reveals somatic piRNAs in flies. Cell 137 (3):509–521. doi:S0092–8674(09)00452–8 [pii]. 10.1016/j.cell.2009.04.027

Lingel A, Simon B, Izaurralde E, Sattler M (2003) Structure and nucleic-acid binding of the Drosophila Argonaute 2 PAZ domain. Nature 426(6965):465–469. doi:10.1038/nature02123. nature02123 [pii]

Lingel A, Simon B, Izaurralde E, Sattler M (2004) Nucleic acid 3′-end recognition by the Argonaute2 PAZ domain. Nat Struct Mol Biol 11(6):576–577. doi:10.1038/nsmb777. nsmb777 [pii]

Lipardi C, Paterson BM (2009) Identification of an RNA-dependent RNA polymerase in Drosophila involved in RNAi and transposon suppression. Proc Natl Acad Sci USA 106 (37):15645–15650. doi:0904984106 [pii]. 10.1073/pnas.0904984106

Lipardi C, Wei Q, Paterson BM (2001) RNAi as random degradative PCR: siRNA primers convert mRNA into dsRNAs that are degraded to generate new siRNAs. Cell 107(3):297–307. doi: S0092–8674(01)00537–2 [pii]

Liu Q, Rand TA, Kalidas S, Du F, Kim HE, Smith DP, Wang X (2003) R2D2, a bridge between the initiation and effector steps of the Drosophila RNAi pathway. Science 301(5641):1921–1925. doi:10.1126/science.1088710. 301/5641/1921 [pii]

Liu J, Carmell MA, Rivas FV, Marsden CG, Thomson JM, Song JJ, Hammond SM, Joshua-Tor L, Hannon GJ (2004) Argonaute2 is the catalytic engine of mammalian RNAi. Science 305 (5689):1437–1441

Liu X, Jiang F, Kalidas S, Smith D, Liu Q (2006) Dicer-2 and R2D2 coordinately bind siRNA to promote assembly of the siRISC complexes. RNA 12(8):1514–1520. doi:rna.101606 [pii]. 10.1261/rna.101606

Liu Y, Ye X, Jiang F, Liang C, Chen D, Peng J, Kinch LN, Grishin NV, Liu Q (2009) C3PO, an endoribonuclease that promotes RNAi by facilitating RISC activation. Science 325 (5941):750–753. doi:325/5941/750 [pii]. 10.1126/science.1176325

Lohmann JU, Endl I, Bosch TC (1999) Silencing of developmental genes in Hydra. Dev Biol 214 (1):211–214. doi:10.1006/dbio.1999.9407. S0012–1606(99)99407–1 [pii]

Lu S, Cullen BR (2004) Adenovirus VA1 noncoding RNA can inhibit small interfering RNA and MicroRNA biogenesis. J Virol 78(23):12868–12876. doi:78/23/12868 [pii]. 10.1128/JVI.78.23.12868–12876.2004

Lu R, Maduro M, Li F, Li HW, Broitman-Maduro G, Li WX, Ding SW (2005) Animal virus replication and RNAi-mediated antiviral silencing in Caenorhabditis elegans. Nature 436 (7053):1040–1043

Lucchetta EM, Carthew RW, Ismagilov RF (2009) The endo-siRNA pathway is essential for robust development of the Drosophila embryo. PLoS One 4(10):e7576. doi:10.1371/journal. pone.0007576

Ma JB, Ye K, Patel DJ (2004) Structural basis for overhang-specific small interfering RNA recognition by the PAZ domain. Nature 429(6989):318–322. doi:10.1038/nature02519. nature02519 [pii]

Ma JB, Yuan YR, Meister G, Pei Y, Tuschl T, Patel DJ (2005) Structural basis for 5'-end-specific recognition of guide RNA by the A. fulgidus Piwi protein. Nature 434(7033):666–670. doi: nature03514 [pii]. 10.1038/nature03514

Ma E, MacRae IJ, Kirsch JF, Doudna JA (2008) Autoinhibition of human dicer by its internal helicase domain. J Mol Biol 380(1):237–243. doi:S0022–2836(08)00547–0 [pii]. 10.1016/j. jmb.2008.05.005

Ma J, Flemr M, Stein P, Berninger P, Malik R, Zavolan M, Svoboda P, Schultz RM (2010) MicroRNA activity is suppressed in mouse oocytes. Curr Biol 20(3):265–270. doi:S0960–9822 (09)02205–2 [pii]. 10.1016/j.cub.2009.12.042

Macrae IJ, Zhou K, Li F, Repic A, Brooks AN, Cande WZ, Adams PD, Doudna JA (2006) Structural basis for double-stranded RNA processing by Dicer. Science 311(5758):195–198. doi:311/5758/195 [pii]. 10.1126/science.1121638

MacRae IJ, Zhou K, Doudna JA (2007) Structural determinants of RNA recognition and cleavage by Dicer. Nat Struct Mol Biol 14(10):934–940. doi:nsmb1293 [pii]. 10.1038/nsmb1293

MacRae IJ, Ma E, Zhou M, Robinson CV, Doudna JA (2008) In vitro reconstitution of the human RISC-loading complex. Proc Natl Acad Sci USA 105(2):512–517. doi:0710869105 [pii]. 10.1073/pnas.0710869105

Maida Y, Yasukawa M, Furuuchi M, Lassmann T, Possemato R, Okamoto N, Kasim V, Hayashizaki Y, Hahn WC, Masutomi K (2009) An RNA-dependent RNA polymerase formed by TERT and the RMRP RNA. Nature 461(7261):230–235. doi:nature08283 [pii]. 10.1038/ nature08283

Malone CD, Hannon GJ (2009a) Molecular evolution of piRNA and transposon control pathways in Drosophila. Cold Spring Harb Symp Quant Biol 74:225–234. doi:sqb.2009.74.052 [pii]. 10.1101/sqb.2009.74.052

Malone CD, Hannon GJ (2009b) Small RNAs as guardians of the genome. Cell 136(4):656–668. doi:S0092–8674(09)00127–5 [pii]. 10.1016/j.cell.2009.01.045

Malone CD, Brennecke J, Dus M, Stark A, McCombie WR, Sachidanandam R, Hannon GJ (2009) Specialized piRNA pathways act in germline and somatic tissues of the Drosophila ovary. Cell 137(3):522–535. doi:S0092–8674(09)00377–8 [pii]. 10.1016/j.cell.2009.03.040

Maniataki E, Mourelatos Z (2005) A human, ATP-independent, RISC assembly machine fueled by pre-miRNA. Genes Dev 19(24):2979–2990. doi:19/24/2979 [pii]. 10.1101/gad.1384005

Marques JT, Carthew RW (2007) A call to arms: coevolution of animal viruses and host innate immune responses. Trends Genet 23(7):359–364. doi:S0168–9525(07)00152–7 [pii]. 10.1016/ j.tig.2007.04.004

Marques JT, Devosse T, Wang D, Zamanian-Daryoush M, Serbinowski P, Hartmann R, Fujita T, Behlke MA, Williams BR (2006) A structural basis for discriminating between self and nonself double-stranded RNAs in mammalian cells. Nat Biotechnol 24(5):559–565. doi:nbt1205 [pii]. 10.1038/nbt1205

Marques JT, Kim K, Wu PH, Alleyne TM, Jafari N, Carthew RW (2010) Loqs and R2D2 act sequentially in the siRNA pathway in Drosophila. Nat Struct Mol Biol 17(1):24–30. doi: nsmb.1735 [pii]. 10.1038/nsmb.1735

Martinez J, Patkaniowska A, Urlaub H, Luhrmann R, Tuschl T (2002) Single-stranded antisense siRNAs guide target RNA cleavage in RNAi. Cell 110(5):563–574. doi:S009286740200908X [pii]

Matranga C, Tomari Y, Shin C, Bartel DP, Zamore PD (2005) Passenger-strand cleavage facilitates assembly of siRNA into Ago2-containing RNAi enzyme complexes. Cell 123 (4):607–620. doi:S0092–8674(05)00922–0 [pii]. 10.1016/j.cell.2005.08.044

Meins F Jr, Si-Ammour A, Blevins T (2005) RNA silencing systems and their relevance to plant development. Annu Rev Cell Dev Biol 21:297–318. doi:10.1146/annurev. cellbio.21.122303.114706

Meister G, Tuschl T (2004) Mechanisms of gene silencing by double-stranded RNA. Nature 431 (7006):343–349. doi:10.1038/nature02873. nature02873 [pii]

Meister G, Landthaler M, Patkaniowska A, Dorsett Y, Teng G, Tuschl T (2004) Human Argonaute2 mediates RNA cleavage targeted by miRNAs and siRNAs. Mol Cell 15 (2):185–197

Meister G, Landthaler M, Peters L, Chen PY, Urlaub H, Luhrmann R, Tuschl T (2005) Identification of novel argonaute-associated proteins. Curr Biol 15(23):2149–2155. doi:S0960–9822(05) 01301–1 [pii]. 10.1016/j.cub.2005.10.048

Meurs E, Chong K, Galabru J, Thomas NS, Kerr IM, Williams BR, Hovanessian AG (1990) Molecular cloning and characterization of the human double-stranded RNA-activated protein kinase induced by interferon. Cell 62(2):379–390. doi:0092–8674(90)90374-N [pii]

Miyoshi K, Tsukumo H, Nagami T, Siomi H, Siomi MC (2005) Slicer function of Drosophila Argonautes and its involvement in RISC formation. Genes Dev 19(23):2837–2848. doi: gad.1370605 [pii]. 10.1101/gad.1370605

Mourelatos Z, Dostie J, Paushkin S, Sharma A, Charroux B, Abel L, Rappsilber J, Mann M, Dreyfuss G (2002) miRNPs: a novel class of ribonucleoproteins containing numerous microRNAs. Genes Dev 16(6):720–728

Mourrain P, Beclin C, Elmayan T, Feuerbach F, Godon C, Morel JB, Jouette D, Lacombe AM, Nikic S, Picault N, Remoue K, Sanial M, Vo TA, Vaucheret H (2000) Arabidopsis SGS2 and SGS3 genes are required for posttranscriptional gene silencing and natural virus resistance. Cell 101(5):533–542. doi:S0092–8674(00)80863–6 [pii]

Murchison EP, Partridge JF, Tam OH, Cheloufi S, Hannon GJ (2005) Characterization of Dicer-deficient murine embryonic stem cells. Proc Natl Acad Sci USA 102(34):12135–12140. doi:0505479102 [pii]. 10.1073/pnas.0505479102

Murchison EP, Stein P, Xuan Z, Pan H, Zhang MQ, Schultz RM, Hannon GJ (2007) Critical roles for Dicer in the female germline. Genes Dev 21(6):682–693. doi:21/6/682 [pii]. 10.1101/ gad.1521307

Murphy D, Dancis B, Brown JR (2008) The evolution of core proteins involved in microRNA biogenesis. BMC Evol Biol 8:92. doi:1471–2148-8–92 [pii]. 10.1186/1471–2148-8–92

Nayak A, Berry B, Tassetto M, Kunitomi M, Acevedo A, Deng C, Krutchinsky A, Gross J, Antoniewski C, Andino R (2010) Cricket paralysis virus antagonizes Argonaute 2 to modulate antiviral defense in Drosophila. Nat Struct Mol Biol 17(5):547–554. doi:nsmb.1810 [pii]. 10.1038/nsmb.1810

Nejepinska J, Malik R, Filkowski J, Flemr M, Filipowicz W, Svoboda P. dsRNA expression in the mouse elicits RNAi in oocytes and low adenosine deamination in somatic cells. Nucleic Acids Res, published online September 8, 2011 doi:10.1093/nar/gkr702

Newmark PA, Reddien PW, Cebria F, Sanchez Alvarado A (2003) Ingestion of bacterially expressed double-stranded RNA inhibits gene expression in planarians. Proc Natl Acad Sci USA 100(Suppl 1):11861–11865. doi:10.1073/pnas.1834205100. 1834205100 [pii]

Ng KK, Arnold JJ, Cameron CE (2008) Structure–function relationships among RNA-dependent RNA polymerases. Curr Top Microbiol Immunol 320:137–156

Nishikura K (2010) Functions and regulation of RNA editing by ADAR deaminases. Annu Rev Biochem 79:321–349. doi:10.1146/annurev-biochem-060208-105251

Nishikura K, Yoo C, Kim U, Murray JM, Estes PA, Cash FE, Liebhaber SA (1991) Substrate specificity of the dsRNA unwinding/modifying activity. EMBO J 10(11):3523–3532

Nykanen A, Haley B, Zamore PD (2001) ATP requirements and small interfering RNA structure in the RNA interference pathway. Cell 107(3):309–321. doi:S0092-8674(01)00547-5 [pii]

Ohnishi Y, Totoki Y, Toyoda A, Watanabe T, Yamamoto Y, Tokunaga K, Sakaki Y, Sasaki H, Hohjoh H (2010) Small RNA class transition from siRNA/piRNA to miRNA during pre-implantation mouse development. Nucleic Acids Res 38(15):5141–5151. doi:gkq229 [pii]. 10.1093/nar/gkq229

Okamura K, Ishizuka A, Siomi H, Siomi MC (2004) Distinct roles for Argonaute proteins in small RNA-directed RNA cleavage pathways. Genes Dev 18(14):1655–1666. doi:10.1101/gad.1210204. 1210204 [pii]

Okamura K, Chung WJ, Ruby JG, Guo H, Bartel DP, Lai EC (2008) The Drosophila hairpin RNA pathway generates endogenous short interfering RNAs. Nature 453(7196):803–806. doi:nature07015 [pii]. 10.1038/nature07015

Okamura K, Robine N, Liu Y, Liu Q, Lai EC (2011) R2D2 organizes small regulatory RNA pathways in Drosophila. Mol Cell Biol 31(4):884–896. doi:MCB.01141–10 [pii]. 10.1128/MCB.01141–10

Orii H, Mochii M, Watanabe K (2003) A simple "soaking method" for RNA interference in the planarian Dugesia japonica. Dev Genes Evol 213(3):138–141. doi:10.1007/s00427–003–0310–3

Pak J, Fire A (2007) Distinct populations of primary and secondary effectors during RNAi in C. elegans. Science 315(5809):241–244. doi:1132839 [pii]. 10.1126/science.1132839

Palladino MJ, Keegan LP, O'Connell MA, Reenan RA (2000) A-to-I pre-mRNA editing in Drosophila is primarily involved in adult nervous system function and integrity. Cell 102 (4):437–449

Park H, Davies MV, Langland JO, Chang HW, Nam YS, Tartaglia J, Paoletti E, Jacobs BL, Kaufman RJ, Venkatesan S (1994) TAR RNA-binding protein is an inhibitor of the interferon-induced protein kinase PKR. Proc Natl Acad Sci USA 91(11):4713–4717

Parker JS, Roe SM, Barford D (2004) Crystal structure of a PIWI protein suggests mechanisms for siRNA recognition and slicer activity. EMBO J 23(24):4727–4737. doi:7600488 [pii]. 10.1038/sj.emboj.7600488

Parker GS, Eckert DM, Bass BL (2006) RDE-4 preferentially binds long dsRNA and its dimeriza-tion is necessary for cleavage of dsRNA to siRNA. RNA 12(5):807–818. doi:rna.2338706 [pii]. 10.1261/rna.2338706

Parker GS, Maity TS, Bass BL (2008) dsRNA binding properties of RDE-4 and TRBP reflect their distinct roles in RNAi. J Mol Biol 384(4):967–979. doi:S0022–2836(08)01261–8 [pii]. 10.1016/j.jmb.2008.10.002

Parrish S, Fire A (2001) Distinct roles for RDE-1 and RDE-4 during RNA interference in Caenorhabditis elegans. RNA 7(10):1397–1402

Patel RC, Sen GC (1998) PACT, a protein activator of the interferon-induced protein kinase, PKR. EMBO J 17(15):4379–4390. doi:10.1093/emboj/17.15.4379

Pavelec DM, Lachowiec J, Duchaine TF, Smith HE, Kennedy S (2009) Requirement for the ERI/DICER complex in endogenous RNA interference and sperm development in Caenorhabditis elegans. Genetics 183(4):1283–1295. doi:genetics.109.108134 [pii]. 10.1534/genetics.109.108134

Pelczar H, Woisard A, Lemaitre JM, Chachou M, Andeol Y (2010) Evidence for an RNA polymerization activity in axolotl and xenopus egg extracts. PLoS One 5(12):e14411. doi:10.1371/journal.pone.0014411

Peters L, Meister G (2007) Argonaute proteins: mediators of RNA silencing. Mol Cell 26 (5):611–623. doi:S1097–2765(07)00257–2 [pii]. 10.1016/j.molcel.2007.05.001

Pfeffer S, Zavolan M, Grasser FA, Chien M, Russo JJ, Ju J, John B, Enright AJ, Marks D, Sander C, Tuschl T (2004) Identification of virus-encoded microRNAs. Science 304(5671):734–736. doi:10.1126/science.1096781. 304/5671/734 [pii]

Pfeffer S, Sewer A, Lagos-Quintana M, Sheridan R, Sander C, Grasser FA, van Dyk LF, Ho CK, Shuman S, Chien M, Russo JJ, Ju J, Randall G, Lindenbach BD, Rice CM, Simon V, Ho DD, Zavolan M, Tuschl T (2005) Identification of microRNAs of the herpesvirus family. Nat Methods 2(4):269–276. doi:nmeth746 [pii]. 10.1038/nmeth746

Pham JW, Pellino JL, Lee YS, Carthew RW, Sontheimer EJ (2004) A Dicer-2-dependent 80s complex cleaves targeted mRNAs during RNAi in Drosophila. Cell 117(1):83–94. doi: S0092867404002582 [pii]

Polson AG, Bass BL (1994) Preferential selection of adenosines for modification by double-stranded RNA adenosine deaminase. EMBO J 13(23):5701–5711

Provost P, Dishart D, Doucet J, Frendewey D, Samuelsson B, Radmark O (2002) Ribonuclease activity and RNA binding of recombinant human Dicer. EMBO J 21(21):5864–5874

Qin H, Chen F, Huan X, Machida S, Song J, Yuan YA (2010) Structure of the Arabidopsis thaliana DCL4 DUF283 domain reveals a noncanonical double-stranded RNA-binding fold for protein–protein interaction. RNA 16(3):474–481. doi:rna.1965310 [pii]. 10.1261/rna.1965310

Roignant JY, Carre C, Mugat B, Szymczak D, Lepesant JA, Antoniewski C (2003) Absence of transitive and systemic pathways allows cell-specific and isoform-specific RNAi in Drosophila. RNA 9(3):299–308

Ruby JG, Jan C, Player C, Axtell MJ, Lee W, Nusbaum C, Ge H, Bartel DP (2006) Large-scale sequencing reveals 21U-RNAs and additional microRNAs and endogenous siRNAs in C. elegans. Cell 127(6):1193–1207. doi:S0092-8674(06)01468-1 [pii]. 10.1016/j.cell.2006.10.040

Sadler AJ, Williams BR (2007) Structure and function of the protein kinase R. Curr Top Microbiol Immunol 316:253–292

Saito K, Ishizuka A, Siomi H, Siomi MC (2005) Processing of pre-microRNAs by the Dicer-1-Loquacious complex in Drosophila cells. PLoS Biol 3(7):e235. doi:05-PLBI-RA-0198R1 [pii]. 10.1371/journal.pbio.0030235

Saleh MC, Tassetto M, van Rij RP, Goic B, Gausson V, Berry B, Jacquier C, Antoniewski C, Andino R (2009) Antiviral immunity in Drosophila requires systemic RNA interference spread. Nature 458(7236):346–350. doi:nature07712 [pii]. 10.1038/nature07712

Sam M, Wurst W, Kluppel M, Jin O, Heng H, Bernstein A (1998) Aquarius, a novel gene isolated by gene trapping with an RNA-dependent RNA polymerase motif. Dev Dyn 212(2):304–317. doi:10.1002/(SICI)1097–0177(199806)212:2<304::AID-AJA15>3.0.CO;2–3 [pii]. 10.1002/(SICI)1097–0177(199806)212:2<304::AID-AJA15>3.0.CO;2–3

Sanchez Alvarado A, Newmark PA (1999) Double-stranded RNA specifically disrupts gene expression during planarian regeneration. Proc Natl Acad Sci USA 96(9):5049–5054

Sanchez-Vargas I, Scott JC, Poole-Smith BK, Franz AW, Barbosa-Solomieu V, Wilusz J, Olson KE, Blair CD (2009) Dengue virus type 2 infections of aedes aegypti are modulated by the mosquito's RNA interference pathway. PLoS Pathog 5(2):e1000299. doi:10.1371/journal.ppat.1000299

Scadden AD (2005) The RISC subunit Tudor-SN binds to hyper-edited double-stranded RNA and promotes its cleavage. Nat Struct Mol Biol 12(6):489–496

Scadden AD, Smith CW (2001) RNAi is antagonized by A→I hyper-editing. EMBO Rep 2(12):1107–1111. doi:10.1093/embo-reports/kve244. kve244 [pii]

Schott DH, Cureton DK, Whelan SP, Hunter CP (2005) An antiviral role for the RNA interference machinery in Caenorhabditis elegans. Proc Natl Acad Sci USA 102(51):18420–18424. doi:0507123102 [pii]. 10.1073/pnas.0507123102

Schwarz DS, Hutvagner G, Haley B, Zamore PD (2002) Evidence that siRNAs function as guides, not primers, in the Drosophila and human RNAi pathways. Mol Cell 10(3):537–548

Schwarz DS, Tomari Y, Zamore PD (2004) The RNA-induced silencing complex is a Mg2+-dependent endonuclease. Curr Biol 14(9):787–791. doi:10.1016/j.cub.2004.03.008. S0960982204001769 [pii]

Shih JD, Fitzgerald MC, Sutherlin M, Hunter CP (2009) The SID-1 double-stranded RNA transporter is not selective for dsRNA length. RNA 15(3):384–390. doi:rna.1286409 [pii]. 10.1261/rna.1286409

Shinagawa T, Ishii S (2003) Generation of Ski-knockdown mice by expressing a long double-strand RNA from an RNA polymerase II promoter. Genes Dev 17(11):1340–1345. doi:10.1101/gad.1073003. 17/11/1340 [pii]

Sijen T, Plasterk RH (2003) Transposon silencing in the Caenorhabditis elegans germ line by natural RNAi. Nature 426(6964):310–314. doi:10.1038/nature02107. nature02107 [pii]

Sijen T, Fleenor J, Simmer F, Thijssen KL, Parrish S, Timmons L, Plasterk RH, Fire A (2001) On the role of RNA amplification in dsRNA-triggered gene silencing. Cell 107(4):465–476

Sijen T, Steiner FA, Thijssen KL, Plasterk RH (2007) Secondary siRNAs result from unprimed RNA synthesis and form a distinct class. Science 315(5809):244–247. doi:1136699 [pii]. 10.1126/science.1136699

Sinkkonen L, Hugenschmidt T, Filipowicz W, Svoboda P (2010) Dicer is associated with ribosomal DNA chromatin in mammalian cells. PLoS One 5(8):e12175. doi:10.1371/journal. pone.0012175

Smalheiser NR, Lugli G, Thimmapuram J, Cook EH, Larson J (2011) Endogenous siRNAs and noncoding RNA-derived small RNAs are expressed in adult mouse hippocampus and are up-regulated in olfactory discrimination training. RNA 17(1):166–181. doi:rna.2123811 [pii]. 10.1261/rna.2123811

Smardon A, Spoerke JM, Stacey SC, Klein ME, Mackin N, Maine EM (2000) EGO-1 is related to RNA-directed RNA polymerase and functions in germ-line development and RNA interference in C. elegans. Curr Biol 10(4):169–178, S0960–9822(00)00323–7 [pii]

Soifer HS, Sano M, Sakurai K, Chomchan P, Saetrom P, Sherman MA, Collingwood MA, Behlke MA, Rossi JJ (2008) A role for the Dicer helicase domain in the processing of thermodynamically unstable hairpin RNAs. Nucleic Acids Res 36(20):6511–6522. doi:gkn687 [pii]. 10.1093/nar/gkn687

Song JJ, Liu J, Tolia NH, Schneiderman J, Smith SK, Martienssen RA, Hannon GJ, Joshua-Tor L (2003) The crystal structure of the Argonaute2 PAZ domain reveals an RNA binding motif in RNAi effector complexes. Nat Struct Biol 10(12):1026–1032. doi:10.1038/nsb1016. nsb1016 [pii]

Song JJ, Smith SK, Hannon GJ, Joshua-Tor L (2004) Crystal structure of Argonaute and its implications for RISC slicer activity. Science 305(5689):1434–1437

Sontheimer EJ (2005) Assembly and function of RNA silencing complexes. Nat Rev Mol Cell Biol 6(2):127–138

Stein P, Svoboda P, Anger M, Schultz RM (2003) RNAi: mammalian oocytes do it without RNA-dependent RNA polymerase. RNA 9(2):187–192

Stein P, Zeng F, Pan H, Schultz RM (2005) Absence of non-specific effects of RNA interference triggered by long double-stranded RNA in mouse oocytes. Dev Biol 286(2):464–471

Suh N, Baehner L, Moltzahn F, Melton C, Shenoy A, Chen J, Blelloch R (2010) MicroRNA function is globally suppressed in mouse oocytes and early embryos. Curr Biol 20(3):271–277. doi:S0960–9822(09)02207–6 [pii]. 10.1016/j.cub.2009.12.044

Sullivan CS, Grundhoff AT, Tevethia S, Pipas JM, Ganem D (2005) SV40-encoded microRNAs regulate viral gene expression and reduce susceptibility to cytotoxic T cells. Nature 435 (7042):682–686. doi:nature03576 [pii]. 10.1038/nature03576

Svoboda P, Cara AD (2006) Hairpin RNA: a secondary structure of primary importance. Cell Mol Life Sci 63(7–8):901–908

Svoboda P, Stein P, Hayashi H, Schultz RM (2000) Selective reduction of dormant maternal mRNAs in mouse oocytes by RNA interference. Development 127(19):4147–4156

Svoboda P, Stein P, Anger M, Bernstein E, Hannon GJ, Schultz RM (2004) RNAi and expression of retrotransposons MuERV-L and IAP in preimplantation mouse embryos. Dev Biol 269 (1):276–285

Tabara H, Grishok A, Mello CC (1998) RNAi in C. elegans: soaking in the genome sequence. Science 282(5388):430–431

Tabara H, Sarkissian M, Kelly WG, Fleenor J, Grishok A, Timmons L, Fire A, Mello CC (1999) The rde-1 gene, RNA interference, and transposon silencing in *C. elegans*. Cell 99 (2):123–132

Tabara H, Yigit E, Siomi H, Mello CC (2002) The dsRNA binding protein RDE-4 interacts with RDE-1, DCR-1, and a DExH-box helicase to direct RNAi in *C. elegans*. Cell 109(7):861–871

Tam OH, Aravin AA, Stein P, Girard A, Murchison EP, Cheloufi S, Hodges E, Anger M, Sachidanandam R, Schultz RM, Hannon GJ (2008) Pseudogene-derived small interfering RNAs regulate gene expression in mouse oocytes. Nature 453(7194):534–538. doi: nature06904 [pii]. 10.1038/nature06904

Tang F, Kaneda M, O'Carroll D, Hajkova P, Barton SC, Sun YA, Lee C, Tarakhovsky A, Lao K, Surani MA (2007) Maternal microRNAs are essential for mouse zygotic development. Genes Dev 21(6):644–648. doi:21/6/644 [pii]. 10.1101/gad.418707

Tang F, Barbacioru C, Wang Y, Nordman E, Lee C, Xu N, Wang X, Bodeau J, Tuch BB, Siddiqui A, Lao K, Surani MA (2009) mRNA-Seq whole-transcriptome analysis of a single cell. Nat Methods 6(5):377–382. doi:nmeth.1315 [pii]. 10.1038/nmeth.1315

Tian B, Bevilacqua PC, Diegelman-Parente A, Mathews MB (2004) The double-stranded-RNA-binding motif: interference and much more. Nat Rev Mol Cell Biol 5(12):1013–1023. doi: nrm1528 [pii]. 10.1038/nrm1528

Timmons L, Fire A (1998) Specific interference by ingested dsRNA. Nature 395(6705):854. doi:10.1038/27579

Tomari Y, Zamore PD (2005) Perspective: machines for RNAi. Genes Dev 19(5):517–529. doi:19/5/517 [pii]. 10.1101/gad.1284105

Tomari Y, Du T, Haley B, Schwarz DS, Bennett R, Cook HA, Koppetsch BS, Theurkauf WE, Zamore PD (2004a) RISC assembly defects in the Drosophila RNAi mutant armitage. Cell 116 (6):831–841. doi:S0092867404002181 [pii]

Tomari Y, Matranga C, Haley B, Martinez N, Zamore PD (2004b) A protein sensor for siRNA asymmetry. Science 306(5700):1377–1380. doi:306/5700/1377 [pii]. 10.1126/science.1102755

Tomari Y, Du T, Zamore PD (2007) Sorting of Drosophila small silencing RNAs. Cell 130 (2):299–308. doi:S0092-8674(07)00761-1 [pii]. 10.1016/j.cell.2007.05.057

Tomoyasu Y, Miller SC, Tomita S, Schoppmeier M, Grossmann D, Bucher G (2008) Exploring systemic RNA interference in insects: a genome-wide survey for RNAi genes in Tribolium. Genome Biol 9(1):R10. doi:gb-2008-9-1-r10 [pii]. 10.1186/gb-2008-9-1-r10

Tonkin LA, Bass BL (2003) Mutations in RNAi rescue aberrant chemotaxis of ADAR mutants. Science 302(5651):1725

Tonkin LA, Saccomanno L, Morse DP, Brodigan T, Krause M, Bass BL (2002) RNA editing by ADARs is important for normal behavior in *Caenorhabditis elegans*. Embo J 21 (22):6025–6035

Tran N, Raponi M, Dawes IW, Arndt GM (2004) Control of specific gene expression in mammalian cells by co-expression of long complementary RNAs. FEBS Lett 573(1–3):127–134. doi:10.1016/j.febslet.2004.07.075. S0014579304009573 [pii]

Tuschl T, Zamore PD, Lehmann R, Bartel DP, Sharp PA (1999) Targeted mRNA degradation by double-stranded RNA in vitro. Genes Dev 13(24):3191–3197

Vasale JJ, Gu W, Thivierge C, Batista PJ, Claycomb JM, Youngman EM, Duchaine TF, Mello CC, Conte D Jr (2010) Sequential rounds of RNA-dependent RNA transcription drive endogenous small-RNA biogenesis in the ERGO-1/Argonaute pathway. Proc Natl Acad Sci USA 107 (8):3582–3587. doi:0911908107 [pii]. 10.1073/pnas.0911908107

Vastenhouw NL, Fischer SE, Robert VJ, Thijssen KL, Fraser AG, Kamath RS, Ahringer J, Plasterk RH (2003) A genome-wide screen identifies 27 genes involved in transposon silencing in *C. elegans*. Curr Biol 13(15):1311–1316. doi:S0960982203005396 [pii]

Voinnet O, Pinto YM, Baulcombe DC (1999) Suppression of gene silencing: a general strategy used by diverse DNA and RNA viruses of plants. Proc Natl Acad Sci USA 96 (24):14147–14152

Volpe TA, Kidner C, Hall IM, Teng G, Grewal SI, Martienssen RA (2002) Regulation of heterochromatic silencing and histone H3 lysine-9 methylation by RNAi. Science 297 (5588):1833–1837

Wang XH, Aliyari R, Li WX, Li HW, Kim K, Carthew R, Atkinson P, Ding SW (2006) RNA interference directs innate immunity against viruses in adult Drosophila. Science 312 (5772):452–454. doi:1125694 [pii]. 10.1126/science.1125694

Wang Y, Medvid R, Melton C, Jaenisch R, Blelloch R (2007) DGCR8 is essential for microRNA biogenesis and silencing of embryonic stem cell self-renewal. Nat Genet 39(3):380–385. doi: ng1969 [pii]. 10.1038/ng1969

Wang HW, Noland C, Siridechadilok B, Taylor DW, Ma E, Federer K, Doudna JA, Nogales E (2009a) Structural insights into RNA processing by the human RISC-loading complex. Nat Struct Mol Biol 16(11):1148–1153. doi:nsmb.1673 [pii]. 10.1038/nsmb.1673

Wang Y, Juranek S, Li H, Sheng G, Wardle GS, Tuschl T, Patel DJ (2009b) Nucleation, propagation and cleavage of target RNAs in Ago silencing complexes. Nature 461 (7265):754–761. doi:nature08434 [pii]. 10.1038/nature08434

Watanabe T, Takeda A, Tsukiyama T, Mise K, Okuno T, Sasaki H, Minami N, Imai H (2006) Identification and characterization of two novel classes of small RNAs in the mouse germline: retrotransposon-derived siRNAs in oocytes and germline small RNAs in testes. Genes Dev 20 (13):1732–1743. doi:gad.1425706 [pii]. 10.1101/gad.1425706

Watanabe T, Totoki Y, Toyoda A, Kaneda M, Kuramochi-Miyagawa S, Obata Y, Chiba H, Kohara Y, Kono T, Nakano T, Surani MA, Sakaki Y, Sasaki H (2008) Endogenous siRNAs from naturally formed dsRNAs regulate transcripts in mouse oocytes. Nature 453(7194):539–543. doi: nature06908 [pii]. 10.1038/nature06908

Whangbo JS, Hunter CP (2008) Environmental RNA interference. Trends Genet 24(6):297–305. doi:S0168–9525(08)00126–1 [pii]. 10.1016/j.tig.2008.03.007

Wianny F, Zernicka-Goetz M (2000) Specific interference with gene function by double-stranded RNA in early mouse development. Nat Cell Biol 2(2):70–75

Wilkins C, Dishongh R, Moore SC, Whitt MA, Chow M, Machaca K (2005) RNA interference is an antiviral defence mechanism in *Caenorhabditis elegans*. Nature 436(7053):1044–1047

Winston WM, Molodowitch C, Hunter CP (2002) Systemic RNAi in *C. elegans* requires the putative transmembrane protein SID-1. Science 295(5564):2456–2459. doi:10.1126/science.1068836. 1068836 [pii]

Winston WM, Sutherlin M, Wright AJ, Feinberg EH, Hunter CP (2007) Caenorhabditis elegans SID-2 is required for environmental RNA interference. Proc Natl Acad Sci USA 104 (25):10565–10570. doi:0611282104 [pii]. 10.1073/pnas.0611282104

Wolfrum C, Shi S, Jayaprakash KN, Jayaraman M, Wang G, Pandey RK, Rajeev KG, Nakayama T, Charrise K, Ndungo EM, Zimmermann T, Koteliansky V, Manoharan M, Stoffel M (2007) Mechanisms and optimization of in vivo delivery of lipophilic siRNAs. Nat Biotechnol 25 (10):1149–1157. doi:nbt1339 [pii]. 10.1038/nbt1339

Xie Q, Guo HS (2006) Systemic antiviral silencing in plants. Virus Res 118(1–2):1–6. doi:S0168–1702(05)00351–5 [pii]. 10.1016/j.virusres.2005.11.012

Xu W, Han Z (2008) Cloning and phylogenetic analysis of sid-1-like genes from aphids. J Insect Sci 8:1–6. doi:10.1673/031.008.3001

Yan KS, Yan S, Farooq A, Han A, Zeng L, Zhou MM (2003) Structure and conserved RNA binding of the PAZ domain. Nature 426(6965):468–474. doi:10.1038/nature02129. nature02129 [pii]

Yang N, Kazazian HH Jr (2006) L1 retrotransposition is suppressed by endogenously encoded small interfering RNAs in human cultured cells. Nat Struct Mol Biol 13(9):763–771. doi: nsmb1141 [pii]. 10.1038/nsmb1141

Yang S, Tutton S, Pierce E, Yoon K (2001) Specific double-stranded RNA interference in undifferentiated mouse embryonic stem cells. Mol Cell Biol 21(22):7807–7816

Yang W, Wang Q, Howell KL, Lee JT, Cho DS, Murray JM, Nishikura K (2005) ADAR1 RNA deaminase limits short interfering RNA efficacy in mammalian cells. J Biol Chem 280 (5):3946–3953

Yang W, Chendrimada TP, Wang Q, Higuchi M, Seeburg PH, Shiekhattar R, Nishikura K (2006) Modulation of microRNA processing and expression through RNA editing by ADAR deaminases. Nat Struct Mol Biol 13(1):13–21

Yi CE, Bekker JM, Miller G, Hill KL, Crosbie RH (2003) Specific and potent RNA interference in terminally differentiated myotubes. J Biol Chem 278(2):934–939. doi:10.1074/jbc. M205946200. M205946200 [pii]

Yigit E, Batista PJ, Bei Y, Pang KM, Chen CC, Tolia NH, Joshua-Tor L, Mitani S, Simard MJ, Mello CC (2006) Analysis of the *C. elegans* Argonaute family reveals that distinct Argonautes act sequentially during RNAi. Cell 127(4):747–757. doi:S0092--8674(06)01293--1 [pii]. 10.1016/j.cell.2006.09.033

Yoda M, Kawamata T, Paroo Z, Ye X, Iwasaki S, Liu Q, Tomari Y (2010) ATP-dependent human RISC assembly pathways. Nat Struct Mol Biol 17(1):17–23. doi:nsmb.1733 [pii]. 10.1038/ nsmb.1733

Yoneyama M, Kikuchi M, Natsukawa T, Shinobu N, Imaizumi T, Miyagishi M, Taira K, Akira S, Fujita T (2004) The RNA helicase RIG-I has an essential function in double-stranded RNA-induced innate antiviral responses. Nat Immunol 5(7):730–737. doi:10.1038/ni1087. ni1087 [pii]

Yuan YR, Pei Y, Ma JB, Kuryavyi V, Zhadina M, Meister G, Chen HY, Dauter Z, Tuschl T, Patel DJ (2005) Crystal structure of *A. aeolicus* argonaute, a site-specific DNA-guided endoribonuclease, provides insights into RISC-mediated mRNA cleavage. Mol Cell 19(3):405–419. doi:S1097--2765(05)01475--9 [pii]. 10.1016/j.molcel.2005.07.011

Zambon RA, Vakharia VN, Wu LP (2006) RNAi is an antiviral immune response against a dsRNA virus in Drosophila melanogaster. Cell Microbiol 8(5):880–889. doi:CMI688 [pii]. 10.1111/ j.1462–5822.2006.00688.x

Zamore PD, Tuschl T, Sharp PA, Bartel DP (2000) RNAi: double-stranded RNA directs the ATP-dependent cleavage of mRNA at 21 to 23 nucleotide intervals. Cell 101(1):25–33

Zhang X, Wang C, Schook LB, Hawken RJ, Rutherford MS (2000) An RNA helicase, RHIV-1, induced by porcine reproductive and respiratory syndrome virus (PRRSV) is mapped on porcine chromosome 10q13. Microb Pathog 28(5):267–278. doi:10.1006/mpat.1999.0349. S0882–4010(99)90349-2 [pii]

Zhang H, Kolb FA, Brondani V, Billy E, Filipowicz W (2002) Human Dicer preferentially cleaves dsRNAs at their termini without a requirement for ATP. EMBO J 21(21):5875–5885

Zhang H, Kolb FA, Jaskiewicz L, Westhof E, Filipowicz W (2004) Single processing center models for human Dicer and bacterial RNase III. Cell 118(1):57–68. doi:10.1016/j. cell.2004.06.017. S009286740400618X [pii]

Zhou R, Czech B, Brennecke J, Sachidanandam R, Wohlschlegel JA, Perrimon N, Hannon GJ (2009) Processing of Drosophila endo-siRNAs depends on a specific Loquacious isoform. RNA 15(10):1886–1895. doi:rna.1611309 [pii]. 10.1261/rna.1611309

Chapter 6
Generation of Functional Long Noncoding RNA Through Transcription and Natural Selection

Riki Kurokawa

Abstract The human genome has been found to generate enormous numbers of transcripts, much more than expected classically. A majority of the transcripts appears to be noncoding RNAs (ncRNAs) that do not encode any protein sequence information. Most of the ncRNAs have been shown to be long ncRNAs (lncRNAs) with lengths of more than 200 nucleotides. Therefore, strong attention on lncRNAs has been emerging. However, knowledge of lncRNAs is far less extensive than microRNAs. Emerging evidence suggests distinct roles of lncRNAs in regulation of gene expression, raising the central questions of how these lncRNAs are generated and selected for specific functions. For an attempt to solve these elusive questions, my major focus in this review is on transcription of the lncRNAs. Examination of the data regarding the transcription of the lncRNAs raises a hypothesis about the origin of the functional lncRNAs in which pervasive transcription serves to generate pools of divergent lncRNAs and that selection of specific lncRNAs by criteria of biological potency supplies functional lncRNAs in living cells. In this review, I explore recent and previous papers regarding lncRNAs and discuss the hypothesis.

Keywords Chemical evolution • long noncoding RNA • nucleosome • SELEX • transcription

6.1 Introduction

Analyzing the transcriptome of the human genome has demonstrated that more than 90% of the genome generates divergent transcripts including a few percent of coding messenger RNAs and mostly noncoding RNAs (ncRNAs). Many of these

R. Kurokawa (✉)
Division of Gene Structure and Function, Research Center for Genomic Medicine,
Saitama Medical University, Saitama, Japan
e-mail: rkurokaw@saitama-med.ac.jp

B. Mallick (eds.), *Regulatory RNAs*, DOI 10.1007/978-3-642-22517-8_6,   151
© Springer-Verlag Berlin Heidelberg (outside the USA) 2012

ncRNAs are defined as long ncRNAs (lncRNAs) of which length is more than 200 nucleotides (nts) (Carninci et al. 2005; Kapranov et al. 2007a; Kapranov et al. 2007b). The focus of the ncRNA studies is on small RNAs like microRNA and piRNA, while issues concerning long ncRNAs are emerging recently. Here, I focus on the lncRNAs in this chapter. There is still pretty much skeptical view of the biological significance of these lncRNAs. Due to very low levels of expression, some lncRNAs might represent a bona fide noise of transcription and be made through random or pervasive transcription. However, other lncRNAs are highly expressed and/or are transcribed in a signal-regulated manner. In order to understand the biological significance of lncRNAs, we need to find biological functions of the lncRNAs from a pool of divergent RNA molecules transcribed from the human genome (see details in Chap. 13). In this review article, I search for mechanisms of generating such numerous numbers of lncRNAs, focusing on their transcriptions. Mechanism of eukaryotic transcription has a key role in generating divergent species of lncRNAs.

Early studies on RNA polymerase II showed that the enzyme catalyzes ribonucleic acid polymers effectively from bare DNA templates (Shenkin and Burdon 1966). Upon incorporation of DNA into chromatin structure, the transcription of the DNA is suppressed. However, once the interaction between DNA and histones in the chromatin is disrupted by genotoxic stimuli including UV, DNA-damaging chemicals, and ionizing irradiation, RNA polymerases are forced to recruit on the site and catalyze RNA synthesis there (Wang et al. 2008). This event might induce the random transcription initiation everywhere in the human genome and generate heterogeneous species of lncRNAs to produce a pool of the random lncRNAs. According to their chemical property, these lncRNAs are expected to be selected for suitable biological processes, like the X chromosome inactivation and transcriptional repression (Lee 2010; Wang et al. 2008). Regarding the process of generating lncRNAs, I would present the hypothesis of a natural selection of functional lncRNAs from the pool of divergent RNA molecules transcribed from the noncoding regions of the human genome. In this review, I explore the papers related to properties and functions of lncRNAs, discuss origin of lncRNAs, and trace a pathway to the hypothesis of generating lncRNAs from the human genome during the biological evolution.

6.2 Diversity of lncRNAs

It is not so easy to categorize divergent species of lncRNAs. Let us observe the diversity of lncRNAs. This is a good topic to start to see chaos of the lncRNAs.

6.2.1 Definition of lncRNA

It has been reported that more than 34,000 possible lncRNAs are revealed by genome-wide analysis of transcriptomes with computational approaches

(Carninci et al. 2005; ENCODE-consortium 2004). Functional annotations of these lncRNAs have been far less established than those of coding mRNAs. Recent challenge to find out functional lncRNAs has predicted functions of 340 lncRNAs (Liao et al. 2011). An attempt to construct a database of known lncRNAs was reported with 160 entries of the published lncRNAs (Amaral et al. 2011). A prevailing definition of lncRNA is the ncRNA with nucleotide length more than 200. The tentative number of 200 nts is practically defined by the technical limitations of separation technologies of RNA molecules. It has been presented a definition of lncRNA as the noncoding RNAs that have been spliced and unspliced RNAs which are not known classes of small RNA.

With this definition, the database was developed and has entries now more than 160 lncRNAs identified from literatures in around 60 different organisms. Examples of the lncRNAs in the database are divergent like Kncqlotl, HOTAIR, and Neat1 (Clemson et al. 2009; Mohammad et al. 2008; Pandey et al. 2008; Rinn et al. 2007). To date, the huge number of lncRNAs has shown to be transcribed from the genome, but only small fractions of the lncRNAs have been annotated to have their biological functions. This raises a question whether some of lncRNAs might be generated as a bona fide noise of transcription. For biological studies, lncRNAs need to be inspected for their functional significance.

6.2.2   Functional lncRNA

Expression levels of lncRNAs are generally low compared to protein-coding mRNAs with a few exceptions like Malat1 (Neat2) and Gomafu (Sone et al. 2007; Tripathi et al. 2010). This suggests that lncRNAs work mostly as regulatory molecules rather than cellular structures (see details in Chap. 8). Some of lncRNAs are involved in transcriptional regulation. At this section, I describe various lncRNAs, focusing on functions of transcriptional regulations.

Recently, we found a transcriptional regulatory lncRNA that acts through an RNA binding protein TLS. The cyclin D1 promoter-associated lncRNA (CCND1-lncRNA) has been found to be transcribed from the CCND1 promoter region upon genotoxic stimuli like ionizing irradiation (Wang et al. 2008).The CCND1 binds to RNA binding protein TLS and exerts inhibitory effect on transcription through inhibition of histone acetyltransferase (HAT) activity of CBP/p300 (Fig. 6.1a).

Khps is an antisense RNA transcribed from T-DMR (tissue-dependent differentially methylated region) of Sphk1 (sphingosine kinase-1) (Imamura et al. 2004). Over expression of Khps1 induces demethylation of the CpG island of T-DMR but methylation of its non-CG region (Imamura et al. 2004). The modulation of the methylation of Sphk1 locus was shown to regulate expression of this locus. This is one example where the antisense transcript of the gene promoter regulates gene expression through methylation status of the gene.

The ncRNA of the dihydrofolate reductase (DHFR) minor promoter has other mechanisms to repress transcription. In quiescent mammalian cells, expression of

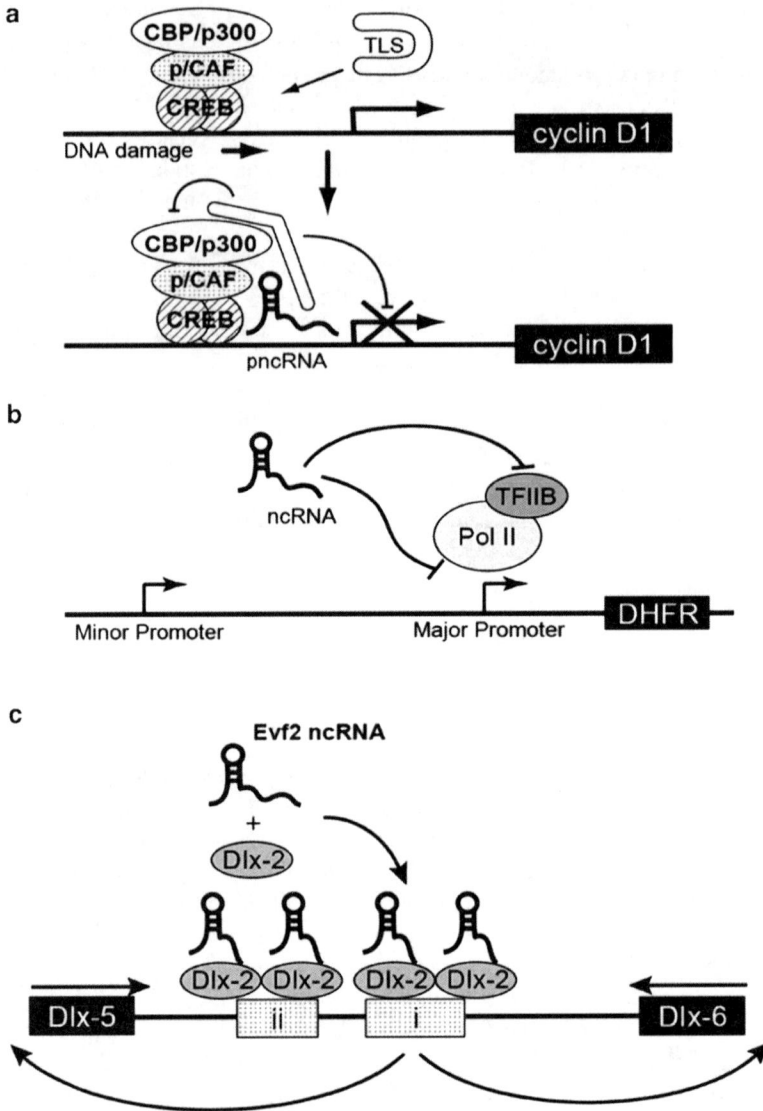

Fig. 6.1 Divergent functions of lncRNAs. (a) Transcription repression by the cyclin D1 promoter-associated lncRNA (CCND1-lncRNA) through RNA binding protein TLS. The DNA damaging-signal like ionizing irradiation induces expression of the CCND1-lncRNA. The TLS-CCND1-lncRNA complex is effectively bound to CBP/p300 and inhibits its histone acetyltransferase activity to repress the transcription of the CCND1 gene. (b) The dihydrofolate reductase (DHFR) minor promoter ncRNA represses transcription of the DHFR gene. This lncRNA directly targets general transcription factor TFIIB and the RNA polymerase II complex to repress the transcription. (c) The Evf2 ncRNA activates transcription. The Evf2 ncRNA is bound to Dlx2 protein and targets the intergenic regions (i and ii) between the loci Dlx5 and Dlx6 to activate transcription of these genes

DHFR is repressed. A transcript of a minor promoter located upstream of a major promoter of DHFR is involved in the repression of this gene (Martianov et al. 2007). In the quiescent cells, the minor promoter transcript inhibited transcriptional initiation from the major promoter through direct binding to TFIIB of the preinitiation complex (Fig. 6.1b). The alternative promoters within the same gene have been observed in various loci. This could be a general mechanism that the transcripts from the alternative promoters have a regulatory role in the transcription of the promoter.

Some lncRNAs also have a function in transcriptional activation. The Evf2 ncRNA is transcribed from an intergenic region between Dlx5 and Dlx6. These two genes belong to the homeodomain protein family. The Evf2 ncRNA is interacted with a protein Dlx2 and bound to the intergenic regions i and ii located between Dlx5 and Dlx6, and activates expression of these loci (Feng et al. 2006). This is another example to regulate transcription in a positive manner (Fig. 6.1c).

X-chromosome inactivation is executed by a well-analyzed lncRNA, 17 kb-length Xist (Zhao et al. 2008). X-chromosome inactivation also employs the ncRNA, the 1.6-kb RepA that is transcribed from the fragment of the Xist locus as an antisense RNA (Zhao et al. 2008). The reduction of expression of Tsix that is a full-length antisense RNA of Xist has a function as a signal. RepA as the sensor receives the reduction of the Tsix expression as the signal, recruiting PRC2 (polycomb repressive complex 2) containing histone methyltransferase activity to the Xist locus, and induces X-chromosome inactivation.

Another example of the lncRNA sensor is HOTAIR (Rinn et al. 2007). During embryonic development, HOTAIR also works as a sensor and exerts gene-silencing effect through recruitment of PRC2 (Rinn et al. 2007). Intriguingly, the lncRNAs with function of the sensors have been shown to need histone-modifying enzymes as an effector molecule. These data suggest that lncRNAs function as sensors for divergent biological signals and regulate gene expression through histone modification.

NEAT1, an abundant 4-kb lncRNA, plays a role in assembly of specific nuclear organelle, paraspeckle (Bond and Fox 2009; Chen and Carmichael 2009). The functions of paraspeckle have been thought to be involved in transcription, pre-mRNA splicing, and nuclear retention of RNA (Bond and Fox 2009; Fox et al. 2002). Depletion of NEAT1 eliminated paraspeckles in many cell lines, including 293, HeLa, and HT-1080, indicating that the NEAT1 is required for the paraspeckle formation. These data show that abundant lncRNAs exert functional roles in assembly of subcellular structures.

Generally speaking, relatively many species of lncRNAs function as a transcriptional regulator, while still others have divergent functions. Each group of lncRNA, for example, HOTAIR and the CCND1-lncRNA, display no significant homology. A question is emerging why so divergent species of lncRNAs have been generated in human and other mammalian cells. To answer the question, analyzing intrinsic mechanisms of transcription is expected to present a radical solution.

6.3 Mechanism of Eukaryotic Transcription

It is no doubt that only transcription makes lncRNAs. Then, we need to analyze the basic core transcriptional machinery extensively. Most of lncRNAs are generated through RNA polymerase II, although some lncRNAs have been shown to be transcribed through RNA polymerase III (Dieci et al. 2007; Kurokawa 2011; Kurokawa et al. 2009). Therefore, this section focuses on the function of RNA polymerase II. This tells us a hint to know how diverse species of lncRNAs are transcribed from the human genome.

6.3.1 RNA Polymerase II and General Transcription Factors

Extensive biochemical and molecular biological studies have demonstrated that the RNA polymerase II complex comprises multiple components: TFIIB, TFIID including TBP and TAFs, TFIIE, TFIIF, and TFIIH (Fig. 6.2), and the precise initiation of the transcription requires the RNA polymerase II with its essential components shown above to form the complex, that is, the holoenzyme of RNA

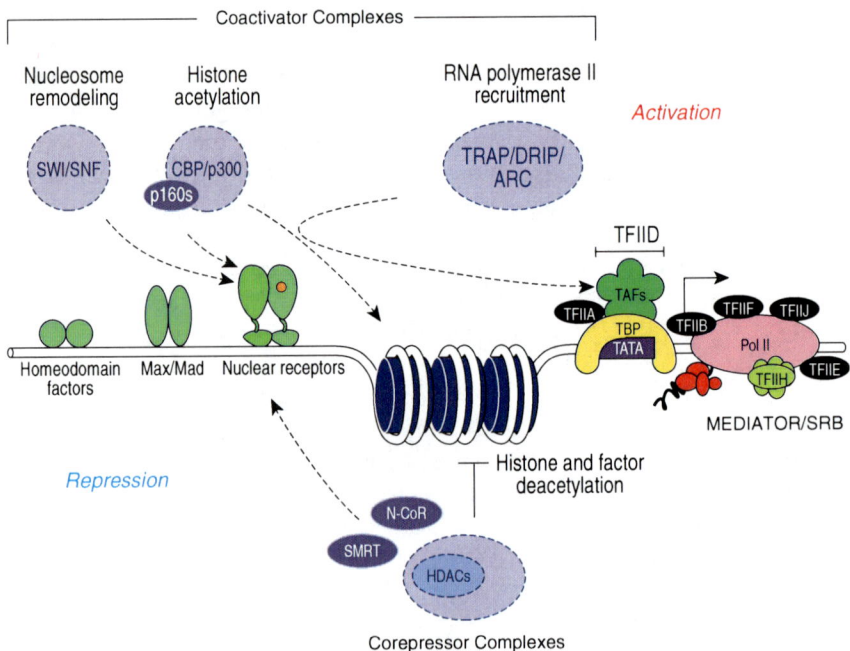

Fig. 6.2 Mechanism of eukaryotic transcription. Simplified scheme of the eukaryotic transcription is shown. There is a growing list of factors involved in this process

polymerase II (Roeder 1991; Weake and Workman 2010). These data indicate that RNA polymerase II alone could not initiate specific and precise transcription, and needs to form the holoenzyme for specific transcription initiation. Contrarily, RNA polymerase II alone could catalyze a random transcription reaction with induction by some protein fractions as described in the next section.

6.3.2   Random or Pervasive Transcription

The fractions of Ehrlich ascites tumor cells (SII) and of HeLa cell (TFIIS) were shown to stimulate nonspecific transcription by RNA polymerase II (Reinberg and Roeder 1987; Sekimizu et al. 1979). These data facilitate understanding randomly initiated transcription of lncRNAs from divergent sites of the human genome. Biochemical analyses with nuclei of the mouse ascitic carcinoma Krebs II cells and RNA polymerase II with endogenous DNA as templates revealed strong activity of the transcription (Shenkin and Burdon 1966). Indeed, the incubation of 0.83 mL of the nuclear fraction of Krebs II cells with [^{3}H]uridine at 37°C for 30 min generated the amount of [^{3}H]RNA ranging from 0.175 to 0.50 mg, showing that significant percentage of the mouse genome is potentially transcribed. The nuclear fraction contained enzymes and substrates that were sufficient for this active transcription. Taken together, these data indicate that the genome has the potential to be transcribed to create divergent RNA species, although these data are all results by in vitro experiments.

6.4   Chromatin Structure and Transcription

The human genome is indeed assembled into chromatin, which consists of genomic DNA, core histones, and related proteins. Chromatin is an essential apparatus to accommodate transcription and related events in nuclei. At default mode, transcription remains suppressed in the chromatin structure. At certain circumstances, transcription is activated by stimuli. Then, nucleosome positioning and nucleosome removal (eviction) play a substantial role in regulation of transcription of lncRNAs. Therefore, we have a look at these issues at the next two sections.

6.4.1   Nucleosome Positioning and Transcription

Nucleosomes comprise of 146 base pairs of DNA wrapped with a histone octamer consisted of histones H2A, H2B, H3, and H4, and function as a fundamental unit of gene-expression process. Recently, it has been shown that various kinds of the histone modification enzyme function at many types of human cell lines.

Nucleosome positions have been published regarding yeast (Lee et al. 2007), *C. elegans* (Johnson et al. 2006), and the human genomes (Schones et al. 2008). The positioning of nucleosomes in the human genome of both resting and activated human CD4[+] T cells has been analyzed by direct sequencing of the nucleosome ends using the high-throughput deep sequencing technology (Schones et al. 2008). Schones et al. have sequenced the ends of the mononucleosome-protected DNA fragments isolated from micrococcal nuclease (MNase)-digested chromatin and presented the genome-wide maps of nucleosome positions in both resting and activated human T cells. They found that nucleosomes are highly "phased (periodically altered positioning of nucleosomes)" relative to the transcription start sites (TSSs) of expressed genes, but this phasing disappeared for unexpressed genes. It has shown that promoters with stalled or posed RNA polymerase II exhibited a nucleosome phasing similar to promoters of transcriptionally active genes, suggesting that the posing of RNA polymerase is one of regulatory mechanisms of transcription.

Gene activation by T-cell receptor signaling is accompanied by nucleosome reorganization in promoters and enhancers as well. Moreover, deposition of variant histone H2A.Z and the H3K4me3 (trimethylation of lysine 4 of histone H3) modification have been suggested to facilitate nucleosome eviction or repositioning in promoter regions of the human genome. The genome-wide maps of nucleosome positions could lead to solutions to understand the relationship between transcription and chromatin structures. Activation of transcription needs to place a massive molecular assembly of basic core transcriptional machinery containing RNA polymerase II at an immediate upstream of the TSS (Fig. 6.2). At this situation, the nucleosome structure is a barrier against initiation of transcription and needs to be removed for activation of transcription. Large-scale analyses have shown nucleosome loss surrounding TSSs in both yeast and human genomes (Bernstein et al. 2004; Ozsolak et al. 2007; Yuan et al. 2005). Consistent with these data, level of nucleosomes at an immediate upstream of TSSs is decreased in an RNA polymerase II-dependent manner in the human genome. This suggests that removing the arrays of nucleosomes is required for the transcription initiation and transcription of lncRNAs as well.

For this nucleosome loss, there are two possible mechanisms involved, that is, nucleosome eviction and nucleosome sliding. The nucleosome eviction has been shown to accompany gene activation (Boeger et al. 2003; Reinke and Horz 2003). Similarly, it has also been reported that nucleosome sliding is associated with activation of the gene, for example, the interferon-B gene by viral infection (Lomvardas and Thanos 2002).

Selection of eviction or sliding of nucleosome might be induced depending upon favorable energy status of each situation. It is speculated that the nucleosome positioning at regulatory regions such as promoters may be maintained by multiple factors like chromatin-modifying enzymes and RNA polymerase II machinery assembly, whereas the nucleosome positioning at nonregulatory regions may be mainly controlled by the underlying DNA sequence features (Ioshikhes et al. 2006; Segal et al. 2006).

These data indicate that removal of the nucleosome on TSS is an initial event of transcription. This prompts us to consider roles of the nucleosome free regions on transcription initiation of lncRNAs.

6.4.2   LncRNAs Are Transcribed from Nucleosome-Free Regions of the Genomes

Nucleosome positioning is an essential procedure for transcription regulation as described above. As a particular case of the nucleosome positioning, nucleosome-free regions (NFRs) are generated in the genomes. The NFR at the 5' and 3' ends of genes is a general region of the transcription initiation of mRNAs and also lncRNAs. It has been identified as NFRs within transcriptional regulatory regions like promoters, enhancers, and also the conserved location of TSSs. These data suggest that regulation of NFRs profoundly affects transcription initiation and related events.

Recent genome-wide mappings of nucleosome positions have been performed in a number of organisms, including yeast (Albert et al. 2007; Bernstein et al. 2004) and humans (Barski et al. 2007; Schones et al. 2008). The genomes of these organisms display a characteristic chromatin structure containing gene-coding regions and transcriptional regulatory regions. Gene-coding regions generally have high nucleosome occupancy with arrays of well-phased nucleosomes extending from the 5' end of a gene. In contrast, transcriptional regulatory regions like promoters and enhancers have low nucleosome occupancy and often contain an NFR. NFRs represent regions with an increased accessibility to MNase digestion and also other factors. Thus, the term NFR refers to a deficiency in experimentally determined canonical nucleosomes and does not necessarily imply a complete lack of histones (Fig. 6.3). Recently, multiple factors are shown to facilitate transcription initiation by inducing the formation and size of NFRs in vivo.

There have been characterized predominantly two major classes of NFRs: 5'-NFRs and 3'-NFRs. In yeast, these NFRs are typically from 80 to 300 bp and are flanked by two well-positioned nucleosomes that often contain the histone variant Htz1 (Albert et al. 2007; Raisner et al. 2005). Recently, two additional classes of NRFs have been reported (Yadon et al. 2010): the NRFs located in open reading frames (ORF-NFRs) and far from ORFs (other NFRs). In yeast and other organisms, the transcriptions of many ncRNAs were found to initiate at the upstream edge of 5'-NRFs or at 3'-NFRs. The conserved locations of the ncRNA TSSs around NFRs suggest that NFRs are a general area of transcription initiation, and apparatus controlling NFR accessibility are critical to transcriptional regulation of ncRNAs.

A variety of factors, including the physical and chemical properties of DNA (Kaplan et al. 2009; Zhang et al. 2009), transcription factors (Hartley and Madhani 2009), and chromatin regulators (Hartley and Madhani 2009) are known to

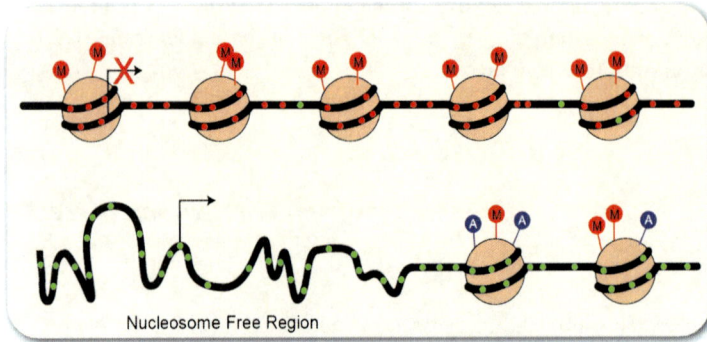

Fig. 6.3 Nucleosome-free regions of the human genome. Upper panel: At the transcription-repressed region, tightly placed nucleosomes form "closed chromatin" with methylated histones (red circle) and methylated DNA mostly at cytosine residues (red dot). Lower panel: At the transcription-activated region, nucleosome-free regions formed at the 5′ and 3′ untranslated regions provide space for the assembly of transcription factors and induce activation of transcription. The NFRs are accompanied with acetylated histones (blue circle) and unmethylated DNAs with cytosine residues (green dot)

positively regulate the formation and size of NFRs in vivo. The activities of these factors in establishing larger NFRs are thought to facilitate the initiation of transcription by allowing transcription factors greater access to DNA. It still remains to be revealed whether there are mechanisms to negatively regulate the size of NRFs in vivo. Recent analysis has shown in yeast that the ATP-dependent chromatin-remodeling enzyme Isw2 functions at the 5′ and 3′ ends of genes to increase nucleosome occupancy within intergenic regions, sliding nucleosomes away from coding regions. Intriguingly, Isw2 is also required to repress noncoding antisense transcripts from the 3′ end of three genes tested (Whitehouse et al. 2007). It was not certain if Isw2-dependent chromatin remodeling generally affects chromatin structure and the ncRNA transcription around NFRs. Yadon et al. have hypothesized that Isw2 might generally repress the ncRNA transcription by negatively regulating the size of NFRs in vivo (Yadon et al. 2010). To examine the hypothesis, they analyzed data from multiple nucleosome-mapping studies to systematically annotate a consensus set of NFRs across the yeast genome (Yadon et al. 2010). Their work identified two additional NFRs above-mentioned across the yeast genome. The Isw2 targets were found to be significantly enriched at all four classes of NFRs. Thus, this identified previously unknown targets of Isw2 at the OFR-NFRs. Furthermore, they employed custom strand-specific tiled microarrays to analyze ncRNA transcripts and found that Isw2 is globally required to repress initiation of cryptic RNA transcripts from NFRs by sliding nucleosomes toward NFRs to restrict their size. Finally, they provided evidence that a potential biological consequence for Isw2-dependent repression of some cryptic transcripts is to prevent transcriptional interference (Yadon et al. 2010). These data present firm evidence to demonstrate that NRF is an area for the transcriptional initiation of ncRNAs and lncRNAs as well.

6.5   Pervasive Transcription Generates lncRNAs

At this section, we discuss involvement of the histone covalent modification in transcriptional regulation of lncRNAs. The histone modification is also one of regulatory elements of transcription of lncRNAs.

6.5.1   Divergent Transcription Makes Antisense RNA

It has been reported that 70% of coding genes have their counterpart of antisense transcripts (Katayama et al. 2005). Thus, antisense RNA occupies a major part of the lncRNA fractions. Then, I describe the transcription of antisense RNA at this section. Our experiments with the CCND1-lncRNA showed that its transcription proceeds in both sense and antisense directions (Wang et al. 2008). The CCND1-lncRNA binds to TLS and exerts inhibitory activity on the HAT activity of CBP/p300, repressing the transcription of the CCND1 gene itself (Fig. 6.1a). Exposure of HeLa cells with ionizing irradiation (IR) resulted in induction of the CCND1-lncRNA expression from the promoter (Fig. 6.4a). The transcripts were found to be a novel class of lncRNAs with 200 and 330 nt (Fig. 6.4b). Intriguingly, the exposure of the cells to IR induced transcription of both sense and antisense of the strands of the promoter (Fig. 6.4c, d). We have not known how this kind of pervasive or bidirectional transcription could occur on lncRNA loci and discuss this mechanism at this section.

Analysis of transcription of lncRNAs requires precise monitoring of nascent transcripts, especially rapidly degraded transcripts. Churchman and Weissman developed an approach, native elongating transcript sequencing (NET-seq), based upon deep sequencing of $3'$ ends of nascent transcripts associated with RNA polymerase, to monitor active transcription at nucleotide resolution (Churchman and Weissman 2011). The NET-seq approach made it possible to determine the precise positions of all active RNA polymerase II complexes by exploiting the extraordinary stability of the RNA polymerase ternary complex with DNA and RNA to capture nascent transcripts directly from living cells without cross-linking. The properties of the $3'$ end of purified transcripts are revealed by deep sequencing, thus providing a quantitative measurement of RNA polymerase II density.

The NET-seq detected the relative amounts of nascent sense and antisense transcripts. Although some divergent promoters were observed, the large majority of promoters had much less antisense transcription than sense transcription. Then, a question arising is why the promoter has a directionality. Intriguingly, the NET-seq indicated a strong positive correlation between antisense transcription levels and previously published measurements of the levels of histone H4 acetylation (Pokholok et al. 2005). The correlation between the antisense transcription and the H4 acetylation indicated that the H4 acetylation may have a causative role in facilitating antisense transcription. To test this end, the effect of deletion of RCO1,

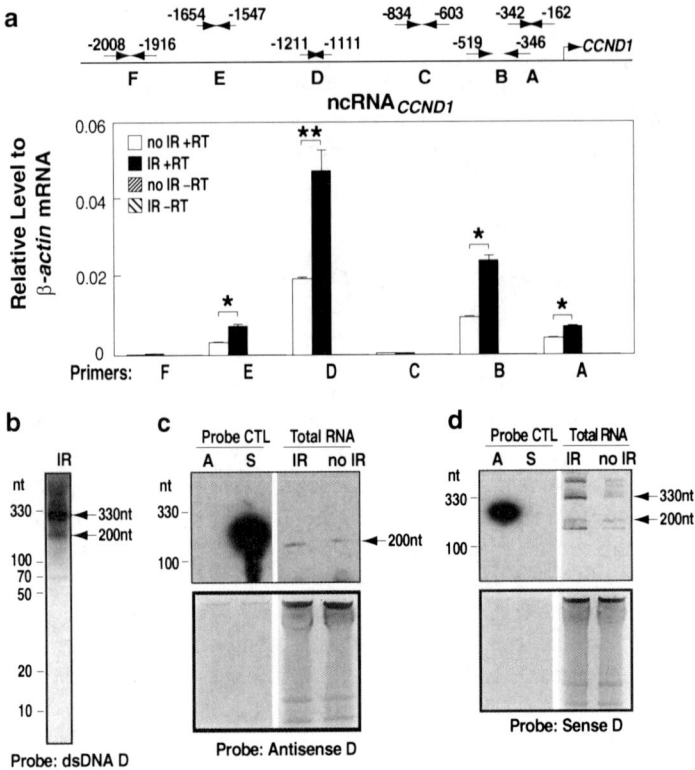

Fig. 6.4 Pervasive transcription of lncRNAs. (**a**) Upper panel: A map of the CCND1 promoter. Lower panel: Relative level of transcription of the CCND1-lncRNAs detected by quantitative PCR. The regions A through F containing GGUG consensus were tested for their transcripts and the expressions of transcription of the regions A, B, D, and E were detected after exposure to ionizing irradiation. (**b**) Northern blotting of the CCND1-lncRNAs transcribed from the region D. (**c–d**) Northern blotting of sense and antisense strands of the region D transcribed. This figure is composed from our data (Wang et al. 2008). According to the Nature Publishing Group, a division of Macmillan Publishers Ltd., the copyright of the Nature paper belongs to the authors

an essential subunit of the Rpd 3 small (Rpd3S) H4 deacetylation complex (Carrozza et al. 2005; Keogh et al. 2005), was examined on the antisense transcription. The experiment revealed a pervasive increase at average fourfold in the unstable antisense transcription. Deletion of EAF3, another essential subunit of Rpd3S, also increased the pervasive transcription, confirming the result with deletion of ECO1. These experiments indicated that the Rpd3S histone complex enforces the promoter directionality. To see how the Rpd3S works in preventing the antisense transcription from promoters, the experiment was performed with the deletion mutants of yeast. These data together with previously published data indicated that the mechanism of Rpd3S action on antisense transcription is involved in the Set2 recruitment to elongation of RNA polymerase II via the Ser2 phosphorylation on its carboxy-terminal domain. This, in turn, through the Set2 methylation activity, allows recruitment of Rpd3S to the 3′ ends of genes, suppressing antisense transcription from downstream nucleosome-free regions (Churchman and Weissman 2011). Although it remains unsolved how the histone H4 acetylation in the body of antisense transcripts can facilitate the initiation of transcription, these data indicate that the histone acetylation has a crucial role in transcription of antisense RNAs and pose a speculation that the histone modification may have a general role in transcriptional regulation of lncRNAs.

Additional NET-seq experiments demonstrated pervasive polymerase pausing and backtracking throughout the body of transcripts. The average pause density showed the prominent peaks at each of the first four nucleosomes, with the peak location occurring in good agreement with in vitro previous biophysical experiments (Churchman and Weissman 2011). Therefore, the nucleosome-induced pausing represents a major barrier to transcriptional elongation in vivo. Taken together, these data demonstrated that the nucleosome is an apparatus to repress transcription through modification of chromatin structure. In the case of the CCND1-lncRNA transcription, the H4 acetylation on antisense region of the promoter could be induced or inhibited by blocking recruitment of Rpd3S. This suggests that epigenetic regulation is also a crucial element for expression of the CCND1-lncRNA or generally lncRNAs as well.

6.5.2   Transcription from Intergenic Regions Generates lncRNAs

We have observed particular histone modifications around the CCND1 loci (Fig. 6.5). The CCND1 locus has been shown to be encompassed between an upstream H3K4me3 region and a downstream H3K36me3 region of the gene body. Guttman et al. recently identified these specific modifications of histone H3 methylation around loci of large intervening noncoding RNAs (lincRNA: hereafter lncRNA), which are related to their expression (Guttman et al. 2009). Their approach to identify lncRNAs using chromatin-state maps indicated discrete transcriptional units intervening known protein-coding loci. The genome-wide chromatin-state mapping with chromatin immunoprecipitation and deep sequencing

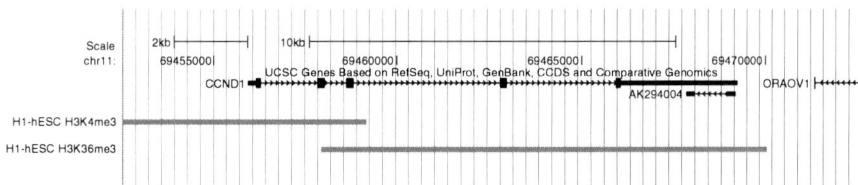

Fig. 6.5   The K4-K36 domain of the CCND1-lncRNA locus. Human embryonic stem cells (ESCs) are tested to detect these histone modifications. The figure was composed from the data from the UCSC Genome Bioinformatics Site: http://genome.ucsc.edu/index.html

(ChIP-Seq) demonstrated that genes actively transcribed by RNA polymerase II are marked byH3K4me3 at their promoter and also by H3K36me3 along the transcribed regions. This distinctive structure was named as a "K4-K36 domain." Therefore, the K4-K36 domain turns out to be a marker for active transcription regions. Then, searching for the K4-K36 structures that reside outside the known protein-coding gene loci makes it possible to systematically discover lncRNAs.

Examining the K4-K36 domains in genome-wide chromatin-state maps was performed in four mouse cell types, that is, mouse embryonic stem cells (ESCs), mouse embryonic fibroblasts (MEFs), mouse lung fibroblasts (MLF), and neural precursor cells (NPCs). This identified the K4-K36 domains of at least 5 kb in length that did not overlap regions containing protein-coding genes as well as known microRNAs and endogenous short interfering RNAs. This revealed the 1,675 spots of the K4-K36 domains that do not overlap with known annotations and identified more than 1,600 large multiexonic RNAs across four mouse cell types. Moreover, these K4-K36 domains display high evolutionary conservation between mouse and human. Together, the results confirm biological significance of these identified lncRNAs in living cells.

Upon identification of a long list of conserved lncRNAs, Guttman et al. developed methods to infer their putative functions that can be tested experimentally (Guttman et al. 2009). To this purpose, the RNA expression profiles of both lncRNAs and protein-coding genes were examined over variety of tissues. The polyadenylated RNA fractions from 16 mouse samples were hybridized to a custom lncRNA array. The samples included the original four cell types (mouse ESCs, NPC, MEF, and MLF), a time course of embryonic development (whole embryo, hind limb, and forelimb at embryonic days 9.5, 10.5, and 13.5), and four normal adult tissues (brain, lung, ovary, and testis). These expression data present important information about biological functions of the lncRNAs. Indeed lncRNAs with an expression pattern was screened for opposite to the known lncRNA HOTAIR. Notably, the most highly anticorrelated lncRNA in the genome was found in the HOXC cluster, in the same euchromatic domain as HOTAIR. Then, this lncRNA was named as "*Frigidair.*" The data imply that *Frigidair* might repress HOTAIR or perhaps activate genes in the HOXD cluster that was shown to be of negative correlation with HOTAIR (Guttman et al. 2009).

Using published data, 118 lncRNAs in which the promoter loci were bound by the transcription factors Oct4 and Nanog have been identified (Loh et al. 2006). Interestingly, more than 70% of these lncRNAs were associated with pluripotency. One of these lncRNAs was found to be only expressed in ESCs and located around 100 kb from Sox2 locus, which encodes a key transcription factor for pluripotency. Transient transfection experiment using the locus of the lncRNA in mouse cells showed that Sox2 and Oct4 were each sufficient to drive expression of this lncRNA promoter, and the expression of both Oct4 and Sox2 together caused synergistic increases in the expression. Transcription factors, Sox2 and Oct4, were found to regulate the expression of the lncRNAs through their binding to the related promoters. Taken together, the data demonstrate that specific sets of functional lncRNAs are highly conserved and implicated in divergent biological significances.

The data examined at this section demonstrate two clues to understand the functions of lncRNA. First one is that the genes near lncRNA loci are strongly biased toward the gene-encoding transcription factors and other transcription-related molecules. The second one with previous data (Rinn et al. 2007) is that the lncRNA like HOTAIR functions with a protein complex bearing chromatin-modifying enzyme activity to regulate transcription. Therefore, these data suggest that distinctive ncRNAs are involved in crucial aspect of regulation of transcription by recruiting chromatin-modifying enzyme to active sites of transcription.

The analysis could divide these lncRNAs into two categories. One category would represent lncRNAs that are associated with mRNAs and that might act in cis to regulate the associated mRNA. Another category would not be associated with mRNAs and would act in trans to regulate the expression of multiple genes nearby or at a distance. These lncRNAs could also include structural RNAs, for example, required to build paraspeckles.

6.6 Chemical Evolution of Functional RNA Molecules from a Pool of Divergent RNAs and Evolution of lncRNAs

Systematic evolution of ligands by exponential enrichment (SELEX) process provides a small oligonucleotide that was coined "aptamers" (derived from the Greek word *aptus*, "to fit") (Tuerk and Gold 1990). The SELEX technology presents a speculation about generation of divergent lncRNAs. Here, we discuss the origin of the lncRNAs.

6.6.1 SELEX Chooses Functional RNA Oligos from an RNA Oligonucleotide Library

Initially, the SELEX technology was developed with experiments regarding the translational operator within the bacteriophage T4 gene 43 mRNA (Tuerk and Gold 1990). These remarkable data of the first SELEX process prompted them to make it more refined technology to select single-stranded nucleic acids, aptamers, and promoted it to target therapeutic applications. Actual SELEX procedure is summarized in Fig. 6.6.

High specificity of aptamers against target molecules stimulated therapeutic interests. It has been started to develop therapeutics with aptamers. Many useful aptamers were identified, some of which are in clinical development nowadays. The first aptamer taken into clinic is Macugen, and many others have been in clinical examinations (Doggrell 2005; Kaiser 2006). These data have proved usefulness of aptamers and also of SELEX technology. Robustness of the SELEX process has raised a hypothesis that lncRNAs have been generated or evolved through a process like SELEX over the biological evolution.

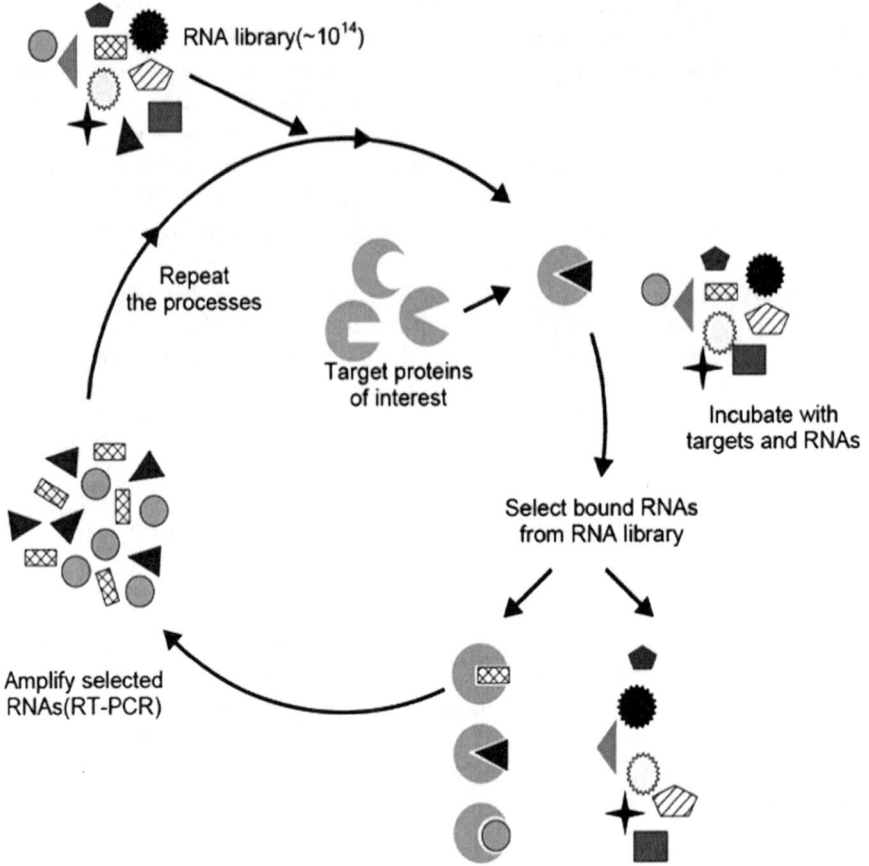

Fig. 6.6 Systematic evolutions of ligands by exponential enrichment (SELEX) procedure. Typical SELEX procedure is summarized

6.6.2 Natural Selection of Functional lncRNAs from a Pool of Divergent lncRNAs

The natural selection of genetic codons through the process like the SELEX procedure is one of intriguing examples of the evolution. We discuss this process at this section. Evolution of the genetic codon displays robustness of SELEX process. The SELEX process with the evolution of the genetic code presents intense implication of evolution of functional RNAs and also lncRNAs. It could be tested if genetic codons that specify a particular amino acid in the canonical genetic code occur more than expectedly at RNA regions that bind to the amino acid. Amazingly, it has been shown that arginine binding sites are predominantly composed of Arg codons which are AGG, AGA, CGG, CGA, CGC, and CGT, even in the aptamers selected in different laboratories using distinctive protocols (Knight and Landweber 1998).

The recent isolation of aptamers to tyrosine turned out to contain actual Tyr codons which are TAT and TAC (Mannironi et al. 2000). Yarus extended this analysis to other amino acids for which aptamers are now available (arginine, isoleucine, and tyrosine) and concluded that the overall probability that the observed codon/binding site association would occur by chance is fairly low (3.3×10^{-7}), suggesting that the interaction of the codons and the corresponding amino acids is selected by specific procedure like SELEX (Yarus 2000). Taken together, analyzing different components of the translation apparatus has been starting to present a reliable scheme of evolution of the genetic code. The primordial code, influenced by direct interactions between bases of primitive tRNA and amino acids, probably dates back to the RNA world. The creation of tRNA and ribozyme-based aminoacyl-tRNA synthetases could make the mechanism of translation more consistent by allowing exchanges of amino acids between codons and achievement of optimal combinations between amino acids and codons. Furthermore, the code probably underwent a process of expansion from relatively a few amino acids into the modern complement of 20 amino acids.

These data indicate that the SELEX has a role in natural selection of functional tRNA during evolutionary process. This piece of information confirms evidence to indicate that the SELEX should work as to select functional lncRNAs from a pool of divergent RNA molecules from the human genome (Fig. 6.7).

6.6.3 Chemical Evolution of Functional RNAs in the Ancient RNA World and Evolution of lncRNAs

Elucidation of the origin of life is one of fundamental problems in natural sciences. Beginning of life on Earth, the chemical evolution is assumed to be an initial event of the biological evolution. The chemical evolution represents that the creation of complex organic compounds originated from simple inorganic chemicals through chemical reactions in a hypothetical shallow pool, a Darwinian pond during early days of the Earth. In the context of the chemical evolution, the SELEX procedure might play a role in accelerating its process. In this section, we discuss possible roles of the SELEX during the chemical evolution and its potency in the biological evolution. At this section, I present a hypothesis to tether the SELEX to natural selection of functional lncRNAs in the biological evolution.

Creating life implies making a hereditary substance, a gene bearing functions to store, replicate, and transfer genetic information to next generations. In the modern life, there are two nucleic acids, DNA and RNA, as a hereditary substance. In ancient time of the primeval Earth, RNA was presumably an initial macromolecule generated in the Darwinian ponds where abiogenic organic compounds could be accumulated. Then, primitive RNA molecules were synthesized and selected to function as a genetic substance, a gene, because of its versatility like enzymatic activity, making three-dimensional structures, and storing information in its

Fig. 6.7 Hypothesis of natural selection of functional lncRNAs. Bare DNA is an efficient template to make random transcripts. The human genome is assembled into chromatin. The transcription of the chromatinized template is repressed. Stimuli like DNA damage induce random or pervasive transcription. The naturally occurring SELEX procedure selects the lncRNAs fitting specific biological actions. The selected ones have been evolved to functional lncRNAs

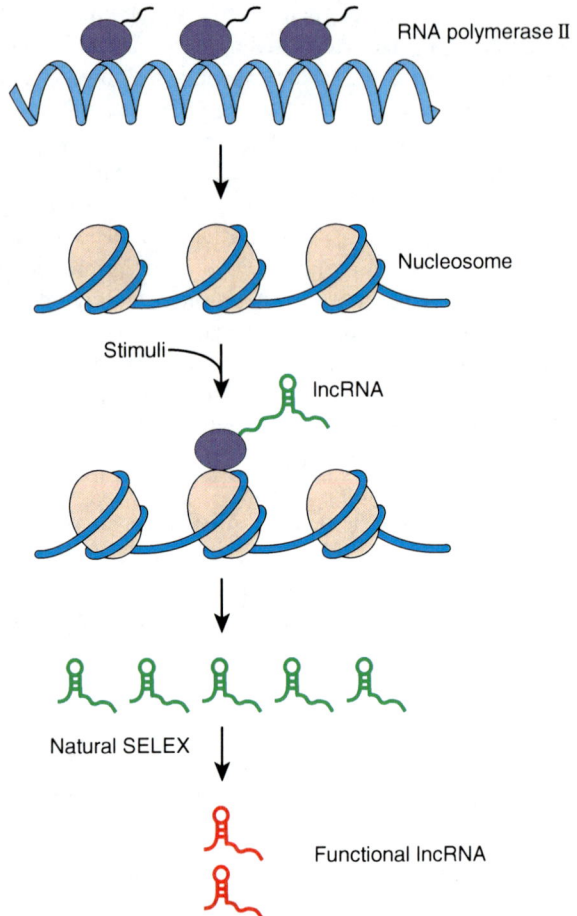

structure. Now, the idea of an ancient RNA world as a probable progenitor of the contemporary living world on Earth is generally recognized (Joyce 1989, 1994). How were primitive RNA molecules self-assembled and evolved into refined molecules with multiple functions? This is still one of central questions regarding the RNA world. One clue comes from selection process of ancient RNA molecules. We discuss a possibility of the SELEX procedure to be involved in the selection of the functional RNA from ancient pools of RNA molecules.

It is also essential to postulate a functional membrane isolating solution containing soluble univalent metal cations (K^+ and Na^+) and Mg^{++} on the ancient Earth, although it is one of major questions how such kinds of the primitive membrane have been developed. However, there is a possible alternative for the membrane. Chetverin and colleagues demonstrated that RNA molecules could form colonies on gels when optimal conditions were attained for their replication (Chetverin et al. 1991). It is conceivable that mixed RNA colonies grown on solid surfaces with no membrane were the earliest cell-free molecular complexes. Some RNA molecules were able to perform genetic functions, while others might produce the structures accommodating the survival of the RNA colonies or function as ribozymes for synthesis and preparation of chemicals for RNA synthesis (Chetverin et al. 1991; Chetverina and Chetverin 1993).This cell-free situation presumably provided conditions for an extremely rapid evolution. These RNA colonies were not isolated from the environment and could readily exchange their molecules as a genetic substance. Instead of gaining a membrane, survival of these alternative cell-free molecular complexes seems plausible because the formation of RNA colonies could have been a natural consequence of the drying of RNA-containing ponds. Then, the naturally occurring SELEX procedure selected the earliest functional RNAs.

The scenario of the natural primordial SELEX was started when mixture of primitive RNA molecules was synthesized in the Darwinian pond. When functionally different RNAs including the RNA-replicating ribozyme were placed in the same pond, the entire RNA population increased in number and, due to spontaneous transesterification (Spirin 2005) and replication errors, increased in diversity of the RNA molecules.

When the pond dried, the RNA molecules were absorbed on the moist surface of clay or other mineral substrate. On the surface of such kinds of clay, mixed RNA colonies containing the RNA-replicating enzyme and several other RNAs were formed and grew. The most successful colonies comprising of the most potent RNA molecules grew more rapidly than others. Subsequent flooding of the dried surface resulted in dissolution of the colony, and the resulting RNA population was again amplified in the shared pond, but starting with the population enriched with "potent" RNA molecules. In this context, the repeating flooding and drying of RNA-containing ponds provided for the systematic enrichment of RNA population with functionally better molecules. This was indeed the natural SELEX and most likely the initial format of development of the ancient RNA world.

The natural SELEX hypothesis is just one possibility. There are other ideas like evolution of pseudogenes and serving as microRNA decoys. The possible problem with the hypothesis is that it needs to have a versatile RNA-replicating ribozyme for multiple cycles of the SELEX at the very beginning of the RNA world. Therefore, proving the natural SELEX hypothesis is still a future homework for modern biology.

The natural SELEX is just a way of selection of the functional lncRNAs. Another question is what selection pressures are needed to generate the functional lncRNAs because the lncRNAs do not encode proteins. A clue for the question might be that the lncRNAs lack strong conservation on the primary sequences. This is major argument for nonfunctionality of the lncRNAs. Contrarily, mRNAs have strong selection pressures to conserve the codon usage and prevent frameshift mutations in a single open reading frame. The lncRNAs might conserve just portions of the primary sequences that are constrained by secondary structures. Indeed, many lncRNAs still contain strongly conserved elements (Ponjavic et al. 2007; Siepel et al. 2005). Comparative genomic analysis predicted that such conserved elements could form secondary structures of the RNA molecules (Torarinsson et al. 2006; Torarinsson et al. 2008). These secondary structures of the lncRNAs could be targeted by specific RNA-binding proteins. Upon binding of the RNA-binding proteins to the lncRNAs through conserved secondary structures, the complexes exert biological functions. Selecting these combinations of the RNA-protein interactions could be mediated through the natural SELEX procedure that I presented above, or any other ways of selection might occur.

The discussion at this section presents a guide to see how the natural SELEX might work on the natural selection in the RNA world. In the Darwinian pond, primordial RNA molecules were passed through rigorous selection of the natural SELEX. The SELEX procedure turns out to be extensively versatile for selecting beneficial molecules from pools of divergent RNA molecules. This postulates us that the similar SELEX procedure also has been playing a role in selecting functional lncRNAs from pools of lncRNAs in living cells.

6.7 Conclusions and Future Perspectives

I have outlined functions of the lncRNA and presented the hypothesis regarding the origin of the lncRNAs at this review. Pervasive and random transcriptions could generate the lncRNA pools. The natural selection might pick the biologically beneficial lncRNAs from the pools. The beneficial lncRNAs could be evolved through the selection process. For instance, Xist was derived from fragments of a pseudogene of a previously active coding gene *Lnx3* and of transposable elements (Duret et al. 2006; Elisaphenko et al. 2008). The prototype of Xist was a sort of nonfunctional lncRNA comprising fragments of the pseudogene and the transposable elements. Unidentified selection step on the prototypic Xist might generate the functional Xist during the biological evolution. For the selection step, the natural SELEX might play a pivotal role in establishment of functional Xist.

There are two crucial events to generate functional lncRNAs in the hypothesis: one is to provide a pool of divergent RNA molecules; second is to select functional lncRNAs from the pool. It is likely that these two events are continuously occurring on the way in the modern living world. Therefore, there are lots of nonfunctional lncRNAs as pools for future selections in living cells. Rapid and small-scale evolution has been progressing in the cells and generating more functional lncRNAs. Then, it is possible to have a quick response to altered environment by selecting novel lncRNAs for survival of cells. Elucidating generation of the lncRNAs will present evidence of the biological and chemical evolutions, and also clue to know fundamental structure of the human genome and transcription mechanisms.

Acknowledgments The author thanks Ms. R. Tanji for her making figures and format of the manuscript. Dr. C.K. Glass gave me valuable advice and also suggestion on writing the manuscript. This work was supported by Takeda Science Foundation, the Naito Foundations, Astellas Foundation for Research on Metabolic Disorders Foundation, and also by Grant-in-Aid for

Scientific Research (B: nos22390057), Grant-in-Aid for Challenging Exploratory Research (nos23659461), and Grant-in-aid for "Support Project of Strategic Research Center in Private Universities" from the Ministry of Education, Culture, Sports, Science and Technology to Saitama Medical University Research Center for Genomic Medicine.

References

Albert I, Mavrich TN, Tomsho LP, Qi J, Zanton SJ, Schuster SC, Pugh BF (2007) Translational and rotational settings of H2A.Z nucleosomes across the *Saccharomyces cerevisiae* genome. Nature 446:572–576

Amaral PP, Clark MB, Gascoigne DK, Dinger ME, Mattick JS (2011) lncRNAdb: a reference database for long noncoding RNAs. Nucleic Acids Res 39:D146–D151

Barski A, Cuddapah S, Cui K, Roh TY, Schones DE, Wang Z, Wei G, Chepelev I, Zhao K (2007) High-resolution profiling of histone methylations in the human genome. Cell 129:823–837

Bernstein BE, Liu CL, Humphrey EL, Perlstein EO, Schreiber SL (2004) Global nucleosome occupancy in yeast. Genome Biol 5:R62

Boeger H, Griesenbeck J, Strattan JS, Kornberg RD (2003) Nucleosomes unfold completely at a transcriptionally active promoter. Mol Cell 11:1587–1598

Bond CS, Fox AH (2009) Paraspeckles: nuclear bodies built on long noncoding RNA. J Cell Biol 186:637–644

Carninci P, Kasukawa T, Katayama S, Gough J, Frith MC, Maeda N, Oyama R, Ravasi T, Lenhard B, Wells C et al (2005) The transcriptional landscape of the mammalian genome. Science 309:1559–1563

Carrozza MJ, Li B, Florens L, Suganuma T, Swanson SK, Lee KK, Shia WJ, Anderson S, Yates J, Washburn MP et al (2005) Histone H3 methylation by Set2 directs deacetylation of coding regions by Rpd3S to suppress spurious intragenic transcription. Cell 123:581–592

Chen LL, Carmichael GG (2009) Altered nuclear retention of mRNAs containing inverted repeats in human embryonic stem cells: functional role of a nuclear noncoding RNA. Mol Cell 35:467–478

Chetverin AB, Chetverina HV, Munishkin AV (1991) On the nature of spontaneous RNA synthesis by Q beta replicase. J Mol Biol 222:3–9

Chetverina HV, Chetverin AB (1993) Cloning of RNA molecules in vitro. Nucleic Acids Res 21:2349–2353

Churchman LS, Weissman JS (2011) Nascent transcript sequencing visualizes transcription at nucleotide resolution. Nature 469:368–373

Clemson CM, Hutchinson JN, Sara SA, Ensminger AW, Fox AH, Chess A, Lawrence JB (2009) An architectural role for a nuclear noncoding RNA: NEAT1 RNA is essential for the structure of paraspeckles. Mol Cell 33:717–726

Dieci G, Fiorino G, Castelnuovo M, Teichmann M, Pagano A (2007) The expanding RNA polymerase III transcriptome. Trends Genet 23:614–622

Doggrell SA (2005) Pegaptanib: the first antiangiogenic agent approved for neovascular macular degeneration. Expert Opin Pharmacother 6:1421–1423

Duret L, Chureau C, Samain S, Weissenbach J, Avner P (2006) The Xist RNA gene evolved in eutherians by pseudogenization of a protein-coding gene. Science 312:1653–1655

Elisaphenko EA, Kolesnikov NN, Shevchenko AI, Rogozin IB, Nesterova TB, Brockdorff N, Zakian SM (2008) A dual origin of the Xist gene from a protein-coding gene and a set of transposable elements. PLoS One 3:e2521

ENCODE-consortium (2004) The ENCODE (ENCyclopedia Of DNA Elements) project. Science 306:636–640

Feng J, Bi C, Clark BS, Mady R, Shah P, Kohtz JD (2006) The Evf-2 noncoding RNA is transcribed from the Dlx-5/6 ultraconserved region and functions as a Dlx-2 transcriptional coactivator. Genes Dev 20:1470–1484

Fox AH, Lam YW, Leung AK, Lyon CE, Andersen J, Mann M, Lamond AI (2002) Paraspeckles: a novel nuclear domain. Curr Biol 12:13–25

Guttman M, Amit I, Garber M, French C, Lin MF, Feldser D, Huarte M, Zuk O, Carey BW, Cassady JP et al (2009) Chromatin signature reveals over a thousand highly conserved large non-coding RNAs in mammals. Nature 458:223–227

Hartley PD, Madhani HD (2009) Mechanisms that specify promoter nucleosome location and identity. Cell 137:445–458

Imamura T, Yamamoto S, Ohgane J, Hattori N, Tanaka S, Shiota K (2004) Non-coding RNA directed DNA demethylation of Sphk1 CpG island. Biochem Biophys Res Commun 322:593–600

Ioshikhes IP, Albert I, Zanton SJ, Pugh BF (2006) Nucleosome positions predicted through comparative genomics. Nat Genet 38:1210–1215

Johnson SM, Tan FJ, McCullough HL, Riordan DP, Fire AZ (2006) Flexibility and constraint in the nucleosome core landscape of Caenorhabditis elegans chromatin. Genome Res 16:1505–1516

Joyce GF (1989) RNA evolution and the origins of life. Nature 338:217–224

Joyce GF (1994) In vitro evolution of nucleic acids. Curr Opin Struct Biol 4:331–336

Kaiser PK (2006) Antivascular endothelial growth factor agents and their development: therapeutic implications in ocular diseases. Am J Ophthalmol 142:660–668

Kaplan N, Moore IK, Fondufe-Mittendorf Y, Gossett AJ, Tillo D, Field Y, LeProust EM, Hughes TR, Lieb JD, Widom J et al (2009) The DNA-encoded nucleosome organization of a eukaryotic genome. Nature 458:362–366

Kapranov P, Cheng J, Dike S, Nix DA, Duttagupta R, Willingham AT, Stadler PF, Hertel J, Hackermuller J, Hofacker IL et al (2007a) RNA maps reveal new RNA classes and a possible function for pervasive transcription. Science 316:1484–1488

Kapranov P, Willingham AT, Gingeras TR (2007b) Genome-wide transcription and the implications for genomic organization. Nat Rev Genet 8:413–423

Katayama S, Tomaru Y, Kasukawa T, Waki K, Nakanishi M, Nakamura M, Nishida H, Yap CC, Suzuki M, Kawai J et al (2005) Antisense transcription in the mammalian transcriptome. Science 309:1564–1566

Keogh MC, Kurdistani SK, Morris SA, Ahn SH, Podolny V, Collins SR, Schuldiner M, Chin K, Punna T, Thompson NJ et al (2005) Cotranscriptional set2 methylation of histone H3 lysine 36 recruits a repressive Rpd3 complex. Cell 123:593–605

Knight RD, Landweber LF (1998) Rhyme or reason: RNA-arginine interactions and the genetic code. Chem Biol 5:R215–220

Kurokawa R (2011) Long noncoding RNA as a regulator for transcription. Prog Mol Subcell Biol 51:29–41

Kurokawa R, Rosenfeld MG, Glass CK (2009) Transcriptional regulation through noncoding RNAs and epigenetic modifications. RNA Biol 6:233–236

Lee JT (2010) The X as model for RNA's niche in epigenomic regulation. Cold Spring Harb Perspect Biol 2:a003749

Lee W, Tillo D, Bray N, Morse RH, Davis RW, Hughes TR, Nislow C (2007) A high-resolution atlas of nucleosome occupancy in yeast. Nat Genet 39:1235–1244

Liao Q, Liu C, Yuan X, Kang S, Miao R, Xiao H, Zhao G, Luo H, Bu D, Zhao H et al (2011) Large-scale prediction of long non-coding RNA functions in a coding-non-coding gene co-expression network. Nucleic Acids Res 39(9):3864–3878

Loh YH, Wu Q, Chew JL, Vega VB, Zhang W, Chen X, Bourque G, George J, Leong B, Liu J et al (2006) The Oct4 and Nanog transcription network regulates pluripotency in mouse embryonic stem cells. Nat Genet 38:431–440

Lomvardas S, Thanos D (2002) Modifying gene expression programs by altering core promoter chromatin architecture. Cell 110:261–271

Mannironi C, Scerch C, Fruscoloni P, Tocchini-Valentini GP (2000) Molecular recognition of amino acids by RNA aptamers: the evolution into an L-tyrosine binder of a dopamine-binding RNA motif. RNA 6:520–527

Martianov I, Ramadass A, Serra Barros A, Chow N, Akoulitchev A (2007) Repression of the human dihydrofolate reductase gene by a non-coding interfering transcript. Nature 445:666–670

Mohammad F, Pandey RR, Nagano T, Chakalova L, Mondal T, Fraser P, Kanduri C (2008) Kcnq1ot1/Lit1 noncoding RNA mediates transcriptional silencing by targeting to the perinucleolar region. Mol Cell Biol 28:3713–3728

Ozsolak F, Song JS, Liu XS, Fisher DE (2007) High-throughput mapping of the chromatin structure of human promoters. Nat Biotechnol 25:244–248

Pandey RR, Mondal T, Mohammad F, Enroth S, Redrup L, Komorowski J, Nagano T, Mancini-Dinardo D, Kanduri C (2008) Kcnq1ot1 antisense noncoding RNA mediates lineage-specific transcriptional silencing through chromatin-level regulation. Mol Cell 32:232–246

Pokholok DK, Harbison CT, Levine S, Cole M, Hannett NM, Lee TI, Bell GW, Walker K, Rolfe PA, Herbolsheimer E et al (2005) Genome-wide map of nucleosome acetylation and methylation in yeast. Cell 122:517–527

Ponjavic J, Ponting CP, Lunter G (2007) Functionality or transcriptional noise? Evidence for selection within long noncoding RNAs. Genome Res 17:556–565

Raisner RM, Hartley PD, Meneghini MD, Bao MZ, Liu CL, Schreiber SL, Rando OJ, Madhani HD (2005) Histone variant H2A.Z marks the 5′ ends of both active and inactive genes in euchromatin. Cell 123:233–248

Reinberg D, Roeder RG (1987) Factors involved in specific transcription by mammalian RNA polymerase II. Transcription factor IIS stimulates elongation of RNA chains. J Biol Chem 262:3331–3337

Reinke H, Horz W (2003) Histones are first hyperacetylated and then lose contact with the activated PHO5 promoter. Mol Cell 11:1599–1607

Rinn JL, Kertesz M, Wang JK, Squazzo SL, Xu X, Brugmann SA, Goodnough LH, Helms JA, Farnham PJ, Segal E et al (2007) Functional demarcation of active and silent chromatin domains in human HOX loci by noncoding RNAs. Cell 129:1311–1323

Roeder RG (1991) The complexities of eukaryotic transcription initiation: regulation of preinitiation complex assembly. Trends Biochem Sci 16:402–408

Schones DE, Cui K, Cuddapah S, Roh TY, Barski A, Wang Z, Wei G, Zhao K (2008) Dynamic regulation of nucleosome positioning in the human genome. Cell 132:887–898

Segal E, Fondufe-Mittendorf Y, Chen L, Thastrom A, Field Y, Moore IK, Wang JP, Widom J (2006) A genomic code for nucleosome positioning. Nature 442:772–778

Sekimizu K, Nakanishi Y, Mizuno D, Natori S (1979) Purification and preparation of antibody to RNA polymerase II stimulatory factors from Ehrlich ascites tumor cells. Biochemistry 18:1582–1588

Shenkin A, Burdon RH (1966) Asymmetric transcription of deoxyribonucleic acid by deoxyribonucleic acid-dependent ribonucleic acid polymerase of Krebs II ascites-tumour cells. Biochem J 98:5C–7C

Siepel A, Bejerano G, Pedersen JS, Hinrichs AS, Hou M, Rosenbloom K, Clawson H, Spieth J, Hillier LW, Richards S et al (2005) Evolutionarily conserved elements in vertebrate, insect, worm, and yeast genomes. Genome Res 15:1034–1050

Sone M, Hayashi T, Tarui H, Agata K, Takeichi M, Nakagawa S (2007) The mRNA-like noncoding RNA Gomafu constitutes a novel nuclear domain in a subset of neurons. J Cell Sci 120:2498–2506

Spirin AS (2005) RNA world and its evolution. Mol Biol (Mosk) 39:550–556

Torarinsson E, Sawera M, Havgaard JH, Fredholm M, Gorodkin J (2006) Thousands of corresponding human and mouse genomic regions unalignable in primary sequence contain common RNA structure. Genome Res 16:885–889

Torarinsson E, Yao Z, Wiklund ED, Bramsen JB, Hansen C, Kjems J, Tommerup N, Ruzzo WL, Gorodkin J (2008) Comparative genomics beyond sequence-based alignments: RNA structures in the ENCODE regions. Genome Res 18:242–251

Tripathi V, Ellis JD, Shen Z, Song DY, Pan Q, Watt AT, Freier SM, Bennett CF, Sharma A, Bubulya PA et al (2010) The nuclear-retained noncoding RNA MALAT1 regulates alternative splicing by modulating SR splicing factor phosphorylation. Mol Cell 39:925–938

Tuerk C, Gold L (1990) Systematic evolution of ligands by exponential enrichment: RNA ligands to bacteriophage T4 DNA polymerase. Science 249:505–510

Wang X, Arai S, Song X, Reichart D, Du K, Pascual G, Tempst P, Rosenfeld MG, Glass CK, Kurokawa R (2008) Induced ncRNAs allosterically modify RNA-binding proteins in cis to inhibit transcription. Nature 454:126–130

Weake VM, Workman JL (2010) Inducible gene expression: diverse regulatory mechanisms. Nat Rev Genet 11:426–437

Whitehouse I, Rando OJ, Delrow J, Tsukiyama T (2007) Chromatin remodelling at promoters suppresses antisense transcription. Nature 450:1031–1035

Yadon AN, Van de Mark D, Basom R, Delrow J, Whitehouse I, Tsukiyama T (2010) Chromatin remodeling around nucleosome-free regions leads to repression of noncoding RNA transcription. Mol Cell Biol 30:5110–5122

Yarus M (2000) RNA-ligand chemistry: a testable source for the genetic code. RNA 6:475–484

Yuan GC, Liu YJ, Dion MF, Slack MD, Wu LF, Altschuler SJ, Rando OJ (2005) Genome-scale identification of nucleosome positions in S. cerevisiae. Science 309:626–630

Zhang Y, Moqtaderi Z, Rattner BP, Euskirchen G, Snyder M, Kadonaga JT, Liu XS, Struhl K (2009) Intrinsic histone-DNA interactions are not the major determinant of nucleosome positions in vivo. Nat Struct Mol Biol 16:847–852

Zhao J, Sun BK, Erwin JA, Song JJ, Lee JT (2008) Polycomb proteins targeted by a short repeat RNA to the mouse X chromosome. Science 322:750–756

Chapter 7
MicroRNA Regulation of Neuronal Differentiation and Plasticity

Christian Barbato and Francesca Ruberti

Abstract MicroRNAs (miRNAs) expressed in the mammalian nervous system exhibit context-dependent functions during different stages of neuronal development, from early neurogenesis and neuronal differentiation to dendritic morphogenesis and neuronal plasticity. miRNAs often act through regulatory networks in specific cellular contexts and at specific times to ensure the progression through each biological state. Crosstalk between miRNAs and RNA-binding proteins introduces an additional layer of regulatory complexity in miRNA-mediated post-transcriptional regulation. Plasticity in localised parts of synapses is necessary for the information storage capacity of the brain. miRNAs and RNA-induced silencing complexes (RISCs) contribute to synaptic plasticity by modulating dendritic mRNA translation and dendritic spines. Specific molecules in neuronal cells may regulate miRNA action at the post-transcriptional and transcriptional level, suggesting that they may be involved in early and late responses underlying synaptic plasticity processes. Studies in animal models show that RISC and specific miRNAs may be recruited in synaptic plasticity processes underpinning learning, memory and cognition. Recent discoveries provide an encouraging starting point to investigate miRNA/RISC involvement in the development, progression and eventual therapeutic treatment of neurological and psychiatric diseases. Here we discuss recent findings that highlight the role of microRNAs in the regulatory networks associated with neuronal differentiation and synaptic plasticity in mammals.

Keywords Behaviour • memory • microRNA • neurons • RISC (RNA-induced silencing complex) • synaptic plasticity

C. Barbato (✉) • F. Ruberti (✉)
Cell Biology and Neurobiology Institute (IBCN), CNR, National Research Council of Italy,
IRCCS Fondazione Santa Lucia, Rome, Italy

Fondazione EBRI – Rita Levi-Montalcini, EBRI – European Brain Research Institute, Rome, Italy
e-mail: francesca.ruberti@inmm.cnr.it

B. Mallick (eds.), *Regulatory RNAs*, DOI 10.1007/978-3-642-22517-8_7,   175
© Springer-Verlag Berlin Heidelberg (outside the USA) 2012

7.1   Introduction

Genome projects have shown that although more than 90% of the human and mouse genome is transcribed, only 1.2% encodes proteins. Indeed, a large number of non-coding RNAs (ncRNAs) have been identified. Among them are microRNAs (miRNAs), which are small non-coding RNA (ncRNAs) that are 20–24 nucleotides in length. They regulate mRNA target expression through base pairing, usually in the 3' untranslated region (3'UTR). In mammals, the recognition of mRNA by miRNA occurs through partial base pairing between the miRNA response element (RE) of the target and usually the 5' end of the miRNA, including nucleotides 2–8. miRNAs derive from transcripts that fold back on themselves to form distinctive hairpin structures (i.e. primay precursor miRNAs) (Bartel 2004), whereas other types of endogenous small RNAs derive either from much longer hairpins that give rise to a greater diversity of small RNAs (e.g. small interfering RNAs, or siRNAs), from two RNA molecules complexed into RNA duplexes (e.g. siRNAs), or from precursors without any suspected double-stranded character (e.g. piwi-interacting RNAs).

miRNA functions are executed by the multi-protein RNA-induced silencing complex (RISC). The core of RISC consists of Argonaute (Ago) proteins, which load only one strand of the miRNA duplex. RISC/miRNA mediates miRNA-dependent mRNA decay and translation repression (reviewed in Bartel 2009).

Each miRNA has been predicted to target dozens to hundreds of genes. As a consequence, an miRNA can regulate several genes in a pathway or even multiple pathways in an additive, synergistic or antagonistic manner. In addition, more than one RE for the same miRNA, as well as several REs for different miRNAs, are present within an mRNA target. Therefore, cooperative regulatory mechanisms may be involved (reviewed in Bartel 2009).

The number of miRNA genes expressed in the nervous system seems to be larger than in other tissues, probably because the nervous system contains many types and subtypes of cells. Some miRNAs are enriched in or unique to brain tissue and neural cells (Lagos-Quintana et al. 2002; Bak et al. 2008). The temporally regulated expression of miRNAs suggests that miRNAs play important roles in the development of the mammalian brain (Miska et al. 2004). Recent studies have demonstrated critical roles for miRNAs in changes in the gene expression programme underlying the transition from progenitor neural cells to mature neurons. miRNAs expressed in the mammalian nervous system exhibit context-dependent functions during different stages of neuronal differentiation, including dendritic morphogenesis, dendritic spine development and neuronal physiology. The diversification of protein synthesis in specific neuronal compartments, such as dendrites and spines, contributes to the formation of neuronal connections during neuronal development and to higher-order cognitive behaviour (Bramham and Wells 2007). Recent studies have shown that miRNAs and RISC may act locally in specific neuronal compartments and also play essential roles in synaptic plasticity and memory processes.

7.2 miRNA Regulation of Neuronal Differentiation

Intricate networks of transcription factors and post-transcriptional regulators, including miRNAs and RNA-binding proteins, are involved in neuronal differentiation. miRNAs often act through regulatory networks in specific cellular contexts and at specific times to ensure the progression through each biological state. Examples of regulatory feedback circuits involving miRNAs, the underlying neuronal cell fates and neuronal differentiation, have been found in invertebrates and mammals.

7.2.1 miRNA Networks and Neuronal Differentiation

In mammals, miR-9 and miR-124 are miRNAs that are highly enriched in the brain. miR-124 was first identified as one of the mouse brain-specific miRNAs, and it has been conserved from *Aplysia*, *Drosophila*, and *C. elegans* to mammals. miR-124 is the most abundant miRNA in the brain, where it accounts for 25–48% of all miRNAs. In mammals, miR-124 is encoded by three genes located on three different chromosomes (Lagos-Quintana et al. 2002). miR-124 is upregulated during neuronal differentiation (Smirnova et al. 2005). It is broadly expressed in all postmitotic neurons in the adult mouse brain. In mammals, miR-124 seems to ensure a switch from non-neuronal to neuronal gene expression. miR-124 is extensively involved in repressing non-neuronal genes, thereby controlling neuronal identity. The overexpression of miR-124 in HeLa cells induces the downregulation of more than 100 non-neuronal mRNAs, producing a neuron-like expression profile and suggesting that it plays a role in neuronal differentiation (Conaco et al. 2006; Lim et al. 2005). In cortical neurons, several non-neuronal mRNA transcripts increase upon miR-124 knockdown (Conaco et al. 2006), and during chick spinal cord development, miR-124 is necessary for the preservation of neuronal identity (Visvanathan et al. 2007). RE1-silencing transcription factor (REST) is a transcriptional repressor of neural genes in non-neuronal tissues, including miR-124 (Conaco et al. 2006). Decreased levels of REST, together with increased miR-124 expression, lead to the terminal differentiation of neuronal cells (Visvanathan et al. 2007). miR-124 downregulates REST activity by silencing non-neuronal mRNAs, including the small C-terminal domain phosphatase-1 (SCP1), an activator of REST that is involved in the anti-neural REST pathway (Visvanathan et al. 2007). In non-neuronal cells and neuronal precursors, REST and SCP1 repress the expression of miR-124 and other neuronal genes, which are derepressed during neural differentiation (Visvanathan et al. 2007; Wu and Xie 2006). These results suggest a negative feedback loop between REST/SCP1 and miR-124 for the rapid transition from neural progenitors to postmitotic neurons. Among the genes controlled by miR-124 is the polypyrimidine tract-binding protein (PTBP1), which is an important splicing regulator (Makeyev et al. 2007).

During neuronal differentiation, PTBP1 expression is substituted by its neuronal homolog, PTBP2. This switch has substantial consequences for the splicing patterns of genes involved in neuronal functions (Makeyev et al. 2007).

A negative feedback interaction controlling neuronal differentiation involves the mutual inhibition of Ephrin-B1 and miR-124 in the developing mouse brain (Arvanitis et al. 2010). Ephrins act as both ligands and receptors that transduce bidirectional signals, reflecting cell–cell contact. Ephrin-B1 and miR-124 show reciprocal expression during mouse cortex differentiation. Ephrin-B1 is expressed in neural progenitor layers, whereas miR-124 expression is elevated in more differentiated cells. Ephrin-B1 signalling reduces miR-124 levels, and reciprocally, miR-124 expression represses ephrin-B1 expression post-transcriptionally. Thus, ephrin-B1 is required for maintaining the progenitor state, and miR-124 promotes differentiation.

miRNAs show defined expression patterns at precise time points during development. In mammals, miR-9 is encoded by three genomic loci: miR-9-1, miR-9-2 and miR-9-3. Their mature forms have identical nucleotide sequences. In the E11.5 mouse brain, miR-9 is most abundantly expressed in the developing medial pallium, but it is also apparent in the ventricular zone of the ganglion eminences. Gain- and loss-of-function studies of miR-9 showed that miR-9 modulates Cajal-Retzius cell differentiation by suppressing the expression of transcription factor of the forkhead family Foxg1 in the mouse medial pallium within the developing telencephalon (Shibata et al. 2008). Where miR-9 is highly expressed, Foxg1 is absent. The defined spatial distribution of miR-9 at precise time points may play a dual role (1) it may regulate cellular differentiation where it is highly expressed, and (2) it may allow a genetic programme to occur in tissues adjacent to where it is absent. Interestingly, studies of miR-9-2 and miR-9-3 in double mutant mice (Shibata et al. 2011) demonstrate that miR-9 controls neural progenitor proliferation and differentiation in the developing telencephalon by regulating the expression of multiple transcription factors. *miR-9-2/3* double mutants exhibit dysregulation of the proliferation and differentiation of pallial and sub-pallial progenitors. Concomitantly, the double mutants exhibit multiple defects in telencephalic structures, possibly due to the dysregulation of Foxg1, nuclear receptor subfamily 2, group E, member 1 (Nr2e1), genomic screened homeobox 2 (Gsh2) and Meis homeobox 2 (Meis2) expression (Shibata et al. 2011).

The simultaneous actions of different miRNAs on the same mRNA target regulate the exit of neural progenitors from the proliferative state and the transition to postmitotic neurons. miR-124 and miR-9* promote neuronal differentiation by repressing BAF53a (Yoo et al. 2009). BAF53a is a chromatin remodelling protein that is essential for neuronal progenitor proliferation. BAF53b is a subunit of the chromatin complex that is essential for postmitotic neuronal development and dendritic morphogenesis. The miRNAs miR-9* and miR-124 have been shown to be regulated by REST (Conaco et al. 2008; Packer et al. 2008). REST-mediated repression of miR-124 and miR-9/9* was found to direct the switch of chromatin regulatory complexes, which distinguish neuronal progenitors from fully differentiated neurons. Therefore, in neuronal precursors, miR-124 and miR-9*

Fig. 7.1 REST-mediated
repression of microRNAs. In
the absence of REST, both
miR-124 and miR-9 promote
neuronal differentiation by
silencing BAF53a

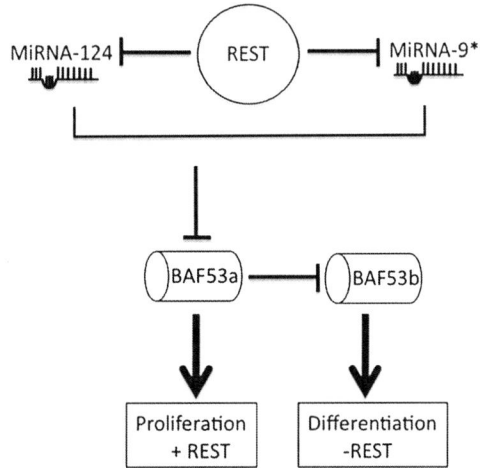

are inhibited by REST and their targets are expressed. Therefore, BAF53a is expressed to promote and sustain the maintenance of the progenitor state. In the absence of active REST, miR-124 and miR-9* repress their targets, which results in a derepression of pro-neural genes, such as BAF53b (Fig. 7.1).

A complex network of transcriptional activators and repressors acting together with miRNAs was recently associated with the neuronal differentiation of human neuroblastoma cells. These studies have shown that, miR-9-2 is expressed by a transcription unit independent from that of the host gene (Laneve et al. 2010). The search for transcription factors involved in miR-9-2 gene regulation has highlighted two antagonistic molecules, the repressor REST and the activator cAMP-response element binding protein (CREB), which function in an opposite and temporally regulated manner (Laneve et al. 2010). Studies in proliferating cells have indicated that REST is bound to the miR-9-2 promoter, which prevents transcriptional activity. Upon retinoic acid treatment, REST depletion and the concomitant phosphorylation of CREB already bound to the promoter activate transcription (Laneve et al. 2010).

The fine regulation of gene expression by miRNAs may occur not only during neuronal differentiation, but also during neuronal maturation processes. KCC2 is a potassium chloride co-transporter that is specifically expressed in neurons, developmentally regulated and abundant in the mature central nervous system (CNS) (Blaesse et al. 2009). The high expression of KCC2 in the mature CNS causes a decrease in the intracellular concentration of chloride ions, resulting in a developmental shift in γ-aminobutyric acid (GABA) action from depolarisation to hyperpolarisation. Recent studies have shown that miRNA-92 is developmentally downregulated during the maturation of rat cerebellar granule neurons in vitro, and that KCC2 is post-transcriptionally regulated by miR-92 (Barbato et al. 2010). Furthermore, the modulation of miR-92 expression levels was shown to regulate KCC2 protein expression and to change the reversal potential of GABA-induced chloride currents in granule neurons (Barbato et al. 2010). miR-92, similar to other miRNAs, might act in concert with a transcriptional control to define the fine-tuned regulation of KCC2 expression. Indeed, it has been shown that KCC2 expression is transcriptionally regulated in the mouse cerebellum during postnatal development by the transcription factor early growth response 4 (Uvarov et al. 2006). Recently, the REST-dual repressor element-1 interaction was identified as a novel mechanism of KCC2 transcriptional regulation that significantly contributes to the developmental switch in neuronal chloride concentration and GABA action in cortical neurons (Yeo et al. 2009). The search for interactions between transcriptional and post-transcriptional mechanisms might further elucidate the developmental regulation of KCC2 expression and the consequent switch in neuronal chloride concentration.

7.2.2   The Interplay Between miRNAs and RNA-Binding Proteins in the Regulation of Neuronal Development and Physiology

Crosstalk between miRNAs and RNA-binding proteins introduces an additional layer of regulatory complexity in miRNA-mediated post-transcriptional regulation. miRNA actions on mRNA targets may be either reduced or enhanced by RNA-binding protein interactions with the same mRNA (reviewed in Krol et al. 2010). Alternatively, a single miRNA may be involved in the regulation of RNA-binding proteins implicated in neuronal differentiation.

Several examples of RNA-binding proteins facilitating miRNA functions have been described (reviewed in Krol et al. 2010). In mammalian neurons, Edbauer et al. (2010) demonstrated a functional association between fragile X mental retardation protein (FMRP), a protein implicated in translation repression, miRNAs and Ago1. In particular, they found that the inhibitory effect of FMRP on translation of the mRNA encoding the N-methyl-D-aspartate (NMDA) receptor subunit NR2A is reinforced by miR-125b. Furthermore, the depletion of FMRP prevents the effect of miR-125 overexpression on spine morphology (Edbauer et al. 2010).

Several RNA-binding proteins counteract the repressive activity of miRISC. An AU-rich RNA-binding protein, Elavl2 [(Embryonic lethal abnormal vision Drosophila)-like 2 or HuB], was found to attenuate the miR-9-2-mediated repression of Foxg1 (Shibata et al. 2011). Studies of telencephalon development in mice have shown that in the ventricular zone of mutant mice lacking *miR-9-2* and *miR-9-3* (*miR-9-2/3*), Foxg1 expression is suppressed by *miR-9* at earlier stages (E14.5), but the suppression is countered by Elavl2 (E16.5), whose expression increases at later stages (Shibata et al. 2011).

Several reports indicate that miRNAs not only act as repressors, but can also act as activators of translation (Vasudevan et al. 2007). Studies performed in miR-9-2/3 double mutant mice and in vitro experiments in P19 cells have shown that two RNA-binding proteins, Elavl1 (HuR) and Msi1 (Musashi homolog 1), together with miRNA-9, target Nr2e1 mRNA 3′UTR to enhance expression (Shibata et al. 2011).

Not only do RNA-binding proteins modulate miRNA activity, but miRNAs may also repress RNA-binding proteins. miR-134 and miR-375 have been found to regulate dendrite density by targeting the translational activator Elavl4 (or HuD) (Abdelmohsen et al. 2010) and the translational repressor Pumilio2 (Pum2) (Fiore et al. 2009), respectively.

The miR-134 gene is clustered together with more than 50 other miRNAs within the Glt2/Dlk1 locus (miR-379-410 cluster). Interestingly, the myocyte enhancer factor 2 (Mef2)-induced expression of miR-134 and at least two other miRNAs of the miR-379-410 cluster are required for the activity-dependent dendritic outgrowth of hippocampal neurons (Fiore et al. 2009). Activity-induced miR-134 promotes dendritic outgrowth by fine tuning Pum2. Recent work has shown that vertebrate Pum2 is significantly upregulated at postnatal day 1, when dendritic outgrowth begins in the hippocampus, and negatively modulates the translation of eIF4E (eukaryotic translation initiation factor 4E) and other transcripts in hippocampal neurons (Vessey et al. 2010).

The levels of miR-375 are highest in early mouse embryos and decrease by E16, whereas HuD levels are low in early embryos and increase markedly by E16, when neurogenesis peaks (Abdelmohsen et al. 2010). miR-375 represses HuD expression through a specific, evolutionarily conserved site on the HuD 3′UTR (Abdelmohsen et al. 2010). The ectopic expression of miR-375 in the mouse hippocampus potently reduces dendrite density. miR-375 overexpression lowers both HuD mRNA stability and translation and recapitulates the effects of HuD silencing, which reduces the levels of target proteins with key functions in neuronal signalling and cytoskeleton organisation (Abdelmohsen et al. 2010).

Besides regulating protein abundance, RNA-binding factors have been implicated in RNA transport and local translation in neuronal dendrites. Neuronal granules are ribonucleoprotein particles that serve to transport mRNAs along microtubules and control local protein synthesis. Components of miRNA machinery, such as Ago proteins, miRNAs and mRNAs repressed by miRNAs, are enriched in evolutionarily conserved cytoplasmic structures called P-bodies (also called Dcp or GW bodies), which function as sites of both mRNA degradation and storage for translationally repressed mRNAs (Liu et al. 2005; Pillai et al. 2005; Bhattacharyya et al. 2006). Recently, P-body-like structures (i.e. dlPbodies) were described in mammalian dendritic neurons. Interestingly, neuronal activity seems to induce the disassembly of dendritically localised P-bodies (Zeitelhofer et al. 2008), or their remodelling, relocalisation to more distant sites and decreased association with Ago2 (Cougot et al. 2008). To date, it is still unknown whether miRNAs/RISC, in combination with RNA-binding proteins, play a role in mRNA dendritic transport.

7.3   miRNAs and Local Protein Synthesis

Plasticity in individual or localised regions of synapses is necessary for the information storage capacity of the brain. Upon the discovery of polyribosomes within dendritic shafts and spines, it was suggested that rapid dendritic protein synthesis, triggered by synaptic activity, serves as a mechanism for long-term plasticity at specific synapses. However, many of the most intriguing questions remain unanswered. For synaptic protein synthesis, the corresponding mRNAs must be transported to the dendritic compartment and translated upon site-specific activation. It is generally thought that dendritic mRNAs are transported in a translationally silenced state within ribonucleoprotein complexes (Kiebler and Bassel 2006). Ongoing efforts are underway to understand post-transcriptional mechanisms regulating gene expression and the consequent abundance of proteins at synapses. The studies described below indicate that miRNAs may contribute to the regulation of dendrites and spine morphology by modulating the expression of dendritic mRNAs.

7.3.1 miRNAs and Dendritic Spines

Recent studies indicate that a new class of small molecules, miRNAs, may participate in the mRNA-specific regulation of local translation by tuning gene expression at the post-transcriptional level. Dendritic spines are specific domains where local protein synthesis occurs. Spine structures are dynamically regulated (Hotulainen and Hoogenraad 2010), and functional and structural changes at spines and synapses are proposed as the basis of learning and memory (Kasai et al. 2010). miRNAs, expressed in a spatially and temporally controlled manner in the brain, are ideal candidates for the modulation of dendritic protein synthesis. miRNAs modulate dendritic morphology by regulating the expression of proteins involved in the actin cytoskeleton (Vo et al. 2005; Schratt et al. 2006; Siegel et al. 2009; Wayman et al. 2008) mRNA transport (Fiore et al. 2009) and neurotransmission (Edbauer et al. 2010).

Schratt and collaborators showed an association between miRNA function and local protein synthesis in mammalian neurons, with the brain-specific miRNA-134 (Schratt et al. 2006). The overexpression of miR-134 causes a significant reduction in dendritic spine size, whereas its inhibition by 2′-O-methyl antisense oligonucleotides induces a slight increase in spine volume. The mRNA target of miR-134 was identified as Lim-domain containing protein kinase 1 (Limk1). The repression of Limk1 translation by miR-134 is mitigated by brain-derived neurotrophic factor (BDNF) stimulation of synaptic activity. In cortical neurons, BDNF induces the translation of the 3′UTR Limk1 mRNA luciferase reporter, but not when neurons are transfected with a reporter in which the miR-134 responsive sequence was mutated. This suggests that BDNF stimulation and the repression of Limk1 translation by miR-134 play a role in synaptic plasticity in the dendritic compartment of hippocampal neurons (Fig. 7.2).

Kosik's group, with the aim of identifying dendritic mRNAs under the control of RISC, individuated several dendritically localised mRNAs. They trapped both known RISC-regulated mRNAs, Limk1 and alpha-isoform of calcium/calmodulin-dependent protein kinase II (αCaMKII) and a novel mRNA, Lysophospholipase

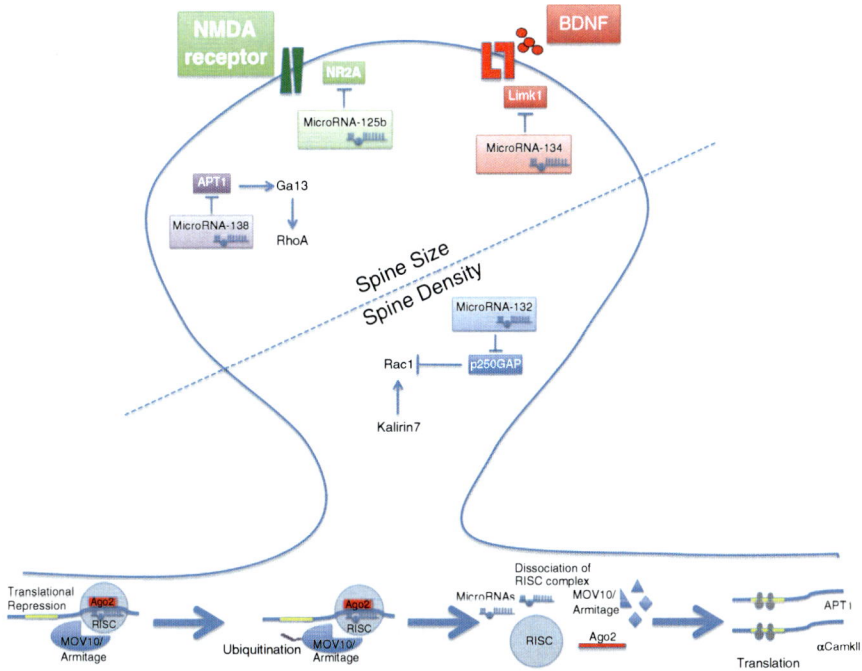

Fig. 7.2 MicroRNAs and RISC pathway regulating synaptic plasticity and dendritic spines. Top part of figure shows several miRNAs and their targets involved in the regulation of dendritic spine size and density. The components of each pathway are indicated with the same colour. miRNAs involved in either spine size or spine density are artificially separated by a *dashed line*. Bottom part of figure depicts the disassembly of the RISC complex after proteasomal degradation of MOV10, a way to allow derepression of APT1 and αCaMKII mRNAs

1 (Lypla1), which is also known as acyl-protein thioesterase (APT) 1, a depalmitoylation enzyme regulated post-transcriptionally by dendritic miR-138 (Banerjee et al. 2009). Previously, Schratt's group showed that in rat hippocampal neurons, miR-138 is enriched at the synapses and negatively modulates spine size through the regulation of APT1 levels (Siegel et al. 2009), followed by the depalmitoylation of Gα13, a downstream target of APT1 that is an activator of Rho downstream of G-protein coupled receptors (Kurose 2003) (Fig. 7.2).

CREB- and activity-regulated miRNA-132 is induced during periods of active synaptogenesis. Gain- and loss-of-function experiments have shown that miR-132 is necessary and sufficient for hippocampal spine formation (Impey et al. 2010). Likewise, the depletion of the miR-132 target Rho family GTPase-Activating Protein, known as p250GAP (Vo et al. 2005; Wayman et al. 2008), increases spine formation, while the introduction of a p250GAP mutant unresponsive to miR-132 attenuates this activity. P250GAP is an inhibitor of Rho family GTPases that may influence spine structure through their ability to regulate actin dynamics (Van Aelst and Cline 2004). Interestingly, the miR-132/p250GAP circuit regulates

Rac1 activity and spine formation by modulating synapse-specific kalirin-7/Rac-1 signalling (Impey et al. 2010) (Fig. 7.2). Consistent with these data, the ablation of the miR-212/132 locus dramatically reduces dendritic length, branching and spine density in newborn hippocampal neurons in young adult mice (Magill et al. 2010). Changes in dendritic arborisation and spine formation in newborn neurons persist for several months after training in a Morris water maze (Dupret et al. 2008).

Alterations in dendrite spine morphology are associated with fragile X syndrome (FXS) (reviewed in Beckel-Mitchener and Greenough 2004). The loss of the Fmr1 gene product FMRP, an mRNA-binding protein involved in translational regulation, leads to FXS. FMRP is thought to repress the synthesis of proteins required for synaptic plasticity. Among the most common symptoms reported in FXS are deficits in attention, inhibitory control and cognitive flexibility (Bassell and Warren 2008). Recent studies have shown that FMRP interacts with the RISC complex through an association with Ago (Edbauer et al. 2010). In hippocampal neurons, FMRP is linked to several specific miRNAs, including miR-132 and miR-125b (Edbauer et al. 2010). miR-132 and miR-125b have been reported to differentially affect dendritic spine morphology. Indeed, miR-132 gain of function increases spine density, while miR-125 gain of function reduces spine width. The effects of miR-132 and miR-125b overexpression on the morphology of mouse hippocampal neurons are abolished in cells with FMRP knockdown. In addition, the negative regulation of the NMDA receptor (NMDAR) subunit NR2A involves both FMRP activity and miR-125b targeting of the NR2A 3'UTR (Fig. 7.2). This indicates that FMRP contributes to miRNA function during synapse development, confirming previous observations that the loss of FMRP modulates NMDAR function in mice (Pfeiffer and Huber 2007).

7.4 RISC and miRNAs in Neuronal Plasticity

The formation of stable memory requires protein synthesis, a feature common to vertebrates and invertebrates. Local protein synthesis at the synapse is partially regulated by miRNAs. By modulating dendritic mRNA translation, miRNAs may contribute to synaptic plasticity. The coordination of different pathways orchestrating protein expression at synapses is fundamental for the control of synaptic plasticity. The assumption that synaptic plasticity underlies memory formation (Morris 2003) and the evidence that certain forms of long-lasting synaptic plasticity depend on protein synthesis (Manahan-Vaughan et al. 2000) suggest that miRNAs may indeed be important for this phenomenon.

7.4.1 RISC, Memory and Behaviour

The first evidence of the involvement of the RISC complex in memory formation was reported by the Kunes laboratory (Ashraf et al. 2006). The regulated disruption of the silencing complex component Armitage leads to the removal of the miRNA-mediated repression of CaMKII, an mRNA involved in synaptic plasticity. Putative miRNA binding sites are present within the 3'UTR of CaMKII, as well as within the 3'UTR of transcripts coding for Staufen and Kinesin Heavy Chain, two dendritic granule-associated proteins. Expression analyses of these genes have indicated that the synaptic translation of CaMKII increases in dicer-, armitage- and aubergine-mutant brains. Indeed, this work suggests an armitage-driven repression of CaMKII expression in the *Drosophila* olfactory system (Table 7.1). Synaptic activation induces a decrease in the levels of Armitage protein and a corresponding increase in CaMKII abundance (Fig. 7.2). The decrease in Armitage protein was due to the activity of the proteasome, which is known to act at the synaptic level to contribute to modulating synaptic protein content (Bingol and Schuman 2005; 2006). Overall, Ashraf et al. (2006) proposed a novel and intriguing regulatory mechanism whereby CaMKII translational repression is driven by miRNAs and, in turn, is relieved by the activity-dependent proteasome-mediated degradation of Armitage. Similarly, recent findings indicate that the mammalian ortholog of Armitage, Moloney

Table 7.1 Summary of studies illustrating microRNAs/RISC involvement in memory and behaviour

miRNA/RISC	Behavioural phenotypes	Target	Reference
Dicer ↓	Enhanced memory in Morris water maze and contextual fear conditioning	?	Konopka et al. (2010)
Ago2 ↓	Reduced cocaine self-administration	?	Schaefer et al. (2010)
	Impaired short-term memory Impaired long-term contextual fear memory		Batassa et al. (2010)
Armitage ↓	Failure in long-term olfactory memory	CaMKII	Ashraf et al. (2006)
miR-124* let-7d* miR-181 *	Altered cocaine conditioned place preference	Genes involved in cocaine-induced plasticity	Chandrasekar and Dreyer (2009, 2011)
miR-134 ↑	Impaired contextual fear conditioning Impaired LTP	CREB	Gao et al. (2010)
miR-132 ↑	Deficits in novel object recognition	MeCP2	Hansen et al. (2010)
miR-212 ↑	Decreased cocaine intake	MeCP2	Im et al. (2010)
miR-219 ↓	Extension of the circadian period		Cheng et al. (2007)
	Attenuation of behavioural response mediated by NMDAR antagonist	CaMKIIγ	Kocerha et al. (2009)
miR-132 ↓	Potentiation of light-induced clock resetting	MeCP2, Ep300, Jarid1a, Btg2, and Paip2a	Alvarez-Saavedra et al. (2011)

* Knock-down of these miRNAs modulates conditioned place preference. The question mark indicates that the miRNA target genes associated to the behavioural phenotype are unknown

leukaemia virus 10 homolog (MOV10), is present at synapses and is rapidly degraded by the proteasome after NMDAR activation. RNA interference-mediated knockdown of MOV10 may release the translational repression of mRNAs regulated by RISC, CamkII and Lypla1 (Banerjee et al. 2009) (Fig. 7.2).

The effect of RISC/Ago2 complex inactivation in the mouse brain was recently investigated (Batassa et al. 2010). In the human and mouse genome, four Ago genes that encode Ago1, Ago2, Ago3 and Ago4 are present. Among Ago proteins, only Ago2 is able to control mRNA expression through the 'slicer activity' of mRNA, which is perfectly complementary to specific miRNAs (Song et al. 2004). Five different plasmids that express siRNA targeting Ago2 mRNA and induce Ago2 downregulation were injected into the dorsal hippocampus of C57BL/6 mice. After surgery, the animals were allowed 1 week of recovery, after which two groups were submitted to hippocampus-related tasks. Ago2 silencing impaired both short-term memory and long-term contextual fear memory (Batassa et al. 2010). These data demonstrate the importance of Ago2 in the hippocampus for both contextual fear conditioning and passive avoidance tasks (Table 7.1). Importantly, when Ago2 expression levels were rescued 3 weeks after injection, memory recovered, indicating that the memory deficit was not due to a broad-spectrum impairment in hippocampal function. This was the first study showing a role for the RISC/Ago2 pathway in mammalian memory formation in vivo. The effects of Ago2 silencing on contextual memory impairment might involve not only alterations in miRNA-mediated post-transcriptional regulation, but also modulations of endogenous siRNAs. Indeed, through deep sequencing technology, endo-siRNAs have been revealed in mammalian cells (Ghildiyal and Zamore 2009). More recently, the expression of endo-siRNAs and other ncRNAs in the adult mouse hippocampus was demonstrated (Smalheiser et al. 2011). Moreover, endo-siRNAs are upregulated during olfactory discrimination training. Among them are several hairpin-derived siRNAs, identified by deep sequencing, which are mapped within genes associated with synaptic plasticity, such as SynGAP1, CaMKIIa and GAP43. These endo-siRNAs can bind to both the sense mRNAs from which they derive, as well as any antisense transcripts that may be expressed on the opposite strand. The studies described above suggest that Ago2 by small inhibitory RNA (siRNAs/miRNAs) might be actively involved in learning and memory processes.

RISC/Ago2 impairment is also associated with other behavioural tasks. Recent data have shown that the regulation of specific genes by cocaine administration may involve miRNAs (Chandrasekar and Dreyer 2009). In addition, Ago2 was found to participate in the maturation of miRNAs from their precursors (O'Carroll et al. 2007). Therefore, the effect of Ago2 deficiency on cocaine addiction was explored (Schaefer et al. 2010). Ago2 knockout in mouse dopamine 2 receptor (Drd2)-expressing neurons reduces animal dependence on cocaine (Table 7.1). Moreover, the acute administration of cocaine leads to an increase in 63 miRNAs in Drd2-expressing neurons in wild-type mice. Of those miRNAs, 23 are Ago2-dependent. Thus, the Ago2/miRNA pathway is an important player in animal behaviour. It will be important to explore whether the other Ago proteins (Ago1, 3 and 4) are involved in the selection of miRNA groups related to other behavioural patterns.

7.4.2 Regulation of miRNA Expression During Synaptic Plasticity

Several observations suggest that the induction of long-term potentiation (LTP) and long-term depression (LTD), two forms of synaptic plasticity, extensively regulates miRNA expression. Park and Tang (2009) performed a temporal expression profile of 60 hippocampal miRNAs following the induction of chemical LTP (C-LTP) and metabotropic glutamate receptor-dependent LTD (mGluR-LTD) in mouse hippocampal slices. They observed that C-LTP or mGluR-LTD evokes changes in the expression levels of most hippocampal miRNAs, suggesting a role for miRNA-mediated translational repression. miRNAs regulated in both experimental paradigms display distinct temporal expression dynamics. Further, many miRNAs are upregulated at specific time points during C-LTP and mGluR-LTD induction, as if to provide an active mechanism for restoring the dormant state of mRNA translation after transient activation (Park and Tang 2009).

Recent studies have demonstrated the fine regulation of mature and precursor miRNA expression by mGluR and NMDAR signalling during LTP induction in the adult dentate gyrus (Wibrand et al. 2010).

The first evidence that miRNA expression is specifically altered during an in vivo learning paradigm in mammals was recently revealed (Smalheiser et al. 2010). In particular, olfactory discrimination training upregulates and reorganises the expression of miRNAs in the adult mouse hippocampus (Smalheiser et al. 2010). Among miRNAs most upregulated by training, miR-10a is particularly intriguing. miR-10a is predicted to target numerous plasticity-related genes, including BDNF, Camk2b, CREB1 and Elavl2. However, miR-10a was reported to produce a positive effect on general protein translation by binding to the 5' terminal oligopyrimidine mRNAs and enhancing their translation (Orom et al. 2008). Further studies are required to evaluate the role of miRNA responses in changes in gene expression associated with this learning paradigm.

7.4.3 miRNAs, Memory and Behavioural Phenotypes

How does the expression of miRNAs affect learning and memory? This was studied in a mouse model with an inducible disruption of the Dicer1 gene in the adult forebrain (Konopka et al. 2010). After the induction of the Dicer1 gene deletion, a progressive loss of a whole set of brain-specific miRNAs was observed. Mice were tested in a battery of both aversively and appetitively motivated cognitive tasks, such as the Morris water maze, IntelliCage system and trace fear conditioning. An enhancement in memory was recorded 12 weeks after the Dicer1 gene mutation (Table 7.1). To date, it is not known whether there is 'better memory with less miRNA' or the reverse, but it is known that miRNAs may modulate memory, that components of LTP require local protein translation, which regulates synaptic plasticity, and that miRNAs have been identified as key regulators of protein synthesis.

The direct involvement of specific miRNAs in establishing long-term synaptic modifications in mammals was recently demonstrated. A novel pathway mediated by Sirtuin 1 (SIRT1) and miR-134 was found to regulate memory and plasticity (Gao et al. 2010). Mammalian SIRT1 is involved in several complex processes relevant to ageing, including DNA repair, genomic stability, neuronal survival and age-dependent neurodegenerative disorders. In the hippocampus of SIRT1 knock-out mice, dendritic spine density in pyramidal neurons is decreased, and associative memory and spatial learning are impaired (Gao et al. 2010). Furthermore, LTP in the CA1 region of acute hippocampal slices is abrogated (Table 7.1). Since CREB protein levels are significantly reduced in the SIRT1-KO hippocampus, but mRNA levels of CREB are not altered, it has been suggested that CREB protein levels are downregulated in SIRT1-KO brains via post-transcriptional mechanisms. Indeed, these effects are mediated via the post-transcriptional regulation of CREB expression by miR-134. Chromatin immunoprecipitation experiments demonstrated that SIRT1 in complex with the transcription factor Ying Yang 1 negatively regulates miR-134 transcription through the association of SIRT1 with DNA sequences upstream of the pre-miR-134 sequence (Gao et al. 2010). In the SIRT1-KO neuronal context, miR-134 is upregulated, with a consequential increase in the translational repression of its target mRNAs (Gao et al. 2010). Furthermore, miR-134 attenuates CREB expression via a specific interaction with miR-134 REs within the 3'UTR of CREB (Gao et al. 2010). Since the inhibition of miR-134 in SIRT1-KO mice rescues LTP and partially rescues memory formation, the synaptic plasticity impairments observed in SIRT1 KO mice are partly due to miR-134 upregulation and the consequent inhibition of miR-134 target genes. These results not only describe a novel pathway regulating memory and plasticity via SIRT1 and miR-134, but also suggest that miRNA-based mechanisms may be involved in other normal or pathological pathways regulated by SIRT1, indicating its value as a potential therapeutic target for the treatment of CNS disorders.

Another miRNA, miR-132, was found to modulate memory processes. A transgenic mouse strain that expresses miR-132 in forebrain neurons shows an increase in dendritic spine density and deficits in novel object recognition (Hansen et al. 2010) (Table 7.1). Consistently, a decrease in MeCP2 expression in the hippocampus of miR-132 transgenic mice was observed. MeCP2 is a target regulated by miR-132 (Klein et al. 2007), and the altered expression of MeCP2 has been associated with the development of Rett syndrome, a neurodevelopmental disorder in which dendrite development and synaptogenesis are affected. Further work is necessary to evaluate whether the dysregulation of miR-132 could contribute to an array of cognitive disorders.

Recent studies have shown that MeCP2 regulates responses to psychostimulants. In particular, MeCP2 was reported to regulate cocaine intake through interactions with miR-212 (Im et al. 2010). The early finding was that MeCP2 levels are increased in the dorsal striatum of rats with extended access to intravenous cocaine self-administration. After MeCP2 downregulation in the rat dorsal striatum, a decrease in cocaine intake is associated with higher expression levels of miR-212 (Im et al. 2010) (Table 7.1). Therefore, the molecular mechanisms underlying behavioural responses to psychostimulants were investigated. First, when miR-212 was inhibited after MeCP2 knockdown, cocaine intake was found to increase in animals with extended periods of access (Im et al. 2010). Second, a negative feedback mechanism between MeCP2 and miR-212 was observed mainly in rats with extended cocaine access (Im et al. 2010). Indeed, MeCP2, which represses miR-212, is repressed in turn by miR-212. Lastly, MeCP2-miR-212 interactions control cocaine's effect on BDNF (Im et al. 2010). Striatal BDNF transmission is known to increase sensitivity to the motivational effects of cocaine (Graham et al. 2007). The work from Im et al. (2010) suggests that repeated cocaine consumption in rats with extended access induces BDNF neosynthesis in the dorsal striatum. Striatal BDNF levels are negatively correlated with miR-212 expression, whereas they are positively correlated with MeCp2 expression (Im et al. 2010). These authors described MeCP2 and miR-212 as being locked in a regulatory loop, demonstrating that they both exert opposite effects on striatal BDNF levels and suggesting that the balance between these two factors may be fundamental in determining vulnerability to cocaine addiction. However, the precise mechanism by which MeCP2-miR-212 interactions affect striatal BDNF levels is still unclear (Im et al. 2010).

Behavioural phenotypes during circadian rhythms have been extensively studied in several organisms. Circadian mutants can exhibit behavioural dysfunctions associated with addiction and human mood disorders (reviewed in Takahashi et al. 2008). In mammals, the circadian rhythm is organised in a hierarchical network of molecular clocks that operate in different tissues. The master clock resides in the suprachiasmatic nucleus (SCN) of the hypothalamus and synchronises the rhythms of the peripheral oscillators. The circadian clock is composed of a cell-autonomous transcription-translation feedback loop. In mammals, the circadian clock is composed of a primary feedback loop involving the genes *Clock*, *Bmal1*, period homologue 1 (*Per1*), *Per2*, cryptochrome 1 (*Cry1*) and *Cry2*. The CLOCK and BMAL1 complex activates the transcription of the *Per* and *Cry* genes. The resulting PER and CRY proteins interact with the CLOCK-BMAL1 complex to inhibit their own transcription. After this phase, the PER-CRY repressor complex is degraded and CLOCK-BMAL1 can then activate a new cycle of transcription. Transcription activation also underlies the resetting, or entrainment, of the clock. CREB has been implicated in light-induced gene transactivation and resetting the circadian clock. Recently, miRNAs were revealed as new players in the landscape of circadian clock regulation. miRNA-132 and miRNA-219 were found to modulate the circadian clock located in the mouse SCN (Cheng et al. 2007). miR-219 is regulated by the transcription factor CLOCK and exhibits a robust circadian rhythm of expression. miR-132 expression is light-inducible via CREB, and miR-132 levels are also rhythmic (Cheng et al. 2007). Loss-of-function experiments showed that miR-219 inhibition extends the circadian period, while miR-132 inactivation potentiates light-induced clock resetting (Table 7.1), suggesting that endogenous miR-132 acts as a negative modulator of light responsiveness (Cheng et al. 2007).

A subset of miR-132-regulated targets in the mouse SCN has been implicated in chromatin remodelling (e.g. *Mecp2, Ep300, Jarid1a*) and in translational control (e. g. *Btg2, Paip2a*) (Alvarez-Saavedra et al. (2011)). The coordinated regulation of these genes has been shown to underlie the miR-132-mediated modulation of period gene expression and fine tuning of clock entrainment (Alvarez-Saavedra et al. 2011).

The direct regulation by miRNAs of core components of the circadian clock has also been observed (Nagel et al. 2009). The miR-192/194 cluster was found to inhibit Per gene family expression through a functional interaction with the 3′UTR of Per mRNA. In addition, miR-192/194 overexpression leads to an altered circadian cycle in cultured cell lines (Nagel et al. 2009).

Interestingly, miR-219 is decreased in the prefrontal cortex of mice after NMDAR signal blockade with dizocilpine (Kocerha et al. 2009). The disruption of NMDA-mediated glutamate signalling has been linked to behavioural deficits observed in psychiatric disorders such as schizophrenia. Dizocilpine is a highly selective phencyclidine-like NMDAR antagonist that can rapidly produce schizophrenia-like behavioural deficits in humans and rodents (Heresco-Levy and Javitt 1998). In particular, dizocilpine treatment in mice produces hyperlocomotor activity and increased stereotypic behaviour. Consistent with a role for miR-219 in NMDAR signalling, calcium/calmodulin-dependent protein kinase II γ subunit, a component of the NMDAR signalling cascade, has been identified as a target of miR-219. The inhibition of miR-219 in vivo attenuates behavioural responses mediated by the block of NMDAR signal transduction (Table 7.1), suggesting that miR-219 is part of a compensatory mechanism that maintains NMDA receptor function (Kocerha et al. 2009).

All of the findings described above underline the importance of the post-transcriptional regulation of miRNAs mediated in several behavioural paradigms, indicating a high level of complexity in transcriptional and post-transcriptional regulatory networks.

7.5   Conclusions and Perspectives

Networks of regulatory mechanisms are involved during the development and differentiation of the nervous system. The examples of REST, miR-124 and miR-9 summarised above suggest that miRNAs can act as reinforcers and backups of specific transcriptional programmes. Thus, an miRNA and its target are oppositely regulated by the same signal, defining a coherent feedback loop. This suggests that an miRNA participates in signalling networks to stabilise fine tissue patterning by repressing its target mRNA in cells where it should be not expressed, or to better control gene expression levels.

Studies of miRNAs and the 3′UTR of mRNAs highlight the complexity and significance of post-transcriptional regulation mediated by the 3′UTR in mammalian gene expression. Multiple cis-elements present within the 3′UTR may act in a synergistic or antagonistic manner to regulate gene expression. Furthermore, miRNA action may be either activated or reduced by RNA-binding proteins. miRNAs may play a crucial role in regulatory networks underlying synaptic plasticity, memory formation and cognitive functions. While this field is still in its infancy, studies are already demonstrating that specific molecules in neuronal cells may modulate miRNA action at the post-transcriptional and transcriptional level. This suggests that they may be involved in early and late responses underlying synaptic plasticity processes.

The analysis of miRNA expression profiles in physiological and pathological conditions, and the individuation of the relationship between the miRNA/RISC pathway and neuronal activity, may elucidate the mechanisms underlying the cellular and molecular basis of neuronal plasticity. A major effort will be needed to define the role of RNA-mediated gene-silencing machinery in neurons, the neuronal miRNA targets, and specific components of RISC that are relevant in neurological and psychiatric diseases.

Despite major challenges that still need to be overcome, miRNAs hold great potential as therapeutic targets (see Chap. 18 by Majer et al., this volume). Future therapies may be directed at specific miRNAs of interest that can affect a multitude of targets involved in modulating the mechanisms of plasticity.

Acknowledgments This work was supported by Compagnia di San Paolo-Programma Neuroscienze-'MicroRNAs in Neurodegenerative Diseases' (to C.B.), and by CNR grant DG. RSTL.059.012 (to F.R.). We thank for graphical assistance—Studio 2CV idee e paesaggi- (www. studio2cv.it).

References

Abdelmohsen K, Hutchison ER, Lee EK, Kuwano Y, Kim MM, Masuda K, Srikantan S, Subaran SS, Marasa BS, Mattson MP, Gorospe M (2010) miR-375 inhibits differentiation of neurites by lowering HuD levels. Mol Cell Biol 30:4197–4210

Alvarez-Saavedra M, Antoun G, Yanagiya A, Oliva-Hernandez R, Cornejo-Palma D, Perez-Iratxeta C, Sonenberg N, Cheng HY (2011) miRNA-132 orchestrates chromatin remodeling and translational control of the circadian clock. Hum Mol Genet 20(4):731–751

Arvanitis DN, Jungas T, Annie Behar A, Alice Davy A (2010) Ephrin-B1 reverse signaling controls a posttranscriptional feedback mechanism via miR-124. Mol Cell Biol 30:2508–2551

Ashraf SI, McLoon AL, Sclarsic SM, Kunes S (2006) Synaptic protein synthesis associated with memory is regulated by the RISC pathway in Drosophila. Cell 124:191–205

Bak M, Silahtaroglu A, Møller M, Christensen M, Rath MF et al (2008) MicroRNA expression in the adult mouse central nervous system. RNA 14(3):432–444

Banerjee S, Neveu P, Kosik KS (2009) A coordinated local translational control point at the synapse involving relief from silencing and MOV10 degradation. Neuron 64:871–884

Barbato C, Ruberti F, Pieri M, Vilardo E, Costanzo M, Ciotti MT, Zona C, Cogoni C (2010) MicroRNA-92 modulates K(+) Cl(−) co-transporter KCC2 expression in cerebellar granule neurons. J Neurochem 113:591–600

Bartel DP (2004) MicroRNAs: genomics, biogenesis, mechanism, and function. Cell 116:281–297

Bartel DP (2009) MicroRNAs: target recognition and regulatory functions. Cell 136:215–233, Review

Bassell GJ, Warren ST (2008) Fragile X syndrome: loss of local mRNA regulation alters synaptic development and function. Neuron 60:201–214

Batassa EM, Costanzi M, Saraulli D, Scardigli R, Barbato C, Cogoni C, Cestari V (2010) RISC activity in hippocampus is essential for contextual memory. Neurosci Lett 471:185–188

Beckel-Mitchener A, Greenough WT (2004) Correlates across the structural, functional, and molecular phenotypes of fragile X syndrome. Ment Retard Dev Disabil Res Rev 10:53–59, Review

Bhattacharyya SN, Habermacher R, Martine U, Closs EI, Filipowicz W (2006) Relief of microRNA-mediated translational repression in human cells subjected to stress. Cell 125:1111–1112

Bingol B, Schuman EM (2005) Synaptic protein degradation by the ubiquitin proteasome system. Curr Opin Neurobiol 15:536–541

Bingol B, Schuman EM (2006) Activity dependent dynamics and sequestration of proteasomes in dendritic spines. Nature 441:1144–1148

Blaesse P, Airaksinen MS, Rivera C, Kaila K (2009) Cation-chloride cotransporters and neuronal function. Neuron 61:820–838

Bramham CR, Wells DG (2007) Dendritic mRNA: transport, translation and function. Nat Rev Neurosci 8:776–789

Chandrasekar V, Dreyer JL (2009) microRNAs miR-124, let-7d and miR-181a regulate cocaine-induced plasticity. Mol Cell Neurosci 42:350–362

Chandrasekar V, Dreyer JL (2011) Regulation of MiR-124, Let-7d, and MiR-181a in the accumbens affects the expression, extinction, and reinstatement of cocaine-induced conditioned place preference. Neuropsychopharmacology 36:1149–1164

Cheng HY, Papp JW, Varlamova O, Dziema H, Russell B, Curfman JP, Nakazawa T, Shimizu K, Okamura H, Impey S, Obrietan K (2007) MicroRNA modulation of circadian-clock period and entrainment. Neuron 54:813–829

Conaco C, Otto S, Han JJ, Mandel G (2006) Reciprocal actions of REST and a microRNA promote neuronal identity. Proc Natl Acad Sci USA 103:2422–2427

Cougot N, Bhattacharyya SN, Tapia-Arancibia L, Bordonné R, Filipowicz W, Bertrand E, Rage F (2008) Dendrites of mammalian neurons contain specialized P-body-like structures that respond to neuronal activation. J Neurosci 28:13793–13804

Dupret D, Revest JM, Koehl M, Ichas F, De Giorgi F, Costet P, Abrous DN, Piazza PV (2008) Spatial relational memory requires hippocampal adult neurogenesis. PLoS One 3(4):e1959

Edbauer D, Neilson JR, Foster KA, Wang CF, Seeburg DP, Batterton MN, Tada T, Dolan BM, Sharp PA, Sheng M (2010) Regulation of synaptic structure and function by FMRP-associated microRNAs miR-125b and miR-132. Neuron 65:373–384

Fiore R, Khudayberdiev S, Christensen M, Siegel G, Flavell SW, Kim TK, Greenberg ME, Schratt G (2009) Mef2-mediated transcription of the miR379–410 cluster regulates activity-dependent dendritogenesis by fine-tuning Pumilio2 protein levels. EMBO J 28:697–710

Gao J, Wang WY, Mao YW, Gräff J, Guan JS, Pan L, Mak G, Kim D, Su SC, Tsai LH (2010) A novel pathway regulates memory and plasticity via SIRT1 and miR-134. Nature 466:1105–1109

Ghildiyal M, Zamore PD (2009) Small silencing RNAs: an expanding universe. Nat Rev Genet 10 (2):94–108

Graham DL, Edwards S, Bachtell RK, DiLeone RJ, Rios M, Self DW (2007) Dynamic BDNF activity in nucleus accumbens with cocaine use increases self-administration and relapse. Nat Neurosci 10:1029–1037

Hansen KF, Sakamoto K, Wayman GA, Impey S, Obrietan K (2010) Transgenic miR132 alters neuronal spine density and impairs novel object recognition memory. PLoS One 5:e15497

Heresco-Levy U, Javitt DC (1998) The role of N-methyl-D-aspartate (NMDA) receptor-mediated neurotransmission in the pathophysiology and therapeutics of psychiatric syndromes. Eur Neuropsychopharmacol 8:141–152

Hotulainen P, Hoogenraad CC (2010) Actin in dendritic spines: connecting dynamics to function. J Cell Biol 189:619–629

Im HI, Hollander JA, Bali P, Kenny PJ (2010) MeCP2 controls BDNF expression and cocaine intake through homeostatic interactions with microRNA-212. Nat Neurosci 13:1120–1127

Impey S, Davare M, Lasiek A, Fortin D, Ando H, Varlamova O, Obrietan K, Soderling TR, Goodman RH, Wayman GA (2010) An activity-induced microRNA controls dendritic spine formation by regulating Rac1-PAK signaling. Mol Cell Neurosci 43:146–156

Kasai H, Fukuda M, Watanabe S, Hayashi-Takagi A, Noguchi J (2010) Structural dynamics of dendritic spines in memory and cognition. Trends Neurosci 33:121–129

Kiebler MA, Bassel GJ (2006) Neuronal RNA granules: movers and makers. Neuron 51:685–690

Klein ME, Lioy DT, Ma L, Impey S, Mandel G, Goodman RH (2007) Homeostatic regulation of MeCP2 expression by a CREB-induced microRNA. Nat Neurosci 10:1513–1514

Kocerha J, Faghihi MA, Lopez-Toledano MA, Huang J, Ramsey AJ, Caron MG, Sales N, Willoughby D, Elmen J, Hansen HF, Orum H, Kauppinen S, Kenny PJ, Wahlestedt C (2009) MicroRNA-219 modulates NMDA receptor-mediated neurobehavioral dysfunction. Proc Natl Acad Sci USA 106:3507–3512

Konopka W, Kiryk A, Novak M, Herwerth M, Parkitna JR, Wawrzyniak M, Kowarsch A, Michaluk P, Dzwonek J, Arnsperger T, Wilczynski G, Merkenschlager M, Theis FJ, Köhr G, Kaczmarek L, Schütz G (2010) MicroRNA loss enhances learning and memory in mice. J Neurosci 30:14835–14842

Krol J, Loedige I, Filipowicz W (2010) The widespread regulation of microRNA biogenesis, function and decay. Nat Rev Genet 11:597–610

Kurosc H (2003) Galpha12 and Galpha13 as key regulatory mediator in signal transduction. Life Sci 74:155–161

Lagos-Quintana M, Rauhut R, Yalcin A, Meyer J, Lendeckel W et al (2002) Identification of tissue-specific microRNAs from mouse. Curr Biol 12:735–739

Laneve P, Gioia U, Andriotto A, Moretti F, Bozzoni I, Caffarelli E (2010) A minicircuitry involving REST and CREB controls miR-9-2 expression during human neuronal differentiation. Nucleic Acids Res 38:6895–905

Lim LP, Lau NC, Garrett-Engele P, Grimson A, Schelter JM et al (2005) Microarray analysis shows that some microRNAs downregulate large numbers of target mRNAs. Nature 433:769–773

Liu J, Rivas FV, Wohlschlegel J, Yates JR 3rd, Parker R, Hannon GJ (2005) A role for the P-body component GW182 in microRNA function. Nat Cell Biol 7:1261–1266

Magill ST, Cambronne XA, Luikart BW, Lioy DT, Leighton BH, Westbrook GL, Mandel G, Goodman RH (2010) MicroRNA-132 regulates dendritic growth and arborization of newborn neurons in the adult hippocampus. Proc Natl Acad Sci USA 107:20382–20387

Majer A, Boese AS, Booth SA (2012) The role of microRNAs in neurodegenerative diseases: implications for early detection and treatment. In: Mallick B (eds) Regulatory RNAs: Springer-Verlag Berlin Heidelberg

Makeyev EV, Zhang J, Carrasco MA, Maniatis T (2007) The microRNA miR-124 promotes neuronal differentiation by triggering brain-specific alternative pre-mRNA splicing. Mol Cell 27:435–448

Manahan-Vaughan D, Kulla A, Frey JU (2000) Requirement of translation but not transcription for the maintenance of long-term depression in the CA1 region of freely moving rats. J Neurosci 20:8572–8576

Miska EA, Alvarez-Saavedra E, Townsend M, Yoshii A, Sestan N et al (2004) Microarray analysis of microRNA expression in the developing mammalian brain. Genome Biol 5(9):R68

Morris RG (2003) Long-term potentiation and memory. Philos Trans R Soc Lond B Biol Sci 358:643–647

Nagel R, Clijsters L, Agami R (2009) The miRNA-192/194 cluster regulates the Period gene family and the circadian clock. FEBS J 276:5447–5455

O'Carroll D, Mecklenbrauker I, Das PP, Santana A, Koenig U, Enright AJ, Miska EA, Tarakhovsky A (2007) A Slicer-independent role for Argonaute 2 in hematopoiesis and the microRNA pathway. Genes Dev 21:1999–2004

Orom UA, Nielsen FC, Lund AH (2008) MicroRNA-10a binds the 5'UTR of ribosomal protein mRNAs and enhances their translation. Mol Cell 30:460–471

Packer AN, Xing Y, Harper SQ, Jones L, Davidson BL (2008) The bifunctional microRNA miR-9/miR-9* regulates REST and CoREST and is downregulated in Huntington's disease. J Neurosci 28:14341–14346

Park CS, Tang SJ (2009) Regulation of microRNA expression by induction of bidirectional synaptic plasticity. J Mol Neurosci 38:50–56

Pfeiffer BE, Huber KM (2007) Fragile X mental retardation protein induces synapse loss through acute postsynaptic translational regulation. J Neurosci 27:3120–3130

Pillai RS, Bhattacharyya SN, Artus CG, Zoller T, Cougot N et al (2005) Inhibition of translational initiation by Let-7 MicroRNA in human cells. Science 309:1573–1576

Schaefer A, Im HI, Venø MT, Fowler CD, Min A, Intrator A, Kjems J, Kenny PJ, O'Carroll D, Greengard P (2010) Argonaute 2 in dopamine 2 receptor-expressing neurons regulates cocaine addiction. J Exp Med 207:1843–1851

Schratt GM, Tuebing F, Nigh EA, Kane CG, Sabatini ME et al (2006) A brain-specific microRNA regulates dendritic spine development. Nature 439:283–289

Shibata M, Kurokawa D, Nakao H, Ohmura T, Aizawa S (2008) MicroRNA-9 modulates Cajal-Retzius cell differentiation by suppressing Foxg1 expression in mouse medial pallium. J Neurosci 28:10415–10421

Shibata M, Nakao H, Kiyonari H, Abe T, Aizawa S (2011) MicroRNA-9 regulates neurogenesis in mouse telencephalon by targeting multiple transcription factors. J Neurosci 31:3407–3422

Siegel G, Obernosterer G, Fiore R, Oehmen M, Bicker S et al (2009) A functional screen implicates microRNA-138-dependent regulation of the depalmitoylation enzyme APT1 in dendritic spine morphogenesis. Nat Cell Biol 11:705–716

Smalheiser NR, Lugli G, Lenon AL, Davis J, Torvik V, Larson J (2010) Olfactory discrimination training up-regulates and reorganizes expression of microRNAs in adult mouse hippocampus. ASN Neuro 2:e00028

Smalheiser NR, Lugli G, Thimmapuram J, Cook EH, Larson J (2011) Endogenous siRNAs and noncoding RNA-derived small RNAs are expressed in adult mouse hippocampus and are upregulated in olfactory discrimination training. RNA 17:166–181

Smirnova L, Grafe A, Seiler A, Schumacher S, Nitsch R, Wulczyn FG (2005) Regulation of miRNA expression during neural cell specification. Eur J Neurosci 21:1469–1477

Song JJ, Smith SK, Hannon GJ, Joshua-Tor L (2004) Crystal structure of Argonaute and its implications for RISC slicer activity. Science 305:1434–1437

Takahashi JS, Hong HK, Ko CH, McDearmon EL (2008) The genetic of mammalian circadian order and disorder: implications for physiology and diseases. Nat Rev Genet 9:764–775

Uvarov P, Ludwig A, Markkanen M, Rivera C, Airaksinen MS (2006) Upregulation of the neuron-specific K+/Cl – cotransporter expression by transcription factor early growth response 4. J Neurosci 26:13463–13473

Van Aelst L, Cline HT (2004) Rho GTPases and activity-dependent dendrite development. Curr Opin Neurobiol 14:297–304

Vasudevan S, Tong Y, Steitz JA (2007) Switching from repression to activation: microRNAs can up-regulate translation. Science 318:1931–1934

Vessey JP, Schoderboeck L, Gingl E, Luzi E, Riefler J, Di Leva F, Karra D, Thomas S, Kiebler MA, Macchi P (2010) Mammalian Pumilio 2 regulates dendrite morphogenesis and synaptic function. Proc Natl Acad Sci USA 107:3222–3227

Visvanathan J, Lee S, Lee B, Lee JW, Lee SK (2007) The microRNA miR-124 antagonizes the anti-neural REST/SCP1 pathway during embryonic CNS development. Genes Dev 21:744–749

Vo N, Klein ME, Varlamova O, Keller DM, Yamamoto T et al (2005) A cAMP-response element binding protein-induced microRNA regulates neuronal morphogenesis. Proc Natl Acad Sci USA 102:16426–16431

Wayman GA, Davare M, Ando H, Fortin D, Varlamova O, Cheng HY, Marks D, Obrietan K, Soderling TR, Goodman RH, Impey S (2008) An activity-regulated microRNA controls dendritic plasticity by down-regulating p250GAP. Proc Natl Acad Sci USA 105:9093–9098

Wibrand K, Panja D, Tiron A, Ofte ML, Skaftnesmo KO et al (2010) Differential regulation of mature and precursor microRNA expression by NMDA and metabotropic glutamate receptor activation during LTP in the adult dentate gyrus in vivo. Eur J Neurosci 31:636–645

Wu J, Xie X (2006) Comparative sequence analysis reveals an intricate network among REST, CREB and miRNA in mediating neuronal gene expression. Genome Biol 7:R85

Yeo M, Berglund K, Augustine G, Liedtke W (2009) Novel repression of Kcc2 transcription by REST-RE-1 controls developmental switch in neuronal chloride. J Neurosci 29:14652–14662

Yoo AS, Staahl BT, Chen L, Crabtree GR (2009) MicroRNA-mediated switching of chromatin-remodelling complexes in neural development. Nature 460:642–646

Zeitelhofer M, Macchi P, Dahm R (2008) Perplexing bodies: the putative roles of P-bodies in neurons. RNA Biol 5:244–248

Chapter 8
Long Noncoding RNA Function and Expression in Cancer

Sally K Abd Ellatif, Tony Gutschner, and Sven Diederichs

Abstract In the last decades, medical research has mainly focused on the 2% of the human genome that serve as blueprint for proteins, assuming that the noncoding sequences were irrelevant and would neither contain significant information nor be of functional importance. However, 70% of the human genome are transcribed into RNA; therefore, the genome contains much more noncoding information than coding, which is present in the cell as noncoding RNA (ncRNA). Some of these ncRNAs are highly expressed, specifically regulated and evolutionarily conserved arguing in favor of their functional significance.

Long ncRNAs (lncRNAs) have been shown to regulate gene expression at various levels including chromatin modification, transcription, and posttranscriptional processing. Here, we review recent advances in ncRNA research for examples such as *XIST*, *H19*, *MALAT1*, *HOTAIR*, and *GAS5*. Many lncRNAs show differential expression patterns that correlate with diagnosis or prognosis in multiple tumor entities and can, thus, serve as an extensive source of new biomarkers. Moreover, these lncRNAs are functionally important and can provide novel insights into the mechanisms underlying tumor development and might serve as new targets in cancer therapy.

Keywords ANRIL • DLEU1 • *GAS5* • *H19* • *HOTAIR* • HULC • lincRNA • Cancer • lncRNA • *MALAT1* • ncRNA • Noncoding RNA • PRINCR1 • Tumor • *XIST*

S. Diederichs (✉)
Molecular RNA Biology & Cancer, German Cancer Research Center (DKFZ), Im Neuenheimer Feld 280 (B150), 69120 Heidelberg, Germany
e-mail: s.diederichs@dkfz.de

B. Mallick (eds.), *Regulatory RNAs*, DOI 10.1007/978-3-642-22517-8_8,
© Springer-Verlag Berlin Heidelberg (outside the USA) 2012

8.1 Introduction

Ribonucleic acids that do not code for proteins are collectively referred to as noncoding RNA (ncRNA). However, this does not imply that such RNAs do not contain information or have no function (Mattick and Makunin 2006). For many years, the central dogma of molecular biology was embraced by scientists that most RNAs mainly serve as messengers between the genetic information stored in the DNA and the functionally important proteins (with the notable exception of RNAs involved in protein biosynthesis itself like rRNA or tRNA). It was assumed that the noncoding sequences were not important, but recent evidence has clearly demonstrated that RNA can be more than a mere messenger. The noncoding parts – previously often perceived as "junk" – contain unrecognized jewels representing a high value for multiple processes in the cell (Habeck 2003; Mattick 2003; Berg 2006). Indeed, the flow of information can end at the RNA stage without proceeding to the protein level.

The complete human transcriptome, i.e., the entire set of all RNA molecules, which are written off the human genome, could be only described and characterized in recent years by two new techniques: ultra deep sequencing and tiling arrays. Both methods have led independently to the same result: a much larger part of the human genome is transcribed into RNA than previously assumed (Fig. 8.1). It is estimated that up to 70% of the sequence of the human genome is transcribed but only up to

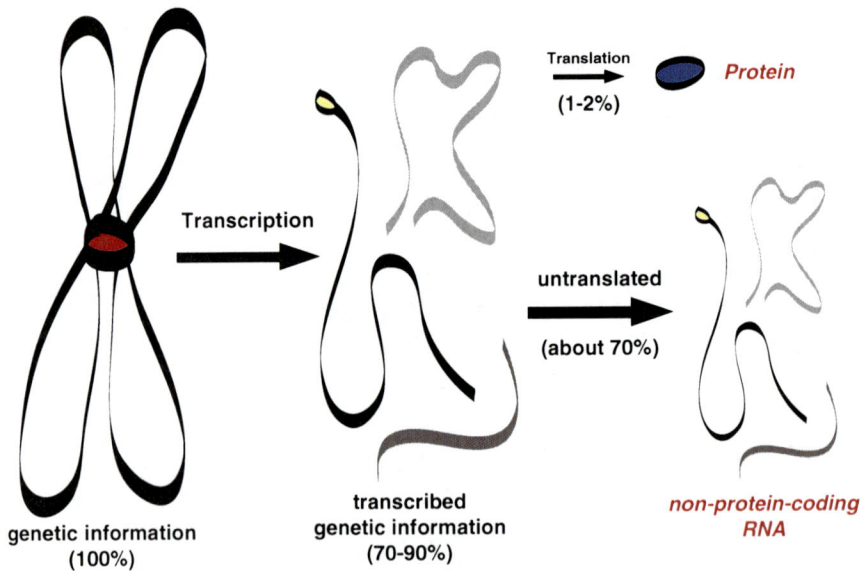

Fig. 8.1 The human genome – much more than protein-coding genes. The human genome contains protein-coding and nonprotein-coding genes. Nearly 70–90% of the genetic information encoded in about three billion base pairs of DNA is transcribed into different classes of RNA, e.g., rRNA, tRNA, mRNA, small, and long ncRNA. However, only a minor subset of the genetic information is converted into proteins (1–2%). The large majority of genomic information is not a template for translation and might function at the RNA level

2% of the human genome two percent serve as blueprints for proteins (Bertone et al. 2004; Carninci et al. 2005; Cheng et al. 2005; Sultan et al. 2008; Diederichs 2010).

Until recently, most of the known ncRNAs were classical "housekeeping" ncRNAs, such as transfer RNAs (tRNAs), ribosomal RNAs (rRNAs), small nuclear RNAs (snRNAs), and small nucleolar RNAs (snoRNAs), which are constitutively expressed and play critical roles in protein biosynthesis (Mattick and Makunin 2006; Chen and Carmichael 2010a; Chen and Carmichael 2010b). In addition to these well-characterized members, new classes have recently enriched the ncRNA world, some of which are very short: microRNA (miRNA) (Winter et al. 2009), small interfering RNA (siRNA), or PIWI-interacting RNA (piRNA); while others are long and termed long ncRNA (lncRNA) or long intergenic ncRNA (lincRNA) (Taft et al. 2010). The short ncRNAs are discussed elsewhere (see Chap. 1) and need to be distinguished from the long ncRNAs, which are the main focus of this chapter.

Although short ncRNAs received most attention in recent years, it has become increasingly clear that mammalian genomes encode numerous *long ncRNAs*. They were first described via the large-scale sequencing of full-length cDNA libraries in the mouse (Okazaki et al. 2002; Mercer et al. 2009). Other names such as large RNA, macroRNA, and long intergenic ncRNA or lincRNA are also used to refer to lncRNA. They are defined as endogenous cellular RNAs of more than 200 nucleotides in length that lack any significant positive-strand open reading frame and are distinct from any known functional RNA classes (including but not limited to ribosomal, transfer, and small nuclear or nucleolar RNAs) (Ponting et al. 2009; Chen and Carmichael 2010b; Lipovich et al. 2010). Thus, this group of ncRNAs is defined by size and lack of protein-coding potential, but does not constitute a homogeneous class of functionally related molecules.

Long ncRNAs often overlap with or are interspersed between coding and noncoding transcripts making their classification a complex task (Carninci et al. 2005; Kapranov et al. 2005; Mercer et al. 2009). For the identification of lncRNA in animals and plants using experimental and computational approaches, see Chap. 13. A preliminary classification places a lncRNA into one or more of five broad categories: (1) sense or (2) antisense, when overlapping one or more exons of another transcript on the same or opposite strand, respectively; (3) bidirectional, when the expression of it and a neighboring coding transcript on the opposite strand is initiated in close genomic proximity; (4) intronic, when it is derived wholly from within an intron of a second transcript (although these sometimes may represent pre-mRNA sequences); or (5) intergenic, when it lies as an independent unit within the genomic interval between two genes (Ponting et al. 2009).

Some lncRNAs show clear evolutionary conservation or strict regulation, implying that they are of functional importance (Huarte et al. 2010). Some transcripts are derived from ultraconserved genomic regions (UCR) (Bejerano et al. 2004). These UCR can be altered in human cancer (Calin et al. 2007). For further information on the transcription and evolution of lncRNAs, see Chap. 6 of this volume.

A recent study identified 5,446 lncRNA genes in the human genome and combined them with lncRNAs from four published sources to derive 6,736 lncRNA genes. The study also examined protein-coding capacity of known genes overlapping with lncRNAs and revealed that 62% of known genes with "hypothetical protein" names lacked protein-coding capacity. This means that the human lncRNA catalog is much larger than previously expected (Jia et al. 2010). While lncRNAs are pervasively transcribed in the genome, their potential involvement in human disease is not yet understood (Gupta et al. 2010).

Several lncRNAs can regulate gene expression at various levels, including chromatin modification, transcription, and posttranscriptional processing (Mercer et al. 2009; Wilusz et al. 2009) (Fig. 8.2). Some lncRNAs, such as the X inactive-specific transcript (*XIST*) or *HOTAIR*, repress the expression of their target genes by interacting with *chromatin remodeling* complexes, which induce the formation of heterochromatin (Rinn et al. 2007; Zhao et al. 2008). Other lncRNAs function by regulating *transcription* through a variety of mechanisms that include interacting with RNA-binding proteins, acting as a coactivator of transcription factors, or repressing a major promoter of their target gene (Feng et al. 2006; Martianov et al. 2007; Wang et al. 2008b).

In addition to chromatin modification and transcriptional regulation, lncRNAs can regulate gene expression at the *posttranscriptional level*. On the one hand, a long antisense transcript can silence a gene posttranscriptionally by annealing to the corresponding sense mRNA transcript forming a duplex that can be cleaved by Dicer into endogenous siRNAs (Ogawa et al. 2008). On the other hand, the lncRNA can induce an alternative splicing pattern of the target gene. For example, the 5′ splice site of the zinc finger HOX mRNA *ZEB2* can be masked by a complementary antisense ncRNA, which prevents the removal of the intron by the spliceosome and leads to the retention of the intronic internal ribosome entry site (IRES) in the mature mRNA. This results in an enhanced translation of the *ZEB2* mRNA (Beltran et al. 2008). The different mechanisms by which lncRNAs perform their functions will be discussed in further detail in the following section. An overview of differentially expressed lncRNAs in cancer is provided in Table 8.1.

8.2 Long Noncoding RNAs in Cancer

8.2.1 XIST: *From X Chromosome Inactivation to Cancer*

XIST is one of the few well-characterized lncRNAs. The *XIST* gene is located on human chromosome Xq13.2 and produces a 17-kb-long ncRNA, best known for its role in dosage compensation/equilibration between the male XY and the female XX gonosomes. This genetic regulatory mechanism involves the silencing (inactivation) of one of the two X chromosomes in female somatic cells to adjust the expression levels of genes present on the X chromosome, so that they are equal to the levels in male cells (Agrelo and Wutz 2010; Lipovich et al. 2010).

XIST was discovered as a transcript associated with the inactive X chromosome (Xi) but not with the active X chromosome (Xa) in human and mouse (Borsani

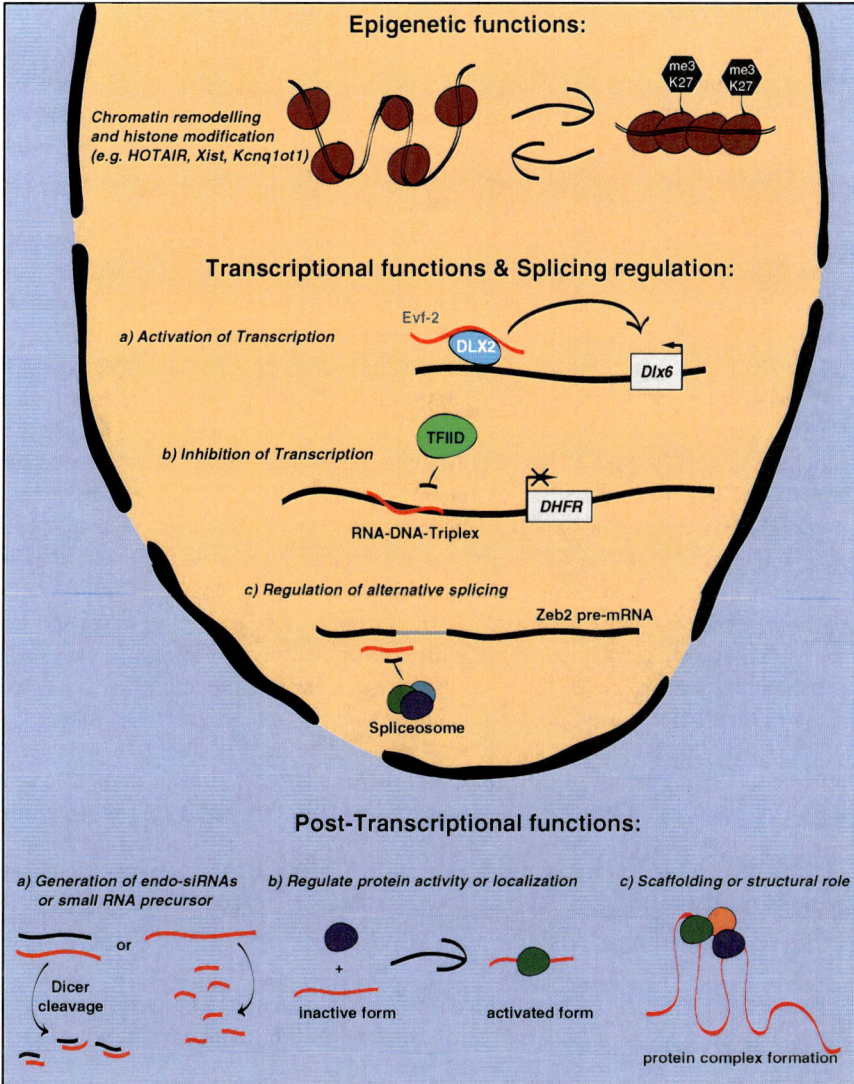

Fig. 8.2 Long noncoding RNA – regulators of gene expression and beyond. Recent studies on long ncRNA function have uncovered a broad range of mechanisms for gene expression control that involve and require these transcripts. At the epigenetic level of expression control, ncRNAs can regulate chromatin remodeling as well as histone modifications. The ncRNA *HOTAIR* recruits the Polycomb repressive complex to the HoxD locus where the lysine 27 residues of histone H3 get trimethylated (H3K27me3). This leads to the formation of heterochromatin and the repression of gene expression. Also, ncRNAs were found to regulate gene expression at the transcriptional level. They can either activate the expression of another gene, e.g., the ncRNA *Evf-2*, which binds to the transcription factor DLX2 and functions as a coactivator for the transcription of the *DLX6* gene. However, long ncRNAs can also inhibit transcription. For the *DHFR* gene, an ncRNA is transcribed from the minor promoter of the *DHFR* gene and forms a triplex structure at the major

et al. 1991; Brockdorff et al. 1991; Brown et al. 1991). *XIST* is transcribed from the X inactivation center (Xic) locus on the X chromosome to be inactivated. It spreads along this chromosome and silences gene expression by epigenetic mechanisms. In detail, *XIST* recruits the chromatin remodeling Polycomb complex that trimethylates the lysine-27 residues of histone H3 leading to heterochromatin formation (Mercer et al. 2009; Agrelo and Wutz 2010).

TSIX, the antisense repressor of *XIST*, is transcribed as a 40-kb-long ncRNA located 15 kb downstream of *XIST* (Lee et al. 1999). *TSIX* is downregulated on the X chromosome destined to be inactivated, allowing the upregulation of *XIST* and its spread along this chromosome. On the other hand, *TSIX* prevents *XIST* upregulation on the active X chromosome (Zhao et al. 2008).

XIST has been implicated in sex and nonsex related cancers (Agrelo and Wutz 2010; Weakley et al. 2011). *XIST* expression is lost in female ovarian, breast, and cervical cancer cell lines (Kawakami et al. 2004b; Benoit et al. 2007). In some epithelial ovarian cancer cell lines, *XIST* expression is undetectable in contrast to normal epithelial ovarian cells (Benoit et al. 2007). When RNA expression is compared between primary and recurrent ovarian tumors from the same patient, *XIST* is the most strongly differentially expressed gene downregulated in the recurrent tumor. *XIST* expression also correlates with the response to chemotherapy and Taxol sensitivity of gynecological cancer cell lines (Huang et al. 2002). Its expression strongly associates with the disease-free survival of Taxol-treated cancer patients (Huang et al. 2002). The loss of *XIST* expression could be due to the loss of Xi that is frequently reported in female tumors and can serve as a marker in certain female tumors (Kawakami et al. 2004b; Agrelo and Wutz 2010).

The relationship between the breast cancer gene *BRCA1*, a well-known tumor suppressor gene, and *XIST* has been controversial: BRCA1 could contribute to the regulation of *XIST* localization on Xi and X chromosome inactivation in somatic cells (Ganesan et al. 2002; Silver et al. 2007). Additionally, BRCA1-deficient breast cancer cell lines have increased *XIST* expression, which is partly attributed to *XIST* expression from Xa. Moreover, *BRCA1*-associated hereditary cancers have significantly higher levels of *XIST* in comparison to sporadic basal-like cancers, suggesting that *XIST* could be used as a marker to distinguish between these two types of tumors (Sirchia et al. 2009). However, some studies did not find a

---

Fig. 8.2 (continued) promoter of this gene. This blocks the binding of TFIID, a general transcription factor, resulting in *DHFR* gene silencing. Long ncRNAs can also regulate alternative splicing of pre-mRNAs. The access of the spliceosome to the 5′ splice site of the *ZEB2* pre-mRNA is hindered by an antisense ncRNA. This leads to intron retention and the formation of an alternative *ZEB2* transcript, which gets translated efficiently due to the presence of an internal ribosome entry site (IRES) in the retained intron. Finally, long ncRNAs function also at the posttranscriptional level. Hybridization of ncRNAs to complementary antisense transcripts leads to RNA-RNA duplex formation. This structure is recognized by Dicer and gets subsequently processed into small endo-siRNAs. Alternatively, long ncRNAs can be precursors for small, single-stranded RNAs like miRNAs or piRNAs. Furthermore, ncRNAs can bind to proteins and regulate their activity, influence their cellular localization, or help larger protein complexes to form by presenting a scaffold or docking platform for the proteins

Table 8.1 LncRNAs differentially expressed in cancer (Recent studies identified long non-coding RNAs as deregulated genes in human cancers. This overview table summarizes specific expression patterns and functional implications of a subset of these cancer-related lncRNAs.)

lncRNA	Size	Characteristics and relations to other genes	Link to cancer	References
XIST	17 kb	Covers Xi and silences gene expression via PRC2 (dosage compensation); TSIX, the antisense repressor of XIST, is transcribed as a 40-kb-long ncRNA located 15 kb downstream of XIST	*Female tumors:* Marker in ovarian and breast cancers *Male tumors:* Hypomethylated XIST promoter can be used as a diagnostic marker in TGCTs and prostate cancer *Nonsex-specific cancers:* Leukemia, gastric, colorectal, intestine, and kidney tumors	Huang et al. (2002), Zhang et al. (2005), Song et al. (2007), Sirchia et al. (2009), Agrelo and Wutz (2010), Weakley et al. (2011)
H19	2.5 kb	Imprinted at the Igf2 locus; contains a microRNA (miR-675) in Exon 1	Universal tumor marker deregulated in TGCTs, HCC, bladder, breast, uterine, ovarian, lung, esophageal, and colorectal cancers	Kondo et al. (1995), Hibi et al. (1996), Verkerk et al. (1997), Adriaenssens et al. (1998), Ariel et al. (1998), Tanos et al. (1999), Ariel et al. (2000), Lottin et al. (2002), Lottin et al. 2005)
			Beckwith-Wiedemann syndrome: biallelic silencing with increased susceptibility to Wilms' tumors	Rump et al. (2005), Riccio et al. (2009)
MALAT1/ NEAT2	8 kb	Intergenic transcript; can be processed into a short tRNA-like ncRNA: mascRNA	Prognostic marker in NSCLC and deregulated in uterine and cervical cancers and HCC	Ji et al. (2003), Yamada et al. (2006), Lin et al. (2007), Guo et al. (2010)
HOTAIR	2.2 kb	Transcribed from a position intergenic and antisense to the HOXC genes	Increased expression in breast cancer and HCC and plays a role in cancer invasiveness and metastasis	Gupta et al. (2010), Yang et al. (2011)
HULC	484 nt	Intergenic two exon transcript Exon 1 mainly consists of a long terminal repeat (MLT1A)	Strongly upregulated gene specifically in HCC	Panzitt et al. (2007)
GAS5	Multiple isoforms	Encodes ten snoRNAs in its introns	Downregulated in breast cancer	Mourtada-Maarabouni et al. (2009)
ANRIL	3,834 nt and splice variants	Antisense noncoding RNA in the INK4 locus	Elevated levels in prostate cancer	Pasmant et al. (2007), Pasmant et al. (2010), Yap et al. (2010)

(continued)

Table 8.1 (continued)

lncRNA	Size	Characteristics and relations to other genes	Link to cancer	References
DLEU2	1.0–1.8 kb	pri-miRNA gives rise to the mature microRNAs miR-16-1 and miR-15	Downregulated in CLL	Migliazza et al. (2001), Mertens et al. (2009), Klein et al. (2010)
PRNCR1	13 kb	Intron-less long noncoding RNA	Upregulated in prostate cancer	(Chung et al. 2011)
TERRA	Heterogeneous sizes (between 100 nt and 9 kb)	Transcribed from several subtelomeric loci toward chromosome ends	Inhibitor of telomerase	Azzalin et al. (2007), Schoeftner and Blasco (2008), Redon et al. (2010)
DD3/PCA3	0.6, 2, 4 kb	Transcribed antisense to BMCC1	Highly specific marker of prostate cancer	Bussemakers et al. (1999), Clarke et al. (2009)
SPRY4-IT1	687 nt	Derived from an intron of the SPRY4 gene	Elevated expression in melanoma	Khaitan et al. (2011)
PCGEM1	1,643 nt	No significant homology to any previously defined gene	Exclusively expressed in prostate tissue and overexpressed in prostate cancer	Srikantan et al. (2000), Ifere and Ananaba (2009b)
MEG3	1.6 kb and splicing isoforms	Maternally expressed from the Dlk1-Gtl2 imprinted locus	Stimulates p53-mediated transactivation and suppresses tumor cell growth, downregulated in brain cancers	Zhang et al. (2003), Zhang et al. (2010)
Zfas1	0.5 kb	Antisense to the 5′-end of the protein-coding gene Znfx1 and hosts three C/D box snoRNAs	Putative tumor suppressor gene, downregulated in breast tumors	Askarian-Amiri et al. (2011)
TUC338	590 nt	Encodes the ultraconserved element 338 which is partially located within the poly(rC)-binding protein 2 (PCBP2) gene	Increased in human HCC	Braconi et al. (2011)
ncRAN	2,186 nt, 2,087 nt	–	Higher expression in neuroblastoma and bladder cancer	Yu et al. (2009), Zhu et al. (2011)
PTENP1	3.9 kb	Pseudogene of PTEN tumor suppressor gene	Reduces the translational repression of the tumor suppressor PTEN	Alimonti et al. (2010), Poliseno et al. (2010)
UCA1	Multiple isoforms 1.4 kb, 2.2 kb, and 2.7 kb	The first two exons mainly consist of a possibly nested LTR element of the ERV1 family (LTR7Y and HERVH)	Sensitive and specific marker for bladder cancer	Wang et al. (2006), Wang et al. (2008a)

correlation between BRCA1 and *XIST* expression or localization (Pageau et al. 2007; Vincent-Salomon et al. 2007). Thus, further analyses are needed to clearly define the role of *XIST* in tumorigenesis of gynecological cancers.

Since male somatic cells have only one X chromosome, they weakly express *XIST*. In contrast, mature testes show strong expression levels of *XIST*. However, absolute levels are much lower than in female somatic cells (Richler et al. 1992). Nevertheless, *XIST* plays a role in male cancers such as testicular germ cell tumors (TGCTs) and prostate cancer. TGCTs, among the most common cancers in young men, show *XIST* expression especially in seminomas (Looijenga et al. 1997; Kawakami et al. 2003).

Normally, the promoter region of *XIST* on the active X chromosome in female and male somatic cells is methylated leading to gene silencing. In contrast, this region is hypomethylated and reactivated in some TGCT patients (mainly seminomas). Therefore, the methylation status of the *XIST* promoter may be used as a diagnostic marker for TGCTs (Kawakami et al. 2004a; Zhang et al. 2005; Lind et al. 2007). Similarly, prostate cancer cells and DNA fragments from the serum of prostate cancer patients show hypomethylated *XIST* promoter regions. This might be used for diagnosis and the identification of the more aggressive cases (Laner et al. 2005; Song et al. 2007).

In addition to sex-specific cancers, *XIST* is potentially involved in various types of gender-independent cancers. In leukemia, *XIST* gene deletion from the Xi is observed. However, this does not lead to reactivation of the Xi (Rack et al. 1994). Gastric fundus cells from normal male mice do not express *XIST*, whereas preneoplastic fundus cells infected with *Helicobacter felis* gain *XIST* expression (Nomura et al. 2004). In addition, microsatellite unstable sporadic colorectal cancers harbor *XIST* gene amplifications suggesting a putative functional role of *XIST* in carcinogenesis (Lassmann et al. 2007). Copy number gains of the *XIST* gene have also been documented in cell lines of collecting duct carcinoma of the kidney (Wu et al. 2009). Moreover, mice that receive oral Benzo[a]pyrene, a polycyclic aromatic hydrocarbon and a common combustion product, develop proximal small intestine tumors with remarkable upregulation of *XIST* expression (Shi et al. 2010).

In summary, *XIST* has proven to be more than a key player in X chromosome inactivation. Different types of cancers show *XIST* deregulation, suggesting its potential value as a diagnostic marker. However, further studies are needed to explore its role and potential functional importance in these cancers.

8.2.2 H19: *Beyond Genetic Imprinting*

H19 is another well-characterized lncRNA. The *H19* gene is located on human chromosome 11p15.5 and gives rise to a 2.5-kb-long, spliced, and polyadenylated cytoplasmic RNA (Bartolomei et al. 1991). *H19* was first identified while screening for genes that are upregulated by α-fetoprotein in murine fetal liver and was identified as the 19th clone in row H, which gave rise to its current name (Pachnis et al. 1984). It was also found in a screen for genes involved in myogenic differentiation (Davis et al. 1987). It is upregulated during embryonic stem cell differentiation, as well (Poirier et al. 1991).

H19 is exclusively expressed from the maternal chromosome. This phenomenon is known as genetic imprinting: imprinted genes are expressed only from either the maternal or the paternal allele. While *H19* is only expressed from the maternal allele, its imprinted neighbor, insulin-like growth factor 2 (*IGF2*), is expressed solely from the paternal allele, and their locus is collectively referred to as the *H19/IGF2* locus (DeChiara et al. 1991; Gabory et al. 2010).

H19 is suggested to play a role in the regulation of *IGF2* at the transcriptional and posttranscriptional level. Maternal deletion of *H19* changes the methylation profiles of *IGF2* on both alleles, which points toward a transcriptional effect of *H19* (Forne et al. 1997). The *IGF2* mRNA-binding protein 1 (IMP1) binds to the $5'$ untranslated region of the *IGF2* mRNA and maintains its stability. *H19* RNA can compete with *IGF2* mRNA for binding to IMP1, interfering with the posttranscriptional regulation of *IGF2* (Runge et al. 2000; Hansen et al. 2004).

In addition, the first exon of *H19* produces a highly conserved microRNA (miR-675) (Cai and Cullen 2007). However, the role of miR-675 remains to be elucidated.

H19 is deregulated in a variety of cancers, and several studies suggest that it can serve as an oncogenic marker in humans. *H19* is increased in TGCTs due to biallelic expression. Moreover, its expression levels depend on the differentiation lineage and maturation stage (Verkerk et al. 1997). *H19* can also be used as a prognostic marker for the early recurrence of bladder cancer: higher *H19* expression is correlated with a shorter disease-free period. Moreover, the number of cells expressing *H19* decreases as tumor grade increases, i.e., cells become less differentiated (Ariel et al. 2000).

In breast adenocarcinomas, *H19* expression is increased as compared to healthy tissues. This upregulation of *H19* significantly correlates with tumor size and estrogen or progesterone receptor expression (Adriaenssens et al. 1998). In addition, *H19* overexpression enhances the tumorigenic properties of breast cancer cells (Lottin et al. 2002). Increased expression of *H19* is also found in hepatocellular carcinoma (HCC), uterine tissue myometrium carcinoma, epithelial ovarian cancer, lung cancer, esophageal carcinoma, and colorectal carcinoma recommending it as universal tumor marker (Kondo et al. 1995; Hibi et al. 1996; Ariel et al. 1998; Tanos et al. 1999; Lottin et al. 2005).

Knockdown studies have been performed to elucidate the function of *H19*. Loss of *H19* in the HCC cell line Hep3B reduces tumor growth after implantation (Matouk et al. 2007). Similar results are found in bladder cancer, suggesting that *H19* acts as an oncogene (Matouk et al. 2007).

In contrast, other studies propose a tumor suppressive role for *H19*. Children affected with the Beckwith-Wiedemann syndrome show biallelic silencing of *H19* and are more susceptible to developing Wilms' tumors (Rump et al. 2005; Riccio et al. 2009). However, these cases also show a biallelic activation of *IGF2* which might be also involved in tumorigenesis. Ectopic *H19* expression causes growth retardation in embryonic tumor cell lines (Hao et al. 1993). In a murine model for colorectal cancer, double mutant mice lacking *H19* and *Apc* show an increased number of polyps compared to *Apc* mutant mice suggesting that *H19* might control the initiation step of tumorigenesis (Colnot et al. 2004). In an HCC mouse model, loss of *H19* causes the tumors to appear earlier (Yoshimizu et al. 2008). Therefore, *H19* may delay the onset of tumor appearance and could act as tumor suppressor.

How can these contrary roles in tumorigenesis be explained? It's oncogenic or tumor suppressive role might depend on the cell type and conditions under investigation. The cellular role of *H19* could also differ at different times in life in embryonic versus adult cells or between human and mouse. Further studies are needed to shed light onto the role of *H19* in cancer and to elucidate whether this role is mediated through the full-length RNA molecule or through the microRNA (Gabory et al. 2010).

8.2.3 MALAT1: *From a Metastasis Marker to a Player in Metastasis?*

Metastasis-associated lung adenocarcinoma transcript 1 (*MALAT1*), also known as *nuclear-enriched abundant transcript 2* (NEAT2), is an ncRNA of about 8 kb expressed from human chromosome 11q13. It was first discovered as a prognostic marker for patient survival and metastasis in non-small-cell lung cancer (NSCLC) (Ji et al. 2003).

MALAT1 is highly conserved across several species indicating its potentially important function. The *MALAT1* transcript is specifically retained in the nucleus and localizes to nuclear speckles, but it is not required to maintain these structures (Hutchinson et al. 2007; Clemson et al. 2009). Nuclear speckles play a role in pre-mRNA processing, and recently, *MALAT1* has been shown to regulate alternative splicing of pre-mRNA by modulating the levels of active serine/arginine (SR) splicing factors. These factors regulate tissue- or cell-type specific alternative splicing in a concentration- and phosphorylation-dependent manner (Zhao et al. 2009; Tripathi et al. 2010). Knockdown of *MALAT1* alters the processing of a subset of pre-mRNAs, which play important roles in cancer biology, e.g., tissue factor or endoglin. This supports the hypothesis that *MALAT1* is a regulator of posttranscriptional RNA processing or modification (Lin et al. 2011).

MALAT1 is expressed in many healthy organs with the highest levels of expression in pancreas and lung (Ji et al. 2003). In several human cancers including lung cancer, uterine endometrial stromal sarcoma, cervical cancer, and HCC, it is deregulated (Ji et al. 2003; Yamada et al. 2006; Lin et al. 2007; Guo et al. 2010).

MALAT1 is significantly associated with metastasis in NSCLC patients. This association with metastasis is stage and histology specific. Therefore, *MALAT1* can serve as an independent prognostic parameter for patient survival in early stage lung adenocarcinoma (Ji et al. 2003). A recent study sheds light onto the role of *MALAT1* in the metastasis process. It suggests that *MALAT1* promotes cell motility of lung cancer cells through transcriptional or posttranscriptional regulation of motility-related genes (Tano et al. 2010). In cervical cancer, depletion of *MALAT1* suppresses proliferation and invasion of CaSki cells. The knockdown of *MALAT1* also leads to an upregulation of caspase-8 and -3 and Bax and downregulation of Bcl-2 and Bcl-xL in these cervical cancer cells, which suggests that *MALAT1* may regulate these genes (Guo et al. 2010). However, since both studies were carried out with individual siRNAs only and lack rescue experiments, further investigations are necessary to exclude unspecific off-target effects and to corroborate the functional importance of *MALAT1* in carcinogenesis or metastasis.

In addition to its potential involvement in lung and cervical cancer, *MALAT1* is upregulated in endometrial stromal sarcoma of the uterus, which is a rare uterine malignancy. Overexpression of *MALAT1* is thought to be a characteristic of endometrial stromal sarcoma and may be correlated with relapse or metastasis (Yamada et al. 2006). Moreover, *MALAT1* is overexpressed in HCC and in other types of cancer including breast and pancreas carcinoma suggesting that it plays a significant role in neoplasia and can be potentially used as a universal marker for carcinomas (Lin et al. 2007).

The *MALAT1* transcript can also be processed into a highly conserved tRNA-like small cytoplasmic RNA of 61 nucleotides that is broadly expressed in human tissues (Wilusz et al. 2008). The function of this so-called mascRNA is unknown.

8.2.4   HOTAIR: *A Chromatin Regulator Goes a Long Way "in trans"*

HOTAIR (HOX antisense intergenic RNA) was discovered while studying the transcriptional landscape of the human *HOX* loci. *HOTAIR* is a 2.2-kb-long ncRNA, transcribed in the antisense direction from the *HOXC* gene cluster. It is the first lncRNA that functions *in trans* repressing transcription across 40 kb along the *HOXD* locus by interacting with the Polycomb repressive complex 2 (PRC2). This interaction leads to the trimethylation of histone H3 lysine-27 (H3K27) at the HOXD locus and consequently its transcriptional silencing (Rinn et al. 2007). In detail, *HOTAIR* executes its function by providing a scaffold for histone modification complexes. Its 5′-domain binds PRC2 while its 3′-domain binds to the LSD1 complex. *HOTAIR* can coordinate the targeting of both complexes to chromatin for a coupled histone H3K27 methylation and H3K4 demethylation, enforcing gene silencing (Tsai et al. 2010).

Given its important role in the epigenetic regulation of gene expression, it is not surprising that *HOTAIR* is deregulated in different types of cancer (Gupta et al. 2010; Yang et al. 2011). *HOTAIR* expression is increased in primary breast tumors and metastases. Its expression level in primary tumors positively correlates with metastasis and poor outcome. Overexpression of *HOTAIR* in epithelial cancer cells targets PRC2 to alter H3K27 methylation and gene expression. This leads to

increased cancer invasiveness and metastasis. On the other hand, *HOTAIR* depletion inhibits cancer invasiveness (Gupta et al. 2010).

HOTAIR expression in hepatocellular carcinoma (HCC) is elevated compared to
noncancerous tissues. In HCC patients, who received a liver transplantation, high
HOTAIR expression levels are an independent prognostic marker for HCC recurrence and shorter survival. *HOTAIR* suppression by siRNAs reduces cell invasion
and cell viability in a liver cancer cell line. In addition, this suppression sensitizes
cancer cells to TNFα-induced apoptosis and increases their sensitivity to the che-
motherapeutic agents cisplatin and doxorubicin (Yang et al. 2011). These studies
suggest that *HOTAIR* may be an important target for cancer diagnosis and therapy.

8.2.5   HULC: *A ncRNA with Strong Regulation in HCC*

Highly upregulated in liver cancer (*HULC*) is an lncRNA named for being the most
strongly upregulated gene in HCC and was first identified by screening an HCC-
specific gene library for deregulated genes. The *HULC* gene locus on chromosome
6p24.3 produces a spliced, 484-nt-long, polyadenylated ncRNA, which localizes to
the cytoplasm. *HULC* might be associated with ribosomes, which could point
toward a potential role in translation or translational regulation (Panzitt et al. 2007).

siRNA-mediated knockdown of *HULC* in HCC cell lines alters the expression of
several genes, some of which are known to be affected in HCC. Therefore, *HULC*
may also have a role in posttranscriptional modulation of gene expression. Since
HULC is specifically upregulated in HCC and can be detected in HCC patients, it
can be used as a tumor marker (Panzitt et al. 2007). *HULC* is also upregulated in
hepatic metastases of colorectal carcinoma. Moreover, it is upregulated in HCC cell
lines that produce hepatitis B virus (HBV) compared to the same parental lines that
do not produce HBV (Matouk et al. 2009). So far, a detailed mechanistic analysis of
the molecular function of *HULC* is lacking but will be necessary to gain deeper
insight into hepatocellular carcinogenesis.

8.2.6   GAS5: *A Host Gene for Small snoRNAs with a Big Impact*

The gene *growth arrest-specific 5* (*GAS5*) was originally identified based on its
increased levels in growth-arrested mouse NIH3T3 fibroblasts (Schneider et al.
1988). In actively growing Friend leukemia or NIH3T3 cells, GAS5 is expressed at
low levels while it increases after density-induced cell cycle arrest (Coccia et al.
1992). *GAS5* RNA levels appear to be regulated by stabilization rather than
transcription (Coccia et al. 1992; Smith and Steitz 1998). Nutrient deprivation in
F9 embryonal carcinoma cells also induces GAS5 expression (Fleming et al. 1998;
Fontanier-Razzaq et al. 2002).

Human *GAS5* is transcribed from chromosome 1q25.1 and is alternatively
spliced. Its exons contain a small and poorly conserved open reading frame

that does not encode a functional protein (Muller et al. 1998; Raho et al. 2000). *GAS5* introns encode multiple small nucleolar RNAs (snoRNAs), which may mediate important biological activities (Smith and Steitz 1998). *GAS5* functions as a "riborepressor": the ncRNA interacts with the DNA-binding domain of the glucocorticoid receptors, thus competing with the glucocorticoid response elements in the genome for binding to these receptors. This suppresses gluco-corticoid-mediated induction of several responsive genes including *cellular inhibitor of apoptosis 2 (cIAP2)* ultimately sensitizing cells to apoptosis (Kino et al. 2010).

The link to apoptosis may explain why *GAS5* is necessary for normal growth arrest in leukemic human T-cell lines as well as human peripheral blood T cells. Overexpression of *GAS5* in these cells causes increased apoptosis and decelerated cell cycle, while downregulation of endogenous *GAS5* inhibits apoptosis and maintains a more rapid cell cycle progression (Mourtada-Maarabouni et al. 2008).

In breast cancer, *GAS5* transcript levels are significantly reduced compared to normal breast epithelial tissues. Moreover, GAS5 expression induces growth arrest and apoptosis independently of other stimuli in some prostate and breast cancer cell lines (Mourtada-Maarabouni et al. 2009). Since effective control of cell survival and proliferation is critical for the prevention of oncogenesis, *GAS5* could play a role in the development and therapy of cancer.

Another link between cancer and *GAS5* is the serine/threonine protein kinase mTOR (mammalian target of rapamycin). mTOR plays a critical role in the control of mammalian cell growth and regulates both cellular protein synthesis and cell proliferation (Fingar et al. 2004; Hay and Sonenberg 2004). The mTOR antagonist rapamycin decreases cell proliferation and is used in cancer therapy of leukemia and other malignancies (Abdel-Karim and Giles 2008; Jiang and Liu 2008). Downregulation of *GAS5* by RNA interference protects both leukemic and primary human T cells from the proliferative inhibition by mTOR antagonists, suggesting that *GAS5* might – directly or indirectly – be required for this inhibitory effect (Mourtada-Maarabouni et al. 2010). Finally, genetic aberrations of the *GAS5* locus have been found in many types of tumors including melanoma, breast, and prostate cancers (Smedley et al. 2000; Nupponen and Carpten 2001; Stange et al. 2006; Morrison et al. 2007; Nakamura et al. 2008). However, their functional significance needs to be established.

8.2.7   ANRIL: *An Oncogene Sharing a Locus with Three Tumor Suppressors*

Antisense noncoding RNA in the INK4 locus (*ANRIL*) was first identified in a study to determine the precise size and end points of a large germline deletion removing the entire *INK4A-ARF-INK4B* gene cluster in a melanoma/neural system tumor syndrome family. *ANRIL* is a 3,834-nt lncRNA with its first exon located in the promoter of the *ARF* gene and overlapping the two exons of *INK4B* in antisense orientation (Pasmant et al. 2007).

This gene cluster codes for three distinct tumor suppressor proteins: p16^{INK4A}, p14^{ARF}, and p15^{INK4B} (Pasmant et al. 2010). Genome-wide association studies identified *ANRIL* as a risk locus for several cancers including breast cancer, nasopharyngeal carcinoma, basal cell carcinoma, and glioma (Shete et al. 2009; Stacey et al. 2009; Bei et al. 2010; Turnbull et al. 2010). *ANRIL* may regulate the *p16^{INK4A}*, *p14^{ARF}*, and *p15^{INK4B}* locus via Polycomb complex-mediated epigenetic silencing and therefore may play a role in cancer. *ANRIL* binds to chromobox7 (CBX7) within the Polycomb repressive complex 1 (PRC1), contributes to CBX7 function, and affects its ability to repress the locus and control senescence. In addition, both *ANRIL* and CBX7 levels are elevated in prostate cancer tissues (Yap et al. 2010). *ANRIL* also binds to and recruits PRC2 to repress the expression of the *p15^{INK4B}* locus. *ANRIL* depletion disrupts the binding of SUZ12, a component of PRC2, to the *p15^{INK4B}* locus. This results in increased expression of *p15^{INK4B}*, but not *p16^{INK4A}* or p14^{ARF}, and inhibits cellular proliferation (Kotake et al. 2011).

8.2.8 DLEU1 *and* DLEU2*: Long Sought Tumor Suppressors Harbor Short MicroRNAs*

The most common genomic aberration in chronic lymphocytic leukemia (CLL) is the loss of a critical region in 13q14.3 [del(13q)] with an occurrence of more than 50% (Bullrich et al. 1996). This led to the assumption that the region might contain tumor suppressor genes (Stilgenbauer et al. 1998; Dohner et al. 2000). By screening primary CLL clones and cell lines, *DLEU1* and *DLEU2*, also known as *LEU1* and *LEU2*, were identified as candidate tumor suppressor genes. *DLEU1* and *DLEU2* are lncRNAs significantly downregulated in CLL cells. *DLEU2* seems to be a pri-miRNA giving rise to the mature microRNAs miR-16-1 and miR-15a that have tumor-suppressive functions (Liu et al. 1997; Wolf et al. 2001; Calin et al. 2002; Mertens et al. 2009; Klein et al. 2010).

8.2.9 PRNCR1*: A Message for Prostate Cancer from 8q24*

Multiple genetic variants in a large region of chromosome 8q24 have been associated with susceptibility to prostate cancer. A mapping and resequencing study focusing on its most centromeric region led to the identification of a 13-kb-long ncRNA, which was named *prostate cancer noncoding RNA 1 (PRNCR1)*. *PRNCR1* may be involved in prostate carcinogenesis, as it is upregulated in prostatic intraepithelial neoplasia and prostate cancer cells. Knockdown of

PRNCR1 decreases the viability of prostate cancer cells and the transactivation activity of androgen receptor (Chung et al. 2011).

8.2.10   TERRA: *Coming to the End of the Chromosome*

Telomeres are heterochromatic structures at the ends of eukaryotic chromosomes and are essential for chromosome stability. Until recently, telomeres have been considered to be transcriptionally silent. However, Northern blot analysis of RNA from a human cervical cancer cell line (HeLa) revealed the existence of *TERRA*. *TERRA* describes a group of lncRNAs transcribed from several subtelomeric loci toward chromosome ends. It localizes to telomeres and is involved in telomeric heterochromatin formation (Azzalin et al. 2007; Schoeftner and Blasco 2008; Deng et al. 2009; Luke and Lingner 2009).

Telomere transcription is an evolutionarily conserved phenomenon in eukaryotic cells suggesting functional importance (Caslini 2010). Critical shortening of telomeres is essential for the limited capacity of normal human cells to divide and the subsequent onset of replicative senescence (Hayflick 1965; Harley et al. 1990). Overcoming the senescence barrier by elongating and maintaining telomeres is a prerequisite for tumor formation (Hanahan and Weinberg 2000). Telomere length maintenance is achieved by activation of telomerase in around 85% of tumors and by the alternative lengthening of telomeres (ALT) mechanism in about 15% of tumors (Bryan et al. 1997; Shay and Bacchetti 1997).

TERRA is a proposed regulator of telomerase, which may act globally or at individual telomeres as a direct inhibitor of the telomerase enzyme (Redon et al. 2010). Reduction of *TERRA* transcription is necessary for telomerase-mediated telomere lengthening. This role may link *TERRA* to cancer. Telomerase-positive cancer cells with high levels of subtelomeric methylation display low levels of *TERRA* compared to matched ALT-positive cancer cells or normal cells (Ng et al. 2009). Moreover, when cell extracts are incubated with an excess of synthetic RNA oligonucleotides mimicking *TERRA*, telomerase activity is inhibited (Schoeftner and Blasco 2008).

Taken together, *TERRA* is the first lncRNA that has not been linked to cancer based on its expression pattern but by virtue of its function. A deregulation, silencing, or mutation of *TERRA* in human cancer remains to be discovered.

8.3   Conclusion and Outlook

The nonprotein-encoding part of the human genome was once referred to as the "dark matter of the genome" (Johnson et al. 2005). However, these parts have now been found to be frequently transcribed into nonprotein-encoding RNAs. These ncRNAs have proven to be important regulators in health and in disease.

Despite the rapid progress in identifying these transcripts, only individual examples have been functionally studied at all, and many important questions remain to be addressed. Molecular studies are needed to explore the functions and mechanisms of action of this novel class of biomolecules. Although the majority of recent literature and current research focuses on short RNAs (miRNAs, siRNAs, piRNAs), the underestimated long ncRNAs have a great potential and show a lot of promise to be important molecules in physiological as well as pathological settings. Therefore, the next step should be genome-scale identification of lncRNAs differentially expressed in a wide variety of human cancers. One way would be to use microarray-based profiling, deep sequencing or RNA-Seq, as well as the careful and detailed validation of the expression and sequencing data by qRT-PCR, Northern blotting, and RACE. lncRNAs specifically expressed or silenced in human cancers could play an important role in these cancer entities, and their functional analysis is required to understand the molecular mechanisms underlying tumorigenesis.

This is where lncRNAs raise a lot of challenges due to their mechanistic heterogeneity that is just beginning to emerge from the first discoveries in this field. Insights into their function can be gained using RNA interference to knock down the lncRNA, genetic loss-of-function models, or overexpression followed by studying the cellular phenotypes associated with tumor development such as proliferation, migration, cell viability, or apoptosis. However, RNAi-mediated loss-of-function studies should always be scrutinized to exclude frequent off-target effects. Essential controls include the use of multiple siRNAs for each gene, nontargeting or scrambled siRNA controls, and a validation of the phenotype specificity in rescue experiments reversing the phenotype by overexpression of the targeted gene. Studying the localization of the lncRNA can also provide valuable insights into its function. Once the cellular fraction, where the lncRNA is normally present, has been established, the lncRNA can be further analyzed. For example, protein interaction partners of the ncRNA could be identified via RNA affinity purification.

RNA immunoprecipitation-sequencing (RIP-Seq) or PAR-CLIP provides other approaches to studying RNA-protein interactions (Hafner et al. 2010; Zhao et al. 2010). Here, all RNAs that bind to a specific protein of interest are pulled down to identify substrate, target, or regulator RNAs – either coding or noncoding. Expanding RIP-based analysis of protein–lncRNA-binding patterns will help to create an experimentally documented lncRNA–protein interactome atlas. Such atlas can be a helpful guide for in-depth studies on the functions of each lncRNA (Lipovich et al. 2010).

We have discussed many lncRNAs with differential expression patterns that could be of diagnostic, prognostic, or predictive value for various types of cancer. lncRNA offer a number of advantages as diagnostic and prognostic markers. First, they can be highly specific; HULC, for example, is highly expressed in primary liver tumors and hepatic metastases of colorectal carcinoma, but not in the primary colon tumors or in nonliver metastases (Matouk et al. 2009). *Prostate-specific gene 1 (PCGEM1), differential display code 3 (DD3), also known as prostate cancer gene 3* (*PCA3*), and *PRNCR1* are three lncRNAs that have been exclusively associated with prostate cancer (de Kok et al. 2002; Ifere and Ananaba 2009a; Chung et al. 2011). LncRNAs can also be used to differentiate between subtypes of the same cancer.

Second, lncRNAs can be detected in biological fluids like blood and urine, which is much less invasive and more convenient than biopsies. A quantitative PCA3 urine test with the potential for general use in clinical settings was developed, the Progensa™ PCA3 urine test. This specific test can help patients who have a first negative biopsy to avoid unnecessary repeated biopsies (Durand et al. 2011).

Third, the ncRNAs can be detected from minute amounts using qRT-PCR amplification. For example, *HULC* can be easily detected in the blood of HCC patients using PCR (Panzitt et al. 2007).

Fourth, lncRNAs expression may potentially correlate with patient response to chemotherapy. As mentioned before, *XIST* expression strongly associates with the disease-free survival of Taxol-treated cancer patients (Huang et al. 2002). lncRNAs are also powerful predictors of patient outcome. For example, *MALAT1* can serve as an independent prognostic parameter for patient survival in early-stage lung adeno-carcinoma because it is significantly associated with metastasis in NSCLC patients (Ji et al. 2003). Also, *HOTAIR* positively correlates with metastasis and poor outcome in primary breast tumors (Gupta et al. 2010).

The great wealth of newly discovered transcripts makes it highly likely that many other lncRNA markers remain to be discovered. The vast amount of cancer genome data becoming rapidly available can only be fully exploited if also the noncoding content of the human cancer genome is studied in great detail – after all, it constitutes the large majority of the genomic information! The more we learn about lncRNA expression patterns in different types of cancer – as well as in healthy cells – the higher the chances for an improved diagnosis and better prognosis will be.

Lastly, uncovering the role of lncRNAs in cancer will not only provide novel insights into the molecular mechanisms in the normal as well as the tumor cell but will also aid in designing novel therapeutic agents. For example, oncogenic lncRNAs could be targeted by RNA interference or an antagolnc (a synthetic RNA that specifically binds a target lncRNA) or tumor suppressive lncRNAs could be induced. Also, designing therapeutic aptamer agents that specifically target (deregulated) lncRNA-protein interactions, modulating the function of lncRNAs, or using lncRNAs to epigenetically silence oncogenes is conceivable and could provide new options in cancer therapy. Here, lncRNAs with increased expression in tumors can reduce the risk of affecting normal tissues during genetic treatment by providing tumor-specific regulatory regions: For example, *H19* expression is increased in a wide range of human cancers. A plasmid BC-819 (DTA-H19) has been developed to make use of this tumor-specific expression of *H19*. The plasmid carries the gene for the A subunit of diphtheria toxin under the regulation of the *H19* gene promoter. It is administered via intratumoral injection and induces the expression of high levels of diphtheria toxin specifically in the tumor resulting in a reduction of tumor size in human trials. While most in vivo studies have investigated BC-819 for the treatment of bladder cancer, recent studies have also yielded encouraging results in NSCLC, colon, pancreatic, and ovarian cancers (Smaldone and Davies 2010; Gibb et al. 2011).

Targeting cancer-specific lncRNAs may provide a way to cancer-specific therapeutics. However, designing molecules to inhibit oncogenic lncRNAs can be challenging due to their extensive secondary structures which underscores the importance of structural studies in designing such therapeutic agents. For example, *HOTAIR* leads to increased cancer invasiveness and metastasis by targeting the chromatin remodeling complex PRC2 to alter H3K27 methylation and gene expression. The administration of an antagolnc against *HOTAIR* would prevent it from binding to PRC2 and may normalize the chromatin state to inhibit cancer cell growth and metastasis (Tsai et al. 2011). Similarly, targeting *MALAT1* with an antagolnc could reduce cell motility of lung cancer cells by affecting *MALAT1*-mediated regulation of motility-related genes. The same principle can be used to target *ANRIL* and prevent its repressive effect on the tumor suppressor locus *INK4*.

GAS5 transcript levels are significantly reduced in breast cancer, and its expression induces growth arrest and apoptosis independently of other stimuli in some prostate and breast cancer cell lines (Mourtada-Maarabouni et al. 2009). Therefore, designing a vector that would induce the expression of *GAS5* when injected into the tumor might provide an attractive therapeutic approach. As mentioned above, *TERRA* is a proposed regulator of telomerase, and telomerase-positive cancer cells display low levels of *TERRA* (Ng et al. 2009). Therefore, a potential therapeutic strategy in such cancer cells would be to enhance *TERRA* expression or administer synthetic *TERRA* mimics.

We are taking our first steps on the road of understanding the role of lncRNAs in cancer, and as we move forward, we are bound to discover new lncRNAs and find out more about their importance in cancer, which will inevitably help us to design better therapeutic agents.

References

Abdel-Karim IA, Giles FJ (2008) Mammalian target of rapamycin as a target in hematological malignancies. Curr Probl Cancer 32(4):161–177

Adriaenssens E, Dumont L, Lottin S, Bolle D, Lepretre A, Delobelle A, Bouali F, Dugimont T, Coll J, Curgy JJ (1998) H19 overexpression in breast adenocarcinoma stromal cells is associated with tumor values and steroid receptor status but independent of p53 and Ki-67 expression. Am J Pathol 153(5):1597–1607

Agrelo R, Wutz A (2010) ConteXt of change–X inactivation and disease. EMBO Mol Med 2 (1):6–15

Alimonti A, Carracedo A, Clohessy JG, Trotman LC, Nardella C, Egia A, Salmena L, Sampieri K, Haveman WJ, Brogi E, Richardson AL, Zhang J, Pandolfi PP (2010) Subtle variations in Pten dose determine cancer susceptibility. Nat Genet 42(5):454–458

Ariel I, Miao HQ, Ji XR, Schneider T, Roll D, de Groot N, Hochberg A, Ayesh S (1998) Imprinted H19 oncofetal RNA is a candidate tumour marker for hepatocellular carcinoma. Mol Pathol 51 (1):21–25

Ariel I, Sughayer M, Fellig Y, Pizov G, Ayesh S, Podeh D, Libdeh BA, Levy C, Birman T, Tykocinski ML, de Groot N, Hochberg A (2000) The imprinted H19 gene is a marker of early recurrence in human bladder carcinoma. Mol Pathol 53(6):320–323

Askarian-Amiri ME, Crawford J, French JD, Smart CE, Smith MA, Clark MB, Ru K, Mercer TR, Thompson ER, Lakhani SR, Vargas AC, Campbell IG, Brown MA, Dinger ME, Mattick JS (2011) SNORD-host RNA Zfas1 is a regulator of mammary development and a potential marker for breast cancer. RNA 17(5):878–891

Azzalin CM, Reichenbach P, Khoriauli L, Giulotto E, Lingner J (2007) Telomeric repeat containing RNA and RNA surveillance factors at mammalian chromosome ends. Science 318(5851):798–801

Bartolomei MS, Zemel S, Tilghman SM (1991) Parental imprinting of the mouse H19 gene. Nature 351(6322):153–155

Bei JX, Li Y, Jia WH, Feng BJ, Zhou G, Chen LZ, Feng QS, Low HQ, Zhang H, He F, Tai ES, Kang T, Liu ET, Liu J, Zeng YX (2010) A genome-wide association study of nasopharyngeal carcinoma identifies three new susceptibility loci. Nat Genet 42(7):599–603

Bejerano G, Pheasant M, Makunin I, Stephen S, Kent WJ, Mattick JS, Haussler D (2004) Ultraconserved elements in the human genome. Science 304(5675):1321–1325

Beltran M, Puig I, Pena C, Garcia JM, Alvarez AB, Pena R, Bonilla F, de Herreros AG (2008) A natural antisense transcript regulates Zeb2/Sip1 gene expression during Snail1-induced epithelial-mesenchymal transition. Genes Dev 22(6):756–769

Benoit MH, Hudson TJ, Maire G, Squire JA, Arcand SL, Provencher D, Mes-Masson AM, Tonin PN (2007) Global analysis of chromosome X gene expression in primary cultures of normal ovarian surface epithelial cells and epithelial ovarian cancer cell lines. Int J Oncol 30(1):5–17

Berg P (2006) Origins of the human genome project: why sequence the human genome when 96% of it is junk? Am J Hum Genet 79(4):603–605

Bertone P, Stolc V, Royce TE, Rozowsky JS, Urban AE, Zhu X, Rinn JL, Tongprasit W, Samanta M, Weissman S, Gerstein M, Snyder M (2004) Global identification of human transcribed sequences with genome tiling arrays. Science 306(5705):2242–2246

Borsani G, Tonlorenzi R, Simmler MC, Dandolo L, Arnaud D, Capra V, Grompe M, Pizzuti A, Muzny D, Lawrence C, Willard HF, Avner P, Ballabio A (1991) Characterization of a murine gene expressed from the inactive X chromosome. Nature 351(6324):325–329

Braconi C, Valeri N, Kogure T, Gasparini P, Huang N, Nuovo GJ, Terracciano L, Croce CM, Patel T (2011) Expression and functional role of a transcribed noncoding RNA with an ultraconserved element in hepatocellular carcinoma. Proc Natl Acad Sci USA 108(2):786–791

Brockdorff N, Ashworth A, Kay GF, Cooper P, Smith S, McCabe VM, Norris DP, Penny GD, Patel D, Rastan S (1991) Conservation of position and exclusive expression of mouse Xist from the inactive X chromosome. Nature 351(6324):329–331

Brown CJ, Ballabio A, Rupert JL, Lafreniere RG, Grompe M, Tonlorenzi R, Willard HF (1991) A gene from the region of the human X inactivation centre is expressed exclusively from the inactive X chromosome. Nature 349(6304):38–44

Bryan TM, Englezou A, Dalla-Pozza L, Dunham MA, Reddel RR (1997) Evidence for an alternative mechanism for maintaining telomere length in human tumors and tumor-derived cell lines. Nat Med 3(11):1271–1274

Bullrich F, Veronese ML, Kitada S, Jurlander J, Caligiuri MA, Reed JC, Croce CM (1996) Minimal region of loss at 13q14 in B-cell chronic lymphocytic leukemia. Blood 88 (8):3109–3115

Bussemakers MJ, van Bokhoven A, Verhaegh GW, Smit FP, Karthaus HF, Schalken JA, Debruyne FM, Ru N, Isaacs WB (1999) DD3: a new prostate-specific gene, highly overexpressed in prostate cancer. Cancer Res 59(23):5975–5979

Cai X, Cullen BR (2007) The imprinted H19 noncoding RNA is a primary microRNA precursor. RNA 13(3):313–316

Calin GA, Dumitru CD, Shimizu M, Bichi R, Zupo S, Noch E, Aldler H, Rattan S, Keating M, Rai K, Rassenti L, Kipps T, Negrini M, Bullrich F, Croce CM (2002) Frequent deletions and downregulation of micro-RNA genes miR15 and miR16 at 13q14 in chronic lymphocytic leukemia. Proc Natl Acad Sci USA 99(24):15524–15529

Calin GA, Liu CG, Ferracin M, Hyslop T, Spizzo R, Sevignani C, Fabbri M, Cimmino A, Lee EJ, Wojcik SE, Shimizu M, Tili E, Rossi S, Taccioli C, Pichiorri F, Liu X, Zupo S, Herlea V, Gramantieri L, Lanza G, Alder H, Rassenti L, Volinia S, Schmittgen TD, Kipps TJ, Negrini M, Croce CM (2007) Ultraconserved regions encoding ncRNAs are altered in human leukemias and carcinomas. Cancer Cell 12(3):215–229

Carninci P, Kasukawa T, Katayama S, Gough J, Frith MC, Maeda N, Oyama R, Ravasi T, Lenhard B, Wells C, Kodzius R, Shimokawa K, Bajic VB, Brenner SE, Batalov S, Forrest AR, Zavolan M, Davis MJ, Wilming LG, Aidinis V, Allen JE, Ambesi-Impiombato A, Apweiler R, Aturaliya RN, Bailey TL, Bansal M, Baxter L, Beisel KW, Bersano T, Bono H, Chalk AM, Chiu KP, Choudhary V, Christoffels A, Clutterbuck DR, Crowe ML, Dalla E, Dalrymple BP, de Bono B, Della GG, di Bernardo D, Down T, Engstrom P, Fagiolini M, Faulkner G, Fletcher CF, Fukushima T, Furuno M, Futaki S, Gariboldi M, Georgii-Hemming P, Gingeras TR, Gojobori T, Green RE, Gustincich S, Harbers M, Hayashi Y, Hensch TK, Hirokawa N, Hill D, Huminiecki L, Iacono M, Ikeo K, Iwama A, Ishikawa T, Jakt M, Kanapin A, Katoh M, Kawasawa Y, Kelso J, Kitamura H, Kitano H, Kollias G, Krishnan SP, Kruger A, Kummerfeld SK, Kurochkin IV, Lareau LF, Lazarevic D, Lipovich L, Liu J, Liuni S, McWilliam S, Madan BM, Madera M, Marchionni L, Matsuda H, Matsuzawa S, Miki H, Mignone F, Miyake S, Morris K, Mottagui-Tabar S, Mulder N, Nakano N, Nakauchi H, Ng P, Nilsson R, Nishiguchi S, Nishikawa S, Nori F, Ohara O, Okazaki Y, Orlando V, Pang KC, Pavan WJ, Pavesi G, Pesole G, Petrovsky N, Piazza S, Reed J, Reid JF, Ring BZ, Ringwald M, Rost B, Ruan Y, Salzberg SL, Sandelin A, Schneider C, Schonbach C, Sekiguchi K, Semple CA, Seno S, Sessa L, Sheng Y, Shibata Y, Shimada H, Shimada K, Silva D, Sinclair B, Sperling S, Stupka E, Sugiura K, Sultana R, Takenaka Y, Taki K, Tammoja K, Tan SL, Tang S, Taylor MS, Tegner J, Teichmann SA, Ueda HR, van Nimwegen E, Verardo R, Wei CL, Yagi K, Yamanishi H, Zabarovsky E, Zhu S, Zimmer A, Hide W, Bult C, Grimmond SM, Teasdale RD, Liu ET, Brusic V, Quackenbush J, Wahlestedt C, Mattick JS, Hume DA, Kai C, Sasaki D, Tomaru Y, Fukuda S, Kanamori-Katayama M, Suzuki M, Aoki J, Arakawa T, Iida J, Imamura K, Itoh M, Kato T, Kawaji H, Kawagashira N, Kawashima T, Kojima M, Kondo S, Konno H, Nakano K, Ninomiya N, Nishio T, Okada M, Plessy C, Shibata K, Shiraki T, Suzuki S, Tagami M, Waki K, Watahiki A, Okamura-Oho Y, Suzuki H, Kawai J, Hayashizaki Y (2005) The transcriptional landscape of the mammalian genome. Science 309(5740):1559–1563

Caslini C (2010) Transcriptional regulation of telomeric non-coding RNA: implications on telomere biology, replicative senescence and cancer. RNA Biol 7(1):18–22

Chen LL, Carmichael GG (2010a) Decoding the function of nuclear long non-coding RNAs. Curr Opin Cell Biol 22(3):357–364

Chen LL, Carmichael GG (2010b) Long noncoding RNAs in mammalian cells: what, where, and why? Wiley Interdiscip Rev RNA 1(1):2–21

Cheng J, Kapranov P, Drenkow J, Dike S, Brubaker S, Patel S, Long J, Stern D, Tammana H, Helt G, Sementchenko V, Piccolboni A, Bekiranov S, Bailey DK, Ganesh M, Ghosh S, Bell I, Gerhard DS, Gingeras TR (2005) Transcriptional maps of 10 human chromosomes at 5-nucleotide resolution. Science 308(5725):1149–1154

Chung S, Nakagawa H, Uemura M, Piao L, Ashikawa K, Hosono N, Takata R, Akamatsu S, Kawaguchi T, Morizono T, Tsunoda T, Daigo Y, Matsuda K, Kamatani N, Nakamura Y, Kubo M (2011) Association of a novel long non-coding RNA in 8q24 with prostate cancer susceptibility. Cancer Sci 102(1):245–252

Clarke RA, Zhao Z, Guo AY, Roper K, Teng L, Fang ZM, Samaratunga H, Lavin MF, Gardiner RA (2009) New genomic structure for prostate cancer specific gene PCA3 within BMCC1: implications for prostate cancer detection and progression. PLoS One 4(3):e4995

Clemson CM, Hutchinson JN, Sara SA, Ensminger AW, Fox AH, Chess A, Lawrence JB (2009) An architectural role for a nuclear noncoding RNA: NEAT1 RNA is essential for the structure of paraspeckles. Mol Cell 33(6):717–726

Coccia EM, Cicala C, Charlesworth A, Ciccarelli C, Rossi GB, Philipson L, Sorrentino V (1992) Regulation and expression of a growth arrest-specific gene (gas5) during growth, differentiation, and development. Mol Cell Biol 12(8):3514–3521

Colnot S, Niwa-Kawakita M, Hamard G, Godard C, Le Plenier S, Houbron C, Romagnolo B, Berrebi D, Giovannini M, Perret C (2004) Colorectal cancers in a new mouse model of familial adenomatous polyposis: influence of genetic and environmental modifiers. Lab Invest 84 (12):1619–1630

Davis RL, Weintraub H, Lassar AB (1987) Expression of a single transfected cDNA converts fibroblasts to myoblasts. Cell 51(6):987–1000

de Kok JB, Verhaegh GW, Roelofs RW, Hessels D, Kiemeney LA, Aalders TW, Swinkels DW, Schalken JA (2002) DD3(PCA3), a very sensitive and specific marker to detect prostate tumors. Cancer Res 62(9):2695–2698

DeChiara TM, Robertson EJ, Efstratiadis A (1991) Parental imprinting of the mouse insulin-like growth factor II gene. Cell 64(4):849–859

Deng Z, Norseen J, Wiedmer A, Riethman H, Lieberman PM (2009) TERRA RNA binding to TRF2 facilitates heterochromatin formation and ORC recruitment at telomeres. Mol Cell 35(4):403–413

Diederichs S (2010) Non-coding RNA in malignant tumors. A new world of tumor biomarkers and target structures in cancer cells. Pathologe 31(Suppl 2):258–262

Dohner H, Stilgenbauer S, Benner A, Leupolt E, Krober A, Bullinger L, Dohner K, Bentz M, Lichter P (2000) Genomic aberrations and survival in chronic lymphocytic leukemia. N Engl J Med 343(26):1910–1916

Durand X, Moutereau S, Xylinas E, de la Taille A (2011) Progensa PCA3 test for prostate cancer. Expert Rev Mol Diagn 11(2):137–144

Feng J, Bi C, Clark BS, Mady R, Shah P, Kohtz JD (2006) The Evf-2 noncoding RNA is transcribed from the Dlx-5/6 ultraconserved region and functions as a Dlx-2 transcriptional coactivator. Genes Dev 20(11):1470–1484

Fingar DC, Richardson CJ, Tee AR, Cheatham L, Tsou C, Blenis J (2004) mTOR controls cell cycle progression through its cell growth effectors S6K1 and 4E-BP1/eukaryotic translation initiation factor 4E. Mol Cell Biol 24(1):200–216

Fleming JV, Hay SM, Harries DN, Rees WD (1998) Effects of nutrient deprivation and differentiation on the expression of growth-arrest genes (gas and gadd) in F9 embryonal carcinoma cells. Biochem J 330(Pt 1):573–579

Fontanier-Razzaq N, Harries DN, Hay SM, Rees WD (2002) Amino acid deficiency upregulates specific mRNAs in murine embryonic cells. J Nutr 132(8):2137–2142

Forne T, Oswald J, Dean W, Saam JR, Bailleul B, Dandolo L, Tilghman SM, Walter J, Reik W (1997) Loss of the maternal H19 gene induces changes in Igf2 methylation in both cis and trans. Proc Natl Acad Sci USA 94(19):10243–10248

Gabory A, Jammes H, Dandolo L (2010) The H19 locus: role of an imprinted non-coding RNA in growth and development. Bioessays 32(6):473–480

Ganesan S, Silver DP, Greenberg RA, Avni D, Drapkin R, Miron A, Mok SC, Randrianarison V, Brodie S, Salstrom J, Rasmussen TP, Klimke A, Marrese C, Marahrens Y, Deng CX, Feunteun J, Livingston DM (2002) BRCA1 supports XIST RNA concentration on the inactive X chromosome. Cell 111(3):393–405

Gibb EA, Brown CJ, Lam WL (2011) The functional role of long non-coding RNA in human carcinomas. Mol Cancer 10:38. doi:1476-4598-10-38 [pii] 10.1186/1476-4598-10-38

Guo F, Li Y, Liu Y, Wang J, Li G (2010) Inhibition of metastasis-associated lung adenocarcinoma transcript 1 in CaSki human cervical cancer cells suppresses cell proliferation and invasion. Acta Biochim Biophys Sin (Shanghai) 42(3):224–229

Gupta RA, Shah N, Wang KC, Kim J, Horlings HM, Wong DJ, Tsai MC, Hung T, Argani P, Rinn JL, Wang Y, Brzoska P, Kong B, Li R, West RB, van de Vijver MJ, Sukumar S, Chang HY (2010) Long non-coding RNA HOTAIR reprograms chromatin state to promote cancer metastasis. Nature 464(7291):1071–1076

Habeck M (2003) Jewels among the junk. Drug Discov Today 8(4):145–146

Hafner M, Landthaler M, Burger L, Khorshid M, Hausser J, Berninger P, Rothballer A, Ascano M Jr, Jungkamp AC, Munschauer M, Ulrich A, Wardle GS, Dewell S, Zavolan M, Tuschl T (2010) Transcriptome-wide identification of RNA-binding protein and microRNA target sites by PAR-CLIP. Cell 141(1):129–141. doi:S0092-8674(10)00245-X [pii] 10.1016/j.cell.2010.03.009

Hanahan D, Weinberg RA (2000) The hallmarks of cancer. Cell 100(1):57–70. doi:S0092-8674 (00), 81683-9 [pii]

Hansen TV, Hammer NA, Nielsen J, Madsen M, Dalbaeck C, Wewer UM, Christiansen J, Nielsen FC (2004) Dwarfism and impaired gut development in insulin-like growth factor II mRNA-binding protein 1-deficient mice. Mol Cell Biol 24(10):4448–4464

Hao Y, Crenshaw T, Moulton T, Newcomb E, Tycko B (1993) Tumour-suppressor activity of H19 RNA. Nature 365(6448):764–767. doi:10.1038/365764a0

Harley CB, Futcher AB, Greider CW (1990) Telomeres shorten during ageing of human fibroblasts. Nature 345(6274):458–460. doi:10.1038/345458a0

Hay N, Sonenberg N (2004) Upstream and downstream of mTOR. Genes Dev 18(16):1926–1945. doi:10.1101/gad.1212704 18/16/1926 [pii]

Hayflick L (1965) The limited in vitro lifetime of human diploid cell strains. Exp Cell Res 37:614–636

Hibi K, Nakamura H, Hirai A, Fujikake Y, Kasai Y, Akiyama S, Ito K, Takagi H (1996) Loss of H19 imprinting in esophageal cancer. Cancer Res 56(3):480–482

Huang KC, Rao PH, Lau CC, Heard E, Ng SK, Brown C, Mok SC, Berkowitz RS, Ng SW (2002) Relationship of XIST expression and responses of ovarian cancer to chemotherapy. Mol Cancer Ther 1(10):769–776

Huarte M, Guttman M, Feldser D, Garber M, Koziol MJ, Kenzelmann-Broz D, Khalil AM, Zuk O, Amit I, Rabani M, Attardi LD, Regev A, Lander ES, Jacks T, Rinn JL (2010) A large intergenic noncoding RNA induced by p53 mediates global gene repression in the p53 response. Cell 142(3):409–419

Hutchinson JN, Ensminger AW, Clemson CM, Lynch CR, Lawrence JB, Chess A (2007) A screen for nuclear transcripts identifies two linked noncoding RNAs associated with SC35 splicing domains. BMC Genomics 8:39. doi:1471-2164-8-39 [pii] 10.1186/1471-2164-8-39

Ifere GO, Ananaba GA (2009a) Prostate cancer gene expression marker 1 (PCGEM1): a patented prostate- specific non-coding gene and regulator of prostate cancer progression. Recent Pat DNA Gene Seq 3(3):151–163. doi:DNAG: 09 [pii]

Ifere GO, Ananaba GA (2009b) Prostate cancer gene expression marker 1 (PCGEM1): a patented prostate- specific non-coding gene and regulator of prostate cancer progression. Recent Pat DNA Gene Seq 3(3):151–163

Ji P, Diederichs S, Wang W, Boing S, Metzger R, Schneider PM, Tidow N, Brandt B, Buerger H, Bulk E, Thomas M, Berdel WE, Serve H, Muller-Tidow C (2003) MALAT-1, a novel noncoding RNA, and thymosin beta4 predict metastasis and survival in early-stage non-small cell lung cancer. Oncogene 22(39):8031–8041

Jia H, Osak M, Bogu GK, Stanton LW, Johnson R, Lipovich L (2010) Genome-wide computational identification and manual annotation of human long noncoding RNA genes. RNA 16 (8):1478–1487. doi:rna.1951310 [pii] 10.1261/rna.1951310

Jiang BH, Liu LZ (2008) Role of mTOR in anticancer drug resistance: perspectives for improved drug treatment. Drug Resist Updat 11(3):63–76. doi:S1368-7646(08), 00018-6 [pii] 10.1016/j.drup.2008.03.001

Johnson JM, Edwards S, Shoemaker D, Schadt EE (2005) Dark matter in the genome: evidence of widespread transcription detected by microarray tiling experiments. Trends Genet 21(2):93–102. doi:S0168-9525(04), 00337-3 [pii] 10.1016/j.tig.2004.12.009

Kapranov P, Drenkow J, Cheng J, Long J, Helt G, Dike S, Gingeras TR (2005) Examples of the complex architecture of the human transcriptome revealed by RACE and high-density tiling arrays. Genome Res 15(7):987–997

Kawakami T, Okamoto K, Sugihara H, Hattori T, Reeve AE, Ogawa O, Okada Y (2003) The roles of supernumerical X chromosomes and XIST expression in testicular germ cell tumors. J Urol 169(4):1546–1552. doi:10.1097/01.ju.0000044927.23323.5a S0022-5347(05), 63816-5 [pii]

Kawakami T, Okamoto K, Ogawa O, Okada Y (2004a) XIST unmethylated DNA fragments in male-derived plasma as a tumour marker for testicular cancer. Lancet 363(9402):40–42. doi: S0140-6736(03), 15170-7 [pii] 10.1016/S0140-6736(03)15170-7

Kawakami T, Zhang C, Taniguchi T, Kim CJ, Okada Y, Sugihara H, Hattori T, Reeve AE, Ogawa O, Okamoto K (2004b) Characterization of loss-of-inactive X in Klinefelter syndrome and female-derived cancer cells. Oncogene 23(36):6163–6169. doi:10.1038/sj.onc.1207808 1207808 [pii]

Khaitan D, Dinger ME, Mazar J, Crawford J, Smith MA, Mattick JS, Perera RJ (2011) The melanoma-upregulated long noncoding RNA SPRY4-IT1 modulates apoptosis and invasion. Cancer Res 71 (11):3852–3862. doi:0008-5472.CAN-10-4460 [pii] 10.1158/0008-5472.CAN-10-4460

Kino T, Hurt DE, Ichijo T, Nader N, Chrousos GP (2010) Noncoding RNA gas5 is a growth arrest-and starvation-associated repressor of the glucocorticoid receptor. Sci Signal 3(107):ra8

Klein U, Lia M, Crespo M, Siegel R, Shen Q, Mo T, Ambesi-Impiombato A, Califano A, Migliazza A, Bhagat G, Dalla-Favera R (2010) The DLEU2/miR-15a/16-1 cluster controls B cell proliferation and its deletion leads to chronic lymphocytic leukemia. Cancer Cell 17 (1):28–40. doi:S1535-6108(09)00419-X [pii] 10.1016/j.ccr.2009.11.019

Kondo M, Suzuki H, Ueda R, Osada H, Takagi K, Takahashi T (1995) Frequent loss of imprinting of the H19 gene is often associated with its overexpression in human lung cancers. Oncogene 10(6):1193–1198

Kotake Y, Nakagawa T, Kitagawa K, Suzuki S, Liu N, Kitagawa M, Xiong Y (2011) Long non-coding RNA ANRIL is required for the PRC2 recruitment to and silencing of p15(INK4B) tumor suppressor gene. Oncogene 30(16):1956–1962

Laner T, Schulz WA, Engers R, Muller M, Florl AR (2005) Hypomethylation of the XIST gene promoter in prostate cancer. Oncol Res 15(5):257–264

Lassmann S, Weis R, Makowiec F, Roth J, Danciu M, Hopt U, Werner M (2007) Array CGH identifies distinct DNA copy number profiles of oncogenes and tumor suppressor genes in chromosomal- and microsatellite-unstable sporadic colorectal carcinomas. J Mol Med 85 (3):293–304. doi:10.1007/s00109-006-0126-5

Lee JT, Davidow LS, Warshawsky D (1999) Tsix, a gene antisense to Xist at the X-inactivation centre. Nat Genet 21(4):400–404. doi:10.1038/7734

Lin R, Maeda S, Liu C, Karin M, Edgington TS (2007) A large noncoding RNA is a marker for murine hepatocellular carcinomas and a spectrum of human carcinomas. Oncogene 26 (6):851–858

Lin R, Roychowdhury-Saha M, Black C, Watt AT, Marcusson EG, Freier SM, Edgington TS (2011) Control of RNA processing by a large non-coding RNA over-expressed in carcinomas. FEBS Lett 585(4):671–676. doi:S0014-5793(11), 00060-3 [pii] 10.1016/j.febslet.2011.01.030

Lind GE, Skotheim RI, Lothe RA (2007) The epigenome of testicular germ cell tumors. APMIS 115(10):1147–1160. doi:APMapm_660.xml [pii] 10.1111/j.1600-0463.2007.apm_660.xml.x

Lipovich L, Johnson R, Lin CY (2010) MacroRNA underdogs in a microRNA world: evolutionary, regulatory, and biomedical significance of mammalian long non-protein-coding RNA. Biochim Biophys Acta 1799(9):597–615

Liu Y, Corcoran M, Rasool O, Ivanova G, Ibbotson R, Grander D, Iyengar A, Baranova A, Kashuba V, Merup M, Wu X, Gardiner A, Mullenbach R, Poltaraus A, Hultstrom AL, Juliusson G, Chapman R, Tiller M, Cotter F, Gahrton G, Yankovsky N, Zabarovsky E, Einhorn S, Oscier D (1997) Cloning of two candidate tumor suppressor genes within a 10 kb region on chromosome 13q14, frequently deleted in chronic lymphocytic leukemia. Oncogene 15(20):2463–2473. doi:10.1038/sj.onc.1201643

Looijenga LH, Gillis AJ, van Gurp RJ, Verkerk AJ, Oosterhuis JW (1997) X inactivation in human testicular tumors. XIST expression and androgen receptor methylation status. Am J Pathol 151 (2):581–590

Lottin S, Adriaenssens E, Dupressoir T, Berteaux N, Montpellier C, Coll J, Dugimont T, Curgy JJ (2002) Overexpression of an ectopic H19 gene enhances the tumorigenic properties of breast cancer cells. Carcinogenesis 23(11):1885–1895

Lottin S, Adriaenssens E, Berteaux N, Lepretre A, Vilain MO, Denhez E, Coll J, Dugimont T, Curgy JJ (2005) The human H19 gene is frequently overexpressed in myometrium and stroma during pathological endometrial proliferative events. Eur J Cancer 41(1):168–177. doi:S0959-8049(04), 00780-4 [pii] 10.1016/j.ejca.2004.09.025

Luke B, Lingner J (2009) TERRA: telomeric repeat-containing RNA. EMBO J 28(17):2503–2510. doi:emboj2009166 [pii] 10.1038/emboj.2009.166

Martianov I, Ramadass A, Serra Barros A, Chow N, Akoulitchev A (2007) Repression of the human dihydrofolate reductase gene by a non-coding interfering transcript. Nature 445 (7128):666–670. doi:nature05519 [pii] 10.1038/nature05519

Matouk IJ, DeGroot N, Mezan S, Ayesh S, Abu-lail R, Hochberg A, Galun E (2007) The H19 non-coding RNA is essential for human tumor growth. PLoS One 2(9):e845. doi:10.1371/journal. pone.0000845

Matouk IJ, Abbasi I, Hochberg A, Galun E, Dweik H, Akkawi M (2009) Highly upregulated in liver cancer noncoding RNA is overexpressed in hepatic colorectal metastasis. Eur J Gastroenterol Hepatol 21(6):688–692

Mattick JS (2003) Challenging the dogma: the hidden layer of non-protein-coding RNAs in complex organisms. Bioessays 25(10):930–939

Mattick JS, Makunin IV (2006) Non-coding RNA. Hum Mol Genet 15(1):R17–R29

Mercer TR, Dinger ME, Mattick JS (2009) Long non-coding RNAs: insights into functions. Nat Rev Genet 10(3):155–159

Mertens D, Philippen A, Ruppel M, Allegra D, Bhattacharya N, Tschuch C, Wolf S, Idler I, Zenz T, Stilgenbauer S (2009) Chronic lymphocytic leukemia and 13q14: miRs and more. Leuk Lymphoma 50(3):502–505

Migliazza A, Bosch F, Komatsu H, Cayanis E, Martinotti S, Toniato E, Guccione E, Qu X, Chien M, Murty VV, Gaidano G, Inghirami G, Zhang P, Fischer S, Kalachikov SM, Russo J, Edelman I, Efstratiadis A, Dalla-Favera R (2001) Nucleotide sequence, transcription map, and mutation analysis of the 13q14 chromosomal region deleted in B-cell chronic lymphocytic leukemia. Blood 97(7):2098–2104

Morrison LE, Jewell SS, Usha L, Blondin BA, Rao RD, Tabesh B, Kemper M, Batus M, Coon JS (2007) Effects of ERBB2 amplicon size and genomic alterations of chromosomes 1, 3, and 10 on patient response to trastuzumab in metastatic breast cancer. Genes Chromosomes Cancer 46 (4):397–405. doi:10.1002/gcc.20419

Mourtada-Maarabouni M, Hedge VL, Kirkham L, Farzaneh F, Williams GT (2008) Growth arrest in human T-cells is controlled by the non-coding RNA growth-arrest-specific transcript 5 (GAS5). J Cell Sci 121(Pt 7):939–946. doi:121/7/939 [pii] 10.1242/jcs.024646

Mourtada-Maarabouni M, Pickard MR, Hedge VL, Farzaneh F, Williams GT (2009) GAS5, a non-protein-coding RNA, controls apoptosis and is downregulated in breast cancer. Oncogene 28 (2):195–208

Mourtada-Maarabouni M, Hasan AM, Farzaneh F, Williams GT (2010) Inhibition of human T-cell proliferation by mammalian target of rapamycin (mTOR) antagonists requires noncoding RNA growth-arrest-specific transcript 5 (GAS5). Mol Pharmacol 78(1):19–28. doi:mol.110.064055 [pii] 10.1124/mol.110.064055

Muller AJ, Chatterjee S, Teresky A, Levine AJ (1998) The gas5 gene is disrupted by a frameshift mutation within its longest open reading frame in several inbred mouse strains and maps to murine chromosome 1. Mamm Genome 9(9):773–774

Nakamura Y, Takahashi N, Kakegawa E, Yoshida K, Ito Y, Kayano H, Niitsu N, Jinnai I, Bessho M (2008) The GAS5 (growth arrest-specific transcript 5) gene fuses to BCL6 as a result of t(1;3)(q25;q27) in a patient with B-cell lymphoma. Cancer Genet Cytogenet 182 (2):144–149. doi:S0165-4608(08), 00035-6 [pii] 10.1016/j.cancergencyto.2008.01.013

Ng LJ, Cropley JE, Pickett HA, Reddel RR, Suter CM (2009) Telomerase activity is associated with an increase in DNA methylation at the proximal subtelomere and a reduction in telomeric transcription. Nucleic Acids Res 37(4):1152–1159. doi:gkn1030 [pii] 10.1093/nar/gkn1030

Nomura S, Baxter T, Yamaguchi H, Leys C, Vartapetian AB, Fox JG, Lee JR, Wang TC, Goldenring JR (2004) Spasmolytic polypeptide expressing metaplasia to preneoplasia in H. felis-infected mice. Gastroenterology 127(2):582–594. doi:doi:S0016508504009217 [pii]

Nupponen NN, Carpten JD (2001) Prostate cancer susceptibility genes: many studies, many results, no answers. Cancer Metastasis Rev 20(3–4):155–164

Ogawa Y, Sun BK, Lee JT (2008) Intersection of the RNA interference and X-inactivation pathways. Science 320(5881):1336–1341. doi:320/5881/1336 [pii] 10.1126/science.1157676

Okazaki Y, Furuno M, Kasukawa T, Adachi J, Bono H, Kondo S, Nikaido I, Osato N, Saito R, Suzuki H, Yamanaka I, Kiyosawa H, Yagi K, Tomaru Y, Hasegawa Y, Nogami A, Schonbach C, Gojobori T, Baldarelli R, Hill DP, Bult C, Hume DA, Quackenbush J, Schriml LM, Kanapin A, Matsuda H, Batalov S, Beisel KW, Blake JA, Bradt D, Brusic V, Chothia C, Corbani LE, Cousins S, Dalla E, Dragani TA, Fletcher CF, Forrest A, Frazer KS, Gaasterland T, Gariboldi M, Gissi C, Godzik A, Gough J, Grimmond S, Gustincich S, Hirokawa N, Jackson IJ, Jarvis ED, Kanai A, Kawaji H, Kawasawa Y, Kedzierski RM, King BL, Konagaya A, Kurochkin IV, Lee Y, Lenhard B, Lyons PA, Maglott DR, Maltais L, Marchionni L, McKenzie L, Miki H, Nagashima T, Numata K, Okido T, Pavan WJ, Pertea G, Pesole G, Petrovsky N, Pillai R, Pontius JU, Qi D, Ramachandran S, Ravasi T, Reed JC, Reed DJ, Reid J, Ring BZ, Ringwald M, Sandelin A, Schneider C, Semple CA, Setou M, Shimada K, Sultana R, Takenaka Y, Taylor MS, Teasdale RD, Tomita M, Verardo R, Wagner L, Wahlestedt C, Wang Y, Watanabe Y, Wells C, Wilming LG, Wynshaw-Boris A, Yanagisawa M, Yang I, Yang L, Yuan Z, Zavolan M, Zhu Y, Zimmer A, Carninci P, Hayatsu N, Hirozane-Kishikawa T, Konno H, Nakamura M, Sakazume N, Sato K, Shiraki T, Waki K, Kawai J, Aizawa K, Arakawa T, Fukuda S, Hara A, Hashizume W, Imotani K, Ishii Y, Itoh M, Kagawa I, Miyazaki A, Sakai K, Sasaki D, Shibata K, Shinagawa A, Yasunishi A, Yoshino M, Waterston R, Lander ES, Rogers J, Birney E, Hayashizaki Y (2002) Analysis of the mouse transcriptome based on functional annotation of 60,770 full-length cDNAs. Nature 420(6915):563–573

Pachnis V, Belayew A, Tilghman SM (1984) Locus unlinked to alpha-fetoprotein under the control of the murine raf and Rif genes. Proc Natl Acad Sci USA 81(17):5523–5527

Pageau GJ, Hall LL, Lawrence JB (2007) BRCA1 does not paint the inactive X to localize XIST RNA but may contribute to broad changes in cancer that impact XIST and Xi heterochromatin. J Cell Biochem 100(4):835–850. doi:10.1002/jcb.21188

Panzitt K, Tschernatsch MM, Guelly C, Moustafa T, Stradner M, Strohmaier HM, Buck CR, Denk H, Schroeder R, Trauner M, Zatloukal K (2007) Characterization of HULC, a novel gene with striking upregulation in hepatocellular carcinoma, as noncoding RNA. Gastroenterology 132(1):330–342

Pasmant E, Laurendeau I, Heron D, Vidaud M, Vidaud D, Bieche I (2007) Characterization of a germ-line deletion, including the entire INK4/ARF locus, in a melanoma-neural system tumor family: identification of ANRIL, an antisense noncoding RNA whose expression coclusters with ARF. Cancer Res 67(8):3963–3969. doi:67/8/3963 [pii] 10.1158/0008-5472.CAN-06-2004

Pasmant E, Sabbagh A, Vidaud M, Bieche I (2010) ANRIL, a long, noncoding RNA, is an unexpected major hotspot in GWAS. FASEB J 25(2):444–8

Poirier F, Chan CT, Timmons PM, Robertson EJ, Evans MJ, Rigby PW (1991) The murine H19 gene is activated during embryonic stem cell differentiation in vitro and at the time of implantation in the developing embryo. Development 113(4):1105–1114

Poliseno L, Salmena L, Zhang J, Carver B, Haveman WJ, Pandolfi PP (2010) A coding-independent function of gene and pseudogene mRNAs regulates tumour biology. Nature 465 (7301):1033–1038. doi:nature09144 [pii] 10.1038/nature09144

Ponting CP, Oliver PL, Reik W (2009) Evolution and functions of long noncoding RNAs. Cell 136 (4):629–641

Rack KA, Chelly J, Gibbons RJ, Rider S, Benjamin D, Lafreniere RG, Oscier D, Hendriks RW, Craig IW, Willard HF et al (1994) Absence of the XIST gene from late-replicating isodicentric X chromosomes in leukaemia. Hum Mol Genet 3(7):1053–1059

Raho G, Barone V, Rossi D, Philipson L, Sorrentino V (2000) The gas 5 gene shows four alternative splicing patterns without coding for a protein. Gene 256(1–2):13–17. doi:S0378-1119(00), 00363-2 [pii]

Redon S, Reichenbach P, Lingner J (2010) The non-coding RNA TERRA is a natural ligand and direct inhibitor of human telomerase. Nucleic Acids Res 38(17):5797–5806. doi:gkq296 [pii] 10.1093/nar/gkq296

Riccio A, Sparago A, Verde G, De Crescenzo A, Citro V, Cubellis MV, Ferrero GB, Silengo MC, Russo S, Larizza L, Cerrato F (2009) Inherited and sporadic epimutations at the IGF2-H19 locus in Beckwith-Wiedemann syndrome and Wilms' tumor. Endocr Dev 14:1–9. doi:000207461 [pii] 10.1159/000207461

Richler C, Soreq H, Wahrman J (1992) X inactivation in mammalian testis is correlated with inactive X-specific transcription. Nat Genet 2(3):192–195. doi:10.1038/ng1192-192

Rinn JL, Kertesz M, Wang JK, Squazzo SL, Xu X, Brugmann SA, Goodnough LH, Helms JA, Farnham PJ, Segal E, Chang HY (2007) Functional demarcation of active and silent chromatin domains in human HOX loci by noncoding RNAs. Cell 129(7):1311–1323. doi:S0092-8674 (07), 00659-9 [pii] 10.1016/j.cell.2007.05.022

Rump P, Zeegers MP, van Essen AJ (2005) Tumor risk in Beckwith-Wiedemann syndrome: a review and meta-analysis. Am J Med Genet A 136(1):95–104. doi:10.1002/ajmg.a.30729

Runge S, Nielsen FC, Nielsen J, Lykke-Andersen J, Wewer UM, Christiansen J (2000) H19 RNA binds four molecules of insulin-like growth factor II mRNA-binding protein. J Biol Chem 275 (38):29562–29569. doi:10.1074/jbc.M001156200 M001156200 [pii]

Schneider C, King RM, Philipson L (1988) Genes specifically expressed at growth arrest of mammalian cells. Cell 54(6):787–793. doi:S0092-8674(88), 91065-3 [pii]

Schoeftner S, Blasco MA (2008) Developmentally regulated transcription of mammalian telomeres by DNA-dependent RNA polymerase II. Nat Cell Biol 10(2):228–236. doi: ncb1685 [pii] 10.1038/ncb1685

Shay JW, Bacchetti S (1997) A survey of telomerase activity in human cancer. Eur J Cancer 33 (5):787–791. doi:S0959-8049(97), 00062-2 [pii] 10.1016/S0959-8049(97)00062-2

Shete S, Hosking FJ, Robertson LB, Dobbins SE, Sanson M, Malmer B, Simon M, Marie Y, Boisselier B, Delattre JY, Hoang-Xuan K, El Hallani S, Idbaih A, Zelenika D, Andersson U, Henriksson R, Bergenheim AT, Feychting M, Lonn S, Ahlbom A, Schramm J, Linnebank M, Hemminki K, Kumar R, Hepworth SJ, Price A, Armstrong G, Liu Y, Gu X, Yu R, Lau C, Schoemaker M, Muir K, Swerdlow A, Lathrop M, Bondy M, Houlston RS (2009) Genome-wide association study identifies five susceptibility loci for glioma. Nat Genet 41(8):899–904. doi:ng.407 [pii] 10.1038/ng.407

Shi Z, Dragin N, Miller ML, Stringer KF, Johansson E, Chen J, Uno S, Gonzalez FJ, Rubio CA, Nebert DW (2010) Oral benzo[a]pyrene-induced cancer: two distinct types in different target organs depend on the mouse Cyp1 genotype. Int J Cancer 127(10):2334–2350. doi:10.1002/ijc.25222

Silver DP, Dimitrov SD, Feunteun J, Gelman R, Drapkin R, Lu SD, Shestakova E, Velmurugan S, Denunzio N, Dragomir S, Mar J, Liu X, Rottenberg S, Jonkers J, Ganesan S, Livingston DM (2007) Further evidence for BRCA1 communication with the inactive X chromosome. Cell 128 (5):991–1002. doi:S0092-8674(07), 00249-8 [pii] 10.1016/j.cell.2007.02.025

Sirchia SM, Tabano S, Monti L, Recalcati MP, Gariboldi M, Grati FR, Porta G, Finelli P, Radice P, Miozzo M (2009) Misbehaviour of XIST RNA in breast cancer cells. PLoS One 4(5):e5559. doi:10.1371/journal.pone.0005559

Smaldone MC, Davies BJ (2010) BC-819, a plasmid comprising the H19 gene regulatory sequences and diphtheria toxin A, for the potential targeted therapy of cancers. Curr Opin Mol Ther 12(5):607–616

Smedley D, Sidhar S, Birdsall S, Bennett D, Herlyn M, Cooper C, Shipley J (2000) Characterization of chromosome 1 abnormalities in malignant melanomas. Genes Chromosomes Cancer 28 (1):121–125. doi:10.1002/(SICI)1098-2264(200005)28:1<121::AID-GCC14>3.0.CO;2-O [pii]

Smith CM, Steitz JA (1998) Classification of gas5 as a multi-small-nucleolar-RNA (snoRNA) host gene and a member of the 5'-terminal oligopyrimidine gene family reveals common features of snoRNA host genes. Mol Cell Biol 18(12):6897–6909

Song MA, Park JH, Jeong KS, Park DS, Kang MS, Lee S (2007) Quantification of CpG methylation at the 5′-region of XIST by pyrosequencing from human serum. Electrophoresis 28 (14):2379–2384. doi:10.1002/elps.200600852

Srikantan V, Zou Z, Petrovics G, Xu L, Augustus M, Davis L, Livezey JR, Connell T, Sesterhenn IA, Yoshino K, Buzard GS, Mostofi FK, McLeod DG, Moul JW, Srivastava S (2000) PCGEM1, a prostate-specific gene, is overexpressed in prostate cancer. Proc Natl Acad Sci USA 97(22):12216–12221. doi:10.1073/pnas.97.22.12216 97/22/12216 [pii]

Stacey SN, Sulem P, Masson G, Gudjonsson SA, Thorleifsson G, Jakobsdottir M, Sigurdsson A, Gudbjartsson DF, Sigurgeirsson B, Benediktsdottir KR, Thorisdottir K, Ragnarsson R, Scherer D, Hemminki K, Rudnai P, Gurzau E, Koppova K, Botella-Estrada R, Soriano V, Juberias P, Saez B, Gilaberte Y, Fuentelsaz V, Corredera C, Grasa M, Hoiom V, Lindblom A, Bonenkamp JJ, van Rossum MM, Aben KK, de Vries E, Santinami M, Di Mauro MG, Maurichi A, Wendt J, Hochleitner P, Pehamberger H, Gudmundsson J, Magnusdottir DN, Gretarsdottir S, Holm H, Steinthorsdottir V, Frigge ML, Blondal T, Saemundsdottir J, Bjarnason H, Kristjansson K, Bjornsdottir G, Okamoto I, Rivoltini L, Rodolfo M, Kiemeney LA, Hansson J, Nagore E, Mayordomo JI, Kumar R, Karagas MR, Nelson HH, Gulcher JR, Rafnar T, Thorsteinsdottir U, Olafsson JH, Kong A, Stefansson K (2009) New common variants affecting susceptibility to basal cell carcinoma. Nat Genet 41(8):909–914. doi:ng.412 [pii] 10.1038/ng.412

Stange DE, Radlwimmer B, Schubert F, Traub F, Pich A, Toedt G, Mendrzyk F, Lehmann U, Eils R, Kreipe H, Lichter P (2006) High-resolution genomic profiling reveals association of chromosomal aberrations on 1q and 16p with histologic and genetic subgroups of invasive breast cancer. Clin Cancer Res 12(2):345–352. doi:12/2/345 [pii] 10.1158/1078-0432.CCR-05-1633

Stilgenbauer S, Nickolenko J, Wilhelm J, Wolf S, Weitz S, Dohner K, Boehm T, Dohner H, Lichter P (1998) Expressed sequences as candidates for a novel tumor suppressor gene at band 13q14 in B-cell chronic lymphocytic leukemia and mantle cell lymphoma. Oncogene 16 (14):1891–1897. doi:10.1038/sj.onc.1201764

Sultan M, Schulz MH, Richard H, Magen A, Klingenhoff A, Scherf M, Seifert M, Borodina T, Soldatov A, Parkhomchuk D, Schmidt D, O'Keeffe S, Haas S, Vingron M, Lehrach H, Yaspo ML (2008) A global view of gene activity and alternative splicing by deep sequencing of the human transcriptome. Science 321(5891):956–960

Taft RJ, Pang KC, Mercer TR, Dinger M, Mattick JS (2010) Non-coding RNAs: regulators of disease. J Pathol 220(2):126–139

Tano K, Mizuno R, Okada T, Rakwal R, Shibato J, Masuo Y, Ijiri K, Akimitsu N (2010) MALAT-1 enhances cell motility of lung adenocarcinoma cells by influencing the expression of motility-related genes. FEBS Lett 584(22):4575–4580

Tanos V, Prus D, Ayesh S, Weinstein D, Tykocinski ML, De-Groot N, Hochberg A, Ariel I (1999) Expression of the imprinted H19 oncofetal RNA in epithelial ovarian cancer. Eur J Obstet Gynecol Reprod Biol 85(1):7–11. doi:S0301211598002759 [pii]

Tripathi V, Ellis JD, Shen Z, Song DY, Pan Q, Watt AT, Freier SM, Bennett CF, Sharma A, Bubulya PA, Blencowe BJ, Prasanth SG, Prasanth KV (2010) The nuclear-retained noncoding RNA MALAT1 regulates alternative splicing by modulating SR splicing factor phosphorylation. Mol Cell 39(6):925–938. doi:S1097-2765(10), 00621-0 [pii] 10.1016/j.molcel.2010.08.011

Tsai MC, Manor O, Wan Y, Mosammaparast N, Wang JK, Lan F, Shi Y, Segal E, Chang HY (2010) Long noncoding RNA as modular scaffold of histone modification complexes. Science 329(5992):689–693. doi:science.1192002 [pii] 10.1126/science.1192002

Tsai MC, Spitale RC, Chang HY (2011) Long intergenic noncoding RNAs: new links in cancer progression. Cancer Res 71(1):3–7. doi:71/1/3 [pii] 10.1158/0008-5472.CAN-10-2483

Turnbull C, Ahmed S, Morrison J, Pernet D, Renwick A, Maranian M, Seal S, Ghoussaini M, Hines S, Healey CS, Hughes D, Warren-Perry M, Tapper W, Eccles D, Evans DG, Hooning M, Schutte M, van den Ouweland A, Houlston R, Ross G, Langford C, Pharoah PD, Stratton MR, Dunning AM, Rahman N, Easton DF (2010) Genome-wide association study identifies five new breast cancer susceptibility loci. Nat Genet 42(6):504–507. doi:ng.586 [pii] 10.1038/ng.586

Verkerk AJ, Ariel I, Dekker MC, Schneider T, van Gurp RJ, de Groot N, Gillis AJ, Oosterhuis JW, Hochberg AA, Looijenga LH (1997) Unique expression patterns of H19 in human testicular cancers of different etiology. Oncogene 14(1):95–107. doi:10.1038/sj.onc.1200802

Vincent-Salomon A, Ganem-Elbaz C, Manie E, Raynal V, Sastre-Garau X, Stoppa-Lyonnet D, Stern MH, Heard E (2007) X inactive-specific transcript RNA coating and genetic instability of the X chromosome in BRCA1 breast tumors. Cancer Res 67(11):5134–5140. doi:67/11/5134 [pii] 10.1158/0008-5472.CAN-07-0465

Wang XS, Zhang Z, Wang HC, Cai JL, Xu QW, Li MQ, Chen YC, Qian XP, Lu TJ, Yu LZ, Zhang Y, Xin DQ, Na YQ, Chen WF (2006) Rapid identification of UCA1 as a very sensitive and specific unique marker for human bladder carcinoma. Clin Cancer Res 12(16):4851–4858. doi:12/16/4851 [pii] 10.1158/1078-0432.CCR-06-0134

Wang F, Li X, Xie X, Zhao L, Chen W (2008a) UCA1, a non-protein-coding RNA upregulated in bladder carcinoma and embryo, influencing cell growth and promoting invasion. FEBS Lett 582(13):1919–1927. doi:S0014-5793(08), 00413-4 [pii] 10.1016/j.febslet.2008.05.012

Wang X, Arai S, Song X, Reichart D, Du K, Pascual G, Tempst P, Rosenfeld MG, Glass CK, Kurokawa R (2008b) Induced ncRNAs allosterically modify RNA-binding proteins in cis to inhibit transcription. Nature 454(7200):126–130. doi:nature06992 [pii] 10.1038/nature06992

Weakley SM, Wang H, Yao Q, Chen C (2011) Expression and function of a large non-coding RNA Gene XIST in human cancer. World J Surg. doi:10.1007/s00268-010-0951-0

Wilusz JE, Freier SM, Spector DL (2008) 3′ end processing of a long nuclear-retained noncoding RNA yields a tRNA-like cytoplasmic RNA. Cell 135(5):919–932. doi:S0092-8674(08), 01303-2 [pii] 10.1016/j.cell.2008.10.012

Wilusz JE, Sunwoo H, Spector DL (2009) Long noncoding RNAs: functional surprises from the RNA world. Genes Dev 23(13):1494–1504. doi:23/13/1494 [pii] 10.1101/gad.1800909

Winter J, Jung S, Keller S, Gregory RI, Diederichs S (2009) Many roads to maturity: microRNA biogenesis pathways and their regulation. Nat Cell Biol 11(3):228–234

Wolf S, Mertens D, Schaffner C, Korz C, Dohner H, Stilgenbauer S, Lichter P (2001) B-cell neoplasia associated gene with multiple splicing (BCMS): the candidate B-CLL gene on 13q14 comprises more than 560 kb covering all critical regions. Hum Mol Genet 10(12):1275–1285

Wu ZS, Lee JH, Kwon JA, Kim SH, Han SH, An JS, Lee ES, Park HR, Kim YS (2009) Genetic alterations and chemosensitivity profile in newly established human renal collecting duct carcinoma cell lines. BJU Int 103(12):1721–1728. doi:BJU8290 [pii] 10.1111/j.1464-410X.2008.08290.x

Yamada K, Kano J, Tsunoda H, Yoshikawa H, Okubo C, Ishiyama T, Noguchi M (2006) Phenotypic characterization of endometrial stromal sarcoma of the uterus. Cancer Sci 97 (2):106–112. doi:CAS [pii] 10.1111/j.1349-7006.2006.00147.x

Yang Z, Zhou L, Wu LM, Lai MC, Xie HY, Zhang F, Zheng SS (2011) Overexpression of long non-coding RNA HOTAIR predicts tumor recurrence in hepatocellular carcinoma patients following liver transplantation. Ann Surg Oncol. doi:10.1245/s10434-011-1581-y

Yap KL, Li S, Munoz-Cabello AM, Raguz S, Zeng L, Mujtaba S, Gil J, Walsh MJ, Zhou MM (2010) Molecular interplay of the noncoding RNA ANRIL and methylated histone H3 lysine 27 by polycomb CBX7 in transcriptional silencing of INK4a. Mol Cell 38(5):662–674

Yoshimizu T, Miroglio A, Ripoche MA, Gabory A, Vernucci M, Riccio A, Colnot S, Godard C, Terris B, Jammes H, Dandolo L (2008) The H19 locus acts in vivo as a tumor suppressor. Proc Natl Acad Sci USA 105(34):12417–12422. doi:0801540105 [pii] 10.1073/pnas.0801540105

Yu M, Ohira M, Li Y, Niizuma H, Oo ML, Zhu Y, Ozaki T, Isogai E, Nakamura Y, Koda T, Oba S, Yu B, Nakagawara A (2009) High expression of ncRAN, a novel non-coding RNA mapped to chromosome 17q25.1, is associated with poor prognosis in neuroblastoma. Int J Oncol 34 (4):931–938

Zhang X, Zhou Y, Mehta KR, Danila DC, Scolavino S, Johnson SR, Klibanski A (2003) A pituitary-derived MEG3 isoform functions as a growth suppressor in tumor cells. J Clin Endocrinol Metab 88(11):5119–5126

Zhang C, Kawakami T, Okada Y, Okamoto K (2005) Distinctive epigenetic phenotype of cancer testis antigen genes among seminomatous and nonseminomatous testicular germ-cell tumors. Genes Chromosomes Cancer 43(1):104–112

Zhang X, Rice K, Wang Y, Chen W, Zhong Y, Nakayama Y, Zhou Y, Klibanski A (2010) Maternally expressed gene 3 (MEG3) noncoding ribonucleic acid: isoform structure, expression, and functions. Endocrinology 151(3):939–947

Zhao J, Sun BK, Erwin JA, Song JJ, Lee JT (2008) Polycomb proteins targeted by a short repeat RNA to the mouse X chromosome. Science 322(5902):750–756

Zhao R, Bodnar MS, Spector DL (2009) Nuclear neighborhoods and gene expression. Curr Opin Genet Dev 19(2):172–179

Zhao J, Ohsumi TK, Kung JT, Ogawa Y, Grau DJ, Sarma K, Song JJ, Kingston RE, Borowsky M, Lee JT (2010) Genome-wide identification of polycomb-associated RNAs by RIP-seq. Mol Cell 40(6):939–953

Zhu Y, Yu M, Li Z, Kong C, Bi J, Li J, Gao Z (2011) ncRAN, a newly identified long noncoding RNA, enhances human bladder tumor growth, invasion, and survival. Urology 77(2):510 e511–515

Part II
Methods

Chapter 9
Expression Profiling of ncRNAs Employing RNP Libraries and Custom LNA/DNA Microarray Analysis

Konstantinia Skreka, Michael Karbiener, Marek Zywicki, Alexander Hüttenhofer, Marcel Scheideler, and Mathieu Rederstorff

Abstract Recently, it has been shown by the ENCODE consortium that more than 90% of the human genome might be transcribed. While only about 1.5% of these transcripts correspond to mRNAs, it was proposed that the majority of them (i.e., 88.5%) might correspond to regulatory noncoding RNAs (ncRNAs). Numerous protocols dedicated to the generation of cDNA libraries coupled to next-generation sequencing (NGS) technologies are currently available to identify novel ncRNA species, and we have recently developed a novel procedure for the generation of ribonucleoprotein (RNP) libraries. To validate differential expression of ncRNAs identified using our or any library generation approach, we describe an innovative ncRNA profiling approach based on microarray technology. Employing LNA probes, dedicated to the analysis of small/microRNAs, and DNA probes, dedicated to the study of longer ncRNAs, our platform enables the study of most ncRNAs independently of their length in a single experiment. Detailed methodological solution description includes the automated design of probes to be spotted on the array, optimization of spotting and labeling of probes, as well as hybridization conditions. All the steps have been improved for the analysis of ncRNAs, which are generally difficult to study owing to their peculiarities in terms of secondary structure or abundance.

Keywords High-throughput sequencing • microarray • ncRNP libraries • noncoding RNAs (ncRNAs) • ribonucleoprotein particles (RNP)

M. Rederstorff (✉)
Biocenter, Innsbruck Medical University, Section for Genomics and RNomics, Innsbruck, Austria

Biopole, Lorraine University, CNRS UMR 7214, Vandoeuvre-les-Nancy, France
e-mail: Mathieu.rederstorff@maem.uhp-nancy.fr

B. Mallick (eds.), *Regulatory RNAs*, DOI 10.1007/978-3-642-22517-8_9,
© Springer-Verlag Berlin Heidelberg (outside the USA) 2012

9.1 Introduction

9.1.1 The ncRNA Transcriptome

Pervasive transcription of eukaryotic genomes has been widely described within the last decade, with the prediction that the majority of transcripts does not code for proteins (Brosius 2005; Carninci et al. 2005; Cheng et al. 2005; Kampa et al. 2004; Mattick and Makunin 2005, 2006; Willingham and Gingeras 2006). High-resolution analysis of about 1% of the human genome by the ENCODE consortium has even shown that more than 90% of the genome might be transcribed, with about 88.5% of the transcripts corresponding to ncRNAs and only 1.5% to mRNAs (Birney et al. 2007). As a consequence, the predicted number of ncRNA transcripts originating from these regions has increased extremely, with the highest estimations corresponding to about 450,000 ncRNAs encoded in the human genome. This suggests that ncRNAs may serve as major regulatory elements in eukaryal genomes (Mattick and Makunin 2005, 2006). However, most of these transcripts remain of unknown function, and it is still a matter of debate which ones represent real, functional ncRNA species (Willingham and Gingeras 2006).

Most, if not all, functional ncRNAs are involved in RNA-protein complexes in the cell, designated as ribonucleoprotein particles (RNPs), with a broad range of possible functions. ncRNAs such as ribosomal RNAs (rRNAs) or transfer RNAs (tRNAs) are involved in protein synthesis, while the more recently identified small interfering or microRNAs (siRNAs, miRNAs) were found to regulate gene expression (Liu et al. 2008; Ghildiyal and Zamore 2009; *see also Chap. 12 of this volume*). New classes of ncRNAs continue to be discovered (Mercer et al. 2008; Guttman et al. 2009; Dinger et al. 2008) and even for already known ncRNAs, a complete understanding of their functions is still lacking.

9.1.2 ncRNA Identification and Profiling: Microarray Versus Next-Generation Sequencing Approaches

To identify novel ncRNA candidates, numerous computational (Kawaji et al. 2009) as well as experimental approaches based on the generation of cDNA libraries (Rederstorff et al. 2010; Huttenhofer and Vogel 2006) have been described. Previously, cDNA libraries were generated using RNA that was size-separated on denaturing gels (Jochl et al. 2008). This led mostly to the cloning and sequencing of RNA species corresponding to ribosomal RNAs (rRNAs) or mRNA degradation products, constituting a major issue. Therefore, to distinguish the biologically relevant ncRNAs from junk transcripts, we developed a novel procedure for the generation of cDNA libraries derived from ncRNAs involved in functional ribonucleoprotein particles (Fig. 9.1) (Rederstorff et al. 2010; Rederstorff and Huttenhofer

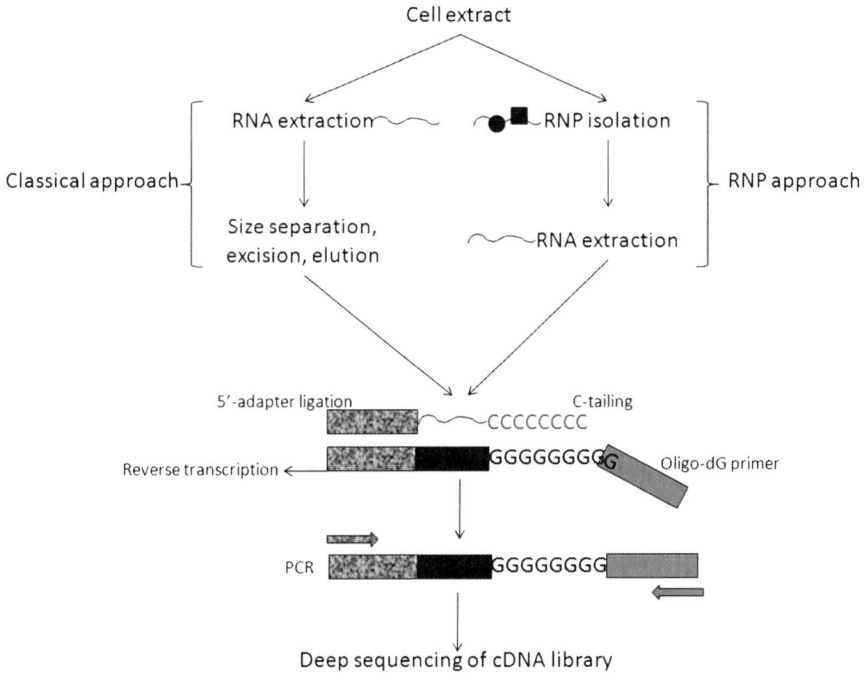

Fig. 9.1 *Generation of ncRNA libraries.* Total RNA or RNP extracts are prepared from the cells or tissues of interest. Classical library generation approaches use naked, phenol-extracted, size-separated RNAs, which mostly leads to sequencing of non functional degradation products. RNP libraries are enriched in functional RNA species, as the ncRNAs they contain are selected upon binding to proteins within functional RNPs: to do so, RNP extracts are sedimentated on 10–30% glycerol gradient before RNAs are extracted from the gradient fractions. Next, since ncRNAs do not contain a poly(A) tail, they are, in the first step of the library generation process, tailed with CTP and poly(A) polymerase. After the addition of C-tails, a primer-adapter is ligated to the 5'-end of ncRNAs, and they are reverse-transcribed into cDNA employing an oligo-d(G) primer. PCR-amplified cDNA libraries are next submitted to high-throughput sequencing

2011a). Thus, RNP libraries constitute an alternative approach for ncRNA transcriptome studies employing RNA deep sequencing (RNA-seq). By employing this approach, we could enrich the libraries with functional ncRNA species compared to size-separated RNA libraries, as well as identify numerous candidates for novel functional ncRNA.

One of the advantages of RNA-seq techniques is that they are open to the identification of novel transcripts. Therefore, such "open" techniques are becoming the standard for high-throughput transcriptome analysis (Hawkins et al. 2010) and will benefit from the variations in library generation protocols to obtain an exhaustive picture of the ncRNA transcriptome within a defined system. On the other hand, microarray technologies belong to the "closed" approach category, which are based on previous knowledge of the transcripts studied and are, thus, not suitable for the identification of previously uncharacterized ncRNA candidates (Table 9.1). However, they remain largely used for various transcriptome analyses and remain a method of choice for the profiling of ncRNAs once their identification has been achieved by RNA-seq approaches.

Therefore, although the recently described "digital gene expression" by next-generation sequencing (NGS) has been introduced as a promising new platform for assessing the copy number of transcripts, thereby providing a digital record of the numerical frequency of a sequence in a sample, the question of whether NGS or microarray technology is better suited for ncRNA profiling is open for debate. A study employing two synthetic ncRNA mixtures consisting of 744 synthetic RNA oligonucleotides enabled a direct comparison of the results obtained from each platform for the known RNA sample content (Willenbrock et al. 2009). Microarray technology appeared to be highly specific and sensitive, surpassing next-generation sequencing for absolute quantification of small ncRNAs (e.g., miRNAs). Nonetheless, sequencing offers other advantages, such as enabling discovery of new sequence variants, although this study indicates that thorough filtering is important in order to avoid overinterpretation of potential sequencing errors. Both technologies deliver highly reproducible expression data and perform well in relative gene expression studies (Willenbrock et al. 2009; Marioni et al. 2008). As for NGS, one has also to be aware of potential inaccuracies in sequencing data that might be introduced due to PCR biases, as most current techniques involve an amplification step of RNA material (Metzker 2010; Carninci et al. 2005). In contrast, such amplification is usually not necessary for microarray analyses.

Other methodologies have been applied to profile ncRNA expression, such as northern blotting (Griffiths-Jones 2006), quantitative reverse transcription PCR-based (qRT-PCR) amplification (Schmittgen et al. 2004) as well as bead-based profiling (Goff et al. 2005) (Table 9.1). However, if a large number of ncRNAs are screened for in a parallel manner, microarray-based profiling has grown in popularity as the primary tool for gene expression analysis (Castoldi et al. 2008). Although microarray-based expression profiling may be less sensitive in detecting low abundant transcripts (compared to qRT-PCR), its application is often less reagent and/or time consuming (compared to northern blotting or cloning approaches) or less expensive (compared to qRT-PCR).

9.1.3   ncRNA Microarrays: The Challenges

Specificity and accuracy of mRNA expression profiling techniques applied to short or long ncRNAs are challenged by (1) the short length of the transcripts that offer little sequence for appending detection molecules (in case of detecting short ncRNAs), (2) a wide range of predicted melting temperatures (T_m) compared to their DNA counterparts, (3) the low copy number of some transcripts, (4) their frequent occurrence in families that in some cases differ by as little as a single

Table 9.1 Comparison of ncRNA profiling approaches

Method	Principle	Major application	Advantages	Drawbacks
Next-generation sequencing (NGS)	cDNA libraries are sequenced	Determination of full transcriptome road map	Discovery of previously unknown transcripts is possible	Highly dependent on libraries generation protocols, which can lead to various biases
		Expression profiling and comparative transcriptomics	Huge amount of data obtained in a single experiment	Huge amount of data obtained, which requires adapted tools for analysis, and space for storage
		Clinical diagnostic applications are emerging		
Microarray	"DNA capture probes" are attached to a solid substrate such as glass. During hybridization, labeled "target" transcripts diffuse passively across the glass surface and will anneal to DNA probes with sequence complementarity and form a nucleic acid duplex. Hybridized targets can be read out by spatial label quantification in order to define the relative abundance of target transcripts	Highly parallel gene expression profiling for thousands of genes (up to genome-wide coverage) for disease diagnostics as well as for the identification of genes that are differentially expressed in a biological process of interest upon a chosen stimulus	Highly parallel analysis of thousands of genes	Novel target transcript sequences cannot be discovered
			As many distinct target sequences can be detected and quantified as DNA capture probes of distinct sequence are attached to the solid substrate	Lower sensitivity compared to NGS and qRT-PCR
			Fast procedure	
			Numerous analysis software tools are available for the analysis of microarray-based gene expression analysis	
			Robust detection of differential expression as thousands of genes that are not differentially expressed serve as normalizers	

(continued)

Table 9.1 (continued)

Method	Principle	Major application	Advantages	Drawbacks
Quantitative reverse transcription polymerase chain reaction (qRT-PCR)	"Target" transcripts are reverse-transcribed to cDNAs. An individual primer pair for each target cDNA is used for the amplification by PCR	Parallel gene expression profiling for up to hundreds of genes for disease diagnostics as well as for the identification of genes that are differentially expressed in a biological process of interest upon a chosen stimulus	Highly parallel analysis of hundreds of genes	As identification of differential expression is strongly dependent on a specific housekeeping gene, this internal reference gene has to be validated for not being responsive to the chosen stimulus in the biological process of interest
		Validation of high-throughput data	The sequence to be detected is technically not limited	Only a limited number of genes can be detected simultaneously
			Fast procedure	Cost
Northern blot (NB)	"Target" transcripts are size-separated by gel electrophoresis and then transferred to a solid substrate such as a membrane. During hybridization, labeled "DNA probes" will anneal to target transcripts with sequence complementarity and form a nucleic acid duplex. Hybridized targets can be read out by spatial label quantification in order to define the relative abundance of target transcripts	Small-scale gene expression profiling to identify genes that are differentially expressed in a biological process of interest upon a chosen stimulus	The sequence to be detected is technically not limited	Compared to the methods above, relatively large amounts of RNA are required as starting material
			Determination of the size of the candidates and of eventual processing products	Semiquantitative
				Genes can only be analyzed sequentially
				In many applications, work with radioactively labeled DNA probes is necessary
				Time consuming

nucleotide (Griffiths-Jones 2006), and, last but not least, (5) secondary structures, particularly in long ncRNAs (Anthony et al. 2003; Chandler et al. 2003). These features complicate the design of suitable capture probes across the complete "RNome" and the optimization of hybridization conditions that are unbiased regarding an accurate detection of all ncRNAs. A combined platform for all ncRNAs independently of their length appears therefore difficult to establish due to the numerous different parameters each class of ncRNAs is presenting. Hybridization temperature gradients (Hutzinger et al. 2010) have been employed to bypass these problems, however, using labeled total RNA with a temperature gradient might also result in higher background due to unspecific hybridization.

To overcome the challenges (1) and (2) in case of short ncRNAs, modified oligonucleotides termed locked nucleic acids (LNA) can be incorporated within the oligonucleotide capture probes immobilized on the array surface. LNA is a synthetic RNA analog characterized by increased thermostability of nucleic acid duplexes by 2–10°C per LNA/RNA hybrid when LNA monomers are introduced into oligonucleotides (Fig. 9.2a). As a consequence of this property, LNA-modified capture probes can be designed such that, despite the limited length of the short ncRNA capture probe, a uniform T_m can be applied to a genome-wide set of probes, allowing the establishment of normalized hybridization conditions. To overcome

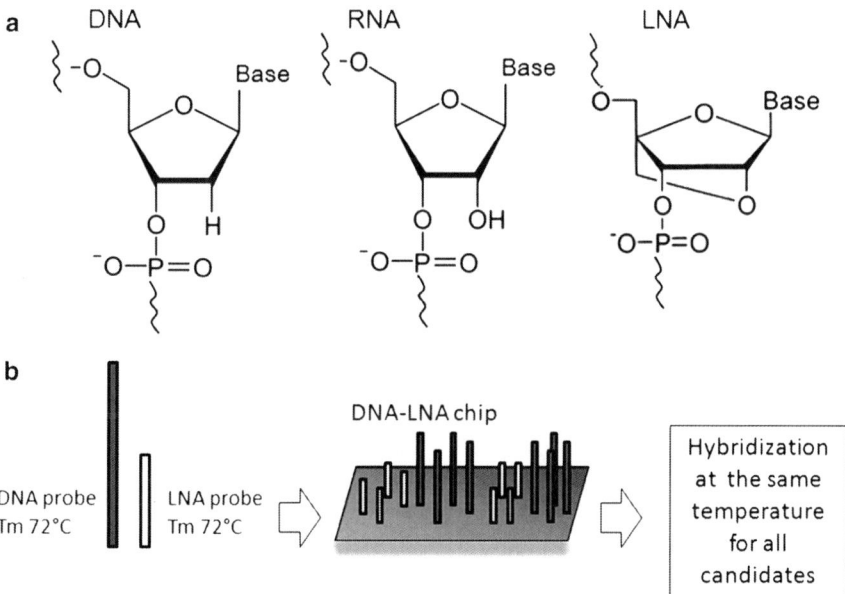

Fig. 9.2 *Hybrid LNA/DNA microarrays.* (**a**) Structures of DNA, RNA, and LNA nucleotide monomers. An LNA is a synthetic RNA analog which maintains a stable 3'-endo conformation due to a bridge connecting the 2' oxygen with the 4' carbon of the ribose. As a consequence, an oligonucleotide containing LNA residues has a higher melting temperature than an RNA or a DNA oligonucleotide of the same sequence. (**b**) Spotting of longer customized DNA probes together with shorter LNA probes featuring identical melting temperature

the challenge (2) in case of long ncRNAs, their longer sequence allows a more flexible capture probe design that meets the requirement of a uniform T_m for the whole capture probe set.

Hence, the combination of LNA probes, dedicated to the analysis of small/microRNAs, together with DNA probes dedicated to longer ncRNAs is described. Such a platform can be functional if the DNA and LNA oligos are designed to all have the same T_m so that they can be used in combination in a single experiment (Fig. 9.2b). As a proof of concept, our study was performed on miRNAs and different longer ncRNAs (LNA probes and custom DNA probes, respectively, Skreka et al., unpublished results, manuscript in preparation).

Concerning low abundant transcripts (3), LNA-modified capture probes yielded a several fold increased sensitivity that was more obvious when lower amounts (2.5–5 μg) of total input RNA were used (Castoldi et al. 2008). Furthermore, the use of microarray slides with reflective optical coating can increase the sensitivity of fluorescence-based detection systems compared to traditional (first-generation) glass slides (Redkar et al. 2006). Moreover, in terms of specificity (4), LNA-modified capture probes hybridize their ncRNA targets in a highly specific manner, as even a single nucleotide mismatch is sufficient to destabilize the heteroduplex. Therefore, LNA-modified capture probes enable efficient discrimination between ncRNA family members, at least when they differ in nucleotides close to the central position (Castoldi et al. 2008).

Secondary structures of long ncRNAs (5) can be addressed by hybridization at higher temperature. A basic prerequisite for that is a higher T_m of the capture probe/RNA hybrid that can be achieved either by incorporation of LNA-modified nucleotides into the capture probe or by increased capture probe sequence length.

9.2   Materials

9.2.1   LNA Probes

The miRCURY LNA microRNA array ready-to-spot probe set 208010 was purchased from Exiqon as an LNA capture probe set for short ncRNAs detection. This set comprises 2,056 capture probes designed to have a uniform T_m of 72°C and covers all miRNAs of miRBase version 9.2.

9.2.2   DNA Probes

The DNA probes used were $5'$-C_6 amino-modified to attach to the slide's surface and their T_m was set at 72°C to comply with the Exiqon LNA set hybridization conditions. DNA probes were desalted and diluted in 3xSSC, 1.5 M Betaine buffer to a final concentration of 20 μM. DNA probes were purchased from Microsynth, Balgach, Switzerland.

9.2.3 ncRNA Chip

The LNA-based capture probe set for short ncRNAs as well as the self-designed DNA-based capture probe set for long ncRNAs were spotted on HiSens epoxy-coated glass slides (Nexterion) using the MicroGrid II Microarrayer (Zinsser Analytic).

9.2.4 Hybridization Station

Hybridizations were performed using the Tecan HS400 hybridization station.

9.2.5 Microarray Scanner

The ncRNA chip was scanned using the Axon instruments GenePix 4000B.

9.2.6 Labeling Kits

The poly(A) labeling kit used was the NCode Rapid miRNA Labeling System (MIRLSRPD-20). This kit contains Alexa Fluor 5 and Alexa Fluor 3 dyes, equivalent to Cy5 and Cy3, respectively.

9.2.7 Poly(A) Tailing Buffer

We found it more efficient to replace the optimized buffers for small RNA poly(A) tailing with a reaction buffer containing 0.05 M Tris-HCl (pH 8.0), 0.25 M NaCl, and 10 mM $MgCl_2$.

9.3   Methods

In this section, after briefly explaining how to generate ncRNA libraries, we will provide hints on how to achieve adequate bioinformatical analysis of NGS data, as this step is of fundamental importance to any subsequent study, and will next focus on the various methodological aspects for the ncRNA profiling using microarray technology (Fig. 9.3).

9.3.1   RNP Library Generation

Library generation approaches aimed at the identification of novel functional ncRNAs have been described extensively (Huttenhofer and Vogel 2006; Jochl et al. 2008; Saxena and Carninci 2010). The original RNP library generation

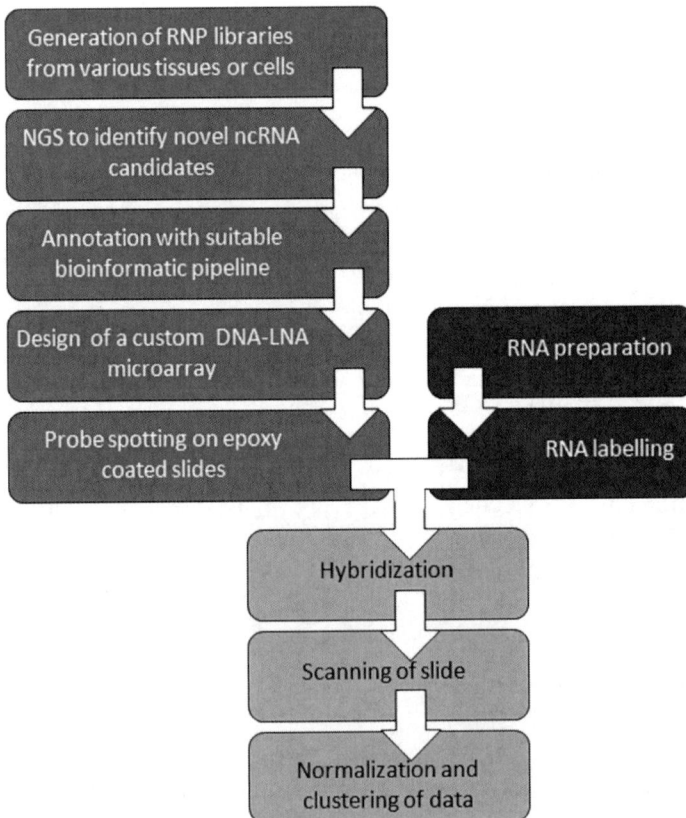

Fig. 9.3 Overview of the experimental strategy for the identification and profiling of ncRNAs

approach enables considerable enrichment of libraries with functional ncRNAs owing to prior isolation of ribonucleoprotein particles (Fig. 9.1) (Rederstorff et al. 2010; Rederstorff and Huttenhofer 2011b; Rederstorff 2011). Briefly, RNP extracts are prepared from the cells or tissues of interest and RNP are sedimented on a 10–30% glycerol gradient. RNAs are extracted from the gradient fractions. Since ncRNAs do not contain a poly(A) tail, they are, in the first step of the library generation process, tailed with CTP and poly(A) polymerase. After the addition of C-tails, a primer-adapter is ligated to the 5′-end of ncRNAs and they are reverse-transcribed into cDNA employing an oligo-d(G) primer. Libraries are next amplified by PCR using specific oligonucleotides. cDNA libraries are then submitted to high-throughput sequencing (Metzker 2010).

9.3.2   NGS Data Analysis

The general concept of the analysis of cDNA libraries containing novel ncRNAs includes similar steps to those used in other types of transcript analysis (*see also Chap. 10 of this volume*). However, due to specific features of ncRNA genes, some aspects vary significantly. The first step consists in removal of any adaptor sequence that has been used during the library preparation from the reads. Sequences of low quality should be discarded at this step as well. Reads are next assembled into contigs. The most appropriate approach for this is based on genome mapping of the reads and assembly of the contigs, based on the overlapping positions within the genomic sequence. An alternative approach, based on de novo contig assembly without any reference genome (Martin et al. 2010), requires relatively large overlaps between the reads and low repetitive sequence content, which is not the case for ncRNAs. During the genome alignment, it is crucial to include all positions of the reads that would map in more than one unique genomic locus, or most of the snoRNAs and other ncRNAs encoded by multiplicated genes would be discarded. The next step is to annotate the existing contigs with the gene information. The most straightforward way for this is to check for overlaps with the genome annotation tables from one of the genome databases, like Ensembl (Hubbard et al. 2009) or UCSC Genome Browser (Fujita et al. 2011). In the case of cDNA libraries containing ncRNAs, such annotation will result in identification of numerous intergenic contigs. To complement this annotation, it is useful to perform a sequence similarity search against noncoding RNA databases, like the Functional RNA Database (fRNAdb) (Mituyama et al. 2009). Finally, prediction of whether the remaining intergenic contigs belong to known ncRNA families, like microRNAs (Friedlander et al. 2008) or snoRNAs (Schattner et al. 2005), can be performed.

Identification of differentially expressed ncRNAs, in other terms, ncRNA profiling, corresponds to the next step in the study. We will next describe in detail a microarray-based approach to do so; however, numerous in silico methods for this purpose also exist, which we will briefly describe here. Since most of these tools have been designed to work with mRNA or microRNA sequencing data, the analysis of putatively novel ncRNAs requires particular precautions. The crucial step in the assessment of differential expression is the correct normalization of the read numbers. The standard method for mRNA transcriptome profiling relies on RPKM values (reads per kilobase of the transcript per million of mapped reads) (Mortazavi et al. 2008), which depends on the length of the transcript. This value reflects the random fragmentation step introduced in RNA-seq protocols for mRNAs, leading to higher representation of fragments originating from longer transcripts. Concerning small ncRNA gene analysis, as no fragmentation is performed, length parameters are not required for standardization. Thus, the best method is to normalize read counts against the total number of reads mapping to the genome. The approach published recently by Robinson and Oshlack (Robinson et al. 2010) is of special recommendation for that purpose. It reduces the library composition bias caused by occupancy of the read space by the most abundant transcripts. Whenever the experiment contains any replicates, the variance normalization should then be performed in the next step. Normalized data can next be used directly for statistical testing of differential expression.

Assembly and annotation of ncRNA-derived cDNA libraries can be performed step-by-step using different tools; however, we recommend the use of the APART pipeline (Automated Pipeline for Analysis of RNA Transcripts) (Zywicki et al. 2011). This novel approach provides fully automated workflow for analysis of ncRNA-containing cDNA libraries. It includes trimming of the reads, genome mapping, as well as contig assembly and annotation. One of the advantages of the APART pipeline is that it can handle nonunique reads, and it offers a convenient way to identify and cluster spurious contigs derived from multiple mapping of a single set of reads. For differential expression analysis, the Bioconductor packages edgeR (Robinson et al. 2010) and DESeq (Anders and Huber 2010) software are recommended. They both provide extensive solutions for normalization and calling of differentially expressed genes. EdgeR additionally supports the library composition normalization by Robinson and Oshlack.

9.3.3 Microarray Probe Design

Design of specific custom oligonucleotide probes for small noncoding RNAs is much more complex than for mRNA arrays. The first task in designing the best probes is to choose the appropriate duplex T_m of the probes. This temperature should be high enough to destabilize structured ncRNAs. However, the increase of T_m requires an extension of the probe length, which is then incompatible with the small sizes of several ncRNA classes (e.g., microRNAs). Employing locked nucleic acid (LNA) during the oligonucleotide synthesis is one way to overcome this limitation. However, their elevated cost is a major drawback, rendering this method unsuitable for large- or genome-scale projects. A good conciliation can be reached by using the combined LNA/DNA platform. However, such an option limits the choice of the T_m to the one of the LNA probe set. The second challenge in the design of ncRNA microarray probes is to verify their specificity. This task, which is relatively easy in case of mRNA probe design, where only poly(A) tailed transcripts (or cDNAs) are used for hybridization, is much more complex here, mainly because of the use of total RNA in the experiment. Based on the pilot ENCODE project findings, one can expect about 90% of the genome to be actively transcribed (Birney et al. 2007); thus, the verification of specificity has to be performed against the whole genome. In the case of ncRNA genes encoded in clusters (e.g., snoRNAs) (Elmen et al. 2005; Nahkuri et al. 2008), a simple search of the probe sequence within the genome can be confusing and may result in multiple genomic hits even for a specific probe. Therefore, we propose to perform the specificity verification in two steps. The first step consists in comparison of the probe sequence with the genomic sequence. Then, the genomic positions of known ncRNAs are obtained from the Functional RNA Database (fRNAdb) (Mituyama et al. 2009), and all the matches of the probe are verified to be annotated as the same ncRNA. One has to keep in mind that, because of their short length, it might be impossible to design any single specific DNA probe of desired T_m for some small AU-rich ncRNAs.

Many tools for designing microarray probes are available at the moment. Concerning ncRNAs, it is important that the tool used provides three major possibilities: (1) calculation of the T_m of RNA-DNA duplexes, (2) optimization of the probes to reach the desired T_m, and (3) verification of the local secondary structure on the RNA target site. One of these tools corresponds to the program OligoWiz (Wernersson and Nielsen 2005). However, for ncRNA probe design, we even suggest to increase the importance of the target RNA structure. Since OligoWiz is only using known genes sequence sets (obtained from the Unigene) for specificity verification, we recommend to additionally filter the results with any local search software (e.g., blast, fasta) using the nucleotide substitution matrix recently released by Eklund et al. (Eklund et al. 2010) on the full genomic sequence.

9.3.4 RNA Preparation and Labeling

9.3.4.1 Total RNA Preparation and Quantification

LNA microarrays can function with small amounts of RNA, starting with as few as 30 ng of total RNA, according to the manufacturer's recommendations (Exiqon), which can be directly labeled. In contrast, DNA microarrays require 10–25 μg of total RNA as starting material for reverse transcription and cDNA labeling (see below). Direct RNA labeling protocols limit the maximum amount of total RNA used as starting material to 5 μg. We recommend using 1–2 μg RNA (total RNA or size-separated RNA) as starting material when employing an LNA-DNA platform. Indeed, we observed that using 5 μg of RNA with this platform increased unspecific hybridization.

Total RNA is extracted from any tissue or cell line by TriReagent (Sigma) treatment. DNase digestion can be applied to ensure that no traces of DNA contaminate the sample. RNA quantity was estimated with a Nanodrop spectrophotometer (Peqlab). RNA quality should be checked with an Agilent 2100 Bioanalyser before labeling.

9.3.4.2 Labeling of RNA

Generally, microarray-based analysis of gene expression employs cyanine-labeled cDNA. During reverse transcription, aminoallyl-modified nucleotides are incorporated in the cDNA. These modified nucleotides react with N-hydroxysuccinimidyl ester (NHS-ester) reactive groups of cyanine dyes (Cy3, Cy5), thus labeling the molecule. This method is also suitable for long ncRNAs but not for small RNA species that are difficult to reverse transcribe. Moreover, reverse transcription is prone to different possible artifacts (Ozsolak et al. 2009). Therefore, alternative methods, such as direct RNA labeling, are required to study ncRNAs of various sizes.

Most ncRNAs present a stable and conserved secondary structure as well as several posttranscriptionally modified nucleotides (Cantara et al. 2011). The most common modifications correspond to the isomerization of uridine residues to pseudouridines (pseudouridilation) or the methylation of the ribose moiety on the 2'-hydroxyl group (2'O-methylation). These modifications enhance stacking interactions and stabilize the RNA structure (Ishitani et al. 2008). However, they might interfere with optimal direct labeling.

Commercial kits aimed at direct labeling of small RNA species are available. Two different methods have been developed, namely, labeling by enzymatic reaction and labeling by ligation of fluorescent dye dendrimers to poly(A) tailed small RNAs, using an oligo-dT bridge oligonucleotide. By testing both methods, we have observed that the latter one, with a slight modification of the poly(A) tailing protocol (see Materials), provides the best results in labeling long structured RNA species until 400 bp (longer ncRNAs were not tested). It was described in the literature that unfolding of some ncRNAs increases efficiency of poly(A) tailing, and that RNAs with a hairpin structure are more difficult to tail (Martin and Keller 1998). We made similar observations, especially regarding long structured ncRNAs such as 7SK RNA. Therefore, we recommend to first heat-denaturate ncRNAs to increase RNA poly(A) tailing yields.

9.3.5 Hybridization

9.3.5.1 Hybridization Temperature

The hybridization temperature suggested for the LNA capture set according to the manufacturer's recommendations (Exiqon) is 56°C, although T_m of the probes is

72°C. DNA capture probes designed with an identical T_m can be combined with the LNA probes, enabling specific results at a high temperature for both short and long ncRNAs in a single experiment, minimizing unspecific hybridization. A different range of temperature is possible, but we recommend a 56°C–65°C range for optimal results.

9.3.5.2 Sensitivity Versus Specificity

Unspecific hybridization, a problem frequently observed in microarray studies, is difficult to quantify (Dufva et al. 2009). Additionally, sensitivity and specificity of the array are important features. In order to increase specificity, high-quality probe design (as previously described), optimal hybridization temperature, as well as stringent enough washing steps are required. Thus, the quantity of RNA used needs to be adjusted so that sensitivity of the array is not compromised under these stringent conditions.

9.3.5.3 Self-Self Hybridization and Dye Swap

One-color array approaches provide an absolute estimation of the expression of a given gene, as the labeled RNA is hybridized on an array spotted with probes corresponding to known sequences. In two-color array approaches, the same array is hybridized simultaneously with RNA deriving usually from two biological samples, each labeled with a different fluorescent dye. The latter approach, thus, provides an estimation of differential expression of the transcripts in each sample.

The dyes usually used for direct RNA labeling, Cy3 and Cy5, have a different dependency on transcript concentration regarding signal emission, thus creating a bias, which can be gene dependent (Cox et al. 2004; Fang et al. 2007). Therefore, when testing one sample for expression, two technical replicates should be produced, labeled each with either Cy3 or Cy5 and hybridized on the same slide (self-self hybridization). When two or more samples are tested for differential expression, an additional hybridization is performed where the dyes are swapped. This means that the technical replicates of each biological sample are hybridized on different arrays together with the technical replicates of the other biological sample, so that, per array, there is a pair of two different fluorescent dyes, e.g., a Cy3–Cy5 pair. Per experiment, 3–5 biological replicates are required to gain statistical significance of the results.

9.3.6 ncRNA Chip Data Analysis

Hybridized slides were scanned with the software GenePix Pro 4.1 (Axon Instruments). After image acquisition and filtering the data for low intensity, heterogeneity, and saturated spots, results were normalized with our in-house developed software ArrayNorm (Pieler et al. 2004). After background correction, the datasets were normalized by global-mean and dye-swap pair normalization. The obtained result files were used for cluster analyses using the Genesis software tool (Sturn et al. 2002).

9.4   Short Protocol

1. *RNA Labeling*
   Start with 1–2 μg RNA (total RNA or size-separated RNA) per labeling reaction. When using high hybridization temperatures (65°C), use 2 μg RNA.
   Denaturation step:

   – Denaturate RNA by heating at 90°C for 3 min.
   – Immediately cool on ice for 2 min to avoid refolding of denatured RNAs.

   Poly(A) tailing and labeling:

   – Proceed with poly(A) tailing according to the labeling kit manufacturer's instructions. Replace the reaction buffer (see Materials) for better results.

2. *Hybridization*

   – A prehybridization step is needed at the same temperature as the hybridization step for 30 min, to reduce unspecific hybridization.
   – Mix labeling reactions when using two-color arrays, whether when performing self-self hybridization or dye swap (see above).
   – The mixed labeled RNA probe is denatured at 90°C for 3 min, spun down, and injected on the slides.
   – Hybridization is preferably performed at stable temperature between 56°C and 65°C for 16 h. We recommend 65°C for more specific results.
   – We recommend using the protocol provided by Exiqon with the LNA capture set for the washing steps.

9.5   Discussion

Microarray technology is currently challenged by other methods such as real-time PCR or next-generation sequencing, for expression profiling as well as for molecular diagnostic (Jordan 2010). However, combining high-throughput sequencing and microarray analysis should be considered, as these methods appear to be very complementary (Coppee 2008). Identification of novel functional regulatory ncRNAs employing RNP library generation approaches coupled to NGS, in combination with differential expression of the candidates in various conditions (disease, development, etc.) employing microarray technology, appears to be an excellent compromise. Indeed, RNA-seq approaches feature numerous biases or drawbacks, generally linked to isolation of RNA protocols or library generation steps. Therefore, the exhaustive transcriptome elucidation would require the generation and sequencing of as many libraries as possible, e.g., from tissues under various stresses or developmental conditions. Employing customized microarray approaches, therefore, enables to check for ncRNA differential expression in various conditions with a limited number of arrays.

Importantly, since several ncRNAs have been associated to different disorders, their use as potent biomarkers for diagnostic purposes is more and more frequent (Gilad et al. 2008). MicroRNAs, for instance, are widely considered as excellent biomarkers for different types of cancers (Lu et al. 2005; Rosenfeld et al. 2008) or neurological diseases (Kocerha et al. 2009). On the other hand, long noncoding RNAs (lincRNAs) were shown to be involved in chromatin regulation (reviewed in Huarte and Rinn 2010) with some of them being differentially expressed in tumors as well (Gupta et al. 2010). For instance, the expression level of the lincRNA HOTAIR in primary breast tumors was shown to be important for the prediction of metastasis or death (Gupta et al. 2010). Other lincRNAs act as antisense transcripts to regulate protein-coding genes expression, which they epigenetically silence (Katayama et al. 2005) or posttranscriptionally regulate (Beltran et al. 2008). Even some snoRNAs have been described as good biomarkers for non-small-cell lung cancer, as they were observed to be differentially expressed in patients (Liao et al. 2010).

Hence, incorporating ncRNAs in diagnostics would provide a more complete picture regarding prognosis and evolution of a disease. The more ncRNA biomarkers will be discovered, the more precise diagnostics will be achieved. ncRNA biomarkers could easily and widely be used employing custom LNA-DNA ncRNA microarrays for diagnostic profiling. Such platforms will enable the analysis of long ncRNAs (e.g., HOTAIR RNA, 2,200 nt) as well as short ncRNAs (e.g., miRNAs, 21 nt) biomarkers in a single, affordable experiment. Importantly enough, amounts of RNA needed for such approaches comply with diagnostic references. Hence, LNA-DNA ncRNA microarrays could become a routinely employed clinical tool for diagnostics or molecular profiling.

Acknowledgments Lorraine University and the Austrian Ministry of Science and Research (GEN-AU project consortium "noncoding RNAs" D-110420-011-015 to A.H. and M.S.) are acknowledged for financial support.

M.Z. was financed on a grant from the Austrian Ministry of Science and Research (D-110420-012-012).

K.S. was financed on a FWF grant (Project Number 2 F012060-03, Signal Processing in Neurons (SPIN) Ph.D. program to A.H.).

M.K. was financed on a GEN-AU grant from the Austrian Ministry of Science and Research (Project Number D-110420-011-015 to M.S.).

References

Anders S, Huber W (2010) Differential expression analysis for sequence count data. Genome Biol 11(10):R106. doi:gb-2010-11-10-r106 [pii] 10.1186/gb-2010-11-10-r106

Anthony RM, Schuitema AR, Chan AB, Boender PJ, Klatser PR, Oskam L (2003) Effect of secondary structure on single nucleotide polymorphism detection with a porous microarray matrix; implications for probe selection. Biotechniques 34(5):1082–1086, 1088–1089

Beltran M, Puig I, Pena C, Garcia JM, Alvarez AB, Pena R, Bonilla F, de Herreros AG (2008) A natural antisense transcript regulates Zeb2/Sip1 gene expression during Snail1-induced epithelial-mesenchymal transition. Genes Dev 22(6):756–769. doi:22/6/756 [pii] 10.1101/gad.455708

Birney E, Stamatoyannopoulos JA, Dutta A, Guigo R, Gingeras TR, Margulies EH, Weng Z, Snyder M, Dermitzakis ET, Thurman RE, Kuehn MS, Taylor CM, Neph S, Koch CM, Asthana S, Malhotra A, Adzhubei I, Greenbaum JA, Andrews RM, Flicek P, Boyle PJ, Cao H, Carter NP, Clelland GK, Davis S, Day N, Dhami P, Dillon SC, Dorschner MO, Fiegler H, Giresi PG, Goldy J, Hawrylycz M, Haydock A, Humbert R, James KD, Johnson BE, Johnson EM, Frum TT, Rosenzweig ER, Karnani N, Lee K, Lefebvre GC, Navas PA, Neri F, Parker SC, Sabo PJ, Sandstrom R, Shafer A, Vetrie D, Weaver M, Wilcox S, Yu M, Collins FS, Dekker J, Lieb JD, Tullius TD, Crawford GE, Sunyaev S, Noble WS, Dunham I, Denoeud F, Reymond A, Kapranov P, Rozowsky J, Zheng D, Castelo R, Frankish A, Harrow J, Ghosh S, Sandelin A, Hofacker IL, Baertsch R, Keefe D, Dike S, Cheng J, Hirsch HA, Sekinger EA, Lagarde J, Abril JF, Shahab A, Flamm C, Fried C, Hackermuller J, Hertel J, Lindemeyer M, Missal K, Tanzer A, Washietl S, Korbel J, Emanuelsson O, Pedersen JS, Holroyd N, Taylor R, Swarbreck D, Matthews N, Dickson MC, Thomas DJ, Weirauch MT, Gilbert J, Drenkow J, Bell I, Zhao X, Srinivasan KG, Sung WK, Ooi HS, Chiu KP, Foissac S, Alioto T, Brent M, Pachter L, Tress ML, Valencia A, Choo SW, Choo CY, Ucla C, Manzano C, Wyss C, Cheung E, Clark TG, Brown JB, Ganesh M, Patel S, Tammana H, Chrast J, Henrichsen CN, Kai C, Kawai J, Nagalakshmi U, Wu J, Lian Z, Lian J, Newburger P, Zhang X, Bickel P, Mattick JS, Carninci P, Hayashizaki Y, Weissman S, Hubbard T, Myers RM, Rogers J, Stadler PF, Lowe TM, Wei CL, Ruan Y, Struhl K, Gerstein M, Antonarakis SE, Fu Y, Green ED, Karaoz U, Siepel A, Taylor J, Liefer LA, Wetterstrand KA, Good PJ, Feingold EA, Guyer MS, Cooper GM, Asimenos G, Dewey CN, Hou M, Nikolaev S, Montoya-Burgos JI, Loytynoja A, Whelan S, Pardi F, Massingham T, Huang H, Zhang NR, Holmes I, Mullikin JC, Ureta-Vidal A, Paten B, Seringhaus M, Church D, Rosenbloom K, Kent WJ, Stone EA, Batzoglou S, Goldman N, Hardison RC, Haussler D, Miller W, Sidow A, Trinklein ND, Zhang ZD, Barrera L, Stuart R, King DC, Ameur A, Enroth S, Bieda MC, Kim J, Bhinge AA, Jiang N, Liu J, Yao F, Vega VB, Lee CW, Ng P, Yang A, Moqtaderi Z, Zhu Z, Xu X, Squazzo S, Oberley MJ, Inman D, Singer MA, Richmond TA, Munn KJ, Rada-Iglesias A, Wallerman O, Komorowski J, Fowler JC, Couttet P, Bruce AW, Dovey OM, Ellis PD, Langford CF, Nix DA, Euskirchen G, Hartman S, Urban AE, Kraus P, Van Calcar S, Heintzman N, Kim TH, Wang K, Qu C, Hon G, Luna R, Glass CK, Rosenfeld MG, Aldred SF, Cooper SJ, Halees A, Lin JM, Shulha HP, Xu M, Haidar JN, Yu Y, Iyer VR, Green RD, Wadelius C, Farnham PJ, Ren B, Harte RA, Hinrichs AS, Trumbower H, Clawson H, Hillman-Jackson J, Zweig AS, Smith K, Thakkapallayil A, Barber G, Kuhn RM, Karolchik D, Armengol L, Bird CP, de Bakker PI, Kern AD, Lopez-Bigas N, Martin JD, Stranger BE, Woodroffe A, Davydov E, Dimas A, Eyras E, Hallgrimsdottir IB, Huppert J, Zody MC, Abecasis GR, Estivill X, Bouffard GG, Guan X, Hansen NF, Idol JR, Maduro VV, Maskeri B, McDowell JC, Park M, Thomas PJ, Young AC, Blakesley RW, Muzny DM, Sodergren E, Wheeler DA, Worley KC, Jiang H, Weinstock GM, Gibbs RA, Graves T, Fulton R, Mardis ER, Wilson RK, Clamp M, Cuff J, Gnerre S, Jaffe DB, Chang JL, Lindblad-Toh K, Lander ES, Koriabine M, Nefedov M, Osoegawa K, Yoshinaga Y, Zhu B, de Jong PJ (2007) Identification and analysis of functional elements in 1% of the human genome by the ENCODE pilot project. Nature 447(7146):799–816. doi:doi:10.1038/nature05874

Brosius J (2005) Waste not, want not–transcript excess in multicellular eukaryotes. Trends Genet 21(5):287–288. doi:S0168-9525(05)00060-0 [pii] 10.1016/j.tig.2005.02.014

Cantara WA, Crain PF, Rozenski J, McCloskey JA, Harris KA, Zhang X, Vendeix FA, Fabris D, Agris PF (2011) The RNA Modification Database, RNAMDB: 2011 update. Nucleic Acids Res 39:D195–201. doi:doi:gkq1028 [pii] 10.1093/nar/gkq1028, (Database issue):D195–201

Carninci P, Kasukawa T, Katayama S, Gough J, Frith MC, Maeda N, Oyama R, Ravasi T, Lenhard B, Wells C, Kodzius R, Shimokawa K, Bajic VB, Brenner SE, Batalov S, Forrest AR, Zavolan M, Davis MJ, Wilming LG, Aidinis V, Allen JE, Ambesi-Impiombato A, Apweiler R, Aturaliya RN, Bailey TL, Bansal M, Baxter L, Beisel KW, Bersano T, Bono H, Chalk AM, Chiu KP, Choudhary V, Christoffels A, Clutterbuck DR, Crowe ML, Dalla E, Dalrymple BP, de Bono B, Della Gatta G, di Bernardo D, Down T, Engstrom P, Fagiolini M, Faulkner G, Fletcher CF, Fukushima T, Furuno M, Futaki S, Gariboldi M, Georgii-Hemming P, Gingeras TR, Gojobori T, Green RE, Gustincich S, Harbers M, Hayashi Y, Hensch TK, Hirokawa N, Hill D, Huminiecki L, Iacono M, Ikeo K, Iwama A, Ishikawa T, Jakt M, Kanapin A, Katoh M, Kawasawa Y, Kelso J, Kitamura H, Kitano H, Kollias G, Krishnan SP, Kruger A, Kummerfeld SK, Kurochkin IV, Lareau LF, Lazarevic D, Lipovich L, Liu J, Liuni S, McWilliam S, Madan Babu M, Madera M, Marchionni L, Matsuda H, Matsuzawa S, Miki H, Mignone F, Miyake S, Morris K, Mottagui-Tabar S, Mulder N, Nakano N, Nakauchi H, Ng P, Nilsson R, Nishiguchi S, Nishikawa S, Nori F, Ohara O, Okazaki Y, Orlando V, Pang KC, Pavan WJ, Pavesi G, Pesole G, Petrovsky N, Piazza S, Reed J, Reid JF, Ring BZ, Ringwald M, Rost B, Ruan Y, Salzberg SL, Sandelin A, Schneider C, Schonbach C, Sekiguchi K, Semple CA, Seno S, Sessa L, Sheng Y, Shibata Y, Shimada H, Shimada K, Silva D, Sinclair B, Sperling S, Stupka E, Sugiura K, Sultana R, Takenaka Y, Taki K, Tammoja K, Tan SL, Tang S, Taylor MS, Tegner J, Teichmann SA, Ueda HR, van Nimwegen E, Verardo R, Wei CL, Yagi K, Yamanishi H, Zabarovsky E, Zhu S, Zimmer A, Hide W, Bult C, Grimmond SM, Teasdale RD, Liu ET, Brusic V, Quackenbush J, Wahlestedt C, Mattick JS, Hume DA, Kai C, Sasaki D, Tomaru Y, Fukuda S, Kanamori-Katayama M, Suzuki M, Aoki J, Arakawa T, Iida J, Imamura K, Itoh M, Kato T, Kawaji H, Kawagashira N, Kawashima T, Kojima M, Kondo S, Konno H, Nakano K, Ninomiya N, Nishio T, Okada M, Plessy C, Shibata K, Shiraki T, Suzuki S, Tagami M, Waki K, Watahiki A, Okamura-Oho Y, Suzuki H, Kawai J, Hayashizaki Y (2005) The transcriptional landscape of the mammalian genome. Science 309(5740):1559–1563. doi:309/5740/1559 [pii] 10.1126/science.1112014

Castoldi M, Schmidt S, Benes V, Hentze MW, Muckenthaler MU (2008) miChip: an array-based method for microRNA expression profiling using locked nucleic acid capture probes. Nat Protoc 3(2):321–329. doi:nprot.2008.4 [pii] 10.1038/nprot.2008.4

Chandler DP, Newton GJ, Small JA, Daly DS (2003) Sequence versus structure for the direct detection of 16S rRNA on planar oligonucleotide microarrays. Appl Environ Microbiol 69 (5):2950–2958

Cheng J, Kapranov P, Drenkow J, Dike S, Brubaker S, Patel S, Long J, Stern D, Tammana H, Helt G, Sementchenko V, Piccolboni A, Bekiranov S, Bailey DK, Ganesh M, Ghosh S, Bell I, Gerhard DS, Gingeras TR (2005) Transcriptional maps of 10 human chromosomes at 5-nucleotide resolution. Science 308(5725):1149–1154. doi:1108625 [pii] 10.1126/science. 1108625

Coppee JY (2008) Do DNA microarrays have their future behind them? Microbes Infect 10 (9):1067–1071. doi:S1286-4579(08)00179-2 [pii] 10.1016/j.micinf.2008.07.003

Cox WG, Beaudet MP, Agnew JY, Ruth JL (2004) Possible sources of dye-related signal correlation bias in two-color DNA microarray assays. Anal Biochem 331(2):243–254. doi:10.1016/j.ab.2004.05.010 S0003269704004166 [pii]

Dinger ME, Amaral PP, Mercer TR, Pang KC, Bruce SJ, Gardiner BB, Askarian-Amiri ME, Ru K, Solda G, Simons C, Sunkin SM, Crowe ML, Grimmond SM, Perkins AC, Mattick JS (2008) Long noncoding RNAs in mouse embryonic stem cell pluripotency and differentiation. Genome Res 18(9):1433–1445. doi:gr.078378.108 [pii] 10.1101/gr.078378.108

Dufva M, Petersen J, Poulsen L (2009) Increasing the specificity and function of DNA microarrays by processing arrays at different stringencies. Anal Bioanal Chem 395(3):669–677. doi:10.1007/s00216-009-2848-z

Eklund AC, Friis P, Wernersson R, Szallasi Z (2010) Optimization of the BLASTN substitution matrix for prediction of non-specific DNA microarray hybridization. Nucleic Acids Res 38(4): e27. doi:gkp1116 [pii] 10.1093/nar/gkp1116

Elmen J, Thonberg H, Ljungberg K, Frieden M, Westergaard M, Xu Y, Wahren B, Liang Z, Orum H, Koch T, Wahlestedt C (2005) Locked nucleic acid (LNA) mediated improvements in siRNA stability and functionality. Nucleic Acids Res 33(1):439–447. doi:33/1/439 [pii] 10.1093/nar/gki193

Fang H, Fan X, Guo L, Shi L, Perkins R, Ge W, Dragan YP, Tong W (2007) Self-self hybridization as an alternative experiment design to dye swap for two-color microarrays. OMICS 11 (1):14–24. doi:10.1089/omi.2006.0002

Friedlander MR, Chen W, Adamidi C, Maaskola J, Einspanier R, Knespel S, Rajewsky N (2008) Discovering microRNAs from deep sequencing data using miRDeep. Nat Biotechnol 26 (4):407–415. doi:nbt1394 [pii] 10.1038/nbt1394

Fujita PA, Rhead B, Zweig AS, Hinrichs AS, Karolchik D, Cline MS, Goldman M, Barber GP, Clawson H, Coelho A, Diekhans M, Dreszer TR, Giardine BM, Harte RA, Hillman-Jackson J, Hsu F, Kirkup V, Kuhn RM, Learned K, Li CH, Meyer LR, Pohl A, Raney BJ, Rosenbloom KR, Smith KE, Haussler D, Kent WJ (2011) The UCSC Genome Browser database: update 2011. Nucleic Acids Res 39:D876–882. doi:doi:gkq963 [pii] 10.1093/nar/gkq963, (Database issue):D876–882

Ghildiyal M, Zamore PD (2009) Small silencing RNAs: an expanding universe. Nat Rev Genet 10 (2):94–108. doi:nrg2504 [pii] 10.1038/nrg2504

Gilad S, Meiri E, Yogev Y, Benjamin S, Lebanony D, Yerushalmi N, Benjamin H, Kushnir M, Cholakh H, Melamed N, Bentwich Z, Hod M, Goren Y, Chajut A (2008) Serum microRNAs are promising novel biomarkers. PLoS One 3(9):e3148. doi:10.1371/journal.pone.0003148

Goff LA, Yang M, Bowers J, Getts RC, Padgett RW, Hart RP (2005) Rational probe optimization and enhanced detection strategy for microRNAs using microarrays. RNA Biol 2(3):93–100. doi:2059 [pii]

Griffiths-Jones S (2006) miRBase: the microRNA sequence database. Methods Mol Biol 342:129–138. doi:1-59745-123-1:129 [pii] 10.1385/1-59745-123-1:129

Gupta RA, Shah N, Wang KC, Kim J, Horlings HM, Wong DJ, Tsai MC, Hung T, Argani P, Rinn JL, Wang Y, Brzoska P, Kong B, Li R, West RB, van de Vijver MJ, Sukumar S, Chang HY (2010) Long non-coding RNA HOTAIR reprograms chromatin state to promote cancer metastasis. Nature 464(7291):1071–1076. doi:nature08975 [pii] 10.1038/nature08975

Guttman M, Amit I, Garber M, French C, Lin MF, Feldser D, Huarte M, Zuk O, Carey BW, Cassady JP, Cabili MN, Jaenisch R, Mikkelsen TS, Jacks T, Hacohen N, Bernstein BE, Kellis M, Regev A, Rinn JL, Lander ES (2009) Chromatin signature reveals over a thousand highly conserved large non-coding RNAs in mammals. Nature 458(7235):223–227. doi:nature07672 [pii] 10.1038/nature07672

Hawkins RD, Hon GC, Ren B (2010) Next-generation genomics: an integrative approach. Nat Rev Genet 11(7):476–486. doi:nrg2795 [pii] 10.1038/nrg2795

Huarte M, Rinn JL (2010) Large non-coding RNAs: missing links in cancer? Hum Mol Genet 19 (R2):R152–161. doi:doi:ddq353 [pii] 10.1093/hmg/ddq353

Hubbard TJ, Aken BL, Ayling S, Ballester B, Beal K, Bragin E, Brent S, Chen Y, Clapham P, Clarke L, Coates G, Fairley S, Fitzgerald S, Fernandez-Banet J, Gordon L, Graf S, Haider S, Hammond M, Holland R, Howe K, Jenkinson A, Johnson N, Kahari A, Keefe D, Keenan S, Kinsella R, Kokocinski F, Kulesha E, Lawson D, Longden I, Megy K, Meidl P, Overduin B, Parker A, Pritchard B, Rios D, Schuster M, Slater G, Smedley D, Spooner W, Spudich G, Trevanion S, Vilella A, Vogel J, White S, Wilder S, Zadissa A, Birney E, Cunningham F, Curwen V, Durbin R, Fernandez-Suarez XM, Herrero J, Kasprzyk A, Proctor G, Smith J,

Searle S, Flicek P (2009) Ensembl 2009. Nucleic Acids Res 37:D690–697. doi:doi:gkn828 [pii] 10.1093/nar/gkn828, (Database issue):D690–697

Huttenhofer A, Vogel J (2006) Experimental approaches to identify non-coding RNAs. Nucleic Acids Res 34(2):635–646. doi:34/2/635 [pii] 10.1093/nar/gkj469

Hutzinger R, Mrazek J, Vorwerk S, Huttenhofer A (2010) NcRNA-microchip analysis: a novel approach to identify differential expression of noncoding RNAs. RNA Biol 7(5):586–595. doi:12971 [pii]

Ishitani R, Yokoyama S, Nureki O (2008) Structure, dynamics, and function of RNA modification enzymes. Curr Opin Struct Biol 18(3):330–339. doi:S0959-440X(08)00069-9 [pii] 10.1016/j.sbi.2008.05.003

Jochl C, Rederstorff M, Hertel J, Stadler PF, Hofacker IL, Schrettl M, Haas H, Huttenhofer A (2008) Small ncRNA transcriptome analysis from Aspergillus fumigatus suggests a novel mechanism for regulation of protein synthesis. Nucleic Acids Res 36(8):2677–2689. doi:gkn123 [pii] 10.1093/nar/gkn123

Jordan BR (2010) Is there a niche for DNA microarrays in molecular diagnostics? Expert Rev Mol Diagn 10(7):875–882. doi:10.1586/erm.10.74

Kampa D, Cheng J, Kapranov P, Yamanaka M, Brubaker S, Cawley S, Drenkow J, Piccolboni A, Bekiranov S, Helt G, Tammana H, Gingeras TR (2004) Novel RNAs identified from an in-depth analysis of the transcriptome of human chromosomes 21 and 22. Genome Res 14 (3):331–342. doi:10.1101/gr.2094104 14/3/331 [pii]

Katayama S, Tomaru Y, Kasukawa T, Waki K, Nakanishi M, Nakamura M, Nishida H, Yap CC, Suzuki M, Kawai J, Suzuki H, Carninci P, Hayashizaki Y, Wells C, Frith M, Ravasi T, Pang KC, Hallinan J, Mattick J, Hume DA, Lipovich L, Batalov S, Engstrom PG, Mizuno Y, Faghihi MA, Sandelin A, Chalk AM, Mottagui-Tabar S, Liang Z, Lenhard B, Wahlestedt C (2005) Antisense transcription in the mammalian transcriptome. Science 309(5740):1564–1566. doi:309/5740/1564 [pii] 10.1126/science.1112009

Kawaji H, Severin J, Lizio M, Waterhouse A, Katayama S, Irvine KM, Hume DA, Forrest AR, Suzuki H, Carninci P, Hayashizaki Y, Daub CO (2009) The FANTOM web resource: from mammalian transcriptional landscape to its dynamic regulation. Genome Biol 10(4):R40. doi:gb-2009-10-4-r40 [pii] 10.1186/gb-2009-10-4-r40

Kocerha J, Kauppinen S, Wahlestedt C (2009) microRNAs in CNS disorders. Neuromolecular Med 11(3):162–172. doi:10.1007/s12017-009-8066-1

Liao J, Yu L, Mei Y, Guarnera M, Shen J, Li R, Liu Z, Jiang F (2010) Small nucleolar RNA signatures as biomarkers for non-small-cell lung cancer. Mol Cancer 9:198. doi:1476-4598-9-198 [pii] 10.1186/1476-4598-9-198

Liu X, Fortin K, Mourelatos Z (2008) MicroRNAs: biogenesis and molecular functions. Brain Pathol 18(1):113–121. doi:BPA121 [pii] 10.1111/j.1750-3639.2007.00121.x

Lu J, Getz G, Miska EA, Alvarez-Saavedra E, Lamb J, Peck D, Sweet-Cordero A, Ebert BL, Mak RH, Ferrando AA, Downing JR, Jacks T, Horvitz HR, Golub TR (2005) MicroRNA expression profiles classify human cancers. Nature 435(7043):834–838. doi:nature03702 [pii] 10.1038/nature03702

Marioni JC, Mason CE, Mane SM, Stephens M, Gilad Y (2008) RNA-seq: an assessment of technical reproducibility and comparison with gene expression arrays. Genome Res 18 (9):1509–1517. doi:gr.079558.108 [pii] 10.1101/gr.079558.108

Martin G, Keller W (1998) Tailing and 3'-end labeling of RNA with yeast poly(A) polymerase and various nucleotides. RNA 4(2):226–230

Martin J, Bruno VM, Fang Z, Meng X, Blow M, Zhang T, Sherlock G, Snyder M, Wang Z (2010) Rnnotator: an automated de novo transcriptome assembly pipeline from stranded RNA-Seq reads. BMC Genomics 11:663. doi:1471-2164-11-663 [pii] 10.1186/1471-2164-11-663

Mattick JS, Makunin IV (2005) Small regulatory RNAs in mammals. Hum Mol Genet 14(1): R121–132. doi:1:R121–132. doi:14/suppl_1/R121 [pii] 10.1093/hmg/ddi101, Spec No 1: R121–132

Mattick JS, Makunin IV (2006) Non-coding RNA. Hum Mol Genet 15 Spec No 1:R17–29. doi:15/suppl_1/R17 [pii] 10.1093/hmg/ddl046

Mercer TR, Dinger ME, Sunkin SM, Mehler MF, Mattick JS (2008) Specific expression of long noncoding RNAs in the mouse brain. Proc Natl Acad Sci USA 105(2):716–721. doi:0706729105 [pii] 10.1073/pnas.0706729105

Metzker ML (2010) Sequencing technologies – the next generation. Nat Rev Genet 11(1):31–46. doi:nrg2626 [pii] 10.1038/nrg2626

Mituyama T, Yamada K, Hattori E, Okida H, Ono Y, Terai G, Yoshizawa A, Komori T, Asai K (2009) The Functional RNA Database 3.0: databases to support mining and annotation of functional RNAs. Nucleic Acids Res 37 (Database issue):D89–92. doi:gkn805 [pii] 10.1093/nar/gkn805

Mortazavi A, Williams BA, McCue K, Schaeffer L, Wold B (2008) Mapping and quantifying mammalian transcriptomes by RNA-Seq. Nat Methods 5(7):621–628. doi:nmeth.1226 [pii] 10.1038/nmeth.1226

Nahkuri S, Taft RJ, Korbie DJ, Mattick JS (2008) Molecular evolution of the HBII-52 snoRNA cluster. J Mol Biol 381(4):810–815. doi:S0022-2836(08)00773-0 [pii] 10.1016/j.jmb.2008.06.057

Ozsolak F, Platt AR, Jones DR, Reifenberger JG, Sass LE, McInerney P, Thompson JF, Bowers J, Jarosz M, Milos PM (2009) Direct RNA sequencing. Nature 461(7265):814–818. doi:nature08390 [pii] 10.1038/nature08390

Pieler R, Sanchez-Cabo F, Hackl H, Thallinger GG, Trajanoski Z (2004) ArrayNorm: comprehensive normalization and analysis of microarray data. Bioinformatics 20(12):1971–1973. doi:10.1093/bioinformatics/bth174 bth174 [pii]

Rederstorff M, Huttenhofer A (2011a) cDNA library generation from ribonucleoprotein particles. Nat Protoc 6(2):166–174. doi:nprot.2010.186 [pii] 10.1038/nprot.2010.186

Rederstorff M, Huttenhofer A (2011) Experimental RNomics, A global approach to identify noncoding RNAs in model organisms, and RNPomics to analyze the non-coding RNP transcriptome. In: Hartman RK, Westhof E (eds) Handbook of RNA biochemistry. Wiley VCH Verlag GmbH, Weinheim

Rederstorff M, Bernhart SH, Tanzer A, Zywicki M, Perfler K, Lukasser M, Hofacker IL, Huttenhofer A (2010) RNPomics: defining the ncRNA transcriptome by cDNA library generation from ribonucleo-protein particles. Nucleic Acids Res 38(10):e113. doi:gkq057 [pii] 10.1093/nar/gkq057

Redkar RJ, Schultz NA, Scheumann V, Burzio LA, Haines DE, Metwalli E, Becker O, Conzone SD (2006) Signal and sensitivity enhancement through optical interference coating for DNA and protein microarray applications. J Biomol Tech 17(2):122–130. doi:17/2/122 [pii]

Robinson MD, McCarthy DJ, Smyth GK (2010) edgeR: a Bioconductor package for differential expression analysis of digital gene expression data. Bioinformatics 26(1):139–140. doi:btp616 [pii] 10.1093/bioinformatics/btp616

Rosenfeld N, Aharonov R, Meiri E, Rosenwald S, Spector Y, Zepeniuk M, Benjamin H, Shabes N, Tabak S, Levy A, Lebanony D, Goren Y, Silberschein E, Targan N, Ben-Ari A, Gilad S, Sion-Vardy N, Tobar A, Feinmesser M, Kharenko O, Nativ O, Nass D, Perelman M, Yosepovich A, Shalmon B, Polak-Charcon S, Fridman E, Avniel A, Bentwich I, Bentwich Z, Cohen D, Chajut A, Barshack I (2008) MicroRNAs accurately identify cancer tissue origin. Nat Biotechnol 26 (4):462–469. doi:nbt1392 [pii] 10.1038/nbt1392

Saxena A, Carninci P (2010) Whole transcriptome analysis: what are we still missing? Wiley Interdiscip Rev Syst Biol Med. doi:10.1002/wsbm.135

Schattner P, Brooks AN, Lowe TM (2005) The tRNAscan-SE, snoscan and snoGPS web servers for the detection of tRNAs and snoRNAs. Nucleic Acids Res 33 (Web Server issue): W686–689. doi:33/suppl_2/W686 [pii] 10.1093/nar/gki366

Schmittgen TD, Jiang J, Liu Q, Yang L (2004) A high-throughput method to monitor the expression of microRNA precursors. Nucleic Acids Res 32(4):e43. doi:10.1093/nar/gnh040 32/4/e43 [pii]

Sturn A, Quackenbush J, Trajanoski Z (2002) Genesis: cluster analysis of microarray data. Bioinformatics 18(1):207–208

Wernersson R, Nielsen HB (2005) OligoWiz 2.0 – integrating sequence feature annotation into the design of microarray probes. Nucleic Acids Res 33 (Web Server issue):W611–615. doi:33/suppl_2/W611 [pii] 10.1093/nar/gki399

Willenbrock H, Salomon J, Sokilde R, Barken KB, Hansen TN, Nielsen FC, Moller S, Litman T (2009) Quantitative miRNA expression analysis: comparing microarrays with next-generation sequencing. RNA 15(11):2028–2034. doi:rna.1699809 [pii] 10.1261/rna.1699809

Willingham AT, Gingeras TR (2006) TUF love for "junk" DNA. Cell 125(7):1215–1220. doi:S0092-8674(06)00767-7 [pii] 10.1016/j.cell.2006.06.009

Zywicki M, Bakowska-Zywicka K, Polacek N (2011) Identification of functional RNA processing products from RNA-seq data. Manuscript in preparation

Chapter 10
Targeted Methods to Improve Small RNA Profiles Generated by Deep Sequencing

Yoshinari Ando, A. Maxwell Burroughs, Mitsuoki Kawano, Michiel Jan Laurens de Hoon, and Yoshihide Hayashizaki

Abstract Several recent reviews expertly address the relative merits of different approaches to preparation and analysis of deep-sequenced small RNA libraries. Here, we focus on an array of protocols and tools with the intention of assisting researchers in improving short RNA profiles constructed with second-generation sequencing. This includes methods and commentaries on the preparation of sequencing-caliber immunoprecipitation RNA libraries, techniques for targeting different populations of RNAs with distinct 5′- and 3′-ends, reduction of adapter dimers in libraries, and dealing with the underappreciated problem of genomic cross-mapping of similar miRNA sequences.

Keywords AGO2 • Deep sequencing • Immunoprecipitation • IP-seq • Library preparation • LNA • miRNA • Second-generation sequencing • small RNA

Abbreviations

AGO	Argonaute
BAP	Bacterial alkaline phosphatase
bp	Base pairs
CAGE	Cap analysis gene expression
cDNA	Complementary DNA
CLIP-seq	Cross-linked immunoprecipitation and deep sequencing of RNAs
HITS-CLIP	High-throughput sequencing of RNAs isolated by cross-linked immunoprecipitation
IP	Immunoprecipitation
LNA	Locked nucleic acid

A.M. Burroughs (✉)
Omics Science Center (OSC), RIKEN Yokohama Institute, Yokohama, Kanagawa, Japan
e-mail: burrough@gsc.riken.jp

B. Mallick (eds.), *Regulatory RNAs*, DOI 10.1007/978-3-642-22517-8_10,
© Springer-Verlag Berlin Heidelberg (outside the USA) 2012

miRNA	microRNA
mRNA	Messenger RNA
nt	Nucleotides
PAR-CLIP	Photoactivatable-ribonucleoside-enhanced cross-linking and immunoprecipitation
PASR	Promoter-associated small RNA
piRNA	Piwi-interacting RNA
qPCR	Quantitative real-time PCR
RISC	RNA-induced silence complex
siRNA	Small interfering RNA
T4 PNK	T4 polynucleotide kinase
TAP	Tobacco acid pyrophosphatase
TASR	Termini-associated small RNA
TEX	Terminator^{TM} 5′-phosphate-dependent exonuclease
UTR	Untranslated region

10.1   Introduction

Increasingly, small RNAs (16–30 nucleotides (nt) in length) in eukaryotes are regarded as crucial contributors to a host of regulatory processes including post- and pretranscriptional gene and transposon silencing, chromatin dynamics, splicing regulation, and transcriptional initiation. The advent of high-throughput, genome-wide analysis platforms have enabled the construction of global small RNA profiles which seek to catalog and quantify the expression of the complete set of small RNAs in desired conditions or backgrounds. Construction of these profiles have changed our appreciation of the diversity and dynamism of eukaryotic small RNA through two primary means: (1) discovery of novel classes of RNA and demonstration of their ubiquity within a transcriptome and (2) offering contextual clues providing the first indications of possible functional roles. Additionally, targeted application of small RNA profiling in efforts to answer specific biological questions has begun to provide new perspective to complex biological problems.

Methods for global small RNA profiling began with availability of microarray and quantitative real-time PCR (qPCR) technologies (Willenbrock et al. 2009; Baker 2010). However, these technologies suffer from several limitations: (1) they are only applicable for measuring the expression of RNAs which have already been characterized; (2) these methods are based on hybridization, and cross-hybridization can result in acquisition of noisy data which, in the case of small RNA, is further compounded by the close sequence similarity within related classes (Kucho et al. 2004); and (3) data is generated by measuring the intensity ratio of the fluorescence label normalized by the background level resulting in difficulties properly quantifying signal at low and high extremes in expression. An additional technology, traditional Sanger sequencing with cloning, has been available for quite some time for use in discovery of novel classes of small RNA. However, the application of this sequencing method to small RNA identification is severely affected by the low expression level of many small RNAs, particularly in restricted cell types in distinct tissues or at very specific developmental stages. Second-generation sequencing technologies provide distinct advantages for small RNA studies compared with the aforementioned technologies because they do not need the sequence information and content of samples beforehand, require no cloning, and enable ultrahigh-throughput sequencing of the compatible length of many distinct classes of regulatory small RNA. Data resulting from these experiments is also comparable as absolute "digital" values which, through careful application of the proper analytical methods, can produce a clear understanding of changes in small RNA expression across different cell types or conditions.

Deep sequencing of small RNAs by second-generation sequencers has revealed a wealth of diverse and distinct classes of small RNA. As these classes are extensively covered elsewhere (Carninci 2010) and also in an introductory chapter in this book (*see Chap. 1*), we will not focus on describing these but will point out that the context of their identification has often provided substantial clues to the functional roles of the small RNAs. For example, the cellular restriction of Piwi-interacting RNAs (piRNAs) in germ line cells hinted at a role controlling transcription in early development and the genome-derived locations of the bidirectionally transcribed 20–100 nt length *p*romoter-/*t*ermini-*a*ssociated *s*mall *R*NAs (PASRs/TASRs) which are located in the upstream/downstream termini of genes (Kapranov et al. 2007; Affymetrix/Cold Spring Harbor Laboratory ENCODE Transcriptome Project 2009) suggested a role in regulating gene expression which, at least for the PASRs, some supporting experimental evidence has been presented (Affymetrix/Cold Spring Harbor Laboratory ENCODE Transcriptome Project 2009). Thus, deep sequencing has contributed to understanding the biological role of these small RNA sequences. Recently, small RNA profiling has increasingly been employed as a tool to assist in answering more specific questions. For example, when applied to microRNA (miRNA), a class of 20–23 nt sequences which posttranscriptionally regulates gene expression, deep sequencing has led to an understanding of the mechanisms behind generation of multiple mature variants (termed isomiRs) from the same miRNA locus (Landgraf et al. 2007; Morin et al. 2008) and their selective expression across distinct cell types (Lee et al. 2010). Additionally, deep sequencing of small RNAs has been utilized to identify enzymes involved in processing of the mature miRNA through $3'$ adenylation and uridylation (Jones et al. 2009; Lehrbach et al. 2009; Burroughs et al. 2010) and editing events by adenosine/cytidine deaminases (de Hoon et al. 2010; Kim et al. 2010).

Second-generation sequencing technology has enabled deep exploration of the small RNA world, increasing our understanding of the vast and varied content of the classes of small RNA in the cell. Careful application of this technology has opened novel avenues of research substantially impacting our understanding of the roles of small RNA. However, care must be taken to apply the proper technique to answer the biological question under investigation. This chapter will focus on a variety of methods, from small RNA library construction to computational data analysis methods that have proven to be useful in the past in assisting with the characterization of small RNA.

10.2   Methods

Small RNA profiling studies with second-generation sequencing technologies are currently performed from the construction of complementary DNA (cDNA) libraries followed by sequencing on a second-generation platform (McCormick et al. 2011). A small RNA library is prepared by enzymatically ligating 5' and 3' adapters to small RNA molecules and is enriched by 10–15 cycles of PCR using primers that contain sequences homologous to these adapters and complementary to the oligos optimal for each sequencing platform. This ligation-based library construction method is popular to most users of second-generation sequencers (Morin et al. 2010; Thomas and Ansel 2010; Lu and Shedge 2011) despite the ligation and PCR amplification bias (Linsen et al. 2009; Tian et al. 2010; McCormick et al. 2011). However, ligation bias can be decreased to some extent by using a truncated version of T4 RNA ligase 2 (T4 Rnl2 truncated) for 3' adapter ligation (Vigneault et al. 2008; Munafo and Robb 2010). This enzyme specifically ligates the preadenylated 5'-end of the 3' adapter to the 3'-end of RNA. Unlike usual RNA ligases, T4 Rnl2 truncated cannot ligate the 5'-phosphorylated RNA or DNA to the 3'-end of RNA and does not circularize 5'-phosphorylated and 3'-hydroxylated RNAs, like miRNAs. Amplification bias is also minimized by optimizing the experimental condition (Aird et al. 2011).

10.2.1   Sequencing Using Second-Generation Sequencers

Comparison of the relative strengths and weaknesses of different second-generation sequencers has been extensively considered elsewhere (Metzker 2010; McCormick et al. 2011; *see Chap. 21*); we will only briefly survey standard library construction approaches here. Relevant to this discussion are recent studies which have shown that the diversity and abundance of small RNAs sequenced in the libraries can be less affected by the sequencing platform than the library preparation method (Linsen et al. 2009; Tian et al. 2010). The primary implication of these findings is that selection of a verified library construction method is important for proper comparison between different samples.

For the Illumina and SOLiD platforms, there are several commercial kits available for small RNA library construction as shown below. These kits have simple, speedy, and verified protocols to make small RNA libraries with or without barcode tags for multiplex sequencing.

Genome Analyzer IIx or *HiSeq2000*: Small RNA Sample Prep Kit v1.5 (Illumina); TruSeq™ Small RNA Sample Prep Kit (Illumina); ScriptMiner™ Small RNA-Seq Library Preparation Kit (EPICENTRE Biotechnologies); NEBNext™ Small RNA Sample Prep Set 1 (New England Biolabs); NEXTflex™ Small RNA Sequencing Kits (Bioo Scientific); Ambion RNA-Seq Library Construction Kit compatible with the Illumina GAII platform (Life Technologies)

SOLiD4hq: SOLiD™ Total RNA-Seq Kit (Life Technologies); NEBNext™ Small RNA Sample Prep Set 3 (New England Biolabs)

While the 454 platform, the first second-generation sequencer to be widely adopted on the market, has been used extensively for small RNA sequencing in the past (Watanabe et al. 2008; Taft et al. 2009), the library construction kit is not commercialized. Due to an average read length of 400 nt (Table 10.1), 454 sequencing is somewhat unwieldy for small RNA sequencing when compared to more recently introduced platforms available on the market. Additionally, the depth of coverage for 454 sequencing is shallower than other platforms. This can be mitigated somewhat through concatemerization of several small RNAs for efficient sequencing. However, this step makes the protocol relatively complicated.

The read length and sequencing depth of the Helicos platform is suitable for small RNA sequencing (Table 10.1), and it has already been used to great effect in analyzing short RNA content (Kapranov et al. 2010). This true single molecule sequencing technology enables dissolution of problems derived from ligation and amplification bias and has been adapted to achieve levels of reproducibility in gene expression not previously observed on other sequencing platforms (Kanamori-Katayama et al. 2011). For small RNA libraries, single molecule sequencing increases the complexity of small RNA libraries enabling the discovery of novel, low-expressed classes of RNA that will fail to effectively amplify using other library preparation techniques (Kapranov et al. 2010). However, the Helicos platform still has a higher raw error rate (substitution 0.2%, insertion 1.5%, and deletion 3.0%) than other platforms, which poses a particular problem when trying to accurately assess expression levels within classes of small RNA with members

Table 10.1 Comparison of second-generation sequencers

Platform	Read length (bases)	Sequencing reaction	Run time (days)	Gb per run
GS FLX/454Life sciences in Roche Diagnostics	500 (400)[a]	Pyrosequencing	10 h	0.4–0.6[b]
Genome Analyzer IIx/Illumina	150	Sequencing by synthesis	14	67[b]
HiSeq2000/Illumina	100	Sequencing by synthesis	10	400–480[b]
SOLiD 4*hq*/Applied biosystems in life technologies	75	Sequencing by ligation	14	300[b]
Heliscope/Helicos Biosciences	25–55 (35)[a]	True single molecule sequencing	8	21–35

[a]Average length
[b]Error-free bases

displaying high levels of sequence similarity (e.g., miRNA). Additionally, a small RNA library construction kit has not been commercialized yet. Despite these difficulties, single molecule sequencing represents the next frontier in exploring small RNA; continued improvements in error rates could improve usefulness of the technology in understanding small RNA expression. In sum, availability of commercial kits or verified protocols is extremely helpful for users in selecting suitable sequencing platforms.

10.2.2 RNA Sample Preparation and Library Production

10.2.2.1 Different 5′-End Structure of Small RNAs

The majority of small RNAs have 5′-monophosphoryl and 3′-hydroxyl termini as products generated by endonucleases like the DICER and argonaute (AGO) proteins. In a ligation-based "standard" library construction method, T4 RNA ligase 1 is essential to ligate the 5′ adaptor sequence to the 5′-end of RNA molecules. T4 RNA ligase 1 requires a 5′-monophosphoryl terminus on the donor sequence in the ligation reaction. Some small RNA species are not detected by standard library preparation techniques because these RNAs have different 5′-end structures (5′-triphosphorylated, 5′-hydroxylated, or 5′-capped). For example, the human hepatitis delta virus produces 5′-capped, 18–25 nt small RNAs using human RNA polymerase II for their replication in human cells (Haussecker et al. 2008). PASRs and TASRs are well known as small RNAs with a 5′-cap structure transcribed by RNA polymerase II (Affymetrix/Cold Spring Harbor Laboratory ENCODE Transcriptome Project 2009). To clone these small RNA species, several treatments (dephosphorylation, rephosphorylation, and decapping reaction) are required before the ligation reaction.

Bacterial alkaline phosphatase (BAP) catalyzes the dephosphorylation of almost all phosphate monoesters from the 5′- and 3′-ends of RNA. T4 polynucleotide kinase (T4 PNK) catalyzes the transfer and exchange of monophosphate from the γ position of ATP to the 5′-hydroxyl terminus of single-stranded RNA. Tobacco acid pyrophosphatase (TAP) hydrolyzes the phosphoric acid anhydride bonds in the triphosphate bridge of the 5′-cap structure, releasing the 7-methylguanosine cap nucleotide (m^7G) and generating a 5′-monophosphorylated terminus. Similarly, TAP digests the triphosphate group at the 5′-end, generating an RNA molecule with a 5′-monophosphate. Terminator™ 5′-phosphate-dependent exonuclease (TEX, EPICENTRE Biotechnologies) digests RNA with a 5′-monophosphate, while the enzyme cannot digest RNA having a 5′-triphosphate, a 5′-cap, or a 5′-hydroxyl group. Using this enzyme, minor small RNA species might be enriched in the library by selectively digesting the miRNA with a 5′-monophosphorylated terminus. Combination of these pretreatments makes it possible to construct various small RNA libraries containing minor RNA species with different 5′-end structures (Fig. 10.1). For example, to construct a library consisting of small RNAs with only

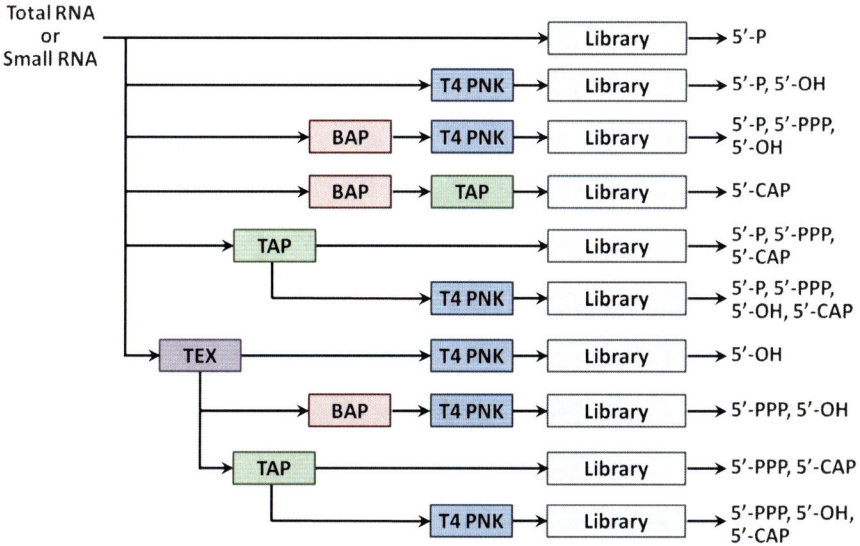

Fig. 10.1 Preparation of small RNA libraries containing various RNAs with different 5′-end structure. T4 PNK: T4 polynucleotide kinase, BAP: bacterial alkaline phosphatase, TAP: tobacco acid pyrophosphatase, TEX: Terminator™ 5′-phosphate-dependent exonuclease, 5′-P: 5′-monophosphorylated RNA, 5′-PPP: 5′-triphosphorylated RNA, 5′-OH: 5′-hydroxylated RNA, 5′-CAP: 5′-capped RNA

5′-cap structures, total RNA or small RNA fraction should be treated by TAP following BAP treatment before library construction.

10.2.2.2 Immunoprecipitated RNAs with Human AGO2 Proteins

Functional miRNAs bind to AGO family proteins forming the *R*NA-*i*nduced *s*ilencing *c*omplex (RISC); RISC-associated miRNA then recognizes and interacts with partially complementary sequences typically located in the 3′ untranslated regions (3′ UTRs) of specific target messenger RNAs (mRNAs), leading to translational repression or mRNA degradation (Hutvagner and Simard 2008; Ender and Meister 2010). High purified fractions of miRNA, which binds with human AGO2 proteins, can be enriched by immunoprecipitation (IP) methods using a high affinity antibody against human AGO2 protein (Azuma-Mukai et al. 2008; Goff et al. 2009; Burroughs et al. 2010; Burroughs et al. 2011). As well as miRNAs, various small RNAs can be enriched in AGO2-IP sequence libraries, suggesting association with a functional RISC (Ender et al. 2008; Burroughs et al. 2011). Although these small RNAs often appear to lack the canonical hairpin structure of miRNA precursors leaving the nature of biogenesis and RISC-incorporation pathways somewhat murky, in at least some cases, a RISC-active functional role has been demonstrated in human cells (Ender et al. 2008; Haussecker et al. 2010; Lee et al. 2009).

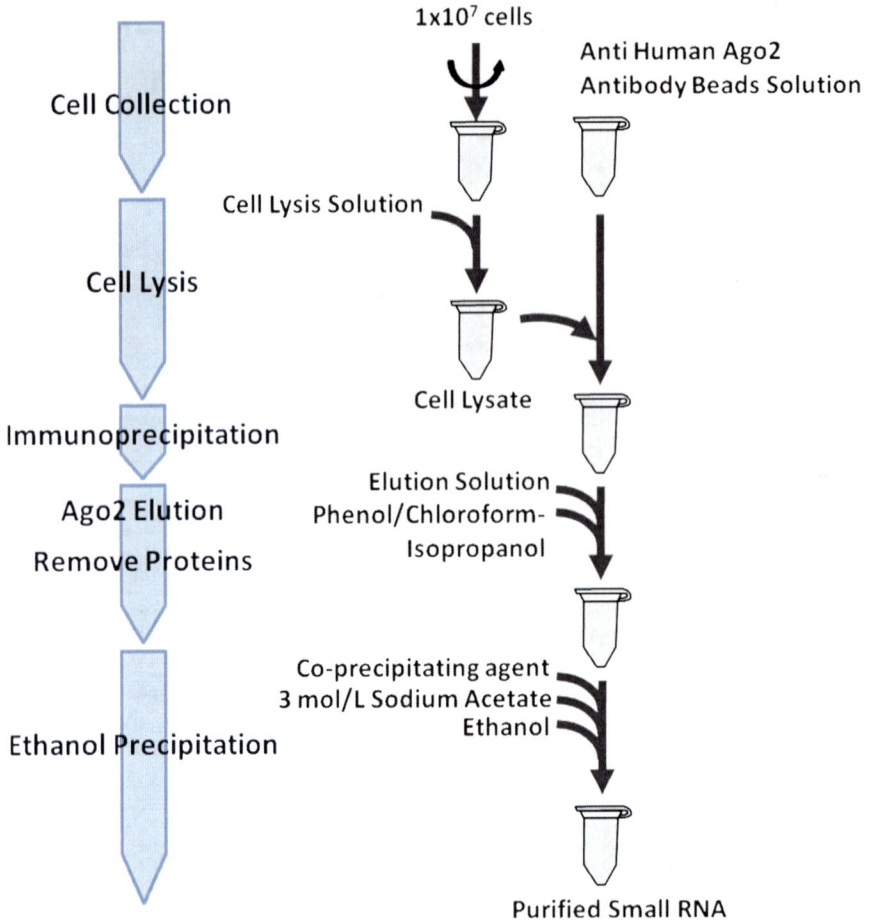

Fig. 10.2 Outline of human AGO2 immunoprecipitation using a commercial kit

The following protocol is designed based on microRNA Isolation Kit Human Ago2 (Wako Pure Chemical Industries) with some modifications (Fig. 10.2). Cultured cells (1×10^7 cells) are collected and incubated in cell lysis solution (20 mM Tris–HCl buffer (pH7.4), 200 mM sodium chloride, and 2.5 mM magnesium chloride) for 10 min on ice. Cell lysates are then cleared by centrifugation at 4°C, and 50 µL of Anti Human Ago2 Antibody Beads Solution in the kit was added to supernatant of cell lysate and mixed by rotation for 2 h at 4°C. Beads are washed twice with cell lysis solution, and each bound AGO2-RNA complex is eluted with elution solution (0.5% SDS). Eluted complexes are extracted with phenol/chloroform, and RNAs are precipitated with ethanol. Purified immunoprecipitated RNAs are used for construction of small RNA libraries.

This basic AGO2 IP-seq protocol has been extended in an effort to decipher the genome-wide microRNA-mRNA interaction map. Using the *high-t*hroughput sequencing of RNAs isolated by *cross-linked immuno*precipitation (HITS-CLIP) (Chi et al. 2009) protocol, Darnell and colleagues identified two distinct RNA populations binding to AGO2 which corresponded to miRNA and its target mRNA sequences. Separate sequencing of the two populations was successful at reducing the size of possible target mRNA sequences for individual miRNAs. A variation of this method which has similarly been applied to AGO2 to identify both miRNA and mRNA targets is referred to as *photoactivatable-ribonucleoside-enhanced cross-linking* and *immuno*precipitation (PAR-CLIP) (Hafner et al. 2010). This protocol is slightly more involving, requiring the insertion of photoreactive ribonucleoside analogs into the RNA of cultured cells. These analogs enhance the RNA-protein cross-linking reaction after exposure to UV light, in theory increasing the efficiency of the detection of interacting RNA species.

10.2.2.3 Dimer Eliminator and Barcode Library

In most methods used to sequence small RNAs, RNA-derived cDNA libraries consisting of cDNA inserts of various sizes ligated between the 5′ and 3′ adapter linker sequences are constructed. A protocol has recently been developed with fewer preparation steps needed to construct small RNA libraries, and its use of preadenylated adapter oligos makes it particularly efficient in capturing small RNAs (Vigneault et al. 2008; Munafo and Robb 2010). However, even when this protocol is used with a commercially available sample preparation kit (Illumina), a large number of sequencing reads without cDNA inserts are often observed, mainly composed of adapter-dimer products. To overcome this problem, we developed a method adding a *locked nucleic acid* (LNA) oligonucleotide, named dimer eliminator-22 (5′-*TACGAGATTT*NN*GATCGTCGGA*-3′; LNA is shown in italics, and NN shows random DNA), that is complementary to the adapter-dimer ligation products: 8 nt of it span the 3′ adapter, while 14 nt span the 5′ adapter during the reverse transcription reaction (Fig. 10.3). LNA treatment reduces adapter dimers which often contaminate standard libraries by as much as threefold, concomitantly increasing the number of non-insert sequence reads (Kawano et al. 2010).

A simplified version of protocol follows. Briefly, a 3′ DNA adapter and a 5′ RNA adapter are sequentially ligated to RNA. The ligation products are reverse-transcribed with 10–20 μM of the dimer eliminator oligo, followed by PCR and polyacrylamide gel purification of insert-containing cDNA products of around 100 base pairs (bp) which corresponds to RNAs of approximately 15–36 nt. The products are subjected to sequencing runs on the Illumina sequencing platform.

We applied this technology for a pooled library construction which requires a barcoding system for multiplex sequencing in order to increase the sequencing performance (Kawano et al. 2010). We designed eight barcode tags (two nucleotide barcode with common AA at the 3′-end): AAAA, GAAA, CAAA, TAAA, AGAA, ACAA, ATAA, and TTAA and incorporated them at the 3′-end of the 5′ RNA adapter oligos. We pooled the samples after ligation to the barcoded adapter and proceeded with the above reactions in a single tube. It is not necessary to prepare

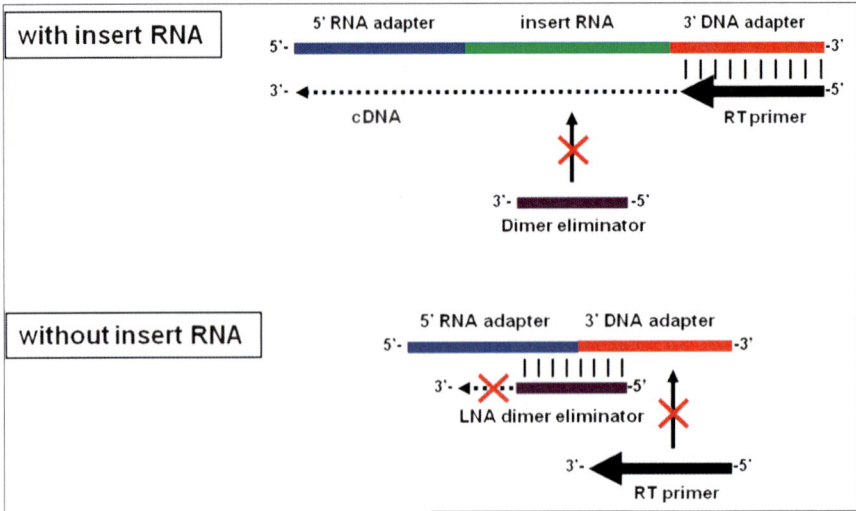

Fig. 10.3 Scheme of adapter dimer reducing technology using LNA dimer eliminator oligo. With an insert RNA between adapters, RT primer can anneal to 3′ DNA adapter and synthesize cDNA product. Without an insert RNA, LNA dimer eliminator oligo can completely anneal to only adapter dimer ligation products. Thus, cDNA synthesis for adapter dimers cannot occur due to interference with the annealing of the RT primer. LNA dimer eliminator oligo cannot be used as a primer in reverse transcription reactions

dimer eliminators corresponding to each of the barcoded adapter sequences because the dimer eliminator-22 contains random sequences at the middle to fit the 5′ RNA adapter sequences and they can be applied to any barcoded adapter.

The barcoding system for high-throughput sequencing has been previously published, but to our knowledge, there has not been any evaluation of bias from the use of barcodes, which would interfere with an accurate expression comparison of the barcoded samples (Hamady et al. 2008; Vigneault et al. 2008; Smith et al. 2009). Therefore, we evaluated the barcoding system by looking at Pearson's correlation values for each of the eight barcodes from two pooled libraries made from the same total RNA by using the dimer eliminator-22. The small RNA expression profiles between the technical replicates showed high correlation: the average Pearson's correlation coefficient was 0.983. Furthermore, the profiles between the distinct barcodes also showed high correlation: 0.894 for all tested barcodes and 0.945 for the following four barcodes: AAAA, GAAA, CAAA, and TAAA. These results show that our protocol reproduces RNA population profiles well with technical replicates and different barcode tags, indicating that little or no bias is generated through LNA treatment (Kawano et al. 2010). We recommend the use of these four barcode tags in comparing samples with more precise small RNA measurements as they show small bias during the complete procedure. This method improves the sequencing yield and efficiency while simplifying library construction and makes it easier to perform large-scale small RNA sequencing.

10.2.3 Analysis of Sequencing Results

Raw tags deep-sequenced in small RNA libraries by second-generation machines are typically subjected to three basic processing steps: (1) removal of adaptor and barcode sequences from raw tags, (2) filtering to remove low-quality tags, and (3) mapping to the appropriate genome assembly. Further downstream analyses can vary depending on the goal of the study, but typically includes annotation of the mapped tags, normalization of tags to enable comparisons of annotated tags across libraries, statistical assessment of the significance of observed expression changes, and identification of novel small RNAs within the generated data. Comparison of different methods which address these issues has been adroitly covered elsewhere (Metzker 2010; McCormick et al. 2011; *see Chap. 9 and Chap. 21*). Instead of revisiting the strengths and weaknesses of these many methods, we will discuss issues facing biologists lacking a strong computational background yet seeking single-solution, comprehensive platforms for second-generation sequencing analysis and focus on solving the underappreciated problem of miRNA genome cross-mapping.

10.2.3.1 Platforms for Analysis of Deep-Sequenced Data

Newcomers to the world of second-generation sequencing are essentially faced with one of two choices when trying to find comprehensive or integrative solutions to sequence data processing and analysis: (1) purchasing a commercial platform or (2) adopting a platform developed in an academic lab. Both options can be fraught with uncertainty. Commercial platforms, like those developed by CLC Bio, Real Time Genomics, and Genomatix, represent a new cost for a laboratory, and while they may perform basic analysis tasks well, they may lack flexibility for adaptation to specialized analysis tasks. On the other hand, while academic platforms (see NGSQC (Dai et al. 2010), SAMS (Bekel et al. 2009), and SEWAL (Pitt et al. 2010) for recent examples) are open source and often more flexible, they may be customized to fit the needs of the research lab constructing the software and may require considerable expertise to limit the impact of portability and performance-related issues. Hence, academic options potentially represent a sizeable personnel and time investment.

As consensus is reached on the best approaches to analyzing second-generation sequencing data, the task of selecting appropriate commercial and open-source efforts will likely become easier, and these approaches will be adopted in lieu of the current, lab-centric approach. Parallels can be drawn to the development of microarray analysis methods and the rise of widely used commercial platforms like GeneSpring GX (Agilent Technologies) and open-source projects like Bioconductor (http://www.bioconductor.org/). A sensible short-term approach, particularly for laboratories involved in smaller-scale sequencing projects, is pooling together freely available programs from different sources most suitable for the necessary tasks. While many methods for mapping and normalization are available, programs targeting a range of additional tasks like managing quality control across multiple sequencing runs (Lassmann et al. 2011), artifact removal (Lassmann et al. 2009), and novel small RNA predictions (Friedlander et al. 2008; Hackenberg et al. 2009) are also becoming easier to acquire and implement.

10.2.3.2   Cross-Mapping of Small RNA Sequences to Multiple miRNA Loci

Many miRNA sequences are closely related at the sequence level and can be grouped into what are termed miRNA "families." Such related miRNA sequences are thought to have descended from a common ancestor, undergoing diversification through gene duplication at pivotal points in animal evolution (Hertel et al. 2006). This sequence similarity, coupled with the short size of mature miRNA sequences (20–23 nt) and the error rates inherent to any sequencing machines, poses difficulties in assessing the true genomic locus for miRNA sequences as often more than one genomic location may appear to be equally suitable for a given tag. Further compounding this cross-mapping problem is the widespread presence of nucleotidyltransferase enzyme–mediated 3' adenine and uridine addition events in mature miRNA (Jones et al. 2009; Katoh et al. 2009; Heo et al. 2009; Burroughs et al. 2010). Since these nucleotides are genuinely present in small RNA libraries, they cannot be filtered using conventional per-nucleotide quality scores (de Hoon et al. 2010).

All of these factors can combine to influence expression measured at individual miRNA loci in deep-sequencing experiments. In a broad sense, this problem is akin to the cross-hybridization in microarray experiments. Perhaps more pressingly, tags with more than one possible mapping location that contain a mismatch to the genome may be improperly assigned to a locus. Inspection of the tags aligning to these loci can lead to the impression that editing sites are present when in fact the mismatch is an artifact of cross-mapping. With an abundance of studies purporting detection of novel editing sites through analysis of deep-sequenced small RNA data (Reid et al. 2008; Su et al. 2010; Schulte et al. 2010; Buchold et al. 2010) concerns about potential, unaddressed effects of cross-mapping prompted the development of a new tool to make corrections (available for download from http://134.160.84.27/osc/english/dataresource/) (de Hoon et al. 2010).

In many ways, the approach of the program resembles efforts to similarly correct for deep-sequenced tags mapping to repetitive regions like promoter-derived *c*ap *a*nalysis *g*ene *e*xpression (CAGE) tags (Faulkner et al. 2008). Ultimately, each small RNA tag that can map to multiple locations in the genome is assigned a weight based on the local expression level of a given position within the same library and the alignment errors at that position. Penalty for alignment errors is determined through comparing the position-specific error profile for each tag against the likelihood of observing an error in the genome-tag mapping location (de Hoon et al. 2010). In general, the effects of cross-mapping correction on the calculated expression at miRNA loci were not widespread; however, for at least 14 loci in the THP-1 cell line expression differences with cross-mapping correction and in the absence of cross-mapping correction was larger than 50% (de Hoon et al. 2010). These effects will vary considerably depending on the composition of the miRNA transcriptome in different cells and conditions.

More dramatic was the impact of cross-mapping correction on apparent editing sites. Applied to the FANTOM4 small RNA sequencing dataset, the ten sites identified as possible editing sites in the absence of cross-mapping correction were reduced to two sites (de Hoon et al. 2010). One of these sites was found to harbor an SNP in the genome sequence, and the other was a previously experimentally characterized editing event (Kawahara et al. 2007). Application of the corrective procedure to another dataset which had shown possible editing events within the *let*-7 miRNA family appeared to show these sites instead resulted from cross-mapping (Reid et al. 2008). These results emphasize the importance of considering the effects of cross-mapping when searching for novel, potential editing sites in mature miRNA sequences derived from second-generation sequencing data. Interestingly, application of the cross-mapping correction to plant miRNA datasets did not appreciably reduce the number of editing sites, suggesting that mature plant miRNAs may be targeted far more frequently by the editing machinery in the cell (Ebhardt et al. 2009; personal observations).

Recently, a genome alignment program named Statmap based on similar principles has been employed for the mapping of CAGE data in the drosophila modENCODE project (Hoskins et al. 2011). A primary difference between the techniques is that instead of being applied as a postmapping corrective procedure, Statmap assigns a mapping likelihood based on mismatch profiles and relative expression of different locations in the genome for every tag in a given library during the mapping procedure (Hoskins et al. 2011). Such a mapping approach has the potential to address cross-mapping issues for short RNA at the genome mapping step.

10.3   Applications: Probing the Extent of 3′ miRNA Addition in Animals

In this chapter, we have attempted to draw attention to and discuss several distinct methodologies which can assist in the design and also improve the yield and analysis of small RNA datasets. As specific examples on how many of these methods can be utilized, the reader is referred to previous research by our group (Burroughs et al. 2010; Burroughs et al. 2011). In this research, miRNA cross-mapping correction was performed on a broad range of deeply sequenced, publicly available datasets across several organisms in an effort to determine the evolutionary depth and global distribution of 3′ miRNA modification events (Burroughs et al. 2010). This revealed not only the conservation in frequency of miRNA 3′ addition events but also a tendency for addition to occur on phylogenetically related miRNA sequences. We also tested the global effects of knockdown of several known and predicted miRNA nucleotidyltransferases on 3′ miRNA addition using library preparation techniques discussed above and again utilizing the cross-mapping corrective procedure to ensure accurate counting of the addition events following sequencing. This global analysis identified the GLD-2 enzyme as a principal miRNA adenyltransferase. Several targeted, miRNA-specific studies indicated a possible role for 3′ adenylation in affecting the association of miRNA with AGO proteins. Deep sequencing of the immunoprecipitated small RNA associating with the AGO1–3 proteins using the protocol described above and employing cross-mapping correction revealed the reduction of adenylated miRNA isomiRs in the AGO2 and AGO3 proteins, possibly supporting a scenario where adenylation interferes with proper AGO association (Burroughs et al. 2010). Further analysis of the cross-mapping corrected isomiRs across the three AGO-derived libraries revealed the presence of a small-scale, intralocus sorting mechanism wherein different isomiRs from the same miRNA locus preferentially associated with distinct AGO proteins in a significantly differential manner (Burroughs et al. 2011), consistent with a previous experimental investigation demonstrating this at a single miRNA locus (Azuma-Mukai et al. 2008).

10.4 Discussion

Small RNA sequencing has proven an extremely effective tool for identifying novel populations of small RNA (*see Chap. 1*) and, in many cases, offering the first clues to their potential biological role. At the same time, the use of small RNA profiling as a standard tool incorporated into more targeted biological experiments is growing, including examining the biogenesis of miRNA (Jones et al. 2009; Lehrbach et al. 2009; Burroughs et al. 2010), the roles of small RNAs in regulating splicing and transcription initiation (Taft et al. 2009; Taft et al. 2010), and the roles of small RNA in regulating heterochromatin formation (Halic and Moazed 2010). While this research is generating exciting new perspectives on fundamental issues in molecular biology, as we have tried to communicate in this chapter, the effectiveness of sequencing as a tool to investigate targeted biological questions depends on its proper application. Several crucial issues need to be considered when planning deep-sequencing experiments. Foremost among these is choice of sequencing platform, which needs to be made after weighing the strengths and weaknesses of each technology in light of the desired experimental outcome; specific issues to consider include tag yield, amplification bias, read length, time for sequencing run, and cost. Similarly, choice of RNA and library preparation technique should be tailored to the class of small RNA that is targeted by the research, and if needed, additional techniques like the LNA treatment outlined above can also be employed to improve yield. Even postsequencing analysis needs to be considered carefully to ensure proper interpretation of the data. One common pitfall in the analysis of miRNA is to ignore potential effects of cross-mapping across closely related miRNA species.

The various methodologies discussed in this chapter are intended to assist researchers seeking to utilize sequencing in various targeted biological contexts. We hope the explanations and discussion provided in this chapter will be of use to investigators both during the project design phase and also during implementation and troubleshooting of projects incorporating second-generation sequencing.

Acknowledgments We thank Takahiro Nishibu, Ryo Ukekawa, Taku Funakoshi, and Tsutomu Kurokawa of Wako Pure Chemical Industries, Ltd. and Carsten Daub of the RIKEN Omics Science Center (OSC) for their collaboration. This work was supported by a research grant for the RIKEN OSC from the Ministry of Education, Culture, Sports, Science, and Technology (MEXT) to YH and a grant for the Innovative Cell Biology by Innovative Technology (Cell Innovation Program) from the MEXT to YH.

References

Affymetrix/Cold Spring Harbor Laboratory ENCODE Transcriptome Project (2009) Post-transcriptional processing generates a diversity of 5′-modified long and short RNAs. Nature 457 (7232):1028–1032. doi:nature07759 [pii] 10.1038/nature07759

Aird D, Ross MG, Chen WS, Danielsson M, Fennell T, Russ C, Jaffe DB, Nusbaum C, Gnirke A (2011) Analyzing and minimizing PCR amplification bias in Illumina sequencing libraries. Genome Biol 12(2):R18. doi:gb-2011-12-2-r18 [pii] 10.1186/gb-2011-12-2-r18

Azuma-Mukai A, Oguri H, Mituyama T, Qian ZR, Asai K, Siomi H, Siomi MC (2008) Characterization of endogenous human Argonautes and their miRNA partners in RNA silencing. Proc Natl Acad Sci USA 105(23):7964–7969. doi:0800334105 [pii] 10.1073/pnas.0800334105

Baker M (2010) MicroRNA profiling: separating signal from noise. Nat Methods 7(9):687–692. doi:nmeth0910-687 [pii] 10.1038/nmeth0910-687

Bekel T, Henckel K, Kuster H, Meyer F, Mittard Runte V, Neuweger H, Paarmann D, Rupp O, Zakrzewski M, Puhler A, Stoye J, Goesmann A (2009) The Sequence Analysis and Management System – SAMS-2.0: data management and sequence analysis adapted to changing requirements from traditional sanger sequencing to ultrafast sequencing technologies. J Biotechnol 140(1–2):3–12 doi: 10.1016/j.jbiotec.2009.01.006

Buchold GM, Coarfa C, Kim J, Milosavljevic A, Gunaratne PH, Matzuk MM (2010) Analysis of microRNA expression in the prepubertal testis. PLoS One 5(12):e15317. doi:10.1371/journal.pone.0015317

Burroughs AM, Ando Y, de Hoon MJ, Tomaru Y, Nishibu T, Ukekawa R, Funakoshi T, Kurokawa T, Suzuki H, Hayashizaki Y, Daub CO (2010) A comprehensive survey of 3′ animal miRNA modification events and a possible role for 3′ adenylation in modulating miRNA targeting effectiveness. Genome Res 20(10):1398–1410. doi:gr.106054.110 [pii] 10.1101/gr.106054.110

Burroughs AM, Ando Y, Hoon ML, Tomaru Y, Suzuki H, Hayashizaki Y, Daub CO (2011) Deep-sequencing of human Argonaute-associated small RNAs provides insight into miRNA sorting and reveals Argonaute association with RNA fragments of diverse origin. RNA Biol 8 (1):158–177. doi:14300 [pii] 10.4161/rna.8.1.14300

Carninci P (2010) RNA dust: where are the genes? DNA Res 17(2):51–59. doi:dsq006 [pii] 10.1093/dnares/dsq006

Chi SW, Zang JB, Mele A, Darnell RB (2009) Argonaute HITS-CLIP decodes microRNA-mRNA interaction maps. Nature 460(7254):479–486. doi:nature08170 [pii] 10.1038/nature08170

Dai M, Thompson RC, Maher C, Contreras-Galindo R, Kaplan MH, Markovitz DM, Omenn G, Meng F (2010) NGSQC: cross-platform quality analysis pipeline for deep sequencing data. BMC Genomics 11(Suppl 4):S7. doi:1471-2164-11-S4-S7 [pii] 10.1186/1471-2164-11-S4-S7

de Hoon MJ, Taft RJ, Hashimoto T, Kanamori-Katayama M, Kawaji H, Kawano M, Kishima M, Lassmann T, Faulkner GJ, Mattick JS, Daub CO, Carninci P, Kawai J, Suzuki H, Hayashizaki Y (2010) Cross-mapping and the identification of editing sites in mature microRNAs in high-throughput sequencing libraries. Genome Res 20(2):257–264. doi:gr.095273.109 [pii] 10.1101/gr.095273.109

Ebhardt HA, Tsang HH, Dai DC, Liu Y, Bostan B, Fahlman RP (2009) Meta-analysis of small RNA-sequencing errors reveals ubiquitous post-transcriptional RNA modifications. Nucleic Acids Res 37(8):2461–2470. doi:gkp093 [pii] 10.1093/nar/gkp093

Ender C, Meister G (2010) Argonaute proteins at a glance. J Cell Sci 123(Pt 11):1819–1823. doi:123/11/1819 [pii] 10.1242/jcs.055210

Ender C, Krek A, Friedlander MR, Beitzinger M, Weinmann L, Chen W, Pfeffer S, Rajewsky N, Meister G (2008) A human snoRNA with microRNA-like functions. Mol Cell 32(4):519–528. doi:S1097-2765(08), 00733-8 [pii] 10.1016/j.molcel.2008.10.017

Faulkner GJ, Forrest AR, Chalk AM, Schroder K, Hayashizaki Y, Carninci P, Hume DA, Grimmond SM (2008) A rescue strategy for multimapping short sequence tags refines surveys of transcriptional activity by CAGE. Genomics 91(3):281–288. doi:S0888-7543(07), 00279-0 [pii] 10.1016/j.ygeno.2007.11.003

Friedlander MR, Chen W, Adamidi C, Maaskola J, Einspanier R, Knespel S, Rajewsky N (2008) Discovering microRNAs from deep sequencing data using miRDeep. Nat Biotechnol 26 (4):407–415. doi:nbt1394 [pii] 10.1038/nbt1394

Goff LA, Davila J, Swerdel MR, Moore JC, Cohen RI, Wu H, Sun YE, Hart RP (2009) Ago2 immunoprecipitation identifies predicted microRNAs in human embryonic stem cells and neural precursors. PLoS One 4(9):e7192. doi:10.1371/journal.pone.0007192

Hackenberg M, Sturm M, Langenberger D, Falcon-Perez JM, Aransay AM (2009) miRanalyzer: a microRNA detection and analysis tool for next-generation sequencing experiments. Nucleic Acids Res 37 (Web Server issue):W68–76. doi:gkp347 [pii] 10.1093/nar/gkp347

Hafner M, Landthaler M, Burger L, Khorshid M, Hausser J, Berninger P, Rothballer A, Ascano M Jr, Jungkamp AC, Munschauer M, Ulrich A, Wardle GS, Dewell S, Zavolan M, Tuschl T (2010) Transcriptome-wide identification of RNA-binding protein and microRNA target sites by PAR-CLIP. Cell 141(1):129–141. doi:S0092-8674(10)00245-X [pii] 10.1016/j.cell.2010.03.009

Halic M, Moazed D (2010) Dicer-independent primal RNAs trigger RNAi and heterochromatin formation. Cell 140(4):504–516. doi:S0092-8674(10), 00020-6 [pii] 10.1016/j.cell.2010.01.019

Hamady M, Walker JJ, Harris JK, Gold NJ, Knight R (2008) Error-correcting barcoded primers for pyrosequencing hundreds of samples in multiplex. Nat Methods 5(3):235–237. doi:nmeth.1184 [pii] 10.1038/nmeth.1184

Haussecker D, Cao D, Huang Y, Parameswaran P, Fire AZ, Kay MA (2008) Capped small RNAs and MOV10 in human hepatitis delta virus replication. Nat Struct Mol Biol 15(7):714–721. doi:nsmb.1440 [pii] 10.1038/nsmb.1440

Haussecker D, Huang Y, Lau A, Parameswaran P, Fire AZ, Kay MA (2010) Human tRNA-derived small RNAs in the global regulation of RNA silencing. RNA 16(4):673–695. doi:rna.2000810 [pii] 10.1261/rna.2000810

Heo I, Joo C, Kim YK, Ha M, Yoon MJ, Cho J, Yeom KH, Han J, Kim VN (2009) TUT4 in concert with Lin28 suppresses microRNA biogenesis through pre-microRNA uridylation. Cell 138 (4):696–708. doi:S0092-8674(09), 00964-7 [pii] 10.1016/j.cell.2009.08.002

Hertel J, Lindemeyer M, Missal K, Fried C, Tanzer A, Flamm C, Hofacker IL, Stadler PF (2006) The expansion of the metazoan microRNA repertoire. BMC Genomics 7:25. doi:1471-2164-7-25 [pii] 10.1186/1471-2164-7-25

Hoskins RA, Landolin JM, Brown JB, Sandler JE, Takahashi H, Lassmann T, Yu C, Booth BW, Zhang D, Wan KH, Yang L, Boley N, Andrews J, Kaufman TC, Graveley BR, Bickel PJ,

Carninci P, Carlson JW, Celniker SE (2011) Genome-wide analysis of promoter architecture in Drosophila melanogaster. Genome Res 21(2):182–192. doi:gr.112466.110 [pii] 10.1101/gr.112466.110

Hutvagner G, Simard MJ (2008) Argonaute proteins: key players in RNA silencing. Nat Rev Mol Cell Biol 9(1):22–32. doi:nrm2321 [pii] 10.1038/nrm2321

Jones MR, Quinton LJ, Blahna MT, Neilson JR, Fu S, Ivanov AR, Wolf DA, Mizgerd JP (2009) Zcchc11-dependent uridylation of microRNA directs cytokine expression. Nat Cell Biol 11 (9):1157–1163. doi:ncb1931 [pii] 10.1038/ncb1931

Kanamori-Katayama M, Itoh M, Kawaji H, Lassmann T, Katayama S, Kojima M, Bertin N, Kaiho A, Ninomiya N, Daub CO, Carninci P, Forrest AR, Hayashizaki Y (2011) Unamplified cap analysis of gene expression on a single-molecule sequencer. Genome Res. doi:gr.115469.110 [pii] 10.1101/gr.115469.110

Kapranov P, Cheng J, Dike S, Nix DA, Duttagupta R, Willingham AT, Stadler PF, Hertel J, Hackermuller J, Hofacker IL, Bell I, Cheung E, Drenkow J, Dumais E, Patel S, Helt G, Ganesh M, Ghosh S, Piccolboni A, Sementchenko V, Tammana H, Gingeras TR (2007) RNA maps reveal new RNA classes and a possible function for pervasive transcription. Science 316(5830):1484–1488. doi:1138341 [pii] 10.1126/science.1138341

Kapranov P, Ozsolak F, Kim SW, Foissac S, Lipson D, Hart C, Roels S, Borel C, Antonarakis SE, Monaghan AP, John B, Milos PM (2010) New class of gene-termini-associated human RNAs suggests a novel RNA copying mechanism. Nature 466(7306):642–646. doi:nature09190 [pii] 10.1038/nature09190

Katoh T, Sakaguchi Y, Miyauchi K, Suzuki T, Kashiwabara S, Baba T (2009) Selective stabilization of mammalian microRNAs by 3′ adenylation mediated by the cytoplasmic poly(A) polymerase GLD-2. Genes Dev 23(4):433–438. doi:23/4/433 [pii] 10.1101/gad.1761509

Kawahara Y, Zinshteyn B, Sethupathy P, Iizasa H, Hatzigeorgiou AG, Nishikura K (2007) Redirection of silencing targets by adenosine-to-inosine editing of miRNAs. Science 315 (5815):1137–1140. doi:315/5815/1137 [pii] 10.1126/science.1138050

Kawano M, Kawazu C, Lizio M, Kawaji H, Carninci P, Suzuki H, Hayashizaki Y (2010) Reduction of non-insert sequence reads by dimer eliminator LNA oligonucleotide for small RNA deep sequencing. Biotechniques 49(4):751–755. doi:000113516 [pii] 10.2144/000113516

Kim YK, Heo I, Kim VN (2010) Modifications of small RNAs and their associated proteins. Cell 143(5):703–709. doi:S0092-8674(10), 01298-5 [pii] 10.1016/j.cell.2010.11.018

Kucho K, Yoneda H, Harada M, Ishiura M (2004) Determinants of sensitivity and specificity in spotted DNA microarrays with unmodified oligonucleotides. Genes Genet Syst 79(4):189–197. doi:JST.JSTAGE/ggs/79.189 [pii] 10.1266/ggs.79.189

Landgraf P, Rusu M, Sheridan R, Sewer A, Iovino N, Aravin A, Pfeffer S, Rice A, Kamphorst AO, Landthaler M, Lin C, Socci ND, Hermida L, Fulci V, Chiaretti S, Foa R, Schliwka J, Fuchs U, Novosel A, Muller RU, Schermer B, Bissels U, Inman J, Phan Q, Chien M, Weir DB, Choksi R, De Vita G, Frezzetti D, Trompeter HI, Hornung V, Teng G, Hartmann G, Palkovits M, Di Lauro R, Wernet P, Macino G, Rogler CE, Nagle JW, Ju J, Papavasiliou FN, Benzing T, Lichter P, Tam W, Brownstein MJ, Bosio A, Borkhardt A, Russo JJ, Sander C, Zavolan M, Tuschl T (2007) A mammalian microRNA expression atlas based on small RNA library sequencing. Cell 129(7):1401–1414. doi:S0092-8674(07), 00604-6 [pii] 10.1016/j.cell.2007.04.040

Lassmann T, Hayashizaki Y, Daub CO (2009) TagDust – a program to eliminate artifacts from next generation sequencing data. Bioinformatics 25(21):2839–2840. doi:btp527 [pii] 10.1093/bioinformatics/btp527

Lassmann T, Hayashizaki Y, Daub CO (2011) SAMStat: monitoring biases in next generation sequencing data. Bioinformatics 27(1):130–131. doi:btq614 [pii] 10.1093/bioinformatics/btq614

Lee YS, Shibata Y, Malhotra A, Dutta A (2009) A novel class of small RNAs: tRNA-derived RNA fragments (tRFs). Genes Dev 23(22):2639–2649. doi:23/22/2639 [pii] 10.1101/gad.1837609

Lee LW, Zhang S, Etheridge A, Ma L, Martin D, Galas D, Wang K (2010) Complexity of the microRNA repertoire revealed by next-generation sequencing. RNA 16(11):2170–2180. doi: rna.2225110 [pii] 10.1261/rna.2225110

Lehrbach NJ, Armisen J, Lightfoot HL, Murfitt KJ, Bugaut A, Balasubramanian S, Miska EA (2009) LIN-28 and the poly(U) polymerase PUP-2 regulate let-7 microRNA processing in Caenorhabditis elegans. Nat Struct Mol Biol 16(10):1016–1020. doi:nsmb.1675 [pii] 10.1038/nsmb.1675

Linsen SE, de Wit E, Janssens G, Heater S, Chapman L, Parkin RK, Fritz B, Wyman SK, de Bruijn E, Voest EE, Kuersten S, Tewari M, Cuppen E (2009) Limitations and possibilities of small RNA digital gene expression profiling. Nat Methods 6(7):474–476. doi:nmeth0709-474 [pii] 10.1038/nmeth0709-474

Lu C, Shedge V (2011) Construction of small RNA cDNA libraries for high-throughput sequencing. Methods Mol Biol 729:141–152. doi:10.1007/978-1-61779-065-2_9

McCormick KP, Willmann MR, Meyers BC (2011) Experimental design, preprocessing, normalization and differential expression analysis of small RNA sequencing experiments. Silence 2 (1):2. doi:1758-907X-2-2 [pii] 10.1186/1758-907X-2-2

Metzker ML (2010) Sequencing technologies – the next generation. Nat Rev Genet 11(1):31–46. doi:nrg2626 [pii] 10.1038/nrg2626

Morin RD, O'Connor MD, Griffith M, Kuchenbauer F, Delaney A, Prabhu AL, Zhao Y, McDonald H, Zeng T, Hirst M, Eaves CJ, Marra MA (2008) Application of massively parallel sequencing to microRNA profiling and discovery in human embryonic stem cells. Genome Res 18(4):610–621. doi:gr.7179508 [pii] 10.1101/gr.7179508

Morin RD, Zhao Y, Prabhu AL, Dhalla N, McDonald H, Pandoh P, Tam A, Zeng T, Hirst M, Marra M (2010) Preparation and analysis of microRNA libraries using the Illumina massively parallel sequencing technology. Methods Mol Biol 650:173–199. doi:10.1007/978-1-60761-769-3_14

Munafo DB, Robb GB (2010) Optimization of enzymatic reaction conditions for generating representative pools of cDNA from small RNA. RNA 16(12):2537–2552. doi:rna.2242610 [pii] 10.1261/rna.2242610

Pitt JN, Rajapakse I, Ferre-D'Amare AR (2010) SEWAL: an open-source platform for next-generation sequence analysis and visualization. Nucleic Acids Res 38(22):7908–7915. doi: gkq661 [pii] 10.1093/nar/gkq661

Reid JG, Nagaraja AK, Lynn FC, Drabek RB, Muzny DM, Shaw CA, Weiss MK, Naghavi AO, Khan M, Zhu H, Tennakoon J, Gunaratne GH, Corry DB, Miller J, McManus MT, German MS, Gibbs RA, Matzuk MM, Gunaratne PH (2008) Mouse let-7 miRNA populations exhibit RNA editing that is constrained in the 5'-seed/cleavage/anchor regions and stabilize predicted mmu-let-7a:mRNA duplexes. Genome Res 18(10):1571–1581. doi:gr.078246.108 [pii] 10.1101/gr.078246.108

Schulte JH, Marschall T, Martin M, Rosenstiel P, Mestdagh P, Schlierf S, Thor T, Vandesompele J, Eggert A, Schreiber S, Rahmann S, Schramm A (2010) Deep sequencing reveals differential expression of microRNAs in favorable versus unfavorable neuroblastoma. Nucleic Acids Res 38(17):5919–5928. doi:gkq342 [pii] 10.1093/nar/gkq342

Smith AM, Heisler LE, Mellor J, Kaper F, Thompson MJ, Chee M, Roth FP, Giaever G, Nislow C (2009) Quantitative phenotyping via deep barcode sequencing. Genome Res 19(10): 1836–1842. doi:gr.093955.109 [pii] 10.1101/gr.093955.109

Su RW, Lei W, Liu JL, Zhang ZR, Jia B, Feng XH, Ren G, Hu SJ, Yang ZM (2010) The integrative analysis of microRNA and mRNA expression in mouse uterus under delayed implantation and activation. PLoS One 5(11):e15513. doi:10.1371/journal.pone.0015513

Taft RJ, Glazov EA, Cloonan N, Simons C, Stephen S, Faulkner GJ, Lassmann T, Forrest AR, Grimmond SM, Schroder K, Irvine K, Arakawa T, Nakamura M, Kubosaki A, Hayashida K, Kawazu C, Murata M, Nishiyori H, Fukuda S, Kawai J, Daub CO, Hume DA, Suzuki H, Orlando V, Carninci P, Hayashizaki Y, Mattick JS (2009) Tiny RNAs associated with transcription start sites in animals. Nat Genet 41(5):572–578. doi:ng.312 [pii] 10.1038/ng.312

Taft RJ, Simons C, Nahkuri S, Oey H, Korbie DJ, Mercer TR, Holst J, Ritchie W, Wong JJ, Rasko JE, Rokhsar DS, Degnan BM, Mattick JS (2010) Nuclear-localized tiny RNAs are associated with transcription initiation and splice sites in metazoans. Nat Struct Mol Biol 17(8):1030–1034. doi:nsmb.1841 [pii] 10.1038/nsmb.1841

Thomas MF, Ansel KM (2010) Construction of small RNA cDNA libraries for deep sequencing. Methods Mol Biol 667:93–111. doi:10.1007/978-1-60761-811-9_7

Tian G, Yin X, Luo H, Xu X, Bolund L, Zhang X (2010) Sequencing bias: comparison of different protocols of microRNA library construction. BMC Biotechnol 10:64. doi:1472-6750-10-64 [pii] 10.1186/1472-6750-10-64

Vigneault F, Sismour AM, Church GM (2008) Efficient microRNA capture and bar-coding via enzymatic oligonucleotide adenylation. Nat Methods 5(9):777–779. doi:10.1038/nmeth.1244

Watanabe T, Totoki Y, Toyoda A, Kaneda M, Kuramochi-Miyagawa S, Obata Y, Chiba H, Kohara Y, Kono T, Nakano T, Surani MA, Sakaki Y, Sasaki H (2008) Endogenous siRNAs from naturally formed dsRNAs regulate transcripts in mouse oocytes. Nature 453(7194):539–543. doi: nature06908 [pii] 10.1038/nature06908

Willenbrock H, Salomon J, Sokilde R, Barken KB, Hansen TN, Nielsen FC, Moller S, Litman T (2009) Quantitative miRNA expression analysis: comparing microarrays with next-generation sequencing. RNA 15(11):2028–2034. doi:rna.1699809 [pii] 10.1261/rna.1699809

Chapter 11
Biocomputational Identification of Bacterial Small RNAs and Their Target Binding Sites

Brian Tjaden

Abstract Over the past decade, small regulatory RNAs (sRNAs) have been found to be widespread among bacteria. A major class of these sRNA genes act as posttranscriptional regulators of messenger RNAs via base-pairing interactions. Members of this class of bacterial sRNA are typically noncoding and 50–300 nucleotides in length. Because these sRNAs do not have open reading frames with distinctive statistical biases nor are they broadly conserved in most cases, bioinformatics approaches for identifying these RNA genes on a genome-wide scale have met with only moderate success. Similarly, computational approaches for predicting message targets of sRNA regulation are still emerging. In this chapter, we survey the state of the field, first in computational identification of sRNAs throughout bacterial genomes and second in computational identification of regulatory targets of sRNA action. We present different classes of techniques that have proven effective at identifying sRNAs or their regulatory targets as well as specific implementations of these bioinformatic techniques along with their strengths and weaknesses.

Keywords Regulatory targets • small noncoding RNA • target prediction

11.1 Introduction

Small regulatory RNA (sRNA) genes pervade bacterial genomes. These sRNAs act by a variety of mechanisms to regulate a range of processes (reviewed in Liu and Camilli 2010; Waters and Storz 2009; Gottesman and Storz 2010). One family of sRNAs interacts with proteins and modifies their activity. Members of this family include 6S (Hindley 1967; Wassarman and Storz 2000), GlmY (Urban et al. 2007),

B. Tjaden (✉)
Computer Science Department, Wellesley College, Wellesley, MA, USA
e-mail: btjaden@wellesley.edu

B. Mallick (eds.), *Regulatory RNAs*, DOI 10.1007/978-3-642-22517-8_11,   273
© Springer-Verlag Berlin Heidelberg (outside the USA) 2012

CsrB/CsrC/CsrD (Liu et al. 1997), 4.5S (Ribes et al. 1990), RNase P (Stark et al. 1978), and tmRNA (Ray and Apirion 1979). Another family is comprised of riboswitches, which are structured elements that bind to small metabolites and are part of the mRNA they regulate (Tucker and Breaker 2005). Clusters of regularly interspaced short palindromic repeat (CRISPRs) represent a class of sRNAs characterized by short tandem repeats separated by distinct spacer sequences. Together with associated proteins, CRISPRs are thought to act as a type of immune system via an RNA interference mechanism (Sorek et al. 2008).

The largest family of sRNAs corresponds to sRNAs that regulate target mRNAs via base pairing (see Chap. 4). Many of these base-pairing sRNAs are *cis*-acting RNAs in that they are transcribed opposite to their target RNA. Because these sRNAs are antisense, at least in part, to their target, they share an extended region of complementarity to their target. Examples of *cis*-acting sRNAs include RNA antitoxins (Gerdes and Wagner 2007), sRNAs such as GadY that direct cleavage of their target (Opdyke et al. 2004), and sRNAs such as IsrA that repress expression of their target's product (Duhring et al. 2006). The best studied group of sRNAs corresponds to *trans*-acting regulators that bind via base pairing to their message targets. In contrast to *cis*-acting sRNAs, the *trans*-acting sRNAs typically have limited complementarity to their targets. For example, Fig. 11.1 shows the

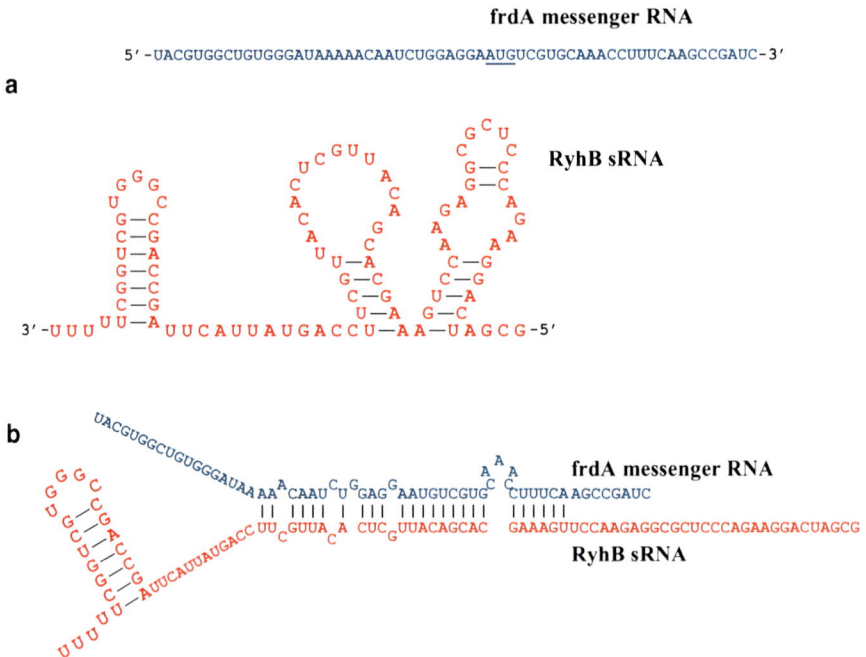

Fig. 11.1 (a) A region around the translation initiation site of the *frdA* mRNA sequence is shown in blue with the *frdA* start codon underlined. The putative secondary structure of the sRNA RyhB is shown in red. (b) The partial complementarity between the *frdA* mRNA and the sRNA RyhB is illustrated in the interaction

putative interaction in *Escherichia coli* for the *trans*-acting sRNA RyhB and its target *frdA*, which codes for a subunit of fumarate reductase. While there are few universal properties of these *trans*-acting sRNAs, some features are shared by the majority. For example, *trans*-acting sRNAs are generally transcribed independently, often approximately 100 nucleotides in length; they most often negatively regulate their targets through translational repression or destabilization of the target RNA, though there are also examples of positive regulation. Many bind the RNA chaperone Hfq at least in enteric bacteria. They typically bind the 5′ UTR of their targets in the neighborhood of the ribosome binding site though binding can also occur far upstream of the ribosome binding site or downstream in the coding sequence; they are often synthesized under specific growth conditions, and many regulate multiple targets. In *E. coli*, where they are perhaps best understood, there are approximately 100 such *trans*-acting sRNAs. The remainder of this chapter focuses primarily on these *trans*-acting sRNAs that act by limited base pairing to their mRNA targets.

11.2 Computational Identification of sRNAs

In this section, we restrict our consideration to *in silico* methods for identifying sRNA genes on a genome-wide scale. We do not consider experimental methods either for detecting individual sRNAs or for genome-wide screens though such experimental approaches are critical for validating predictions from *in silico* methods (see Chap. 14). For review of experimental methods for characterizing sRNAs, including genome-tiling microarrays, high-throughput sequencing, shotgun cloning and RNomics, direct labeling and sequencing, functional genetic screens, and isolation of sRNAs through co-purification with proteins, see Altuvia (2007) and Sharma and Vogel (2009). Here, we present different classes of computational techniques that have proven effective at identifying sRNAs as well as specific implementations of these bioinformatics techniques along with their strengths and weaknesses.

While computational methods that predict protein-coding genes rely heavily on codon usage statistics to identify coding sequences within a genome, in contrast, sRNAs have no open reading frames with distinctive statistical biases so that computational methods for predicting sRNAs must rely on other signals suggestive of a gene. Bioinformatics approaches for predicting sRNAs typically use one (or more) of four types of data: primary sequence data, conservation of primary sequence, secondary structure information, and conservation of secondary structure. We will consider each of the four types of data in turn in the context of biocomputational identification of sRNAs.

11.2.1   Primary Sequence Data

While sRNAs have been identified within protein-coding sequences, antisense to protein-coding sequences, and within the untranslated regions between genes co-transcribed as part of an operon, most sRNAs characterized to date reside in intergenic regions of a genome. As a result, many computational screens for sRNAs restrict their searches to intergenic regions of a genome. Transcription signals such as promoter sequences and Rho-independent terminators are often used as indicators of sRNA genes. TransTermHP is a commonly used tool to predict Rho-independent terminators (Kingsford et al. 2007). TransTermHP predictions can be downloaded for many genomes from the authors' website (http://transterm. cbcb.umd.edu/), or the program can be downloaded and executed on a genome of interest, typically requiring less than a minute when run on a desktop computer. While computational prediction of Rho-independent terminators is fairly reliable, accurate prediction of promoter sequences is a more challenging problem. Many tools that predict promoter sequences suffer from high false positive rates because promoter-like sequences abound in a genome and it is difficult to distinguish spurious sites from sites of RNA polymerase binding that lead to functional transcripts (Haugen et al. 2008). Furthermore, promoter prediction methods cannot easily be applied broadly across bacterial genomes since promoter sequences are specific to various sigma factors. Those sRNA prediction methods that attempt to identify promoters primarily have been applied to *E. coli* and they model, either with a consensus sequence or a position-specific scoring matrix (Staden 1984), the -10 region and -35 region corresponding to σ^{70}.

Regulatory sites corresponding to specific transcription factors have been used to focus searches for sRNAs in intergenic sequences, though more commonly transcription factor binding sites are identified after candidate sRNAs have been determined in order to elucidate possible regulatory mechanisms targeting the sRNA. Under the assumption that sRNAs have a different nucleotide composition than the intergenic sequences in which they reside, GC content and dinucleotide frequencies have been used to distinguish possible sRNAs from background genomic sequence. Finally, in order to limit false positive predictions that may correspond to untranslated regions of neighboring protein-coding genes, computational methods may restrict their searches to those regions of intergenic sequences that are sufficiently far from neighboring protein-coding genes, for example, at least 50 nucleotides from start or stop codons of annotated genes, or to intergenic sequences on the opposite strand from neighboring protein-coding genes.

11.2.1.1   Example Programs Using Primary Sequence Data

One of the seminal studies in computational identification of sRNAs relies on a combination of primary sequence data and primary sequence conservation

(Argaman et al. 2001). Argaman et al. used four criteria to search for candidate sRNAs in *E. coli*: (1) only intergenic sequences were considered, (2) promoter sequences corresponding to the RNA polymerase sigma factor σ^{70} and Rho-independent terminators were identified, (3) only sequences with 50–400 nucleotides between a predicted promoter and terminator were retained, and (4) candidate sRNA sequences were required to demonstrate significant conservation in related genomes as determined by BLAST (Altschul et al. 1990). These criteria led to the prediction of 24 sRNAs in *E. coli*, at a time when only 10 sRNAs had been identified previously. Of the 24 predictions, 23 were tested experimentally by Northern blot, primer extension, and RACE (rapid amplification of cDNA ends), and 14 of these 23 showed evidence of an expressed transcript corresponding to the computational prediction.

Intergenic Sequence Inspector (ISI) uses similar criteria for predicting sRNAs (Pichon and Felden 2003). ISI searches intergenic regions for primary sequence conservation and allows the user to incorporate information about promoters, terminators, and RNA secondary structure, while filtering the results based on size and GC content. ISI was tested for its ability to correctly predict known sRNAs in *E. coli*. Later, ISI together with Northern blots and microarrays were used to identify 12 sRNAs in *Staphylococcus aureus* (Pichon and Felden 2005).

11.2.2 Primary Sequence Conservation

Few, if any, *trans*-acting sRNAs that bind via base pairing with their targets are conserved broadly across bacteria. However, many sRNAs are conserved, particularly in closely related species. As a result, conservation of a sequence, usually detected by BLAST (Altschul et al. 1990) or a similar approach, can indicate the presence of an sRNA. The primary challenge associated with using conservation as a predictor of sRNAs is that sequences may be conserved for a variety of reasons, and it is not trivial to distinguish conservation of an sRNA from that of untranslated regions of genes, regulatory sites, and other functional elements in a genome.

Not only can conservation of a sequence be used as an indicator of an sRNA, but the pattern or profile of conservation across a large number of genomes (hundreds or thousands) offers predictive power. In this vein, nucleic acid phylogenetic profiling has recently been used to identify sRNAs on a genome-wide scale (Marchais et al. 2009). Of course, any approach that relies on conservation will be unable to identify orphan sRNAs, and the extent of orphan sRNA genes in bacteria remains unknown, in part, because many computational approaches for identifying sRNAs rely on conservation and, thus, are bias against discovery of orphans.

11.2.2.1 Example Programs Using Primary Sequence Conservation

sRNAPredict is a computational method for predicting sRNAs based both on primary sequence data and primary sequence conservation (Livny et al. 2005). sRNAPredict searches a genome for sRNAs by considering, for each intergenic sequence, the extent of primary sequence conservation as well as distance to neighboring protein-coding genes and possible promoters and Rho-independent terminators in the intergenic sequence as predicted by other programs. One study suggests that sRNAPredict is among the most effective computational methods for sRNA prediction (Lu et al. 2011). When applied to *Pseudomonas aeruginosa*, 34 novel sRNAs were predicted, 31 of which were experimentally tested via Northern analysis and 17 were found to encode sRNA transcripts (Livny et al. 2006). sRNAPredict's methodology has been applied throughout sequenced bacterial genomes in the SIPHT pipeline (Livny et al. 2008).

Most sRNA prediction methods that employ comparative genomics analyses only incorporate conservation information from a handful of closely related genomes when identifying candidate sRNAs. In contrast, NAPP (nucleic acid phylogenetic profiling) uses the pattern of conservation across a large number (e.g., hundreds) of genomes when identifying candidate sRNAs (Marchais et al. 2009). RNA genes tend to have a pattern of primary sequence conservation similar to that of other RNA genes and less similar to that of protein-coding genes. NAPP exploits this distinguishing feature by clustering similar conservation patterns in order to identify groups of candidate RNA genes. In *Staphylococcus aureus*, NAPP predicted 189 candidate sRNAs. Of these predictions, 24 were tested by Northern blot and 7 showed evidence of sRNA expression (Marchais et al. 2009).

11.2.3 Secondary Structure Information

While it is unclear if, in general, mRNAs have lower folding free energies than background sequences (Workman and Krogh 1999), the thermodynamic stability of an sRNA sequence may help distinguish it from intergenic sequences not encoding RNAs (Washietl et al. 2005). The stability of a sequence can be estimated by folding the sequence into the most energetically favorable secondary structure or into an ensemble of possible secondary structures each with a corresponding measure of stability. When considering a given intergenic sequence, the thermodynamic stability of the sequence's structure can provide some indication as to whether the sequence is likely to correspond to a functional RNA. While thermodynamic stability has not been employed on its own to predict sRNAs, it has been used successfully in collaboration with other features such as structural conservation of a sequence.

11.2.3.1   Example Programs Using Secondary Structure

The BIOPROP program has been used as the basis of a machine learning approach for predicting RNA genes (Carter et al. 2001). The approach uses a combination of neural networks and support vector machines to classify sequences as RNAs based on mono- and dinucleotide composition, thermodynamic stability, and presence of five short nucleotide motifs often found in RNA sequences. The approach was used to predict approximately 370 small RNAs in *E. coli*, though the predictions were not tested experimentally.

11.2.4   Secondary Structure Conservation

Many sequences are conserved in closely related genomes. The pattern of conservation of a sequence can be an effective predictor of the functional role of the sequence. Rivas and Eddy (2001) have developed an approach that classifies a sequence into one of three groups based on its pattern of conservation. Conserved sequences with mutation patterns consistent with synonymous codons may be classified as protein-coding sequences. Conserved sequences whose mutation pattern conserves a secondary structure, such as via compensatory mutations in base-pairing nucleotides in an RNA structure, may be classified as structural RNAs (Fig. 11.2). Conserved sequences with no obvious pattern in mutations may be classified as intergenic or nonfunctional sequences. As a first step in determining conserved structural RNAs, typically a pairwise sequence alignment or multiple sequence alignment is performed to identify related sequences and mutation patterns. Then the mutation patterns are assessed for covarying mutations and their likelihood of corresponding to conserved RNA structural elements. Programs for predicting structural RNAs tend to work best when the input alignments correspond to sequences with approximately 65–85% similarity (Rivas and Eddy 2001). Comparative sequences need to be sufficiently similar so that RNA structures are conserved, yet sufficiently divergent so as to allow for compensatory mutations. Such approaches are effective at identifying RNA elements; however, they generally lack the ability to discriminate sRNAs from other RNA structures that abound in a transcriptome.

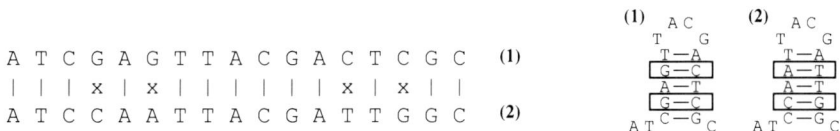

Fig. 11.2 The pairwise alignment of two sequences (1) and (2) is illustrated along with a hairpin secondary structure for each of the two sequences. Four nucleotides differ between the two sequences, yet the structures of the two sequences are conserved since the four differences correspond to compensatory base pair mutations

11.2.4.1   Example Programs Using Secondary Structure Conservation

QRNA is a tool that can use comparative sequence analysis to predict structural RNA genes by identifying patterns of compensatory mutations consistent with some base-paired secondary structure (Rivas and Eddy 2001). QRNA requires an input pairwise sequence alignment and uses a pair stochastic context-free grammar to identify RNA candidates. QRNA was initially applied to *E. coli* (Rivas et al. 2001), hyperthermophiles (Klein et al. 2002), and yeast (McCutcheon and Eddy 2003), but has served more broadly as the prototype for structural RNA gene finding.

Motivated by QRNA, ddbRNA searches for compensatory mutations consistent with secondary structure in conserved sequences but allows input three-way alignments in addition to pairwise alignments (di Bernardo et al. 2003). MSARI extends the search for compensatory mutations with general multiple sequence alignments as input (Coventry et al. 2004).

RNAz, like MSARI, uses multiple sequence alignments as input to identify mutation patterns indicative of structural RNAs (Washietl et al. 2005; Gruber et al. 2010). However, RNAz also incorporates thermodynamic stability information of a sequence when classifying it with a support vector machine as a structural RNA. RNAz was used in conjunction with QRNA in one study (del Val et al. 2007), though both independently have been applied widely. Following the model of RNAz, the Dynalign program has been used as the basis of a support vector machine classifier for structural RNA genes (Uzilov et al. 2006).

sRNAFinder is a biocomputational tool that attempts to integrate in a unified probabilistic framework the various sources of heterogeneous data that evince sRNAs in a genome (Tjaden 2008a). sRNAFinder uses a hidden Markov model to predict sRNAs based on a variety of sources of information such as promoter and terminator signals, conserved structure, and expression data if available. sRNAFinder has been applied to both bacteria and archaea in genome-wide screens for sRNAs (Swiercz et al. 2008; Gvakharia et al. 2010; Babski et al. 2011).

11.2.4.2   Perspectives

As a summary of some of the available tools used for genome-wide screens of sRNAs, Table 11.1 lists the more commonly used computational approaches for sRNA prediction along with which of the four abovementioned data types each approach employs. Despite the abundance of computational methods for predicting sRNAs, there has been a dearth of systematic comparisons of these methods. Many of the abovementioned bioinformatics approaches suffer from low specificity when used for genome-wide screens. Further, there is little practical guidance for users of these tools. Recently, Lu et al. (2011) assessed four of the leading biocomputational tools for sRNA prediction: eQRNA (Rivas and Eddy 2001), RNAz (Washietl et al. 2005; Gruber et al. 2010), sRNAPredict3 and SIPHT (Livny et al. 2005; Livny et al.

Table 11.1 Some common methods for predicting sRNA genes as well as genome-wide screens that employ these methods

Name of method	Type of data used	References
Argaman et al.	1,2	Argaman et al. (2001)
Axmann et al.	2,3	Axmann et al. (2005)
BIOPROP	1,3	Carter et al. (2001)
Chen et al.	1	Chen et al. (2002)
Coenye et al.	3,4	Coenye et al. (2007)
ddbRNA	4	di Bernardo et al. (2003)
Dynalign	3,4	Uzilov et al. (2006)
GMMI	1	Yackie et al. (2006)
ISI	1,2	Pichon and Felden (2003)
Lenz et al.	1,2	Lenz et al. (2004)
MSARI	4	Coventry et al. (2004)
NAPP	2	Marchais et al. (2009)
nocoRNAc	1	Herbig and Nieselt (2011)
Panek et al.	1,2	Panek et al. (2008)
PSoL	1,2,3	Wang et al. (2006)
QRNA	4	Rivas and Eddy (2001)
Saetrom et al.	1,3	Saetrom et al. (2005)
Schattner	1	Schattner (2002)
SIPHT	1,2	Livny et al. (2008)
sRNAFinder	1,4	Tjaden (2008a)
sRNAPredict	1,2	Livny et al. (2005), Livny et al. (2006)
sRNAscanner	1	Sridhar et al. (2010)
RNAz	3,4	Washietl et al. (2005), Gruber et al. (2010)
Tran et al.	3	Tran et al. (2009)
Voss et al.	3,4	Voss et al. (2009)
Xiao et al.	1,2,3	Xiao et al. (2009)

The middle column indicates the type of data used by the method: (1) primary sequence data, (2) primary sequence conservation, (3) secondary structure information, (4) secondary structure conservation

2006; Livny et al. 2008), and NAPP (Marchais et al. 2009). As a benchmark upon which to test the tools, the authors compiled a set of over 700 sRNAs taken from experimentally confirmed sRNAs, RNAs characterized in the Rfam database (Gardner et al. 2008), sRNAs suggested from genome-tiling microarray experiments, and sRNAs suggested by RNA-seq experiments conducted in six different bacteria. The authors found that the mean sensitivities of the tools, i.e., the percentage of putative sRNAs correctly predicted by the tools, ranged from 20% to 49%. The mean precisions of the tools ranged from 4% to 12%. Such low precisions suggest potentially high false positive rates associated with the tools' predictions. Even when predictions from the various tools were combined, the ability of the tools to distinguish sRNAs on a genome-wide scale remained unimpressive when compared, for example, to protein-coding gene prediction. In general, the tools performed well in predicting the strand and length of the sRNAs. Nevertheless, the modest sensitivity and poor precision of the tools suggest that there is much room for improvement in the field of computational sRNA prediction. In light of recent studies using RNA-seq experiments to identify sRNAs in a range of bacteria (Sittka et al. 2008; Liu et al. 2009; Yoder-Himes et al. 2009; Albrecht et al. 2010; Sharma et al. 2010), it is clear that the role of high-throughput sequencing methods will be increasingly important in elucidating sRNAs. As such, the demand for computational tools that integrate data from high-throughput sequencing experiments to identify functional transcripts is likely to grow for the foreseeable future.

11.3 Computational Identification of sRNA Regulatory Targets

As in the case of microRNAs in eukaryotes, *trans*-acting sRNA regulators that bind via base pairing to messages typically affect the translation and stability of their targets. Commonly, these sRNAs inhibit the translation of their mRNA target, for example, by binding in the neighborhood of the translation initiation site and blocking ribosome binding (Fig. 11.1b). Either as an alternative or complement to ribosome occlusion, sRNAs may decrease the stability of the message and target it for degradation by RNase E (Pfeiffer et al. 2009; Desnoyers et al. 2009). Thus, sRNAs often act stoichiometrically where they are degraded along with their targets. Less commonly, sRNAs can activate translation, for example, by freeing translation initiation sites that would otherwise be occluded by an inhibitory secondary structure (Prevost et al. 2007; Urban and Vogel 2008).

In most cases studied to date in enteric bacteria, the chaperone protein Hfq binds both sRNA and mRNA and facilitates interaction. By binding to both sRNA and mRNA, Hfq may increase the local concentrations of both or affect the RNAs' secondary structures (Brennan and Link 2007). However, some sRNAs, particularly in Gram-positive bacteria, do not require Hfq (Boisset et al. 2007; Silvaggi et al. 2006), and the role of an RNA chaperone protein in sRNA:mRNA interactions throughout bacteria more broadly remains unclear.

Base pairing between an sRNA and mRNA occurs over a short region, usually one to two dozen nucleotides, and is imperfect. For these reasons, computational methods for predicting the targets of an sRNA have met with only moderate success, often generating large numbers of false positive predictions. Within the region of interaction between an sRNA and one of its message targets, a few mutational studies have suggested that only about 4–9 nucleotides are essential for regulatory effect (Kawamoto et al. 2006). One sRNA can interact with multiple mRNAs (reviewed in Papenfort and Vogel 2009), enabling sRNAs to be effective participants in bacterial global responses to specific, often stress related, conditions.

A number of experimental approaches have been employed for the purpose of large scale target identification (reviewed in Vogel and Wagner 2007; Sharma and Vogel 2009). Examples of such approaches include genetic screens (Altuvia et al. 1997), co-immunoprecipitation of interaction complexes (Zhang et al. 2003), and sRNAs being used as bait for capturing target mRNAs by affinity purification

(Antal et al. 2005; Douchin et al. 2006). Also, candidates can be identified by investigating differentially expressed targets through transcriptome or proteome screens following deletion or expression of an sRNA (Masse et al. 2005; Papenfort et al. 2006; Rasmussen et al. 2005). One limitation of this approach is that distinguishing primary (physically interacting) targets from secondary downstream regulatory effects is difficult. Consequently, validation of candidate targets, such as through site-specific mutagenesis, is important.

11.3.1 Data Used for Computational Prediction of sRNA Regulatory Targets

Most methods for predicting sRNA:mRNA interactions use one or more of a few different types of data in order to generate predictions. The strength of hybridization between the two RNAs is often used as an indicator of an interaction. Hybridization strength typically refers to inter- but not intramolecular interactions involving the two RNAs, and it can be assessed in a variety of ways. Most simply, the two RNA sequences can be aligned and the (possibly weighted) count of matching base pairs can be used as an indicator of hybridization strength. Using a thermodynamic approach, the energy contribution of stacking base pairs can be aggregated together with the destabilizing effects of bulge and interior loops (Mathews et al. 1999). Hybridization strengths can be assessed for two RNAs either for a specific structured interaction or over an ensemble of different structured interactions between the two RNAs.

Intramolecular structures can also be used to elucidate potential sRNA:mRNA interactions. One approach involves simultaneously computing the joint secondary structure of the two interacting RNAs, for example, by concatenating the two RNA sequences and "folding" the concatenated sequence, while another approach involves computing the accessibility of the interacting region of each RNA by determining its propensity for being unpaired (not part of a structure) in an ensemble of secondary structures. Methods that restrict interactions by disallowing certain complex structures such as pseudoknots and kissing hairpins may increase their computational efficiency at a cost to their sensitivity.

While hybridization, structure, and accessibility are the most common features used by methods to predict targets, other features can also be employed. Since most known interactions occur around the message target's translation initiation site, the search for interactions is often restricted to this region. Comparative genomics may be used to identify interacting regions, as more highly conserved regions may be more likely to act as interacting sites than less highly conserved regions. Since many interactions contain a seed region comprised of a short series of consecutive base pairs between the two RNAs, the existence of such a seed region may be used to hone the search for target candidates. Potential Hfq binding sites can be incorporated into prediction algorithms, though there has been little success thus

Table 11.2 Some common methods for predicting sRNA regulatory targets

Name of method	Type of data used	URL of web server	References
RNAhybrid	1	http://bibiserv.techfak.uni-bielefeld.de/rnahybrid/	Rehmsmeier et al. (2004)
RNAup	1,2	http://rna.tbi.univie.ac.at/cgi-bin/RNAup.cgi	Muckstein et al. (2006)
TargetRNA	1,3	http://snowwhite.wellesley.edu/targetRNA/	Tjaden et al. (2006)
Mandin et al.	1		Mandin et al. (2007)
Boisset et al.	4		Boisset et al. (2007)
RNAplex	1		Tafer and Hofacker (2008)
IntaRNA	1,2,3	http://rna.informatik.uni-freiburg.de:8080/IntaRNA.jsp	Busch et al. (2008)
sRNATarget	3,4	http://ccb.bmi.ac.cn/srnatarget/index.php	Zhao et al. (2008)
RactIP	4		Kato et al. (2010)
Peer and Margalit	2,5		Peer and Margalit (2011)
PETcofold	4,5	http://rth.dk/resources/petcofold/submit.php	Seemann et al. (2011)
ripalign	4,5		Li et al. (2011)

The second column indicates, for each method, the type of data used by the method: (1) hybridization, (2) accessibility, (3) seed region, (4) joint structure, (5) conservation

far in reliable prediction of Hfq binding sites in RNAs. Similarly, RNA architectural modules have yet to be applied to sRNA regulatory target prediction (Westhof 2010). Thus, further research is needed in the field of computational identification of mRNA targets of sRNA action. Table 11.2 summarizes some of the more popular computational methods and the type of data they employ when making their predictions.

11.3.2 Example Programs That Predict Regulatory Targets of sRNA Action

Some of the early bioinformatics methods for predicting sRNA:mRNA interactions stem from methods designed to predict microRNA targets in eukaryotic genomes. RNAhybrid, for example, predicts targets of microRNA action and was initially applied to *Drosophila* (Rehmsmeier et al. 2004). To predict the likelihood of interaction of two RNAs, RNAhybrid simplifies the classic approach for folding RNA sequences (Zuker and Stiegler 1981) by disallowing intramolecular interactions and restricting loop sizes to be at most 15 nucleotides in length. Similar to RNAhybrid, RNAup considers the intermolecular energy of two interacting RNAs (Muckstein et al. 2006). However, RNAup combines this hybridization energy with the energy associated with the interacting regions being unpaired (i.e., accessible) in the intramolecular structures. The probability of two RNAs interacting at a given site is determined as the sum of the two energies over all possible types of binding. RNAup was evaluated by comparing its RNA interaction predictions to data from RNA interference experiments. RNAup is distributed as part of the Vienna RNA package and can be used via a Web interface (Hofacker 2003). Because of the computational costs associated with RNAup, it takes days to execute when applied to genome-wide screens.

TargetRNA was one of the first biocomputational tools designed specifically for predicting targets of sRNAs in bacteria (Tjaden et al. 2006). TargetRNA offers two approaches for predicting the hybridization of an sRNA with a candidate target: an individual base pair model that is an extension of the Smith-Waterman dynamic program (Smith and Waterman 1981) and a stacked base pair model that is similar to RNAhybrid (Rehmsmeier et al. 2004). TargetRNA was evaluated on its ability to correctly predict previously published sRNA:mRNA interactions as well as on targets showing differential expression in microarray assays and Northern blot experiments following ectopic expression of several sRNAs in *E. coli*. TargetRNA demonstrated variable results – for some sRNAs the predictions corresponded well with putative targets and for some sRNAs the predictions did not correspond well with putative targets. Some useful features of TargetRNA include its calculation of p values enabling the significance of a prediction to be estimated, its incorporation of a seed region of consecutive interacting nucleotides, its speed (genome-wide searches take a few seconds), its ability to identify orthologous sRNA:mRNA interactions in other genomes, and its user-friendly Web interface. The primary limitation of TargetRNA is the significant false positive rate of its predictions. An extension of TargetRNA, RNATarget, addresses the inverse problem of identifying an sRNA in a genome that may regulate a given mRNA target (Tjaden 2008b). Like TargetRNA, RNATarget suffers from a significant number of false positive predictions.

Mandin et al. developed a computational method to scan a genome for mRNA targets of an sRNA based on thermodynamic pairing energies (Mandin et al. 2007). The method combines favorably contributing stacking energies of predicted base-pairing nucleotides with unfavorably contributing energies of bulge and internal loops. The contribution of loop energies to a predicted interaction are determined from a training set of three known sRNA:mRNA interactions in *E. coli* and one in *Staphylococcus aureus*. Unlike other methods that restrict their searches to the 5' end of sRNAs, both 5' and 3' regions were screened for possible targets. The computational approach was applied in *Listeria monocytogenes*, where target predictions for three of nine sRNAs were validated experimentally.

As part of their investigation of RNAIII in *Staphylococcus aureus*, Boisset et al. (2007) use a computational approach to predict the sRNA's targets that can be viewed as an instantiation of the more general tool PairFold (Andronescu et al. 2003). The authors concatenate part of the 5' UTR sequence around the ribosome binding site of each mRNA with an eight nucleotide linker sequence and with a hypothesized interacting region of RNAIII. The concatenated sequence is then folded with RNAfold (Hofacker et al. 1994) to identify joint structures with free energy below a chosen threshold. Two of the predicted RNAIII targets were then validated experimentally.

RNAplex uses an approach comparable to that of RNAhybrid, which is to say that intramolecular interactions are ignored when computing a hybridization energy for two interacting RNAs (Tafer and Hofacker 2008). RNAplex uses a simplified energy model, i.e., an affine function for loop size rather than a logarithmic function, in order to achieve a 10–27 speedup as compared to RNAhybrid. When evaluating RNAplex target predictions for eight sRNAs in *E. coli*, on average, RNAplex scores more than 100 mRNAs higher than the known target, suggesting a potentially substantial false positive rate.

Similar to RNAup, IntaRNA is a tool that uses a combination of hybridization energy (intermolecular base pairings) and accessibility of an interacting region (unpaired in the intramolecular structure) to predict the interaction of two RNAs (Busch et al. 2008). As in the case of TargetRNA, IntaRNA can restrict predictions to those containing a seed region composed of consecutive base-pairing nucleotides in the interaction. On a test set of 18 sRNA:mRNA interactions, IntaRNA's accuracy was found to be comparable to or better than that of other approaches while requiring less time to execute than other approaches that incorporate accessibility information. IntaRNA is available for use via a Web interface (Smith et al. 2010) and typically takes minutes to hours to execute for a genome-wide screen.

sRNATarget is a tool that uses either a Naive Bayes classifier or a support vector machine to discriminate sRNA:mRNA interactions (Zhao et al. 2008). The classifiers are based on ten features, seven of which are derived from minimum free energy structures of the RNAs, one of which corresponds to a seed region, and two of which correspond to A/U rich sequences in unstructured regions of minimum free energy structures. The latter two features are meant to model possible Hfq binding regions in the two RNAs. This is one of the few examples of a biocomputational tool that incorporates information about possible Hfq binding sites. Of the ten features, the difference between the minimum free energy of the joint sRNA:mRNA structure and the minimum free energies of the individual sRNA and mRNA structures are found to offer the best discriminatory power in classifying interactions. The tool was evaluated on a set of previously reported sRNA:mRNA interactions in *E. coli* and later made available through a Web server (Cao et al. 2009).

RactIP is a method that uses integer linear programming to estimate the joint secondary structure of two interacting RNAs (Kato et al. 2010). Aspects of feasible structures are modeled as a series of linear constraints. The objective function, which is the sum of weighted base pairs in the joint secondary structures, is maximized using an integer programming solver. RactIP does not use the classic thermodynamic parameters (Mathews et al. 1999) for modeling stacked base pairs or various loops within the structures. RactIP's primary contribution is a more efficient method for estimating joint secondary structure as compared to more computationally expensive methods for computing the interaction partition function (McCaskill 1990). RactIP was initially evaluated on 18 sRNA:mRNA examples in *E. coli* for its ability to accurately predict the interacting nucleotides given a known sRNA:mRNA interaction as input. However, it was not applied to the more general problem of predicting targets of a given sRNA. A recent study evaluating its ability to predict mRNA targets for a given sRNA found its specificity to be prohibitively low (Tjaden 2011, unpublished).

Peer and Margalit approach the problem of predicting sRNA:mRNA interactions from a different perspective – rather than focusing on the mRNA targets, they investigate the target-binding regions of sRNAs (Peer and Margalit 2011). The authors consider different features that may be used to identify sRNA interaction sites. In particular, they find that sRNA interaction sites are better conserved, as determined by primary sequence conservation, than other regions of the sRNA and that sRNA interactions sites are more accessible, i.e., less likely to participate in structured regions, than other regions of the sRNA. As a result, the authors offer evidence that false positive predictions can be reduced by focusing searches for sRNA targets to messages that interact with probable sRNA interaction sites as determined by conservation and accessibility. This finding is an important advance as it addresses, at least in part, the main limitation of sRNA target prediction methods – their high false positive rate.

11.3.3   Comparison of Tools

For three of the abovementioned tools, IntaRNA, TargetRNA, and RNAplex, we assessed their performance on a test set of 73 validated sRNA:mRNA interactions involving 24 sRNAs in *Escherichia coli* (Peer and Margalit 2011). Each tool was used to predict message targets in *E. coli* for the 24 sRNAs. Only regions of messages within 30 nucleotides upstream of the start of translation and within 20 nucleotides downstream of the start of translation were considered. For IntaRNA, a seed of 7 was used and predictions at or below 32 energy scores ranging from -25.0 to -9.5 were considered. For TargetRNA, a seed of 7 was used and predictions at or below 14 p values ranging from 0.0001 to 0.12 were considered. For RNAplex, predictions at or below 17 energy scores ranging from -25.0 to -9.0 were considered. Figure 11.3 illustrates the sensitivity of the three tools in identifying the 73 sRNA:mRNA interactions in the test set. At larger score thresholds, as the tools make more predictions for each sRNA, the sensitivities of their predictions rise. However, when making as many as 200 predictions per sRNA, no tool identifies even half of the 73 sRNA:mRNA interactions, and when making a more reasonable 30 predictions per sRNA, no tool identifies even a quarter of the 73 interactions.

11.3.4   Perspectives

Taken together, the abovementioned methods offer a number of insights into the field of computational prediction of sRNA targets. First, most approaches for predicting

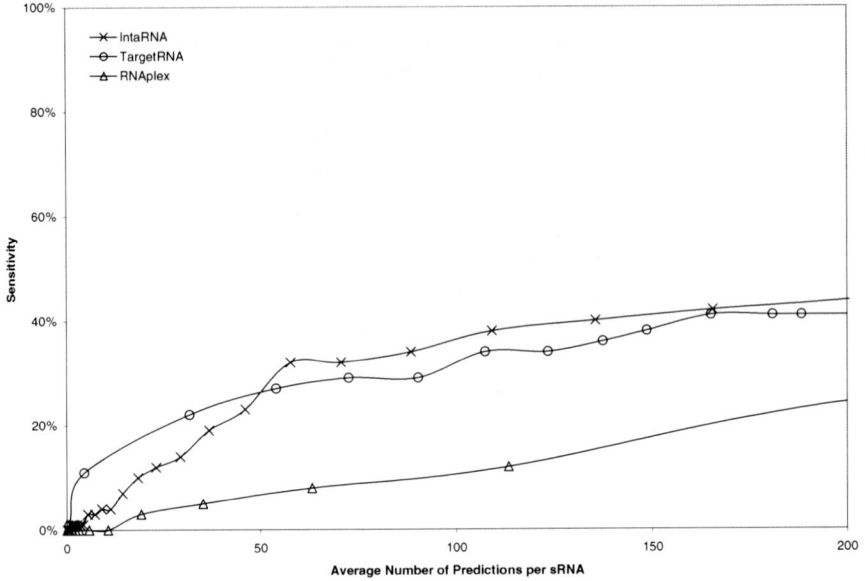

Fig. 11.3 The performance of three computational tools (IntaRNA, TargetRNA, and RNAplex) is shown on a test set of 73 validated sRNA:mRNA interactions involving 24 sRNAs. The ordinate axis shows the sensitivity of each tool, i.e., the percent of the 73 interactions correctly predicted by the tool. The abscissa axis indicates the number of predictions made by a tool averaged over the 24 sRNAs. Each of the three lines corresponds to the performance of a tool when predictions from the tool are considered at various score thresholds specific to the tool

sRNA:mRNA interactions use some subset of the same few features such as the existence of a seed region, thermodynamically favorable hybridization and structures, and conservation in other genomes. When applied to entire genomes, a number of approaches demonstrate a trade-off between faster execution (seconds) with less accurate results and slower execution (hours or days) with more accurate results. Two of the main challenges for the field of computational identification of sRNA regulatory targets are limiting the number of false positive predictions and training methods on a small number of verified sRNA:mRNA interactions, especially outside of *Escherichia coli*. As more interactions are verified experimentally, benchmark data sets can be developed and systematic comparisons of the various tools will offer more compelling assessments of the tools and better practical guidance for users of the tools. Thus, accurate prediction of sRNA regulatory targets remains a challenging problem, with many opportunities for further investigation and advancement.

References

Albrecht M, Sharma CM, Reinhardt R, Vogel J, Rudel T (2010) Deep sequencing-based discovery of the *Chlamydia trachomatis* transcriptome. Nucleic Acids Res 38(3):868–877

Altschul SF, Gish W, Miller W, Myers EW, Lipman DJ (1990) Basic local alignment search tool. J Mol Biol 215:403–410

Altuvia S (2007) Identification of bacterial small non-coding RNAs: experimental approaches. Curr Opin Microbiol 10(3):257–261

Altuvia S, Weinstein-Fischer D, Zhang A, Postow L, Storz G (1997) A small, stable RNA induced by oxidative stress: role as a pleiotropic regulator and antimutator. Cell 90(1):43–53

Andronescu M, Aguirre-Hernandez R, Condon A, Hoos HH (2003) RNAsoft: a suite of RNA secondary structure prediction and design software tools. Nucleic Acids Res 31 (13):3416–3422

Antal M, Bourdeau V, Douchin V, Felden B (2005) A small bacterial RNA regulates a putative ABC transporter. J Biol Chem 280:7901–7908

Argaman L, Hershberg R, Vogel J, Bejerano G, Wagner EG, Margalit H, Altuvia S (2001) Novel small RNA-encoding genes in the intergenic regions of *Escherichia coli*. Curr Biol 11 (12):941–950

Axmann IM, Kensche P, Vogel J, Kohl S, Herzel H, Hess WR (2005) Identification of cyanobacterial non-coding RNAs by comparative genome analysis. Genome Biol 6:R73

Babski J, Tjaden B, Voss B, Jellen-Ritter A, Marchfelder A, Hess WR, Soppa J (2011) Bioinformatics prediction and experimental verification of sRNAs in the haloarchaeon *Haloferax volcanii*. RNA Biol 8(5):806–816

Boisset S, Geissmann T, Huntzinger E, Fechter P, Bendridi N, Possedko M, Chevalier C, Helfer AC, Benito Y, Jacquier A, Gaspin C, Vandenesch F, Romby P (2007) *Staphylococcus aureus* RNAIII coordinately represses the synthesis of virulence factors and the transcription regulator Rot by an antisense mechanism. Genes Dev 21:1353–1366

Brennan RG, Link TM (2007) Hfq structure, function and ligand binding. Curr Opin Microbiol 10(2):125–133

Busch A, Richter AS, Backofen R (2008) IntaRNA: efficient prediction of bacterial sRNA targets incorporating target site accessibility and seed regions. Bioinformatics 24(24):2849–2856

Cao Y, Zhao Y, Cha L, Ying X, Wang L, Shao N, Li W (2009) sRNATarget: a web server for prediction of bacterial sRNA targets. Bioinformation 3(8):364–366

Carter RJ, Dubchak I, Holbrook SR (2001) A computational approach to identify genes for functional RNAs in genomic sequences. Nucleic Acids Res 29(19):3928–3938

Chen S, Lesnik EA, Hall TA, Sampath R, Griffey RH, Ecker DJ, Blyn LB (2002) A bioinformatics based approach to discover small RNA genes in the *Escherichia coli* genome. Biosystems 65:157–177

Coenye T, Drevinek P, Mahenthiralingam E, Shah SA, Gill RT, Vandamme P, Ussery DW (2007) Identification of putative noncoding RNA genes in the *Burkholderia cenocepacia* J2315 genome. FEMS Microbiol Lett 276(1):83–92

Coventry A, Kleitman DJ, Berger B (2004) MSARI: multiple sequence alignments for statistical detection of RNA secondary structure. Proc Natl Acad Sci USA 101(33):12102–12107

del Val C, Rivas E, Torres-Quesada O, Toro N, Jimenez-Zurdo JI (2007) Identification of differentially expressed small non-coding RNAs in the legume endosymbiont *Sinorhizobium meliloti* by comparative genomics. Mol Microbiol 66(5):1080–1091

Desnoyers G, Morissette A, Prevost K, Masse E (2009) Small RNA-induced differential degradation of the polycistronic mRNA *iscRSUA*. EMBO J 28:1551–1561

di Bernardo D, Down T, Hubbard T (2003) ddbRNA: detection of conserved secondary structures in multiple alignments. Bioinformatics 19(13):1606–1611

Douchin V, Bohn C, Bouloc P (2006) Down-regulation of porins by a small RNA bypasses the essentiality of the regulated intramembrane proteolysis protease RseP in *Escherichia coli*. J Biol Chem 281:12253–12259

Duhring U, Axmann IM, Hess WR, Wilde A (2006) An internal antisense RNA regulates expression of the photosynthesis gene *isiA*. Proc Natl Acad Sci USA 103:7054–7058

Gardner PP, Daub J, Tate JG, Nawrocki EP, Kolbe DL, Lindgreen S, Wilkinson AC, Finn RD, Griffiths-Jones S, Eddy SR, Bateman A (2008) Rfam: updates to the RNA families database. Nucleic Acids Res 37(S1):D136–D140

Gerdes K, Wagner EGH (2007) RNA antitoxins. Curr Opin Microbiol 10:117–124

Gottesman S, Storz G (2010) Bacterial small RNA regulators: versatile roles and rapidly evolving mechanisms. Cold Spring Harbor Perspectives in Biology, Editors John F. Atkins, Raymond F. Gesteland, and Thomas R. Cech, Cold Spring Harbor Press

Gruber AR, Findeiss S, Washietl S, Hofacker IL, Stadler PF (2010) RNAZ 2.0: improved noncoding RNA detection. Pac Symp Biocomput 15:69–79

Gvakharia BO, Tjaden B, Vajrala N, Sayavedra-Soto LA, Arp D (2010) Computational prediction and transcriptional analysis of sRNAs in *Nitrosomonas europaea*. FEMS Microbiol Lett 312 (1):46–54

Haugen SP, Ross W, Gourse RL (2008) Advances in bacterial promoter recognition and its control by factors that do not bind DNA. Nat Rev Microbiol 6:507–519

Herbig A, Nieselt K (2011) nocoRNAc: characterization of non-coding RNAs in prokaryotes. BMC Bioinformatics 12:40

Hindley J (1967) Fractionation of ^{32}P-labeled ribonucleic acids on polyacrylamide gels and their characterization by fingerprinting. J Mol Biol 30:125–136

Hofacker IL (2003) Vienna RNA secondary structure server. Nucleic Acids Res 31 (13):3429–3431

Hofacker IL, Fontana W, Stadler PF, Bonhoeffer LS, Tacker M, Schuster P (1994) Fast folding and comparison of RNA secondary structure. Monatshefte Chem 125(2):167–188

Kato Y, Sato K, Hamada M, Watanabe Y, Asai K, Akutsu T (2010) RactIP: fast and accurate prediction of RNA–RNA interaction using interaction programming. Bioinformatics 26: i460–i466

Kawamoto H, Koide Y, Morita T, Aiba H (2006) Base-pairing requirement for RNA silencing by a bacterial small RNA and acceleration of duplex formation by Hfq. Mol Microbiol 61 (4):1013–1022

Kingsford CL, Ayanbule K, Salzberg SL (2007) Rapid, accurate, computational discovery of Rho-independent transcription terminators illuminates their relationship to DNA uptake. Genome Biol 8:R22

Klein RJ, Misulovin Z, Eddy SR (2002) Noncoding RNA genes identified in AT-rich hyperther-mophiles. Proc Natl Acad Sci USA 99(11):7542–7547

Lenz DH, Mok KC, Lilley BN, Kulkarni RV, Wingreen NS, Bassler BL (2004) The small RNA chaperone Hfq and multiple small RNAs control quorum sensing in *Vibrio harveyi* and *Vibrio cholerae*. Cell 118:69–82

Li AX, Marz M, Qin J, Reidys CM (2011) RNA–RNA interaction prediction based on multiple sequence alignments. Bioinformatics 27(4):456–463

Liu JM, Camilli A (2010) A broadening world of bacterial small RNAs. Curr Opin Microbiol 13:18–23

Liu MY, Gui G, Wei B, Preston JF III, Oakford L, Yuksel U, Geidroc DP, Romeo T (1997) The RNA molecule CsrB binds to the global regulatory protein CsrA and antagonizes its activity in *Escherichia coli*. J Biol Chem 272:17502–17510

Liu JM, Livny J, Lawrence MS, Kimball MD, Waldor MK, Camilli A (2009) Experimental discovery of sRNAs in *Vibrio cholerae* by direct cloning, 5S/tRNA depletion and parallel sequencing. Nucleic Acids Res 37(6):e46

Livny J, Fogel MA, Davis BM, Waldor MK (2005) sRNAPredict: an integrative computational approach to identify sRNAs in bacterial genomes. Nucleic Acids Res 33(13):4096–4105

Livny J, Brencic A, Lory S, Waldor MK (2006) Identification of 17 *Pseudomonas aeruginosa* sRNAs and prediction of sRNA-encoding genes in 10 diverse pathogens using the bioinfor-matics tool sRNAPredict2. Nucleic Acids Res 34(12):3484–3493

Livny J, Teonadi H, Livny M, Waldor MK (2008) High-throughput, kingdom-wide prediction and annotation of bacterial non-coding RNAs. PLoS One 3(9):e3197

Lu X, Goodrich-Blair H, Tjaden B (2011) Assessing computational tools for the discovery of small RNA genes in bacteria. RNA 17:1635–1647

Mandin P, Repoila F, Vergassola M, Geissmann T, Cossart P (2007) Identification of new noncoding RNAs in *Listeria monocytogenes* and prediction of mRNA targets. Nucleic Acids Res 35(3):962–974

Marchais A, Naville M, Bohn C, Bouloc P, Gautheret D (2009) Single-pass classification of all noncoding sequences in a bacterial genome using phylogenetic profiles. Genome Res 19:1084–1092

Masse E, Vanderpool CK, Gottesman S (2005) Effect of RyhB small RNA on global iron use in *Escherichia coli*. J Bacteriol 187(20):6870–6873

Mathews DH, Sabina J, Zuker M, Turner DH (1999) Expanded sequence dependence of thermodynamic parameters provides robust prediction of RNA secondary structure. J Mol Biol 288:910–940

McCaskill JS (1990) The equilibrium partition function and base pair binding probabilities for RNA secondary structure. Biopolymers 29:1105–1119

McCutcheon JP, Eddy SR (2003) Computational identification of non-coding RNAs in *Saccharomyces cerevisiae* by comparative genomics. Nucleic Acids Res 31(14):4119–4128

Muckstein U, Tafer H, Hackermuller J, Bernhart SH, Stadler PF, Hofacker IL (2006) Thermodynamics of RNA–RNA binding. Bioinformatics 22(10):1177–1182

Opdyke J, Kang JG, Storz G (2004) GadY, a small RNA regulator of acid response genes in *Escherichia coli*. J Bacteriol 186(20):6698–6705

Panek J, Bobek J, Mikulik K, Basler M, Vohradsky J (2008) Biocomputational prediction of small non-coding RNAs in *Streptomyces*. BMC Genomics 9:217

Papenfort K, Vogel J (2009) Multiple target regulation by small noncoding RNAs rewires gene expression at the post-transcriptional level. Res Microbiol 160(4):278–287

Papenfort K, Pfeiffer V, Mika F, Lucchini S, Hinton JCD, Vogel J (2006) Sigma(E)-dependent small RNAs of *Salmonella* respond to membrane stress by accelerating global *omp* mRNA decay. Mol Microbiol 62(6):1674–1688

Peer A, Margalit H (2011) Accessibility and evolutionary conservation mark bacterial small-RNA target-binding regions. J Bacteriol 193(7):1690–1701

Pfeiffer V, Papenfort K, Lucchini S, Hinton JCD, Vogel J (2009) Coding sequence targeting by MicC RNA reveals bacterial mRNA silencing downstream of translational initiation. Nat Struct Mol Biol 16:840–846

Pichon C, Felden B (2003) Intergenic sequence inspector: searching and identifying bacterial RNAs. Bioinformatics 19(13):1707–1709

Pichon C, Felden B (2005) Small RNA genes expressed from *Staphylococcus aureus* genomic and pathogenicity islands with specific expression among pathogenic strains. Proc Natl Acad Sci USA 102(40):14249–14254

Prevost K, Salvail H, Desnoyers G, Jacques JF, Phaneuf E, Masse E (2007) The small RNA RyhB activates the translation of *shiA* mRNA encoding a permease of shikimate, a compound involved in siderophore synthesis. Mol Microbiol 64(5):1260–1273

Rasmussen AA, Eriksen M, Gilany K, Udesen C, Franch T, Petersen C, Valentin-Hansen P (2005) Regulation of *ompA* mRNA stability: the role of a small regulatory RNA in growth phase-dependent control. Mol Microbiol 58(5):1421–1429

Ray BK, Apirion D (1979) Characterization of 10S RNA: a new stable rna molecule from *Escherichia coli*. Mol Gen Genet 174(1):25–32

Rehmsmeier M, Steffen P, Hochsmann M, Giegerich R (2004) Fast and effective prediction of microRNA/target duplexes. RNA 10:1507–1517

Ribes V, Romisch K, Giner A, Dobberstein B, Tollervey D (1990) *E. coli* 4.5S RNA is part of a ribonucleoprotein particle that has properties related to signal recognition particle. Cell 63 (3):591–600

Rivas E, Eddy SR (2001) Noncoding RNA gene detection using comparative sequence analysis. BMC Bioinformatics 2(1):8

Rivas E, Klein RJ, Jones TA, Eddy SR (2001) Computational identification of noncoding RNAs in *E. coli* by comparative genomics. Curr Biol 11(17):1369–1373

Saetrom P, Sneve R, Kristiansen KI, Snove O Jr, Grunfeld T, Rognes T, Seeberg E (2005) Predicting non-coding RNA genes in *Escherichia coli* with boosted genetic programming. Nucleic Acids Res 33(10):3263–3270

Schattner P (2002) Searching for RNA genes using base-composition statistics. Nucleic Acids Res 30(9):2076–2082

Seemann SE, Richter AS, Gesell T, Backofen R, Gorodkin J (2011) PETcofold: predicting conserved interactions and structures of two multiple alignments of RNA sequences. Bioinformatics 27(2):211–219

Sharma CM, Vogel J (2009) Experimental approaches for the discovery and characterization of regulatory small RNAs. Curr Opin Microbiol 12(5):536–546

Sharma CM, Hoffmann S, Darfeuille F, Reignier J, Findeiss S, Sittka A, Chabas S, Reiche K, Hackermuller J, Reinhardt R, Stadler PF, Vogel J (2010) The primary transcriptome of the major human pathogen *Helicobacter pylori*. Nature 464:250–255

Silvaggi JM, Perkins JB, Losick R (2006) Genes for small, noncoding RNAs under sporulation control in *Bacillus subtilis*. J Bacteriol 188(2):532–541

Sittka A, Lucchini S, Papenfort K, Charma CM, Rolle K, Binnewies TT, Hinton JCD, Vogel J (2008) Deep sequencing analysis of small noncoding RNA and mRNA targets of the global post-transcriptional regulator, Hfq. PLoS Genet 4(8):e1000163

Smith TF, Waterman MS (1981) Identification of common molecular subsequences. J Mol Biol 147:195–197

Smith C, Heyne S, Richter AS, Will S, Backofen R (2010) Freiburg RNA tools: a web server integrating IntaRNA, ExpaRNA and LocARNA. Nucleic Acids Res 38(S2):W373–W377

Sorek R, Kunin V, Hugenholtz P (2008) CRISPR – a widespread system that provides acquired resistance against phages in bacteria and archaea. Nat Rev Microbiol 6:181–186

Sridhar J, Narmada SR, Sabarinathan R, Ou HY, Deng Z, Sekar K, Rafi ZA, Rajakumar K (2010) sRNAscanner: a computational tool for intergenic small RNA detection in bacterial genomes. PLoS One 5(8):e11970

Staden R (1984) Computer methods to locate signals in nucleic acid sequences. Nucleic Acids Res 12:505–519

Stark BC, Kole R, Bowman EJ, Altman S (1978) Ribonuclease P: an enzyme with an essential RNA component. Proc Natl Acad Sci USA 75(8):3717–3721

Swiercz JP, Hindra BJ, Haiser HJ, Di Berardo C, Tjaden B, Elliot MA (2008) Small non-coding RNAs in *Streptomyces coelicolor*. Nucleic Acids Res 36(22):7240–7251

Tafer H, Hofacker IL (2008) RNAplex: a fast tool for RNA–RNA interaction search. Bioinformatics 24(22):2657–2663

Tjaden B (2008a) Prediction of small, noncoding RNAs in bacteria using heterogeneous data. J Math Biol 56(1–2):183–200

Tjaden B (2008b) TargetRNA: a tool for predicting targets of small RNA action in bacteria. Nucleic Acids Res 36:W109–W113

Tjaden B, Goodwin SS, Opdyke JA, Guillier M, Fu DX, Gottesman S, Storz G (2006) Target prediction for small, noncoding RNAs in bacteria. Nucleic Acids Res 34(9):2791–2802

Tran TT, Zhou F, Marshburn S, Stead M, Kushner SR, Xu Y (2009) *De novo* computational prediction of non-coding RNA genes in prokaryotic genomes. Bioinformatics 25(22):2897–2905

Tucker BJ, Breaker RR (2005) Riboswitches as versatile gene control elements. Curr Opin Struct Biol 15(3):342–348

Urban JH, Vogel J (2008) Two seemingly homologous noncoding RNAs act hierarchically to activate *glmS* mRNA translation. PLoS Biol 6(3):e64

Urban JH, Papenfort K, Thomsen J, Schmitz RA, Vogel J (2007) A conserved small RNA promotes discoordinate expression of the *glmUS* operon mRNA to activate GlmS synthesis. J Mol Biol 373:521–528

Uzilov AV, Keegan JM, Mathews DH (2006) Detection of non-coding RNAs on the basis of predicted secondary structure formation free energy change. BMC Bioinformatics 7:173

Vogel J, Wagner EGH (2007) Target identification of small noncoding RNAs in bacteria. Curr Opin Microbiol 10(3):262–270

Voss B, Georg J, Schon V, Ude S, Hess WR (2009) Biocomputational prediction of non-coding RNAs in model cyanobacteria. BMC Genomics 10:123

Wang C, Ding C, Meraz RF, Holbrook SR (2006) PSoL: a positive sample only learning algorithm for finding non-coding RNA genes. Bioinformatics 22(21):2590–2596

Washietl S, Hofacker IL, Stadler PF (2005) Fast and reliable prediction of noncoding RNAs. Proc Natl Acad Sci USA 102(7):2454–2459

Wassarman KM, Storz G (2000) 6S RNA regulates *E. coli* RNA polymerase activity. Cell 101:613–623

Waters LS, Storz G (2009) Regulatory RNAs in bacteria. Cell 136:615–628

Westhof E (2010) The amazing world of bacterial structured RNAs. Genome Biol 11(3):108

Workman C, Krogh A (1999) No evidence that mRNAs have lower folding free energies than random sequences with the same dinucleotide distribution. Nucleic Acids Res 27(24):4816–4822

Xiao B, Li W, Guo G, Li B, Liu Z, Jia K, Guo Y, Mao X, Zou Q (2009) Identification of small noncoding RNAs in *Helicobacter pylori* by a bioinformatics-based approach. Curr Microbiol 58(3):258–263

Yackie N, Numata K, Saito R, Kanai A, Tomita M (2006) Prediction of non-coding and antisense RNA genes in *Escherichia coli* with gapped Markov model. Gene 372:171–181

Yoder-Himes DR, Chain PSG, Zhu Y, Wurtzel O, Rubin EM, Tiedje JM, Sorek R (2009) Mapping the *Burkholderia cenocepacia* niche response via high-throughput sequencing. Proc Natl Acad Sci USA 106(10):3976–3981

Zhang A, Wassarman KM, Rosenow C, Tjaden B, Storz G, Gottesman S (2003) Global analysis of small RNA and mRNA targets of Hfq. Mol Microbiol 50(4):1111–1124

Zhao Y, Li H, Hou Y, Cha L, Cao Y, Wang L, Ying X, Li W (2008) Construction of two mathematical models for predictions of bacterial sRNA targets. Biochem Biophys Res Commun 372:346–350

Zuker M, Stiegler P (1981) Optimal computer folding of large RNA sequences using thermodynamics and auxiliary information. Nucleic Acids Res 9(1):133–148

Chapter 12
Identifying Functional miRNA Targets Using Overexpression and Knockdown Methods

Elizabeth L. Johnson, Eric J. Suh, Talia R. Chapman, and Hilary A. Coller

Abstract MicroRNAs are important regulators of gene expression that can posttranscriptionally regulate transcript abundance. Misregulation of miRNA expression has been associated with cell cycle abnormalities and disease states, thus reinforcing the importance of understanding the functional role of miRNAs. Genomic approaches are particularly valuable for studying miRNA function because they can address the ability of a single miRNA to direct the regulation of multiple mRNA targets. This chapter details methods for monitoring differentially expressed miRNAs using expression profiling, monitoring miRNA-related cell cycle phenotypes using flow cytometry, identifying miRNA targets using overexpression and knockdown methods, and validating target sites in 3′UTRs using luciferase reporters.

Keywords 3′UTR luciferase assay • cell cycle flow cytometry • expression profiling • LNA • miRNAs • overexpression • sponge

12.1 Introduction

MicroRNAs (miRNAs) are short single-stranded sequences of RNA, around 22 nt in length, that have been shown to be important regulators of mammalian gene expression. Altered miRNA expression has been linked to abnormal cell cycle control, and perturbing miRNA levels has allowed for the identification of specific and functional mRNA targets. In a growing number of instances, miRNA targets have been directly linked with cell cycle control (Johnson et al. 2007;

H.A. Coller (✉)
Department of Molecular Biology, Princeton University, Princeton, NJ, USA
e-mail: hcoller@princeton.edu

B. Mallick (eds.), *Regulatory RNAs*, DOI 10.1007/978-3-642-22517-8_12,
© Springer-Verlag Berlin Heidelberg (outside the USA) 2012

Linsley et al. 2007). Two representative examples of cell cycle control by miRNAs include a study in which overexpression of the miRNA *let-7* led to G_2/M arrest through the downregulation of Cdc34 and a study in which depletion of miR-221/222 in cancer cell lines was shown to cause an increased fraction of cells in the G_0/G_1 state by alleviating the negative regulation of the cell cycle inhibitor p27 (Legesse-Miller et al. 2009; le Sage et al. 2007). miRNA misregulation has also been associated with poor prognosis in cancer patients along with increased incidence of certain cancers (Takamizawa et al. 2004; He et al. 2005; Calin et al. 2008). Understanding how individual miRNAs affect the gene expression landscape of a cell can provide insight into how miRNA misregulation contributes to the development of proliferative diseases and other associated pathologies.

Complications to understanding the regulatory function of even a single miRNA arise from the ability of one miRNA to repress the expression of a multitude of mRNA targets. In order to repress target genes, miRNAs are loaded into the RNA-induced silencing complex (RISC), which recognizes complementary sequences in the $3'$ untranslated region ($3'$UTR) of the target mRNA. This interaction with RISC is primarily thought to cause transcript destabilization but can also cause translational inhibition through a variety of mechanisms (Guo et al. 2010). Because decreases in mRNA stability often correlate with decreases in gene expression, gene expression levels can serve as an indicator of miRNA effectiveness. Therefore, identifying transcripts that decrease in abundance upon overexpression of an miRNA and increase in expression when miRNA levels are depleted allows for a better understanding of the specific functional role of miRNAs in a biological system.

To begin understanding how miRNAs are regulating a system of interest, one can perform a genome-wide analysis of miRNA expression changes between distinct cellular states. Such an approach offers an unbiased assessment of the global changes in miRNA levels in a given system. Once an individual miRNA of interest is identified from miRNA expression profiling, cell cycle flow can be used to determine miRNA-associated cell cycle phenotypes. Then changes in target expression can be monitored by microarray in response to overexpression and knockdown of the miRNA. Targets of interest can be identified based on their responses to changes in the miRNA's expression levels and can be subsequently validated using luciferase-based $3'$UTR assays. Some of the major advantages of this genome-wide approach are the identification of multiple miRNA targets of interest in one experiment and the potential to organize related transcripts into their associated pathways.

Multiple challenges arise in this analysis ranging from experimental design to data analysis. Approaches to these challenges are addressed in this chapter. We will focus on an example in which the targets of the *let-7* miRNA are identified using microarray analysis, and the functional effects on the cell cycle are determined for *let-7* and its targets.

12.2   Materials

12.2.1   miRNA Expression Profiling to Identify miRNAs That Change in Abundance Between Cell Cycle States

1. miRNA Complete Labeling and Hyb Kit (Agilent 5190-0456)
2. Gene Expression Wash Buffer Kit (Agilent 5188-5327)
3. MicroRNA Spike-In Kit (Agilent 5190-1934)
4. *mir*Vana™ miRNA Isolation Kit (Applied Biosystems AM1560)
5. Micro Bio-Spin 6 Columns, in Tris Buffer, 6 kD Limit (Bio-Rad 732-6221)
6. RNase-free water
7. RNA 6000 Nano Kit (Agilent 5067-1511)
8. 2100 Bioanalyzer (Agilent)
9. Microarray Scanner (Agilent)
10. Hybridization Chamber, Stainless (Agilent)
11. Hybridization Chamber Gasket Slides, 8 Arrays/Slide (Agilent)
12. Hybridization oven, 20 rpm, set at 55°C (Agilent)
13. TruSeq Small RNA Sample Prep Kit (Illumina RS-200-0012)

12.2.2   Reverse Transfection of miRNAs into Primary Human Foreskin Fibroblasts

1. Human Foreskin Fibroblasts (ATCC CRL-2522™)
2. Pre-miR™ miRNA Precursor Molecule (hsa-let-7b) (Ambion AM17100)
3. Locked Nucleic Acid (LNA™) (hsa-let-7b) (Exiqon)
4. Pre-miR Negative Control #1 RNA (Ambion AM17100)
5. Oligofectamine™ Transfection Reagent (Invitrogen 12252-011)
6. Opti-MEM® Reduced Serum Media (Invitrogen 11058-021)
7. Dulbecco's Phosphate-Buffered Saline (D-PBS) (Invitrogen 14190144)
8. Trypsin, 0.5% (10×) with EDTA 4Na (Invitrogen 15400-054) – diluted to 1× in PBS
9. Twenty percent Fetal Bovine Serum (FBS) in Dulbecco's Modified Eagle Media (DMEM) (Invitrogen)
10. Trypsin Inhibitors – 0.25–0.5 mg/ml in PBS (Invitrogen 17075-029)

12.2.3   Monitoring Changes in Cell Cycle Status

12.2.3.1   Monitoring DNA Synthesis Using Flow Cytometry

1. 5-Ethynyl-2′-deoxyuridine (EdU) (Invitrogen A10044)
2. Permeabilization buffer – 0.2% Triton X-100 in PBS

3. Wash buffer – 1% BSA, 0.2% Triton X-100 in PBS
4. Blocking solution – 5% BSA, 0.2% Triton X-100 in PBS
5. 4′,6-Diamidino-2-phenylindole (DAPI) Solution (Sigma D9542-10MG) – 1 μg/ml DAPI solution diluted 1:1000 from 1 mg/ml DAPI stock solution in 0.1% Triton X-100 in PBS
6. Reaction cocktail (per sample) – 50 μl 1 M Tris pH 8.5, 10 μl 100 mM $CuSO_4$ in water, 2.5 μl 2 mM Alexa Fluor® 488 azide (Invitrogen A10266), 50 μl 1 M ascorbic acid, 387.5 μl water
7. Polystyrene Round-Bottom Tube (5 mL) (BD Falcon 352053)

12.2.3.2   Cell Cycle Analysis Using Pyronin Y–Hoechst 33342 Staining to Quantify Cells in the G_0 Cell Cycle State

1. Hanks Buffered Saline Solution (1×) (HBSS) (Invitrogen 14170112)
2. Pyronin Y (Sigma-Aldrich P9172-1 G)
3. Hoechst 33342 (Invitrogen H1399)
4. Ethanol (70%) – chilled at −20°C
5. Polystyrene Round-Bottom Tube (5 mL) (BD Falcon 352053)

12.2.4   Monitoring Changes in Gene Expression Using Microarray Analysis in Response to miRNA Overexpression or Knockdown

1. TRIzol® Reagent (Invitrogen 15596026)
2. Chloroform Isoamyl Alcohol 24:1 (Sigma 25666-100ML)
3. Ethanol (100%) (Sigma E7023)
4. Nuclease-free water
5. 15-ml Heavy Phase-lock Gel Tubes (PLG) (5 PRIME 2302850)
6. Quick Amp Microarray Labeling Kit, Two-Color (Agilent 5190-0444)
7. RNA Spike-in control, Two-Color (Agilent 5188-5279)
8. Whole Human Genome Microarray Kit, 4 × 44 K (Agilent G4112F)
9. Gene Expression Hybridization Kit (Agilent 5188-5242)
10. Gene Expression Wash Buffer 1 and 2 (Agilent 5188-5325, 5188-5326)
11. Microarray Wash Buffer Additive (Agilent 5190-0401)
12. Prehybridization Buffer (Agilent 5190-0402)
13. RNeasy Mini Kit (Qiagen 74104)
14. Hybridization Chamber Gasket Slides, 4 Arrays/Slide (Agilent)

12.2.5 Identifying miRNA Targets and Using 3'UTR Luciferase Assays to Validate Target Responsiveness to miRNAs

1. HEK 293 cells (ATCC CRL-1573™)
2. psiCHECK™-1 Vector (Promega C8011)
3. pRL-CMV Renilla Luciferase Control Plasmid (Promega E2261)
4. Pre-miR™ miRNA Precursor Molecule (hsa-let-7b) (Ambion AM17100)
5. Lipofectamine™ 2000 Transfection Reagent (Invitrogen 11668-019)
6. Opti-MEM® I Reduced-Serum Medium (1×) (Invitrogen 31985-070)
7. Firefly & Renilla Luciferase Assay Kit (Biotium 30005-1)
8. OptiPlate-96, White Opaque 96-well Microplate (PerkinElmer 6005290)
9. GloMax® 96 Microplate Luminometer (Promega E6501)

12.3 Methods

The following subsections describe methods that will guide the reader in techniques that include choosing an miRNA to study, identifying cell cycle effects of the chosen miRNA, and using expression profiling to identify then validate targets regulated by the miRNA. This complete analysis can lead to the identification of specific transcripts and pathways that are involved in miRNA-mediated cell cycle control. A schematic of the workflow recommended to complete this analysis is detailed in Fig. 12.1.

12.3.1 Expression Profiling to Identify miRNAs That Change in Abundance Between Cell Cycle States

miRNA expression levels are measured using a variety of standard molecular biology techniques of varying throughput, from low-throughput northern blots to high-throughput next-generation sequencing. Many of the sample preparation procedures have been simplified and packaged in commercially available kits. Here we compare two methods to quantify miRNA expression: miRNA microarrays and high-throughput sequencing of small RNAs (refer to Chap. 9 of this volume for more information on expression profiling of ncRNAs). Expression profiling of miRNAs offers an unbiased method to determine miRNAs that may have functional importance in a particular biological system. The application of miRNA microarray technology assumes that miRNAs that change in abundance between two states are important for the proper regulation of cell state transitions. Disadvantages of this approach include that it will not detect changes unrelated to miRNA levels, for instance, changes in the accessibility of the miRNA binding site. Another possible disadvantage is that miRNA profiling can result in a multitude of promising

ACTION	Methods	
IDENTIFY miRNA(s) TO STUDY →	Total RNA isolation ·····>	miRNA microarray; miRNA deep sequencing
TEST THE miRNA(s) FUNCTIONAL EFFECT ON THE CELL CYCLE →	Reverse Transfection of miRNA, LNA, or sponge into cells ·····>	Cell cycle analysis by flow cytometry
IDENTIFY miRNA RESPONSIVE mRNA TARGETS	Reverse transfection of miRNA, LNA, or sponge into cells ·····>	mRNA microarray and bioinformatics to identify interesting targets
	Reverse transfection of miRNA and reporter vector into cells ·····>	Validate mRNA target responsiveness using luciferase 3′UTR reporter assay

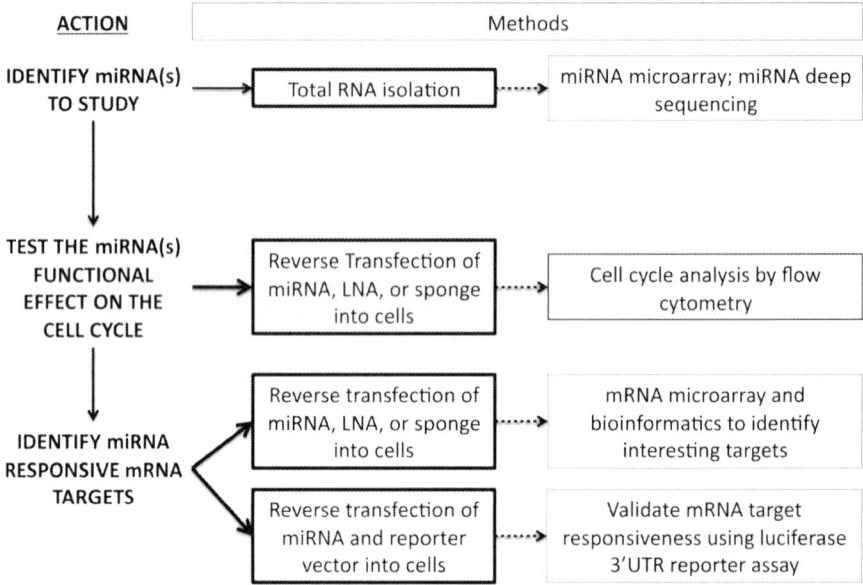

Fig. 12.1 Schematic providing an overview of the techniques used in this chapter

candidates and it may not be clear which miRNAs are most important. It is therefore important to use supporting methods to select specific miRNAs for further analysis. Possible filtering criteria include the strength and consistency of the change in miRNA abundance, information in the literature, and the extent to which the miRNA's targets correlate with the biological process under consideration. Among these, the most straightforward way to select miRNAs for continued experimentation is to consider the magnitude of the abundance change while keeping in mind that abundance does not always perfectly correlate with how effective an miRNA is at regulating predicted targets in a particular condition.

12.3.1.1   miRNA Microarrays

The advantages of miRNA microarrays for quantifying miRNA expression are the simple, established workflow and the ease of data processing when using readily available computational packages and methods. Agilent miRNA microarrays, for example, are easily used to quantify the differences in miRNA levels between proliferating and quiescent fibroblasts. For Agilent microarrays, a cyanine 3-conjugated pCp (Cy3-pCp) is ligated to the 3′ end of the RNA by T4 RNA ligase, followed by purification and hybridization to the array probes (Wang et al. 2007). We have had success using Agilent's miRNA Complete Labeling and Hyb Kit to complete these steps.

There are also drawbacks to using microarrays. The only miRNAs that can be monitored with miRNA microarrays are those that are present on the microarray, and thus, there is no opportunity for de novo discovery. Microarrays also have a compressed dynamic range compared to digital counting expression analysis for sequencing. Finally, microarray results can be confounded by cross-hybridization artifacts, which can lead to bias between isoforms as well as bias from miRNA precursors.

We suggest at least two separate labeling and hybridization replicates and at least three biological replicates for each experimental condition in order to accurately distinguish technical variation from the biological variation of interest.

Brief Overview of miRNA Microarray Methods

Total RNA can be isolated from 10^6 primary human fibroblasts per condition using the mirVana miRNA isolation kit and eluted into nuclease-free water. Total RNA isolation is recommended over methods that attempt to isolate only small RNAs, as the latter can introduce biases and loss of sample compared to the former.

To assess the quantity and quality of the RNA, the RNA can be analyzed both on a spectrophotometer such as the Nanodrop 2100 and Agilent's analytical microfluidic Bioanalyzer system. In general, we suggest quantifying the RNA on the former and assessing the quality on the latter. Microarrays require extremely pure and nondegraded RNA samples, as data from profiling the expression of mature miRNAs can be confounded when degradation causes a change in miRNA precursor chemistry and size. This RNA degradation may result in the unwanted hybridization of miRNA precursors to probes designed to specifically bind to short mature miRNA sequences. In general, we only use total RNA samples with a Bioanalyzer RNA integrity number greater than 9.5.

Following the Agilent protocol, dilute the total RNA sample to 25 ng/µl in RNase-free water, and use 4 µl (100 ng) along with the properly diluted spike-in reagents in the initial calf intestinal alkaline phosphatase (CIP) reaction. After denaturing the reaction with DMSO and heating the reaction, the total RNA samples are labeled at the $3'$ end with the Cy3-pCp. These labeled RNA samples are desalted using the Micro Bio-Spin 6 columns, eluted in RNase-free water, dried in a warm SpeedVac, and then hybridized with a spike-in control to the Agilent array using the methods described in Agilent's protocol.

The arrays are scanned in Agilent's microarray scanner using the standard settings described in the miRNA microarray protocol. The $\log_2$-normalized data can then be processed by multiple linear regression. Some experimental artifacts (i.e., "hidden" experimental variables) may only affect a subset of genes, thus creating widespread biases that would be difficult to remove with standard linear regression. We find surrogate variable analysis (SVA) to be particularly useful in this circumstance (Leek and Storey 2007); SVA is a technique to estimate the effects of such unknown experimental variables by searching for statistically significant patterns in the regression residuals matrix. The combination of thorough regression along all experimental and technical variables with SVA generally produces results that correlate well with quantitative real-time polymerase chain reaction (qRT-PCR) data. Artificially stringent normalization criteria can lead to loss of biological signal and inadequate characterization of biologically driven changes.

12.3.1.2   High-Throughput Sequencing of Small RNAs

The advantages of using high-throughput sequencing to profile miRNA expression include the ability to discover and profile miRNAs de novo without a reference genome sequence, the ability to detect modifications and alternative forms of miRNAs (such as uridylated miRNAs or processing variants), greater dynamic ranges for expression changes, and the ability to distinguish between mature and precursor forms of miRNAs (Wang et al. 2009). The drawbacks consist of the slightly more difficult sample preparation (compared to a microarray or qRT-PCR platform), the relatively more complicated data processing steps, and the lack of mature ways to handle normalization and analysis of digital counting data.

In general, we have found microarray results to be satisfactory for initial experiments and much simpler than high-throughput sequencing of small RNAs. We have chosen to focus on the former platform based on those tradeoffs, but for truly accurate digital count quantification, small RNA sequencing can be multiplexed using a bar-coding strategy, thus allowing for larger dynamic range of detection along with accurate internal controls for normalization. We have used and included in the recommended materials sections reagents for Illumina Inc.'s small RNA sequencing platform. However, multiple sequencing platforms offer small RNA quantification by sequencing including Roche's 454 sequencer, Applied Biosystems/Life Technologies' SOLiD system, and Life Technologies' Ion Torrent Personal Genome Machine. As the data for digital sequencing quantification is likely to follow a discrete statistical distribution rather than a continuous, normal distribution, care must be taken in choosing a method of statistical analysis.

Both microarrays and high-throughput sequencing require careful, reproducible pipetting techniques along with enough replicates and control samples to accurately partition the variation in the data using statistical methods such as ANOVA. From this perspective, sequencing technologies have an advantage because of the ability to use multiplexed bar code strategies to allow for internal normalization.

Once microarray or sequencing analysis of the sample has been completed, a small set of miRNAs of interest can be selected based on the most striking abundance changes between the two states. Other criteria such as information in the literature, relevance to the biological system of interest, and correlation to other supporting datasets (previously generated microarray, proteomics, metabolomics, etc., data) can be useful when singling out a potential miRNA candidate to characterize functionally. The next section of detailed protocols explains how to use overexpression and knockdown analysis to analyze the effects of the chosen miRNA on cell cycle progression.

12.3.2 Reverse Transfection of miRNAs into Primary Human Foreskin Fibroblasts

A cell's response to perturbations of miRNA levels can be measured in a multitude of ways, some of which are described in the methods below. In order to identify important phenotypic effects of a particular miRNA, it is possible to overexpress or deplete levels through the transfection of interacting oligonucleotides into the cell. The reagent oligofectamine is used to introduce RNA oligonucleotides into a liposome that can readily penetrate a cell's plasma membrane. The procedure below introduces the oligofectamine/miRNA transfection complexes to cells while they are not adherent to a tissue culture plate, allowing for faster and more efficient uptake of exogenous RNAs (Chesnoy and Huang 2000):

1. Plate cells at 60% confluency the day before transfection.
2. Dilute *let-7* pre-miR (overexpression) or *let-7* LNA (knockdown) to 50 nM in Opti-MEM from a 50-μM stock.
3. Add 25 μl of Oligofectamine reagent to 350 μl of Opti-MEM for each plate.
4. Add 375 μl of diluted Oligofectamine to 625 μl of diluted miRNA for each plate and incubate at room temperature for 15 min.
5. During the 15-min incubation, trypsinize and resuspend cells at a concentration of 375,000 cells/ml in Opti-MEM.
6. Add transfection reaction from step 4 to 4 ml of cells from step 5.
7. Add transfection reaction/cell mixture to a 10-cm tissue culture plate.
8. Incubate at 37°C for 4 h.
9. Supplement the cells with 5 ml of media containing 20% FBS with no antibiotics.
10. Change to full serum 24 h after transfection.
11. Collect samples for phenotypic analysis 12 h after serum addition. RNA is needed for microarray analysis (Sect. 12.3.3), or cells can be fixed for flow cytometry analysis (Sect. 12.3.5).

Common challenges with this procedure involve low cell survival after transfection and varied transfection efficiency. To address excessive cell death after transfection, it may be necessary to empirically determine the concentration of nucleic acid needed for a successful transfection. To determine the effectiveness of the overexpression or knockdown, it is possible to do real-time PCR (qRT-PCR) targeting the mature form of the miRNA.

12.3.3 Monitoring Changes in Cell Cycle Status

Any phenotypic changes related to proper cell cycle progression in response to the transfection can be readily identified by cell cycle flow cytometry. The major advantage of flow cytometry is that it allows an analysis of the cell cycle status for each of tens of thousands of cells simultaneously based on the intracellular levels of specific fluorescent dyes. The methods below address identifying defects in cell cycle progression (monitoring DNA synthesis using labeled nucleotides and flow cytometry) and cell cycle exit defects (pyronin Y–Hoechst staining).

12.3.3.1 Monitoring DNA Synthesis Using Flow Cytometry

Monitoring DNA synthesis using flow cytometry is a preferred method of measuring cell cycle status because of the ability to clearly identify cells progressing through S phase. This approach allows for a determination of whether cells have stalled or accelerated their cell division cycles. The method described provides information on both DNA content and DNA synthesis to give a clear picture of how fast cells are progressing through S phase in relation to the control. Other advantages of this method include reproducibility and ease of execution. In this method, DAPI staining allows for the identification of DNA content, while fluorescent monitoring of the thymidine analogue 5-ethynyl-2'-deoxyuridine (EdU) indicates the amount of new DNA synthesis during a defined labeling period (Fig. 12.2). A fluorescent azide (Alexa Fluor® 488-azide) is added to newly incorporated EdU molecules by a copper-catalyzed reaction. The Alexa Fluor® 488 intensity indicates the amount of new DNA synthesis (Salic and Mitchison 2008). This method provides clear data on the fraction of cells in G_0/G_1, early S phase, late S phase, and G_2/M:

1. After transfection (Sect. 12.3.2), dilute 10 mM EdU in cell culture media for a final concentration of 10 μM and incubate for 2 h at 37°C.
2. Wash cells with PBS, add 1× trypsin, and incubate for 5 min.
3. Inactivate the trypsin by adding an equal volume of trypsin inhibitors to the cell suspension in a conical tube.
4. Spin the conical tube at 1,600 rpm for 5 min to pellet cells.
5. Decant supernatant and suspend the cells in 1% BSA in PBS at a concentration of 1×10^7 cells/ml.
6. Add 100 μl 4% paraformaldehyde and incubate the conical tube in the dark for 15 min.
7. Add 2 ml of 1% BSA and PBS and spin the conical tube at 1,600 rpm for 5 min.
8. Decant supernatant and suspend cells in 100 μl 1% BSA in PBS (note: safe stopping point – cells can stay at 4°C for up to 1 week).
9. Permeabilize cells with 100 μl of 0.2% Triton X in PBS and protect from light.
10. Incubate in reaction cocktail for 30 min.
11. Add 2 ml of wash buffer and centrifuge cells at 1,600 rpm for 5 min.
12. Remove supernatant, add 2 ml of wash buffer, and centrifuge cells at 1,600 rpm for 5 min.
13. Remove supernatant and resuspend pellet in 500 μl of DAPI solution.
14. Transfer sample to a flow tube, vortex, and incubate in the dark on ice for 10 min.

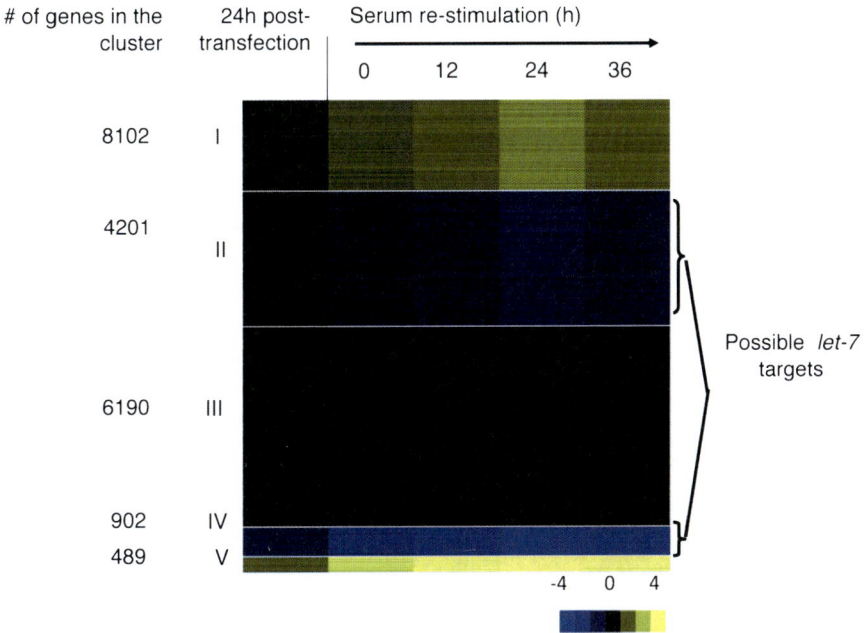

Fig. 12.2 Identification of *let-7* targets by microarray analysis. Primary human diploid fibroblasts were transfected with a noncoding microRNA or *let-7* microRNA. Twenty-four hours after transfection, samples were serum-starved for 36 h and then stimulated by readdition of serum. Samples were collected 24 h post transfection and at time-points 0, 12, 24, and 36 h after serum stimulation. RNA was isolated and microarrays were performed such that samples transfected with the noncoding control microRNA were compared with samples collected from the *let-7* transfected cells at each time-point. Genes repressed in the *let-7* transfected samples compared to control samples are indicated in blue, while genes present at higher levels in *let-7* transfected samples are in yellow. K-means clustering analysis of expression intensities resulted in five clusters that are shown in heat map format. Further motif and pathway analysis of this data identified Cdc34 as an important *let-7* target showcasing the power of this method. This research was originally published in the Journal of Biological Chemistry (Legesse-Miller et al. (2009). © the American Society for Biochemistry and Molecular Biology)

15. Analyze fluorescence using flow cytometry.
16. Excitation is UV (max 350 nm) for DAPI and 495 nm for Alexa Fluor® 488. Emission is measured at 461 nm for DAPI and 518 nm for Alexa Fluor® 488.

One common difficulty with this protocol is incomplete DNA staining with the DAPI solution. Precise cell counting is necessary for even DAPI staining across conditions, and in situations of incomplete staining, it is appropriate to double the DAPI concentration for a clearer determination of DNA content. Other staining concerns involve the extra wash in step 12, which is critical for eliminating nonspecific EdU staining. It is important to achieve specific and high-resolution staining in order to accurately classify cells into their respective cell cycle states. Additionally, it is important to note the condition of the EdU and ascorbic acid stocks. Diluted EdU should be used within a year and the ascorbic acid in solution should be discarded when it has oxidized (when the clear solution starts to turn orange). Another minor technical issue to consider is the importance of using a consistent 2-h labeling period in step 1 in order to ensure the reproducibility of the data. It is also important to stain the cells with the reaction cocktail within 1 week of the fixation in paraformaldehyde.

12.3.3.2 Pyronin Y–Hoechst 33342 Staining

To further delineate cells that have reversibly exited the cell cycle into G_0/quiescence from cells in G_1, RNA content can be monitored with the use of the fluorescent dye pyronin Y (PY) (Darzynkiewicz et al. 2004). Low PY levels represent the characteristic depletion of ribosomal RNA and lowered translational potential of quiescent cells. Hence, quiescent cells are identified as cells that have 2N DNA content (monitored by Hoechst staining) and PY levels lower than the levels in S phase cells (Fig. 12.3):

1. Make PY–Hoechst 33342 staining solution – 1 mg Hoechst 33342, 2 mg pyronin Y in 500 ml of room temperature HBSS. Make sure to mix the PY–Hoechst 33342 solution well and allow enough time before staining for the solution components to completely dissolve into the HBSS.

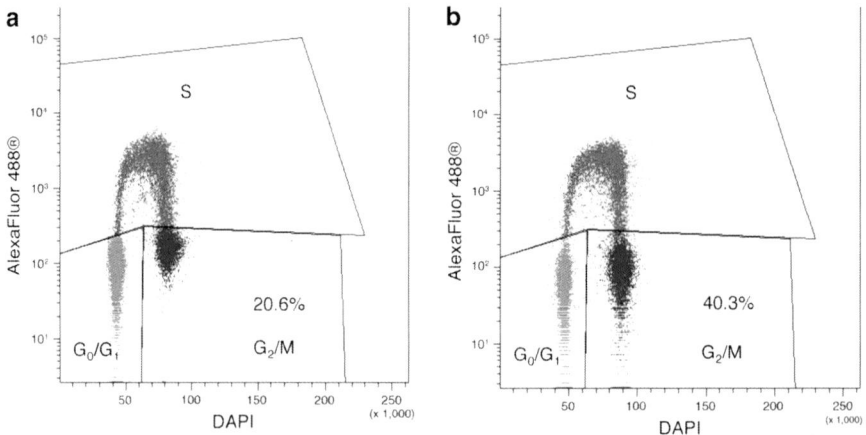

Fig. 12.3 Flow cytometry to detect new DNA synthesis in cells transfected with noncoding miRNA or *let-7* miRNA. Fibroblasts were transfected with noncoding miRNA or *let-7* miRNA, allowed to recover for 24 h, serum-starved, and restimulated for 32 h. Cells were pulsed with the nucleotide analogue EdU and collected for cell cycle analysis. Cells were labeled with DAPI for DNA content and a fluorescent molecule that covalently attached to the EdU incorporated into the DNA. Cells were analyzed by flow cytometry, and the dot plots are shown. DNA content is plotted on the x-axis, and EdU incorporation is shown on the y-axis. Cell cycle status was determined in cells transfected with a control (**a**) and cells transfected with *let-7b* pre-miR (**b**). There is an increased fraction of cells in the G_2/M cell cycle state upon overexpression of *let-7b* in primary human fibroblast as compared to the control

2. After transfection (Sect. 12.3.2), trypsinize, spin down cells at 1,000 rpm, and resuspend cells in 1 ml of cold PBS at a density of 2×10^6 cells/ml.
3. Add 10 ml of cold 70% ethanol to a 15-ml conical tube.
4. Transfer 1 ml of cell suspension from step 1 into ethanol from step 2 and incubate on ice for at least 2 h (note: cells can stay in ethanol fixative for up to 1 week).
5. Centrifuge tubes with fixed cells in ethanol at $300 \times g$ at $4°C$ for 5 min.
6. Remove supernatant and resuspend the cell pellet in 2 ml of cold HBSS.
7. Centrifuge tubes, remove supernatant, and resuspend cells in 500 μl of cold HBSS.
8. Transfer 500 μl of cell suspension to flow tube and place tube on ice.
9. Add 500 μl cold PY–Hoechst 33342 staining solution to flow tube.
10. Incubate cells for 20 min before measuring cell fluorescence on flow cytometer.
11. Excitation is UV (355 nm) for Hoechst and 488 nm for pyronin Y. Emission is measured at 450 ± 50 nm for Hoechst and 576 ± 26 nm for pyronin Y.

Inconsistent staining can confound results. In order to avoid this challenge, which mainly stems from imprecise cell counting and improper dilution of the PY–Hoechst 33342 staining solution, much care should be taken with steps 1 and 2 of this protocol. Since there is spillover of the Hoechst 33342 signal into the pyronin Y channel, it may be necessary to use compensation to compose cell cycle dot plots as seen in Fig. 12.3. Samples that are to be compared should be run on the same day as changes in cytometer settings from one day to the next can introduce variability that make the data difficult to compare.

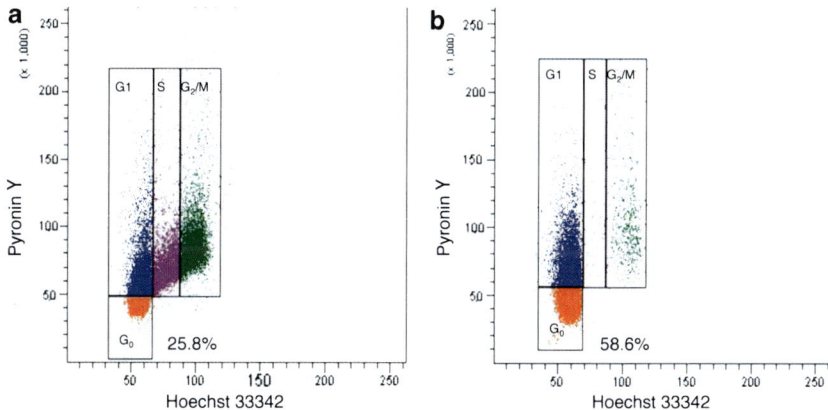

Fig. 12.4 Flow cytometry dot plots of Pyronin Y–Hoechst 33342 (PY–H) staining to determine G_0 cell population. Cycling (**a**) or quiescent (**b**) fibroblasts were fixed and stained with a PY–H solution. Quiescence was induced by maintaining the cells in a contact-inhibited state for 7 days. Cell cycle status was determined using DNA content (x-axis) and RNA content (y-axis) information. In the sample of cells induced into quiescence, more cells are classified to the G_0 cell cycle state than in the sample taken from cycling cells. G_0 = orange, G_1 = blue, S = magenta, G_2/M = green (Lemons et al. 2010)

Deviations from the control in a particular cell cycle state indicate that the miRNA of interest is involved in a process that regulated the indicated cell cycle transition. To further identify the exact transcripts and pathways involved in the cell cycle phenotype, it is necessary to look at global changes in transcript abundance in response to miRNA overexpression or knockdown using gene expression microarrays (Sect. 12.3.4).

12.3.4 Monitoring Changes in Gene Expression Using Microarray Analysis in Response to Overexpression or Knockdown to Identify miRNA Targets

Since a single miRNA can affect a large number of mRNA targets, it is convenient to use microarrays to identify transcripts that are responsive to perturbations in miRNA levels. Microarrays give an investigator the ability to monitor expression levels of many genes in a single experiment, and Agilent's two-color microarray format further allows for the direct identification of expression changes between control and experimental conditions.

The commentary below goes into detail on two of the most important aspects of microarray analysis: Performing a quality RNA isolation and experimental design. An overview of downstream array processing steps follows.

12.3.4.1 Total RNA Isolation from Transfection in Sect. 12.3.2

RNA is isolated from other cellular components while maintaining transcript integrity using the TRIzol reagent (Chomczynski and Sacchi 1987). TRIzol contains guanidinium thiocyanate, a potent RNase inhibitor, that allows for the deactivation of cellular RNase activity in the cell lysate. Upon the addition of chloroform to the cell lysate, phase-lock gel tubes (PLG) are used to create a gel barrier between aqueous and organic layers of the extraction mixture after centrifugation. The RNA is present in the aqueous layer. To recover clean RNA, the RNA is precipitated out of solution by incubating the aqueous layer in cold isopropanol. Centrifuging this mixture results in a small pellet of RNA on the side of the centrifuge tube. Ethanol washes allow the recovery of RNA free from protein and salt contamination before resuspending the pellet in nuclease-free water. Special precautions must be observed when working with RNA to avoid degradation from RNases. RNases are ubiquitous; therefore, a 0.1% SDS solution should be used to clean materials such as gloves that come into close proximity of RNA samples. All reagents/containers should be RNase free. A 0.1% concentration of diethylpyrocarbonate (DEPC) in water can deactivate RNases. Autoclaving for 15 min per liter should inactivate DEPC allowing for downstream analysis. Additionally, extra caution should be taken when working with TRIzol® as it contains harmful toxins and irritants:

1. Add 3 ml of TRIzol® reagent to one 10-cm tissue culture plate of Human Foreskin Fibroblasts.
2. Incubate plate at room temperature for 5 min (safe stopping point – at this point, the lysate can be stored at 80°C).
3. While the plate is incubating, spin down phase-lock gel (PLG) tubes at 1500 × g to get gel to bottom of tube.
4. Transfer TRIzol/cell mixture to a 15-mL PLG tube and add 600 uL of chloroform.
5. Shake tube by hand for 15 s and avoid mixing by vortexing.
6. Spin at 3000 × g for 10 min at 4°C.
7. Pour clear supernatant into a clean 15-ml conical tube. The red organic layer should stay separated by the gel. If there is red liquid in the top layer, add 200 μl of chloroform and spin at 4°C for another 10 min.
8. Add 1.5 ml of isopropanol. Invert 10 times and incubate at room temperature for 10 min.
9. Spin tube at 11000 × g for 10 min at 4°C.
10. Remove supernatant with a pipet, being careful not to disturb the pellet. Note: Pellet may be loose and easily disturbed. If this is the case, spin tube at 3000 × g for 5 min at 4°C for a more compact pellet. Repeat this at any subsequent step to avoid losing the pellet.
11. Add 3 ml 75% ethanol to conical tube.
12. Spin tube at 7500 × g for 5 min at 4°C.
13. Remove supernatant with a pipet, being careful not to disturb the pellet.
14. Spin down residual ethanol at 2000 × g for 5 min at 4°C and use a pipet to carefully remove supernatant.
15. Air dry pellet under a fume hood for 5 min.
16. Resuspend pellet in 50–200 μl of nuclease-free water.

Insufficient yields and the isolation of low-quality (degraded) RNA are major concerns when performing this protocol. If it is difficult to see a compact pellet or the pellet is loose and it is difficult to decant supernatants without losing the pellet, it can be helpful to do the spins in steps 9 and 12 at 13,000 g. At this speed, it is prudent to use VWR's SuperClear™ Gatefree™ centrifuge tubes to avoid cracking, which can occur when using a conventional conical tube at such high speeds. Without practice, isolating RNA from a small amount of cells using this method can result in abnormally low yields. In this case, it may be useful to use a carrier such as glycogen or to purchase a column-based RNA isolation kit such as Ambion's PureLink RNA mini kit. This kit allows RNA isolation with a lower risk of losing the RNA during the wash steps.

As previously mentioned in Sect. 12.3.1.1.1, the quality of RNA used for microarray analysis is one of the major determinants of the quality of the final gene expression analysis. If contamination from salts, proteins, and buffers results in a low 260/280 or 260/230 ratio (<2), it is possible to purify the RNA by doing another isopropanol precipitation or using a commercial cleanup kit such as Qiagen's RNeasy kit for RNA cleanup. Note that if there is Qiagen RLT buffer on the column after the final spin before elution, the RNA sample's 260/230 ratio will be lower than usual, but the RNA is fine for further analysis.

For two-color arrays, it is important to choose a reference sample that represents an appropriate control. For an miRNA overexpression/knockdown experiment, it is appropriate to use total RNA from the negative control RNA transfection as the reference sample. Any changes observed in the experimental condition are then in reference to the negative control sample and all the information needed to evaluate transcript abundance changes is contained on a single array. This design controls for experimental noise caused by array effects. The negative control RNA should be as closely matched to the experimental RNA without actually targeting genes as possible. It is also important when looking at miRNAs that cause cell cycle defects to analyze genes that are direct targets of the miRNA. Other gene expression changes can be expected if the fraction of cells in different phases of the cell cycle changes as a result of the transfection, but these changes will not necessarily reflect the direct action of the overexpressed miRNA. In addition, swamping out the RISC complex may result in upregulation of miRNA targets if other miRNAs can no longer access RISC. Using knockdown methods in addition to overexpression and ensuring the chosen target has an miRNA binding site will help control for off target miRNA effects. Once RNA quality and experimental design are satisfactory, it is appropriate to proceed to RNA labeling, hybridization, and array scanning.

12.3.4.2   RNA Labeling, Hybridization, and Array Scanning

Agilent two-color arrays are printed with gene-specific probes. To determine the relative abundance of transcripts between experimental and reference conditions, the RNA is first reverse transcribed by Moloney Murine Leukemia Virus Reverse Transcriptase (M-MLV RT) to cDNA. cDNA from the experimental condition (miRNA overexpression/knockdown transfection) is labeled using the fluorescent nucleotide cyanine 3-dCTP (Cy3-dCTP), while total RNA from the control condition (negative control RNA transfection) is labeled using the fluorescent nucleotide cyanine 5-dCTP (Cy5-dCTP) by T7 RNA polymerase to make cRNA. Note that Cy5 is sensitive to high ozone levels and should be handled in an ozone-free environment in areas with characteristically high ozone levels.

The cRNA is then incubated with the microarray slide resulting in the hybridization of labeled cRNA with probes on the array surface. After the hybridization is complete, the array is scanned to measure fluorescence intensity of all the features on the array. The gene associated with a probe is readily identified by the gridded position on the array slide. The relative fluorescence intensity at that gridded position represents the expression of that transcript in the sample. The $\log_2$ transformed ratio between experimental and reference samples is used to determine the fold change in expression of a gene between the experimental and control conditions.

Special attention should be taken to understand the specific software settings when using Agilent's feature extractor to determine fluorescence intensities. Using the default settings may normalize out important biological variance instead of just experimental abnormalities.

12.3.4.3 Data Processing and Pathway Analysis

Microarray analysis provides a wealth of information about the experimental condition. Determining patterns in expression levels can help an investigator identify interesting biological processes that are associated with the miRNA of interest. Clustering of genes based on their expression patterns groups together genes that change similarly in response to the overexpression and knockdown of the miRNA of interest. There are multiple clustering algorithms that will allow for the visualization of data in the form of a heat map. Source code and user-friendly interfaces for clustering and visualizing this data can be found at http://rana.lbl.gov/EisenSoftware.htm (Eisen et al. 1998).

Once genes that decrease or increase in expression in response to miRNA overexpression or knockdown, respectively, have been identified, it is important to determine the subset of these genes that actually contain recognition sites for the miRNA of interest in their 3'UTR. This correlation suggests that the miRNA has the potential to regulate the transcript. Direct targeting of the transcript by the miRNA of interest can be further validated by 3'UTR luciferase reporter assays using methods detailed in Sect. 12.3.5. To identify miRNA targets, popular algorithms include TargetScan (http://www.targetscan.org/), PicTar (http://pictar.mdc-berlin.de/), and miRanda (http://www.microrna.org/microrna/home.do). We suggest compiling a comprehensive list that incorporates the results from searching all these sites. Possible miRNA targets can be selected from the list among genes that display the appropriate expression changes in the overexpression/knockdown microarray data.

Genes that have been clustered into coregulated groups can be introduced into pathway analysis software to identify biological pathways associated with the miRNA. Table 12.1 lists a few pathway analysis tools that we have found to be valuable in our studies.

12.3.5 Identifying miRNA Targets and Validating Responsiveness to miRNAs Using 3'UTR Luciferase Assays

Integration of data from miRNA microarray and overexpression/knockdown microarray experiments can help to narrow down a truly responsive miRNA target from a large list of candidates. Below are three criteria to choose a classically responsive miRNA target:

Table 12.1 Pathway analysis tools

	Details	Website	Advantages	Disadvantages
Gene ontology term finder (Ashburner et al. 2000)	Determines if a list of potential genes has overrepresentation of particular functional categories using hypergeometric statistics	http://go.princeton.edu/ – a collection of GO tools	Ease of use, free	Does not take into account GO hierarchy. Some GO terms are not specific and uninformative. Only uses discrete gene cluster information
Gene set enrichment analysis (Subramanian et al. 2005)	Rigorously determines if gene expression data or a gene rank-order list has enrichment of particular gene sets	http://www.broadinstitute.org/gsea/index.jsp	Quantitative and free, well documented, many available gene sets including computationally predicted sets	Complex interface and data presentation. Gene sets can be confusing or redundant. Only uses continuous gene expression data or rank-order lists
Ingenuity	Determines pathways in which genes are significantly upregulated or downregulated	http://www.ingenuity.com/	Well-annotated pathways, extensive database, good data visualization	Proprietary algorithm and database, only uses continuous expression data
iPAGE (Goodarzi et al. 2009)	Uses mutual information techniques to identify pathways that are most informative about the experimentally determined gene clusters	https://iget.princeton.edu/	Quantitative, free, easy to use Web page and GUI, can take discrete gene clusters or continuous expression data as an input	Can be less sensitive than other techniques

1. Potential target goes down in expression with overexpression of miRNA.
2. Potential target increases in expression when the miRNA is depleted.
3. Potential target has one or more miRNA motif(s) in its 3′UTR.

Once a target is chosen, the 3′UTR sequence of the transcript can be cloned downstream of luciferase to create a reporter for miRNA activity. The luciferase gene makes a preferable reporter because changes in its expression level are easily monitored using a luminometer after cell lysates are exposed to the luciferase substrate D-luciferin. Decreased luminescence in relation to a negative control small RNA is evidence that the miRNA can regulate the expression of the transcript containing that 3′UTR sequence.

12.3.5.1 Creation of Luciferase Reporter Vector

Several commercial nonviral luciferase reporter vectors are available to assess 3′UTR response to miRNA activity. These vectors can be transfected using chemical transfection, electroporation, or nucleofection using the Lonza Nucleofector device. They generally come in one of two varieties: with only one luciferase enzyme in front of a multiple cloning site, or with two orthogonal luciferase enzymes, with one upstream of a multiple cloning site and the other used as an internal control. Three commonly used vectors include pMIR-REPORT (Ambion) and psiCHECK-1 (Promega), which are of the first variety, and psiCHECK-2 (Promega), which is of the second variety. For many in vitro cell culture applications, a pMIR-REPORT derivative cotransfected with an orthogonal luciferase vector (e.g., pRL-TK) as a transfection and input control is generally a very accurate and efficient way to assay the repression of gene expression by a cotransfected miRNA. A neomycin resistance cassette in pMIR-REPORT allows for selecting immortal stable cell lines.

However, for cells in which selection of nonviral vectors is not feasible, having two separate vectors can lead to differential loss of one vector or the other and therefore more experimental noise and bias. In these experiments, the two-luciferase system vectors are more useful as they are internally controlled on the same plasmid. psiCHECK-2, however, has the drawback that it does not have a selectable marker, making stable lines difficult to generate even in immortalized cell lines.

For stable cell line production, instead of luciferase expression, a viral GFP fusion vector can be used. We have developed a retroviral vector from pRetroQ-AcGFP-C1, in which the 3′UTR of GFP contains a multiple cloning site to allow for insertion of a variety of either 3′UTR gene sequences, miRNA target sites, or sponge cassettes. To convert the pRetroQ-AcGFP-C1 vector to a pRetroQ-UTR vector, an in-frame stop codon must be inserted in the 5′ end of the multiple cloning site. We inserted the sequence AACTGAGCCTTAATTAAGTCAT into the BglII site, leaving a number of restriction sites 3′ of the stop codon. This inserted sequence has a PacI site to cut out concatemeric ligation products. In brief, the

oligos GATCTAACTGAGCCTTAATTAAGTCATA and GATCTATGACT-
TAATTAAGGCTCAGTTA are 5′ phosphorylated with T4 PNK, pRetroQ-
AcGFP-C1 is digested with BglII and then dephosphorylated with calf intestinal
alkaline phosphatase, and then the oligos are annealed and ligated into the plasmid.
Finally, the plasmid is digested with PacI and then religated to obtain the final
product.

This vector can be used similarly to a luciferase reporter vector, if an endogenous gene's 3′UTR sequence is ligated into the cloning site. By cloning sites
complementary to miRNA seed sequences, pRetroQ-UTR derivatives can be used
as a reporter for miRNA levels. In addition, with 15 or more concatemeric binding
sites (especially binding sites with noncomplementary sequences against
nucleotides 9–12 of the miRNA), one can construct an miRNA sponge. In practice,
however, stable tranfections of miRNA sponges lead to less repression of the
miRNA compared to transient transfection of the sponge vector (Ebert et al.
2007). The sponge vector will repress miRNA expression by sequestering the
miRNA when it binds to complementary 3′UTR sites. Transient transfection of
this sponge vector can be used instead of the LNA to knockdown miRNA expression levels in Sect. 12.3.2. miRNA sponges can also be used to test other cell
phenotypes when the miRNA of interest is repressed.

12.3.5.2   Transfection of Luciferase Reporter Vectors into HEK 293 Cells

psiCHECK-1 and pRL-TK are used in the following protocol to measure luciferase
activity. Transfections can be done in duplicate to estimate technical error. Note
that lipofectamine instead of oligofectamine is used when transfecting HEK 293
because HEK 293 cells are more robust to the toxicity of the lipofectamine than
fibroblasts. This makes it possible to take advantage of lipofectamine's greater
transfection efficiency. The renilla control plasmid is not under specific posttranscriptional regulation and serves as a control for transfection efficiency of the
reporter construct. The following transfection conditions will allow determination
of whether an miRNA decreases luciferase expression more than a nonspecific
control RNA:

Reporter Construct + Renilla Control Construct (plasmid endogenous control) + pre-miR

Reporter Construct + Renilla Control Construct (plasmid endogenous control) + negative control

1. Plate HEK 293 cells in 6-well plates at 60% confluency the day before starting
   transfection.
2. Make 1% lipofectamine working solution in Opti-MEM and incubate for
   15 min.
3. Dilute pre-miR or negative control RNA to 50 nM in Opti-MEM.
4. Add 0.5 μg each of reporter plasmid and renilla control plasmid to pre-miR and
   negative control tubes.

5. Add 300 µl of lipofectamine working solution to DNA/RNA tubes and incubate at room temperature for 5 min.
6. Aspirate media off of cells and wash with 1 ml of PBS.
7. Add 200 µl of lipofectamine/DNA/RNA complex and 800 µl of Opti-MEM to cells.
8. Incubate for 4 h at 37°C.
9. Add 1 ml of DMEM + 20% FBS to wells with no antibiotic for recovery.
10. Aspirate media from wells, add 200 µl of lysis buffer to each well and rock at room temperature for 1 h (plates can be stored at −20°C at this point).

12.3.5.3 Measuring Luciferase Activity

Luciferase activity is measured by monitoring the emission of light associated with the luciferase-catalyzed oxidation of D-luciferin into oxyluciferin using a luminometer. Renilla-luciferase activity is measured in a similar manner except coelenterazine is used as the substrate. Luciferase activity should be measured in triplicate for each condition to estimate experimental error. We have found a 96-well plate format to be convenient for these assays. Commercial kits provide a simple workflow in which cells from the transfection are lysed in a buffer and then exposed to solutions containing either the substrate for firefly or renilla luciferase. The buffers in these kits are optimized to sustain a detectable signal while minimizing substrate autoluminescence (Sambrook and Russell 2001). Final luciferase expression levels are calculated by subtracting the respective luciferase and renilla-luciferase negative control values from the experimental conditions then dividing the luciferase values by the renilla luciferase values. This will result in a normalized luminescence value that is comparable between transfection conditions. Equation 12.1 details this normalization scheme.

$$luciferase_{\text{normalized}} = \frac{luciferase_{\text{experimental}} - luciferase_{\text{background}}}{renilla_{\text{experimental}} - renilla_{\text{experimental}}} \qquad (12.1)$$

A two-tailed student's t test is used to determine if normalized luciferase values from the technical replicates are significantly different in their means between pre-miR and negative control transfection conditions.

12.4 Conclusions

The analyses that we have described allow for the identification of an miRNA involved in cell cycle regulation. The protocols can further identify specific targets of the miRNA and evaluate if the target of choice is posttranscriptionally regulated by the miRNA using 3′UTR luciferase reporter assays. The studies can be extended by perturbing the levels of the target identified by this analysis to determine if the target has the same effect on cell cycle progression as the miRNA. Taken together, the protocols described here will allow for a deeper understanding of the mechanisms of miRNA regulation of the cell cycle.

Acknowledgments H.A.C. is the Milton E. Cassel scholar of the Rita Allen Foundation. E.J.S. and E.L.J. are supported in part by a National Science Foundation Graduate Research Fellowship DGE-0646086. E.L.J. acknowledges support from NIH Training Grant 2T32 CA009528. This work was funded by NIGMS Center of Excellence grant P50 GM071508, PhRMA Foundation grant 2007RSGl9572, NIH/NIGMS 1R01 GM081686, and NIH/NIGMS 1R01 GM086465. The authors wish to acknowledge the members of the Coller Laboratory (Princeton University), Jamol Pender (Princeton University), Tina deCoste (Princeton University), Donna Storton (Princeton University), Alison Gammie (Princeton University), and David Lofton (Stanford University) for technical assistance and helpful discussions.

References

Ashburner M, Ball CA, Blake JA, Botstein D, Butler H, Cherry JM, Davis AP, Dolinski K, Dwight SS, Eppig JT, Harris MA, Hill DP, Issel-Tarver L, Kasarskis A, Lewis S, Matese JC, Richardson JE, Ringwald M, Rubin GM, Sherlock G (2000) Gene ontology: tool for the unification of biology. The gene ontology consortium. Nat Genet 25(1):25–29. doi:10.1038/75556

Calin GA, Cimmino A, Fabbri M, Ferracin M, Wojcik SE, Shimizu M, Taccioli C, Zanesi N, Garzon R, Aqeilan RI, Alder H, Volinia S, Rassenti L, Liu X, Liu CG, Kipps TJ, Negrini M, Croce CM (2008) MiR-15a and miR-16-1 cluster functions in human leukemia. Proc Natl Acad Sci USA 105(13):5166–5171. doi:0800121105 [pii]10.1073/pnas.0800121105

Chesnoy S, Huang L (2000) Structure and function of lipid-DNA complexes for gene delivery. Annu Rev Biophys Biomol Struct 29:27–47. doi:29/1/27 [pii] 10.1146/annurev. biophys.29.1.27

Chomczynski P, Sacchi N (1987) Single-step method of RNA isolation by acid guanidinium thiocyanate-phenol-chloroform extraction. Anal Biochem 162(1):156–159. doi:10.1006/abio.1987.99990003-2697(87)90021-2 [pii]

Darzynkiewicz Z, Juan G, Srour EF (2004) Differential staining of DNA and RNA. Curr Protoc Cytom. doi:10.1002/0471142956.cy0703s30 Chapter 7: Unit 7 3

Ebert MS, Neilson JR, Sharp PA (2007) MicroRNA sponges: competitive inhibitors of small RNAs in mammalian cells. Nat Methods 4(9):721–726. doi:nmeth1079 [pii]10.1038/nmeth1079

Eisen MB, Spellman PT, Brown PO, Botstein D (1998) Cluster analysis and display of genome-wide expression patterns. Proc Natl Acad Sci USA 95(25):14863–14868

Goodarzi H, Elemento O, Tavazoie S (2009) Revealing global regulatory perturbations across human cancers. Mol Cell 36(5):900–911. doi:S1097-2765(09)00857-0 [pii]10.1016/j.molcel.2009.11.016

Guo H, Ingolia NT, Weissman JS, Bartel DP (2010) Mammalian microRNAs predominantly act to decrease target mRNA levels. Nature 466(7308):835–840. doi:nature09267 [pii]10.1038/nature09267

He H, Jazdzewski K, Li W, Liyanarachchi S, Nagy R, Volinia S, Calin GA, Liu CG, Franssila K, Suster S, Kloos RT, Croce CM, de la Chapelle A (2005) The role of microRNA genes in papillary thyroid carcinoma. Proc Natl Acad Sci USA 102(52):19075–19080. doi:0509603102 [pii]10.1073/pnas.0509603102

Johnson CD, Esquela-Kerscher A, Stefani G, Byrom M, Kelnar K, Ovcharenko D, Wilson M, Wang X, Shelton J, Shingara J, Chin L, Brown D, Slack FJ (2007) The let-7 microRNA represses cell proliferation pathways in human cells. Cancer Res 67(16):7713–7722. doi:67/16/7713 [pii]10.1158/0008-5472.CAN-07-1083

le Sage C, Nagel R, Egan DA, Schrier M, Mesman E, Mangiola A, Anile C, Maira G, Mercatelli N, Ciafre SA, Farace MG, Agami R (2007) Regulation of the p27(Kip1) tumor suppressor by miR-221 and miR-222 promotes cancer cell proliferation. EMBO J 26(15):3699–3708. doi:7601790 [pii]10.1038/sj.emboj.7601790

Leek JT, Storey JD (2007) Capturing heterogeneity in gene expression studies by surrogate variable analysis. PLoS Genet 3(9):1724–1735. doi:07-PLGE-RA-0237 [pii]10.1371/journal.pgen.0030161

Legesse-Miller A, Elemento O, Pfau SJ, Forman JJ, Tavazoie S, Coller HA (2009) let-7 Overexpression leads to an increased fraction of cells in G2/M, direct down-regulation of Cdc34, and stabilization of Wee1 kinase in primary fibroblasts. J Biol Chem 284 (11):6605–6609. doi:C900002200 [pii]10.1074/jbc.C900002200

Lemons JM, Feng XJ, Bennett BD, Legesse-Miller A, Johnson EL, Raitman I, Pollina EA, Rabitz HA, Rabinowitz JD, Coller HA (2010) Quiescent fibroblasts exhibit high metabolic activity. PLoS Biol 8(10):e1000514. doi:10.1371/journal.pbio.1000514

Linsley PS, Schelter J, Burchard J, Kibukawa M, Martin MM, Bartz SR, Johnson JM, Cummins JM, Raymond CK, Dai H, Chau N, Cleary M, Jackson AL, Carleton M, Lim L (2007) Transcripts targeted by the microRNA-16 family cooperatively regulate cell cycle progression. Mol Cell Biol 27(6):2240–2252. doi:MCB.02005-06 [pii]10.1128/MCB.02005-06

Salic A, Mitchison TJ (2008) A chemical method for fast and sensitive detection of DNA synthesis in vivo. Proc Natl Acad Sci USA 105(7):2415–2420. doi:0712168105 [pii]10.1073/pnas.0712168105

Sambrook J, Russell DW (2001) Molecular cloning: a laboratory manual, 3rd edn. Cold Spring Harbor Laboratory Press, Cold Spring Harbor

Subramanian A, Tamayo P, Mootha VK, Mukherjee S, Ebert BL, Gillette MA, Paulovich A, Pomeroy SL, Golub TR, Lander ES, Mesirov JP (2005) Gene set enrichment analysis: a knowledge-based approach for interpreting genome-wide expression profiles. Proc Natl Acad Sci USA 102(43):15545–15550. doi:0506580102 [pii]10.1073/pnas.0506580102

Takamizawa J, Konishi H, Yanagisawa K, Tomida S, Osada H, Endoh H, Harano T, Yatabe Y, Nagino M, Nimura Y, Mitsudomi T, Takahashi T (2004) Reduced expression of the let-7 microRNAs in human lung cancers in association with shortened postoperative survival. Cancer Res 64(11):3753–3756. doi:10.1158/0008-5472.CAN-04-063764/11/3753 [pii]

Wang H, Ach RA, Curry B (2007) Direct and sensitive miRNA profiling from low-input total RNA. RNA 13(1):151–159. doi:rna.234507 [pii]10.1261/rna.234507

Wang Z, Gerstein M, Snyder M (2009) RNA-Seq: a revolutionary tool for transcriptomics. Nat Rev Genet 10(1):57–63. doi:nrg2484 [pii]10.1038/nrg2484

Chapter 13
Identification of lncRNAs Using Computational and Experimental Approaches

Phil Chi Khang Au and Qian-Hao Zhu

Abstract Over the last decade, we have been illuminated by the startling discovery that many long noncoding RNAs (lncRNAs) are implicated in diverse and substantial biological processes. The identification of most lncRNAs to date has been unintentional and was mainly from subtractive hybridization or mutagenesis screening, initially aimed to identify protein-coding genes of interest. However, the characterization of lncRNAs and their acceptance as important regulators of many developmental and biological pathways have led to strategies specific to their isolation. Experimental methodologies to identify lncRNAs include RNA sequencing, lncRNA-specific microarray, and RNA-immunoprecipitation by taking advantage of their association with known RNA-binding proteins. The past decade has also generated a significant number of EST (expressed sequence tag)-based transcriptome databases which enabled computational methodologies to target their isolation. This chapter discusses several current and powerful computational and experimental approaches to identify lncRNAs.

Keywords cDNA library preparation • *in silico* identification of lncRNA • long noncoding RNA identification • RNA-immunoprecipitation

Abbreviations

ChIP   Chromatin immunoprecipitation
COOLAIR   Cold-induced long antisense intragenic RNA
DHFR   Dihydrofolate reductase
DRS   Direct RNA sequencing

P.C.K. Au (✉)
Commonwealth Scientific and Industrial Research Organisation Plant Industry,
Canberra, ACT, Australia
e-mail: Phil.au@csiro.au

B. Mallick (eds.), *Regulatory RNAs*, DOI 10.1007/978-3-642-22517-8_13,   319
© Springer-Verlag Berlin Heidelberg (outside the USA) 2012

FLC	Flowering locus C
Gas5	Growth arrest–specific 5
H3K4me3	Trimethylation of lysine 4 of histone H3
H3K36me3	Trimethylation of lysine 36 of histone H3
lincRNAs	Large intervening noncoding RNAs
lncRNA	Long noncoding RNA
PRC2	Polycomb repressive complex 2
RNA-IP	RNA-immunoprecipitation
SRA	Steroid receptor RNA activator
TUs	Transcript units
Xist	X-inactive specific transcript

13.1 Introduction

The past decade saw the characterization and implication of many long noncoding RNAs (lncRNAs) in diverse biological processes independent to those known to play specific roles in protein synthesis such as ribosomal RNAs, which were the first lncRNAs discovered in the 1950s (Palade 1955, 1958). These bona fide lncRNAs participate in diverse gene regulatory pathways including imprinting (Nagano et al. 2008; Rinn et al. 2007), X-chromosome inactivation (Ogawa et al. 2008; Zhao et al. 2008), transcriptional (Feng et al. 2006; Martianov et al. 2007; Wang et al. 2008) and posttranscriptional regulation (Beltran et al. 2008; He et al. 2008; Ogawa et al. 2008), and changes in gene expression resulting in tumor progression as discussed in the previous chapter (see Chap. 8, this volume). In addition, while different orders of eukaryotes have approximately the same number of protein-coding genes but vastly different phenotypic complexity (Mattick 2004; Taft et al. 2007), the number of noncoding genes increases proportionally with increasing developmental complexity with 98% of the human transcriptome represented as noncoding RNAs (Mattick and Makunin 2005; Taft et al. 2007), suggesting that functional ncRNAs including lncRNAs may represent the key to understanding the mechanisms that gave rise to species complexity. Following these discoveries, efforts in the isolation of lncRNAs in numerous species have commenced using experimental and computational approaches.

However, previous to our knowledge of the regulatory significance of these novel lncRNAs, approaches used in the identification and study of lncRNAs in animals and plants did not differ to those used in the identification of functional mRNA as lncRNAs were initially identified serendipitously through screening methodologies aimed to identify functional protein-coding RNAs. Perturbation conditions can be used to identify up-regulated transcripts of interest that can be captured through subtractive hybridization screening, while mutagenesis in forward genetics studies can be used to identity genes responsible for interesting phenotypes. These studies have allowed for the isolation of genes that encode both coding and noncoding RNA transcripts, with the latter determined to possess no protein-coding characteristics (e.g., no open reading frames or ORFs) through sequence analyses or when mapped to fully or partially nongenic regions of an available genome. lncRNAs identified through these methodologies include *Xist* (Brown et al. 1991), *Gas5* (Schneider et al. 1988), lncRNA from upstream of the *Pho5* gene (Kramer and Andersen 1980), *MeiRNA* (Watanabe and Yamamoto 1994), and plant *Zm401* (Li et al. 2001).

Subsequently, specific approaches were developed for their identification. Experimental approaches employed for the identification of lncRNAs include but are not limited to next-generation deep sequencing, lncRNA-specific microarray, and RNA-immunoprecipitation. Notably, the generation of many EST-based transcriptome databases has facilitated the development of computational approaches to identify lncRNAs. In this chapter, we will discuss these selected but powerful approaches specific for lncRNA identification.

13.2   Computational Identification of lncRNAs

13.2.1   *De Novo Prediction from Genomic Sequences*

Protein-coding genes exhibit clear evolutionary signatures that can be exploited in their computational predication by comparative genomics methods (Solovyev et al. 2006; Stark et al. 2007). However, lncRNAs generally lack common sequence patterns and characteristic evolutionary patterns, making their de novo detection in genomic DNA sequences more difficult than protein-coding genes (see Chap. 6, this volume). Additionally, as a heterogeneous group, lncRNAs usually do not exhibit conserved secondary structures observed in other types of ncRNAs, such as rRNAs, tRNAs, snoRNA, snRNAs, and miRNAs (Pang et al. 2006), which further complicates their computational identification. Nevertheless, it has been noted that the splice sites and intron positions are generally conserved and under purifying selection in both protein-coding genes and lncRNAs (Ponjavic et al. 2007; Rodriguez-Trelles et al. 2006). Based on these observations, bioinformatics tools have been developed to computationally identify lncRNAs in *Drosophila melanogaster* (Hiller et al. 2009) and human (Rose et al. 2011). The assumption underlying these approaches is that a functional pair of donor (5′) and acceptor (3′) splice sites will be retained over long evolutionary time scales only if (1) the locus is transcribed into a functional transcript and (2) accurate intron removal is necessary to produce a functional transcript (Hiller et al. 2009; Rose et al. 2011). Application of the conserved-intron-based approach in 15 *Drosophila* genomes identified 129 novel lncRNAs that are largely unstructured and not associated with significant sequence conservation (Hiller et al. 2009). This strategy, however, is not suitable for discovery of lncRNAs in vertebrates due to their much longer introns and more variable intron sizes. To circumvent this issue, the same group introduced a new strategy by combining comparative genomics and machine learning approach to predict conserved novel splice donor and acceptor sites in human genome based on conserved exon/intron structure in 44 vertebrate genomes (Rose et al. 2011). This new approach uncovered both coding and noncoding exons, but only a small portion of the predicted exons showed homology to protein-coding exons because features of protein-coding genes were deliberately neglected by the program.

Similar to programs for prediction of protein-coding genes and other types of ncRNAs (e.g., miRNA), these de novo lncRNA identification programs still heavily rely on conserved features of lncRNAs (such as exon/intron structure and intron positions), and conservation of these features across different species, making them unfeasible for prediction of lncRNAs in species without comparative genome information. Notably, de novo identification of lncRNA genes on a genome scale is still a challenging task, and more efforts should be devoted to this area because discovery of certain tissue-specific or rare transcripts that are difficult to be isolated even by the cutting-edge high-throughput sequencing technology may have to rely on genome-wide ab initio computational prediction programs.

13.2.2   In Silico *Identification of lncRNAs Using cDNAs and ESTs*

The rationale for *in silico* identification rests on the notion that the majority of lncRNAs are transcribed by RNA polymerase II; therefore, they are capped, polyadenylated, and often spliced just like protein-coding mRNAs but lack discernible open reading frames (Erdmann et al. 1999). Thus, lncRNAs can be distinguished from protein-coding mRNAs based on their potential coding capacity.

The starting data for *in silico* identification can be cDNAs or ESTs available in public databases, such as GenBank and FANTOM, or novel transcripts generated by high-throughput experiments, e.g., full-length cDNA cloning, tiling arrays, and deep sequencing. Generally, RNA sequences are compared against genomic sequences to remove those that overlap with predicted protein-coding genes; the remaining RNA sequences are then screened for ORFs. There is no defined ORF length for lncRNAs, but 70 or 100 amino acids are commonly used as a threshold. A number of ORF prediction programs are available, including GeneMark.hmm (Lukashin and Borodovsky 1998), GenScan (Burge and Karlin 1998), ESTScan2 (Lottaz et al. 2003), ANGLE (Shimizu et al. 2006), and ORF-Predictor (Jia et al. 2010). More sophisticated bioinformatics tools have also been developed to estimate the protein-coding potential of an RNA sequence, such as CRITICA (Badger and Olsen 1999), DIANA-EST (Hatzigeorgiou et al. 2001), CSTminer (Mignone et al. 2003), and RNAcode (Washietl et al. 2011). In addition, programs, such as CONC (Liu et al. 2006) and Coding Potential Calculator (CPC, Kong et al. 2007), developed based on support vector machines (SVM), have been used to assess the coding potential of putative lncRNAs. These SVM algorithms exploited multiple distinct features of mRNAs in the machine learning methods to distinguish lncRNAs from protein-coding mRNAs. More recently, integrated ncRNA finder (incRNA), an integrative machine learning method which combines a large amount of expression data, RNA secondary structure stability, and evolutionary conservation at the protein and nucleic-acid level, has been applied in prediction and characterization of ncRNAs in *C. elegans* (Lu et al. 2011). This program, however, was not designed specifically for lncRNA identification. Hyperlinks of these programs along with databases accommodating lncRNA information are listed in Table 13.1.

Reliable *in silico* identification of lncRNAs depends on the completeness of the full-length status of the input sequences. It should also be noted that approaches used in the assessment of the protein-coding potential of a transcript are based on the assumption that an RNA can be unequivocally annotated as protein-coding or noncoding. In some cases, however, RNAs might be bifunctional, i.e., they can be translated into proteins but also work independently as regulatory RNAs (Dinger et al. 2008; Au et al. 2011). For instance, most mammalian cells have both coding and noncoding RNA isoforms of steroid receptor RNA activator (*SRA*). The function of the protein (SRAP) encoded by *SRA* remains to be deciphered, but the noncoding isoform of *SRA* has been shown to be the coactivator of many nuclear receptors (Chooniedass-Kothari et al. 2004). Meanwhile, RNAs without protein-coding capacity but with short ORFs may encode small peptides (Hanada et al. 2007; Rohrig et al. 2002; Xu and Ganem 2010). One example of such RNA is *Enod40* which is identified in leguminous plants. *Enod40* RNA contains a conserved secondary structure that is important for interaction with the nuclear

Table 13.1 Resources for long noncoding RNA research

Type	Tool	Source	Reference
Protein-coding potential assessment and lncRNA discovery	CONC	http://cubic.bioc.columbia.edu/_liu/conc/	Liu et al. (2006)
	CPC	http://cpc.cbi.pku.edu.cn/	Kong et al. (2007)
	ORF Finder	http://www.ncbi.nlm.nih.gov/gorf/gorf.html	NCBI
	CRITICA	http://www.ttaxus.com/software.html	Badger and Olsen (1999)
	CSTminer	http://t.caspur.it/CSTminer/	Mignone et al. (2003)
	ESTScan	http://www.ch.embnet.org/software/ESTScan.html	Lottaz et al. (2003)
	GeneMark.hmm	http://exon.biology.gatech.edu/	Lukashin and Borodovsky (1998)
	GenScan	http://genes.mit.edu/GENSCAN.html	Burge and Karlin (1998)
	incRNA	http://incrna.gersteinlab.org/	Lu et al. (2011)
	RNAcode	http://wash.github.com/rnacode/	Washietl et al. (2011)
lncRNA database	lncRNA	http://www.lncrnadb.org/	Amaral et al. (2011)
	NONCODE	http://www.noncode.org/	Liu et al. (2005)
	NRED	http://jsm-research.imb.uq.edu.au/nred/cgi-bin/ncrnadb.pl	Dinger et al. (2009)
	RNAdb	http://research.imb.uq.edu.au/rnadb/	Pang et al. (2007)

RNA-binding protein MtRBP1. It also contains two short overlapping ORFs encoding two peptides of 10-25 amino acids that could be involved in nodule formation by regulation of sucrose utilization in nodules (Rohrig et al. 2002). Therefore, when it comes to the functional characterization of single transcripts, the presence of an ORF should not exclude a priori the existence of additional regulatory functions at the RNA level, and vice versa (Solda et al. 2009).

13.3   Experimental Identification of lncRNAs

13.3.1   Identification of lncRNAs by Whole-Genome Tiling Array and RNA Sequencing

In both plants and animals, a number of lncRNAs have been identified using full-length cDNAs and ESTs that were generated using the traditional Sanger sequencing approach (Ben Amor et al. 2009; Hirsch et al. 2006; Jia et al. 2010; Khachane and Harrison 2010; MacIntosh et al. 2001; Maeda et al. 2006). Full-length cDNA sequencing is the golden standard as it provides the full-length sequence required to determine the exonic structure and confirm noncoding potential; however, the main limitation of this approach is that it is time consuming and expensive. The disadvantage of the cDNA sequencing approach can be well overcome by whole-genome tiling microarray and RNA sequencing (RNA-seq).

Tiling DNA microarray is designed to probe the expression pattern of the whole transcriptome of interested tissues or developmental stages at a high resolution. Using this technology, it has been found that the whole genome of most eukaryotes is pervasively transcribed (Li et al. 2006; Shoemaker et al. 2001; Stolc et al. 2005). For example, the analysis of 1% of the human genome by the ENCODE Consortium suggested that 93% of the genome is transcribed (ENCODE Project Consortium 2007). In rice, more than 5,000 uniquely transcribed intergenic regions were found using whole-genome tiling array (Li et al. 2006). These unannotated transcripts identified by tiling arrays provided a rich source for lncRNA discovery. However, tiling arrays rely on existing knowledge of a reference genome, without which tiling arrays will only be able to identify novel exons but will not be able to provide their connections. Furthermore, they suffer from a lack of sensitivity in detecting rare transcripts due to high level of background, cross-hybridization of related sequences, and saturation of signals.

Transcriptome sequencing or RNA-seq has emerged as a new technology for tackling the complexity of eukaryotic transcription in an unbiased manner. The technique has a wide dynamic range spanning at least four to five orders of magnitude (Cloonan et al. 2008; Mortazavi et al. 2008; Wilhelm et al. 2008) and allows accurate quantification of expression levels of transcripts. These characteristics make RNA-seq the most suitable approach so far for lncRNA discovery. To perform RNA-seq, RNAs are first converted into a cDNA library through either RNA fragmentation or DNA fragmentation. Sequencing adaptors are subsequently ligated to 5′ and 3′ ends of each cDNA fragment, and a short sequence is generated from each cDNA using a next-generation high-throughput sequencing technology, such as 454, Illumina, or SOLiD. A single cDNA preparation can yield over hundreds of millions of short reads that are then computationally aligned to a reference genome. Sequences mapping to a single exon can be generally unambiguously assigned to the corresponding gene, although RNAs that are produced by highly similar members of paralogous genes present an alignment challenge. With improvement in read length, it should become easier to exactly align these sequence reads.

Results from RNA-seq suggest the existence of a large number of novel transcribed regions in every genome investigated, even for the well-annotated genomes, such as *Drosophila melanogaster* and *Arabidopsis thaliana*. For instance, by combining tiling arrays and RNA-seq, the modENCODE project has explored the *Drosophila melanogaster* transcriptome at unprecedented depth throughout various developmental stages of both males and females (Graveley et al. 2011). From this comprehensive study, ~2,000 novel transcribed regions that do not link to any annotated gene models were identified, and about two-thirds of these novel transcripts have an ORF of less than 100 amino acids, including a multiexon lncRNA in the well-studied Bithorax complex, which is expressed in embryos and adult males but not females (Graveley et al. 2011). In another study, more than 7,000 novel transcript units (TUs) were identified in eight rice tissues, about half of these novel TUs contain multiple exons, the majority of which lack protein-coding capacity (Zhang et al. 2010). With cost reduction and increase in read length, RNA-seq is expected to provide more applications in the discovery of novel RNA species, and with further increase in sequencing depth, chances are good that even rare regulatory lncRNAs expressed only in certain tissues can be identified.

All current RNA-seq protocols require RNA to be converted into cDNA. This cDNA conversion process introduces biases and artifacts at various steps, such as priming with random hexamers, cDNA synthesis, ligation, and amplification, which can all interfere with the proper characterization and quantitation of transcripts. The recently developed direct RNA sequencing (DRS) approach, in which RNA is sequenced directly without prior conversion to cDNA, is able to overcome these biases and therefore allows more accurate quantitation of transcripts. More importantly, DRS requires only femtomole or attomole levels of input RNA and involves relatively simple sample preparation (Ozsolak and Milos 2011; Ozsolak et al. 2009), which should be more feasible to discover very lowly expressed lncRNAs. So far, a key challenge for DRS is to generate the multimillion-level read quantities that are required for many RNA applications and to further reduce error rates and input RNA quantities (Ozsolak and Milos 2011).

13.3.2 Identification of lncRNA Using lncRNA-Specific Microarray

Microarrays or tiling arrays have been developed for transcriptome analyses as well as quantitative and comparative analyses of mRNA expression between one or two samples of different origins. Microarrays are glass or silicon slides, whose surface are printed with DNA probes in a grid-like arrangement. These probes represent the entire level of cellular transcripts of an organism and are typically single stranded with variable lengths (predominantly 25-70 nucleotides). Samples used in microarrays are prepared from extracted RNA, converted to cDNA, and generally labeled with Cy3 or Cy5 fluorescent dyes (CyScribe™ first-strand cDNA labeling kit). These labeled cDNAs are applied to the microarray slide, allowing complementary cDNA and probe to hybridize resulting in a spot on the slide. The slide is subjected to a scanner that displays the result as a red or green dot, depending on the fluorescent dye used (Fig. 13.1). The intensity of the fluorescence is a quantitative measure of transcript abundance. If two samples are used and labeled with different dyes, the appearance of a red, green, yellow, or orange color dot is a reflection on the relative abundance of the transcript between the two RNA pools.

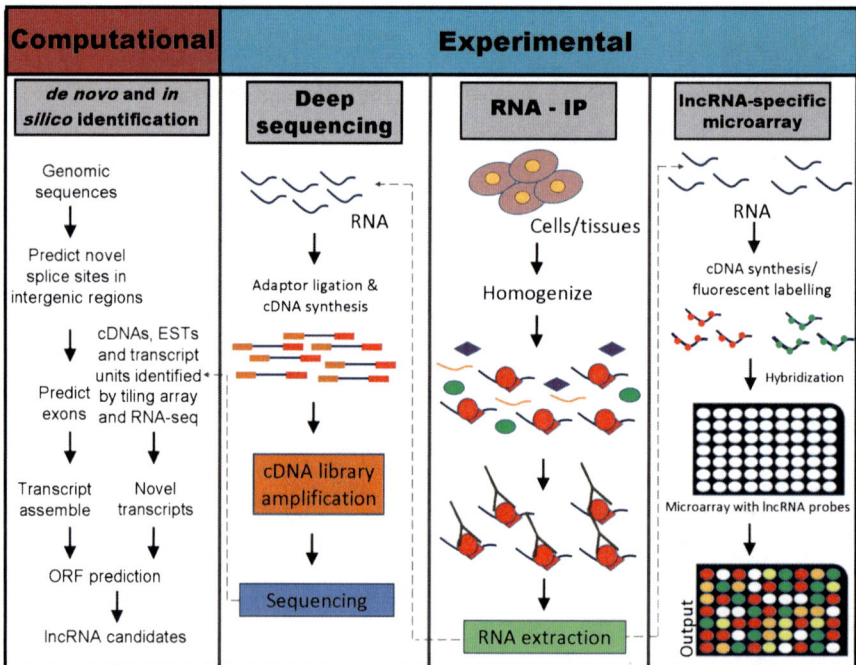

Fig. 13.1 Schematic illustration of computational and experimental approaches currently used in the identification of lncRNA

The same principle can be applied to the identification and expression analysis of lncRNAs. The limitation is related to the fact that most commercially available microarray slides contain probes that represent mRNA populations. However, with new advances in microarray technology, the ability to create custom oligonucleotide microarray is now possible. Babak et al. (2005) searched for functional lncRNAs using a microarray containing 3,478 intergenic and intronic sequences of both sense and antisense orientation that are conserved between human, mouse, and rat genomes. They identified 55 highly expressed novel lncRNAs, of which eight were confirmed to be expressed in mouse tissues by northern blot analyses. More recently, Hung et al. (2011) identified promoter-derived lncRNA transcripts using a high-density array containing probes that tile for the regulatory region of 56 human cell cycle genes. This was used to identify differentially induced lncRNAs from a large range of tissues or cells which allowed for the subsequent identification and validation of lncRNAs induced in human cancers, during specific oncogenic stimuli and stem cell differentiation as well as under DNA damage condition.

In plants, a microarray covering both the sense and antisense strands, and 50 kb upstream and downstream of the *FLC* (*Flowering Locus C*) gene was used to identify *COOLAIR* (Swiezewski et al. 2009), a lncRNA shown to be involved in the repression of the *FLC* gene during cold treatment.

Following microarray analyses, validation of lncRNA expression typically employs qRT-PCR and northern blot hybridization. Custom oligonucleotide microarrays can be purchased from a range of manufacturers including Roche NimbleGen, MYcroarray, Aligent Technologies, and Biosynthesis.

Limitations in the use of lncRNA-specific microarray to identify lncRNAs are that the methodology requires existing knowledge of an available genome for probe design and it highly relies on the quality of the designed probes to cover predicted regions that potentially transcribe lncRNAs and therefore omits regions that do transcribe lncRNAs but are not incorporated into the microarray as probes. Of the 3,478 sequences analyzed, only 55 novel lncRNAs have been isolated in the study by Babak et al. (2005). Lowly expressed transcripts as determined by microarray are usually discarded as they may potentially represent artifacts. However, in some cases, the biological significance of a lncRNA is not correlated to its abundance as one or few transcripts can adequately function. These lncRNAs include *Xist* (Brockdorff et al. 1992; Penny et al. 1996), *Air* (Nagano et al. 2008), and lncRNA transcribed from an upstream region of the *DHFR* (*dihydrofolate reductase*) locus (Martianov et al. 2007), whose functions are to induce gene silencing by physically associating with target gene promoters. Under these regulatory mechanisms, potentially one to several stably expressed transcripts per nucleus will suffice.

13.3.3 Chromatin Signature–Based Approach

Actively transcribed genes are marked by a distinctive chromatin signature that consists of a short region with trimethylation of lysine 4 of histone H3 (H3K4me3, mark of active promoter) and a long region with trimethylation of lysine 36 of histone H3 (H3K36me3, mark of transcribed region). After generation of genome-wide chromatin-state maps of H3K4me3 and H3K36me3 using ChIP-seq, Guttman et al. (2009) tested the idea of whether intergenic lncRNAs (lincRNAs) can be identified by searching for actively transcribed intergenic regions defined by a K4–K36 domain. Using this strategy, Guttman et al. (2009) identified 1,250 unannotated intergenic regions (at least 5 kb in size) in four mouse cell types. Expression of these RNA transcripts was then confirmed by hybridizing DNA microarray containing oligonucleotides that tile across the corresponding K4–K36 domain regions with poly(A)$^+$-selected RNA (Guttman et al. 2009). These lincRNAs showed similar expression levels as protein-coding genes but lacked protein-coding capacity. The same strategy has been later used to isolate ~3,300 lincRNAs from six human cell types. Approximately 38% of such identified human lincRNAs were found to be associated with chromatin modification complex, such as PRC2 and CoREST, suggesting a role for these lincRNAs in regulating gene expression through chromatin modification mechanisms (Khalil et al. 2009).

13.3.4   Identification of lncRNA Using RNA-Immunoprecipitation and Template-Switch cDNA Library Preparation

13.3.4.1   Principles

RNA-immunoprecipitation (RNA-IP) is a more function-orientated method to identify lncRNAs of interest. This method may require some established knowledge of characterized lncRNA-protein interaction as it employs antibodies against a specific RNA-binding protein in order to isolate the entire population of lncRNA associated. This is generally followed by cDNA library construction and deep sequencing to identify low abundant lncRNAs. Therefore, RNA is immunoprecipitated based on their interaction with a protein of known function, giving us predetermined clues to their functional roles. In order to preserve in vivo associations and prevent nonspecific in vitro associations during immunoprecipitation, cross-linking using either formaldehyde (generally used in plants) or UV-irradiation (in mammalian cells) is usually recommended.

Depending on the subcellular localization of the RNA-protein complex of interest, one may wish to perform nuclear RNA-IP in which a nuclear isolation step is performed prior to lysis. Such method is used to remove all cytoplasmic components that can form nonspecific association with the antibodies, reducing the quality of the immunoprecipitated sample. This method is feasible provided that an RNA amplification step is incorporated as nuclear RNA-IP tend to give minimal amount of RNA which may not be adequate for cDNA preparation using standard cDNA construction kits, designed for total or poly(A)$^+$ RNA pools. This is not ideal as with any amplification steps, the quality of the resulting data is reduced.

In other cases, whole-cell RNA-IP is suitable, particularly if the subcellular localization of the RNA-protein complex is unclear. However, one must note that whole-cell RNA-IP tends to give more background noise associated with nonspecific interaction between the antibodies and ribosomal/mitochondrial RNA including those derived from chloroplasts in plant tissues. Therefore, stringent bioinformatics analyses must follow to filter these confounding RNA from the true RNA population. Following RNA-IP, the pool of RNAs can be converted to cDNA using cDNA library preparation protocols such as template-switch cDNA library preparation (Fig. 13.2) and their identities interrogated by using high-throughput deep sequencing (see Sect. 13.3.1) or alternatively through the use of microarray (see Sect. 13.3.2) by using fluorescent-labeled cDNA. While RNA-IP combined with lncRNA-specific microarray has yet to be utilized in the identification of lncRNA, it is a powerful approach to consider. Landthaler et al. (2008) identified human Argonaute–associated mRNA via RNA-IP followed by mRNA microarray analyses.

RNA-IP followed by template-switch cDNA library preparation was recently employed by Zhao et al. (2008) to successfully identify lncRNAs associated with the mammalian Polycomb complex. This library construction protocol is in-house, cost-effective, and can preserve strand specificity via addition of adaptors in a directional manner (Fig. 13.2). The adaptor and primer sequences used in this cDNA library construction protocol are compatible with and suitable for high-throughput single-end Illumina® GAII sequencing. We have adapted this library construction methodology in our laboratory to successfully study lncRNAs in plants. In this section, we will describe whole-cell RNA-IP and template-switch cDNA library preparation in detail for lncRNA identification in animals and plants. Protocols on nuclear RNA-IP in animals (Zhao et al. 2008) and plants (Wierzbicki et al. 2008) can be found in the corresponding publication.

As nonspecific association between RNA molecules and antibody is common, negative controls must be included such that the RNA pools isolated can be comparatively analyzed and these nonspecific RNAs can be omitted from the resulting data. Controls can include cells/tissues derived from an organism that does not express the RNA-binding protein and/or from an immunoprecipitation without antibodies. Since nonspecific RNA and antibody association is common, the former control is recommended.

13.3.4.2 Materials

- Mortar and pestle
- Rotator for 1.5-mL microcentrifuge tubes
- Heat block or water bath set to 65°C
- PCR cycler
- Stratalinker (for animal cells)
- Microcentrifuge set to 4°C
- Liquid nitrogen

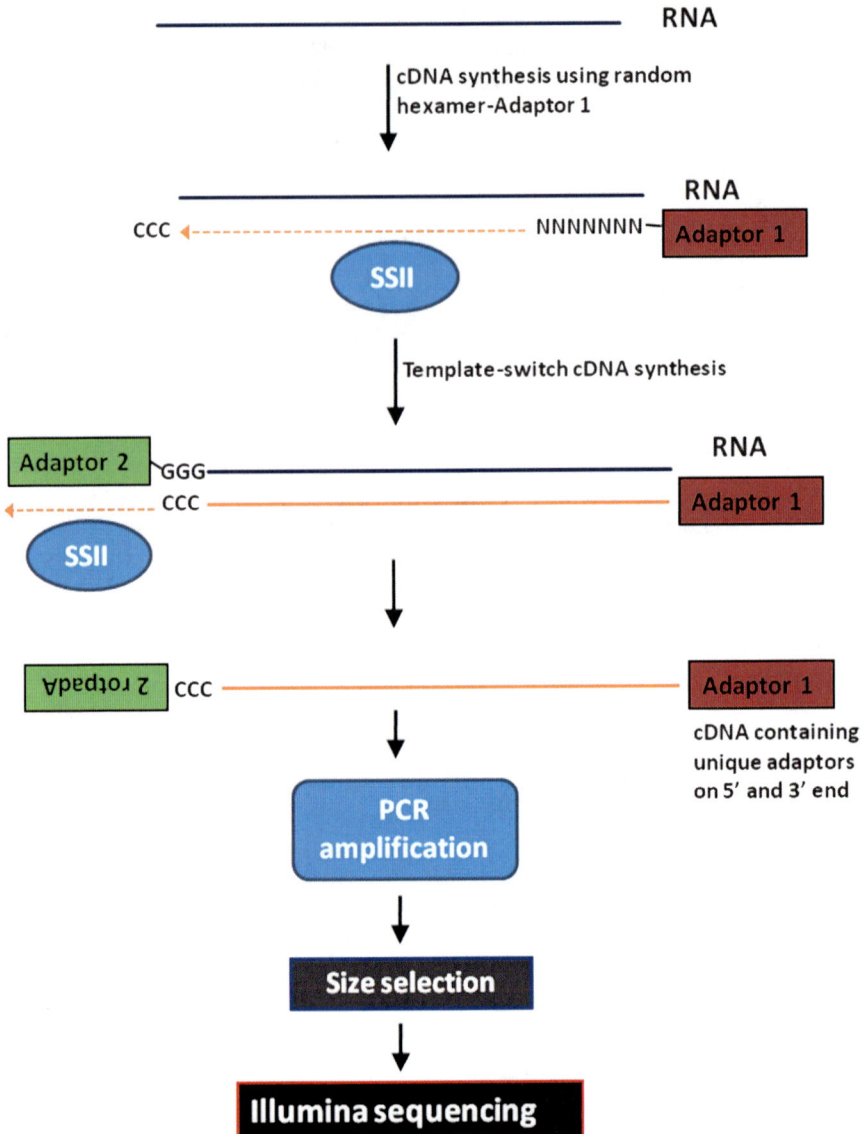

Fig. 13.2 Schematic illustration of template-switch cDNA library preparation

- Absolute ethanol
- DEPC-treated H_2O
- Sodium acetate, 3 M, pH 5.2
- Acid phenol/chloroform pH 4.5 (Ambion)
- Formaldehyde solution 37 wt% in H_2O (Sigma-Aldrich)
- Antibodies of choice
- Salmon sperm DNA/Protein A/G agarose (Millipore)

- Glycogen 5 mg/mL (Ambion)
- RNase out (Invitrogen)
- SuperScript™ II Reverse Transcriptase (Invitrogen)
- RQ1 Rnase-free DNase (Promega)
- Platinum® *Taq* high-fidelity DNA polymerase
- Set of dNTPs (Fisher Biotec)
- Proteinase K (Progen)
- NuSieve® 3:1 agarose (Lonza)
- Tris-Borate-EDTA buffer
- Adaptor/primer sequences:
  - Adaptor 1: 5′ CTTTCCCTACACGACGCTCTTCCGATCTNNNNNN 3′
  - Adaptor 2: 5′CAAGCAGAAGACGGCATACGAGCTCTTCCGATCTggg3′ – lowercase nucleotides are ribonucleic acid
  - Illumina forward primer: 5′AATGATACGGCGACCACCGAGATCTA-CACTCTTTCCCTACACGACGCTCTTCCGATCT 3′
  - Illumina reverse primer: 5′ CAAGCAGAAGACGGCATACGAGCTC-TTCCGATCT 3′
- Extraction buffer:
  - Plant: 20 mM Tris–HCl pH 7.5, 4 mM $MgCl_2$, 5 mM DTT, 0.1% SDS, 100 µL/10 mL Protease inhibitor cocktail for plants* (Sigma-Aldrich), 1 mM PMSF*, 40 u/mL RNase out* (Invitrogen)
  - Polysome lysis buffer for animal cells: 100 mM KCl, 5 mM $MgCl_2$, 10 mM HEPES, 0.5% nonidet P-40, 1 mM DTT, 100 u/mL RNase out* (Invitrogen), 2 mM vanadyl ribonucleoside complexes solution* (Sigma-Aldrich), 25 µL/mL protease inhibitor cocktail for mammalian tissues* (Sigma-Aldrich)
- Wash buffer: 150 mM NaCl, 20 mM Tris–HCl pH 8.0, 2 mM EDTA, 1% Triton X-100, 0.1% SDS, 1 mM PMSF, 40 U/ml RNase out* (Invitrogen).
- RNA-IP elution buffer: 100 mM Tris–HCl pH 8.0, 10 mM EDTA, 1% SDS, 40 U/ml RNase out* (Invitrogen).

*Add just prior to use

13.3.4.3 Methods

General aseptic pipetting techniques, sterile and/or autoclaved consumables, and buffers must be used in all steps to prevent RNA degradation. In addition, all steps of RNA-IP are performed in mild denaturing conditions and include enzymes that inhibit RNases. Centrifugation steps and tubes containing samples are maintained at 4°C unless indicated.

Tissue/cell preparation and extraction:

1. Plants:

    (a) Collect 3 g of plant tissues from each experimental sample and negative control.
    (b) Vacuum infiltrate plant tissues with 0.5% formaldehyde for 2 min, repeat for another 8 min. Stop reaction with addition of glycine to a final concentration of 70 mM and vacuum infiltrate for 1 min with another 4 min repeat. Wash tissues at least four times with DEPC-treated H_2O. Freeze tissues in liquid nitrogen and store at $-80°C$ or continue to extraction.
    (c) Grind tissues into fine powder under liquid nitrogen and then homogenize in 10 mL of plant extraction buffer. Transfer each homogenate into a 50-mL Falcon tube on ice.
    (d) Divide homogenate into 1 mL aliquots in 1.5-mL microcentrifuge tubes.

2. Mammalian cells:

    (a) Grow cells in 10-cm dish until confluent ($\sim 5.0 \times 10^6$ cells).
    (b) Place dish on ice in Stratalinker with lid off and UV-irradiate once for 150 mJ/cm^2.
    (c) Harvest each dish of cells with 1 mL of polysome lysis buffer, scrape cells with a cell scraper to detach cells, and transfer to microcentrifuge tubes on ice.

3. Remove cell debris by centrifugation at maximum speed for 20 min at 4°C and collect lysate (supernatant) into new microcentrifuge tubes.
    Preclearing:
4. In order to remove lysate components that form nonspecific association with the protein agarose beads, a preclearing step should be included. Incubate each lysate aliquot on a rotator at 4°C for 1 h with a desired amount of protein A/G agarose (generally 20 μL/mL lysate will suffice). Centrifuge at 4,000 rpm for 1 min and transfer precleared lysate (supernatant) into new microcentrifuge tubes.
    Immunoprecipitation:
5. Perform immunoprecipitation on each precleared lysate aliquot with 20 μL of protein A/G agarose beads and antibodies (optimization required, generally 5 μg of antibody per lysate aliquot is sufficient) for 2–4 h on a rotator at 4°C. Overnight incubation is not recommended as this may exhaust the RNase inhibitors and lead to RNA degradation.
    Wash:
6. Centrifuge at 4,000 rpm for 1 min at 4°C and discard supernatant.
7. Wash the immunoprecipitated complexes (now attached to the beads) three times with 1 mL wash buffer/tube at 4°C, 5–10 min on a rotator per wash. Centrifuge at 4,000 rpm for 1 min at 4°C between each wash. The length and number of washes can be optimized to improve the quality of the resulting RNA extract.
    Elution:

8. Following the last wash and centrifugation, discard supernatant. Optional validation at this step (see Sect. 13.3.4.4).
9. Add 50 µL of RNA-IP elution buffer to the beads in each tube and rotate at room temperature for 10 min. The high concentration of SDS and RNase inhibitor in the elution buffer prevents RNA degradation at room temperature.
10. Centrifuge at 4,000 rpm for 1 min at 4°C, collect the eluate (supernatant) into a new microcentrifuge tube and save on ice.
11. Repeat elution by adding another 50 µL of RNA-IP elution buffer to the beads and incubate at 65°C with gentle frequent mixing to denature and release the protein complexes.
12. Combine two eluates, add 20 µg of proteinase K, and incubate at 65°C for 1 h to reverse cross-link. Proteinase K inactivates any DNase and RNase and digests the protein complexes to allow the release of the RNA.
    RNA extraction:
13. Extract RNA from each tube with 100 µL of acidic phenol/chloroform (Ambion), followed by overnight ethanol precipitation in the presence of acidic sodium acetate and 30 µg glycogen at −80°C.
14. Centrifuge at maximum speed for 20 min at 4°C.
15. Wash pellet with 70% ethanol and air dry pellet for 2 min.
16. Resuspend RNA pellet in desired amount of DEPC-treated water containing 40 units of RNase inhibitor. Optional validation at this step (see Sect. 13.3.4.4).
    DNase treatment:
17. Remove any contaminating DNA with 1 u RQ1 DNase/tube for 15 min at 37°C.
18. Extract RNA with an equal volume of acid phenol/chloroform, pool all supernatants into one single tube, ethanol precipitate, and wash as previously described.
19. Dissolve and pool all RNA pellets with a desired amount of DEPC-treated water containing 40 units of RNase inhibitor.
20. Determine RNA concentration and store RNA at −80°C.
    Template-switch cDNA library preparation:
21. Concentrate 200–300 ng of each RNA sample to a volume of 1.5 µL by vacuum centrifugation if necessary. Note: Lower amounts of RNA is not recommended as this changes the ratio of RNA to adaptors/primers which can result in adaptor primer and/or primer dimer formation at subsequent steps preventing efficient cDNA synthesis and amplification. Include a negative control using water as template.
22. Random prime each RNA/water sample by adding 0.5 µL of 2.4 µM Adaptor 1 to the RNA and incubate for 10 min at 72°C in a PCR cycler.
23. Synthesize first-strand cDNA using SuperScript II Reverse Transcriptase in a 5-µL reaction by adding on ice: 1 µL 5× first-strand buffer, 0.5 µL 20 mM DTT, 0.5 µL 10 mM dNTPs, 0.5 µL RNase out (40 u/µL), and 0.5 µL Super-Script II (200 u/µL). Incubate the tubes at the following conditions: 20°C for 10 min followed by 37°C for 10 min and 42°C for 45 min.
24. Denature required amount of 10 µM Adaptor 2 at 72°C for 5 min.

25. To each tube, add 0.5 μL denatured 10 μM Adaptor 2 and 0.5 μL SuperScript II. Incubate at 42°C for 30 min followed by 72°C for 15 min.
26. PCR amplify 1 μL of the first-strand cDNA template using Platinum *Taq* high-fidelity DNA polymerase in a 25-μL reaction containing: 1 μL of template, 2.5 μL 10× high-fidelity PCR buffer, 1 μL 10 mM dNTPs, 1 μL 50 mM MgSO$_4$, 2.5 μL 2.5 μM Illumina forward primer, 2.5 μL 10 μM Illumina reverse primer, 0.5 μL Platinum *Taq* high-fidelity polymerase (5 u/μL), and 14 μL RNase/DNase–free H$_2$O. Perform PCR as follows: 94°C for 2 min, 24 cycles of [94°C for 30 s, 65°C for 30 s, 72°C for 30 s], and 72°C for 5 min. Optional validation at this step (see Sect. 13.3.4.4).
27. Run 10 μL of PCR products on a 3% NuSieve/1× TBE gel. A smear should be observed.
28. Size select the range of PCR products desired for analyses (200–500 bp is preferred for Illumina sequencing applications) by gel extraction using QIAquick Gel extraction kit (Qiagen).

13.3.4.4 Validation

Validation of RNA-IP and cDNA library preparation can be performed at the protein, RNA, or cDNA level. Following RNA-IP immediately after the last wash but before elution, the immunoprecipitated complex can be validated for the presence of the specific RNA-binding protein of interest between the experimental and control samples using Western blot analyses. This can be done by denaturing the pull-down beads/products at 95°C with addition of Western blot loading buffer followed by SDS polyacrylamide gel electrophoresis and blotting.

Following RNA extraction and DNase treatment, one may perform gene-specific RT-PCR to validate the quality of the RNA using primers against transcripts that are known to associate with the RNA-binding protein in question and against unrelated transcripts as negative controls.

Libraries can also be validated by PCR for transcripts known to associate with the RNA-binding protein of interest and for unrelated transcripts as negative controls as above. One may also choose to subclone libraries into routinely used cloning and sequencing vectors such as pGEM®-T Easy (Promega) for validation using standard DNA sequencing. Unlike Phusion high-fidelity DNA polymerase, Platinum *Taq* high-fidelity DNA polymerase adds a single deoxyadenosine to the 3′ end of PCR products allowing subsequent cloning into pGEM®-T Easy vectors. However, this may not be suitable for whole-cell RNA-IP as the overrepresentation of ribosomal RNA will limit subcloning and sequencing/identification of the lower abundant but relevant sequences present in the library.

13.4   Conclusions and Future Views

We have come a long way from unintentionally identifying lncRNAs to the development of specific computational and experimental strategies specific for their identification. While computational approaches are increasingly popular and have been successful in identifying lncRNAs in both animals and plants, they will continue to prove challenging due to the intrinsic nature of lncRNA being mostly derived from unconserved regions and lacking common signatures and structural characteristics that have facilitated the identification and functional analyses of protein-coding RNA. Without doubt, experimental approaches must always follow to validate and confirm the noncoding functional capabilities of a lncRNA identified using computational approaches.

Experimental approaches have proven more successful as evidenced by their direct identification and characterization of lncRNA, particularly through the use of lncRNA-specific microarray and RNA-IP. Undoubtedly, lncRNA-specific microarray is a powerful method for predicting, identifying, and characterizing functional lncRNA of high abundance. For the identification of lncRNA of low abundance, next-generation deep sequencing is a more effective method, and combined with RNA-IP can achieve the purpose of identifying low abundant novel lncRNA with specific functions. To circumvent the issue associated with low RNA concentration from nuclear RNA-IP which limits cDNA library preparation, one may consider using nuclear RNA-IP followed by direct RNA sequencing.

Our understanding of lncRNA is only at its primordial phase but will be expected to grow in an exponential way with improvement in computational programs and their ability to analyze and generate infinite amount of data and the increasing read length and depth from our deep sequencing technology. We expect lncRNA to become the focus of many more regulatory pathways and ultimately our unique origin.

Acknowledgments This work was partly supported by an Australian Research Council Future Fellowship (FT0991956). We thank Dr. Ming-Bo Wang for his assistance in the revision of this manuscript.

References

Amaral PP, Clark MB, Gascoigne DK, Dinger ME, Mattick JS (2011) lncRNAdb: a reference database for long noncoding RNAs. Nucleic Acids Res 39 (Database issue):D146-151. doi: gkq1138 [pii] 10.1093/nar/gkq1138

Au PC, Zhu QH, Dennis ES, Wang MB (2011) Long non-coding RNA-mediated mechanisms independent of the RNAi pathway in animals and plants. RNA Biol 8(3):14382 [pii] 10.1093/nar/gkq1138

Babak T, Blencowe BJ, Hughes TR (2005) A systematic search for new mammalian noncoding RNAs indicates little conserved intergenic transcription. BMC Genomics 6:104. doi:1471-2164-6-104 [pii] 10.1186/1471-2164-6-104

Badger JH, Olsen GJ (1999) CRITICA: coding region identification tool invoking comparative analysis. Mol Biol Evol 16(4):512–524

Beltran M, Puig I, Pena C, Garcia JM, Alvarez AB, Pena R, Bonilla F, de Herreros AG (2008) A natural antisense transcript regulates Zeb2/Sip1 gene expression during Snail1-induced epithelial-mesenchymal transition. Genes Dev 22(6):756–769. doi:22/6/756 [pii] 10.1101/gad.455708

Ben Amor B, Wirth S, Merchan F, Laporte P, d'Aubenton-Carafa Y, Hirsch J, Maizel A, Mallory A, Lucas A, Deragon JM, Vaucheret H, Thermes C, Crespi M (2009) Novel long non-protein coding RNAs involved in Arabidopsis differentiation and stress responses. Genome Res 19(1):57–69. doi:gr.080275.108 [pii] 10.1101/gr.080275.108

Brockdorff N, Ashworth A, Kay GF, McCabe VM, Norris DP, Cooper PJ, Swift S, Rastan S (1992) The product of the mouse Xist gene is a 15 kb inactive X-specific transcript containing no conserved ORF and located in the nucleus. Cell 71(3):515–526. doi:0092-8674(92)90519-I [pii]

Brown CJ, Ballabio A, Rupert JL, Lafreniere RG, Grompe M, Tonlorenzi R, Willard HF (1991) A gene from the region of the human X inactivation centre is expressed exclusively from the inactive X chromosome. Nature 349(6304):38–44. doi:10.1038/349038a0

Burge CB, Karlin S (1998) Finding the genes in genomic DNA. Curr Opin Struct Biol 8 (3):346–354. doi:S0959-440X(98), 80069-9 [pii]

Chooniedass-Kothari S, Emberley E, Hamedani MK, Troup S, Wang X, Czosnek A, Hube F, Mutawe M, Watson PH, Leygue E (2004) The steroid receptor RNA activator is the first functional RNA encoding a protein. FEBS Lett 566 (1–3):43–47. doi:10.1016/j.febslet.2004.03.104, S0014579304004387 [pii]

Cloonan N, Forrest AR, Kolle G, Gardiner BB, Faulkner GJ, Brown MK, Taylor DF, Steptoe AL, Wani S, Bethel G, Robertson AJ, Perkins AC, Bruce SJ, Lee CC, Ranade SS, Peckham HE, Manning JM, McKernan KJ, Grimmond SM (2008) Stem cell transcriptome profiling via massive-scale mRNA sequencing. Nat Methods 5(7):613–619. doi:nmeth.1223 [pii] 10.1038/nmeth.1223

Dinger ME, Pang KC, Mercer TR, Mattick JS (2008) Differentiating protein-coding and noncoding RNA: challenges and ambiguities. PLoS Comput Biol 4(11):e1000176. doi:10.1371/journal.pcbi.1000176

Dinger ME, Pang KC, Mercer TR, Crowe ML, Grimmond SM, Mattick JS (2009) NRED: a database of long noncoding RNA expression. Nucleic Acids Res 37 (Database issue):D122-126. doi:gkn617 [pii], 10.1093/nar/gkn617

ENCODE Project Consortium (2007) Identification and analysis of functional elements in 1% of the human genome by the ENCODE pilot project. Nature 447(7146):799–816. doi:10.1038/nature05874

Erdmann VA, Szymanski M, Hochberg A, de Groot N, Barciszewski J (1999) Collection of mRNA-like non-coding RNAs. Nucleic Acids Res 27(1):192–195. doi:gkc101 [pii]

Feng J, Bi C, Clark BS, Mady R, Shah P, Kohtz JD (2006) The Evf-2 noncoding RNA is transcribed from the Dlx-5/6 ultraconserved region and functions as a Dlx-2 transcriptional coactivator. Genes Dev 20(11):1470–1484. doi:gad.1416106 [pii] 10.1101/gad.1416106

Graveley BR, Brooks AN, Carlson JW, Duff MO, Landolin JM, Yang L, Artieri CG, van Baren MJ, Boley N, Booth BW, Brown JB, Cherbas L, Davis CA, Dobin A, Li R, Lin W, Malone JH, Mattiuzzo NR, Miller D, Sturgill D, Tuch BB, Zaleski C, Zhang D, Blanchette M, Dudoit S, Eads B, Green RE, Hammonds A, Jiang L, Kapranov P, Langton L, Perrimon N, Sandler JE, Wan KH, Willingham A, Zhang Y, Zou Y, Andrews J, Bickel PJ, Brenner SE, Brent MR, Cherbas P, Gingeras TR, Hoskins RA, Kaufman TC, Oliver B, Celniker SE (2011) The developmental transcriptome of Drosophila melanogaster. Nature 471(7339):473–479. doi:nature09715 [pii] 10.1038/nature09715

Guttman M, Amit I, Garber M, French C, Lin MF, Feldser D, Huarte M, Zuk O, Carey BW, Cassady JP, Cabili MN, Jaenisch R, Mikkelsen TS, Jacks T, Hacohen N, Bernstein BE, Kellis M, Regev A, Rinn JL, Lander ES (2009) Chromatin signature reveals over a thousand highly conserved large non-coding RNAs in mammals. Nature 458(7235):223–227. doi:nature07672 [pii] 10.1038/nature07672

Hanada K, Zhang X, Borevitz JO, Li WH, Shiu SH (2007) A large number of novel coding small open reading frames in the intergenic regions of the Arabidopsis thaliana genome are transcribed and/or under purifying selection. Genome Res 17(5):632–640. doi:gr.5836207 [pii] 10.1101/gr.5836207

Hatzigeorgiou AG, Fiziev P, Reczko M (2001) DIANA-EST: a statistical analysis. Bioinformatics 17(10):913–919

He Y, Vogelstein B, Velculescu VE, Papadopoulos N, Kinzler KW (2008) The antisense transcriptomes of human cells. Science 322(5909):1855–1857. doi:1163853 [pii] 10.1126/science.1163853

Hiller M, Findeiss S, Lein S, Marz M, Nickel C, Rose D, Schulz C, Backofen R, Prohaska SJ, Reuter G, Stadler PF (2009) Conserved introns reveal novel transcripts in Drosophila melanogaster. Genome Res 19(7):1289–1300. doi:gr.090050.108 [pii] 10.1101/gr.090050.108

Hirsch J, Lefort V, Vankersschaver M, Boualem A, Lucas A, Thermes C, d'Aubenton-Carafa Y, Crespi M (2006) Characterization of 43 non-protein-coding mRNA genes in Arabidopsis, including the MIR162a-derived transcripts. Plant Physiol 140(4):1192–1204. doi: pp.105.073817 [pii] 10.1104/pp.105.073817

Hung T, Wang Y, Lin MF, Koegel AK, Kotake Y, Grant GD, Horlings HM, Shah N, Umbricht C, Wang P, Kong B, Langerod A, Borresen-Dale AL, Kim SK, van de Vijver M, Sukumar S, Whitfield ML, Kellis M, Xiong Y, Wong DJ, Chang HY (2011) Extensive and coordinated transcription of noncoding RNAs within cell-cycle promoters. Nat Genet. doi:ng.848 [pii] 10.1038/ng.848

Jia H, Osak M, Bogu GK, Stanton LW, Johnson R, Lipovich L (2010) Genome-wide computational identification and manual annotation of human long noncoding RNA genes. RNA 16 (8):1478–1487. doi:rna.1951310 [pii] 10.1261/rna.1951310

Khachane AN, Harrison PM (2010) Mining mammalian transcript data for functional long noncoding RNAs. PLoS One 5(4):e10316. doi:10.1371/journal.pone.0010316

Khalil AM, Guttman M, Huarte M, Garber M, Raj A, Rivea Morales D, Thomas K, Presser A, Bernstein BE, van Oudenaarden A, Regev A, Lander ES, Rinn JL (2009) Many human large intergenic noncoding RNAs associate with chromatin-modifying complexes and affect gene expression. Proc Natl Acad Sci USA 106(28):11667–11672. doi:0904715106 [pii] 10.1073/pnas.0904715106

Kong L, Zhang Y, Ye ZQ, Liu XQ, Zhao SQ, Wei L, Gao G (2007) CPC: assess the protein-coding potential of transcripts using sequence features and support vector machine. Nucleic Acids Res 35(Web Server issue):W345–349. doi:35/suppl_2/W345 [pii] 10.1093/nar/gkm391

Kramer RA, Andersen N (1980) Isolation of yeast genes with mRNA levels controlled by phosphate concentration. Proc Natl Acad Sci USA 77(11):6541–6545

Landthaler M, Gaidatzis D, Rothballer A, Chen PY, Soll SJ, Dinic L, Ojo T, Hafner M, Zavolan M, Tuschl T (2008) Molecular characterization of human Argonaute-containing ribonucleoprotein complexes and their bound target mRNAs. RNA 14(12):2580–2596. doi:rna.1351608 [pii] 10.1261/rna.1351608

Li CX, Liu JQ, Yu JJ, Zhao Q, Ao GM (2001) Cloning and expression analysis of pollen-specific cDNA ZM401 from *Zea mays*. J Agr Biotechnol 9(4):374–377

Li L, Wang X, Stolc V, Li X, Zhang D, Su N, Tongprasit W, Li S, Cheng Z, Wang J, Deng XW (2006) Genome-wide transcription analyses in rice using tiling microarrays. Nat Genet 38 (1):124–129. doi:ng1704 [pii] 10.1038/ng1704

Liu J, Gough J, Rost B (2006) Distinguishing protein-coding from non-coding RNAs through support vector machines. PLoS Genet 2(4):e29. doi:10.1371/journal.pgen.0020029

Liu C, Bai B, Skogerbo G, Cai L, Deng W, Zhang Y, Bu D, Zhao Y, Chen R (2005) NONCODE: an integrated knowledge database of non-coding RNAs. Nucleic Acids Res 33 (Database issue):D112-115. doi:33/suppl_1/D112 [pii] 10.1093/nar/gki041

Lottaz C, Iseli C, Jongeneel CV, Bucher P (2003) Modeling sequencing errors by combining Hidden Markov models. Bioinformatics 19(Suppl 2):103–112

Lu ZJ, Yip KY, Wang G, Shou C, Hillier LW, Khurana E, Agarwal A, Auerbach R, Rozowsky J, Cheng C, Kato M, Miller DM, Slack F, Snyder M, Waterston RH, Reinke V, Gerstein MB (2011) Prediction and characterization of noncoding RNAs in C. elegans by integrating conservation, secondary structure, and high-throughput sequencing and array data. Genome Res 21(2):276–285. doi:gr.110189.110 [pii] 10.1101/gr.110189.110

Lukashin AV, Borodovsky M (1998) GeneMark.hmm: new solutions for gene finding. Nucleic Acids Res 26(4):1107–1115. doi:gkb200 [pii]

MacIntosh GC, Wilkerson C, Green PJ (2001) Identification and analysis of Arabidopsis expressed sequence tags characteristic of non-coding RNAs. Plant Physiol 127(3):765–776

Maeda N, Kasukawa T, Oyama R, Gough J, Frith M, Engstrom PG, Lenhard B, Aturaliya RN, Batalov S, Beisel KW, Bult CJ, Fletcher CF, Forrest AR, Furuno M, Hill D, Itoh M, Kanamori-Katayama M, Katayama S, Katoh M, Kawashima T, Quackenbush J, Ravasi T, Ring BZ, Shibata K, Sugiura K, Takenaka Y, Teasdale RD, Wells CA, Zhu Y, Kai C, Kawai J, Hume DA, Carninci P, Hayashizaki Y (2006) Transcript annotation in FANTOM3: mouse gene catalog based on physical cDNAs. PLoS Genet 2(4):e62. doi:10.1371/journal.pgen.0020062

Martianov I, Ramadass A, Serra Barros A, Chow N, Akoulitchev A (2007) Repression of the human dihydrofolate reductase gene by a non-coding interfering transcript. Nature 445 (7128):666–670. doi:nature05519 [pii] 10.1038/nature05519

Mattick JS (2004) RNA regulation: a new genetics? Nat Rev Genet 5(4):316–323. doi:10.1038/ nrg1321 nrg1321 [pii]

Mattick JS, Makunin IV (2005) Small regulatory RNAs in mammals. Hum Mol Genet 14(1): R121–132. doi:14/suppl_1/R121 [pii] 10.1093/hmg/ddi101

Mignone F, Grillo G, Liuni S, Pesole G (2003) Computational identification of protein coding potential of conserved sequence tags through cross-species evolutionary analysis. Nucleic Acids Res 31(15):4639–4645

Mortazavi A, Williams BA, McCue K, Schaeffer L, Wold B (2008) Mapping and quantifying mammalian transcriptomes by RNA-Seq. Nat Methods 5(7):621–628. doi:nmeth.1226 [pii] 10.1038/nmeth.1226

Nagano T, Mitchell JA, Sanz LA, Pauler FM, Ferguson-Smith AC, Feil R, Fraser P (2008) The Air noncoding RNA epigenetically silences transcription by targeting G9a to chromatin. Science 322(5908):1717–1720. doi:1163802 [pii] 10.1126/science.1163802

Ogawa Y, Sun BK, Lee JT (2008) Intersection of the RNA interference and X-inactivation pathways. Science 320(5881):1336–1341. doi:320/5881/1336 [pii] 10.1126/science.1157676

Ozsolak F, Milos PM (2011) RNA sequencing: advances, challenges and opportunities. Nat Rev Genet 12(2):87–98. doi:nrg2934 [pii] 10.1038/nrg2934

Ozsolak F, Platt AR, Jones DR, Reifenberger JG, Sass LE, McInerney P, Thompson JF, Bowers J, Jarosz M, Milos PM (2009) Direct RNA sequencing. Nature 461(7265):814–818. doi: nature08390 [pii] 10.1038/nature08390

Palade GE (1955) A small particulate component of the cytoplasm. J Biophys Biochem Cytol 1 (1):59–68

Palade GE (1958) Microsomal particles and protein synthesis. Pergamon Press, London

Pang KC, Frith MC, Mattick JS (2006) Rapid evolution of noncoding RNAs: lack of conservation does not mean lack of function. Trends Genet 22(1):1–5. doi:S0168-9525(05), 00322-7 [pii] 10.1016/j.tig.2005.10.003

Pang KC, Stephen S, Dinger ME, Engstrom PG, Lenhard B, Mattick JS (2007) RNAdb 2.0–an expanded database of mammalian non-coding RNAs. Nucleic Acids Res 35 (Database issue): D178-182. doi:gkl926 [pii] 10.1093/nar/gkl926

Penny GD, Kay GF, Sheardown SA, Rastan S, Brockdorff N (1996) Requirement for Xist in X chromosome inactivation. Nature 379(6561):131–137. doi:10.1038/379131a0

Ponjavic J, Ponting CP, Lunter G (2007) Functionality or transcriptional noise? Evidence for selection within long noncoding RNAs. Genome Res 17(5):556–565. doi:gr.6036807 [pii] 10.1101/gr.6036807

Rinn JL, Kertesz M, Wang JK, Squazzo SL, Xu X, Brugmann SA, Goodnough LH, Helms JA, Farnham PJ, Segal E, Chang HY (2007) Functional demarcation of active and silent chromatin domains in human HOX loci by noncoding RNAs. Cell 129(7):1311–1323. doi:S0092-8674 (07), 00659-9 [pii] 10.1016/j.cell.2007.05.022

Rodriguez-Trelles F, Tarrio R, Ayala FJ (2006) Origins and evolution of spliceosomal introns. Annu Rev Genet 40:47–76. doi:10.1146/annurev.genet.40.110405.090625

Rohrig H, Schmidt J, Miklashevichs E, Schell J, John M (2002) Soybean ENOD40 encodes two peptides that bind to sucrose synthase. Proc Natl Acad Sci USA 99(4):1915–1920. doi:10.1073/pnas.022664799 022664799 [pii]

Rose D, Hiller M, Schutt K, Hackermuller J, Backofen R, Stadler PF (2011) Computational discovery of human coding and non-coding transcripts with conserved splice sites. Bioinformatics. doi:btr314 [pii] 10.1093/bioinformatics/btr314

Schneider C, King RM, Philipson L (1988) Genes specifically expressed at growth arrest of mammalian cells. Cell 54(6):787–793. doi:S0092-8674(88), 91065-3 [pii]

Shimizu K, Adachi J, Muraoka Y (2006) ANGLE: a sequencing errors resistant program for predicting protein coding regions in unfinished cDNA. J Bioinform Comput Biol 4 (3):649–664. doi:S0219720006002260 [pii]

Shoemaker DD, Schadt EE, Armour CD, He YD, Garrett-Engele P, McDonagh PD, Loerch PM, Leonardson A, Lum PY, Cavet G, Wu LF, Altschuler SJ, Edwards S, King J, Tsang JS, Schimmack G, Schelter JM, Koch J, Ziman M, Marton MJ, Li B, Cundiff P, Ward T, Castle J, Krolewski M, Meyer MR, Mao M, Burchard J, Kidd MJ, Dai H, Phillips JW, Linsley PS, Stoughton R, Scherer S, Boguski MS (2001) Experimental annotation of the human genome using microarray technology. Nature 409(6822):922–927. doi:10.1038/35057141

Solda G, Makunin IV, Sezerman OU, Corradin A, Corti G, Guffanti A (2009) An Ariadne's thread to the identification and annotation of noncoding RNAs in eukaryotes. Brief Bioinform 10 (5):475–489. doi:bbp022 [pii] 10.1093/bib/bbp022

Solovyev V, Kosarev P, Seledsov I, Vorobyev D (2006) Automatic annotation of eukaryotic genes, pseudogenes and promoters. Genome Biol 7(Suppl 1):S10 11–12. doi:gb-2006–7-s1-s10 [pii] 10.1186/gb-2006–7-s1-s10

Stark A, Lin MF, Kheradpour P, Pedersen JS, Parts L, Carlson JW, Crosby MA, Rasmussen MD, Roy S, Deoras AN, Ruby JG, Brennecke J, Hodges E, Hinrichs AS, Caspi A, Paten B, Park SW, Han MV, Maeder ML, Polansky BJ, Robson BE, Aerts S, van Helden J, Hassan B, Gilbert DG, Eastman DA, Rice M, Weir M, Hahn MW, Park Y, Dewey CN, Pachter L, Kent WJ, Haussler D, Lai EC, Bartel DP, Hannon GJ, Kaufman TC, Eisen MB, Clark AG, Smith D, Celniker SE, Gelbart WM, Kellis M (2007) Discovery of functional elements in 12 Drosophila genomes using evolutionary signatures. Nature 450(7167):219–232. doi:nature06340 [pii] 10.1038/nature06340

Stolc V, Samanta MP, Tongprasit W, Sethi H, Liang S, Nelson DC, Hegeman A, Nelson C, Rancour D, Bednarek S, Ulrich EL, Zhao Q, Wrobel RL, Newman CS, Fox BG, Phillips GN Jr, Markley JL, Sussman MR (2005) Identification of transcribed sequences in *Arabidopsis thaliana* by using high-resolution genome tiling arrays. Proc Natl Acad Sci USA 102 (12):4453–4458. doi:0408203102 [pii] 10.1073/pnas.0408203102

Swiezewski S, Liu F, Magusin A, Dean C (2009) Cold-induced silencing by long antisense transcripts of an Arabidopsis Polycomb target. Nature 462(7274):799–802. doi:nature08618 [pii] 10.1038/nature08618

Taft RJ, Pheasant M, Mattick JS (2007) The relationship between non-protein-coding DNA and eukaryotic complexity. Bioessays 29(3):288–299. doi:10.1002/bies.20544

Wang X, Arai S, Song X, Reichart D, Du K, Pascual G, Tempst P, Rosenfeld MG, Glass CK, Kurokawa R (2008) Induced ncRNAs allosterically modify RNA-binding proteins in cis to inhibit transcription. Nature 454(7200):126–130. doi:nature06992 [pii] 10.1038/nature06992

Washietl S, Findeiss S, Muller SA, Kalkhof S, von Bergen M, Hofacker IL, Stadler PF, Goldman N (2011) RNAcode: robust discrimination of coding and noncoding regions in comparative sequence data. RNA 17(4):578–594. doi:rna.2536111 [pii] 10.1261/rna.2536111

Watanabe Y, Yamamoto M (1994) S. pombe mei2+ encodes an RNA-binding protein essential for premeiotic DNA synthesis and meiosis I, which cooperates with a novel RNA species meiRNA. Cell 78(3):487–498. doi:0092–8674(94)90426-X [pii]

Wierzbicki AT, Haag JR, Pikaard CS (2008) Noncoding transcription by RNA polymerase Pol IVb/Pol V mediates transcriptional silencing of overlapping and adjacent genes. Cell 135 (4):635–648. doi:S0092-8674(08), 01192-6 [pii] 10.1016/j.cell.2008.09.035

Wilhelm BT, Marguerat S, Watt S, Schubert F, Wood V, Goodhead I, Penkett CJ, Rogers J, Bahler J (2008) Dynamic repertoire of a eukaryotic transcriptome surveyed at single-nucleotide resolution. Nature 453(7199):1239–1243. doi:nature07002 [pii] 10.1038/nature07002

Xu Y, Ganem D (2010) Making sense of antisense: seemingly noncoding RNAs antisense to the master regulator of Kaposi's sarcoma-associated herpesvirus lytic replication do not regulate that transcript but serve as mRNAs encoding small peptides. J Virol 84(11):5465–5475. doi: JVI.02705-09 [pii] 10.1128/JVI.02705-09

Zhang G, Guo G, Hu X, Zhang Y, Li Q, Li R, Zhuang R, Lu Z, He Z, Fang X, Chen L, Tian W, Tao Y, Kristiansen K, Zhang X, Li S, Yang H, Wang J (2010) Deep RNA sequencing at single base-pair resolution reveals high complexity of the rice transcriptome. Genome Res 20(5):646–654. doi:gr.100677.109 [pii] 10.1101/gr.100677.109

Zhao J, Sun BK, Erwin JA, Song JJ, Lee JT (2008) Polycomb proteins targeted by a short repeat RNA to the mouse X chromosome. Science 322(5902):750–756. doi:322/5902/750 [pii] 10.1126/science.1163045

Chapter 14
Experimental Analyses of RNA-Based Regulations in Bacteria

Marc Hallier, Svetlana Chabelskaya, and Brice Felden

Abstract The paradigm of small, usually noncoding, RNAs (sRNAs) as major performers in gene regulation in bacteria is soundly accepted. The number of sRNAs identified in bacteria has markedly increased. Most sRNAs exert regulatory functions by pairing with mRNAs and/or interacting with dedicated proteins. Apart from these *trans*-acting sRNAs, many mRNA leaders sense environmental signals or intracellular concentrations of metabolites and thus adopt structures that prevent or activate their expression. Recent studies on various bacteria indicate that antisense transcription is widespread. All these sRNAs are components of regulatory circuits involved in stress adaptation, metabolism, and virulence. Although still a new field, the study of sRNAs has already extended our understanding of numerous regulatory circuits in bacteria. In this chapter, we focus on the experimental methods that have allowed the inventory of the sRNAs and their mRNA targets expressed in bacteria, with emphasis on the technologies that unravel the mechanisms of the target gene regulations at molecular level.

Keywords Antisense RNAs • *cis*-encoded sRNAs • computational predictions • EMSA • eubacteria • genetics • high-throughput methods • microarray • mRNA stability • phylogeny • prokaryotes • proteomics • riboregulator • RNA-protein complexes • small regulatory RNAs • sRNA • structure probing • *trans*-encoded sRNAs • translational regulation

B. Felden (✉)
Inserm U835-Upres EA2311'Biochimie Pharmaceutique, Université de Rennes, Rennes, France
e-mail: bfelden@univ-rennes1.fr

B. Mallick (eds.), *Regulatory RNAs*, DOI 10.1007/978-3-642-22517-8_14,
© Springer-Verlag Berlin Heidelberg (outside the USA) 2012

14.1   Introduction

Small, usually noncoding (there are exceptions) regulatory RNA molecules that guide the expression of other genes are conjointly referred to as sRNAs. The number of predicted and experimentally confirmed prokaryotic sRNAs has grown significantly in recent years due, in large part, to the development and utilization of computational methods for predicting sRNA-encoding loci. Most bacterial sRNAs characterized up to now act as intermediate genetic elements of signal transduction pathways that are themselves initiated by an assortment of external stimuli. A wealth of experimental evidence suggests that sRNA-based regulation of gene expression is an archetype common to all domains of life.

Bacterial sRNAs can modulate DNA replication, gene transcription, mRNA stability, and translation (see Gottesman and Storz (2010) for a review). They fulfill these activities through several mechanisms. Detailed information is available in the Chap. 4 of this volume. The sRNAs can be categorized in the following classes based on their mechanisms of action. Many bacterial sRNAs act on target genes by base parings, having either extensive or more limited complementarities with their mRNA targets. The most prevailing mechanism implies antisense pairing between the regulatory sRNA and mRNA target(s) (Wagner et al. 2002). In many instances, a single sRNA mediates regulatory effects on different mRNA targets. Others modulate the activity of proteins (protein targets) by mimicking structures of other nucleic acids. In those cases, sRNAs offer single or multiple binding sites to dedicated proteins to competitively alleviate protein-mediated regulation of target mRNAs. One additional and wide class of regulatory sRNAs comprises riboswitches, which are part of the mRNA they regulate (Breaker 2010). A newly discovered class of sRNAs includes CRISPR (*c*lustered *r*egulatory *i*nterspaced *s*hort *p*alindromic *r*epeats) RNAs which are central to defense mechanisms against foreign DNAs in many bacteria that are involved in DNA maintenance or silencing (Horvath and Barrangou 2010).

Since their initial, fortuitous discovery in the late 1960s, the bacterial sRNA world has greatly expanded, especially during the last decade. The objectives of this chapter are to provide aids and tips to scientists interested in listing and studying the RNome of their favorite bacterium, together with their mRNA and/or protein target candidates. Once the RNome and predicted targets are identified, procedures are described to examine the molecular bases of the detected regulations that utilize RNAs. Also, it summarizes our current experimental knowledge on the various "state of the art" technologies that can be employed for analyzing RNA-based regulatory mechanisms in living cells. All of these strategies will evolve rapidly, as was the case for the spectacular developments of the high-throughput methods for sRNA findings.

14.2 Bacterial sRNA Identification

14.2.1 Computational Methods

14.2.1.1 Bioinformatic Screens

The lack of genome-wide annotations for sRNA-encoding genes compared to those for genes encoding proteins, tRNAs, and rRNAs is due to the difficulty of identifying sRNA-encoding loci by computational methods. Indeed, parameters used in conventional genome annotation are meaningless for the prediction of sRNA genes. Therefore, there is no universal method for their detection (see Backofen and Hess (2010) for a recent review). The use of computational methods to identify the bacterial sRNAs is a difficult task because they usually do not contain recurrent nucleotide motifs such as ribosome binding sites or open reading frames, they are generally small (~50–600 nt-long sRNAs), most are only conserved among closely related bacterial species and sometimes only expressed in "pathotype-specific" strains. Bacterial sRNA gene prediction can be achieved by various methods with variable efficacy, frequently relying on comparative genome analysis. Detailed information is available in the Chap. 11 of this volume.

Some methods rely on primary genomic sequence analysis. As specific examples, the GC content can be useful since sRNA genes usually contain a higher GC content than the remaining portions of the genome; the identification of transcription signals including promoters and rho-independent terminators (Pichon and Felden 2005) and/or the presence of transcription factor binding sites can be used to detect novel sRNA genes. Other methods rely on RNA secondary structure information, such as RNA detection based on thermodynamic stability and minimum free energy of RNA folding. Using multiple sequence alignments, conserved secondary structures can be detected based on compensatory base pair mutations. Some computational approaches use a combination of primary sequence and secondary structure information to increase their predictive capability (Pichon and Felden 2003). All the computational studies predicting the existence of novel sRNA genes have to be challenged experimentally. Once the computational inventory of the putative sRNAs for a bacterium has been performed, their distributions among the bacterial phylogeny can be assessed. Also, within the intergenic regions, primary sequences can be conserved in phylogenetically related species, allowing the identification of novel sRNA genes by comparative genomics (Wassarman et al. 2001).

14.2.1.2 Phylogenetic Analyses

Phylogenetic comparison of gene sequences is an elegant way to hypothesize the course of evolution. It was pioneered in the 1970s by analyzing 16S ribosomal RNA nucleotide sequences and structures (Woese et al. 1975), and later applied to various sRNAs (Felden et al. 2001). In the case of sRNAs, the approach is to analyze the pattern of nucleotide substitutions detected in a pairwise alignment of two homologous sRNA gene sequences. Pairwise alignments of sRNA genes are constrained by structural RNA evolution. A conserved coding region shows an arrangement of synonymous substitutions whereas a conserved structural sRNA reveals a pattern of compensatory mutations consistent with base-paired secondary structures. This strategy allows detection of structurally conserved sRNAs. Some sRNA genes, however, do not have well-conserved intramolecular secondary structures, and therefore, their identification is unattainable with such an approach.

SRNAs detected in *E. coli* can usually be identified, by sequence comparison, with closely related enterobacteria, but additional approaches are necessary to find the equivalent sRNAs in other bacterial species (Gottesman et al. 2006). For the 4.5S RNA, the RNase P RNA, and tmRNA that are housekeeping RNAs present in all prokaryotes (tmRNA genes are detected in all eubacteria but not in archaea) (Felden et al. 1999), primary sequence is sufficiently conserved between bacteria to detect them by sequence comparison. For the 6S housekeeping RNA gene, however, its primary sequence is highly divergent between bacteria and therefore difficult to detect at first glance (Pichon and Felden 2005), despite possessing a consensus secondary structure that mimics an open promoter for transcription initiation (Wassarman and Storz 2000). Interestingly, a recent procedure that uses suboptimal structures is capable of identifying homologous sRNAs in strongly divergent bacterial species, including the 6S RNA (Panek et al. 2011).

So far, all the identified bacterial sRNA genes are located in the core genome, sometimes also in mobile accessory elements, and some are detected in multiple copies (Felden et al. 2011). Besides the housekeeping RNAs such as 4.5S, RNase P, tmRNA, and the 6S RNA, conservation of most sRNAs is restricted to a few bacterial genera, and among them, a substantial fraction is only detected within a single bacterial species. Therefore, each bacterium possesses a "genus-specific" subset of sRNAs, and in that case, the phylogenetic approach is useless and bench work is essential.

14.2.2 Biochemical Approaches

14.2.2.1 Direct Methods

Initially, highly expressed bacterial RNAs other than rRNAs and tRNAs were discovered casually through metabolic [^{32}P]-labeling of total RNAs, followed by direct analysis by fractionations. The 4.5S (part of the secretion machinery) and 6S RNAs (RNA polymerase modulator) were first identified (Griffin 1971; Hindley 1967), followed by tmRNA, responsible for the release of stalled ribosomes and the RNase P RNA that is the catalytic module of a ribozyme responsible for 5′-end pre-tRNA maturation (Ray and Apirion 1979). Two-dimensional gel electrophoresis led to the discovery of the Spot 42 RNA (Ikemura and Dahlberg 1973) that mediates discoordinate expression of the *E. coli* galactose operon (Moller et al. 2002). Then, natural antisense RNAs (asRNA) were identified in plasmids controlling their copy numbers (Stougaard et al. 1981). Even bacterial sRNAs expressed at lower yields, compared to those mentioned above, could be detected directly from cell extraction and RNA sequence determination (e.g., SprD; Pichon and Felden 2005). The major limitation of the direct methods is that they can only detect the highly expressed RNAs, usually transcribed under specific environmental conditions that, for many RNAs, are still unknown. In bacteria, 10–20% of the genes are predicted to encode regulatory RNAs (Romby and Charpentier 2010). Therefore, to identify more bacterial RNAs, specific purification strategies were implemented.

14.2.2.2 Purification and Detection of sRNAs Forming Complexes with Proteins

Several sRNAs interrelate with cellular proteins. As a specific example, the *E. coli* CsrA protein forms a stable ribonucleoprotein complex with the CsrB RNA (Liu et al. 1997). Therefore, bacterial cellular extracts can be incubated with polyclonal antisera against an RNA-binding protein to enrich RNAs by co-immunoprecipitation (Co-IP) followed by hybridization to tiling arrays. An RNA-binding protein that was used for such experiments in various bacteria (Christiansen et al. 2006; Sonnleitner et al. 2008; Zhang et al. 2003) is the Sm-like Hfq protein. During the "Hfq-coIP," however, there are several incubation steps that can favor the detection of those having the higher affinity for the protein but can lead to the loss of the RNAs interacting with a lower affinity with Hfq. Hfq is required for both intracellular stability and target mRNA pairing for many sRNAs and was used as a lure to identify novel Hfq-associated sRNAs in bacteria expressing high levels of Hfq. An improvement came from the use of an antibody specific to DNA/RNA hybrids to detect on the microarrays of the RNAs previously coIP with Hfq (Hu et al. 2006). The problem of the species specificity of the Hfq antibodies can be overcome by adding a tag (triple FLAG) at the chromosomal *hfq* gene, performing coIP and converting the extracted RNAs into cDNAs that are pyrosequenced (Sittka et al. 2008). The sRNAs can be determined by RNA sequencing, RNomics (gel extraction of 50–500 nt size range RNAs, reverse transcription, cloning and sequencing; Huttenhofer et al. 2004), or by microarray analysis.

14.2.2.3 High-Throughput Methods

In recent years, genome-wide global approaches for sRNA identifications have been implemented, based on microarrays (Perez et al. 2009), shotgun cloning and Sanger sequencing of small-sized cDNAs (RNomics; Vogel et al. 2003), 454 sequencing (the first of the so-called next generation sequencing methods to become commercially available in 2004; for a recent application on finding bacterial sRNAs, see Bohn et al. (2010)), or Illumina high-throughput sequencing (HTS; Beaume et al. 2010). Microarrays are designed either with oligonucleotides or

"PCR-derived" DNA probes for a selected set of RNAs (low-density; Pichon and Felden 2005) or tiling arrays containing up to thousands of oligonucleotides covering both the sense and antisense DNA strands encompassing the intergenic regions (high-density; Rasmussen et al. 2009). Tiling arrays have allowed the identification of sRNAs in many bacteria (Toledo-Arana et al. 2009; for the RNAs expressed by *Listeria*), and their high probe density allows reasonable estimations of 5′ and 3′ RNA extremities. A major disadvantage of the tiling arrays, however, is that they are provided by commercials and therefore are very expensive. RNomics consist in conventional Sanger sequencing of cloned cDNAs produced from size-fractionated (gel separation and extraction) or total RNAs (Huttenhofer et al. 2004).

High-throughput sequencing (HTS) technologies have now replaced both the cloning and Sanger procedures, using either 454 GS FLX pyrosequencing, SOLiD (massive parallel sequencing based on oligonucleotides ligations), Solexa GA or Heliscope sequencing, or the Pacific Biosciences sequencing method (MacLean et al. 2009). For all high-throughput sequencers, sRNA cDNA library preparation is based on few basic steps including RNA extraction and isolation, 3′ and 5′ linker ligation, reverse transcription, and PCR amplification (Fig. 14.1). The RNAs are reverse transcribed into millions of cDNAs that are sequenced, such analysis being usually named "RNA-Seq." It allows the detection of RNA transcripts that include sRNAs at the genome-wide level (Perkins et al. 2009). Using HTS, sRNAs expressed at a given time and under selected experimental conditions are identified and quantified. HTS reveals an exhaustive inventory of the RNAs expressed in a sample, the detection sensitivity being only limited by the sequencing depth. To improve the finding of novel sRNAs by high-throughput sequencing, alternative sRNA library preparation methods using ligation, extension, and circularization have been recently developed. They are faster and simpler than the widely used procedures, and the constructed libraries are compatible with high-level multiplex analysis (Kwon 2011).

Most current RNA-seq (cDNA sequencing at massive scale) methods rely on cDNA synthesis and an array of subsequent manipulation steps, which places limitations on the current strategies. Another limitation imposed by cDNA synthesis is template switching. During reverse transcription, the nascent cDNA can sometimes dissociate from the template RNA and reanneal to a different stretch of RNA with a sequence similar to the initial template, generating artefactual chimeric cDNAs. In addition to their requirement for cDNA synthesis, RNA-seq approaches present other difficulties: the RNA-seq signal across transcripts tends to show nonuniformity of coverage, which may be a result of biases introduced during priming with random hexamers, cDNA synthesis, ligation, amplification, or sequencing. RNA-seq can result in transcript-length bias because of the multiple fragmentations and the RNA/cDNA size-selection steps. RNA-seq often involves a poly(A) mRNA enrichment step that could increase the RNA degradation products. The existing approaches are not sufficient to detect certain transcripts and/or cover their entire length. Therefore, the length normalization step is a source of errors for quantitative applications.

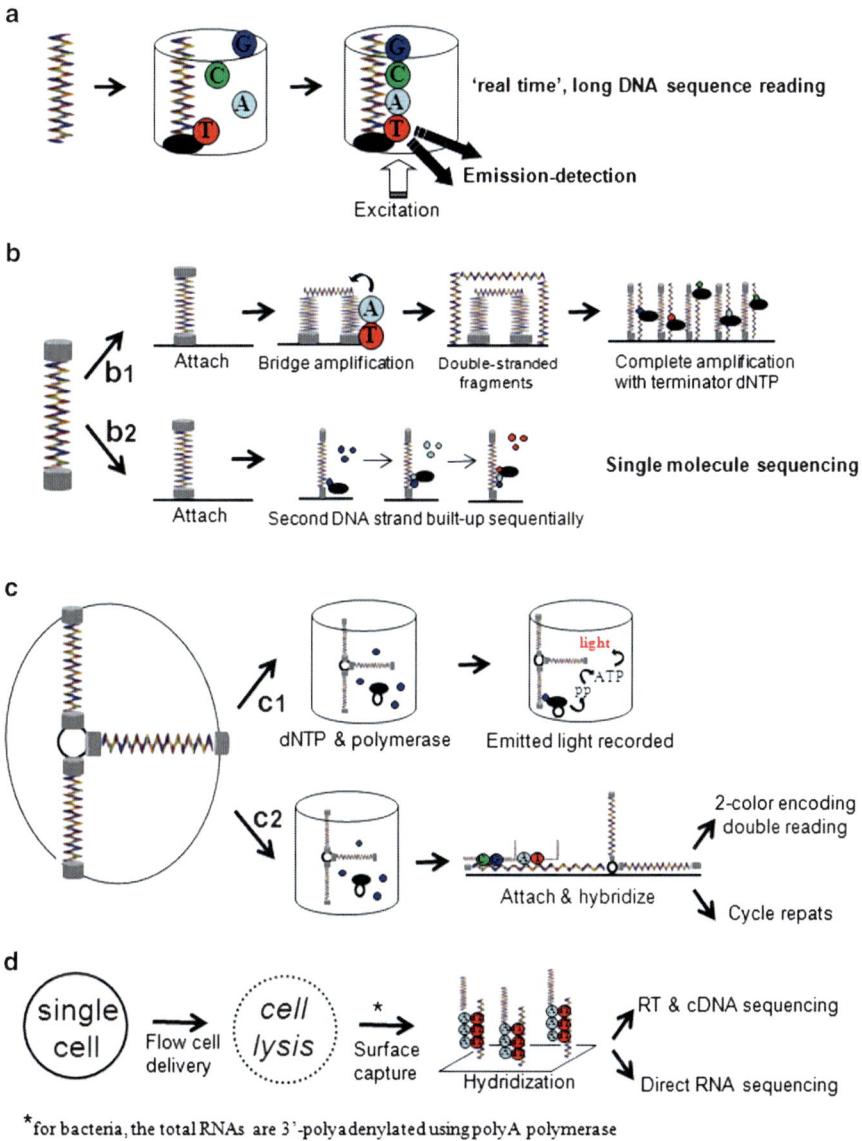

Fig. 14.1 *High-throughput sequencing engineering.* The various technologies can be carried out on (**a**) naked (*colored strings*), (**b**) adaptor ligated (*gray cylinders*) single-stranded DNA fragments that can, in some cases, (**c**) be immobilized on beads (*empty black circles*) and amplified in water-oil emulsion (*large black oval*). (**d**) Emerging techniques are available for single-cell applications. (**a**) The sequencing method developed by Pacific Biosciences in wells encompassing a trapped DNA polymerase (*black and filled oval*) and dNTPs (*colored circles*; A: dATP, C: dCTP, G: dGTP, and T: dTTP). Fluorophores are cleaved as the complementary DNA strand expands. Uninterrupted fluorescence detection combined with an elevated dNTP concentration authorizes rapid and extended reading. (b1) Solexa GA sequencing. DNAs with ligated adaptors are attached and immobilized on substrate followed by "solid-phase" bridge amplification with unlabeled

These limitations can be alleviated by emerging RNA analysis technologies that modify the method of RNA characterization. Massive direct RNA sequencing (DRS) approaches were recently developed by the Helicos approach (Ozsolak et al. 2009) relying on hybridization of femtomoles of 3'-polyadenylated RNA templates to single channels of poly(dT)-coated sequencing surfaces, followed by sequencing by synthesis. DRS requires only femtomole or attomole levels of input RNA and involves relatively simple sample preparation. DRS sample preparation involving polyadenylation can be applied to any RNA species, allowing both short and long RNAs to be observed in a single run. Methods for high-quality RNA isolation from small quantities of cells are available. The main limitation preventing reliable, global profiling of minute RNA quantities has been the incompatibility of high-throughput RNA profiling approaches with low-quantity RNA samples. A number of both hybridization- and sequencing-based technologies are now allowing reliable transcriptome profiles to be obtained from minute cell quantities. Amplification-free RNA-seq approaches have recently been developed that minimize the quantity of input RNA required. One approach involves the sequencing of first-strand cDNA products from as little as ~500 picograms of RNA, with priming carried out in solution with oligo(dT) or random hexamers. Another strategy uses poly(dT) primers on sequencing surfaces to select poly (A) + mRNA from cellular lysates, followed by on-surface first-strand cDNA synthesis and sequencing (Fig. 14.1). Microfluidic capabilities could be combined with DRS for single-cell applications. Hybridization-based methodologies are providing promise for working with very low quantities of RNA. The NanoString

---

Fig. 14.1 (continued) nucleotides and enzymes. The double-stranded DNA fragments are denatured and PCR amplified in sufficient amounts so that the accumulated fluorophore is perceived. The use of terminator dNTPs and DNA polymerase results in synthesis of the complementary DNA strands. (b2) Heliscope sequencing captures DNAs with ligated adaptors to a substrate. Each fluorescent dNTP is successively used to build a complementary DNA strand, the used fluorophores being removed at the beginning of each round, diminishing background signal. (c1) The 454 pyrosequencing method in which the ligated DNA fragments are immobilized to the outside of microscopic beads prior to PCR amplification in a water-oil emulsion. The beads are isolated in wells, a cDNA strand is constructed, and each nucleotide incorporation releases pyrophosphate (pp) that allows ATP production used for a chemiluminescent reaction, the light produced being recorded and analyzed. (c2) SOLiD sample making up is as for the 454 pyrosequencing. After amplification, the beads are attached to a glass slide and subjected to sequential hybridization starting with an oligonucleotide complementary to the adaptor sequence linked to short random oligonucleotides bearing known 3' dinucleotides and a corresponding fluorophore. After 5 cycles, cDNAs are melted away from their templates and the process is reiterated. During a second cycle, synthesis reinitiates at the position immediately upstream from where synthesis began initially, generating reads of 30–50 nucleotides. Repeats of these cycles ensure that nucleotides in the gap between the known dinucleotides are read. (**d**) Single-molecule DNA and RNA sequencing technologies can be adapted for single-cell applications. Cells are delivered to flow cells using fluidic systems, followed by cell lysis and mRNA capture by hybridization on poly(dT)-coated sequencing surfaces. For bacteria, the sRNAs have to be gel purified (please refer to the RNomic subheading) based on their sizes followed by 3'-polyadenylation using poly A polymerase. Sequence analysis can be performed by direct RNA sequencing or by "on-surface" cDNA synthesis followed by single-molecule DNA sequencing

nCounter System provides RNA quantification without cDNA synthesis and relies on the generation of target-specific probes (Fig. 14.1). This approach requires ~100 ng of RNA or 2,000–5,000 cells, but optimization of the probe hybridization and surface immobilization steps may reduce input RNA quantity. Recent advances in RNA-seq have provided a powerful toolbox for transcriptome characterization and quantification. These technologies will allow the building of an exhaustive catalog of transcripts expressed from genomes ranging from bacteria to mammalian cells.

14.2.3 Conclusions

Several *in-silico* and data-based techniques were progressively implemented to identify and verify the expression of a number of sRNAs in numerous bacteria including human and animal pathogens. An indisputable prerequisite to the upcoming sRNA functional and structural studies is to determine their 5′ and 3′ boundaries by primer extension (5′-ends) or by RACE (*r*apid *a*mplification of *c*DNA *e*nds) analyses, as well as to monitor their expression profiles during growth. A few elegant genetic studies have identified novel sRNAs starting from their mRNA targets (Mandin and Gottesman 2009; Yamamoto et al. 2011). They have screened for posttranscriptional regulators, including sRNAs, regulating genes of interest by constructing translational fusions of the 5′ ends of genes of interest to reporter genes in the chromosome, and they have scanned a plasmid library of the bacterial genome to isolate clones that can affect the activity of the fusion. In most cases, however, the mRNA and/or protein targets of the sRNAs are unknown and their identifications remain a challenging task. This is because the mechanisms underlying the pairing of sRNAs to their mRNA targets are not yet well understood, rendering the prediction of mRNA targets difficult.

14.3 Target Identification

14.3.1 Introduction

The discovery of new sRNAs raises the question about their functional roles in bacterial physiology. The analysis of the biological effects due to the inactivation or overexpression of sRNAs in bacteria can provide some information about the affected metabolism they are involved in. Additional data can be obtained by the characterization of the environmental growth conditions affecting their expression levels. The true challenge is to identify their primary target(s) which corresponds to those interacting directly with the sRNA. Most sRNAs of known functions regulate gene expression by binding to their mRNAs or to protein targets with specificity.

For a few sRNAs, their targets were discovered casually. In the case of 6S sRNA, the identification of its protein target results from the characterization of macromolecules which copurify with the sRNA. The 6S RNA from *E. coli* cell extracts migrates as an 11S particle in a sucrose density gradient separation (Lee et al. 1978). Analysis of protein composition of the 11S fraction revealed that the 6S RNA interacts with the sigma 70 subunit of RNA polymerase (Wassarman and Storz 2000). Other sRNAs were discovered by copurifications with RNA-binding proteins (Heeb et al. 2002; Liu et al. 1997). For example, the carbon storage regulator CsrA is an RNA-binding factor involved in the mRNA decay of *glgCAP* transcript which regulates the glycogen metabolism of *E. coli*. The CsrB sRNA was found associated to a histidine-tagged recombinant CsrA protein during the purification of the protein and then identified as a regulator of CrsA (Liu et al. 1997). For other sRNAs such as *cis*-encoded sRNAs, their direct targets are easily identified since sRNAs are transcribed from the opposite strand of their target genes. It cannot be excluded, however, that they possess additional targets encoded in *trans*. The targets for *trans*-encoded sRNAs are trickier to detect. We expose here diverse approaches that have been successfully used to identify targets of several *trans*-encoded sRNAs. These include "large-scale" screens using biocomputational prediction algorithms, microarrays, proteomic analyses, genomic methods, and capture by affinity chromatography.

14.3.2 "Large-Scale" Screens

14.3.2.1 Computational Methods

Identification of target mRNAs of sRNAs by computational methods requires the availability of bacterial genomic sequences. Since sRNAs are able to interact with all domains of a target mRNA, including the sequences within the open reading frame and those encompassing the translation start site as well as their 3′-UTRs (*un*translated *r*egions), an accurate annotation of the genomes is essential for the identification of their mRNA targets. The first step is the search for complementary regions between sRNA and putative mRNA targets. In rare cases, when base-pairing interactions between sRNA and its mRNA target are long, contiguous, and perfect, targets could be detected by simple BlastN or Fasta3 searches. This approach was successfully applied to the identification of the *ompC* and *tisAB* mRNAs as targets of MicC and IstR-1 sRNA, respectively (Chen et al. 2004; Vogel et al. 2004).

In most cases, the pairing sequences are short and lead to the formation of partial and imperfect duplexes. The interaction between the two RNAs can also imply multiple binding sites. Moreover, "sRNA-mRNA" complexes can involve internal loops which complicate the prediction of sRNA targets (Huntzinger et al. 2005; Gottesman 2004; Argaman and Altuvia 2000). Computational methods were developed to search for complementarities between sRNAs and mRNAs. In most cases, the algorithms scan the genomic regions that have the highest probability of pairing with the sRNA. These regions include genomic sequences encompassing the translational start site of mRNA and the sequences upstream and downstream of the annotated ORFs (*open reading frames*). Algorithms can predict the secondary structures of two interacting RNA molecules by means of free energy minimization (Alkan et al. 2006), or they calculate hybridization and/or thermodynamic score(s) for "sRNA-mRNA" duplex formations (Target RNA, RNA-up, RNA-hybrid, intaRNA are examples of such available software) (Tjaden et al. 2006; Tjaden 2008; Muckstein et al. 2006; Rehmsmeier et al. 2004; Busch et al. 2008). The programs constantly evolve to take into account the RNA secondary structures, the GU wobble base pairs, and the intrinsic specificity of some bacterial genomes such as those possessing low GC contents (e.g., *Staphylococcus* or *Listeria*) (Mandin et al. 2007). The "sRNA-mRNA target" binding regions are substantially conserved between the species and are usually more accessible than random nucleotide sequences. Algorithms based on the evolutionary conservations and surface accessibilities of sRNAs allow the prediction of the "sRNA-mRNA target" binding domains (Peer and Margalit 2011). The progressive inclusion of these parameters in the algorithms designed for target prediction is expected to reduce the false positive hits. All computational approaches, however, do not include the protein chaperones, such as Hfq, that are required for a number of base-pairing interactions to happen in several bacteria. The level of false positive predictions is high but provides a fairly good starting point for the use of experimental data required to confirm or rebut the computational predictions.

14.3.2.2 Genetic Methods

A popular genetic approach is based on the design, selection, and analysis of bacterial mutants that are genetically engineered in the particular function under study. A genetic approach using random insertions of a bacteriophage carrying a reporter gene into the bacterial genome was performed to identify mRNAs targets of an sRNA. The λplacMu bacteriophage allows the random insertion of a *lacZ* reporter gene that is deleted for its promoter and for the translation signals. The integration of this phage into a gene in the correct orientation and reading frame leads to a gene fusion in which the expression of *lacZ* depends on the promoter and the translation start site of the target gene. The expression level of the fusion proteins containing the N-terminal sequences encoded by the target gene and the β-galactosidase protein is analyzed by a colorimetric assay with 5-bromo-4-chloro-3-indolyl-β-D-galactoside (X-Gal). The random insertion of λplacMu in a strain carrying an inducible sRNA leads to the detection of colonies containing fusion genes which are either upregulated (blue colonies) or downregulated (white colonies), according to the sRNA expression levels. Target genes are identified by cloning and sequencing. Since the reporter gene contains the translational start site of the target mRNA, an antisense effect induced by an sRNA onto translational initiation can be detected. However, this method excludes the identification of target genes which are regulated by an interaction of the sRNA with its 3′-UTR and within the ORF and do not discriminate between a direct or indirect regulation of the expression of the target gene. sRNAs can modify the expression of a set of genes involved in signal transductions or genes encoding general transcription factors which, in turn, modulate the expression of several genes. By this approach, the Storz team showed that the expression of five genes was regulated by the OxyS sRNA (Altuvia et al. 1997). The translation of only one mRNA target, the *fhlA* mRNA that encodes a transcriptional regulator of the formate hydrogenase, is directly repressed by OxyS through base pairing that blocks the Shine-Dalgarno (SD) sequence (Altuvia et al. 1998). Bacteriophage Mu can be used to randomly insert LacZ reporter genes in other gram-negative bacteria, as in *Pseudomonas syringae* and *Legionella pneumophilia* (Choi and Kim 2009). A similar genetic approach could also be performed in gram-positive bacteria by using modified Tn917 transposons (Choi and Kim 2009).

14.3.2.3 Microarray Analysis

Global searches of the mRNA targets of sRNAs can be performed by microarrays which allow the comparison of the mRNA profiles of strains expressing various levels of an sRNA using differential fluorescence labeling of two cDNA pools. Although some sRNAs regulate the expression of genes by modulating translation, they can also stabilize or destabilize the target mRNA secondary structure and/or increase/decrease protection against ribonucleases which leads to substantial variation of target mRNA stability. Microarray analysis depends on the expression levels of target mRNAs upon pairing. It allows the identification of target genes regulated at the transcriptional and posttranscriptional levels, but also a part of those regulated at translational levels. Generally, microarray analysis compares a deleted sRNA strain, a wild-type strain, and/or an sRNA-overexpressing strain. Total RNAs extracted from bacteria are reverse transcribed into cDNAs which are labeled by fluorescent dyes and hybridized to whole-genome microarrays (cDNA or oligo-arrays). Analysis requires the availability of whole-genome microarray chips for the studied bacterium. For many bacteria, microarrays can be purchased from commercial sources or are custom-made with the help of dedicated firms.

In *E. coli*, 80% of the mRNA half-lives are between 3 and 8 min (Bernstein et al. 2002). So, RNA extraction procedures that require additional steps such as an enzymatic digestion of the cell wall lead to reduced quantities and qualities of the mRNAs. The success of the microarray analyses depends upon the detection of target mRNA expression and requires the characterization of the expression profile of the sRNA. The sRNA expression profiles usually vary according to the growth phases, and many sRNAs require specific experimental conditions to be induced. For example, the expression of *ryhB* is regulated by iron through the fur "iron-dependent" transcription factor. For the microarray analysis, wild-type and "*ryhB*-deleted" strains were grown in "iron-depleted" minimal medium to induce RyhB in order to maximize the differential expression of the genes regulated by RhyB

(Davis et al. 2005). Microarray analysis allows the identification of a multitude of up- or downregulated genes according to the presence or absence of a given sRNA. The largest fraction of these variations in genetic expression does not result from direct regulation by an sRNA but reports the overall effect of sRNA inactivation or overproduction. The deregulation of the expression of primary targets leads to numerous variations of secondary targets. If the primary target is a general regulator, changing its expression levels can deregulate many secondary targets. The best examples come from the gram-positive pathogen *Staphylococcus aureus*. RNAIII is a multifunctional RNA that encodes δ-hemolysin at its 5′-domain while also acting as a regulatory RNA in controlling the expression of Rot (*repressor of toxins*), a transcriptional factor, at the posttranscriptional level (Boisset et al. 2007; Geisinger et al. 2006; Novick et al. 1993). Thus, a major drawback of this technique is that it leads to the detection of indirect rather than direct effects on mRNA expression. One possibility to favor the identification of primary targets is to perform a "short-term" expression of the sRNA. The analysis of the mRNA profiles after this short (5–15 min) sRNA induction enhances the identification of primary transcripts whose expression levels are modulated by the sRNA. Plasmids containing a controlled promoter, as an "IPTG-inducible" *lac* promoter, an "arabinose-inducible" *araBAD* promoter, or a *tet*-inducible promoter (anhydrotetracycline, aTc), have been successfully used to identify mRNA targets through a transient expression of the sRNA under study (Bohn et al. 2010; Boysen et al. 2010; Guillier and Gottesman 2006; Massé et al. 2005).

14.3.2.4 Proteomic Analysis

Quantitative proteomic approaches are used to compare the protein expression profiles in the presence or absence of an sRNA expressed in a bacterial strain. Proteins are resolved on 1D or 2D gels, and the bands or spots corresponding to differentially expressed proteins are excised from the gels and identified by mass spectrometry. Unlike the transcriptomic analyses by microarrays, proteomic studies allow the detection of the translational regulations triggered by the sRNAs. While the results obtained by the microarrays are strongly influenced by the expression level of the mRNAs and/or by their stability, proteomic analysis is mainly hampered by sample preparation difficulties. Whereas some bacteria are easily lysed by mechanical disruption in specific buffers, others require enzymatic digestions to get rid of the cell wall. Sample preparation is also critical. For 2D gel- and mass spectrometry-based proteomic analysis, proteins are separated based on their net charges in the first dimension and also on their molecular masses in the second dimension. The proteins from the samples must be brought into a state allowing their isoelectric focusing separations on immobilized pH gradients. Nucleic acids cause streaking in the first dimension and therefore have to be removed from the protein samples.

Other problems come from the difficulty in identifying proteins that are fused into a single spot on 2D gels ("spot crowding") or that are intractable by mass spectrometry analysis. Moreover, the analysis of poorly soluble proteins, such as membrane proteins and multiprotein complexes, and proteins of low abundance or aberrant pIs is problematic. In summary, only 10% of all proteins can be visualized by proteomic studies. On the other hand, proteomic analysis performed on subproteomic fractions of a bacterium, such as cytosolic, membrane, cell surface-associated, and extracellular proteins, can provide attractive information on the biological signaling pathway regulated by the sRNA. Due to the relatively low number of individual protein spots that can be resolved, however, and because of the incompleteness of protein identification by mass spectrometry, proteomics provides only partial information about the pattern of protein expressions.

In some cases, a single 1D separation gel of the "whole-cell" proteins allows the identification of a target gene by mass spectrometry. As specific examples, the OmpD outer membrane protein was identified as a target of the InvR sRNA encoded by a *Salmonella* pathogenicity island (Pfeiffer et al. 2007). Prefractionation of protein samples through subcellular isolation is useful to perform 1D gel electrophoresis. The reduction in the number of proteins in the sample increases the probability of detecting only one band on the gel which is linked to the variation of expression of a single protein. For example, the analysis of the extracellular protein expression profile from *Staphylococcus aureus* strains expressing various amounts of the sRNA SprD allowed the identification of a target, the Sbi immune-evasion molecule (Chabelskaya et al. 2010). The expression levels of a given protein are either reduced or increased when the sRNA is lacking or, conversely, is overexpressed. The regulation can be either direct on the mRNA encoding the protein or indirect via additional regulators. Additional strategies such as microarray analyses, after a short-term sRNA expression, should be used to identify its primary targets (Boysen et al. 2010).

14.3.3 Capture by Affinity Chromatography

The main problems encountered with "large-scale" screens using bioinformatics, microarray, and proteomic approaches are either in detecting many false positive targets or the difficulty in identifying the primary targets of the sRNAs. In order to favor the identification of the primary targets, experiments based on affinity chromatography were developed. The sRNAs are immobilized covalently, or not, onto a column and can be used as baits for capturing target mRNAs or target proteins by affinity purifications from bacterial extracts. The native sequence of the sRNA can be covalently linked to Sepharose using cyanogen bromide activation. One limitation of this strategy is that the reaction between the sRNA and the matrix occurs at random location, which can limit the number of accessible protein/RNA binding sites. The direct coupling of the sRNA to adipic acid dihydrazide agarose beads could also be performed, but this immobilization procedure requires sRNA oxidation with sodium periodate, limiting the attachment of the sRNA 3′-end onto the matrix (Hovhannisyan and Carstens 2009).

A covalent immobilization approach was successfully used to identify the protein target of the Rcd sRNA. Rcd is transcribed from *cer* which stabilizes the Col E1 multicopy plasmid in *E. coli*. The in vitro transcribed, unmodified Rcd RNA was covalently linked to "CNBR-activated" sepharose beads and incubated with the total proteins from *E. coli* cell lysates. Proteins retained on the column were eluted by a NaCl step-gradient and resolved by 1D gel electrophoresis. Maldi spectroscopy fingerprint analyses of the eluted proteins have identified the tryptophanase enzyme (TnaA) as a target of Rcd (Chant and Summers 2007). The binding of Rcd increases the affinity of TnaA for tryptophan, stimulating indole production which induces a cell division delay.

Noncovalent immobilizations of sRNAs require their specific in vitro chemical modifications. After in vitro biotinylation, the sRNAs are immobilized on "streptavidin-bearing" matrices in accordance with the high-affinity interaction between streptavidin and biotin. This noncovalent binding approach was used to identify the mRNA targets of RseX, an sRNA expressed in enterobacteria. Biotinylated RseX immobilized on streptavidin beads was incubated with total RNAs extracted from *E. coli* strains. The eluted RNAs were reverse transcribed into complementary DNAs before hybridization on a pan-genome *E. coli* DNA chip. Affinity captures coupled to microarray analyses led to the identification of two direct targets of RseX, the mRNAs coding for the outer membrane proteins OmpA and OmpC (Douchin et al. 2006).

For some sRNAs, auxiliary factors must be incorporated during the immobilization process to obtain a functional chromatography column. The presence of the RNA chaperone Hfq protein can be required for the base pairing between sRNAs and their mRNA targets. This is the case for the RydC sRNA from enterobacteria that adopts a conformational change when it binds to Hfq. Preformed "RydC-Hfq" complexes using a His-tagged recombinant Hfq and an in vitro RydC transcript were immobilized noncovalently on nickel-sepharose beads by the chaperone protein. This affinity chromatography column devoid of free, nonfunctional, RydC molecules was used to capture a target mRNA, the polycistronic *yejABEF* mRNA encoding an ABC transporter which confers resistance to antimicrobial peptides and contributes to *Salmonella* virulence (Antal et al. 2005).

The nonspecific binding of proteins and/or RNAs to the sRNAs is a major drawback of the in vitro affinity chromatography methods. Indeed, affinity chromatography using Rcd has captured the EF-Tu protein, an abundant tRNA binding macromolecule involved in translation, which is highly expressed in bacteria (representing 5% of the total proteins in the cells). This nonspecific purification is partially due to the large amount of the bait immobilized on a column, in comparison with the available protein targets. A similar experimental pitfall was observed with the unspecific binding of 16S or 23S ribosomal RNA fragments during target mRNA purification.

An interesting approach to maximize binding specificity uses RNA aptamers to isolate in vivo-assembled "RNA-protein" or "RNA–RNA" complexes (Fig. 14.2). RNA aptamers are highly structured molecules that exhibit high affinity and specificity for their ligands. Aptamer-tagged sRNAs can be expressed in vivo and linked noncovalently to Sepharose that contains the specific ligand of the aptamer (Fig. 14.2). The Vogel team showed that engineering sRNAs with various aptamers did not affect their stability when overexpressed in *Salmonella*. Aptamer-tagged GcvB, InvR, and RybB sRNAs are active and repress translation of their respective mRNA targets by "Hfq-dependent" base-pairing mechanisms (Said et al. 2009). The in vivo overexpression of tagged sRNAs does not only allow their purification according to their respective affinities with their dedicated aptamers but also the copurification of their associated proteins, as Hfq. Moreover, chromosomal integration of the tag (the aptamer) into the sRNA gene can be performed to increase the specificity of the protein interactions by restoring endogenous levels of sRNA expression (Fig. 14.2). This promising approach could be used to identify protein ligands of sRNAs but also of their mRNA targets (Said et al. 2009).

14.3.4   Conclusion

Diverse experimental and biocomputational approaches have been developed to identify sRNA target(s). Some are aimed at identifying sRNA targets directly by bioinformatic methods that predict mRNA targets on the basis of direct interactions with the sRNAs. Also, protein and mRNA targets can be identified directly by copurification with a given sRNA (Chant and Summers 2007; Windbichler et al. 2008). In the case of the proteomic and transcriptomic approaches, they provide a list of putative targets for an sRNA. The regulation can be either direct through a physical interaction between the sRNA and the mRNA, or indirect via numerous intermediates. The "pulse expression" of a given sRNA under the control of an inducible promoter should prevent the pleiotropic effects that can result from the constitutive expression of the sRNA. It should mainly allow the detection of the direct targets and avoid putative downstream effects. Even in that case, however, the expression of several genes can be modulated. The discrimination between the primary and secondary targets is even more puzzling because one sRNA can regulate multiple primary targets. Indeed, in addition to Rot, RNAIII has many other direct targets encoding bacterial virulence factors such as the alpha-hemolysin (hla), the protein A (SpA), the staphylocoagulase (Coa) (Morfeldt et al. 1995; Huntzinger et al. 2005; Chevalier et al. 2010; Boisset et al. 2007). Thus, once the targets of an sRNA are identified, the next challenge is to understand the mechanisms of each of the regulations at molecular level as well as their integration in the overall regulatory networks of gene expression regulations.

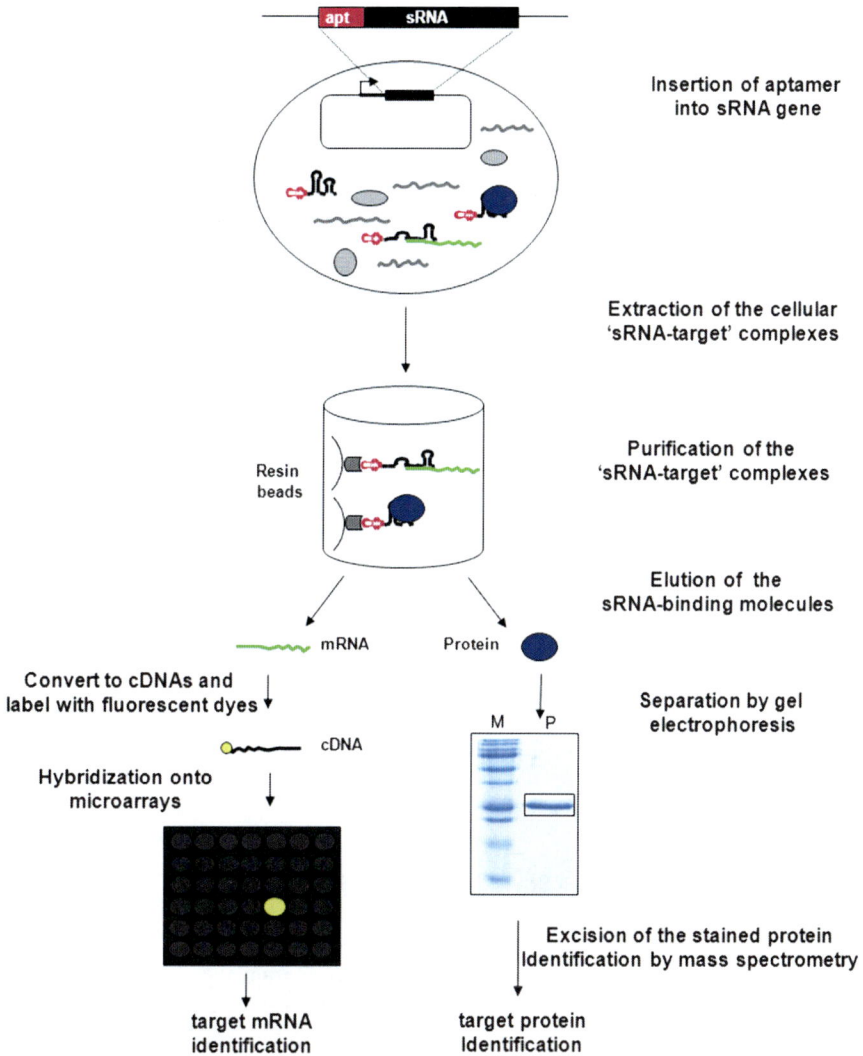

Fig. 14.2 *sRNA target identification using "aptamer-tagged" sRNAs expressed in vivo.* An aptamer module is genetically engineered and inserted within the *sRNA* gene in the bacterial chromosome. The in vivo expressed "tagged-sRNA" contains an additional sequence which allows the "high-affinity" purification of the modified sRNA. In vivo preformed "sRNA-mRNA" and/or "sRNA-protein" complexes are purified on sepharose beads containing the aptamer ligand (strep-tomycin or tobramycin). Specifically associated mRNAs and proteins are recovered from the "aptamer-tagged" sRNA resin. sRNA binding proteins are resolved on PAGE, and the proteins are excised from the gel before mass spectrometry identification (p: protein sample, M: molecular weight markers). For the identification of the mRNAs that interact specifically with the sRNAs, the mRNAs are converted into cDNAs and labeled with fluorescent dyes. The probe is hybridized to a "whole-genome" microarray to identify the target genes

14.4 Elucidation of sRNA-Based Regulatory Mechanisms

The different types of sRNAs and their mechanisms of action were previously discussed in details in several excellent reviews (Deveau et al. 2010; Gottesman and Storz 2010; Repoila and Darfeuille 2009; Waters and Storz 2009) and in the chapter "*Small regulatory RNAs (sRNAs) – key players in prokaryotic metabolism, stress response and virulence*" of this book. Here, we propose an overview of a general experimental strategy to elucidate the mechanisms of actions of sRNAs with special attention to mRNA and protein targets.

The identification of putative targets is only the first step in the elucidation of the role(s) of an sRNA. To determine the function(s) of sRNAs, further experimental manipulations are indispensable consisting in (1) the validation of the target and (2) the subsequent explanation of the regulatory mechanisms. This can be achieved by several approaches. Strategies that will be discussed hereafter will concern only primary mRNAs and protein targets.

14.4.1 Target Validation

Generally, RNA-mediated effects can be monitored in vivo in strains in which the RNA regulator gene of interest has been deleted or overexpressed. "sRNA-based" repression or activation of mRNA targets often results in changes in protein levels that may be analyzed by Western blot. If no specific antibodies are available, chromosomal or plasmid constructions in which regulatory regions of putative mRNA targets have been fused with a reporter gene may be used. Several systems have been developed in both gram-positive and gram-negative bacteria. They use different reporter genes such as *lacZ*, encoding β-galactosidase (Boisset et al. 2007; Huntzinger et al. 2005) or *gfp* (*green fluorescent protein*) (Pfeiffer et al. 2007; Urban and Vogel 2007). Such constructions often contain the translational regulatory regions of genes of interest fused with a reporter gene and placed under the control of a promoter independent of regulation by sRNAs. This allows the investigation of direct regulation of the mRNA targets at translational level without possible outcomes at transcriptional level.

In addition to the translational regulation, the sRNA/mRNA pairing can induce changes in mRNA stability. Changes in mRNA target levels often accompany the sRNA-mediated regulation of translation (Boisset et al. 2007; Morita et al. 2005; reviewed in Kaberdin and Blasi 2006). This can be due to recruitment of RNases such as RNase E or RNase III upon sRNA-mRNA duplex formation. It can also be due to increased accessibility of less translated and, therefore, less protected mRNAs for RNases. However, sRNAs can promote changes in mRNA target levels without direct translational repression or activation (Desnoyers et al. 2009; Opdyke et al. 2011; Pfeiffer et al. 2007). So, it is important to verify if sRNA binding leads to changes at the mRNA levels (degradation and/or processing). Several approaches such as Northern blot, RT-qPCR, and RNase protection assays can be used to monitor the changes in target RNA levels and prove or disprove the "RNA-based" regulation.

These in vivo analyses are usually the first step in studying the regulatory mechanism because they confirm the target regulation in the cellular context. Furthermore, they could also give the first indications about the mechanism of regulation.

14.4.2   Elucidation of the Regulatory Mechanisms at Molecular Level

Unraveling the precise regulatory mechanism(s) is necessary to study the function (s) of an sRNA. This can be achieved by several approaches, both in vitro and in vivo.

14.4.2.1   In Vitro Approaches

In vitro experiments are required to provide insight into the mechanism(s) of action of "sRNA-mediated" regulation. They can help in validating the direct interaction (s) between an sRNA and its protein or mRNA targets. For both the protein and mRNA targets, similar strategies can be used. These biochemical experiments include *electrophoretic mobility shift assay* (EMSA) that allows visualization of the complex between sRNA with its target(s) and confirmation of direct binding. Complex formation between an sRNA and its mRNA or protein target(s) usually induces structural changes of the sRNA and the target mRNA or protein. This can be analyzed by enzymatic or chemical probes in solution which allows the mapping of the interaction sites and structural changes that accompany complex formation. Additionally, "RNA-mediated" regulation of translation initiation can be monitored by ribosome toeprint assays. These allow the examination of the formation of translation initiation complexes on an mRNA in vitro in the absence or presence of an sRNA.

Analysis of Complex Formation by EMSA

This analysis can be applied to study the binding of an sRNA with its mRNA or protein targets. EMSA experiments are performed with in vitro synthesized RNAs and purified native proteins. As for most in vitro applications, RNAs are transcribed in vitro with T7 or SP6 RNA polymerases from a PCR product or a linearized plasmid template carrying the T7 or SP6 promoter fused with the gene of interest. To avoid the presence of the short transcripts, RNAs are purified by electrophoresis from denaturing "polyacrylamide-urea" gels. Labeling of RNAs can be achieved by incorporating [α-32P] UTP during transcription or by terminal labeling with polynucleotide kinase (5'-ends) or RNA ligase (3'-ends). For complex formation, a fixed concentration of labeled RNA (either the sRNA or the mRNA target) is mixed with an increasing concentration of the unlabeled second RNA or purified protein (Fig. 14.3a). A large concentration range is usually used in the first set of experiments, generally from 10 nM to 10 μM. The complexes are

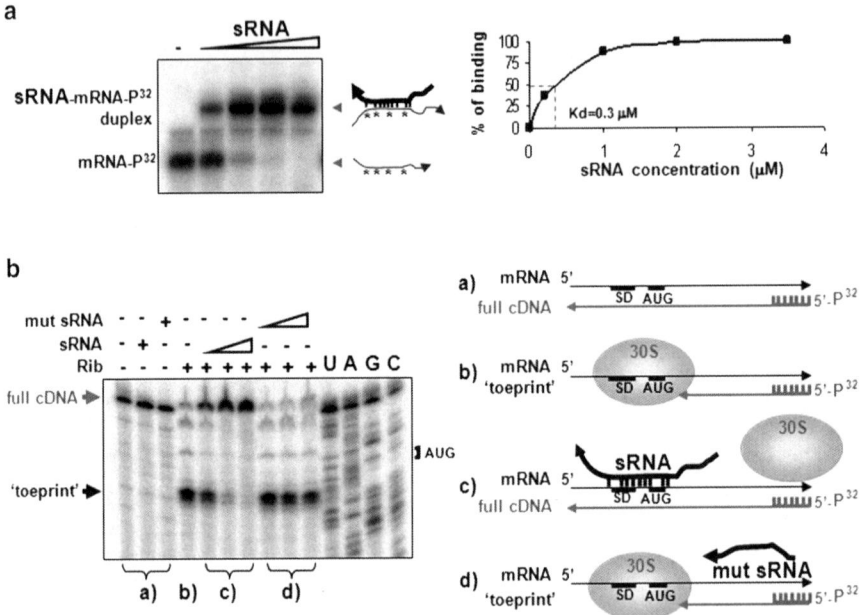

Fig. 14.3 *Experimental approaches for studying the mechanisms of regulation mediated by an sRNA.* (**a**) Complex formation between an sRNA and its mRNA target analyzed by EMSA. The purified labeled *sbi* mRNA (mRNA-P^{32}) was incubated in the presence of increasing amounts of unlabeled SprD (sRNA). Complexes were separated on polyacrylamide gels under native conditions. The autoradiogram obtained after gel electrophoresis is presented. The dissociation constant value (K_d) of the "sRNA-mRNA" complexes (*right panel*) is experimentally determined as the concentration of unlabeled sRNA for which 50% of labeled mRNA target is in complex. (**b**) Toeprint assays illustrating the inhibitory effect of an sRNA (here: SprD) on ribosome loading and translation initiation of *sbi* mRNA. "+/−" indicates the presence of purified ribosomes, sRNA (SprD), or an sRNA mutant lacking the domain of interaction with its mRNA target (mut SprD). U, A, G, and C refer to the mRNA sequencing ladders. (a) In the absence of ribosomes, the reverse transcriptase progresses until the 5' end of the mRNA is reached (*arrowhead*); (b) 30S ribosome loading onto the mRNA that prematurely stops cDNA elongation by the reverse transcriptase and leads to the appearance of a "toeprint" located ~17 nt downstream on the initiation codon (*arrow*); (c) the binding of the sRNA onto the mRNA reduces ribosome loading onto the mRNA in a "concentration-dependent" manner, leading to the progressive decrease of the "toeprint"; (d) an sRNA (mut sRNA) incapable of pairing with the mRNA target fails to interfere with ribosome loading onto the mRNA and, consequently, does not inhibit "toeprint" formation. It demonstrates that the sRNA inhibits mRNA translation initiation

separated on native polyacrylamide gels and visualized by autoradiography or phosphorimaging. This analysis can provide further information about the binding affinity (Fig. 14.3a), the conditions that favor the interactions between the molecules (temperature, pH, buffer compositions), and also if the interaction requires additional helper molecules, as for example, the RNA protein chaperone Hfq (Antal et al. 2005; Geissmann and Touati 2004; Heidrich et al. 2007). Furthermore, in conjunction with a mutational analysis, this approach allows the identification of the binding sites (Kawamoto et al. 2006).

Probing RNA Structures in Solution

This approach allows the study of the secondary structures of RNAs in solution, either free or interacting with ligands (for example, proteins, nucleic acids, and small molecules). The tools for probing RNA structures under statistical conditions can be nucleases or chemical agents (reviewed in Chevalier et al. 2009; Felden 2007). Most of the enzymes induce cleavages of the sRNA in unpaired regions in native buffers: RNase T1 cleaves $3'$ of unpaired Gs, U2 at unpaired As and to a lesser extent at Gs, and the S1 nuclease at unpaired residues, whereas RNase V1 is the only probe which cleaves at paired or stacked residues. RNase T2 cleaves in unpaired regions with a slight preference for As. The limitations of enzymes as probes for RNA structures are due to their size which could prevent the RNA sequences from being cleaved. Therefore, smaller probes can be used: RNase mimics (imidazole conjugates which cleave at unpaired or flexible ribonucleotides), lead (II) acetate (cleaves phosphodiester bonds in unpaired or flexible regions), or base-specific probes (the carbodiimide CMCT modifies atomic positions of nucleotides N3U and N1G; dimethyl sulfate [DMS] methylates N1A, N3C, and N7G; diethylpyrocarbonate [DEPC] monitors the reactivity of N7A; and kethoxal reacts with N1G and N2G) (reviewed in Chevalier et al. 2009; Felden 2007). By using probes with different specificities toward accessible or buried residues, it is possible to determine the reactivity of each nucleotide of RNAs for enzymes or chemical probes. This allows the discrimination of unpaired from paired regions as well as the detection of structural changes resulting from complex formation (Boisset et al. 2007; Bordeau and Felden 2002; Darfeuille et al. 2007; Heidrich et al. 2007; Pfeiffer et al. 2007). The detection of the cleavages can be performed by several methods, depending essentially on the lengths of the studied RNAs. One uses $5'$ or $3'$ end-labeled RNAs, and another involves primer extension by reverse transcriptase on the unlabeled modified RNA with $5'$ labeled DNA primers. Probing studies of RNA conformations have the advantage of being performed under native, physiological conditions. Chemicals such as DMS and lead (II) acetate can be also used to monitor RNA structures in living cells (Altuvia et al. 1997; Benito et al. 2000; Lindell et al. 2002). RNA structure models based on these chemical and enzymatic probes can be refined by using site-directed mutagenesis of selected residues predicted to be critical for RNA folding. These models are useful to identify the nucleotides participating in ligand interactions as well as to infer new functional hypotheses. Such analysis can also be applied to map the cleavage positions of bacterial RNase E or RNase III within the mRNA sequence, in the presence or absence of the sRNAs (Boisset et al. 2007; Viegas et al. 2011).

Testing the Effects of sRNAs on Translation Initiation

A primer extension inhibition (toeprints) assay was developed to study the formation of bacterial ribosomal initiation complexes in vitro (Hartz et al. 1988). The bacterial 30S, or rarely 70S, purified ribosomal subunits are utilized to form translation initiation complexes on the mRNAs to be translated. This method is based on the inhibition of reverse transcriptase elongation by the ribosomes loaded on the mRNA (Fig. 14.3b). The location of the pause ("toeprint") during reverse transcriptase elongation corresponds to the position at the $3'$ edge of the mRNA sequences covered by the loaded ribosome. In the presence of an initiator fMet-tRNA^{fMet}, the "toeprint" is situated on the mRNA at position +16 or + 17 on the initiation codon. A modification due to the binding of purified ribosomes can be visualized by the variation in the intensity of the "toeprints." Initially, this approach was used to monitor the effects of ribosomal components and translation factors on the formation of active ribosomal initiation complexes. Since the discovery of bacterial sRNAs, this method has been largely used to test the effects of sRNAs on mRNA target translation initiation in vitro (reviewed in Fechter et al. 2009). When the "RNA-based" regulation is accompanied by the fast degradation of the mRNA targets in vivo, it is problematic to discriminate between transcriptional and posttranscriptional regulation, as for example, in the case of the specific RNAIII-based regulation in S. aureus. So, the toeprint assays are a simple in vitro way to provide evidence for translational repression or activation by an sRNA. The binding of the sRNA to the ribosome binding site (RBS) of an mRNA leads to RBS masking and to the competition between the sRNA and the ribosome, thus preventing the formation of the ribosomal initiation complex (Boisset et al. 2007; Chabelskaya et al. 2010; Holmqvist et al. 2010). Several recent works using toeprint analyses have reported regulation by sRNA binding outside of the mRNAs RBS. A number of cases of translational repression by sRNAs interacting with the 5′-UTRs of target mRNAs (Darfeuille et al. 2007; Sharma et al. 2007) and also translation inhibition by an sRNA that interacts within the coding sequence of a target mRNA (Bouvier et al. 2008; Heidrich et al. 2007) have been reported. In addition to toeprint analysis, another in vitro method can be used for testing potential translational regulators. The commercially available "mRNA decay-free 70S ribosome translation system" (PURESYSTEM) was used to demonstrate the effect of sRNAs on target mRNA translation in vitro (Maki et al. 2008; Sharma et al. 2007).

14.4.2.2 In Vivo Approaches

Independent of the strategy chosen to study the biological roles of the sRNAs, the confirmation of their mechanisms of action has to be collected in vivo. If the sRNA acts by pairing with mRNA target(s), mutations or deletions of the sRNA and/or of its mRNA target(s) in the area of the interaction should affect the regulation (Boisset et al. 2007; Chabelskaya et al. 2010). A strong validation of a direct "sRNA-mRNA" interaction comes from compensatory base pair exchanges in vivo. Mutations in the interaction sites of sRNA and target mRNA should abolish the regulation, whereas compensatory mutations at both sites on the two interacting RNAs should restore the regulation (Bouvier et al. 2008; Papenfort et al. 2008; Pfeiffer et al. 2007; Vogel et al. 2004).

Also, as for target validation (paragraph 4.1), the specificity of "RNA–RNA" interactions can be assayed in vivo using a two-plasmid *E. coli*-based system (Urban and Vogel 2007). In this approach, the $5'$ regulatory sequences of an mRNA target are cloned as translational fusions into a reporter gene (GFP) and coexpressed with the sRNA under the control of constitutive promoters to avoid transcriptional regulations. Since the regulation can occur within the coding sequence of a target mRNA, this has to be taken into account during the construction of translational fusions. The direct interaction of the sRNA with its mRNA target can be challenged experimentally by introducing point mutations that abolish or restore regulation. Such "two-plasmid" systems are applicable to monitor the interactions between the sRNA and its mRNA target(s) in various bacteria.

In addition to regulation at translational level, the most frequent outcome of the "sRNA-mRNA" binding is to modify mRNA stability. The pairings can result in mRNA degradation (Afonyushkin et al. 2005; Huntzinger et al. 2005; Pfeiffer et al. 2009) or, sometimes, in protecting the mRNA from degradation (McCullen et al. 2010). The role of various RNases in the degradation process of the RNA duplex can be tested in vivo. Thus, monitoring the RNA levels in strains lacking specific RNase(s) can provide information about the implication of an RNase into the regulatory process and to test whether or not the regulation is irreversible (Afonyushkin et al. 2005; Boisset et al. 2007; Masse et al. 2003; McCullen et al. 2010; Opdyke et al. 2011; Viegas et al. 2011). Moreover, if the target mRNA is not rapidly hydrolyzed, the primer extension analysis can be applied to study the "sRNA-mRNA" interactions and to map the $5'/3'$ ends of the cleaved mRNAs. In the case of the bicistronic *gadX-gadW* mRNA, the $5'$-ends of the processed mRNAs resulting from GadY (the sRNA) binding were determined by this method (Opdyke et al. 2011). Likewise, the identification of $5'$ termini of processed mRNA transcripts can be performed by $5'$ *r*apid *a*mplification of *c*DNA *e*nds by PCR (RACE) (Vogel et al. 2004). $3'$ RACE analysis can also be used to estimate the "sRNA-mRNA" pairing domains (Pfeiffer et al. 2009).

Overall, there is a variety of strategies to validate putative mRNA targets and to understand the role(s) of the sRNAs. Because of space limitations, only selected experimental approaches are described in this chapter, and therefore, we apologize for the authors and methods that are not mentioned and discussed in same. Moreover, new approaches are emerging, as for example, a "RNA walk" method to investigate the "RNA–RNA" interactions between an sRNA and its target (Lustig et al. 2010). Combining in vitro and in vivo approaches is a reasonable line of attack to elucidate the mechanism(s) of action of the sRNAs. It is essential to remember that the in vitro systems are very sensitive to experimental conditions. Therefore, in order to minimize the experimental artifacts linked to the in vitro methods (biases), one should select the physiologically relevant conditions for the in vitro experiments. Certainly, the suggested mechanism(s) based on in vitro data must always be confirmed in vivo.

14.5 General Conclusions

A significant number of bacterial sRNAs were discovered during the last decade by computational and biochemical approaches. Given the expeditious ongoing technological developments, as illustrated recently with the high-throughput sequencing and their upcoming availability at low costs, we anticipate a considerable increase of the sRNA catalog in many bacteria, especially in the human pathogens considering that several sRNAs have been shown to be involved in bacterial virulence. The forthcoming challenge is to characterize their primary, direct targets. Indeed, sRNA target identification is the limiting tread for the functional investigation of many sRNAs. The task is further complicated by the multiplicity of the targets per sRNA and the multiplicity of mechanisms of action for a given sRNA. Most of the current "large-scale" technologies favor the identification of secondary targets. Approaches based on affinity captures of sRNA targets focus on identifying the primary targets but are confronted with multiple nonspecific ligands. Thus, a combination of these different methods is required not only to detect the primary targets but also to provide further information on their regulatory processes. Numerous sophisticated tools are currently available to characterize the mechanism(s) of action of the sRNAs both in vitro and in vivo. Whereas in vitro analysis can be performed for any sRNA from any bacterial species, the in vivo experiments require specific experimental strategies according to the bacterium studied. Classical genetic approaches easily set up in *E. coli*, as site-directed mutagenesis, are much more complicated to accomplish in other bacteria such as gram-positive bacteria. As for target identification, an association of in vitro and in vivo complementary approaches is required to elucidate the mechanisms underlying target regulations.

In this chapter, strategies are described that can be applied to identify sRNAs and their primary targets and to study their mechanisms of regulation. The proper use of these methods is mandatory to integrate the sRNAs into bacterial physiology and especially to integrate them into the various signaling pathways triggered by external stimuli. This upcoming challenge will allow understanding of their involvement in elaborate gene regulatory networks in bacteria. One should keep in mind that an sRNA is able to regulate multiple target genes and that the expression of one gene can be controlled by several sRNAs and also by additional regulators, such as the general transcription factors.

References

Afonyushkin T, Vecerek B, Moll I, Blasi U, Kaberdin VR (2005) Both RNase E and RNase III control the stability of sodB mRNA upon translational inhibition by the small regulatory RNA RyhB. Nucleic Acids Res 33(5):1678–1689. doi:10.1093/nar/gki313

Alkan C, Karakoc E, Nadeau JH, Sahinalp SC, Zhang K (2006) RNA–RNA interaction prediction and antisense RNA target search. J Comput Biol 13(2):267–282

Altuvia S, Weinstein-Fischer D, Zhang A, Postow L, Storz G (1997) A small, stable RNA induced by oxidative stress: role as a pleiotropic regulator and antimutator. Cell 90(1):43–53

Altuvia S, Zhang A, Argaman L, Tiwari A, Storz G (1998) The *Escherichia coli* OxyS regulatory RNA represses fhlA translation by blocking ribosome binding. EMBO J 17(20):6069–6075

Antal M, Bordeau V, Douchin V, Felden B (2005) A small bacterial RNA regulates a putative ABC transporter. J Biol Chem 280(9):7901–7908

Argaman L, Altuvia S (2000) fhlA repression by OxyS RNA: kissing complex formation at two sites results in a stable antisense-target RNA complex. J Mol Biol 300(5):1101–1112

Backofen R, Hess WR (2010) Computational prediction of sRNAs and their targets in bacteria. RNA Biol 7(1):33–42

Beaume M, Hernandez D, Farinelli L, Deluen C, Linder P, Gaspin C, Romby P, Schrenzel J, Francois P (2010) Cartography of methicillin-resistant *S. aureus* transcripts: detection, orientation and temporal expression during growth phase and stress conditions. PLoS One 5(5): e10725. doi:10.1371/journal.pone.0010725

Benito Y, Kolb FA, Romby P, Lina G, Etienne J, Vandenesch F (2000) Probing the structure of RNAIII, the *Staphylococcus aureus* agr regulatory RNA, and identification of the RNA domain involved in repression of protein A expression. RNA 6(5):668–679

Bernstein JA, Khodursky AB, Lin PH, Lin-Chao S, Cohen SN (2002) Global analysis of mRNA decay and abundance in *Escherichia coli* at single-gene resolution using two-color fluorescent DNA microarrays. Proc Natl Acad Sci USA 99(15):9697–9702

Bohn C, Rigoulay C, Chabelskaya S, Sharma CM, Marchais A, Skorski P, Borezee-Durant E, Barbet R, Jacquet E, Jacq A, Gautheret D, Felden B, Vogel J, Bouloc P (2010) Experimental discovery of small RNAs in *Staphylococcus aureus* reveals a riboregulator of central metabolism. Nucleic Acids Res 38(19):6620–6636

Boisset S, Geissmann T, Huntzinger E, Fechter P, Bendridi N, Possedko M, Chevalier C, Helfer AC, Benito Y, Jacquier A, Gaspin C, Vandenesch F, Romby P (2007) *Staphylococcus aureus* RNAIII coordinately represses the synthesis of virulence factors and the transcription regulator Rot by an antisense mechanism. Genes Dev 21(11):1353–1366

Bordeau V, Felden B (2002) Ribosomal protein S1 induces a conformational change of tmRNA; more than one protein S1 per molecule of tmRNA. Biochimie 84(8):723–729

Bouvier M, Sharma CM, Mika F, Nierhaus KH, Vogel J (2008) Small RNA binding to 5′ mRNA coding region inhibits translational initiation. Mol Cell 32(6):827–837. doi:10.1016/j.molcel.2008.10.027

Boysen A, Moller-Jensen J, Kallipolitis B, Valentin-Hansen P, Overgaard M (2010) Translational regulation of gene expression by an anaerobically induced small non-coding RNA in *Escherichia coli*. J Biol Chem 285(14):10690–10702

Breaker RR (2010) Riboswitches and the RNA World. Cold Spring Harb Perspect Biol. doi:10.1101/cshperspect.a003566

Busch A, Richter AS, Backofen R (2008) IntaRNA: efficient prediction of bacterial sRNA targets incorporating target site accessibility and seed regions. Bioinformatics 24(24):2849–2856

Chabelskaya S, Gaillot O, Felden B (2010) A *Staphylococcus aureus* small RNA is required for bacterial virulence and regulates the expression of an immune-evasion molecule. PLoS Pathog 6(6):e1000927

Chant EL, Summers DK (2007) Indole signalling contributes to the stable maintenance of *Escherichia coli* multicopy plasmids. Mol Microbiol 63(1):35–43

Chen S, Zhang A, Blyn LB, Storz G (2004) MicC, a second small-RNA regulator of Omp protein expression in *Escherichia coli*. J Bacteriol 186(20):6689–6697

Chevalier C, Geissmann T, Helfer AC, Romby P (2009) Probing mRNA structure and sRNA–mRNA interactions in bacteria using enzymes and lead(II). Methods Mol Biol 540:215–232. doi:10.1007/978-1-59745-558-9_16

Chevalier C, Boisset S, Romilly C, Masquida B, Fechter P, Geissmann T, Vandenesch F, Romby P (2010) *Staphylococcus aureus* RNAIII binds to two distant regions of coa mRNA to arrest translation and promote mRNA degradation. PLoS Pathog 6(3):e1000809

Choi KH, Kim KJ (2009) Applications of transposon-based gene delivery system in bacteria. J Microbiol Biotechnol 19(3):217–228

Christiansen JK, Nielsen JS, Ebersbach T, Valentin-Hansen P, Sogaard-Andersen L, Kallipolitis BH (2006) Identification of small Hfq-binding RNAs in *Listeria monocytogenes*. RNA 12 (7):1383–1396. doi:10.1261/rna.49706

Darfeuille F, Unoson C, Vogel J, Wagner EG (2007) An antisense RNA inhibits translation by competing with standby ribosomes. Mol Cell 26(3):381–392. doi:10.1016/j.molcel.2007.04.003

Davis BM, Quinones M, Pratt J, Ding Y, Waldor MK (2005) Characterization of the small untranslated RNA RyhB and its regulon in *Vibrio cholerae*. J Bacteriol 187(12):4005–4014

Desnoyers G, Morissette A, Prevost K, Masse E (2009) Small RNA-induced differential degradation of the polycistronic mRNA iscRSUA. EMBO J 28(11):1551–1561. doi:10.1038/emboj.2009.116

Deveau H, Garneau JE, Moineau S (2010) CRISPR/Cas system and its role in phage-bacteria interactions. Annu Rev Microbiol 64:475–493. doi:10.1146/annurev.micro.112408.134123

Douchin V, Bohn C, Bouloc P (2006) Down-regulation of porins by a small RNA bypasses the essentiality of the regulated intramembrane proteolysis protease RseP in *Escherichia coli*. J Biol Chem 281(18):12253–12259

Fechter P, Chevalier C, Yusupova G, Yusupov M, Romby P, Marzi S (2009) Ribosomal initiation complexes probed by toeprinting and effect of trans-acting translational regulators in bacteria. Methods Mol Biol 540:247–263. doi:10.1007/978-1-59745-558-9_18

Felden B (2007) RNA structure: experimental analysis. Curr Opin Microbiol 10(3):286–291. doi:10.1016/j.mib.2007.05.001

Felden B, Gesteland RF, Atkins JF (1999) Eubacterial tmRNAs: everywhere except the alpha-proteobacteria? Biochim Biophys Acta 1446(1–2):145–148

Felden B, Massire C, Westhof E, Atkins JF, Gesteland RF (2001) Phylogenetic analysis of tmRNA genes within a bacterial subgroup reveals a specific structural signature. Nucleic Acids Res 29 (7):1602–1607

Felden B, Vandenesch F, Bouloc P, Romby P (2011) The *Staphylococcus aureus* RNome and its commitment to virulence. PLoS Pathog 7(3):e1002006. doi:10.1371/journal.ppat.1002006

Geisinger E, Adhikari RP, Jin R, Ross HF, Novick RP (2006) Inhibition of rot translation by RNAIII, a key feature of agr function. Mol Microbiol 61(4):1038–1048

Geissmann TA, Touati D (2004) Hfq, a new chaperoning role: binding to messenger RNA determines access for small RNA regulator. EMBO J 23(2):396–405. doi:10.1038/sj.emboj.7600058

Gottesman S (2004) The small RNA regulators of *Escherichia coli*: roles and mechanisms. Annu Rev Microbiol 58:303–328

Gottesman S, Storz G (2010) Bacterial small RNA regulators: versatile roles and rapidly evolving variations. Cold Spring Harb Perspect Biol. doi:10.1101/cshperspect.a003798

Gottesman S, McCullen CA, Guillier M, Vanderpool CK, Majdalani N, Benhammou J, Thompson KM, FitzGerald PC, Sowa NA, FitzGerald DJ (2006) Small RNA regulators and the bacterial response to stress. Cold Spring Harb Symp Quant Biol 71:1–11. doi:10.1101/sqb.2006.71.016

Griffin BE (1971) Separation of 32P-labelled ribonucleic acid components. The use of polyethy-lenimine-cellulose (TLC) as a second dimension in separating oligoribonucleotides of '4.5S' and 5S from *E. coli*. FEBS Lett 15(3):165–168

Guillier M, Gottesman S (2006) Remodelling of the *Escherichia coli* outer membrane by two small regulatory RNAs. Mol Microbiol 59(1):231–247

Hartz D, McPheeters DS, Traut R, Gold L (1988) Extension inhibition analysis of translation initiation complexes. Methods Enzymol 164:419–425

Heeb S, Blumer C, Haas D (2002) Regulatory RNA as mediator in GacA/RsmA-dependent global control of exoproduct formation in *Pseudomonas fluorescens* CHA0. J Bacteriol 184 (4):1046–1056

Heidrich N, Moll I, Brantl S (2007) In vitro analysis of the interaction between the small RNA SR1 and its primary target ahrC mRNA. Nucleic Acids Res 35(13):4331–4346. doi:10.1093/nar/gkm439

Hindley J (1967) Fractionation of 32P-labelled ribonucleic acids on polyacrylamide gels and their characterization by fingerprinting. J Mol Biol 30(1):125–136

Holmqvist E, Reimegard J, Sterk M, Grantcharova N, Romling U, Wagner EG (2010) Two antisense RNAs target the transcriptional regulator CsgD to inhibit curli synthesis. EMBO J 29(11):1840–1850. doi:10.1038/emboj.2010.73

Horvath P, Barrangou R (2010) CRISPR/Cas, the immune system of bacteria and archaea. Science 327(5962):167–170. doi:10.1126/science.1179555

Hovhannisyan R, Carstens R (2009) Affinity chromatography using 2′ fluoro-substituted RNAs for detection of RNA-protein interactions in RNase-rich or RNase-treated extracts. Biotechniques 46(2):95–98

Hu Z, Zhang A, Storz G, Gottesman S, Leppla SH (2006) An antibody-based microarray assay for small RNA detection. Nucleic Acids Res 34(7):e52. doi:10.1093/nar/gkl142

Huntzinger E, Boisset S, Saveanu C, Benito Y, Geissmann T, Namane A, Lina G, Etienne J, Ehresmann B, Ehresmann C, Jacquier A, Vandenesch F, Romby P (2005) *Staphylococcus aureus* RNAIII and the endoribonuclease III coordinately regulate spa gene expression. EMBO J 24(4):824–835

Huttenhofer A, Cavaille J, Bachellerie JP (2004) Experimental RNomics: a global approach to identifying small nuclear RNAs and their targets in different model organisms. Methods Mol Biol 265:409–428. doi:10.1385/1-59259-775-0:409

Ikemura T, Dahlberg JE (1973) Small ribonucleic acids of *Escherichia coli*. I. Characterization by polyacrylamide gel electrophoresis and fingerprint analysis. J Biol Chem 248(14):5024–5032

Kaberdin VR, Blasi U (2006) Translation initiation and the fate of bacterial mRNAs. FEMS Microbiol Rev 30(6):967–979. doi:10.1111/j.1574-6976.2006.00043.x

Kawamoto H, Koide Y, Morita T, Aiba H (2006) Base-pairing requirement for RNA silencing by a bacterial small RNA and acceleration of duplex formation by Hfq. Mol Microbiol 61 (4):1013–1022. doi:10.1111/j.1365-2958.2006.05288.x

Kwon YS (2011) Small RNA library preparation for next-generation sequencing by single ligation, extension and circularization technology. Biotechnol Lett. doi:10.1007/s10529-011-0611-y

Lee SY, Bailey SC, Apirion D (1978) Small stable RNAs from *Escherichia coli*: evidence for the existence of new molecules and for a new ribonucleoprotein particle containing 6S RNA. J Bacteriol 133(2):1015–1023

Lindell M, Romby P, Wagner EG (2002) Lead(II) as a probe for investigating RNA structure in vivo. RNA 8(4):534–541

Liu MY, Gui G, Wei B, Preston JF III, Oakford L, Yuksel U, Giedroc DP, Romeo T (1997) The RNA molecule CsrB binds to the global regulatory protein CsrA and antagonizes its activity in *Escherichia coli*. J Biol Chem 272(28):17502–17510

Lustig Y, Wachtel C, Safro M, Liu L, Michaeli S (2010) 'RNA walk' a novel approach to study RNA–RNA interactions between a small RNA and its target. Nucleic Acids Res 38(1):e5. doi:10.1093/nar/gkp872

MacLean D, Jones JD, Studholme DJ (2009) Application of 'next-generation' sequencing technologies to microbial genetics. Nat Rev Microbiol 7(4):287–296. doi:10.1038/nrmicro2122

Maki K, Uno K, Morita T, Aiba H (2008) RNA, but not protein partners, is directly responsible for translational silencing by a bacterial Hfq-binding small RNA. Proc Natl Acad Sci USA 105 (30):10332–10337. doi:10.1073/pnas.0803106105

Mandin P, Gottesman S (2009) A genetic approach for finding small RNAs regulators of genes of interest identifies RybC as regulating the DpiA/DpiB two-component system. Mol Microbiol 72(3):551–565. doi:10.1111/j.1365-2958.2009.06665.x

Mandin P, Repoila F, Vergassola M, Geissmann T, Cossart P (2007) Identification of new noncoding RNAs in *Listeria monocytogenes* and prediction of mRNA targets. Nucleic Acids Res 35(3):962–974

Masse E, Escorcia FE, Gottesman S (2003) Coupled degradation of a small regulatory RNA and its mRNA targets in *Escherichia coli*. Genes Dev 17(19):2374–2383. doi:10.1101/gad.1127103

Massé E, Vanderpool CK, Gottesman S (2005) Effect of RyhB small RNA on global iron use in *Escherichia coli*. J Bacteriol 187(20):6962–6971

McCullen CA, Benhammou JN, Majdalani N, Gottesman S (2010) Mechanism of positive regulation by DsrA and RprA small noncoding RNAs: pairing increases translation and protects rpoS mRNA from degradation. J Bacteriol 192(21):5559–5571. doi:10.1128/jb.00464-10

Moller T, Franch T, Udesen C, Gerdes K, Valentin-Hansen P (2002) Spot 42 RNA mediates discoordinate expression of the *E. coli* galactose operon. Genes Dev 16(13):1696–1706. doi:10.1101/gad.231702

Morfeldt E, Taylor D, von Gabain A, Arvidson S (1995) Activation of alpha-toxin translation in *Staphylococcus aureus* by the trans-encoded antisense RNA, RNAIII. EMBO J 14 (18):4569–4577

Morita T, Maki K, Aiba H (2005) RNase E-based ribonucleoprotein complexes: mechanical basis of mRNA destabilization mediated by bacterial noncoding RNAs. Genes Dev 19 (18):2176–2186. doi:10.1101/gad.1330405

Muckstein U, Tafer H, Hackermuller J, Bernhart SH, Stadler PF, Hofacker IL (2006) Thermodynamics of RNA–RNA binding. Bioinformatics 22(10):1177–1182

Novick RP, Ross HF, Projan SJ, Kornblum J, Kreiswirth B, Moghazeh S (1993) Synthesis of staphylococcal virulence factors is controlled by a regulatory RNA molecule. EMBO J 12 (10):3967–3975

Opdyke JA, Fozo EM, Hemm MR, Storz G (2011) RNase III participates in GadY-dependent cleavage of the gadX-gadW mRNA. J Mol Biol 406(1):29–43. doi:10.1016/j.jmb.2010.12.009

Ozsolak F, Platt AR, Jones DR, Reifenberger JG, Sass LE, McInerney P, Thompson JF, Bowers J, Jarosz M, Milos PM (2009) Direct RNA sequencing. Nature 461(7265):814–818. doi:10.1038/nature08390

Panek J, Krasny L, Bobek J, Jezkova E, Korelusova J, Vohradsky J (2011) The suboptimal structures find the optimal RNAs: homology search for bacterial non-coding RNAs using suboptimal RNA structures. Nucleic Acids Res 39(8):3418–3426. doi:10.1093/nar/gkq1186

Papenfort K, Pfeiffer V, Lucchini S, Sonawane A, Hinton JC, Vogel J (2008) Systematic deletion of Salmonella small RNA genes identifies CyaR, a conserved CRP-dependent riboregulator of OmpX synthesis. Mol Microbiol 68(4):890–906. doi:10.1111/j.1365-2958.2008.06189.x

Peer A, Margalit H (2011) Accessibility and evolutionary conservation mark bacterial small-RNA target-binding regions. J Bacteriol 193(7):1690–1701

Perez N, Trevino J, Liu Z, Ho SC, Babitzke P, Sumby P (2009) A genome-wide analysis of small regulatory RNAs in the human pathogen group A Streptococcus. PLoS One 4(11):e7668. doi:10.1371/journal.pone.0007668

Perkins TT, Kingsley RA, Fookes MC, Gardner PP, James KD, Yu L, Assefa SA, He M, Croucher NJ, Pickard DJ, Maskell DJ, Parkhill J, Choudhary J, Thomson NR, Dougan G (2009) A strand-specific RNA-Seq analysis of the transcriptome of the typhoid bacillus *Salmonella typhi*. PLoS Genet 5(7):e1000569. doi:10.1371/journal.pgen.1000569

Pfeiffer V, Sittka A, Tomer R, Tedin K, Brinkmann V, Vogel J (2007) A small non-coding RNA of the invasion gene island (SPI-1) represses outer membrane protein synthesis from the *Salmonella core* genome. Mol Microbiol 66(5):1174–1191

Pfeiffer V, Papenfort K, Lucchini S, Hinton JC, Vogel J (2009) Coding sequence targeting by MicC RNA reveals bacterial mRNA silencing downstream of translational initiation. Nat Struct Mol Biol 16(8):840–846. doi:10.1038/nsmb.1631

Pichon C, Felden B (2003) Intergenic sequence inspector: searching and identifying bacterial RNAs. Bioinformatics 19(13):1707–1709

Pichon C, Felden B (2005) Small RNA genes expressed from *Staphylococcus aureus* genomic and pathogenicity islands with specific expression among pathogenic strains. Proc Natl Acad Sci USA 102(40):14249–14254. doi:10.1073/pnas.0503838102

Rasmussen S, Nielsen HB, Jarmer H (2009) The transcriptionally active regions in the genome of *Bacillus subtilis*. Mol Microbiol 73(6):1043–1057. doi:10.1111/j.1365-2958.2009.06830.x

Ray BK, Apirion D (1979) Characterization of 10S RNA: a new stable rna molecule from *Escherichia coli*. Mol Gen Genet 174(1):25–32

Rehmsmeier M, Steffen P, Hochsmann M, Giegerich R (2004) Fast and effective prediction of microRNA/target duplexes. RNA 10(10):1507–1517

Repoila F, Darfeuille F (2009) Small regulatory non-coding RNAs in bacteria: physiology and mechanistic aspects. Biol Cell 101(2):117–131. doi:10.1042/bc20070137

Romby P, Charpentier E (2010) An overview of RNAs with regulatory functions in gram-positive bacteria. Cell Mol Life Sci 67(2):217–237. doi:10.1007/s00018-009-0162-8

Said N, Rieder R, Hurwitz R, Deckert J, Urlaub H, Vogel J (2009) In vivo expression and purification of aptamer-tagged small RNA regulators. Nucleic Acids Res 37(20):e133

Sharma CM, Darfeuille F, Plantinga TH, Vogel J (2007) A small RNA regulates multiple ABC transporter mRNAs by targeting C/A-rich elements inside and upstream of ribosome-binding sites. Genes Dev 21(21):2804–2817. doi:10.1101/gad.447207

Sittka A, Lucchini S, Papenfort K, Sharma CM, Rolle K, Binnewies TT, Hinton JC, Vogel J (2008) Deep sequencing analysis of small noncoding RNA and mRNA targets of the global post-transcriptional regulator, Hfq. PLoS Genet 4(8):e1000163. doi:10.1371/journal.pgen.1000163

Sonnleitner E, Sorger-Domenigg T, Madej MJ, Findeiss S, Hackermuller J, Huttenhofer A, Stadler PF, Blasi U, Moll I (2008) Detection of small RNAs in Pseudomonas aeruginosa by RNomics and structure-based bioinformatic tools. Microbiology 154(Pt 10):3175–3187. doi:10.1099/mic.0.2008/019703-0

Stougaard P, Molin S, Nordstrom K (1981) RNAs involved in copy-number control and incompatibility of plasmid R1. Proc Natl Acad Sci USA 78(10):6008–6012

Tjaden B (2008) TargetRNA: a tool for predicting targets of small RNA action in bacteria. Nucleic Acids Res 36(Web Server issue):W109–W113

Tjaden B, Goodwin SS, Opdyke JA, Guillier M, Fu DX, Gottesman S, Storz G (2006) Target prediction for small, noncoding RNAs in bacteria. Nucleic Acids Res 34(9):2791–2802

Toledo-Arana A, Dussurget O, Nikitas G, Sesto N, Guet-Revillet H, Balestrino D, Loh E, Gripenland J, Tiensuu T, Vaitkevicius K, Barthelemy M, Vergassola M, Nahori MA, Soubigou G, Regnault B, Coppee JY, Lecuit M, Johansson J, Cossart P (2009) The *Listeria* transcriptional landscape from saprophytism to virulence. Nature 459(7249):950–956. doi:10.1038/nature08080

Urban JH, Vogel J (2007) Translational control and target recognition by *Escherichia coli* small RNAs in vivo. Nucleic Acids Res 35(3):1018–1037. doi:10.1093/nar/gkl1040

Viegas SC, Silva IJ, Saramago M, Domingues S, Arraiano CM (2011) Regulation of the small regulatory RNA MicA by ribonuclease III: a target-dependent pathway. Nucleic Acids Res 39 (7):2918–2930. doi:10.1093/nar/gkq1239

Vogel J, Bartels V, Tang TH, Churakov G, Slagter-Jager JG, Huttenhofer A, Wagner EG (2003) RNomics in *Escherichia coli* detects new sRNA species and indicates parallel transcriptional output in bacteria. Nucleic Acids Res 31(22):6435–6443

Vogel J, Argaman L, Wagner EG, Altuvia S (2004) The small RNA IstR inhibits synthesis of an SOS-induced toxic peptide. Curr Biol 14(24):2271–2276

Wagner EG, Altuvia S, Romby P (2002) Antisense RNAs in bacteria and their genetic elements. Adv Genet 46:361–398

Wassarman KM, Storz G (2000) 6S RNA regulates *E. coli* RNA polymerase activity. Cell 101 (6):613–623

Wassarman KM, Repoila F, Rosenow C, Storz G, Gottesman S (2001) Identification of novel small RNAs using comparative genomics and microarrays. Genes Dev 15(13):1637–1651. doi:10.1101/gad.901001

Waters LS, Storz G (2009) Regulatory RNAs in bacteria. Cell 136(4):615–628. doi:10.1016/j. cell.2009.01.043

Windbichler N, von Pelchrzim F, Mayer O, Csaszar E, Schroeder R (2008) Isolation of small RNA-binding proteins from *E. coli*: evidence for frequent interaction of RNAs with RNA polymerase. RNA Biol 5(1):30–40

Woese CR, Fox GE, Zablen L, Uchida T, Bonen L, Pechman K, Lewis BJ, Stahl D (1975) Conservation of primary structure in 16S ribosomal RNA. Nature 254(5495):83–86

Yamamoto S, Izumiya H, Mitobe J, Morita M, Arakawa E, Ohnishi M, Watanabe H (2011) Identification of a chitin-induced small RNA that regulates translation of the tfoX gene, encoding a positive regulator of natural competence in *Vibrio cholerae*. J Bacteriol 193 (8):1953–1965. doi:10.1128/jb.01340-10

Zhang A, Wassarman KM, Rosenow C, Tjaden BC, Storz G, Gottesman S (2003) Global analysis of small RNA and mRNA targets of Hfq. Mol Microbiol 50(4):1111–1124

Chapter 15
Microregulators Ruling Over Pluripotent Stem Cells

Shijun Hu and Andrew Stephen Lee

Abstract Embryonic stem cells (ESCs) and induced pluripotent stem cells (iPSCs) have very promising clinical applications in regenerative medicine due to their ability for unlimited self-renewal and potential to differentiate into every cell type of the adult human body. microRNAs (miRNAs) are a novel class of single-stranded noncoding RNA that have been found to play a significant role in epigenetic regulation of ESCs and iPSCs. Understanding the miRNA regulatory pathways underlying the induction and maintenance of pluripotency in ESCs and iPSCs will facilitate basic research and stem cell–based regenerative therapies.

Keywords Embryonic stem cells • induced pluripotent stem cells • microRNAs

15.1 Pluripotent Stem Cells and Micromanagers Building up a Synchronized Network

Embryonic stem cells (ESCs) are derived from the inner cell mass of pre-implantation blastocysts and are technically defined as cells which can both self-renew indefinitely and differentiate into all adult cell types. These potential for self-renewal and differentiation are considered the primary defining features of ESCs. The regulatory mechanism of self-renewal and differentiation in ESCs is a topic that has been intensely investigated since the end of the last century. Much progress has been

S. Hu (✉)
Department of Medicine, Division of Cardiology, Stanford University School of Medicine, Stanford, CA, USA

Department of Radiology, Stanford University School of Medicine, Stanford, CA, USA

Institute for Stem Cell Biology and Regenerative Medicine, Stanford University, Stanford, CA, USA
e-mail: shijunhu@stanford.edu

B. Mallick (eds.), *Regulatory RNAs*, DOI 10.1007/978-3-642-22517-8_15,
© Springer-Verlag Berlin Heidelberg (outside the USA) 2012

made in this field, including characterization of transcriptional regulatory networks controlling pluripotency. A number of transcriptional factors and networks that are critical for "stemness" of ESC, including Oct4, Sox2, Nanog, and Tcf3, have been identified in recent years (Takahashi et al. 2007). These proteins have emerged as key regulators of pluripotency, working together to maintain of self-renewal and capacity for differentiation. However, much remains unknown about the essential nature of "stemness" and the regulation of proliferation and differentiation of ESCs at the molecular and genomic levels.

Induced pluripotent stem cells (iPSCs) generated by de-differentiation of adult somatic cells are a potential solution for the ethical issues surrounding ESCs as well as the potential for ESCS to be rejected immunologically after cellular transplantation. Reprogramming somatic cells to a pluripotent state is typically accomplished through the induction of different combinations of transcription factors such as Oct4, Sox2, Klf4, and C-myc (Takahashi et al. 2007) or Oct4, Sox2, Nanog, and Lin28 (Yu et al. 2007). Reports reveal that iPSCs have similar gene expression and chromatin patterns to those of ESCs. Further, iPSCs spontaneously differentiate to form teratoma and possess self-renewal property as ESCs (Chin et al. 2009).

Both ESCs and iPSCs have significant potential for clinical therapies due to their potential to differentiate into a wide range of cell types. Much progress has been made toward understanding the core regulatory circuitry for self-renewal and differentiation of these pluripotent cells, which is fundamental for their clinical applications, however our current understanding is not complete. Although pluripotency and self-renewal are regulated by a core network of transcription factors – Oct4, Sox2, Nanog, and Klf4 – that are integrated into complex molecular circuits, scientists do not yet have significant control over direction of cell fate. To figure out this missing link, other elements must be considered in order to understand the process of self-renewal and differentiation of ESCs and iPSCs. Recently, miRNAs have been shown to act as microswitches that reveal certain hidden dimensions of the ES cell cycle (Ganga raju and Lin 2009; Mallick et al. 2011).

miRNAs are small noncoding single-stranded RNAs, about 22–24 nucleotides in length that silence messenger RNA (mRNA) expression by binding the 3′UTR of target mRNAs and mediate RNA destruction and translational suppression (Bartel 2009). miRNAs have been shown to be powerful regulatory elements as a single miRNA can silence hundreds of genes. In the process of stem cell self-renewal and differentiation, miRNAs potentially target a great number of genes involved in the regulation of pluripotency and interact with core transcriptional factors (Tiscornia and Izpisua Belmonte 2010). Refer to Chap. 17 (this volume) for details on role of miRNAs in development and stem cell differentiation.

A number of miRNAs have been identified as important regulators of pluripotency in ESCs and iPSCs, including let-7 (Melton et al. 2010), miR-302s (Card et al. 2008), miR-371–373 (Stadler et al. 2010), miR-145 (Xu et al. 2009), among others (Table 15.1). Nucleotides 2–7 of mature miRNAs form the key region for target mRNA recognition and hybridize nearly perfectly with complementary sequences on the 3′ UTR of target mRNAs for loading into the Argonaute

Table 15.1 Important miRNAs involved in stem cell maintenance

Stem cell process	miRNAs involved	Potential targets
Self-renewal	miR-302 cluster	Cyclin D1, Lefty, Nr2f2
	miR-290 cluster	P21, Rbl2, Lats2, Wee1, Fbx15, caspase 2, El24
	miR-195	Cdkn1a
	miR-92b	P57
Differentiation	Let-7	C-myc, Sal14, Lin28
	miR-145	Klf4, Sox2, Oct4
	miR-1	Dll-1
	miR-9	Stathmin

protein–containing RNA-induced silencing complex (RISC). Therefore, target recognition is very important for the elucidation of miRNA functions. Many ESC-specific miRNAs are cotranscribed as polycistronic transcripts, indicating common upstream regulation and coordinated expression patterns. These ESC-specific miRNAs are regulated by a core set of ESC-specific pluripotent transcriptional factors (Marson et al. 2008) and are not present or highly expressed in differentiated cells and somatic cells. Conversely, a great number of miRNAs such as the let-7 family are expressed in differentiated cell types at high levels and in ESCs at low levels (Viswanathan et al. 2008).

15.2 Role of Dicer and Dgcr8 in Stem Cell Maintenance

In mammals, the loss of entire miRNA expression gives critical clues to the importance of miRNAs in early development and pluripotency. For instance, Dicer−/− mice which lack the miRNA processing and regulatory pathways display embryonic lethality and complete loss of stem cell compartments (Bernstein et al. 2003). The loss of Dicer in mouse embryonic stem cells (mESCs) leads to an acute loss of proliferative potential. Moreover, Dicer−/− mESCs display an altered cell cycle profile, with an increase in the number of cells in phase G1-G0. Dicer is not the only key factor involved in miRNA processing. Dgcr8 is another protein that has also been implicated in maintenance of pluripotency in mESCs. Like Dicer−/− embryos, Dgcr8−/− embryos arrest in early development, though this phenotype is less severe than Dicer−/− counterparts (Wang et al. 2007). Dgcr8−/− mESCs demonstrate a distinct impairment in cell proliferation and cannot efficiently silence pathways that maintain pluripotency even under conditions of differentiation. Although the morphology of Dicer−/− and Dgcr8−/− in human ESCs (hESCs) is similar to their wild-type counterparts, these cells are not able to downregulate stem cell–specific marker genes and show defects in self-renewal. The cell cycles of Dicer−/− and Dgcr8−/− hESCs exhibit delays in G1-S and G2-M transition (Melton et al. 2010). These data demonstrate the importance of miRNA-based regulatory pathways as a whole to the regulation of ESC self-renewal and pluripotency.

15.3 Distinct miRNA Signatures in Pluripotent Stem Cells

15.3.1 miR-302s/367 Cluster

In both human and mouse, the miR-302s/367 cluster overlaps with the LA-related protein 7 (LARP7) gene. The human miR-302 cluster consists of several different miRNAs cotranscribed in a polycistronic manner: miR-302a, miR-302a*, miR-302b, miR-302b*, miR-302c, miR-302c*, miR-302d, miR-302d*, miR-367, and miR-367* with a highly conserved 5′ region (Landgraf et al. 2007). Recently, miRNA homologues to the miR-302 family have been reported and registered as miR-302e and miR-302f which are from different chromosomes (Morin et al. 2008) but are separate from the miR-302s/367 cluster. The miR-302s/367 cluster is exclusively expressed at high levels in ESCs and iPSCs but not in somatic stem cells or differentiated cells (Wilson et al. 2009).

ESC expression of the miR-302s/367 cluster is dependent on binding of its promoter region by the pluripotency factors Oct4, Nanog, Sox2, and Rex1 (Card et al. 2008; Barroso-delJesus et al. 2008), indicating that expression of ESC-specific miRNAs is governed by a regulatory network controlled by the ESC-specific transcriptional program.

Many key molecules involved in self-renewal and differentiation of pluripotent stem cells have been proven as targets of miR-302s. The cell cycle regulator, cyclin D1, has been recently found to be post-transcriptionally regulated by the miR-302 cluster in hESCs (Card et al. 2008). Several studies have indicated that during ESC differentiation, the Nodal inhibitor Lefty is post-transcriptionally targeted by the miR-302 cluster for suppression (Barroso-delJesus et al. 2011; Rosa et al. 2009). Recently, Nr2f2 (nuclear receptor subfamily 2, group F, member 2) that can inhibit Oct4 expression has been proven to be a target of the miR-302 cluster. The transcriptional factors Oct4, Nr2f2, and miR-302 are linked in a regulatory circuit that governs both pluripotency and differentiation in hESCs (Rosa and Brivanlou 2011).

In human fibroblasts, miR-302b coupled with miR-372 can promote reprogramming into iPSCs by targeting several genes which are involved in cell cycle, epithelial-mesenchymal transition (EMT), and epigenetic regulation, including Tgfbr2 and Rhoc (Subramanyam et al. 2011). Impressively, the miR-302 cluster alone may drive human skin cancer cells or hair follicle cells to iPSC fates (Lin et al. 2008; Lin et al. 2011). A recent study has indicated overexpression of the miR-302s/367 cluster and addition of valproic acid (VPA) can rapidly and efficiently reprogram mouse and human fibroblasts to iPSCs without exogenous transcription factors (Anokye-Danso et al. 2011). It seems that miR-367 plays a critical role in the process of miR-302s/367-mediated reprogramming, which can activate the expression of Oct4.

15.3.2 miR-290 Cluster

The mouse miR-290 cluster includes 14 miRNAs (miR-290-5p, 290-3p, 291a-5p, 291a-3p, 292-5p, 292-3p, 291b-5p, 291b-3p, 293, 293*, 294, 294*, 295, and 295*) coded by 2.2-kb DNA fragment (Landgraf et al. 2007). The miRNA cluster homologous to miR-290 in human is the miR-371 cluster, which includes miR-371-5p, 371-3p, 371b-5p, 371b-3p, 372, 373, and 373*. The majority of mice deficient in the miR-290 cluster die during embryogenesis. Surviving females are infertile due to the absence of germ cells. In Dicer−/− ESCs, reintroduction of the miR-290 cluster can partially rescue the proliferation defects of Dicer-deficient ESCs. Similarly, in Dgcr8−/− ESCs, the members of the miR-290 family can rescue cell proliferation defects by suppression of several key regulators of the G1/S checkpoint (Wang et al. 2008). G1/S restriction in ESCs is largely absent, allowing the cells to move through the G1 to S phase rapidly. The miR-290 cluster regulates this transition in ESCs through mediating cell cycle regulators directly and indirectly, including targeting of P21, Rbl2, Lats2, Wee1, and Fbx15 (Wang and Blelloch 2009; Lichner et al. 2011; Sinkkonen et al. 2008). Importantly, both miR-290 and miR-371 clusters are regulated by ESC-specific transcription factors, including Oct4, Sox2, Nanog, and Tcf3 (Marson et al. 2008).

Furthermore, the miR-290 cluster has been implicated in promotion of cell survival in murine ESCs by targeting two critical apoptotic regulators directly: caspase 2, the most highly conserved mammalian caspase, and El24, a p53 transcriptional target (Zheng et al. 2011). Overall, the miR-290 cluster can be classified as a regulator of pluripotency and cell survival in ESCs.

15.3.3 Let-7 Family

During differentiation, ESCs must shift to alternative molecular programs that inhibit self-renewal and promote differentiation into specialized cell types. The let-7 family miRNAs are undetectable in ESCs but are highly expressed in differentiated cell types and tightly regulated during ESC differentiation. Let-7 is considered as an "anti-stemness" miRNA and functions as a pro-differentiation factor during ESC differentiation (Melton et al. 2010).

The RNA-binding protein Lin28 has been identified as an inhibitor of let-7 expression through binding of let-7 precursor RNAs and inhibition of Dicer cleavage activity (Viswanathan et al. 2008). Other studies have indicated that Zcchc11 functions with Lin28 to catalyze the addition of a short stretch of uridines to the 3′-end of the Lin28-bound let-7 precursor RNAs (Heo et al. 2009; Hagan et al. 2009).

This has put forth a model in which Lin28 recruits Zcchc11 to let-7 precursors, and promotes the addition of uridines to induce blockade of let-7 maturation in ESCs.

In Dgcr8−/− ESCs, the reintroduction of let-7 can rescue silencing of the self-renewal pathways by directly suppressing several ESC-specific transcription factors, including c-Myc, Sall4, and Lin28 (Melton et al. 2010). Inhibition of let-7 family via antisense oligos enhances reprogramming of mouse fibroblast to iPSCs, highlighting the potential importance of the let-7 family in maintenance of the differentiated state (Melton et al. 2010).

15.3.4 Other miRNAs Regulating the Pluripotent Stem Cell Network

Many other miRNAs have been investigated to be involved in the regulation of self-renewal and differentiation of pluripotent cells. In human ESCs, miR-195 and miR-372 have been found to partially rescue cell cycle impairment in Dicer−/− hESCs. miR-195 has been shown to regulate the G1/M checkpoint inhibitory kinase Wee1, and miR-372 has been shown to regulate the G1/S checkpoint inhibitor Cdkn1a (Qi et al. 2009). miR-92b has also been identified as a regulator of G1/S transition in human ESCs by repression of the P57 checkpoint gene (Sengupta et al. 2009).

miR-145 functions by promoting cell differentiation from the pluripotent state through repression of key pluripotent transcriptional factors including Klf4, Sox2, and Oct4 (Xu et al. 2009). In human ESCs, overexpression of miR-145 blocks cell self-renewal and induces differentiation. Conversely, inhibition of miR-145 has been shown to increase the capacity of hESC self-renewal. Expression of miR-145 is negatively regulated by binding of Oct4 in a negative feedback circuit.

Muscle-specific miRNAs miR-1 and miR-133 promote mesodermal differentiation in ESCs when ectopically expressed in murine ESCs through direct repression of Dll-1, a key ligand in the Notch signaling pathway (Ivey et al. 2008). During retinoic acid–induced smooth muscle cell (SMC) differentiation, overexpression of miR-1 promotes SMC differentiation by repressing Klf4 (Xie et al. 2011). During cardiac differentiation of ESCs, inhibition of miR-296-3p and miR-200c* decreases, while inhibition of miR-465-5p increases (Sun et al. 2011). These miRNAs are regulated by neuregulin 1 signaling. Identification of new microRNAs that are important for ESC cardiac differentiation and regulated by neuregulin 1 is an important field of continued research.

During neuronal progenitor cell (NPC) differentiation from human ESCs, inhibition of miR-9 activity through addition of antisense oligos has been shown to restrict hNPC proliferation (Delaloy et al. 2010). Stathmin is a target of miR-9 and mediates its effects in early NPCs. This observation indicates tissue-specific miRNAs can narrow the gene expression profile of cells to a defined lineage and thereby promote differentiation into a specific cell type.

In mouse ESCs, a recent study discovered miR-134, miR-296, and miR-470 regulate pluripotent transcription factors Nanog, Oct4, and Sox2 (Tay et al. 2008). Another study by Wellner et al. has shown that introduction of miR-200c, miR-203, and miR-183 into ESCs can upregulate markers of differentiation and reduce capacity for self-renewal in mESCs through repression of Klf4 and Sox2 (Wellner et al. 2009).

Interestingly, it was recently found that expression of a large cluster of miRNAs encoded in the Dlk1-Dio3 gene cluster on mouse chromosome 12 uniquely distinguishes some miPSCs and mESCs. This gene cluster is expressed in ESCs and iPSCs but is silenced in incompletely reprogrammed iPSCs (Liu et al. 2010; Stadtfeld et al. 2010). Determination of the miRNAs that are critical in this cluster for maintenance of pluripotency in ESCs and iPSCs and the corresponding targets of these miRNAs should yield significant insights to the importance of miRNA regulatory pathways in ESCs and iPSCs. Determination of a homologous miRNA cluster in human ESCs and iPSCs is also critical to future clinical applications of pluripotent stem cells.

15.4   miRNA-Based Cellular Reprogramming

miRNAs modulate target genes in a tissue specific manner, and hence participate in cellular reprogramming by modulating proteins that function in reprogramming. University of Pennsylvania researchers have shown that miRNAs can be used to reprogram mouse and human fibroblasts into iPS cells without the use of inefficient exogenous transcription factors (Anokye-Danso et al. 2011). Below is a schematic for miRNA-based reprogramming in adult somatic cells. Figure 15.1 shows the flowchart for miRNA-based cellular reprogramming of IMR-90 human fibroblast cells.

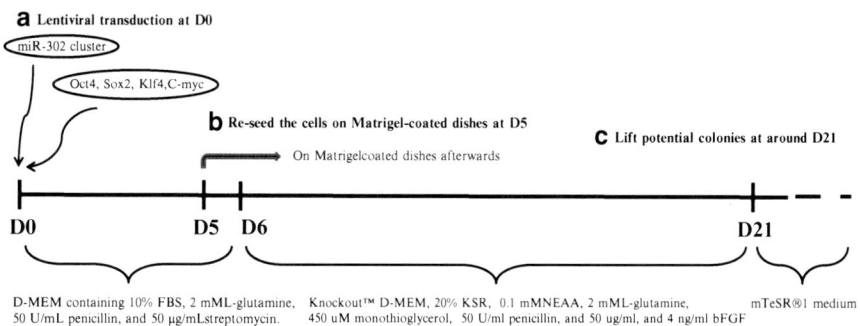

Fig. 15.1 Flowchart shows the iPSC' generation with traditional transcription factors (Klf4, Oct4, Sox2, and C-Myc) and the miR-302 cluster. Typically, the whole procedure includes three steps: (**a**) lentiviral transduction, (**b**) cell reseeding on Matrigel-coated dishes, and (**c**) lifting potential colonies. "D0" represents day 0

15.4.1 Steps to Reprogram IMR-90 Human Fibroblast with miR-302 Cluster

15.4.1.1 Materials

A. Lentiviral Transduction

IMR-90 human fibroblast (ATCC, Manassas, VA, USA)
D-MEM (Invitrogen, Carlsbad, CA, USA)
Fetal bovine serum (FBS) (Invitrogen, Carlsbad, CA, USA)
L-Glutamine (Invitrogen, Carlsbad, CA, USA)
Polybrene (Sigma-Aldrich, USA)

B. Coating Tissue Culture Dishes with BD Matrigel™ hESC-Qualified Matrix

BD Matrigel™ hESC-qualified Matrix (BD Biosciences, San Jose, CA, USA)
KnockOut™ D-MEM (Invitrogen, Carlsbad, CA, USA)
Tissue culture plate (BD Falcon™, San Jose, CA, USA)
BD Falcon™ conical tubes (BD Biosciences, San Jose, CA, USA)
Parafilm® M (Pechiney Plastic Packaging Company, Chicago, IL, USA)

C. Transfer Infected IMR90 Cells on Matrigel™-Coated Dish

TrypLE™ (Invitrogen, Carlsbad, CA, USA)
Countess® Automated Cell Counter (Invitrogen, Carlsbad, CA, USA)
KnockOut™ Serum Replacement (KSR) (Invitrogen, Carlsbad, CA, USA)
Nonessential amino acids (NEAA) (Invitrogen, Carlsbad, CA, USA)
Monothioglycerol (Sigma-Aldrich, USA)
bFGF (R&D Systems, Minneapolis, MN, USA)

D. Lift Single Colonies to Establish the Potential iPSC Line

mTeSR®1 medium for maintenance of human ESCs and iPSCs (STEMCELL
   Technologies, Vancouver, BC, Canada)

15.4.1.2 Methods

One of major hindrances in iPSC derivation and therapeutic usage is low reprogramming efficiency, about from 0.01% to 0.2%. Several approaches have been applied to improve reprogramming efficiency, including small molecule–based methods. miRNAs provide a valuable tool to improve reprogramming efficiency. The hESC-specific miR-302 cluster has been considered as substantial inducer of iPSC generation.

A. Lentiviral Transduction

Very high gene expression can be obtained by lentiviral transduction. Therefore, in this protocol, we employ the lentivirus to overexpress the reprogramming transcription factors and miR-302 cluster.

1. IMR90 human fibroblast cells are maintained with D-MEM containing 10% FBS, 2 mM L-glutamine, 50 U/mL penicillin, and 50 μg/mL streptomycin.
2. When IMR90 were plated at about 80% confluence per well of six-well plate, the cells are infected with 1×10^7 TU titer of individual lentivirus (Oct4, Sox2, Klf4, C-Myc, and miR-302 cluster including miR-302a, miR-302b, miR302c, and miR-302d), with 6 μg/mL polybrene, in 1 mL of fresh medium.
3. After 24 h of incubation, change fresh medium. Change medium every day until day 5 posttransduction.

B. Coating Tissue Culture Dishes with BD Matrigel™ hESC-Qualified Matrix

To avoid the mouse feeder cells due to the potential usage of human iPSC line for regenerative medicine, we derive the iPSCs on feeder-free surfaces using Matrigel-coated tissue culture dishes.

1. Take out an aliquot of frozen BD Matrigel™ hESC-qualified Matrix from −80°C refrigerator. Thaw it on ice until liquid.
2. Dispense 25 mL of cold KnockOut™ D-MEM into a 50-mL conical tube and keep on ice.
3. Add the thawed BD Matrigel™ Matrix into the cold KnockOut™ D-MEM medium and mix thoroughly.
4. Coat tissue culture dishes with the diluted BD Matrigel™ solution. For one well of six-well plate, use 1 mL of diluted BD Matrigel™ solution. Swirl the dish to spread the BD Matrigel™ solution evenly.
5. Coated dish should be left at room temperature for at least 1 h. If not used immediately, the coated plate must be sealed by Parafilm® M and can be stored at 4°C for at most 1 week.
6. Remove the diluted BD Matrigel™ solution by aspiration before you use the plates.

C. Transfer the Transduced IMR90 Cells on Matrigel™-Coated Dish

The iPSCs can be generated on Matrigel-coated dishes to avoid the usage of murine feeder cells.

1. Lift the cells at day 5 post-transduction with TrypLE™ Express.
2. Count the cells with Countess® Automated Cell Counter.
3. Seed 5×10^4 transduced cells into 1 well of Matrigel™-coated six-well plate. Keep culture the cells with previous medium for another 24 h.
4. Change medium to hESC medium containing KnockOut™ D-MEM, 20% KnockOut™ Serum Replacement (KSR), 0.1 mM NEAA, 2 mM L-glutamine, 450 uM monothioglycerol, 50 U/mL penicillin, and 50 μg/mL, supplemental with 4 ng/mL bFGF.
5. Change fresh hESC medium every day afterward.

D. Lifting the Single Colonies to Establish the Potential iPSC Lines

Single potential iPSC colonies can be obtained through lifting and reseeding iPSCs physically. The iPSC cell lines can be maintained in mTeSR®1 medium under the feeder-free condition.

1. Typically, after day 21 transduction, the colonies with hESC morphology can be observed.
2. Change to fresh medium and locate undifferentiated colonies.
3. Use 200-μL tip, deattach one single colony physically, and transfer it into a 1.5-mL tube.
4. Pipette up and down strongly several times to break the colony into several small pieces.
5. Reseed the colony pieces onto another new Matrigel™-coated dish.
6. Maintain and split these picked potential iPSCs in mTeSR®1 medium afterward.

Typically, over expression of the miR-302 cluster can increase the reprogramming efficiency of iPSC generation by at least five times as compared to traditional 4-factor reprogramming. iPSCs derived with miR-302s have similar pluripotency to human embryonic stem cells, and form derivatives in vitro and in vivo of three germ layers.

15.5 Future Perspectives

Taken together, miRNAs are becoming crucial players in stem cell biology, opening up avenues to unravel the cellular and molecular mechanisms underlying self-renewal and pluripotency of ESCs and iPSCs. While a number of studies have demonstrated that miRNAs play a significant role in the regulation of ESC pluripotency, functional characterization of miRNAs in pluripotent cells is still in its infancy.

Recently, it has been demonstrated that mouse and human cells can be reprogrammed to iPSCs by direct transfection of cells by mature double-stranded miRNAs, including the combination of miR-200c, miR-302s, and miR-369s family miRNAs (Ambasudhan et al. 2011). This strategy bypasses the usage of vectors for gene delivery and provides a very promising method for deriving iPSCs. For the purposes of regenerative therapy, pluripotent cells can be differentiated into various therapeutic cell types such as neuronal and endothelial cells prior to transplantation to reduce their capacity for tumorigenicity, although they are not completely devoid of this potential (Ohm et al. 2010; Ghosh et al. 2011). These examples demonstrate the significance of miRNA function in the regulation of multiple signaling networks involved in iPSC generation.

How does miR-302s maintain "stemness" of ESCs and iPSCs? How do different miRNAs interact with each other directly or indirectly to maintain stem cell signatures? How do miRNAs regulate stem cell differentiation globally? There are still plenty of questions to be answered in this field. Further investigation is needed to uncover the complicated miRNA regulatory networks and cross-networks which underlie the self-renewal and differentiation of ESCs and iPSCs. Researchers can employ a number of techniques including in vitro cell culture, knockout mice, as well as transgenic animal models to address the roles of miRNAs in pluripotent cells and development. Hopefully in the near future, continued research into this field will not only provide broad insights into the function of miRNAs but also identify novel miRNA-based networks that can be used to improve clinical applications of pluripotent cells.

Acknowledgment We thank the grant supports of AHA postdoctoral fellowship (SH).

References

Ambasudhan R, Talantova M, Coleman R, Yuan X, Zhu S, Lipton SA, Ding S (2011) Direct reprogramming of adult human fibroblasts to functional neurons under defined conditions. Cell Stem Cell 9(2):113–118. doi:S1934-5909(11)00332-8 [pii] 10.1016/j.stem.2011.07.002

Anokye-Danso F, Trivedi CM, Juhr D, Gupta M, Cui Z, Tian Y, Zhang Y, Yang W, Gruber PJ, Epstein JA, Morrisey EE (2011) Highly efficient miRNA-mediated reprogramming of mouse and human somatic cells to pluripotency. Cell Stem Cell 8(4):376–388. doi:S1934-5909(11) 00111-1 [pii] 10.1016/j.stem.2011.03.001

Barroso-delJesus A, Romero-Lopez C, Lucena-Aguilar G, Melen GJ, Sanchez L, Ligero G, Berzal-Herranz A, Menendez P (2008) Embryonic stem cell-specific miR302-367 cluster: human gene structure and functional characterization of its core promoter. Mol Cell Biol 28 (21):6609–6619. doi:MCB.00398-08 [pii] 10.1128/MCB.00398-08

Barroso-delJesus A, Lucena-Aguilar G, Sanchez L, Ligero G, Gutierrez-Aranda I, Menendez P (2011) The nodal inhibitor lefty is negatively modulated by the microRNA miR-302 in human embryonic stem cells. FASEB J 25(5):1497–1508. doi:fj.10-172221 [pii] 10.1096/fj.10-172221

Bartel DP (2009) MicroRNAs: target recognition and regulatory functions. Cell 136(2):215–233. doi:S0092-8674(09)00008-7 [pii] 10.1016/j.cell.2009.01.002

Bernstein E, Kim SY, Carmell MA, Murchison EP, Alcorn H, Li MZ, Mills AA, Elledge SJ, Anderson KV, Hannon GJ (2003) Dicer is essential for mouse development. Nat Genet 35 (3):215–217. doi:10.1038/ng1253 ng1253 [pii]

Card DA, Hebbar PB, Li L, Trotter KW, Komatsu Y, Mishina Y, Archer TK (2008) Oct4/Sox2-regulated miR-302 targets cyclin D1 in human embryonic stem cells. Mol Cell Biol 28 (20):6426–6438. doi:MCB.00359-08 [pii] 10.1128/MCB.00359-08

Chin MH, Mason MJ, Xie W, Volinia S, Singer M, Peterson C, Ambartsumyan G, Aimiuwu O, Richter L, Zhang J, Khvorostov I, Ott V, Grunstein M, Lavon N, Benvenisty N, Croce CM, Clark AT, Baxter T, Pyle AD, Teitell MA, Pelegrini M, Plath K, Lowry WE (2009) Induced pluripotent stem cells and embryonic stem cells are distinguished by gene expression signatures. Cell Stem Cell 5(1):111–123. doi:S1934-5909(09)00292-6 [pii] 10.1016/j.stem.2009.06.008

Delaloy C, Liu L, Lee JA, Su H, Shen F, Yang GY, Young WL, Ivey KN, Gao FB (2010) MicroRNA-9 coordinates proliferation and migration of human embryonic stem cell-derived neural progenitors. Cell Stem Cell 6(4):323–335. doi:S1934-5909(10)00092-5 [pii] 10.1016/j.stem.2010.02.015

Gangaraju VK, Lin H (2009) MicroRNAs: key regulators of stem cells. Nat Rev Mol Cell Biol 10 (2):116–125. doi:nrm2621 [pii] 10.1038/nrm2621

Ghosh Z, Huang M, Hu S, Wilson KD, Dey D, Wu JC (2011) Dissecting the oncogenic and tumorigenic potential of differentiated human induced pluripotent stem cells and human embryonic stem cells. Cancer Res 71(14):5030–5039. doi:0008-5472.CAN-10-4402 [pii] 10.1158/0008-5472.CAN-10-4402

Hagan JP, Piskounova E, Gregory RI (2009) Lin28 recruits the TUTase Zcchc11 to inhibit let-7 maturation in mouse embryonic stem cells. Nat Struct Mol Biol 16(10):1021–1025. doi: nsmb.1676 [pii] 10.1038/nsmb.1676

Heo I, Joo C, Kim YK, Ha M, Yoon MJ, Cho J, Yeom KH, Han J, Kim VN (2009) TUT4 in concert with Lin28 suppresses microRNA biogenesis through pre-microRNA uridylation. Cell 138 (4):696–708. doi:S0092-8674(09)00964-7 [pii] 10.1016/j.cell.2009.08.002

Ivey KN, Muth A, Arnold J, King FW, Yeh RF, Fish JE, Hsiao EC, Schwartz RJ, Conklin BR, Bernstein HS, Srivastava D (2008) MicroRNA regulation of cell lineages in mouse and human embryonic stem cells. Cell Stem Cell 2(3):219–229. doi:S1934-5909(08)00057-X [pii] 10.1016/j.stem.2008.01.016

Landgraf P, Rusu M, Sheridan R, Sewer A, Iovino N, Aravin A, Pfeffer S, Rice A, Kamphorst AO, Landthaler M, Lin C, Socci ND, Hermida L, Fulci V, Chiaretti S, Foa R, Schliwka J, Fuchs U, Novosel A, Muller RU, Schermer B, Bissels U, Inman J, Phan Q, Chien M, Weir DB, Choksi R, De Vita G, Frezzetti D, Trompeter HI, Hornung V, Teng G, Hartmann G, Palkovits M, Di Lauro R, Wernet P, Macino G, Rogler CE, Nagle JW, Ju J, Papavasiliou FN, Benzing T, Lichter P, Tam W, Brownstein MJ, Bosio A, Borkhardt A, Russo JJ, Sander C, Zavolan M, Tuschl T (2007) A mammalian microRNA expression atlas based on small RNA library sequencing. Cell 129(7):1401–1414. doi:S0092-8674(07)00604-6 [pii] 10.1016/j.cell.2007.04.040

Lichner Z, Pall E, Kerekes A, Pallinger E, Maraghechi P, Bosze Z, Gocza E (2011) The miR-290-295 cluster promotes pluripotency maintenance by regulating cell cycle phase distribution in mouse embryonic stem cells. Differentiation 81(1):11–24. doi:S0301-4681(10)00076-9 [pii] 10.1016/j.diff.2010.08.002

Lin SL, Chang DC, Chang-Lin S, Lin CH, Wu DT, Chen DT, Ying SY (2008) Mir-302 reprograms human skin cancer cells into a pluripotent ES-cell-like state. RNA 14(10):2115–2124. doi: rna.1162708 [pii] 10.1261/rna.1162708

Lin SL, Chang DC, Lin CH, Ying SY, Leu D, Wu DT (2011) Regulation of somatic cell reprogramming through inducible mir-302 expression. Nucleic Acids Res 39(3):1054–1065. doi:gkq850 [pii] 10.1093/nar/gkq850

Liu L, Luo GZ, Yang W, Zhao X, Zheng Q, Lv Z, Li W, Wu HJ, Wang L, Wang XJ, Zhou Q (2010) Activation of the imprinted Dlk1-Dio3 region correlates with pluripotency levels of mouse stem cells. J Biol Chem 285(25):19483–19490. doi:M110.131995 [pii] 10.1074/jbc. M110.131995

Mallick B, Chakrabarti J, Ghosh Z (2011) MicroRNA reins in embryonic and cancer stem cells. RNA Biol 8(3):415–426. doi:14497 [pii]

Marson A, Levine SS, Cole MF, Frampton GM, Brambrink T, Johnstone S, Guenther MG, Johnston WK, Wernig M, Newman J, Calabrese JM, Dennis LM, Volkert TL, Gupta S, Love J, Hannett N, Sharp PA, Bartel DP, Jaenisch R, Young RA (2008) Connecting microRNA genes to the core transcriptional regulatory circuitry of embryonic stem cells. Cell 134 (3):521–533. doi:S0092-8674(08)00938-0 [pii] 10.1016/j.cell.2008.07.020

Melton C, Judson RL, Blelloch R (2010) Opposing microRNA families regulate self-renewal in mouse embryonic stem cells. Nature 463(7281):621–626. doi:nature08725 [pii] 10.1038/nature08725

Morin RD, O'Connor MD, Griffith M, Kuchenbauer F, Delaney A, Prabhu AL, Zhao Y, McDonald H, Zeng T, Hirst M, Eaves CJ, Marra MA (2008) Application of massively parallel sequencing to microRNA profiling and discovery in human embryonic stem cells. Genome Res 18(4):610–621. doi:gr.7179508 [pii] 10.1101/gr.7179508

Ohm JE, Mali P, Van Neste L, Berman DM, Liang L, Pandiyan K, Briggs KJ, Zhang W, Argani P, Simons B, Yu W, Matsui W, Van Criekinge W, Rassool FV, Zambidis E, Schuebel KE, Cope L, Yen J, Mohammad HP, Cheng L, Baylin SB (2010) Cancer-related epigenome changes associated with reprogramming to induced pluripotent stem cells. Cancer Res 70 (19):7662–7673. doi:0008-5472.CAN-10-1361 [pii] 10.1158/0008-5472.CAN-10-1361

Qi J, Yu JY, Shcherbata HR, Mathieu J, Wang AJ, Seal S, Zhou W, Stadler BM, Bourgin D, Wang L, Nelson A, Ware C, Raymond C, Lim LP, Magnus J, Ivanovska I, Diaz R, Ball A, Cleary MA, Ruohola-Baker H (2009) microRNAs regulate human embryonic stem cell division. Cell Cycle 8(22):3729–3741. doi:10033 [pii]

Rosa A, Brivanlou AH (2011) A regulatory circuitry comprised of miR-302 and the transcription factors OCT4 and NR2F2 regulates human embryonic stem cell differentiation. EMBO J 30 (2):237–248. doi:emboj2010319 [pii] 10.1038/emboj.2010.319

Rosa A, Spagnoli FM, Brivanlou AH (2009) The miR-430/427/302 family controls mesendodermal fate specification via species-specific target selection. Dev Cell 16 (4):517–527. doi:S1534-5807(09)00081-1 [pii] 10.1016/j.devcel.2009.02.007

Sengupta S, Nie J, Wagner RJ, Yang C, Stewart R, Thomson JA (2009) MicroRNA 92b controls the G1/S checkpoint gene p57 in human embryonic stem cells. Stem Cells 27(7):1524–1528. doi:10.1002/stem.84

Sinkkonen L, Hugenschmidt T, Berninger P, Gaidatzis D, Mohn F, Artus-Revel CG, Zavolan M, Svoboda P, Filipowicz W (2008) MicroRNAs control de novo DNA methylation through regulation of transcriptional repressors in mouse embryonic stem cells. Nat Struct Mol Biol 15(3):259–267. doi:nsmb.1391 [pii] 10.1038/nsmb.1391

Stadler B, Ivanovska I, Mehta K, Song S, Nelson A, Tan Y, Mathieu J, Darby C, Blau CA, Ware C, Peters G, Miller DG, Shen L, Cleary MA, Ruohola-Baker H (2010) Characterization of microRNAs involved in embryonic stem cell states. Stem Cells Dev 19(7):935–950. doi:10.1089/scd.2009.0426

Stadtfeld M, Apostolou E, Akutsu H, Fukuda A, Follett P, Natesan S, Kono T, Shioda T, Hochedlinger K (2010) Aberrant silencing of imprinted genes on chromosome 12qF1 in mouse induced pluripotent stem cells. Nature 465(7295):175–181. doi:nature09017 [pii] 10.1038/nature09017

Subramanyam D, Lamouille S, Judson RL, Liu JY, Bucay N, Derynck R, Blelloch R (2011) Multiple targets of miR-302 and miR-372 promote reprogramming of human fibroblasts to induced pluripotent stem cells. Nat Biotechnol. doi:nbt.1862 [pii] 10.1038/nbt.1862

Sun M, Yan X, Bian Y, Caggiano AO, Morgan JP (2011) Improving murine embryonic stem cell differentiation into cardiomyocytes with neuregulin1: differential expression of microRNA. Am J Physiol Cell Physiol. doi:ajpcell.00141.2010 [pii] 10.1152/ajpcell.00141.2010

Takahashi K, Tanabe K, Ohnuki M, Narita M, Ichisaka T, Tomoda K, Yamanaka S (2007) Induction of pluripotent stem cells from adult human fibroblasts by defined factors. Cell 131 (5):861–872. doi:S0092-8674(07)01471-7 [pii] 10.1016/j.cell.2007.11.019

Tay Y, Zhang J, Thomson AM, Lim B, Rigoutsos I (2008) MicroRNAs to Nanog, Oct4 and Sox2 coding regions modulate embryonic stem cell differentiation. Nature 455(7216):1124–1128. doi:nature07299 [pii] 10.1038/nature07299

Tiscornia G, Izpisua Belmonte JC (2010) MicroRNAs in embryonic stem cell function and fate. Genes Dev 24(24):2732–2741. doi:24/24/2732 [pii] 10.1101/gad.1982910

Viswanathan SR, Daley GQ, Gregory RI (2008) Selective blockade of microRNA processing by Lin28. Science 320(5872):97–100. doi:1154040 [pii] 10.1126/science.1154040

Wang Y, Blelloch R (2009) Cell cycle regulation by microRNAs in embryonic stem cells. Cancer Res 69(10):4093–4096. doi:0008-5472.CAN-09-0309 [pii] 10.1158/0008-5472.CAN-09-0309

Wang Y, Medvid R, Melton C, Jaenisch R, Blelloch R (2007) DGCR8 is essential for microRNA biogenesis and silencing of embryonic stem cell self-renewal. Nat Genet 39(3):380–385. doi: ng1969 [pii] 10.1038/ng1969

Wang Y, Baskerville S, Shenoy A, Babiarz JE, Baehner L, Blelloch R (2008) Embryonic stem cell-specific microRNAs regulate the G1-S transition and promote rapid proliferation. Nat Genet 40(12):1478–1483. doi:ng.250 [pii] 10.1038/ng.250

Wellner U, Schubert J, Burk UC, Schmalhofer O, Zhu F, Sonntag A, Waldvogel B, Vannier C, Darling D, zur Hausen A, Brunton VG, Morton J, Sansom O, Schuler J, Stemmler MP, Herzberger C, Hopt U, Keck T, Brabletz S, Brabletz T (2009) The EMT-activator ZEB1 promotes tumorigenicity by repressing stemness-inhibiting microRNAs. Nat Cell Biol 11 (12):1487–1495. doi:ncb1998 [pii] 10.1038/ncb1998

Wilson KD, Venkatasubrahmanyam S, Jia F, Sun N, Butte AJ, Wu JC (2009) MicroRNA profiling of human-induced pluripotent stem cells. Stem Cells Dev 18(5):749–758. doi:10.1089/scd.2008.0247

Xie C, Huang H, Sun X, Guo Y, Hamblin M, Ritchie RP, Garcia-Barrio MT, Zhang J, Chen YE (2011) MicroRNA-1 regulates smooth muscle cell differentiation by repressing Kruppel-like factor 4. Stem Cells Dev 20(2):205–210. doi:10.1089/scd.2010.0283

Xu N, Papagiannakopoulos T, Pan G, Thomson JA, Kosik KS (2009) MicroRNA-145 regulates OCT4, SOX2, and KLF4 and represses pluripotency in human embryonic stem cells. Cell 137 (4):647–658. doi:S0092-8674(09)00252-9 [pii] 10.1016/j.cell.2009.02.038

Yu J, Vodyanik MA, Smuga-Otto K, Antosiewicz-Bourget J, Frane JL, Tian S, Nie J, Jonsdottir GA, Ruotti V, Stewart R, Slukvin II, Thomson JA (2007) Induced pluripotent stem cell lines derived from human somatic cells. Science 318(5858):1917–1920. doi:1151526 [pii] 10.1126/science.1151526

Zheng GX, Ravi A, Calabrese JM, Medeiros LA, Kirak O, Dennis LM, Jaenisch R, Burge CB, Sharp PA (2011) A latent pro-survival function for the mir-290-295 cluster in mouse embryonic stem cells. PLoS Genet 7(5):e1002054. doi:10.1371/journal.pgen.1002054 PGENETICS-D-11-00242 [pii]

Part III
Applications

Chapter 16
The Role of RNA Interference in Targeting the Cancer Stem Cell and Clinical Trials for Cancer

Russell C. Langan*, John Mullinax*, Manish Raiji*, and Itzhak Avital

Abstract Recently, compelling evidence has emerged in support of the cancer stem cell (CSC) theory for solid organ cancers. The CSC theory postulates that CSCs account for tumor initiation, tumor propagation, therapeutic resistance, and relapse following surgery or therapy. CSCs are able to do this through traits including (1) self-renewal, either through symmetric or asymmetric cell division via nonrandom chromosomal cosegregation; (2) the capacity for differentiation, which allows for the recapitulation of all cell types of the original tumor; and (3) tumor initiating capacity, which is the ability to propagate tumors when transplanted into a separate environment. The CSC theory provides better understanding of neoplastic formation and tumor propagation which may lead to novel exciting therapies for advanced cancer. Clear identification CSC-specific surface markers, genes (Nanog, Oct3/4, STAT3), and pathways such as TGF-β, hedgehog, and Wnt-β-catenin are crucial to the development of CSC-targeted treatments. RNAi provides a unique opportunity to silence cancer-causing stem cell genes at the pretranslation level, which is otherwise not possible with conventional therapies such as cytotoxic chemotherapy, small molecule inhibitors, or monoclonal antibodies. Owing to the explosion of knowledge generated by a growing understanding of the human genome and the development of high-throughput gene expression profiling of tissue stem cells, a plethora of genes that contribute to tumor initiation and the metastatic cascade are being discovered. RNAi therapy against multidrug resistance genes and CSC genes may provide exceptional benefit and herald a paradigm shift in the treatment of deadly diseases.

Keywords Cancer stem cell • cancer stem cell therapeutic • pluripotency pathways • polo-like kinase-1 (PLK1)

---

*These authors have equally contributed.

I. Avital (✉)
Surgery Branch, National Cancer Institute, National Institutes of Health, Bethesda, MD, USA
e-mail: Avitali@mail.nih.gov

B. Mallick (eds.), *Regulatory RNAs*, DOI 10.1007/978-3-642-22517-8_16,
© Springer-Verlag Berlin Heidelberg (outside the USA) 2012

16.1 Introduction to the Cancer Stem Cell

16.1.1 The Cancer Stem Cell Theory

Current nonsurgical therapy for patients with locally advanced or metastatic disease continues to fall short of providing durable survival benefit. In the United States, a total of 1.6 million new cases of primary cancers and 572,000 deaths are projected to occur in 2011 (Siegel et al. 2011). Factors contributing to this abysmal prognosis include delayed diagnosis, resistance to conventional therapies, and the lack of understanding of neoplastic formation and tumor propagation. Furthermore, many drugs considered to have efficacy against cancer were discovered decades ago and without a clear understanding of the mechanism of action (Haney 2007). Though the molecular changes involved in malignant transformation are better understood today, systemic therapy continues to fall short of providing long-term survival for patients with advanced disease.

Recently, compelling evidence has emerged in support of the cancer stem cell (CSC) theory for many solid organ cancers. Traditionally, the vast majority of all cancer drugs were crude cytotoxic agents that discriminated poorly between cancer cells and normal cells in a given tissue. These agents provided minimal benefit because they preferentially affect the actively proliferating cells of a tumor mass and in many instances recurrence and/or progression occurred via the minority of tumor cells that were left unscathed. It is this particular cell population which has been proposed as the CSC population.

The CSC theory was first developed after noting similarities between embryonic tissue and cancer with respect to their enormous capacity for proliferation and differentiation. This observation led to the "Embryonal Rest" hypothesis which postulated that the stem cells in adult tissue acquire mutations leading to cancer (Pal et al. 2010). This hypothesis holds that because many mutations are necessary for a cell to become tumorigenic, it is unlikely that well-differentiated cells within a given organ live long enough to form cancer. Due to their long life span, however, tissue stem cells are able to accumulate the number of mutations required for carcinogenesis (Pal et al. 2010). Validation of this hypothesis was first established in acute myelogenous leukemia (AML) (Lapidot et al. 1994), and Bonnet et al. found that AML is clonally derived, hierarchically organized, and can be serially passaged in murine models (Bonnet and Dick 1997).

The CSC theory postulates that CSCs account for tumor initiation, tumor propagation, therapeutic resistance, and relapse following surgery or therapy. CSCs are able to do this through traits including (1) self-renewal, either through symmetric or asymmetric cell division via nonrandom chromosomal cosegregation; (2) the capacity for differentiation, which allows for the recapitulation of all cell types of the original tumor; and (3) tumor initiating capacity, which is the ability to propagate tumors when transplanted into a separate environment. Furthermore, traits of motility, invasiveness, and self-renewal are central to cancer and are reflections of the malignant stem cell subpopulations located within solid organ cancers. Identification and proper characterization is imperative to understand the biology of the CSCs. The CSC theory provides better understanding of neoplastic formation and tumor propagation which may lead to novel exciting therapies for advanced cancer.

16.1.2   Potential Role of Cancer Therapeutics Targeting the Cancer Stem Cell

RNA interference (RNAi) harnesses the capability to target intricate CSC or pluripotency pathways associated with tumor initiation and propagation and the metastatic cascade. The ability of RNAi to suppress the translation of any gene expands the universe of drug targets to the entire genome. It will be possible to inhibit expression of not only tyrosine kinase receptors but also any anti-apoptotic gene, survival factor, and tumor-promoting growth factor, the key participants in their signaling pathways or genes essential for cell proliferation. Currently, improved in vitro culture methods to maintain CSC activity from primary tumor samples open the arena of high-throughput screening of RNAi libraries. These studies can therefore provide novel agents to target this critical cell population via RNAi.

RNAi targeting a CSC provides intelligently designed anticancer therapies to silence cancer-causing genes which would otherwise not be amenable to conventional therapies such as cytotoxic chemotherapy, small molecule inhibitors, or monoclonal antibodies. This chapter will focus on the role of RNAi in targeting the CSC and CSC pluripotency pathways and discuss current clinical trials implementing RNAi targeting these cells. Specifically, we will describe the use of RNAi screens in stem cells and discuss RNAi approaches to target CSC function, multidrug resistance genes, and pluripotency pathways. By analyzing the gene expression of CSCs, a novel RNAi therapy will be developed to target the population of cells that is responsible for recurrence and metastasis.

16.2   Targeting the Cancer Stem Cell: RNAi Techniques

16.2.1   RNAi Screening in Human Stem Cells and Cancer Stem Cells

Prior to understanding the role RNAi plays in targeting the CSC, we must first review its role in normal human stem cells. Stem cells discern themselves from other cells by two paramount characteristics, self-renewal and differentiation (Chap. 17, this volume). Embryonic stem cells (ESCs) are pluripotent stem cells derived from the inner cell mass of a blastocyst. ESCs are therefore attractive cells to study molecular regulation of cell lineage commitment and differentiation since they can give rise to all three germ layers: endoderm, mesoderm, and ectoderm (Zou 2010). The ESC provides a possible strategy for high-throughput functional screening of genes that are required for cell lineage differentiation of ESC (Zou 2010) (Fig. 16.1). RNAi screening expands knowledge of stem cell function for potential drug discovery since it systematically eliminates gene function (Haney 2007).

Recent advances have allowed for superb growth of stem cell RNAi libraries. Screening and the creation of such libraries is a promising weapon to explore regulatory mechanisms and may allow for identification and validation of proper genetic targets (Karlsson et al. 2010). These advances in high-throughput technologies which measure changes of gene expression at the mRNA level have enabled extensive global transcriptome profiling of stem cells including ESC and HSC along with their differentiated progenies (Birney et al. 2007). Currently, technological platforms such as subtractive cDNA library analysis, expressed sequence tags (ESTs) sequencing, serial analysis of gene expression (SAGE), DNA microarrays, and massive parallel signature sequencing (MPSS) transcriptomic efforts are being initiated to identify specific human stem cell genes (Birney et al. 2007).

Early studies have shown that RNAi is an effective strategy to regulate gene expression for a plethora of human stem cell types including the ESC. Work of Yang et al. showed that long dsRNA represses the target gene expression in undifferentiated ESC population by mRNA degradation (Yang et al. 2001). They conclude that this serves as an important model to study ESC differentiation. In another study, Zou et al. transfected differentiated ESC with small interfering RNA (siRNA) and found diminished expression of PU.1 and c-EBPα genes (Zou 2010). Similarly, knockdown of Shp-2 expression in differentiated ESC resulted in reduction of hemangioblast development (Zou 2010).

In an evaluation of pluripotency, Velkey et al. used a knockdown model involving Oct-4, a gene associated with ESC pluripotency (Fig. 16.1). They transfected ESC with plasmids containing an independently expressed reporter gene and an RNA polymerase type III promoter to constitutively express small stem-loop RNA transcripts corresponding to Oct-4 mRNA. Cells transfected with Oct-4 shRNA demonstrated reduced levels of Oct-4 mRNA and exhibited characteristics of trophectodermal differentiation (Velkey and O'Shea 2003; Zou 2010) Furthermore, Oct-4 has been knocked down via siRNA in both murine and human models (Velkey and O'Shea 2003; Zou et al. 2010).

It has also been shown that siRNA has the ability to knock down gene expression in HSCs. Scherr et al. showed that HSCs are sensitive to genetic interference by transfecting human CD34+ HSC and CD34+ chronic myeloid leukemia (CML) cells with siRNA targeting bcr-abl (Zou 2010). Results showed an 87% reduction of bcr-abl genetic expression which essentially demonstrates that siRNA can specifically and efficiently interfere with the expression of an oncogenic fusion gene in HSC (Scherr et al. 2003; Zou 2010). In a similar model, researchers evaluated the Abelson helper integration site (*ahi-1*), an oncogene that is dysregulated in CML

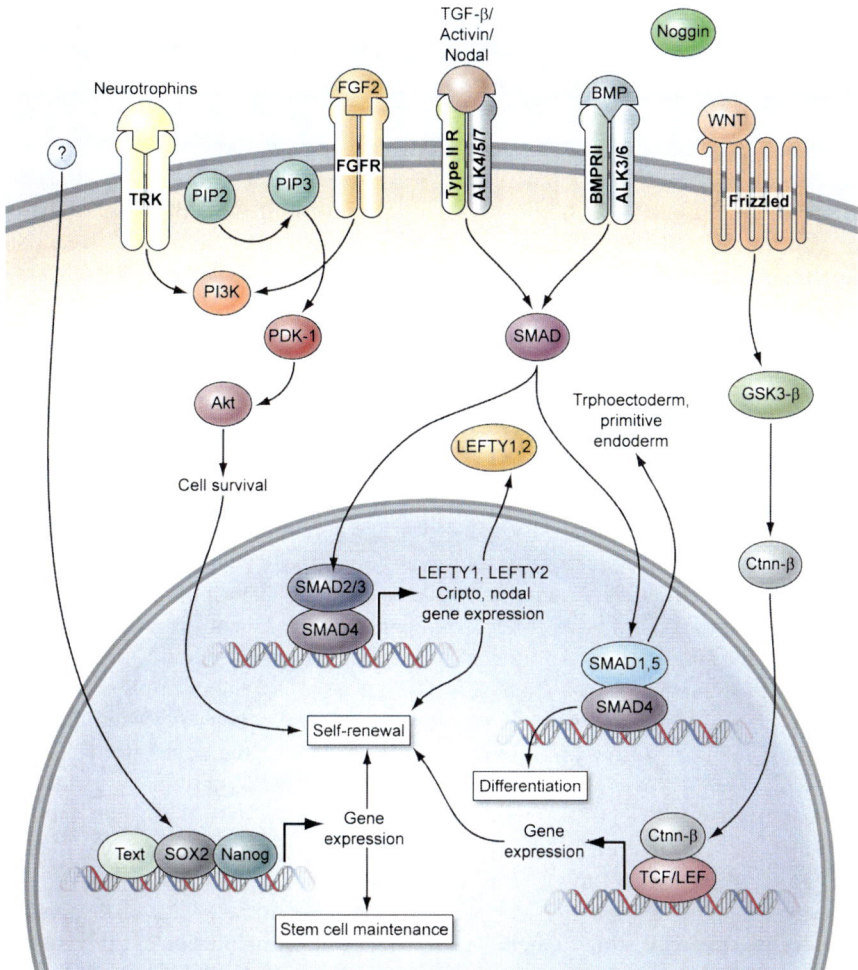

Fig. 16.1 Embryonic stem cell pluripotency pathways. Key: TGF-β, transforming growth factor-β; SOX2, sex determining region Y-box 2; TRK, tropomyosin-receptor-kinase; GSK-3β, glycogen synthase kinase 3 beta; FGFR, fibroblast growth factor receptors; ALK, anaplastic lymphoma kinase. BMPR, bone morphogenetic protein receptor

stem cells which have elevated levels of bcr-abl transcripts (Zou 2010). Overexpression of *ahi-1* in HSC confers growth advantages in vitro and causes leukemia in vivo. It was found that RNAi therapy against *ahi-1* in human cord blood bcr-abl-transduced lin (−), CD34 (+) cells, and CML stem cells reduced their growth autonomy in vitro (Zou 2010). This experiment therefore alludes to *ahi-1* as a potential therapeutic target in CML.

In order to ascertain a stem cell gene signature, Sperger et al. compared mRNA expression patterns via cDNA microarrays in five human ESC lines to a panel of 69 other different human cell lines (Sperger et al. 2003). Results showed that the genes

highly expressed in the ESCs were previously proven ESC-associated genes –
transcription factors Oct3/4, FoxD3, and Sox2 and a DNA methyltransferase
DNMT3B (Sperger et al. 2003). In addition, genes involved in the Wnt-β-catenin
signaling pathway, such as FZD7, FZD8, and Tcf3, were also highly expressed.
This data is consistent with previous mouse genetics data showing that different
dosage of β-catenin could modulate ESC differentiation (Sperger et al. 2003;
Birney et al. 2007). Validation of this data came with the results of Bhattacharya
et al. who used DNA microarrays to identify 92 genes that were enriched in six
different human ESC lines (Bhattacharya et al. 2004; Birney et al. 2007). Results
showed upregulation of these known ESC markers such as Oct3/4, Nanog, GTCM-1,
Connexin 43/GJA1, TDGF1, and Galanin (Bhattacharya et al. 2004; Birney et al.
2007). A cross-comparison of the gene lists generated by these experiments showed
that Oct3/4, Nanog, Sox2, Rex1, DNMT3B, Lin28, TDGF1, and GDF3 are commonly expressed in all human ESCs (Birney et al. 2007) (Fig. 16.1).

HSCs can fully reconstitute all blood cell elements because in addition to their
ability to self-renew, they give rise to both lymphoid and myeloid lineages. This
occurs through progressive restriction of lineage potential, and HSCs subsequently
acquire the characteristics of mature, fully differentiated cells. Several groups have
used high-throughput sequencing strategies to uncover the transcripts that specify
HSCs (Birney et al. 2007). These approaches require construction of cDNA
libraries from murine HSCs followed by the subtraction of mature housekeeping
genes prior to sequencing and analysis (Birney et al. 2007). A cross-comparison
among these efforts demonstrates that the evolutionarily conserved and developmental regulatory pathways prominent in HSCs include the genes from the Wnt
pathway (Lef1, Tcf4, Dsh), the TGF-β pathway (BMP4, activin C, serine and
therine kinases NIK and Ski), the hedgehog pathway (SMO), the Notch pathway
(Notch1 and manic fringe), the homeobox regulatory cascade (Hoxa9, Meis-1,
TGIF, and Enx-1), and Bmi-1 (Birney et al. 2007).

Mounting evidence indicates that some of these intracellular pathways are
involved in the regulation of stem cell self-renewal or maintenance. For instance,
activation of Wnt pathway by expressing β-catenin in HSCs results in growth
factor–independent growth and enhanced self-renewal properties (Birney et al.
2007). Expression of β-catenin in the granulocyte-monocyte progenitor (GMP)
confers enhanced self-renewal activity and leukemic potential of these cells in
chronic myelogenous leukemia (CML) patients (Birney et al. 2007). Bmi-1, a
suppressor of the Ink4 locus (encoding p16 and p19 cell growth inhibitors), is
essential for the determination of the proliferation potential of HSCs. BMP signaling controls the number of HSCs by regulating the size of HSC niche which is the
proposed hematopoietic microenvironment in bone marrow supporting HSC self-
renewal via the BMP receptor BMPRIA (Birney et al. 2007).

Recently, a shared collection of several hundred genes has been identified in
different populations of stem cells (Zou 2010). Further elucidation is required to
determine whether these genes are specifically required for the maintenance of stem
cell function. RNAi may not only provide an approach to knock down these genes
but also expand our understanding of the molecular regulation of function in CSCs

(Zou 2010). Further investigations are required to improve the sensitivity of current genomic and proteomic technologies (Birney et al. 2007) through rigorous functional analysis of selected candidate genes using in vivo and in vitro assays. The development and implementation of better computational algorithms are required (Birney et al. 2007). This will allow for an integrated systems level analysis and modeling of the collective data from the combined genomic, transcriptomic, and proteomic efforts (Birney et al. 2007).

16.2.2   Cancer Stem Cell Pathways Amenable to RNAi Therapy

Systemic chemotherapy for cancer can mediate a response in terms of disease burden and, in some cases, lead to complete regression of disease. Sometimes, the response conveyed by conventional cytotoxic chemotherapy is not durable and patients have local recurrence or develop distant metastasis. The CSC theory hypothesizes that a small group of cells – the putative CSC population – is responsible for relapse or progression of the disease at a distant site. These CSCs are theorized to harbor specific mutations leading to altered cellular pathways that enable them to escape the cytotoxicity of conventional chemotherapy. Like bone marrow–derived stem cells following myeloablation, CSCs are able to regenerate following current chemotherapy regimens. To translate the basic science behind CSC research into the clinic, these pathways unique to CSC must be exploited, and RNAi offers a unique opportunity.

Human cancers are classified by the embryonic tissue from which they are derived. Nearly 90% of neoplasms are of epithelial origin, approximately 5% are of mesenchymal origin, and approximately 5% are liquid cancers derived from components in the peripheral blood (Harless 2011). Each of these malignancies is treated differently in the clinical setting, but similarities exist at the cellular level in terms of altered or overactive pathways that are amenable to RNAi therapy. These similarities allow for broad application of therapies targeting aberrant pathways even though differences exist in the clinical applications in terms of delivery and timing of therapy (i.e., during or after a surgical resection).

Cellular pathways preferentially altered in CSC offer targets for RNAi therapy (Marquardt et al. 2011; Zou 2010; McDermott and Wicha 2010) (Table 16.1). Many of these pathways regulate gene expression through downstream transcription

Table 16.1 CSC-associated pathways amenable to RNAi therapy

Pathway	Key components amenable to RNAi therapy
Hedgehog	SMO, GLi transcription factors
Notch	Notch receptor subtypes 1–4
Wnt	Lef1, Tcf4, Dsh, JNK
TGF-β	MEK, ERK, SMAD4
IL-6	STAT3, mTOR

factors in tissue progenitor cells, making them fundamental to embryogenesis, organogenesis, and tissue homeostasis (Merchant and Matsui 2010). They are generally well conserved among species, allowing for basic research along a wide spectrum of experimental animal models. There is great potential to abrogate recurrence and metastasis after initial local therapy by preferentially targeting these CSC-associated pathways.

A thorough understanding of the pathways responsible for the unique properties of CSC is paramount to planning therapy based on RNAi. The CSC theory holds that migration and self-renewal are central to the function of the CSC and the pathways associated with these two characteristics are most intriguing from a therapeutic perspective. This section will explore the pathways associated with the fundamental CSC function in more depth in an effort to characterize how RNAi therapy might be directed.

16.2.2.1 The Hedgehog Pathway

The number of genes required to be activated or suppressed in the process of organogenesis requires a hierarchical control of gene expression. Individual intracellular pathways leading to the activation or suppression of a single gene would no doubt overwhelm a cell in terms of energy expenditure. The hierarchical model of gene expression allows a single signal to effect a significant change within a cell. The hedgehog (Hh) pathway is an important example of a highly conserved pathway that ends with the transcriptional regulation of large sets of genes.

The Hh pathway occurs in tissues of both mesenchymal and epithelial origin (McDermott and Wicha 2010; Peacock et al. 2007). Initiation occurs when one of the three ligands – sonic hedgehog (SHh), desert hedgehog (DHh), and Indian hedgehog (IHh) – binds at the cell surface with patched (Ptch), a cell surface receptor (Fig. 16.2). Ligand binding displaces Ptch from the cilia, allowing smoothened (Smo) to move in into place, thereby activating any of three GLi transcription factors. Once activated, these factors move into the nucleus of the cell and effect downstream expression of target genes (Pal et al. 2010). Each of the three forms of the GLi transcription factors – GLi1, GLi2, and GLi3 – acts differently at the level of gene expression, with GLi1 being responsible for activation of gene expression, GLi2 suppressing gene expression, and GLi3 alternating between the two depending on posttranslational modification (Merchant and Matsui 2010).

At each level, there is an opportunity for interference in the pathway, and there are currently several phase I trials of small molecule inhibitors against Smo (Pal et al. 2010). Specifically delivering therapy only to malignant cells is a hallmark of optimal cancer therapy. As the Hh pathway is required for physiologic tissue homeostasis, inhibition must be preferential in the malignant cells of a given tissue and not the normal tissue to avoid potentially catastrophic adverse events. Since CSCs are known to have vastly increased levels of Hh expression (Li et al. 2007), therapy against the Hh pathway should preferentially treat CSCs.

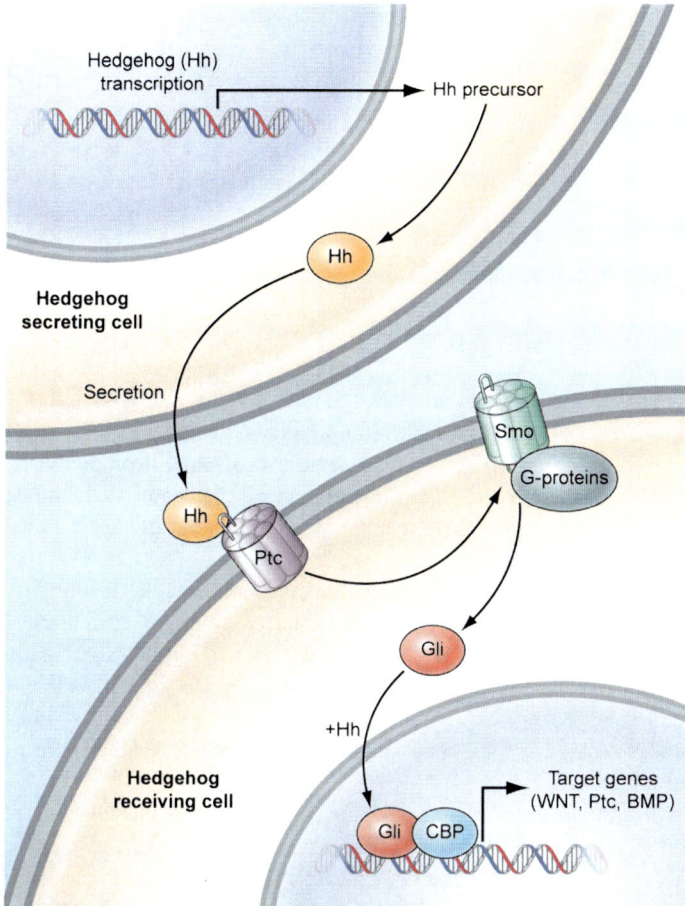

Fig. 16.2 The hedgehog signaling pathway. Key: Hh, Hedgehog; Ptc, Patch transmembrane receptor; Smo, Smoothened transmembrane receptor

Using RNAi therapy, the gene for downstream target such as a cell surface receptor may be suppressed. In the case of the Hh pathway, suppressing the Ptch receptor would potentially not be as beneficial since there is some evidence that a mutated KRAS can have constitutively active GLi1 transcription factor (Ji et al. 2007). It is more likely RNAi toward the transcription factors GLi1 or GLi2 would allow for more therapeutic potential with greater specificity as these factors have opposite functions. In vitro models have shown to increase apoptosis in cells transfected with siRNA against GLi1 and GLi2 and eliminate metastatic potential when transfected with shRNA against Smo (Merchant and Matsui 2010).

The goal of RNAi therapy against the Hh pathway is not necessarily cell death. When used to preferentially treat CSCs, the greatest benefit of this therapy may be in eliminating the ability to self-renew, thereby enhancing the ability of cytotoxic agents to eradicate tumor. By combining RNAi therapy with more conventional therapies, the opportunities are much greater in terms of specific therapeutic goals. As CSCs are hypothesized to be responsible for the recurrence of previously eradicated disease, eliminating this phenotype from the malignant lesion may be just as important as complete regression alone.

16.2.2.2 The Notch Signaling Pathway

The Notch signaling pathway is responsible for self-renewal and differentiation and has been studied most extensively in mammary stem cells (McDermott and Wicha 2010; Pal et al. 2010). There are four Notch receptors (Notch 1–4) with four ligands (Jagged1, Jagged 2, Delta, and Delta-like). Ligand binding initiates proteolysis of the transmembrane Notch receptor by γ-secretase. Having been cleaved, the intracellular domain of the Notch cell surface receptor translocates to the nucleus where it combines with two other factors to induce expression of target genes such as cyclin D1 and c-Myc.

The Notch pathway is important in the differentiation of the precursor mammary cell. It has been shown to determine the polarity of breast epithelium in mouse mammary cell lines (Bouras et al. 2008). By demonstrating the Notch pathway to be most active in the luminal cells of the breast tissue, its role in cell polarity and differentiation is clearer. The Notch1 plays a greater role than the other subtype, and it appears to be a repressor of stem cell proliferation in the mouse model by permitting differentiation.

Studies in human breast cancer cell lines have further clarified the role of the Notch pathway in the differentiation of breast stem cells. One recent report shows a differential expression of the Notch receptor subtype in breast CSCs (Harrison et al. 2010). In contrast to the previous report in the mouse model, the authors show that inhibition of Notch1 and Notch4 led to a decrease in proliferation of the CSC. Interestingly, the Notch4 receptor is more highly expressed in breast CSC as compared to bulk breast tumor cells. Therefore, it reasons that a therapy targeting the Notch4 subtype specifically could eliminate the off-target effects of therapy to the normal breast stem cells.

Current therapies developed against the Notch pathway are largely centered on the inhibition of the γ-secretase enzyme. As this enzyme is responsible for cleaving the receptor into the active Notch intracellular domain (NICD), it has been postulated that a γ-secretase inhibitor (GSI) would have substantial antitumor activity against those lesions with increased expression of the Notch pathway. While a highly sophisticated and targeted therapy, the small molecule inhibitors currently being studied do nothing to protect the physiologic Notch expression in tissue stem cells throughout the body. Based on the data reported in breast CSC, a therapy targeting Notch4 could potentially achieve this goal. RNAi therapy is ideally suited for this as the sequence difference between the Notch receptor subtypes could be exploited in the form of either siRNA or shRNA. Targeted inhibition of the Notch4 subtype may well achieve preferential suppression of CSCs without harming tissue stem cells of the same organ.

16.2.2.3   The Wnt Signaling Pathway

Another fundamental pathway involved in organogenesis and normal tissue renewal is the Wnt signaling pathway. In the β-catenin-dependent pathway, a Wnt ligand binds to the Frizzled transmembrane receptor. This interaction leads to the stabilization of β-catenin and its dissociation from the GSK-3β protein. The free β-catenin thus translocates to the nucleus where it serves as a transcription factor for downstream genes such as cyclin D1 and c-Myc (Smalley and Dale 1999; Pal et al. 2010; de Sousa et al. 2011).

Increased activity of this Wnt-β-catenin pathway has been implicated in several malignancies, most notably colon cancer (de Sousa et al. 2011; Zeki et al. 2011). One of the components of the complex that binds β-catenin in the cytoplasm is the adenomatous polyposis coli (APC) protein. Germline mutations in the APC gene are well documented and lead to the hereditary syndrome of familial adenomatous polyposis (FAP), a condition which confers a near 100% risk of colon cancer by the age of 50. Acquired mutations in the same gene are implicated early in the progression from adenoma to carcinoma in the well-described sequence (Vogelstein et al. 1988). The APC gene is described as a tumor suppressor gene because mutations allow Wnt-independent translocation of β-catenin with therefore uncontrolled transcription of downstream genes.

The mortality associated with cancer is generally due to recurrence, and the morbidity many times is secondary to metastasis (i.e., liver failure from replacement with metastatic deposits). Metastasis has been historically thought to follow a "seed and soil" method whereby a single tumor cell breaks off from the primary lesion and implants at a distant site. While there is certainly an element of truth in terms of the physical translocation of cells, it is not likely that all cells in a given tumor mass have the ability to recapitulate the primary lesion in a distant organ. Only cells capable of the recently described phenomenon of epithelial to mesenchymal transition (EMT) are likely capable of tumorigenesis in distant organs. Mutations in the Wnt-β-catenin pathway are closely associated with the ability to undergo EMT (Li and Zhou 2011), and suppression of β-catenin nuclear translocation is fundamental to stemming the overexpression of Wnt-associated genes. Since CSCs have been shown to have increased expression of genes in the Wnt-β-catenin pathway, targeting cells with increased free intracytoplasmic β-catenin could mitigate metastases.

In order to evaluate the hypotheses associated with the CSC theory, the ability to identify the CSCs is of utmost importance. There have been a variety of methods employed such as labeling DNA with fluorescent nucleotides (Hari et al. 2011), but the most commonly employed method involves cell surface markers (Li et al. 2007; Lapidot et al. 1994; Al-Hajj et al. 2003; Hermann et al. 2007). There are a variety of putative cell surface markers reported to be specific to CSC, and one specific marker, leucine-rich repeat-containing G-protein-coupled receptor 5 (Lgr5), is associated with the Wnt-β-catenin pathway. As a Wnt target gene, Lgr5 is exclusively expressed on the columnar cells of the intestinal crypt (Barker et al. 2007). The receptor is an intriguing marker for stem cells, specifically CSCs, as it denotes activation of the Wnt signaling pathway.

The ability to convert a CSC to a more well-differentiated cell through expression-level knockdown of a CSC-associated gene is something offered only by RNAi therapy. As described above, the Wnt signaling pathway offers several sites of interest for development of such a therapy. For example, the use of siRNA against the Lgr5 receptor might convert an otherwise self-renewing tumor cell into one more sensitive to cytotoxic therapy or even destined for apoptosis. The ability to finely regulate these intracellular, even intranuclear, processes makes RNAi therapy exciting as a CSC-specific treatment.

16.2.3   Delivery of RNAi

There are significant hurdles that exist in the field of RNAi targeting the CSC. The first includes the difficulty of delivering RNA to its point of action within the target cellular cytoplasm while avoiding off-target gene silencing effects (see Chap. 19, this volume). Secondly, progress on isolating the CSC and determining specific CSC gene signatures is underway, but the precise gene targets for RNAi are not yet clear.

The placement of interfering RNA oligonucleotides into the cytoplasm of target cells is hampered by an innate immune response to dsRNA via the IFN pathway (Sen 2001). Innate immune cells are capable of recognizing pathogen-associated molecular patterns (PAMPs), which are generally not present in host tissues (Janeway and Medzhitov 2002). The discovery that long dsRNA is processed into short-nucleotide sequences that do not trigger an IFN response provided the basis of RNAi as an in vivo therapeutic strategy (Elbashir et al. 2001). Another issue arising in the delivery of RNAi involves off-target gene silencing effects (Jackson et al. 2003). Since targeting a CSC is truly novel, off-target effects and toxicity profiles at this point are largely unknown. Current work on chemical modification of synthetic siRNA, including 2'-uridine modification (Cekaite et al. 2007) and 2'-O-methyl substitutions (Jackson et al. 2006), holds promising results for minimizing these off-target effects of RNAi.

In order to translate in vitro studies of CSC RNAi to in vivo therapeutics, the biological barriers to the systemic administration of nucleic acids must be overcome. First-pass metabolism and renal excretion occurs rapidly, with up to 30% of delivered nucleic acid accumulating in the kidneys and a substantial portion accumulating within hepatocytes (Higuchi et al. 2010). Investigation of chemical modifications to protect oligonucleotides from the effects of 3'-endonuclease, the primary enzyme responsible for in vivo nucleic acid degradation (Kennedy et al. 2004), has shown promise in extending the bioavailability of naked RNAi oligonucleotides. The relatively small proportion of nucleic acid that remains intact after first-pass metabolism and serum degradation then must cross the cellular membrane to reach its site of action in the cellular cytoplasm. The negatively charged nucleic acid is repelled by the negatively charged outer layer of the cellular phospholipid bilayer, while the hydrophobic inner membrane repels the hydrophilic RNA. Liposomal RNA encapsulation may solve this problem, and local administration of RNAi may prove to be invaluable for a variety of benign and malignant skin and ocular diseases (Geusens et al. 2009). Unfortunately, this does not address the lack of anatomical access to the vast majority of tumors and, thus, fails to fully address the biological barriers to systemic administration of naked RNA oligonucleotides. That being said, we will discuss a trial of a locally administered siRNA targeting a CSC-associated pathway later in this chapter.

Recently, advances in nanotechnology hold promise as a stable delivery system for nucleic acids. The use of polymer-based nanoparticles has emerged as a leading area of research into direct drug delivery systems (Duncan 2003). Traditional cationic polymers, including *poly*-ethylenimines, interact with negative charges on siRNA to provide excellent packaging of the nucleic acid core. However, these cationic polymers have significant in vivo toxicities. Attempts at using polyethylene glycol (PEG) as a shield provide a possible method of packaging siRNA while protecting the host from cationic polymer toxicity (Malek et al. 2008).

Targeting of RNAi molecules provides a promising method of increasing the proportion of RNAi nucleic acids that reach their therapeutic targets after systemic administration. This can be accomplished either by directly tagging siRNA with molecules that are able to target cell surface molecules or encapsulating siRNA into nanoparticles with targeting molecules bound to its surface (Jeong et al. 2009). These molecules can include ligands to cell surface receptors (Ikeda and Taira 2006), aptamers specific for cell surface proteins (Chu et al. 2006), and targeted antibodies (Toloue and Ford 2011). Research into RNAi therapeutics for solid tumors using targeted delivery systems is currently underway, with a trial involving anti-RRM2 siRNA targeted via a nanoparticle targeted toward human transferring protein (TF), which is known to be upregulated on the surface of solid tumor cells (Davis et al. 2010). The use of targeted RNAi against CSC holds promise for a major advancement in oncologic medicine.

16.2.4   Advances in Cancer Stem Cell Laboratory Practices

The stochastic model of tumorigenesis suggests that the population of cells most susceptible to modern therapeutics is the terminally differentiated population of cells within a tumor (Sell 2004). However, the CSC theory holds that a subpopulation of CSCs accumulate the necessary mutations to become tumorigenic while retaining the stem-like nature that allows them to repopulate the tumor cellular burden. The presence of CSCs in hematologic malignancy has been well established (Dick 1996; Lapidot et al. 1994), and the use of RNAi against these stem cells is actively being researched (Zuber et al. 2011). Currently, improved in vitro culture methods to maintain CSC activity from primary tumor samples open the arena of high-throughput screening of siRNA libraries. These studies can therefore provide novel agents to target this critical cell population.

In order to properly target a CSC, one must first identify and characterize the CSC. Much of the current research into CSCs has focused on cell surface markers as a method of identifying cells capable of initiating tumors in immunodeficient mice (Al-Hajj and Clarke 2004). The identification of cell surface patterns unique to a stem population has been found in leukemia (Lapidot et al. 1994), breast (Al-Hajj et al. 2003), prostate (Collins et al. 2005), and a variety of other tumors. A more functional approach of CSC isolation involves isolating the side population (SP) of tumor cells that demonstrate low uptake of Hoescht dye 33342. This is based on studies showing that these SP cells are consistent with ESCs (Goodell et al. 1997). However, it is clear that the SP is a heterogenous group of cells and that only a small subset of SP cells are consistent with stem cells (Pearce et al. 2004).

Stem cells in tissues from epithelial origins have a subpopulation of cells that retain genetic material through a process of asymmetric cell division (Bickenbach and Mackenzie 1984). These so-called label-retaining cells have been investigated in several solid tumors and have been shown to have increased tumor-initiating capacity in melanoma (Roesch et al. 2010) and breast cancer (Pece et al. 2010). A technique to isolate live label-retaining cells has been developed which allows for further experimentation after identification, not just prior as in most other studies (Hari et al. 2011). As advancements continue to be made in CSC laboratory practices, we will see further RNAi options to target the CSC.

16.3 Current Clinical Trials of RNAi Therapy

16.3.1 A Clinical Trial of siRNA Targeting PLK1

RNAi offers new hope in the treatment of many deadly neoplastic diseases. Due to current advances in CSC laboratory practices and advances in gene libraries, novel therapies have been developed utilizing RNAi. Our group is beginning a clinical trial involving the intra-arterial administration of lipid nanoparticles containing siRNA targeting the PLK1 gene. Eligible patients include those with primary liver cancer or patients with hepatic metastases from other malignancies. Historic controls have shown that patients with unresectable metastatic liver disease hold a 5-year survival less than 5%. For selected histologies, the 5-year survival for resectable hepatic metastases ranges from 20% to 60%, suggesting that control of liver metastases could result in prolonged survival (Kemeny 2010). Although phase I trials of hepatic arterial infusion therapy have shown it to be a safe delivery mechanism, agents used to date have limited efficacy (Callahan and Kemeny 2010). Our strategy differs in that we are targeting a gene associated with the CSC.

The polo-like kinase (PLK) family is characterized by serine/threonine kinase activity and is fundamental to progression through the cell. The four mammalian PLK family members (PLK1, 2, 3, and 4) are characterized by their unique phosphopeptide-binding polo-box domains and have been shown to play nonredundant roles in cell cycle regulation (Barr et al. 2004). In mammalian cells, PLK1 acts to phosphorylate a diverse array of cell cycle proteins and is known to be critical for mitotic progression and cytokinesis. PLK1 expression is temporally controlled in proliferating cells, becoming upregulated as a cell approaches and enters mitosis.

PLK1 is overexpressed in many human tumor types, and its overexpression is a negative prognostic indicator of patient outcome in a variety of cancers (Strebhardt and Ullrich 2006). Inhibition of PLK1 activity in proliferating cancer cells rapidly induces mitotic arrest and apoptosis (Steegmaier et al. 2007). Partial inhibition of PLK1 can also sensitize cancer cells to the cytotoxic effects of conventional chemotherapeutics, likely due to the functional role of PLK1 in the DNA damage and spindle assembly checkpoints (Spankuch et al. 2007). More recently, it has been found that cancer cells harboring mutations in the *KRAS* oncogene or the gene encoding the p53 tumor suppressor protein have increased dependency on PLK1 to support their dysregulated growth and survival (Luo and Zhou 2009).

All of these features combine to make PLK1 a promising therapeutic target in oncology (Strebhardt and Ullrich 2006). Recent evidence also suggests that PLK1 may have implications in CSC pathways (Grinshtein et al. 2011). Grinshtein et al. demonstrated the role of PLK1 in the CSC using a neuroblastoma model (Grinshtein et al. 2011). Andrisani et al. showed that hepatic CSCs have a gene signature which includes genes shown to be involved in cell proliferation (Andrisani et al. 2011). Interestingly, patients with tumors expressing that gene signature tended to have a poor prognosis. This data suggests a mechanistic link between these two gene clusters involving the proliferation cluster gene PLK1 (Andrisani et al. 2011).

16.3.1.1 Rationale for an RNAi Approach to PLK1

A number of small molecule PLK1 inhibitors are currently in development, but siRNA therapy has several advantages over these novel drugs. Because it acts at the level of gene expression, RNAi is highly specific and allows for the more selective inhibition of closely related proteins compared to the relative promiscuity of kinase inhibitors. Current PLK1 inhibitors, for example, also inhibit PLK2 and PLK3 kinase activity (Steegmaier et al. 2007). The cellular effect of PLK1 depletion by RNAi also differs from its functional inhibition by small molecules, likely due to the loss of both kinase and polo-box functionality (McInnes et al. 2006). The duration of drug effect with siRNA is another attractive advantage. Once RNAi is established within mammalian cells, gene silencing can persist for many days due to the relative stability of activated RNA-induced silencing complex (RISC) in the presence of its complementary mRNA. Therefore, the maintenance of drug activity in the tissues can be uncoupled from the requirement to maintain an effective drug concentration in the blood.

There are several obstacles to the delivery of siRNA therapy which include evasion of the immune response, penetration of tumor tissue, and intracellular deposit of nucleic acid. To evade the immune response, the siPLK-1 siRNA in TKM-080301 contains 2'-O-methyl (2'-OMe)-modified ribonucleotide bases at selected positions in both the antisense and sense strands. These modifications are intended to reduce or eliminate the potential for nonspecific pharmacologic effects related to activation of the innate immune response, which can occur when using native, unmodified DNA and RNA (Judge et al. 2006; Judge et al. 2009). The delivery of siRNA into the cytoplasm of target cells is an absolute requirement for drug activity and is a key technological hurdle to developing siRNA therapies (de Fougerolles 2008). Nonformulated siRNAs are very sensitive to nuclease degradation and are also rapidly eliminated, thereby limiting distribution to the intended target tissue(s). TKM-080301 was designed for intravenous delivery of siRNA to solid tumors and has been prepared in the form of stable nucleic acid lipid particles (SNALP). These lipid nanoparticles have a diameter of approximately 75–90 nm.

To ensure penetration of the tumor tissue, SNALP delivery takes advantage of the "enhanced permeation and retention" effect. This effect, also referred to as passive disease site targeting, involves charge-neutral carriers of suitable size (≤ 100 nm diameter) that pass through the fenestrated walls (gaps less than 400 nm) of the neovascular blood vessels typically found within tumors (Seymour 1992; Yuan et al. 1995). To ensure intracellular deposit of nucleic acid, the lipid components of TKM-080301 protect the siRNA from degradation by plasma and tissue nucleases, prevent rapid clearance of the siRNA in the kidneys, and enable effective intracellular uptake into cancer cells (Judge et al. 2009). Once in the tumor tissue, intact lipid particles are believed to enter cells via endocytosis, followed by fusion with the endosomal membrane and release of siRNA into the cytoplasm, wherein it can interact with the RISC to facilitate RNAi.

16.3.1.2   Pharmacology of TKM-080301

The oligonucleotide sequence of siPLK-1, the active drug substance loaded into the SNALP, has been selected from a panel of approximately 50 siRNA duplexes designed to target human PLK1. Selection was based on a series of in vitro activity screens in human cancer cell lines, the results of which consistently showed that siPLK-1 has the greatest potency in silencing human PLK1 expression and inhibiting cancer cell growth. TKM-080301, the formulated drug product that includes the siPLK-1 molecule, does not induce activation of the innate immune response in vitro using mouse dendritic cells (DC) or human peripheral blood mononuclear cells (PBMC). Also, in murine models, this compound elicited no immune response by the animal. However, mild increases in IL-6 and MCP-1 have been observed in monkeys, as well as IL-6, TNF-α, IL-1β, and IL-8 in human whole blood culture assays. Importantly, these cytokine inductions by TKM-080301 can be abrogated in human whole blood culture assays when the cells are preexposed to clinically relevant concentrations of dexamethasone.

16.3.2   *Future Multidisciplinary Use of RNAi in Cancer Stem Cell Therapeutics*

Clear identification CSC-specific surface markers, genes (Nanog, Oct3/4, STAT3), and pathways such as TGF-β, hedgehog, and Wnt-β-catenin are crucial to the development of CSC-targeted treatments. RNAi provides a unique opportunity to silence cancer-causing stem cell genes at the pretranslation level, which is otherwise not possible with conventional therapies such as cytotoxic chemotherapy, small molecule inhibitors, or monoclonal antibodies. Owing to the explosion of knowledge generated by a growing understanding of the human genome and the development of high-throughput gene expression profiling of CSCs, a plethora of genes that contribute to tumor initiation and the metastatic cascade are being discovered. RNAi therapy against multidrug resistance genes and CSC genes may provide exceptional benefit and herald a paradigm shift in the treatment of deadly diseases.

References

Al-Hajj M, Clarke MF (2004) Self-renewal and solid tumor stem cells. Oncogene 23 (43):7274–7282. doi:10.1038/sj.onc.1207947 1207947 [pii]

Al-Hajj M, Wicha MS, Benito-Hernandez A, Morrison SJ, Clarke MF (2003) Prospective identification of tumorigenic breast cancer cells. Proc Natl Acad Sci USA 100(7):3983–3988. doi:10.1073/pnas.0530291100 0530291100 [pii]

Andrisani OM, Studach L, Merle P (2011) Gene signatures in hepatocellular carcinoma (HCC). Semin Cancer Biol 21(1):4–9. doi:S1044-579X(10), 00080-5 [pii] 10.1016/j.semcancer.2010.09.002

Barker N, van Es JH, Kuipers J, Kujala P, van den Born M, Cozijnsen M, Haegebarth A, Korving J, Begthel H, Peters PJ, Clevers H (2007) Identification of stem cells in small intestine and colon by marker gene Lgr5. Nature 449(7165):1003–1007. doi:nature06196 [pii] 10.1038/nature06196

Barr FA, Sillje HH, Nigg EA (2004) Polo-like kinases and the orchestration of cell division. Nat Rev Mol Cell Biol 5(6):429–440. doi:10.1038/nrm1401 nrm1401 [pii]

Bhattacharya B, Miura T, Brandenberger R, Mejido J, Luo Y, Yang AX, Joshi BH, Ginis I, Thies RS, Amit M, Lyons I, Condie BG, Itskovitz-Eldor J, Rao MS, Puri RK (2004) Gene expression in human embryonic stem cell lines: unique molecular signature. Blood 103(8):2956–2964. doi:10.1182/blood-2003-09-3314 2003-09-3314 [pii]

Bickenbach JR, Mackenzie IC (1984) Identification and localization of label-retaining cells in hamster epithelia. J Invest Dermatol 82(6):618–622

Birney E, Stamatoyannopoulos JA, Dutta A, Guigo R, Gingeras TR, Margulies EH, Weng Z, Snyder M, Dermitzakis ET, Thurman RE, Kuehn MS, Taylor CM, Neph S, Koch CM, Asthana S, Malhotra A, Adzhubei I, Greenbaum JA, Andrews RM, Flicek P, Boyle PJ, Cao H, Carter

NP, Clelland GK, Davis S, Day N, Dhami P, Dillon SC, Dorschner MO, Fiegler H, Giresi PG, Goldy J, Hawrylycz M, Haydock A, Humbert R, James KD, Johnson BE, Johnson EM, Frum TT, Rosenzweig ER, Karnani N, Lee K, Lefebvre GC, Navas PA, Neri F, Parker SC, Sabo PJ, Sandstrom R, Shafer A, Vetrie D, Weaver M, Wilcox S, Yu M, Collins FS, Dekker J, Lieb JD, Tullius TD, Crawford GE, Sunyaev S, Noble WS, Dunham I, Denoeud F, Reymond A, Kapranov P, Rozowsky J, Zheng D, Castelo R, Frankish A, Harrow J, Ghosh S, Sandelin A, Hofacker IL, Baertsch R, Keefe D, Dike S, Cheng J, Hirsch HA, Sekinger EA, Lagarde J, Abril JF, Shahab A, Flamm C, Fried C, Hackermuller J, Hertel J, Lindemeyer M, Missal K, Tanzer A, Washietl S, Korbel J, Emanuelsson O, Pedersen JS, Holroyd N, Taylor R, Swarbreck D, Matthews N, Dickson MC, Thomas DJ, Weirauch MT, Gilbert J, Drenkow J, Bell I, Zhao X, Srinivasan KG, Sung WK, Ooi HS, Chiu KP, Foissac S, Alioto T, Brent M, Pachter L, Tress ML, Valencia A, Choo SW, Choo CY, Ucla C, Manzano C, Wyss C, Cheung E, Clark TG, Brown JB, Ganesh M, Patel S, Tammana H, Chrast J, Henrichsen CN, Kai C, Kawai J, Nagalakshmi U, Wu J, Lian Z, Lian J, Newburger P, Zhang X, Bickel P, Mattick JS, Carninci P, Hayashizaki Y, Weissman S, Hubbard T, Myers RM, Rogers J, Stadler PF, Lowe TM, Wei CL, Ruan Y, Struhl K, Gerstein M, Antonarakis SE, Fu Y, Green ED, Karaoz U, Siepel A, Taylor J, Liefer LA, Wetterstrand KA, Good PJ, Feingold EA, Guyer MS, Cooper GM, Asimenos G, Dewey CN, Hou M, Nikolaev S, Montoya-Burgos JI, Loytynoja A, Whelan S, Pardi F, Massingham T, Huang H, Zhang NR, Holmes I, Mullikin JC, Ureta-Vidal A, Paten B, Seringhaus M, Church D, Rosenbloom K, Kent WJ, Stone EA, Batzoglou S, Goldman N, Hardison RC, Haussler D, Miller W, Sidow A, Trinklein ND, Zhang ZD, Barrera L, Stuart R, King DC, Ameur A, Enroth S, Bieda MC, Kim J, Bhinge AA, Jiang N, Liu J, Yao F, Vega VB, Lee CW, Ng P, Yang A, Moqtaderi Z, Zhu Z, Xu X, Squazzo S, Oberley MJ, Inman D, Singer MA, Richmond TA, Munn KJ, Rada-Iglesias A, Wallerman O, Komorowski J, Fowler JC, Couttet P, Bruce AW, Dovey OM, Ellis PD, Langford CF, Nix DA, Euskirchen G, Hartman S, Urban AE, Kraus P, Van Calcar S, Heintzman N, Kim TH, Wang K, Qu C, Hon G, Luna R, Glass CK, Rosenfeld MG, Aldred SF, Cooper SJ, Halees A, Lin JM, Shulha HP, Xu M, Haidar JN, Yu Y, Iyer VR, Green RD, Wadelius C, Farnham PJ, Ren B, Harte RA, Hinrichs AS, Trumbower H, Clawson H, Hillman-Jackson J, Zweig AS, Smith K, Thakkapallayil A, Barber G, Kuhn RM, Karolchik D, Armengol L, Bird CP, de Bakker PI, Kern AD, Lopez-Bigas N, Martin JD, Stranger BE, Woodroffe A, Davydov E, Dimas A, Eyras E, Hallgrimsdottir IB, Huppert J, Zody MC, Abecasis GR, Estivill X, Bouffard GG, Guan X, Hansen NF, Idol JR, Maduro VV, Maskeri B, McDowell JC, Park M, Thomas PJ, Young AC, Blakesley RW, Muzny DM, Sodergren E, Wheeler DA, Worley KC, Jiang H, Weinstock GM, Gibbs RA, Graves T, Fulton R, Mardis ER, Wilson RK, Clamp M, Cuff J, Gnerre S, Jaffe DB, Chang JL, Lindblad-Toh K, Lander ES, Koriabine M, Nefedov M, Osoegawa K, Yoshinaga Y, Zhu B, de Jong PJ (2007) Identification and analysis of functional elements in 1% of the human genome by the ENCODE pilot project. Nature 447 (7146):799–816. doi:10.1038/nature05874

Bonnet D, Dick JE (1997) Human acute myeloid leukemia is organized as a hierarchy that originates from a primitive hematopoietic cell. Nat Med 3(7):730–737

Bouras T, Pal B, Vaillant F, Harburg G, Asselin-Labat ML, Oakes SR, Lindeman GJ, Visvader JE (2008) Notch signaling regulates mammary stem cell function and luminal cell-fate commitment. Cell Stem Cell 3(4):429–441. doi:S1934-5909(08), 00399-8 [pii] 10.1016/j.stem.2008.08.001

Callahan MK, Kemeny NE (2010) Implanted hepatic arterial infusion pumps. Cancer J 16 (2):142–149. doi:10.1097/PPO.0b013e3181d7ea51 00130404-201003000-00009 [pii]

Cekaite L, Furset G, Hovig E, Sioud M (2007) Gene expression analysis in blood cells in response to unmodified and 2'-modified siRNAs reveals TLR-dependent and independent effects. J Mol Biol 365(1):90–108. doi:S0022-2836(06), 01225-3 [pii] 10.1016/j.jmb.2006.09.034

Chu TC, Twu KY, Ellington AD, Levy M (2006) Aptamer mediated siRNA delivery. Nucleic Acids Res 34(10):e73. doi:34/10/e73 [pii] 10.1093/nar/gkl388

Collins AT, Berry PA, Hyde C, Stower MJ, Maitland NJ (2005) Prospective identification of tumorigenic prostate cancer stem cells. Cancer Res 65(23):10946–10951. doi:65/23/10946 [pii] 10.1158/0008-5472.CAN-05-2018

Davis ME, Zuckerman JE, Choi CH, Seligson D, Tolcher A, Alabi CA, Yen Y, Heidel JD, Ribas A (2010) Evidence of RNAi in humans from systemically administered siRNA via targeted nanoparticles. Nature 464(7291):1067–1070. doi:nature08956 [pii] 10.1038/nature08956

de Fougerolles AR (2008) Delivery vehicles for small interfering RNA in vivo. Hum Gene Ther 19 (2):125–132. doi:10.1089/hum.2008.928

de Sousa MF, Vermeulen L, Richel D, Medema JP (2011) Targeting Wnt signaling in colon cancer stem cells. Clin Cancer Res 17:647–653. doi:10.1158/1078-0432.CCR-10-1204

Dick JE (1996) Normal and leukemic human stem cells assayed in SCID mice. Semin Immunol 8 (4):197–206. doi:S1044-5323(96), 90025-1 [pii] 10.1006/smim.1996.0025

Duncan R (2003) The dawning era of polymer therapeutics. Nat Rev Drug Discov 2(5):347–360. doi:10.1038/nrd1088 nrd1088 [pii]

Elbashir SM, Harborth J, Lendeckel W, Yalcin A, Weber K, Tuschl T (2001) Duplexes of 21-nucleotide RNAs mediate RNA interference in cultured mammalian cells. Nature 411 (6836):494–498. doi:10.1038/35078107 35078107 [pii]

Geusens B, Sanders N, Prow T, Van Gele M, Lambert J (2009) Cutaneous short-interfering RNA therapy. Expert Opin Drug Deliv 6(12):1333–1349. doi:10.1517/17425240903304032

Zou GM (2010) RNAi in stem cells: current status and future perspectives. Methods Mol Biol 650:3–14. doi:10.1007/978-1-60761-769-3

Goodell MA, Rosenzweig M, Kim H, Marks DF, DeMaria M, Paradis G, Grupp SA, Sieff CA, Mulligan RC, Johnson RP (1997) Dye efflux studies suggest that hematopoietic stem cells expressing low or undetectable levels of CD34 antigen exist in multiple species. Nat Med 3 (12):1337–1345

Grinshtein N, Datti A, Fujitani M, Uehling D, Prakesch M, Isaac M, Irwin MS, Wrana JL, Al-Awar R, Kaplan DR (2011) Small molecule kinase inhibitor screen identifies polo-like kinase 1 as a target for neuroblastoma tumor-initiating cells. Cancer Res 71(4):1385–1395. doi:0008-5472. CAN-10-2484 [pii] 10.1158/0008-5472.CAN-10-2484

Haney SA (2007) Expanding the repertoire of RNA interference screens for developing new anticancer drug targets. Expert Opin Ther Targets 11(11):1429–1441. doi:10.1517/14728222.11.11.1429

Hari D, Xin HW, Jaiswal K, Wiegand G, Kim BK, Ambe C, Burka D, Koizumi T, Ray S, Garfield S, Thorgeirsson S, Avital I (2011) Isolation of live label-retaining cells and cells undergoing asymmetric cell division via nonrandom chromosomal cosegregation from human cancers. Stem Cells Dev. doi:10.1089/scd.2010.0455

Harless WW (2011) Cancer treatments transform residual cancer cell phenotype. Cancer Cell Int 11(1):1. doi:1475-2867-11-1 [pii] 10.1186/1475-2867-11-1

Harrison H, Farnie G, Howell SJ, Rock RE, Stylianou S, Brennan KR, Bundred NJ, Clarke RB (2010) Regulation of breast cancer stem cell activity by signaling through the Notch4 receptor. Cancer Res 70(2):709–718. doi:0008-5472.CAN-09-1681 [pii] 10.1158/0008-5472.CAN-09-1681

Hermann PC, Huber SL, Herrler T, Aicher A, Ellwart JW, Guba M, Bruns CJ, Heeschen C (2007) Distinct populations of cancer stem cells determine tumor growth and metastatic activity in human pancreatic cancer. Cell Stem Cell 1(3):313–323. doi:S1934-5909(07), 00066-5 [pii] 10.1016/j.stem.2007.06.002

Higuchi Y, Kawakami S, Hashida M (2010) Strategies for in vivo delivery of siRNAs: recent progress. BioDrugs 24(3):195–205. doi:4 [pii] 10.2165/11534450-000000000-00000

Ikeda Y, Taira K (2006) Ligand-targeted delivery of therapeutic siRNA. Pharm Res 23 (8):1631–1640. doi:10.1007/s11095-006-9001-x

Jackson AL, Bartz SR, Schelter J, Kobayashi SV, Burchard J, Mao M, Li B, Cavet G, Linsley PS (2003) Expression profiling reveals off-target gene regulation by RNAi. Nat Biotechnol 21 (6):635–637. doi:10.1038/nbt831 nbt831 [pii]

Jackson AL, Burchard J, Leake D, Reynolds A, Schelter J, Guo J, Johnson JM, Lim L, Karpilow J, Nichols K, Marshall W, Khvorova A, Linsley PS (2006) Position-specific chemical modification of siRNAs reduces "off-target" transcript silencing. RNA 12(7):1197–1205. doi:rna.30706 [pii] 10.1261/rna.30706

Janeway CA Jr, Medzhitov R (2002) Innate immune recognition. Annu Rev Immunol 20:197–216. doi:10.1146/annurev.immunol.20.083001.084359 083001.084359 [pii]

Jeong JH, Mok H, Oh YK, Park TG (2009) siRNA conjugate delivery systems. Bioconjug Chem 20(1):5–14. doi:10.1021/bc800278e 10.1021/bc800278e [pii]

Ji Z, Mei FC, Xie J, Cheng X (2007) Oncogenic KRAS activates hedgehog signaling pathway in pancreatic cancer cells. J Biol Chem 282(19):14048–14055. doi:M611089200 [pii] 10.1074/jbc.M611089200

Judge AD, Bola G, Lee AC, MacLachlan I (2006) Design of noninflammatory synthetic siRNA mediating potent gene silencing in vivo. Mol Ther 13(3):494–505. doi:S1525-0016(05), 01674-6 [pii] 10.1016/j.ymthe.2005.11.002

Judge AD, Robbins M, Tavakoli I, Levi J, Hu L, Fronda A, Ambegia E, McClintock K, MacLachlan I (2009) Confirming the RNAi-mediated mechanism of action of siRNA-based cancer therapeutics in mice. J Clin Invest 119(3):661–673. doi:37515 [pii] 10.1172/JCI37515

Karlsson C, Larsson J, Baudet A (2010) Forward RNAi screens in human stem cells. Methods Mol Biol 650:29–43. doi:10.1007/978-1-60761-769-3

Kemeny N (2010) The management of resectable and unresectable liver metastases from colorectal cancer. Curr Opin Oncol 22(4):364–373. doi:10.1097/CCO.0b013e32833a6c8a

Kennedy S, Wang D, Ruvkun G (2004) A conserved siRNA-degrading RNase negatively regulates RNA interference in C. elegans. Nature 427(6975):645–649. doi:10.1038/nature02302 nature02302 [pii]

Lapidot T, Sirard C, Vormoor J, Murdoch B, Hoang T, Caceres-Cortes J, Minden M, Paterson B, Caligiuri MA, Dick JE (1994) A cell initiating human acute myeloid leukaemia after transplantation into SCID mice. Nature 367(6464):645–648. doi:10.1038/367645a0

Li J, Zhou BP (2011) Activation of beta-catenin and Akt pathways by twist are critical for the maintenance of EMT associated cancer stem cell-like characters. BMC Cancer 11:49. doi:1471-2407-11-49 [pii] 10.1186/1471-2407-11-49

Li C, Heidt DG, Dalerba P, Burant CF, Zhang L, Adsay V, Wicha M, Clarke MF, Simeone DM (2007) Identification of pancreatic cancer stem cells. Cancer Res 67(3):1030–1037. doi:67/3/1030 [pii] 10.1158/0008-5472.CAN-06-2030

Luo L, Zhou A (2009) Antifibrotic activity of anisodamine in vivo is associated with changed intrahepatic levels of matrix metalloproteinase-2 and its inhibitor tissue inhibitors of metalloproteinases-2 and transforming growth factor beta1 in rats with carbon tetrachloride-induced liver injury. J Gastroenterol Hepatol 24(6):1070–1076. doi:JGH5756 [pii] 10.1111/j.1440-1746.2008.05756.x

Malek A, Czubayko F, Aigner A (2008) PEG grafting of polyethylenimine (PEI) exerts different effects on DNA transfection and siRNA-induced gene targeting efficacy. J Drug Target 16(2):124–139. doi:790598183 [pii] 10.1080/10611860701849058

Marquardt JU, Raggi C, Andersen JB, Seo D, Avital I, Geller D, Lee YH, Kitade M, Holczbauer A, Gillen MC, Conner EA, Factor VM, Thorgeirsson SS (2011) Human hepatic cancer stem cells are characterized by common stemness traits and diverse oncogenic pathways. Hepatology doi:10.1002/hep.24454

McDermott SP, Wicha MS (2010) Targeting breast cancer stem cells. Mol Oncol 4:404–419. doi:10.1016/j.molonc.2010.06.005

McInnes C, Mazumdar A, Mezna M, Meades C, Midgley C, Scaerou F, Carpenter L, Mackenzie M, Taylor P, Walkinshaw M, Fischer PM, Glover D (2006) Inhibitors of Polo-like kinase reveal roles in spindle-pole maintenance. Nat Chem Biol 2(11):608–617. doi:nchembio825 [pii] 10.1038/nchembio825

Merchant AA, Matsui W (2010) Targeting Hedgehog–a cancer stem cell pathway. Clin Cancer Res 16:3130–3140. doi:10.1158/1078-0432.CCR-09-2846

Pal A, Valdez KE, Carletti MZ, Behbod F (2010) Targeting the perpetrator: breast cancer stem cell therapeutics. Curr Drug Targets 11:1147–1156

Peacock CD, Wang Q, Gesell GS, Corcoran-Schwartz IM, Jones E, Kim J, Devereux WL, Rhodes JT, Ca H, Pa B, Watkins DN, Matsui W (2007) Hedgehog signaling maintains a tumor stem cell compartment in multiple myeloma. Proc Natl Acad Sci USA 104:4048–4053. doi:10.1073/pnas.0611682104

Pearce DJ, Ridler CM, Simpson C, Bonnet D (2004) Multiparameter analysis of murine bone marrow side population cells. Blood 103(7):2541–2546. doi:10.1182/blood-2003-09-3281 2003-09-3281 [pii]

Pece S, Tosoni D, Confalonieri S, Mazzarol G, Vecchi M, Ronzoni S, Bernard L, Viale G, Pelicci PG, Di Fiore PP (2010) Biological and molecular heterogeneity of breast cancers correlates with their cancer stem cell content. Cell 140(1):62–73. doi:S0092-8674(09), 01554-2 [pii] 10.1016/j.cell.2009.12.007

Roesch A, Fukunaga-Kalabis M, Schmidt EC, Zabierowski SE, Brafford PA, Vultur A, Basu D, Gimotty P, Vogt T, Herlyn M (2010) A temporarily distinct subpopulation of slow-cycling melanoma cells is required for continuous tumor growth. Cell 141(4):583–594. doi:S0092-8674(10)00437-X [pii] 10.1016/j.cell.2010.04.020

Scherr M, Battmer K, Winkler T, Heidenreich O, Ganser A, Eder M (2003) Specific inhibition of bcr-abl gene expression by small interfering RNA. Blood 101(4):1566–1569. doi:10.1182/blood-2002-06-1685 2002-06-1685 [pii]

Sell S (2004) Stem cell origin of cancer and differentiation therapy. Crit Rev Oncol Hematol 51 (1):1–28. doi:10.1016/j.critrevonc.2004.04.007 S104084280400068X [pii]

Sen GC (2001) Viruses and interferons. Annu Rev Microbiol 55:255–281. doi:10.1146/annurev.micro.55.1.255

Seymour LW (1992) Passive tumor targeting of soluble macromolecules and drug conjugates. Crit Rev Ther Drug Carrier Syst 9(2):135–187

Siegel R, Ward E, Brawley O, Jemal A (2011) Cancer statistics, 2011: the impact of eliminating socioeconomic and racial disparities on premature cancer deaths. CA Cancer J Clin 61 (4):212–236. doi:caac.20121 [pii] 10.3322/caac.20121

Smalley MJ, Dale TC (1999) Wnt signalling in mammalian development and cancer. Cancer Metastasis Rev 18(2):215–230

Spankuch B, Kurunci-Csacsko E, Kaufmann M, Strebhardt K (2007) Rational combinations of siRNAs targeting Plk1 with breast cancer drugs. Oncogene 26(39):5793–5807. doi:1210355 [pii] 10.1038/sj.onc.1210355

Sperger JM, Chen X, Draper JS, Antosiewicz JE, Chon CH, Jones SB, Brooks JD, Andrews PW, Brown PO, Thomson JA (2003) Gene expression patterns in human embryonic stem cells and human pluripotent germ cell tumors. Proc Natl Acad Sci USA 100(23):13350–13355. doi:10.1073/pnas.2235735100 2235735100 [pii]

Steegmaier M, Hoffmann M, Baum A, Lenart P, Petronczki M, Krssak M, Gurtler U, Garin-Chesa P, Lieb S, Quant J, Grauert M, Adolf GR, Kraut N, Peters JM, Rettig WJ (2007) BI 2536, a potent and selective inhibitor of polo-like kinase 1, inhibits tumor growth in vivo. Curr Biol 17 (4):316–322. doi:S0960-9822(06), 02671-6 [pii] 10.1016/j.cub.2006.12.037

Strebhardt K, Ullrich A (2006) Targeting polo-like kinase 1 for cancer therapy. Nat Rev Cancer 6 (4):321–330. doi:nrc1841 [pii] 10.1038/nrc1841

Toloue MM, Ford LP (2011) Antibody targeted siRNA delivery. Methods Mol Biol 764:123–139. doi:10.1007/978-1-61779-188-8_8

Velkey JM, O'Shea KS (2003) Oct4 RNA interference induces trophectoderm differentiation in mouse embryonic stem cells. Genesis 37(1):18–24. doi:10.1002/gene.10218

Vogelstein B, Fearon ER, Hamilton SR, Kern SE, Preisinger AC, Leppert M, Nakamura Y, White R, Smits AM, Bos JL (1988) Genetic alterations during colorectal-tumor development. N Engl J Med 319(9):525–532. doi:10.1056/NEJM198809013190901

Yang S, Tutton S, Pierce E, Yoon K (2001) Specific double-stranded RNA interference in undifferentiated mouse embryonic stem cells. Mol Cell Biol 21(22):7807–7816. doi:10.1128/MCB.21.22.7807-7816.2001

Yuan F, Dellian M, Fukumura D, Leunig M, Berk DA, Torchilin VP, Jain RK (1995) Vascular permeability in a human tumor xenograft: molecular size dependence and cutoff size. Cancer Res 55(17):3752–3756

Zeki SS, Graham TA, Wright NA (2011) Stem cells and their implications for colorectal cancer. Nat Rev Gastroenterol Hepatol 8(2):90–100. doi:nrgastro.2010.211 [pii] 10.1038/nrgastro.2010.211

Zou GM, Lebron C, Fu Y (2010) RNAi knockdown of redox signaling protein Ape1 in the differentiation of mouse embryonic stem cells. Methods Mol Biol 650:121–128. doi:10.1007/978-1-60761-769-3_10

Zuber J, Shi J, Wang E, Rappaport AR, Herrmann H, Sison Ea, Magoon D, Qi J, Blatt K, Wunderlich M, Taylor MJ, Johns C, Chicas A, Mulloy JC, Kogan SC, Brown P, Valent P, Bradner JE, Lowe SW, Vakoc CR (2011) RNAi screen identifies Brd4 as a therapeutic target in acute myeloid leukaemia. Nature:1–7. doi:10.1038/nature10334

Chapter 17
MicroRNAs in Development, Stem Cell Differentiation, and Regenerative Medicine

Betty Chang, Ihor R. Lemischka, and Christoph Schaniel

Abstract Mammalian development and cellular differentiation are robust but tightly controlled processes. MicroRNAs have emerged as key players in posttranscriptional regulation of gene expression during development and cellular differentiation. As analytical tools advance from cloning techniques to microarrays and most recently to massively parallel deep sequencing technologies, the space of known microRNAs and their target mRNAs is better defined and is leading to a comprehensive catalog combined with functional characterization. Several tissue- and cell-lineage-specific microRNAs have been identified, some of which are associated with distinct stages of cell identity from stem to progenitor to terminally differentiated cells. We describe the important functional roles of some of these microRNAs as exemplified by the ability of their exogenous expression to elicit changes in cell fate and discuss how, with this knowledge, we can dispense with genetic manipulation and begin to harness the advantage of microRNAs, microRNA mimics, microRNA antagonists (antagomirs), antisense RNA, siRNA, and alike molecules as tools for regenerative medicine and therapy.

Keywords microRNAs • stem cells • development • differentiation • regenerative medicine

Abbreviations

AML	Acute myeloid leukemia
AS (C)	Adipose tissue–derived stem (cell)
asRNA	Antisense RNA

C. Schaniel (✉)
Department of Developmental and Regenerative Biology, Mount Sinai School of Medicine,
The Black Family Stem Cell Institute, New York, NY, USA
e-mail: christoph.schaniel@mssm.edu

B. Mallick (eds.), *Regulatory RNAs*, DOI 10.1007/978-3-642-22517-8_17,
© Springer-Verlag Berlin Heidelberg (outside the USA) 2012

BP	B-cell progenitor
C. elegans	*Caenorhabditis elegans*
CLP	Common lymphoid progenitor
CMP	Common myeloid progenitor
CMV	Cytomegalovirus
Cre	Causes recombination
Dgcr8	DiGeorge syndrome critical region gene 8
DNA	Deoxyribonucleic acid
E	Embryonic (stage/day postconception)
EB	Embryoid body
EC	Embryo(nic) carcinoma
EMT	Epithelial-mesenchymal transition
ES	Embryonic stem
GFP	Green fluorescent protein
GMP	Granulocyte macrophage progenitor
h	Human
HIV	Human immunodeficiency virus
HSC	Hematopoietic stem cell
ICM	Inner cell mass
iPS(C)	Induced pluripotent stem (cell)
lacZ	Gene encoding bacterial β-galactosidase
LNA	Locked nucleic acid
loxP	Locus of chromosomal crossover in the bacteriophage P1
LV	Lentivirus
m	Mouse
MEP	Megakaryocyte erythrocyte progenitor
miR	MicroRNA
miRISC	MicroRNA-induced silencing complex
MPP	Multipotent progenitor
ncRNA	Noncoding RNA
NK	Natural killer (cell)
NKP	NK cell progenitor
NSC	Neural stem cell
P	Postnatal (day)
PCR	Polymerase chain reaction
RA	Retinoic acid
RCME	Recombination-mediated cassette exchange
RNA	Ribonucleic acid
shRNA	Short hairpin RNA
siRNA	Small interfering RNA
SNALP	Stable nucleic-acid-lipid particles
SRF	Serum response factor
Th	T helper type
TK	Thymidine kinase

TP	T-cell progenitor
TSS	Transcription/transcript start site
UTR	Untranslated region

17.1   Introduction

In the nematode *C. elegans*, defects in developmental timing led Ambros and colleagues to discover the first microRNA in 1993 (Lee et al. 1993). Aberrations within the gene *lin-4* in *C. elegans* caused animals to display early larval stage phenotypes late in development. The gene encoding *lin-4* was localized within an intron and found to generate two small RNAs of 61 and 22 nucleotides. These small RNAs harbored sequences complementary to the 3′untranslated region (3′UTR) of the mRNA encoding *lin-14*, a protein specific to early larval stages that accumulates in the *lin-4* mutants. Sequences for the small RNAs, when introduced into the *lin-4* mutants, were able to rescue the developmental timing defects, and this correlated with the downregulation of *lin-14* protein levels, suggesting an antisense targeting of the *lin-14* mRNA 3′UTR to inhibit protein expression. This discovery intimately linked these small RNAs, which we now call microRNAs, with developmental processes and highlights their importance in developmental regulation.

The biogenesis from primary to precursor to functional mature microRNA requires the coordinated effort of several RNA-binding proteins and enzymes. Disruption of genes involved in microRNA biogenesis by either targeted deletion or knockdown results in developmental defects in vertebrates (Bernstein et al. 2003; Wienholds et al. 2003; Giraldez et al. 2005; Harfe 2005b; Kanellopoulou 2005; Murchison 2005; Wang et al. 2007). These studies characterizing members of the microRNA biogenesis pathway, including Drosha, DiGeorge syndrome critical region gene 8 (Dgcr8), and Dicer, outlined in detail in previous chapters, demonstrate the essential roles of microRNAs in vertebrate development. *Danio rerio*, zebra fish, appear to progress through early developmental events normally in the absence of *Dicer1*, mutated by targeted gene inactivation (Wienholds et al. 2003); however, growth is arrested 8–10 days postfertilization. Normal growth in the early zebra fish embryo was further dampened by the introduction of morpholinos, antisense RNA, against *Dicer1* to target maternal mRNA contribution. Maternal-zygotic disruption of *Dicer* (Giraldez et al. 2005), also performed in zebra fish, showed the ability of animals to undergo early fate specification events such as axis formation and generation of embryonic cell lineages; however, the fish have impairments in morphogenesis and neural development during gastrulation. In these fish, the addition of a single microRNA, miR-430, was able to, in part, rescue neural-developmental defects. The loss of *Dicer1* in mice halted development at embryonic stage 7.5, with small embryos lacking expression of the embryonic stem cell marker, Oct4, and the mesodermal lineage marker, Brachyury, and eventually succumbing to lethality before embryonic stage 8.5 (Bernstein et al. 2003). The lack of functional Dicer more specifically leads to deficits in proliferation and differentiation within mouse embryonic stem (mES) cells and depletion of stem cells from the embryo (Kanellopoulou 2005; Murchison 2005; Bernstein et al. 2003). Similar impairment is seen in mES cells lacking functional Dgcr8, a Drosha cofactor, where mES cells display defects in cell proliferation rates and are unable to suppress pluripotency markers and thus are unable to exit from the self-renewal state (Wang et al. 2007; Wang and Blelloch 2011). Conditional deletion of *Dicer*, by Cre recombinase–mediated excision in homozygous mice harboring loxP sites flanking the exon encoding the RNase III domain in the Dicer gene, within the developing limb, led to malformations and morphogenesis defects (Harfe 2005b) consistent with studies in zebra fish. Disruption of *Dicer1*, also by tissue-specific conditional ablation, leads to defects in skin morphogenesis and epithelial stratification in the mouse (Yi et al. 2006). Interruptions in the microRNA biogenesis pathway cause an ablation of mature microRNAs. This suggests that microRNAs are vital to proper development in vertebrates. The more pronounced defects observed in mice over zebra fish lacking proper microRNA biogenesis imply an evolutionarily enhanced role for microRNAs in higher vertebrates.

The diversity of cell fate changes during vertebrate development is often characterized by alterations in gene expression regulated by key transcription factors. The POU-family transcription factor Oct4 defines the inner cell mass (ICM) from where ES cells are derived, and the loss of Oct4 denotes the departure from the pluripotent state toward cellular differentiation. Transcription factors display temporal- and cell-specific expression and serve to mark and shepherd the progression in development toward specified cell fates. During gastrulation and early development, transcription factors establish cell programs to define morphogenesis and patterning in the developing organism. MicroRNAs exhibit temporal- and cell-type-specific expression and act as a posttranscriptional control unit to provide an additional level of regulation. Cataloging of microRNAs across several organisms has defined the milieu of microRNAs expressed at different stages of cellular differentiation and tissue specification. Bioinformatics' predictions in conjunction with cloning, microarray, and next-generation sequencing techniques have defined, identified, and verified a growing list of mature microRNAs.

MicroRNAs demonstrate spatiotemporal expression patterns from *C. elegans* to vertebrates throughout development and in adult organisms. The coordinated expression and activity of microRNAs within *C. elegans* ensures the correct progression among larval stages and ultimately to adulthood (Lee et al. 1993; Reinhart et al. 2000; Rougvie 2001; Großhans et al. 2005). Heterochronic microRNAs *lin-4* and *let-7*, differentially expressed during larval development, act to promote stage progression and serve as switches in developmental timing. *Lin-4* and *let-7* are evolutionarily conserved microRNAs within vertebrates. The mammalian counterparts of *lin-4* are known as miR-125a and miR-125b, and *let-7* has several homologues encoded at several loci that are known collectively as the *let-7* family. Both of these microRNA families target the 3′UTR of the ES cell marker *Lin-28*, an RNA-binding protein that recognizes and sequesters the stem loop of the *let-7* precursor microRNA preventing Dicer-mediated conversion to mature *let-7*, thus forming a regulatory feedback loop. The mature forms of miR-

125 and *let-7* are expressed widely in somatic cells and tissues and serve various functions within specific cell types. The microRNA miR-125b is expressed in the brain, hematopoietic cells, skin, and muscle, among other cell types, and is linked to proliferation and, in different tissues, the advancement or blockage of cellular differentiation. Inappropriate expression and regulation of these microRNAs are also associated with many cancers and other disorders.

Selective expression of individual microRNAs during development plays an important role in cell specification at different stages of embryogenesis and animal maturation. For example, in mice, miR-1 specifies cardiac tissue; however, overexpression leads to growth arrest due to accelerated differentiation leading to depletion of progenitor cells (Zhao et al. 2005). Single microRNAs, by design, have a predicted potential to target a multitude of mRNAs (Lewis et al. 2003). The cellular and coordinated gene expression context surrounding the microRNA influences its activity and functionality. Aberrant microRNA expression is linked to disease – exhibiting developmental defects and cancer phenotypes – since the ability to target multiple genes leads to improper global regulation. To understand and predict potential inappropriate regulation, there is an imperative for better characterization of microRNA targets by computational and biochemical techniques. Target prediction algorithms are beyond the scope of this chapter but incorporate mining mRNA transcripts for sequences complementary to microRNA (Lewis et al. 2003). Biochemical techniques to elucidate microRNA-mediated regulation include ascertaining targeting by luciferase or microRNA-sensor activity or methods examining miRISC incorporation of microRNAs together with their mRNA targets using Ago2-imunoprecipitation followed by parallel sequencing (Landthaler et al. 2008; Chi et al. 2009; Hafner et al. 2010).

The growing body of evidence uncovering microRNAs playing significant roles in posttranscriptional regulation of gene expression during development, differentiation, and in diseased cells makes microRNAs attractive targets for use in regenerative medicine. Several studies demonstrate the ability of ectopic overexpression of microRNAs to drive cellular differentiation, transforming microRNAs into a valuable resource in directed differentiation for research and future cell transplantation therapy. The discovery of small RNAs has burgeoned into the flourishing field of small interfering RNAs and short hairpin RNAs vital to current advances in research and therapeutic applications (see Chap. 19). Antisense RNAs are currently in clinical trials in, for example, diabetes and cancer therapy (Knowling and Morris 2011). Reprogramming to induced pluripotency and trans-differentiation using microRNAs (Anokye-Danso et al. 2011; Yoo et al. 2011; Ambasudhan et al. 2011) illustrates the efficacy of microRNAs to promote dramatic changes in cell identity and presents them as promising factors in regenerative medicine. Transcriptional regulation of microRNAs, microRNA moieties as mimics or other vehicles, and their complementary and inhibitory sequences may be studied for the development of therapeutic tools. In this chapter, we describe the current knowledge on the role of microRNAs in development and differentiation and their application to either direct differentiation of embryonic and adult stem cells or to prevent differentiation to undesirable or diseased cells for regenerative medicine.

17.2 Spatiotemporal MicroRNA Expression Patterns in Vertebrates

Specialization of tissues within the body is shaped by differential gene expression and is typically defined by measurements at the RNA and protein levels. These studies described below define tissue-specific microRNA expression in mouse and human. Within these tissues are arrays of cellular subtypes all of which contribute to the functionality of the larger tissue. Closer examination of each cell type and, in particular, progenitor and terminally differentiated cells identifies microRNAs to distinguish both location-specific as well as cell-fate-specific and identity-specific characterization and function to further define supporting roles for microRNAs in cellular differentiation and development. Early techniques in tissues and specific cell types consisted of cloning small RNAs and northern blotting. This was followed by oligonucleotide arrays and sequence-specific reverse transcription coupled with PCR amplification. Today, discovery and interrogation of mature microRNA species is achieved using massive parallel sequencing platforms. In animals, lacZ-based sensors harboring microRNA target sites map functional activity of mature microRNAs by lineage tracing along embryonic maturation. Moving to cellular systems, overexpression of individual microRNAs by transgenic or mimic introduction allows for gain of function analyses. Antisense molecules or antagomirs can also be utilized, such as locked nucleic acid (LNA) oligonucleotides, for loss of function experiments. Cell-based analyses also allow for luciferase reporter-mediated functional characterization of microRNA target sites. Here we outline some examples of microRNAs distinguishing tissue and cell states with a focus on the diversity of expression.

17.2.1 Tissue-Specific Expression Patterns

A survey for microRNAs was done on 18.5-week-old mice by cloning small, about 21-nucleotide RNA species from heart, liver, spleen, small intestine, colon tissue, and the cortex, cerebellum, and midsection of the brain (Lagos-Quintana et al. 2002). This study uncovered that although a given microRNA may be expressed in several tissues, expression levels of specific microRNAs dominate within specific tissues. For example, miR-1 accounted for over 45% of the microRNA clones derived from the heart. Sempere and colleagues also probed tissues of the mouse and human by northern blotting against 119 microRNAs (Sempere et al. 2004) and defined 30 microRNAs as specific to a single organ or tissue among those profiled.

The majority of the microRNAs found in the brain, liver, heart, and skeletal muscle were conserved between mouse and human. Several microRNAs, *let-7a*, *let7b*, miR-30b, and miR-30c, were also found to have abundant tissue-wide expression. MicroRNAs encoded within the same loci shared coordinate expression patterns. For example, miR-194 and miR-215 were found in the mouse kidney and miR-1 and miR-133 were detected in both the heart and skeletal muscle in both mouse and human. Other microRNAs arising from the same genomic region show differential expression patterns. For example, miR-132 was detected in mouse and human brain but miR-212 was not. Oligonucleotide microarrays were also used to interrogate microRNA expression in mouse tissues (Strauss et al. 2006; Chen et al. 2007). All of these studies identify miR-124 as brain-specific microRNA and miR-122 specific to the liver. Mouse fetal liver cells were collected at embryonic stages E16.5, E17.5, perinatal day P1, and from the adult mouse and profiled by microarray for microRNA expression (Rogler et al. 2009). The miR-23b cluster was upregulated in late fetal development within fetal hepatocytes beginning at E17.5 through adulthood. During hepatocyte differentiation to bile duct cells, the miR-23b cluster family of microRNAs was downregulated and when overexpressed blocked this differentiation step.

In another approach, Mansfield and colleagues developed sensors for in vivo microRNA target validation (Mansfield et al. 2004). Transgenic insertions of the lacZ gene engineered with microRNA target elements within its 3′UTR served as a tracking system for microRNA activity in specific tissues, cells, or developmental stages during embryonic patterning of the mouse. Interestingly, *let-7* family members displayed differing expression patterns in the developing limb, where *let-7c* was expressed in the anterior limb and *let-7e* in the limb ectoderm. The expression of miR-1 in the heart was detected within the myocardium chamber, ventricles, and the atrioventricular canal. MicroRNAs encoded within Hox gene clusters (Krumlauf 1994), essential transcriptional regulators that specify segment identity and accurate body patterning, also demonstrate temporal expression patterns similar to their host Hox genes. MiR-10a is encoded upstream of Hoxb4, and identical mature microRNAs miR-196a-1 and miR-196a-2 lie upstream of the Hoxb9 and Hoxc9 genes, respectively. MiR-10a and Hoxb4 display coordinate expression patterns. Near-perfect sequence complementarity to miR-196 in the Hoxb8 3′UTR (Mansfield et al. 2004; Yekta 2004) creates mutually exclusive spatial expression patterning of miR-196/Hoxb8 during limb development.

In situ hybridization studies in the zebra fish brain using LNA probes for selected microRNAs show tissue-wide and restricted expression patterns. MiR-92b was found to be specific to neural precursor stem cells, whereas miR-124 defined cells that were transitioning from a proliferative stage to a mature differentiated neuronal stage, and miR-218a was only detected in motor neurons (Kapsimali et al. 2007). Temporal examination of mouse cerebellum, cortex, and midbrain (Smirnova et al. 2005) from embryonic stage 12 (E12) to birth (P0) and after birth at days 2 (P2) and 14 (P14) characterizes expression of miR-26 in early brain development at E12 and miR-29 late in brain development at P14. Three microRNAs, miR-124, miR-125, and miR-128, accrued at increasing expression levels in parallel with the maturation of neuronal cells.

17.2.2 Cell-Specific Expression Patterns

17.2.2.1 Embryonic Stem Cells

mES and hES cells derived from the ICM and cultured in vitro have the ability to continually self-renew and retain their pluripotent potential to differentiate into cells of the three primary germ layers, endoderm, mesoderm, and ectoderm (Fig. 17.1a). Several studies defining microRNAs expressed in mES and hES cells (Houbaviy et al. 2003; Suh et al. 2004; Wang et al. 2007; Sempere et al. 2004; Strauss et al. 2006; Chen et al. 2007; Wang et al. 2008) identify sets of microRNA families from distinct clusters with similar seed sequences that are both specific to ES cells and downregulated upon loss of self-renewal and pluripotency.

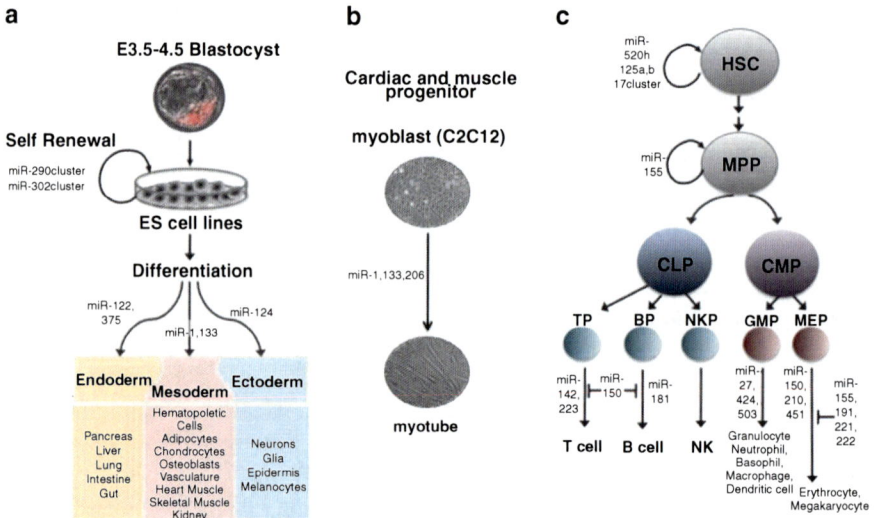

Fig. 17.1 *MicroRNAs associated with cell fate and cell fate decisions, selected examples.* (a) Embryonic stem (ES) cells derived from the inner cell mass (ICM) (*red* depicts Nanog-RFP reporter expression); cells of the blastocyst differentiate to primary germ layers endoderm, mesoderm, and ectoderm. (b) Myogenesis of cardiac and muscle progenitor cells; microRNAs are able to push the myoblast cell line C2C12 toward myotube formation. (c) A simplification of hematopoietic cell hierarchy; microRNAs are implicated in self-renewal of hematopoietic stem cells (HSC), in multipotent progenitors (MPP), and in cell progression from common lymphoid progenitors (CLP) to T-cell progenitors (TP), B-cell progenitors (BP), and natural killer progenitors (NKP) and their terminal cells; and from common myeloid progenitors (CMP) to granulocyte-macrophage progenitor (GMP) and megakaryocyte-erythroid progenitor (MEP) to their terminal cell lineages

The miR-290 in the mouse and the miR-302 family in both the mouse and human are examples of ES cell–specific microRNAs that promote self-renewal by modulating cell cycle proteins.

17.2.2.2   Adipogenic Lineage

MiR-143 was observed in cultured adipocytes (Esau et al. 2004) and is enriched in mouse and human adipose tissues. Using antisense oligonucleotides (ASO) to miR-143 prevented expression of mature adipocyte traits such as expression of GLUT4, ASL, PPARγ2, and triglyceride accumulation. The predicted target of miR-143, ERK5 exhibited diminished expression by western blot analysis in cells overexpressing miR-143, and ERK5 increased with miR-143-ASO treatment. MiR-143 regulation via binding to the ERK5 3′UTR was recently confirmed (Noguchi et al. 2011). Closer examination of microRNA specificity in mouse adipose tissue revealed miR-143 localization to mature white adipocytes and a reduction in brown fat, where miR-455 was highly expressed (Walden et al. 2009). Walden and colleagues also identify miR-1, miR-133a, and miR-206 expression in premature and mature brown adipocytes exclusive of any expression in white adipocytes.

17.2.2.3   Neuronal Lineage

In neural specification by induction of mES cells to defined neuronal cell types using RA treatment (Smirnova et al. 2005), expression of miR-124 and miR-128 were raised in neurons and miR-23 and miR-29 were elevated in astrocytes. Neuronal progenitors and terminally differentiated neuronal cells derived from embryoid bodies (EB) were compared to primary cortical neurons and showed common expression of miR-124a, miR-9/9*, miR-22, and miR-125b (Krichevsky et al. 2006).

17.2.2.4   Pancreatic Lineage

Mir-375 and miR-376 appear to be exclusively expressed in pancreatic cells and when overexpressed decrease insulin secretion (Poy et al. 2004; Harfe 2005a). Mouse α- and β-cell lines also express miR-375. Within the developing mouse pancreas, miR-124a was differentially upregulated at E18.5 relative to E14.5 and was found to target Foxa2 in pancreatic β-cells (Baroukh et al. 2007). Pancreatic islets isolated from human fetal pancreas also show distinct expression of miR-375 and miR-376 along with miR-7 and miR-9 (Joglekar et al. 2009).

17.2.2.5   Hematopoietic Lineage

Mouse hematopoietic tissues were profiled via northern blotting by Chen and colleagues to demonstrate expression of miR-223 in the bone marrow and miR-142s in all hematopoietic cells (Chen 2004). This study also found miR-181 expression in the thymus, brain, lung, bone marrow, and spleen. Ectopic expression of these individual microRNAs in lineage-depleted cells isolated from mouse bone marrow promoted B lymphocyte expansion by miR-181 and T-lymphoid lineage expansion by miR-142s and miR-223. Expression of microRNAs in seven cell types of the mouse hematopoietic system, representing a hierarchy of cellular differentiation, was profiled by microarray (Monticelli et al. 2005). Levels of miR-150 increased in both B- and T-cell progenitors but declined during the differentiation of T lymphocytes to T helper type 1 (Th1) and Th2 cells. In addition, miR-146 was exclusively upregulated in Th1 cells. MiR-150 was identified as specific to several hematopoietic progenitor cells (Monticelli et al. 2005; Lu et al. 2008) and when aberrantly overexpressed, blocks c-Myb expression to impair lymphocyte maturation and dramatically depletes animals of the mature B-cell population (Xiao et al. 2007). In contrast, miR-150 promotes differentiation to the megakaryocyte and erythroid lineage (Lu et al. 2008).

CD34-positive hematopoietic stem cells (HSCs) isolated from human bone marrow express miR-155, which is able to block in vitro differentiation to myeloid and erythroid cells (Georgantas et al. 2007). Expression of miR-155 was decreased in erythroid progenitor cells (Masaki et al. 2007), while miR-451 increased in expression and was able to induce differentiation (Zhan et al. 2007) in in vitro culture.

17.2.2.6   Cardiac and Skeletal Muscle Lineages

The mesoderm lineage gives rise to both cardiac and skeletal muscle cells, so it is not surprising that they share similar microRNA expression profiles. MiR-1 and miR-133 are detected in both cardiac and skeletal muscle cell in the mouse and human, in the embryo, as well as in adult cells, but there are some clear differences during adult cell specification. In skeletal muscle, miR-1 and miR-206 target sites are found in connexin-43 mRNA whose downregulation is concurrent with myotube differentiation; interestingly, miR-206 is absent from the heart (Anderson et al. 2006).

17.2.2.7   Epithelial Lineage

MiR-203 is expressed in the skin and hair follicle cells in mice (Yi et al. 2006; Yi et al. 2008). HNF1alpha transcriptionally activates miR-194 to induce differentiation in the Caco-2 cell line, cells derived from the human intestinal epithelium, and marks primary human intestinal epithelium cells (Hino et al. 2007; Hino et al. 2008).

The epithelial-mesenchymal transition (EMT) is another shift in development with coordinated gene expression, where the transcription factor Twist is able to induce this conversion and activate expression of miR-10b and miR-21. These microRNAs, however, are not able to solely elicit an EMT response (Bracken et al. 2009). The miR-200 cluster of microRNAs is highly expressed in epithelial tissues, and they are able to block the EMT in mice.

MicroRNAs exhibiting either tissue- or cell-specific expression or enriched levels are summarized in Table 17.1.

17.2.3   Transcriptional Regulation of MicroRNA Genes

MicroRNA genes are predominately transcribed by RNA polymerase II and are subject to the same regulation by transcription factors as protein coding genes. Not surprisingly, transcription factors that define cell identity and cell fate have been shown to regulate microRNA expression.

ES cell–specific transcription factors Nanog, Oct4, and Sox2 were mapped by chromatin immunoprecipitation to several putative microRNA transcript start sites (TSS) loci (Marson et al. 2008). Among those microRNA TSS bound by all three transcription factors are the ES cell–specific clusters encoding the miR-290 family in the mouse and the miR-302 family in both the mouse and human. Interestingly, Nanog, Oct4, and Sox2 also reside at several loci regulating differentiation-associated microRNAs, for example, let-7, miR-124, and miR-451, suggesting both positive and negative transcription factor regulation of microRNAs in ES cells.

Muscle-specific transcription factors serum response factor (SRF), MyoD, and MEF2 positively regulate miR-1, miR-133, and miR-206 expression in the developing muscle and during differentiation of C2C12 mouse myoblasts to myotubes. MEF2 binds enhancer regions within the locus, between sequences encoding miR-1-2 and miR-133a-1 in the mouse to induce expression in somite myotubes, skeletal muscle fibers and activates these microRNAs in the developing heart tube during embryogenesis (Liu et al. 2007). MyoD induces miR-206 expression to inhibit follistatin-like 1 (Fstl1) and Utrn in skeletal muscle (Rosenberg et al. 2006). TGFβ1 in conjunction with Smad3 negatively regulates transcription of miR-24 which is able to enhance myogenic differentiation in mice (Sun et al. 2008).

Transcription of the noncoding RNA (ncRNA), bic (B-cell integration cluster), was determined to have a vital impact on the immune response of mice, where bic loss caused immunodeficiency with lineage commitment of T-lymphocyte progenitors skewed to Th2 cells and exhibit B-cell deficiency (Turner and Vigorito 2008). Interestingly, miR-155, outlined earlier with expression specific to hematopoietic progenitors, is encoded within an exon of bic, and overexpression studies of miR-155 in mice cause B-cell malignancies. Gfi1 transcriptionally inhibits miR-21 and miR-196b to maintain hematopoietic progenitor cells, and

Table 17.1 Tissue- and cell-enriched distribution of microRNAs in mouse and human

MicroRNA	Organism (m)-mouse, (h)-human	Tissue(s)	Cell type(s)	Reference(s)
let-7	h		Neural stem cell[a]	Liu et al. (2009)
miR-1	m, h	Heart, skeletal muscle, brown adipose	Cardiomyocytes, muscle satellite cells, brown adipocytes	Lagos-Quintana et al. (2002), Sempere et al. (2004), Mansfield et al. (2004), Zhao et al. (2005), Chen et al. (2005), Anderson et al. (2006), Chen et al. (2007), Walden et al. (2009)
miR-101	m, h	Brain	Chondroblast	Lagos-Quintana et al. (2002), Suomi et al. (2008)
miR-103	h	Bone marrow	Mesenchymal stem cell	Liu et al. (2009)
miR-107	h	Bone marrow	Mesenchymal stem cell	Liu et al. (2009)
miR-10a	m	Fetal liver	Hepatocytes	Rogler et al. (2009)
miR-10b	m	Kidney		Chen et al. (2007)
miR-122	m, h	Liver	Hepatocyte	Lagos-Quintana et al. (2002), Sempere et al. (2004), Chen et al. (2007)
miR-124	m, h	Brain, pancreas	Neuronal progenitor cells, neurons, beta-cell	Sempere et al. (2004), Smirnova et al. (2005), Baroukh et al. (2007), Chen et al. (2007)
miR-125a,b	m, h	Brain	Neural stem cells, hematopoietic stem cell	Lagos-Quintana et al. (2002), Liu et al. (2009), O'Connell et al. (2010), Ooi et al. (2010)
miR-126	m, h	Bone marrow	Hematopoietic stem cell, embryonic stem cell	Shen et al. (2008), Liu et al. (2009)
miR-127	m, h	Brain		Lagos-Quintana et al. (2002)
miR-127	h	Brain	Hematopoietic stem cell	Liao et al. (2008)
miR-128	m, h	Brain	Neurons	Lagos-Quintana et al. (2002), Smirnova et al. (2005), Chen et al. (2007)
miR-129-3p	m	Brain		Chen et al. (2007)
miR-130	m	Lung		Sempere et al. (2004)
miR-131	m, h	Brain		Lagos-Quintana et al. (2002)
miR-132	m, h	Brain		Lagos-Quintana et al. (2002), Chen et al. (2007)
miR-133	m	Skeletal muscle, heart, brown adipose	Brown adipocytes	Sempere et al. (2004), Chen et al. (2005), Chen et al. (2007), Walden et al. (2009)

miRNA	Species	Tissue	Cell type	References
miR-135	m, h	Brain		Sempere et al. (2004), Chen et al. (2007)
miR-137	m	Brain		Chen et al. (2007)
miR-140	h	Bone marrow	Mesenchymal stem cell	Liu et al. (2009)
miR-140	m		Chondrocyte	Nakamura et al. (2011)
miR-141	m	Hair follicles	Pre-osteoblast	Yi et al. (2006), Itoh et al. (2009)
miR-142	m, h	Colon, hematopoietic	Hematopoietic	Lagos-Quintana et al. (2002), Chen (2004)
miR-143	m	Spleen, adipose, heart	Pre-adipocyte, white adipocytes, mesenchymal stem cell, cardiac progenitors	Lagos-Quintana et al. (2002), Esau et al. (2004), Walden et al. (2009), Liu et al. (2009), Cordes et al. (2009)
miR-145	m	Heart	Cardiac progenitors	Cordes et al. (2009)
miR-146	m		Th1 lymphocyte	Monticelli et al. (2005)
miR-150	m		Pre-B, pre-T lymphocyte, megakaryocyte-erythrocyte progenitors	Monticelli et al. (2005), Lu et al. (2008)
miR-153	m	Brain		Chen et al. (2007)
miR-155	h	Bone marrow	Hematopoietic stem cell	Georgantas et al. (2007), Romania et al. (2008)
miR-18	m	Lung		Sempere et al. (2004)
miR-181	m	Thymus, brain, lung, bone marrow, spleen	Myogenic	Chen (2004), Naguibneva et al. (2006)
miR-182	m		Sensory neurons, hair cells	Weston et al. (2011)
miR-183	m	Brain	Embryonic stem cell, sensory neurons, hair cells	Sempere et al. (2004), Chen et al. (2007), Weston et al. (2011)
miR-189	m	Spleen		Sempere et al. (2004)
miR-194	h		Intestinal epithelial cell	Hino et al. (2007), Hino et al. (2008)
miR-199a,b	m	Hair follicles	Chondroblast, osteoblast	Yi et al. (2006), Suomi et al. (2008)
miR-19a	m	Lung		Sempere et al. (2004)
miR-19b	m	Epidermis		Yi et al. (2006)
miR-20	m	Epidermis		Yi et al. (2006)
miR-200a,b,c	m	Hair follicles	Pre-osteoblast	Yi et al. (2006), Itoh et al. (2009)

(continued)

Table 17.1 (continued)

MicroRNA	Organism (m)-mouse, (h)-human	Tissue(s)	Cell type(s)	Reference(s)
miR-203	m	Epidermis		Yi et al. (2008)
miR-204	m, h	Brain	Mesenchymal stem cells	Chen et al. (2007), Huang et al. (2010)
miR-206	m	Skeletal muscle, brown adipose	Myotubes, brown adipocyte	Sempere et al. (2004), Anderson et al. (2006), Yuasa et al. (2008), Walden et al. (2009)
miR-208	m, h	Heart		Sempere et al. (2004)
miR-211	m, h		Mesenchymal stem cells	Huang et al. (2010)
miR-212	m	Spleen		Sempere et al. (2004)
miR-213	m	Lung		Sempere et al. (2004)
miR-219	m, h	Brain		Sempere et al. (2004)
miR-221	m		Erythrocyte	Felli (2005)
miR-222	m		Erythrocyte	Felli (2005)
miR-223	m	Bone marrow	Orthoclase	Chen (2004), Sugatani and Hruska (2007)
miR-23a	m		Astrocyte	Smirnova et al. (2005)
miR-23b	m	Fetal liver	Hepatocytes	Rogler et al. (2009)
miR-24	m	Lung	Osteoblast, chondroblast	Sempere et al. (2004), Suomi et al. (2008)
miR-24-1	m	Fetal liver	Hepatocytes	Rogler et al. (2009)
miR-27a	h	Bone marrow	Mesenchymal stem cells	Schoolmeesters et al. (2009)
miR-27b	m	Fetal liver	Hepatocytes	Rogler et al. (2009)
miR-29	m		Astrocyte	Smirnova et al. (2005)
miR-290, 291-3p,5p, 292-3p,5p, 293, 294, 295	m		Embryonic stem cell	Houbaviy et al. (2003), Strauss et al. (2006), Chen et al. (2007)
miR-299-5p	h		Megakaryocytes	Tenedini et al. (2010)
miR-302a,b,c,d	m, h		Embryonic stem cell	Houbaviy et al. (2003), Suh et al. (2004), Strauss et al. (2006), Chen et al. (2007)

microRNA	Species	Tissue	Cell type	Reference
miR-30b,c	m	Kidney[a]		Sempere et al. (2004)
miR-32	m	Lung		Sempere et al. (2004)
miR-365	h		Hematopoietic stem cell	Liao et al. (2008)
miR-367	m, h		Embryonic stem cell	Houbaviy et al. (2003), Suh et al. (2004), Strauss et al. (2006), Chen et al. (2007)
miR-375	m, h	Pancreas	Pancreatic α- and β-islet cell	Poy et al. (2004)
miR-376	h		Pancreatic β-islet cell	Joglekar et al. (2009), Poy et al. (2004)
miR-383	m	Brain		Chen et al. (2007)
miR-424	h		Macrophage	Rosa et al. (2007)
miR-429	m	Hair follicles		Yi et al. (2006)
miR-451	m, h		Erythrocyte	Zhan et al. (2007)
miR-452	m, h	Hematopoietic	Hematopoietic stem cell, neural crest cells	Liao et al. (2008), Sheehy et al. (2010)
miR-455	m	Adipose	Brown adipocytes	Walden et al. (2009)
miR-489	h	Bone marrow	Mesenchymal stem cells	Schoolmeesters et al. (2009)
miR-499	m	Fetal heart	Cardiomyocyte	Wilson et al. (2010)
miR-520 h	h		Hematopoietic stem cell	Liao et al. (2008), Liu et al. (2009)
miR-526b	h		Hematopoietic stem cell	Liao et al. (2008)
miR-638	h	Bone marrow	Mesenchymal stem cell	Liu et al. (2009)
miR-642a-3p	h	Adipose		Zaragosi et al. (2011)
miR-663	h	Bone marrow	Mesenchymal stem cell	Liu et al. (2009)
miR-7	h		Pancreatic islet cells	Joglekar et al. (2009)
miR-9/9*	m, h	Brain	Neural stem cell	Sempere et al. (2004), Chen et al. (2007)
miR-92b	h		Neural stem cell	Liu et al. (2009)
miR-96	m		Sensory neurons, hair cells	Weston et al. (2011)
miR-99a	m	Spleen		Sempere et al. (2004)

[a] Indicates microRNA detected at elevated levels but not exclusive to specified tissue or cell type

expression of these microRNAs is essential for myelopoiesis toward granulocytes (Velu et al. 2009).

17.3 MicroRNAs in Directed Differentiation

A concert of techniques encompassing cytokine and small molecule exposure and varying formulations of cell culture media have been established to direct differentiation of mES and hES cells toward defined lineages. Gene knockdowns causing loss of pluripotency and self-renewal in ES cells also promote differentiation to a diversity of cell types. The advantage of microRNAs in directed differentiation is the ability to simultaneously target multiple genes, eliminating the need for individual gene targeting. However, unwanted off-target effects may occur and the consequences of this are discussed later in the chapter. Cell fate–determining microRNAs can be defined by studies identifying mature microRNAs that increase in expression using conventional directed differentiation methods. Table 17.2 identifies some of these characterized effects of microRNA-mediated induction or inhibition of cell fate progression and their impacts on cell identity with selected examples illustrated in Fig. 17.1.

17.3.1 MicroRNA Differential Expression upon Differentiation

Levels of mature members of the miR-12 cluster decline in expression prior to the downregulation of Oct4 RNA, a marker of pluripotency and ES cells in differentiating hES cells (Suh et al. 2004). Studies aimed to study global microRNA changes by expression profiling include mouse and human embryo carcinoma (EC) and ES cells upon leukemia inhibitory factor withdrawal, retinoic acid (RA) induction, and EB formation (Houbaviy et al. 2003; Sempere et al. 2004; Smirnova et al. 2005; Suh et al. 2004; Wu and Belasco 2005; Krichevsky et al. 2006; Strauss et al. 2006; Chen et al. 2007; Tzur et al. 2008; Marson et al. 2008). Chen and colleagues describe increased expression of miR-152, miR-193, miR-197, and miR-206 during EB differentiation at 3, 6, and 9 days. Of these, miR-206 has demonstrated effects on enhancing skeletal muscle differentiation.

In addition to tissue profiling, Sempere and colleagues examined mEC and hEC cells treated with RA, an inducer of neuronal differentiation, by northern blotting for brain-specific microRNAs identified from tissue profiling, and found 19 microRNAs to be enriched in the RA-treated mEC and hEC cells. These include miR-124a, miR-124b, miR-9 and miR-9* as recurring neuronal specific microRNAs which have potent effects on cell fate, described later for their role in trans-differentiation.

Differentiation of hES cells toward an endoderm lineage (Fig. 17.1a) promotes the elevation of miR-24, miR-10a, miR-122, and miR-192 levels (Tzur et al. 2008).

Table 17.2 Examples of microRNAs regulating differentiation

MicroRNA(s)	Organism (m)-mouse, (h)-human	Mode of regulation (drive, block)	Progenitor cell	Terminal cell	Known target(s)	Reference(s)
miR-1, miR-133, miR-206	m, h	Drives	Myoblast	Myotube	HSP60, HSP70, Casp9, Connexin43, Fstl1, Utrn	Anderson et al. (2006), Chen et al. (2005), Chen et al. (2010), Kim (2006), Nakajima et al. (2006), Zhao et al. (2005), Lim et al. (2005), Xu et al. (2007)
miR-124	h	Drives	HeLa	Neuronal lineage	PTBP1	Lim et al. (2005), Makeyev et al. (2007)
miR-125b	m	Blocks	MSC	Osteoblast		
miR-150	m	Blocks	HSC	Mature lymphocyte	c-Myb	Xiao et al. (2007)
miR-150	m	Drives	HSC	Megakaryocyte, erythrocyte		Lu et al. (2008)
miR-155	h	Blocks	HSC, bone marrow CD34+ cells	Myeloid, erythroid		Georgantas et al. (2007)
miR-196	h	Drives	HL60	Myeloid lineage	HoxB8	Kawasaki and Taira (2004)
miR-196a	h	Drives	ASC	Osteoblast		Kim et al. (2009)
miR-203	m	Drives	Epidermal stem cell	Terminal epithelial cells	P63	Yi et al. (2008)
miR-221, miR-222	h	Blocks	Hematopoietic progenitor	Erythrocyte	kit	Felli (2005)
miR-223	m	Blocks	MSC	Osteoblast		Sugatani and Hruska (2007)
miR-24	m	Drives	Myoblast	Myotube		Sun et al. (2008)
miR-302a,b,c,d, miR-367	m, h	Reprograms	Fibroblast	iPS		Anokye-Danso et al. (2011)
miR-34a	m	Drives	NSC	Neurons, neuron elongation	SIRT1	Aranha et al. (2011)

(continued)

Table 17.2 (continued)

MicroRNA(s)	Organism (m)-mouse, (h)-human	Mode of regulation (drive, block)	Progenitor cell	Terminal cell	Known target(s)	Reference(s)
miR-424	h	Drives differentiation	HSC/CMP	Monocyte, megakaryocyte	NFI-A	Rosa et al. (2007)
miR-9/9*, miR-124	h	Drives trans-differentiation	Neonatal foreskin, adult fibroblast	Neuron	BAF53a, REST, CoREST, PTBP1	Yoo et al. (2011)

ASC adipose tissue derived stem cell, *CMP* common myeloid progenitor, *HSC* hematopoietic stem cell, *iPSC* induced pluripotent stem cell, *MSC* mesenchymal stem cell, *NSC* neural stem cell

Mouse pre-adipocyte-derived small RNA cloning libraries generated at days 1 and 9 after induction of differentiation using defined media reveal several microRNAs increasing during adipogenesis: miR-10b, miR-15a, miR-26a, miR-99a, miR-101, miR-101b, miR-143, miR-151*, miR-152, miR-183, miR-185, miR-423, and let7b (Kajimoto 2006). Individual antisense inhibition of any of these microRNAs does not block change in cell fate suggesting that these microRNAs do not elicit a switch of cell fate but arise as a consequence of differentiation. Two microRNAs, miR-181a and miR-182b, were observed to decline across adipocyte differentiation.

Another model of adult stem cells is adipose tissue–derived stem (AS) cells of mesenchymal progenitor status. When hAS cells are differentiated to osteoblasts, miR-26 increases and targets SMAD1 to facilitate terminal differentiation (Luzi et al. 2007). Overexpression of miR-196a in hAS cells reduced proliferation rates and enhanced osteogenic differentiation.

The mouse C2C12 cell line is a widely used and accepted cellular model for myoblast differentiation into myotubes. MiR-1 and miR-133 are upregulated during C2C12 myoblast serum starvation–driven differentiation (Chen et al. 2005) (Fig. 17.1b). MiR-181 also was found to increase in C2C12 myotubes during differentiation and targets HoxA12 expression. Inhibition of miR-181 in these cells impedes differentiation, but overexpression does not trigger terminal differentiation (Naguibneva et al. 2006). Also increased during C2C12 differentiation is miR-206 and is also found to be elevated in newly formed muscle fibers and myotubes in mice (Yuasa et al. 2008). MiR-26a also increases in myotubes differentiated from C2C12 cells (Wong and Tellam 2008).

17.3.2 MicroRNAs Inducing Differentiation

Examples of differentiation-inducing microRNAs are demonstrated in the differentiation of many adult progenitor and stem cells. In the hematopoietic system, miR-196 induces myeloid differentiation by directly blocking Hoxb8 in HL60 cells (Kawasaki and Taira 2004). Interestingly, this microRNA-target pair is also involved in murine limb bud formation as discussed earlier (Mansfield et al. 2004), indicating that conserved targeting interactions of microRNAs are dependent on the spatiotemporal context. Expression of miR-150 drives differentiation of mouse megakaryocyte-erythrocyte progenitor cell within in vitro and in vivo systems (Lu et al. 2008) (Fig. 17.1c). Ectopic expression of miR-424 in myeloid progenitor cells isolated from human bone marrow stimulates monocyte and macrophage differentiation and diminishes NFI-A protein levels by direct inhibiting via 3′UTR elements (Rosa et al. 2007).

MiR-1 and miR-133 are able to promote mesoderm marker expression in mES and hES cells, and in one study only, miR-1 is able to push ES cells toward a cardiac fate, whereas miR-133 blocks myogenic terminal differentiation (Ivey et al. 2008; Ivey and Srivastava 2010). It is unclear if the contrast in miR-133 function in cells derived from ES cells versus the C2C12 cell line is due to differences in their differentiation potential. Similar observations noted in vivo may also arise from the presence of similar yet distinct progenitor cells. Mouse ES cells transfected with a construct conferring constitutive expression of miR-1 and transplanted into infarcted mouse hearts were able to repair cardiac function (Glass and Singla 2011). MiR-1 and miR-133 induce myogenic differentiation in C2C12 cells, and miR-206 does likewise even in the presence of serum, implying a dominant effector microRNA activity (Kim 2006). Also in C2C12 mouse myoblasts, transfection with a miR-1 overexpression cassette led to enhanced differentiation and increased cellular fusion, a hallmark of myotube formation (Nakajima et al. 2006) although this augmentation did not detract from the ability for osteoblast or adipocyte formation given appropriate differentiation conditions.

17.3.3 MicroRNAs Shifting Cellular Programs

Selective inhibition of microRNAs is also able to drive differentiation. Many microRNAs with primary expression in stem and progenitor cells function to maintain a proliferative state, perhaps enabling expansion of progenitor populations in adult organisms. MicroRNAs may function to inhibit differentiation and support proliferation of progenitor populations. MiR-223 is a bone marrow–specific microRNA which in mouse osteoclast precursors prevents differentiation, which is reversed when miR-223 is inhibited (Sugatani and Hruska 2007). Decreasing expression of miR-223 also leads to erythroid differentiation (Felli et al. 2009).

The potency of microRNAs in altering cellular programs is another area of interest. MiR-1 when transfected into human HeLa cells drives expression of a muscle-specific genetic program (Lim et al. 2005), and the transfection of miR-124 into the same cell line shifts expression toward a brain-like pattern as measured by gene expression microarrays. Incidentally, miR-124a targets Foxa2 (Baroukh et al. 2007), an early marker for endoderm lineage differentiation.

Remarkably, lentiviruses overexpressing the neuronal lineage–specific miR-9/9* and miR-124 are able to trans-differentiate human neonatal foreskin and adult fibroblasts to a neuronal cell fate (Yoo et al. 2011) with the ability to elicit functional action potentials. This conversion was facilitated by NEUROD2 expression and was enhanced by addition of the transcription factors ASCL1 and MTY1L. The transformational competency of miR-124 is corroborated by the coupling of miR-124 expression with two transcription factors MYT1L and POU3F2 (BRN2) for sufficient reprogramming of human primary dermal fibroblasts originated from the mesoderm to neurons capable of forming action potentials and functional synapses (Ambasudhan et al. 2011).

MicroRNAs also participate in dedifferentiation mechanisms, as several microRNAs demonstrate the ability to inhibit differentiation progression. These microRNAs are candidates for reprogramming studies. The miR-290 family has been demonstrated to orchestrate the enhanced proliferation rates of ES cells by modulating the cell cycle (Wang et al. 2008; Wang and Blelloch 2011) and, additionally, enhances cellular reprogramming (Judson et al. 2009). Both the mES- and hES-specific miR-302 family microRNA cluster alone is able to reprogram somatic cells to induced pluripotency and generate germ line–competent iPS cells (Anokye-Danso et al. 2011). Refer to the Chap. 15 (this volume) for details about miRNA-based reprogramming approach and associated protocols.

17.3.4   Therapeutic and Research Advantages of Small RNAs

The ubiquitous expression of miR-30 throughout all cell types and tissues has turned it into a widely utilized tool in the laboratory. The miR-30 stem loop sequence is commonly used for short hairpin RNA (shRNA) generation owing to the efficient expression and processing of miR-30 (Zeng et al. 2002; Zhou 2005). ShRNA enables continual and stable expression of siRNAs under control of a defined promoter; however, this generally requires vector integration into the host cell genome. The advantage and potential of microRNAs for future therapeutic applications is to administer a microRNA mimic, antagomir, or siRNA without incurring genetic manipulation of target cells, much like a small molecule drug. ES and iPS cells can be manipulated in vitro by these small RNAs and induced to differentiate away from a pluripotent state toward terminally differentiated states (Fig. 17.2a) without altering genomic integrity. Small RNAs can also be delivered in vivo akin to a pharmaceutical product to promote in situ target cell differentiation or to expand progenitor cell populations (Fig. 17.2b).

A clever application benefiting from knowledge of differential microRNA expression patterns enables segregation of cell populations (Brown et al. 2007) where a gene encoding thymidine kinase (TK) is transgenically introduced as a "suicide" lentivirus harboring microRNA targeting sites within the TK 3′UTR. Upon ganciclovir drug treatment, only cells that have downregulated TK are able to survive. The use of cells derived from differentiating ES and iPS cells poses a danger for cellular transplantation, for the remaining pluripotent cells cause tumor formation as evidenced by teratoma formation competence being a defining characteristic of these cells. Use of this "suicide" TK harboring differentiation-specific microRNA target sites for use in cellular transplantation would eradicate trace residual undifferentiated cells. This method can be adapted to deplete other progenitor cells or undesired cell lineages generated during differentiation steps to develop pure and segregated cell populations (Fig. 17.2c).

Prosser and colleagues have recently introduced a set of tools for in vivo functional analysis of microRNAs (Prosser et al. 2011). They provide a total of 428 different targeting vectors for the genomic loci of 476 distinct microRNAs and have established 392 mouse ES cell lines with conditional targeting capacity to target individual microRNAs by recombination-mediated cassette exchange (RCME). These mES cells are able to generate chimeric mice and are germline competent. Crossing such mice containing conditionally targeting individual

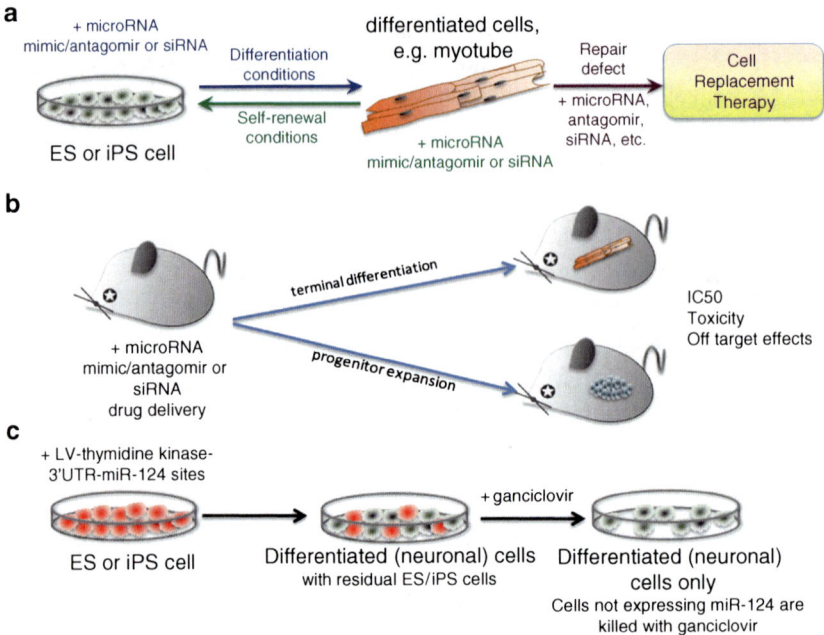

Fig. 17.2 *MicroRNA applications to enhance regenerative medicine.* (**a**) MicroRNA mimics, antagomirs, and siRNAs repress or relieve repression of target gene expression without genetic manipulation for in vitro differentiation, reprogramming, and treatment of, for example, embryonic stem (ES) or induced pluripotent stem (iPS) cells, progenitor cells, adult somatic cells, and terminally differentiated cells for use in cell transplantation. (**b**) MicroRNA mimics, antagomirs, and siRNAs can be administered in vivo as therapeutics to alter tissue- and cell-specific differentiation. (**c**) A "suicide" gene carrier (Brown et al. 2007), for example, lentivirus (LV), for thymidine kinase (TK) harboring 3′UTR microRNA target sites for cell-specific microRNA integrated into progenitor cells. Cell populations expressing microRNAs of interest, for example, neural-specific miR-124, treated with ganciclovir survive, and those with intact TK expression will die

microRNA loci with tissue- or cell-lineage-specific Cre recombinase–expressing animals will enable in vivo characterization of microRNA function during development and within specialized cell compartments. Fluorescent proteins and other traceable markers can also be introduced by RMCE for lineage tracing of individual microRNA loci. This will not only provide a resource for microRNA and microRNA-target characterization during development in vivo but will also create a multitude of mouse models and derived cell types to characterize microRNA function for differentiation, trans-differentiation, and reprogramming studies.

Researchers are also developing microRNA "sponges" as alternatives to designing siRNAs to target misregulated microRNAs (Brown and Naldini 2009). These microRNA sponges function by offering alternative binding sites for sequestration of microRNAs, thus preventing them from binding and suppressing their natural targets. This approach typically relies on strong promoters, active in distinct tissues or cell types to achieve target specificity and to drive expression of an engineered sequence encoding multiple microRNA target sites, often downstream of a reporter gene. The multiple copies act as targets for binding of the natural microRNAs. This technique can be readily coupled with the microRNA loci transgenic models developed by Prosser and colleagues.

Currently, antisense-RNAs (asRNA) are being developed as therapeutics for diabetes, cancer, and coronary artery disease, among other indications (Knowling and Morris 2011). The most advanced asRNA drug marketed as Vitravene by Isis Pharmaceuticals, Inc. and approved by the FDA in 1998 is being used in the treatment of cytomegaloviral infection of the retina (CMV retinitis) by targeting CMV mRNA in order to prevent translation and thereby relieve inflammation. Santaris Pharma A/S began a Phase 2 study of miravirsen, a selective inhibitor of miR-122, a liver-specific microRNA, in 2010 for the treatment of hepatitis C infection. The hepatitis C virus is able to usurp miR-122 in the liver to promote its own replication to produce additional viral particles. Miravirsen is based on LNA technology, and its antisense targeting attenuates the exploitation of miR-122 by hepatitis C virus to inhibit viral infection.

17.4   MicroRNAs in Regenerative Medicine

Cells isolated from the developing mouse blastocyst and cultured in vitro were established in 1981 as mouse ES cells (Martin 1981; Evans and Kaufman 1981). These cultured mES cells maintain continual self-renewal and pluripotency and the ability to give rise to all cells of the primary germ layers, the endoderm, mesoderm, and ectoderm (Fig. 17.1a), as well as the germ line. In 1998, James Thomson established the first human ES cell lines from human embryos (Thomson 1998) and with it a new potential for regenerative medicine. The ability to culture hES cells and to derive a multitude of cell lineages in vitro gives hope of advancing cell transplantation therapy to multiple disease areas and provides a tremendous resource for studying disease etiology and developmental processes in the laboratory.

The advent of reprogramming of somatic cells in mouse and human (Takahashi and Yamanaka 2006; Takahashi et al. 2007; Yu et al. 2007) to induced pluripotent stem (iPS) cells breathed new life into the fields of stem cell biology and regenerative medicine. We can now routinely reprogram patient-specific somatic cells to iPS cells and subsequently differentiate them to defined lineages for in vitro study of disease progression and, perhaps eventually, for in vivo transplantation.

MicroRNA expression profile comparison between ES cells and iPS cells has identified sets of microRNAs in both mouse and human that are not elevated or downregulated during or following reprogramming (Wilson et al. 2009; Stadtfeld et al. 2010; Mallanna and Rizzino 2010). A better understanding of the differences in both mRNA and microRNA expression between ES cells and iPS cells and the functional impact of these differences on their maintenance, homeostasis, and developmental potential is vital for realizing their biochemical promise.

Many microRNAs are distinctly expressed during differentiation and may promote either differentiation or the maintenance and proliferation of progenitor cells. These microRNAs (Table 17.2) are candidates for directed differentiation or for antisense targeting in both in vitro and in vivo systems. MicroRNA or antagomir influx to these cells for directed differentiation is typically performed by transgenic methods; the advantage of small RNAs is that they can also be introduced into cells and tissues as synthetically generated oligonucleotides (Fig. 17.2a, b), thus averting genetic manipulation and its associated complications.

Several cell replacement therapy options employing microRNAs exist to differentiate or treat cells in vitro or in vivo (Yang and Wu 2007). MiR-375 can be used to differentiate pancreatic islet cells from diabetic patient-derived iPS cells. MiR-1, miR-133, etc., are prime candidates in cases of muscle injury in cardiac muscle (as recently demonstrated in mice (Glass and Singla 2011)) or skeletal muscle where miR-1 and miR-133 can in vivo induce satellite cell differentiation.

Careful monitoring and extensive studies of microRNAs and small RNA utilization in regenerative medicine need to be performed before implementation. The potential for a given microRNA to have hundreds or thousands of targets poses a hurdle for proper cell targeting when these are administered in vivo. A study of how each therapeutic small RNA effects areas of the body beyond their intended target and therefore causes possible abnormal differentiation or dedifferentiation and/or proliferation is crucial to determine if a given potentially therapeutic small RNA will function as an oncogene or cause other pathological states.

MiR-195 increases upon cardiac hypertrophy, whereas miR-133 declines (van Rooij et al. 2006). Upon loss of miR-208, a heart-specific microRNA, the heart is protected against hypertrophy (Callis et al. 2008). These microRNAs are potential candidates for in vivo therapeutic treatment. Several congenital disorders manifest in part as hypertrophic cardiac cells to impair heart function. These include LEOPARD syndrome where we have previously shown by patient-specific iPS cell differentiation that enlarged and hypertrophic cardiomyocytes can be generated, thus recapitulating the disease in vitro (Carvajal-Vergara et al. 2010). These cardiomyocytes are prime candidates for in vitro microRNA treatment toward employment in cell replacement therapy. miRagen Therapeutics is currently conducting a preclinical study to investigate the effect of a miR-208/499 antagomir in chronic heart failure.

17.5 Discussion

The discovery of microRNAs in 1993 and the isolation of mES cells in 1981, hES cells in 1998, and invention of miPS and hiPS cells in 2006 and 2007, respectively, have created two very dynamically growing fields. Research discoveries in both areas have burgeoned as demonstrated by the microRNA studies in ES and iPS biology discussed in this chapter. We summarize a growing body of knowledge on microRNA expression in development with respect to spatial and temporal expression and, most importantly, their expression in progenitor and terminally differentiated cells as well as their roles in maintaining cell identity or mediating a switch in cell fate and how they may impact regenerative medicine (Tables 17.1 and 17.2). The repertoire lists of microRNAs and their targets in spatiotemporal expression patterns during development are increasingly comprehensive, and these microRNA regulatory networks will inform all areas of developmental control. This will lead to advances in mechanistic insights of developmental disorders and defects in lineage progression to ultimately inform regenerative medicine applications. Although we aim to cover all aspects of microRNAs involved in differentiation, our summary is by no means complete, and we expect the increasing interest in integrating microRNAs and regenerative medicine to contribute much more knowledge. The clear potential advantages of microRNA and other small RNA therapeutics are the absence of genetic manipulation and ability to deliver them in vivo. The clinical utility of siRNA-, shRNA-, asRNA-, or microRNA-based therapeutics has not yet been realized; however, ongoing clinical trials provide hope for success (Knowling and Morris 2011).

Once microRNA targets are better defined within a given cell- or tissue-specific state or program, they can begin to be applied for therapeutic use. Many research programs, not covered here, aim to elucidate microRNA targeting of the transcriptome during changes in cell fate by RNA foot printing of miRISC-associated microRNAs and their mRNA targets. Target definition can define the course of therapy for both safety and efficacy where cocktails of siRNAs or small molecules for inhibition of a set of mRNAs or proteins may prevent undesired off-target effects stemming from a microRNA. Although a valid concern, it is worth noting that if a microRNA is administered to induce differentiation or a change in cell identity to a fate where it is typically observed, aberrant off-target effects will be minimal and, in fact, as yet unknown mRNAs maybe essential targets in the cell fate switch. This may not hold true when eliciting a transition between two cell fates that do not fall within the same lineage tree in vertebrate development.

Before we can take full advantage of RNA-based therapeutic applications, several obstacles need to be addressed. One of the challenges in any RNA-based therapy continues to be the delivery system. Viruses allow long-term delivery of vectors encoding the RNA therapeutic. However, of great concern is the potential for integration-mediated mutagenesis and consequent possible development of cancer or other pathologies. More recently, nanoparticles have become an attractive delivery vehicle. Lipid-based nanoparticles such as the stable nucleic-acid-lipid particles (SNALP) consist of a lipid bilayer containing a mixture of cationic and fusogenic lipids coated with diffusible polyethylene glycol (Morrissey et al. 2005). The lipid combination not only protects the RNA therapeutic from serum nucleases but also enhances cellular endosomal uptake followed by cytoplasmic release. Nanoparticles consisting of biodegradable polymers that allow more precise pharmacokinetic release are another exciting and promising delivery system. Another challenge, namely how to steer the RNA therapeutic to the appropriate target cell, has been overcome, for example, by engineering the nanoparticle surface to incorporate a cell-type-specific ligand for targeted delivery (Davis et al. 2010).

An early example of a small RNA therapeutic approach was the intravenous delivery of synthetic microRNA mimics, antagomirs (Pedersen et al. 2007). In a primate model of chronic hepatitis C viral infection, LNA-modified oligonucleotides complementary to the liver-specific miR-122 provided long-term suppression of viral replication (Lanford et al. 2010). A version of this LNA targeting miR-122 for therapy in humans is currently in clinical trials by Santaris Pharma A/S. Plasmid- and virus-based approaches are also being used for reducing endogenous microRNA levels (Brown and Naldini 2009). Exploiting and antagonizing microRNA regulation using inducible microRNA sponges is another exciting avenue for use in therapeutic as well as experimental applications (Ebert et al. 2007).

A currently ongoing clinical trial employs tripled RNA-based gene therapy to inhibit HIV infection and replication. The strategy is based on transplantation of autologous $CD34^+$ cells lentivirally transduced to express three RNA-based anti-HIV components, namely an shRNA targeting the tat/rev genes, which enhance viral replication, a decoy of the viral transactivating region mRNA hairpin to sequester the viral Tat protein, and lastly a ribozyme that targets the chemokine receptor CCR5. The pilot feasibility study demonstrated that the treated patients tolerated the therapy well and that the observed toxicity was strictly related to and in the typical range for hematopoietic stem/progenitor cell transplantation (DiGiusto et al. 2010). The reader is directed to refer to the Chap. 16 and the following reviews on the broad range of potential applications of targeted gene silencing for therapy and the recent status of clinical trials using siRNA/shRNA and microRNA (Wahid et al. 2010; Czech et al. 2011; Burnett et al. 2011; Davidson and McCray 2011).

Caution is warranted in using microRNAs to induce differentiation in vivo; although demonstrated recently to aid in the repair of the infarcted mouse heart by miR-1 transfected mES cells (Glass and Singla 2011), transgenic overexpression in mice of miR-1 in cardiac and skeletal muscle progenitors led to developmental arrest at E13.5. This was probably due to the early differentiation and depletion of the progenitor populations (Chen et al. 2005).

Future advances in microRNA-target identification and prediction, as well as a more comprehensive understanding of the roles of microRNAs in specific cell fate transitions from stem to progenitor to specialized mature cell will help define and refine microRNA-based directed cellular differentiation strategies for use in regenerative medicine.

Acknowledgments The authors thank the members of the Lemischka and Moore labs for helpful discussions and feedback. Images of C2C12 myoblast differentiation were kindly provided by Shao-En Ong.

References

Ambasudhan R, Talantova M, Coleman R, Yuan X, Zhu S, Lipton S, Ding S (2011) Direct reprogramming of adult human fibroblasts to functional neurons under defined conditions. Cell Stem Cell 9(2):113–118. doi:10.1016/j.stem.2011.07.002

Anderson C, Catoe H, Werner R (2006) MIR-206 regulates connexin43 expression during skeletal muscle development. Nucleic Acids Res 34(20):5863–5871. doi:10.1093/nar/gkl743

Anokye-Danso F, Trivedi CM, Juhr D, Gupta M, Cui Z, Tian Y, Zhang Y, Yang W, Gruber PJ, Epstein JA, Morrisey EE (2011) Highly efficient miRNA-mediated reprogramming of mouse and human somatic cells to pluripotency. Cell Stem Cell 8(4):376–388. doi:10.1073/pnas.0506216102

Aranha MM, Santos DM, Sola S, Steer CJ, Rodrigues CM (2011) miR-34a regulates mouse neural stem cell differentiation. PLoS One 6(8):e21396. doi:10.1371/journal.pone.0021396

Baroukh N, Ravier MA, Loder MK, Hill EV, Bounacer A, Scharfmann R, Rutter GA, Van Obberghen E (2007) MicroRNA-124a regulates Foxa2 expression and intracellular signaling in pancreatic – cell lines. J Biol Chem 282(27):19575–19588. doi:10.1074/jbc.M611841200

Bernstein E, Kim SY, Carmell MA, Murchison EP, Alcorn H, Li MZ, Mills AA, Elledge SJ, Anderson KV, Hannon GJ (2003) Dicer is essential for mouse development. Nat Genet 35(3):215–217. doi:10.1038/ng1253

Bracken CP, Gregory PA, Khew-Goodall Y, Goodall GJ (2009) The role of microRNAs in metastasis and epithelial-mesenchymal transition. Cell Mol Life Sci 66(10):1682–1699. doi:10.1007/s00018-009-8750-1

Brown BD, Naldini L (2009) Exploiting and antagonizing microRNA regulation for therapeutic and experimental applications. Nat Rev Genet 10(8):578–585. doi:10.1038/nrg2628

Brown BD, Gentner B, Cantore A, Colleoni S, Amendola M, Zingale A, Alessia B, Lazzari G, Galli C, Naldini L (2007) Endogenous microRNA can be broadly exploited to regulate transgene expression according to tissue, lineage and differentiation state. Nat Biotechnol 25(12):1457–1467. doi:10.1038/nbt1372

Burnett JC, Rossi JJ, Tiemann K (2011) Current progress of siRNA/shRNA therapeutics in clinical trials. Biotechnol J. doi:10.1002/biot.201100054

Callis TE, Deng Z, Chen JF, Wang DZ (2008) Muscling through the microRNA world. Exp Biol Med 233(2):131–138. doi:10.3181/0709-mr-237

Carvajal-Vergara X, Sevilla A, D'Souza SL, Ang Y-S, Schaniel C, Lee D-F, Yang L, Kaplan AD, Adler ED, Rozov R, Ge Y, Cohen N, Edelmann LJ, Chang B, Waghray A, Su J, Pardo S, Lichtenbelt KD, Tartaglia M, Gelb BD, Lemischka IR (2010) Patient-specific induced pluripotent stem-cell-derived models of LEOPARD syndrome. Nature 465(7299):808–812. doi:10.1038/nature09005

Chen CZ (2004) MicroRNAs modulate hematopoietic lineage differentiation. Science 303(5654):83–86. doi:10.1126/science.1091903

Chen J-F, Mandel EM, Thomson JM, Wu Q, Callis TE, Hammond SM, Conlon FL, Wang D-Z (2005) The role of microRNA-1 and microRNA-133 in skeletal muscle proliferation and differentiation. Nat Genet 38(2):228–233. doi:10.1038/ng1725

Chen C, Ridzon D, Lee C-T, Blake J, Sun Y, Strauss WM (2007) Defining embryonic stem cell identity using differentiation-related microRNAs and their potential targets. Mamm Genome 18(5):316–327. doi:10.1007/s00335-007-9032-6

Chen JF, Tao Y, Li J, Deng Z, Yan Z, Xiao X, Wang DZ (2010) microRNA-1 and microRNA-206 regulate skeletal muscle satellite cell proliferation and differentiation by repressing Pax7. J Cell Biol 190(5):867–879. doi:10.1083/jcb.200911036

Chi SW, Zang JB, Mele A, Darnell RB (2009) Argonaute HITS-CLIP decodes microRNA–mRNA interaction maps. Nature. doi:10.1038/nature08170

Cordes KR, Sheehy NT, White MP, Berry EC, Morton SU, Muth AN, Lee T-H, Miano JM, Ivey KN, Srivastava D (2009) miR-145 and miR-143 regulate smooth muscle cell fate and plasticity. Nature. doi:10.1038/nature08195

Czech MP, Aouadi M, Tesz GJ (2011) RNAi-based therapeutic strategies for metabolic disease. Nat Rev Endocrinol 7(8):473–484. doi:10.1038/nrendo.2011.57

Davidson BL, McCray PB Jr (2011) Current prospects for RNA interference-based therapies. Nat Rev Genet 12(5):329–340. doi:10.1038/nrg2968

Davis ME, Zuckerman JE, Choi CH, Seligson D, Tolcher A, Alabi CA, Yen Y, Heidel JD, Ribas A (2010) Evidence of RNAi in humans from systemically administered siRNA via targeted nanoparticles. Nature 464(7291):1067–1070. doi:10.1038/nature08956

DiGiusto DL, Krishnan A, Li L, Li H, Li S, Rao A, Mi S, Yam P, Stinson S, Kalos M, Alvarnas J, Lacey SF, Yee JK, Li M, Couture L, Hsu D, Forman SJ, Rossi JJ, Zaia JA (2010) RNA-based gene therapy for HIV with lentiviral vector-modified CD34(+) cells in patients undergoing transplantation for AIDS-related lymphoma. Sci Transl Med 2(36):36–43. doi:10.1126/scitranslmed.3000931

Ebert MS, Neilson JR, Sharp PA (2007) MicroRNA sponges: competitive inhibitors of small RNAs in mammalian cells. Nat Methods 4(9):721–726. doi:10.1038/nmeth1079

Esau C, Kang X, Peralta E, Hanson E, Marcusson EG, Ravichandran LV, Sun Y, Koo S, Perera RJ, Jain R, Dean NM, Freier SM, Bennett CF, Lollo B, Griffey R (2004) MicroRNA-143 regulates adipocyte differentiation. J Biol Chem 279(50):52361–52365. doi:10.1074/jbc.C400438200

Evans M, Kaufman M (1981) Establishment in culture of pluripotential cells from mouse embryos. Nature 292(5819):154–156. doi:10.1038/292154a0

Felli N (2005) MicroRNAs 221 and 222 inhibit normal erythropoiesis and erythroleukemic cell growth via kit receptor down-modulation. Proc Natl Acad Sci 102(50):18081–18086. doi:10.1073/pnas.0506216102

Felli N, Pedini F, Romania P, Biffoni M, Morsilli O, Castelli G, Santoro S, Chicarella S, Sorrentino A, Peschle C, Marziali G (2009) MicroRNA 223-dependent expression of LMO2 regulates normal erythropoiesis. Haematologica 94(4):479–486. doi:10.3324/haematol.2008.002345

Georgantas RW, Hildreth R, Morisot S, Alder J, Liu C-G, Heimfeld S, Calin GA, Croce CM, Civin CI (2007) CD34+ hematopoietic stem-progenitor cell microRNA expression and function: a circuit diagram of differentiation control. Proc Natl Acad Sci 104(8):2750–2755. doi:10.1073/pnas.0610983104

Giraldez AJ, Cinalli RM, Glasner ME, Enright AJ, Thomson JM, Baskerville S, Hammond SM, Bartel DP, Schier AF (2005) MicroRNAs regulate brain morphogenesis in zebrafish. Science 308(5723):833–838. doi:10.1016/j.gde.2010.04.003

Glass C, Singla DK (2011) microRNA-1 transfected embryonic stem cells enhance cardiac myocyte differentiation and inhibit apoptosis by modulating PTEN/Akt pathway in the infarcted heart. Am J Physiol Heart Circ Physiol. doi:10.1152/ajpheart.00271.2011

Großhans H, Johnson T, Reinert KL, Gerstein M, Slack FJ (2005) The temporal patterning microRNA *let-7* regulates several transcription factors at the larval to adult transition in *C. elegans*. Dev Cell 8(3):321–330. doi:10.1016/j.devcel.2004.12.019

Hafner M, Landthaler M, Burger L, Khorshid M, Hausser J, Berninger P, Rothballer A, Ascano M, Jungkamp A-C, Munschauer M, Ulrich A, Wardle GS, Dewell S, Zavolan M, Tuschl T (2010) Transcriptome-wide identification of RNA-binding protein and microRNA target sites by PAR-CLIP. Cell 141(1):129–141. doi:10.1016/j.cell.2010.03.009

Harfe BD (2005a) MicroRNAs in vertebrate development. Curr Opin Genet Dev 15(4):410–415. doi:10.1016/j.gde.2005.06.012

Harfe BD (2005b) The RNaseIII enzyme Dicer is required for morphogenesis but not patterning of the vertebrate limb. Proc Natl Acad Sci 102(31):10898–10903. doi:10.1073/pnas.0504834102

Hino K, Fukao T, Watanabe M (2007) Regulatory interaction of HNF1 to microRNA194 gene during intestinal epithelial cell differentiation. Nucleic Acids Symp Ser 51(1):415–416. doi:10.1093/nass/nrm208

Hino K, Tsuchiya K, Fukao T, Kiga K, Okamoto R, Kanai T, Watanabe M (2008) Inducible expression of microRNA-194 is regulated by HNF-1 during intestinal epithelial cell differentiation. RNA 14(7):1433–1442. doi:10.1261/rna.810208

Houbaviy HB, Murray MF, Sharp PA (2003) Embryonic stem cell-specific microRNAs. Dev Cell 5(2):351–358. doi:10.1016/s1534-5807(03), 00227-2

Huang J, Zhao L, Xing L, Chen D (2010) MicroRNA-204 regulates Runx2 protein expression and mesenchymal progenitor cell differentiation. Stem Cells 28(2):357–364. doi:10.1002/stem.288

Itoh T, Nozawa Y, Akao Y (2009) MicroRNA-141 and -200a are involved in bone morphogenetic protein-2-induced mouse pre-osteoblast differentiation by targeting distal-less homeobox 5. J Biol Chem 284(29):19272–19279. doi:10.1074/jbc.M109.014001

Ivey KN, Srivastava D (2010) MicroRNAs as regulators of differentiation and cell fate decisions. Cell Stem Cell 7(1):36–41. doi:10.1016/j.stem.2010.06.012

Ivey KN, Muth A, Arnold J, King FW, Yeh R-F, Fish JE, Hsiao EC, Schwartz RJ, Conklin BR, Bernstein HS, Srivastava D (2008) MicroRNA regulation of cell lineages in mouse and human embryonic stem cells. Cell Stem Cell 2(3):219–229. doi:10.1016/j.stem.2008.01.016

Joglekar M, Joglekar V, Hardikar A (2009) Expression of islet-specific microRNAs during human pancreatic development. Gene Expr Patterns 9(2):109–113. doi:10.1016/j.gep. 2008.10.001

Judson RL, Babiarz JE, Venere M, Blelloch R (2009) Embryonic stem cell–specific microRNAs promote induced pluripotency. Nat Biotechnol 27(5):459–461. doi:10.1038/nbt.1535

Kajimoto K (2006) MicroRNA and 3T3-L1 pre-adipocyte differentiation. RNA 12(9):1626–1632. doi:10.1261/rna.7228806

Kanellopoulou C (2005) Dicer-deficient mouse embryonic stem cells are defective in differentiation and centromeric silencing. Genes Dev 19(4):489–501. doi:10.1101/gad.1248505

Kapsimali M, Kloosterman WP, de Bruijn E, Rosa F, Plasterk RHA, Wilson SW (2007) MicroRNAs show a wide diversity of expression profiles in the developing and mature central nervous system. Genome Biol 8(8):R173. doi:10.1186/gb-2007-8-8-r173

Kawasaki H, Taira K (2004) MicroRNA-196 inhibits HOXB8 expression in myeloid differentiation of HL60 cells. Nucleic Acids Symp Ser 48:211–212. doi:10.1093/nass/48.1

Kim HK (2006) Muscle-specific microRNA miR-206 promotes muscle differentiation. J Cell Biol 174(5):677–687. doi:10.1083/jcb.200603008

Kim YJ, Bae SW, Yu SS, Bae YC, Jung JS (2009) miR-196a regulates proliferation and osteogenic differentiation in mesenchymal stem cells derived from human adipose tissue. J Bone Miner Res 24(5):816–825. doi:10.1359/jbmr.081230

Knowling S, Morris KV (2011) Non-coding RNA. and antisense RNA Nature's trash or treasure? Biochimie. doi:10.1016/j.biochi.2011.07.031

Krichevsky AM, Sonntag KC, Isacson O, Kosik KS (2006) Specific microRNAs modulate embryonic stem cell-derived neurogenesis. Stem Cells 24(4):857–864. doi:10.1634/stemcells.2005-0441

Krumlauf R (1994) Hox genes in vertebrate development. Cell 78(2):191–201. doi:10.1016/0092-8674(94), 90290-9

Lagos-Quintana M, Rauhut R, Yalcin A, Meyer J, Lendeckel W, Tuschl T (2002) Identification of tissue-specific microRNAs from mouse. Curr Biol 12(9):735–739. doi:10.1016/S0960-9822 (02), 00809-6

Landthaler M, Gaidatzis D, Rothballer A, Chen PY, Soll SJ, Dinic L, Ojo T, Hafner M, Zavolan M, Tuschl T (2008) Molecular characterization of human Argonaute-containing ribonucleoprotein complexes and their bound target mRNAs. RNA 14(12):2580–2596. doi:10.1261/rna.1351608

Lanford RE, Hildebrandt-Eriksen ES, Petri A, Persson R, Lindow M, Munk ME, Kauppinen S, Orum H (2010) Therapeutic silencing of microRNA-122 in primates with chronic hepatitis C virus infection. Science 327(5962):198–201. doi:10.1126/science.1178178

Lee RC, Feinbaum RL, Ambros V (1993) The *C. elegans* heterochronic gene lin-4 encodes small RNAs with antisense complementarity to *lin-14*. Cell 75:843–854. doi:10.1016/0092-8674(93) 90529-Y

Lewis BP, Shih IH, Jones-Rhoades MW, Bartel DP, Burge CB (2003) Prediction of mammalian microRNA targets. Cell 115(7):787–798. doi:10.1016/S0092-8674(03), 01018-3

Liao R, Sun J, Zhang L, Lou G, Chen M, Zhou D, Chen Z, Zhang S (2008) MicroRNAs play a role in the development of human hematopoietic stem cells. J Cell Biochem 104(3):805–817. doi:10.1002/jcb.21668

Lim LP, Lau NC, Garrett-Engle P, Grimson A, Schetter JM, Castle J, Bartel DP, Linsley PS, Johnson JM (2005) Microarray analysis shows that some microRNAs downregulate large numbers of target mRNAs. Nature 433(7027):769–773. doi:10.1038/nature03315

Liu N, Williams AH, Kim Y, McAnally J, Bezprozvannaya S, Sutherland LB, Richardson JA, Bassel-Duby R, Olson EN (2007) An intragenic MEF2-dependent enhancer directs muscle-specific expression of microRNAs 1 and 133. Proc Natl Acad Sci 104(52):20844–20849. doi:10.1073/pnas.0710558105

Liu S-P, Fu R-H, Yu H-H, Li K-W, Tsai C-H, Shyu W-C, Lin S-Z (2009) MicroRNAs regulation modulated self-renewal and lineage differentiation of stem cells. Cell Transplant 18 (9):1039–1045. doi:10.3727/096368909x471224

Lu J, Guo S, Ebert BL, Zhang H, Peng X, Bosco J, Pretz J, Schlanger R, Wang JY, Mak RH (2008) MicroRNA-mediated control of cell fate in megakaryocyte-erythrocyte progenitors. Dev Cell 14(6):843–853. doi:10.1016/j.devcel.2008.03.012

Luzi E, Marini F, Sala SC, Tognarini I, Galli G, Brandi ML (2007) Osteogenic differentiation of human adipose tissue-derived stem cells is modulated by the miR-26a targeting of the SMAD1 transcription factor. J Bone Miner Res 23(2):287–295. doi:10.1359/jbmr.071011

Makeyev EV, Zhang J, Carrasco MA, Maniatis T (2007) The microRNA miR-124 promotes neuronal differentiation by triggering brain-specific alternative pre-mRNA splicing. Mol Cell 27(3):435–448. doi:10.1016/j.molcel.2007.07.015

Mallanna SK, Rizzino A (2010) Emerging roles of microRNAs in the control of embryonic stem cells and the generation of induced pluripotent stem cells. Dev Biol 344(1):16–25. doi:10.1016/j.ydbio.2010.05.014

Mansfield JH, Harfe BD, Nissen R, Obenauer J, Srineel J, Chaudhuri A, Farzan-Kashani R, Zuker M, Pasquinelli AE, Ruvkun G, Sharp PA, Tabin CJ, McManus MT (2004) MicroRNA-responsive 'sensor' transgenes uncover Hox-like and other developmentally regulated patterns of vertebrate microRNA expression. Nat Genet 36(10):1079–1083. doi:10.1038/ng1421

Marson A, Levine SS, Cole MF, Frampton GM, Brambrink T, Johnstone S, Guenther MG, Johnston WK, Wernig M, Newman J, Calabrese JM, Dennis LM, Volkert TL, Gupta S, Love J, Hannett N, Sharp PA, Bartel DP, Jaenisch R, Young RA (2008) Connecting microRNA genes to the core transcriptional regulatory circuitry of embryonic stem cells. Cell 134 (3):521–533. doi:10.1016/j.cell.2008.07.020

Martin G (1981) Isolation of a pluripotent cell line from early mouse embryos cultured in medium conditioned by teratocarcinoma stem cells. Proc Natl Acad Sci 78(12):7634–7638. doi:10.1073/pnas.78.12.7634

Masaki S, Ohtsuka R, Abe Y, Muta K, Umemura T (2007) Expression patterns of microRNAs 155 and 451 during normal human erythropoiesis. Biochem Biophys Res Commun 364 (3):509–514. doi:10.1016/j.bbrc.2007.10.077

Monticelli S, Ansel KM, Xiao C, Socci ND, Krichevsky AM, Thai T-H, Rajewsky N, Marks DS, Sander C, Rajewsky K, Rao A, Kosik KS (2005) MicroRNA profiling of the murine hematopoietic system. Genome Biol 6(8):R71. doi:10.1186/gb-2005-6-8-r71

Morrissey DV, Lockridge JA, Shaw L, Blanchard K, Jensen K, Breen W, Hartsough K, Machemer L, Radka S, Jadhav V, Vaish N, Zinnen S, Vargeese C, Bowman K, Shaffer CS, Jeffs LB, Judge A, MacLachlan I, Polisky B (2005) Potent and persistent in vivo anti-HBV activity of chemically modified siRNAs. Nat Biotechnol 23(8):1002–1007. doi:10.1038/nbt1122

Murchison EP (2005) Characterization of Dicer-deficient murine embryonic stem cells. Proc Natl Acad Sci 102(34):12135–12140. doi:10.1073/pnas.0505479102

Naguibneva I, Ameyar-Zazoua M, Polesskaya A, Ait-Si-Ali S, Groisman R, Souidi M, Cuvellier S, Harel-Bellan A et al (2006) The microRNA miR-181 targets the homeobox protein Hox-A11 during mammalian myoblast differentiation. Nat Cell Biol 8(3):278–284. doi:10.1038/ncb1373

Nakajima N, Takahashi T, Kitamura R, Isodono K, Asada S, Ueyama T, Matsubara H, Oh H (2006) MicroRNA-1 facilitates skeletal myogenic differentiation without affecting osteoblastic and adipogenic differentiation. Biochem Biophys Res Commun 350(4):1006–1012. doi:10.1016/j.bbrc.2006.09.153

Nakamura Y, Inloes JB, Katagiri T, Kobayashi T (2011) Chondrocyte-specific microRNA-140 regulates endochondral bone development and targets Dnpep to modulate bone morphogenetic protein signaling. Mol Cell Biol 31(14):3019–3028. doi:10.1128/mcb.05178-11

Noguchi S, Mori T, Hoshino Y, Maruo K, Yamada N, Kitade Y, Naoe T, Akao Y (2011) MicroRNA-143 functions as a tumor suppressor in human bladder cancer T24 cells. Cancer Lett 307(2):211–220. doi:10.1016/j.canlet.2011.04.005

O'Connell RM, Chaudhuri AA, Rao DS, Gibson WSJ, Balazs AB, Baltimore D (2010) MicroRNAs enriched in hematopoietic stem cells differentially regulate long-term hematopoietic output. Proc Natl Acad Sci 107(32):14235–14240. doi:10.1073/pnas.1009798107

Ooi AGL, Sahoo D, Adorno M, Wang Y, Weissman IL, Park CY (2010) MicroRNA-125b expands hematopoietic stem cells and enriches for the lymphoid-balanced and lymphoid-biased subsets. Proc Natl Acad Sci 107(50):21505–21510. doi:10.1073/pnas.1016218107

Pedersen IM, Cheng G, Wieland S, Volinia S, Croce CM, Chisari FV, David M (2007) Interferon modulation of cellular microRNAs as an antiviral mechanism. Nature 449(7164):919–922. doi:10.1038/nature06205

Poy MN, Eliasson L, Krutzfeldt J, Kuwajima S, Ma X, MacDonald PE, Pfeffer S, Tuschl T, Rajewsky N, Rorsman P, Stoffel M (2004) A pancreatic islet-specific microRNA regulates insulin secretion. Nature 432(7014):226–230. doi:10.1038/nature03076

Prosser HM, Koike-Yusa H, Cooper JD, Law FC, Bradley A (2011) A resource of vectors and ES cells for targeted deletion of microRNAs in mice. Nat Biotechnol. doi:10.1038/nbt.1929

Reinhart BJ, Slack FJ, Basson M, Pasquinelli AE, Bettinger JC, Fougvie AE, Horvitz HR, Ruvkun G (2000) The 21-nucleotide *let-7* RNA regulates developmental timing in *Caenorhabditis elegans*. Nature 403(6772):901–906. doi:10.1038/35002607

Rogler CE, LeVoci L, Ader T, Massimi A, Tchaikovskaya T, Norel R, Rogler LE (2009) MicroRNA-23b cluster microRNAs regulate transforming growth factor-beta/bone morphogenetic protein signaling and liver stem cell differentiation by targeting Smads. Hepatology 50 (2):575–584. doi:10.1002/hep. 22982

Romania P, Lulli V, Pelosi E, Biffoni M, Peschle C, Marziali G (2008) MicroRNA 155 modulates megakaryopoiesis at progenitor and precursor level by targeting Ets-1 and Meis1 transcription factors. Br J Haematol. doi:10.1111/j.1365-2141.2008.07382.x

Rosa A, Ballarino M, Sorrentino A, Sthandier O, De Angelis FG, Marchioni M, Masella B, Guarini A, Fatica A, Peschle C, Bozzoni I (2007) The interplay between the master transcription factor PU1 and miR-424 regulates human monocyte/macrophage differentiation. Proc Natl Acad Sci 104(50):19849–19854. doi:10.1073/pnas.0706963104

Rosenberg MI, Georges SA, Asawachaicharn A, Analau E, Tapscott SJ (2006) MyoD inhibits Fstl1 and Utrn expression by inducing transcription of miR-206. J Cell Biol 175(1):77–85. doi:10.1083/jcb.200603039

Rougvie AE (2001) Control of developmental timing in animals. Nature Rev Genet 2(9):690–701. doi:10.1038/35088566

Schoolmeesters A, Eklund T, Leake D, Vermeulen A, Smith Q, Aldred SF, Federov Y (2009) Functional profiling reveals critical role for miRNA in differentiation of human mesenchymal stem cells. PLoS One 4(5):e5605. doi:10.1371/journal.pone.0005605

Sempere LF, Freemantle S, Pitha-Rowe I, Moss EG, Dmitrovsky E, Ambros V (2004) Expression profiling of mammalian microRNAs uncovers a subset of brain-expressed microRNAs with possible roles in murine and human neuronal differentiation. Genome Biol 5(3):R13. doi:10.1186/gb-2004-5-3-r13

Sheehy NT, Cordes KR, White MP, Ivey KN, Srivastava D (2010) The neural crest-enriched microRNA miR-452 regulates epithelial-mesenchymal signaling in the first pharyngeal arch. Development 137(24):4307–4316. doi:10.1242/dev.052647

Shen WF, Hu YL, Uttarwar L, Passegue E, Largman C (2008) MicroRNA-126 regulates HOXA9 by binding to the homeobox. Mol Cell Biol 28(14):4609–4619. doi:10.1128/mcb.01652-07

Smirnova L, Gräfe A, Seiler A, Schumacher S, Nitsch R, Wulczyn FG (2005) Regulation of miRNA expression during neural cell specification. Eur J Neurosci 21(6):1469–1477. doi:10.1111/j.1460-9568.2005.03978.x

Stadtfeld M, Apostolou E, Akutsu H, Fukuda A, Follett P, Natesan S, Kono T, Shioda T, Hochedlinger K (2010) Aberrant silencing of imprinted genes on chromosome 12qF1 in mouse induced pluripotent stem cells. Nature 465(7295):175–181. doi:10.1038/nature09017

Strauss WM, Chen C, Lee C-T, Ridzon D (2006) Nonrestrictive developmental regulation of microRNA gene expression. Mamm Genome 17(8):833–840. doi:10.1007/s00335-006-0025-7

Sugatani T, Hruska KA (2007) MicroRNA-223 is a key factor in osteoclast differentiation. J Cell Biochem 101(4):996–999. doi:10.1002/jcb.21335

Suh M-R, Lee Y, Kim JY, Kim S-K, Moon S-H, Lee JY, Cha K-Y, Chung HM, Yoon HS, Moon SY (2004) Human embryonic stem cells express a unique set of microRNAs. Dev Biol 270 (2):488–498. doi:10.1016/j.ydbio.2004.02.019

Sun Q, Zhang Y, Yang G, Chen X, Cao G, Wang J, Sun Y, Zhang P, Fan M, Shao N, Yang X (2008) Transforming growth factor-β-regulated miR-24 promotes skeletal muscle differentiation. Nucleic Acids Res 36(8):2690–2699. doi:10.1093/nar/gkn032

Suomi S, Taipaleenmaki H, Seppanen A, Ripatti T, Vaananen K, Hentunen T, Saamanen AM, Laitala-Leinonen T (2008) MicroRNAs regulate osteogenesis and chondrogenesis of mouse bone marrow stromal cells. Gene Regul Syst Biol 2:177–191

Takahashi K, Yamanaka S (2006) Induction of pluripotent stem cells from mouse embryonic and adult fibroblast cultures by defined factors. Cell 126(4):663–676. doi:10.1016/j.cell.2006.07.024

Takahashi K, Tanabe K, Ohnuki M, Narita M, Ichisaka T, Tomoda K, Yamanaka S (2007) Induction of pluripotent stem cells from adult human fibroblasts by defined factors. Cell 131 (5):861–872. doi:10.1016/j.cell.2007.11.019

Tenedini E, Roncaglia E, Ferrari F, Orlandi C, Bianchi E, Bicciato S, Tagliafico E, Ferrari S (2010) Integrated analysis of microRNA and mRNA expression profiles in physiological myelopoiesis: role of hsa-mir-299-5p in CD34+ progenitor cells commitment. Cell Death Dis 1(2):e28. doi:10.1038/cddis.2010.5

Thomson JA (1998) Embryonic stem cell lines derived from human blastocysts. Science 282 (5391):1145–1147. doi:10.1126/science.282.5391.1145

Turner M, Vigorito E (2008) Regulation of B- and T-cell differentiation by a single microRNA. Biochem Soc Trans 36(3):531. doi:10.1042/bst0360531

Tzur G, Levy A, Meiri E, Barad O, Spector Y, Bentwich Z, Mizrahi L, Katzenellenbogen M, Ben-Shushan E, Reubinoff BE, Galun E (2008) MicroRNA expression patterns and function in endodermal differentiation of human embryonic stem cells. PLoS One 3(11):e3726. doi:10.1371/journal.pone.0003726

van Rooij E, Sutherland LB, Liu N, Williams AH, McAnally J, Gerard RD, Richardson JA, Olson EN (2006) A signature pattern of stress-responsive microRNAs that can evoke cardiac hypertrophy and heart failure. Proc Natl Acad Sci 103(48):18255–18260. doi:10.1073/pnas.0608791103

Velu CS, Baktula AM, Grimes HL (2009) Gfi1 regulates miR-21 and miR-196b to control myelopoiesis. Blood 113(19):4720–4728. doi:10.1182/blood-2008-11-190215

Wahid F, Shehzad A, Khan T, Kim YY (2010) MicroRNAs: synthesis, mechanism, function, and recent clinical trials. Biochim Biophys Acta 1803(11):1231–1243. doi:10.1016/j.bbamcr.2010.06.013

Walden TB, Timmons JA, Keller P, Nedergaard J, Cannon B (2009) Distinct expression of muscle-specific MicroRNAs (myomirs) in brown adipocytes. J Cell Physiol 218(2):444–449. doi:10.1002/jcp. 21621

Wang Y, Blelloch R (2011) Cell cycle regulation by microRNAs in stem cells. Results Probl Cell Differ 53:459–472. doi:10.1007/978-3-642-19065-0_19

Wang Y, Medvid R, Melton C, Jaenisch R, Blelloch R (2007) DGCR8 is essential for microRNA biogenesis and silencing of embryonic stem cell self-renewal. Nat Genet 39(3):380–385. doi:10.1038/ng1969

Wang Y, Baskerville S, Shenoy A, Babiarz JE, Baehner L, Blelloch R (2008) Embryonic stem cell–specific microRNAs regulate the G1-S transition and promote rapid proliferation. Nat Genet 40(12):1478–1483. doi:10.1038/ng.250

Weston MD, Pierce ML, Jensen-Smith HC, Fritzsch B, Rocha-Sanchez S, Beisel KW, Soukup GA (2011) MicroRNA-183 family expression in hair cell development and requirement of microRNAs for hair cell maintenance and survival. Dev Dyn 240(4):808–819. doi:10.1002/dvdy.22591

Wienholds E, Koudijs MJ, van Eeden FJM, Cuppen E, Plasterk RHA (2003) The microRNA-producing enzyme Dicer1 is essential for zebrafish development. Nat Genet 35(3):217–218. doi:10.1038/ng1251

Wilson KD, Venkatasubrahmanyam S, Jia F, Sun N, Butte AJ, Wu JC (2009) MicroRNA profiling of human-induced pluripotent stem cells. Stem Cells Dev 18(5):749–757. doi:10.1089/scd.2008.0247

Wilson KD, Hu S, Venkatasubrahmanyam S, Fu JD, Sun N, Abilez OJ, Baugh JJA, Jia F, Ghosh Z, Li RA, Butte AJ, Wu JC (2010) Dynamic microRNA expression programs during cardiac differentiation of human embryonic stem cells: role for miR-499. Circ Cardiovasc Genet 3(5):426–435. doi:10.1161/circgenetics.109.934281

Wong CF, Tellam RL (2008) MicroRNA-26a targets the histone methyltransferase enhancer of Zeste homolog 2 during myogenesis. J Biol Chem 283(15):9836–9843. doi:10.1074/jbc.M709614200

Wu L, Belasco JG (2005) Micro-RNA regulation of the mammalian *lin-28* gene during neuronal differentiation of embryonal carcinoma cells. Mol Cell Biol 25(21):9198–9208. doi:10.1128/mcb.25.21.9198-9208.2005

Xiao C, Calado DP, Galler G, Thai T-H, Patterson HC, Wang J, Rajewsky N, Bender TP, Rajewsky K (2007) MiR-150 controls B cell differentiation by targeting the transcription factor c-Myb. Cell 131(1):146–159. doi:10.1016/j.cell.2007.07.021

Xu C, Lu Y, Pan Z, Chu W, Luo X, Lin H, Xiao J, Shan H, Wang Z, Yang B (2007) The muscle-specific microRNAs miR-1 and miR-133 produce opposing effects on apoptosis by targeting HSP60, HSP70 and caspase-9 in cardiomyocytes. J Cell Sci 120(17):3045–3052. doi:10.1242/jcs.010728

Yang Z, Wu J (2007) MicroRNAs and regenerative medicine. DNA Cell Biol 26(4):257–264. doi:10.1089/dna.2006.0548

Yekta S (2004) MicroRNA-directed cleavage of HOXB8 mRNA. Science 304(5670):594–596. doi:10.1126/science.1097434

Yi R, O'Carroll D, Pasolli HA, Zhang Z, Dietrich FS, Tarakhovsky A, Fuchs E (2006) Morphogenesis in skin is governed by discrete sets of differentially expressed microRNAs. Nat Genet 38(3):356–362. doi:10.1038/ng1744

Yi R, Poy MN, Stoffel M, Fuchs E (2008) A skin microRNA promotes differentiation by repressing 'stemness'. Nature 452(7184):225–229. doi:10.1038/nature06642

Yoo AS, Sun AX, Li L, Shcheglovitov A, Portmann T, Li Y, Lee-Messer C, Dolmetsch RE, Tsien RW, Crabtree GR (2011) MicroRNA-mediated conversion of human fibroblasts to neurons. Nature 476(7359):228–231. doi:10.1038/nature10323

Yu J, Vodyanik MA, Smuga-Otto K, Antosiewicz-Bourget J, Frane JL, Tian S, Nie J, Jonsdottir GA, Ruotti V, Stewart R, Slukvin II, Thomson JA (2007) Induced pluripotent stem cell lines derived from human somatic cells. Science 318(5858):1917–1920. doi:10.1126/science.1151526

Yuasa K, Hagiwara Y, Ando M, Nakamura A, Takeda S, Hijikata T (2008) MicroRNA-206 is highly expressed in newly formed muscle fibers implications regarding potential for muscle regeneration and maturation in muscular dystrophy. Cell Struct Funct 33(2):163–169. doi:10.1038/nmeth.1323

Zaragosi L-E, Wdziekonski G, Le Brigand K, Waldmann R, Dani C, Barbry P (2011) Small RNA sequencing reveals miR-642a-3p as a novel adipocyte-specific microRNA and miR-30 as a key regulator of human adipogenesis. Genome Biol 12(7):R64. doi:10.1186/gb-2011-12-7-r64

Zeng Y, Wagner E, Cullen B (2002) Both natural and designed micro RNAs can inhibit the expression of cognate mRNAs when expressed in human cells. Mol Cell 9(6):1327–1333. doi:10.1016/S1097-2765(02), 00541-5

Zhan M, Miller CP, Papayannopoulou T, Stamatoyannopoulos G, Song C-Z (2007) MicroRNA expression dynamics during murine and human erythroid differentiation. Exp Hematol 35 (7):1015–1025. doi:10.1016/j.exphem.2007.03.014

Zhao Y, Samal E, Srivastava D (2005) Serum response factor regulates a muscle-specific microRNA that targets Hand2 during cardiogenesis. Nature 436(7048):214–220. doi:10.1038/nature03817

Zhou H (2005) An RNA polymerase II construct synthesizes short-hairpin RNA with a quantitative indicator and mediates highly efficient RNAi. Nucleic Acids Res 33(6):e62. doi:10.1093/nar/gni061

Chapter 18
The Role of MicroRNAs in Neurodegenerative Diseases: Implications for Early Detection and Treatment

Anna Majer, Amrit S. Boese, and Stephanie A. Booth

Abstract MicroRNAs (miRNAs) are small noncoding RNAs that can posttranscriptionally regulate gene expression in development, differentiation, and in response to various stimuli. Numerous miRNAs are very specifically expressed within the central nervous system suggesting they regulate important brain functions. MiRNAs are also required for the postmitotic survival of neurons, strongly suggesting a crucial role in survival and neuroprotection. The fact that diverse arrays of miRNAs have been reported to be dysregulated in several neurodegenerative diseases implies that they can contribute to pathogenesis. As a group, the global burden of neurodegenerative disease is huge and includes conditions such as Alzheimer's disease and other dementias, for which the numbers are steadily rising with the aging population, as well as communicable diseases caused by prions that are of public health concern. As yet, no drugs to halt or even delay the progression of these diseases are available, and this is a huge focus of global research. The best time for therapeutic intervention would be before significant memory loss and tissue destruction occurs such that interventions to boost cell repair and to promote neuroprotective mechanisms could provide significant health benefits. MicroRNA research promises to further elucidate the pathways, genes, and proteins that contribute to the neurodegenerative process that may serve as potential therapeutic targets. Furthermore, given the evidence of the neuroprotective properties of some miRNAs, these small RNA species may themselves be the focus for drug development. Here, we review recent studies that imply a link between miRNA function and neurodegeneration plus discuss how increased knowledge of miRNAs may be used in diagnosis and treatment of neurodegenerative diseases.

S.A. Booth (✉)
Department of Medical Microbiology and Infectious Diseases, Faculty of Medicine, University of Manitoba, Winnipeg, Manitoba, Canada

Molecular PathoBiology, National Microbiology Laboratory, Canadian Science Center for Human and Animal Health, Public Health Agency of Canada, Winnipeg, Manitoba, Canada
e-mail: Stephanie.Booth@phac-aspc.gc.ca

B. Mallick (eds.), *Regulatory RNAs*, DOI 10.1007/978-3-642-22517-8_18,
© Springer-Verlag Berlin Heidelberg (outside the USA) 2012

Keywords Alzheimer's • amyotrophic lateral sclerosis • biomarkers • central nervous system • disease • fragile X syndrome • Huntington's • miRNAs • multiple sclerosis • neurodegeneration • Parkinson's • prion • therapeutics

18.1 Introduction

Neurological disorders represent one of the greatest present-day threats to public health. According to the World Health Organization, the neurological burden is likely to become an increasingly serious and unmanageable problem affecting the entire world (Ferri et al. 2005). As a group, the global burden of neurological disease is higher than that for malignancies plus digestive and respiratory diseases. These diseases include noncommunicable neurodegenerative conditions such as Alzheimer's, Parkinson's, and Huntington's disease whose numbers are steadily rising with the aging population. Although multiple sclerosis predominantly affects young adults, the number of cases is also on the rise. At present, the complete mechanism involved in the degenerative process associated with these neurological diseases remains largely unknown which, in turn, hinders the development of effective therapies. The best time for therapeutic intervention is before memory loss and tissue destruction occurs; a time when interventions to boost cellular repair and to promote neuroprotective mechanisms provide the most significant health benefits, namely, to improve the symptoms related to disease and to prolong disease progression. Concurrent identification of biomarkers able to identify susceptible individuals and patients with early stages of disease is also required to be able to distinguish between these neurodegenerative diseases and for effective treatment.

The discovery of microRNAs (miRNAs) has unlocked a novel avenue for therapy and biomarker design. MicroRNAs are small noncoding RNAs that are highly evolutionarily conserved. These RNA species bind to unique sites in the regulatory regions of numerous genes within the RNA-induced silencing complex (RISC) resulting in translational repression or degradation. Each miRNA is predicted to target tens to hundreds of genes and, therefore, is able to regulate the expression of multiple and diverse proteins involved in a biological process. This posttranscriptional mechanism of regulation can be simultaneously evoked in a cell-type and context-dependent fashion.

An abundance of miRNAs in the nervous system initially implied their importance in this tissue, and subsequent studies have uncovered pivotal roles for miRNAs in fundamental processes such as neuronal differentiation, development, plasticity, and survival (see Chap. 7, this volume). Many of these studies implicate a general role for CNS-specific miRNAs in these functions, while the explicit identities of the miRNAs involved remain, for the most part, unresolved. Teasing apart the regulatory loops in which miRNAs play such important roles will be an exciting stage in biological research. Not surprisingly, links between miRNA dysfunction and neurodegenerative diseases are also becoming increasingly more apparent. Loss of miRNA expression in the brain leads to neurodegeneration in a number of animal models. There is also evidence from human tissues that the dysregulation of miRNA expression plays a role in the development of neurodegenerative disorders. Studying this novel layer of gene regulation promises to augment our knowledge of brain dysfunction and pathology.

The focus of this chapter is to describe the current contribution of miRNA activity in a number of neurodegenerative conditions. Furthermore, we discuss the potential usefulness of disease-specific microRNAs as biomarkers and tools for targeted therapy.

18.2   The Role of MicroRNAs in Neurodegenerative Disorders

Some of the most compelling evidence for miRNA involvement in neurodegeneration has emerged from investigating Dicer knockout in animal models. Dicer is an essential enzyme in the miRNA biogenesis pathway, and its knockout prevents newly synthesized pre-miRNAs from being processed into mature, functional forms. Dicer ablation in neurogenic progenitors leads to dramatic impairment of neuronal differentiation and subsequent lethality in a number of models (Bernstein et al. 2003; Choi et al. 2008; Kim et al. 2007; Schaefer et al. 2007; Davis et al. 2008; Kawase-Koga et al. 2009). More specifically, miRNAs appear to be essential for the differentiation, survival, and maturation of newborn postmitotic neurons (De Pietri Tonelli et al. 2008) as well as for normal cellular expansion (Kawase-Koga et al. 2010). Conditional Dicer loss in certain cell types such as Purkinje cells in the cerebellum (Schaefer et al. 2007), dopaminergic neurons of the midbrain (Kim et al. 2007), neocortical neurons (De Pietri Tonelli et al. 2008), oligodendrocytes (Shin et al. 2009), and neuronal stem cells (Kawase-Koga et al. 2010) leads to cell death, providing further evidence that the long-term health and survival of differentiated postmitotic neurons is also governed by miRNAs. Degeneration and death in these cells could possibly be attributed to an increase in proapoptotic proteins, and/or a decrease in prosurvival proteins (Kawase-Koga et al. 2010), perhaps in accordance with documented heterochromatin abnormalities (Fukagawa et al. 2004; Kanellopoulou et al. 2005; Kawase-Koga et al. 2010). All of these phenotypes are reminiscent of progressive neurodegeneration in the absence of Dicer and were the first reports to raise the possibility of an involvement of miRNAs in neurodegenerative disorders.

The extent of miRNA dysregulation in neurodegenerative diseases was initially determined using global miRNA screening techniques such as miRNA microarrays (Saba et al. 2008; Otaegui et al. 2009; Nunez-Iglesias et al. 2010; Wang et al. 2011). While global dysregulation of miRNAs in diseased tissues is evident, the contributions of specific miRNAs to the initiation and progression of neurodegenerative disease are just beginning to be understood (see Table 18.1 for a list of miRNAs and validated targets discussed in this chapter).

Table 18.1 List of microRNAs and validated gene targets that are implicated in numerous neurodegenerative diseases.

Disease	Relative expression	MicroRNA	Gene target(s)	Gene function	References
AD	Down	miR-17-5p, miR-20a, miR-101, miR-106a[a], miR-106b	APP	Amyloid precursor protein once cleaved may hasten neurodegeneration	Patel et al. (2008)[a], Hebert et al. (2009), Vilardo et al. (2010), Long and Lahiri (2011)
–	Down	miR-9[a], miR-29a/b-1, miR-107, miR-298, miR-328, miR-29c[b]	BACE1	Produces Aβ 1-40 and Aβ 1-42 products from cleaving APP which perpetuates neurodegeneration	Lukiw (2007)[a], Hebert et al. (2008), Wang et al. (2008), Boissonneault et al. (2009), Zong et al. (2011)[b]
–	–	miR-101	Cox-2	Cyclooxygenase 2 that promotes the formation of amyloids in the brain	Vilardo et al. (2010)
–	Down	miR-103, miR-107	Cofilin	Actin-binding protein that reorganizes actin filaments	Yao et al. (2010)
–	Down	miR-29a	NAV3	Potential regulator of neurite outgrowth and axonal guidance	Shioya et al. (2010)
–	Down	miR-124	PTBP1	Major regulator of APP splicing producing exon 7 and 8 where abundance may lead to Aβ accumulation	Smith et al. (2011)
–	Up early/down late	miR-106b	TGF-B type II receptor	Neurotrophic and neuroprotective roles in growth and survival of neurons	Wang et al. (2010)
–	Up	miR-34a	Bcl2	Serves a neuroprotective role by inhibiting CASP3	Wang et al. (2009)
–	Up	miR-146a	CFH	Important repressor of the inflammatory response in the brain	Lukiw et al. (2008)
–	–	–	IRAK-1	Essential component of toll-like/IL-1 receptor signaling	Cui et al. (2010)
PD	Down	miR-133b	PITX3	Important transcription factor for normal survival of dopaminergic neurons in the substantia nigra	Kim et al. (2007)
–	Down	miR-7, miR-153	α-synuclein	Is a major contributor to PD because it forms insoluble fibrils	Junn et al. (2009), Doxakis (2010)

Disease	miRNA	Expression	Target	Function	Reference
–	miR-433	N/A	FGF20	If not regulated can stimulate α-synuclein production	Wang et al. (2008a)
HD	miR-30a	Up	BDNF	Essential for normal neuronal function and survival of the nervous system	Mellios et al. (2008), Marti et al. (2010)
–	miR-9/miR-9*	Down	REST/CoREST	An essential transcription repressor complex inhibiting expression of many neuronal genes	Packer et al. (2008)
–	miR-124a	Down	SCP-1	Cofactor for REST activity	Visvanathan et al. (2007), Johnson et al. (2008), Packer et al. (2008)
–	miR-132	Down	MeCP2	Interacts with REST/CoREST, further inhibiting transcription such as the production of BDNF	Klein et al. (2007)
Prion	miR-342-3p, miR-494	Up	N/A	N/A	Saba et al. (2008), Montag et al. (2009)
–	miR-191	Up	EGR1	Transcription factor regulating many genes involved in neuronal function	Saba et al. (2008)
ALS	miR-9	Down	NF-H	Important component of the neurofilament that once deregulated may perpetuate disease	Haramati et al. (2010)
–	miR-206	Down	HDAC4	Hinders the repair of neurons	Williams et al. (2009)
MS	miR-17, miR-20a	Down	N/A	Inhibit T-cell activation	Cox et al. (2010)
–	miR-34a, miR-155, miR-326	Up	CD47	A marker of self that inhibits macrophage activation against cells expressing this marker	Junker et al. (2009)
–	miR-326	Up	Est-1	Negative regulator of T_H-17 differentiation	Du et al. (2009)
FXS	miR-125b	N/A	NR2A	Essential subunit of the NMDA receptor	Edbauer et al. (2010)
–	miR-19b, miR-302b*, miR-323-3p	N/A	FMR1	Encodes FMRP and functions in synaptogenesis by suppressing translation	Yi et al. (2010)

The relative expression for each miRNA reflects the expression level in the infected sample as compared to the control.

AD Alzheimer's disease, PD Parkinson's disease, HD Huntington's disease, Prion prion disease, ALS amyotrophic lateral sclerosis, MS multiple sclerosis, FXS fragile x syndrome

N/A no available data

[a]Indicates a discrepancy found between studies where the expression level of the miRNAs is different

[b]miRNA that is upregulated in diseased samples and corresponding reference

18.2.1 Alzheimer's Disease

Alzheimer's disease (AD) is the most common cause of neurodegeneration, accounting for 60–70% of all dementia cases worldwide (Fratiglioni and Qui 2009). Currently, it affects 2% of the population, and the incidence of disease is expected to increase 20- to 40-fold in the next 50 years with the aging population. The cause of Alzheimer's disease is multifaceted and is believed to be due to the accumulation of extracellular deposits of amyloid fibers (amyloid plaques) and intracellular inclusions (neurofibrillary tangles) leading to gliosis (proliferation of glial cells) and degeneration. Amyloid plaques are produced from the sequential cleavage of amyloid precursor protein (APP) by the β-site APP-cleaving enzyme1 (BACE1) into Aβ 1-40 and Aβ 1-42 products, the main component of amyloid. Over time, these fragments aggregate to form extracellular plaques that are toxic to neurons because they disrupt calcium homeostasis, triggering apoptosis (O'Brien and Wong 2010). In addition, APP is a transmembrane protein critical for neuronal growth, survival, and synapse formation and also plays an important role in post-injury repair. Perhaps the loss of normal APP during the disease process results in the impairment of an essential function, and this may also contribute to disease progression.

BACE1 protein abundance is highly correlated with disease, and therefore, the mechanism by which this enzyme is regulated is profoundly important to the understanding of the development of Alzheimer's disease. Modulation of BACE1 expression by miR-9 and a miR-29a/b-1 cluster was the first evidence of the possible involvement of miRNAs in the development of Alzheimer's disease (Hebert et al. 2008). Hebert et al. also found the levels of miR-9 and a miR-29a/b-1 to be downregulated in a pool of sporadic AD brain samples. In contrast, a second study observed an increase of miR-9 levels in AD hippocampus (Lukiw 2007). Focusing further functional experiments on the miR-29a/b-1 cluster, it was determined that BACE1 protein expression could be significantly repressed in cultured cells. MiR-29a and miR-29b-1 are developmentally regulated in mouse brain in a similar fashion to BACE1, and their levels correlate with BACE1 expression in AD (Hebert et al. 2008). Furthermore, miR-107 (Wang et al. 2008b), miR-298, and miR-328 (Boissonneault et al. 2009) have since been found capable of the posttranscriptional regulation of BACE1 and also exhibit decreased expression in AD patients. This is consistent with the existence of a molecular link between miRNA expression in sporadic AD and the amyloid cascade. Recent evidence suggests that miR-29c, part of the miR-29 family, may play a protective role during early AD (Zong et al. 2011). The authors showed that miR-29c levels are increased in 3- and 6-month-old APPswe/PSΔE9 mice and that miR-29c targets BACE1, inhibiting protein translation (Zong et al. 2011). Furthermore, Aβ 1-40 peptide levels were decreased in miR-29c transgenic mice that overexpressed miRNA-29c, which may be due to miRNA regulation of BACE1 and subsequent effects of this regulation on APP processing (Zong et al. 2011). Further investigation into the potential involvement of miR-29c and other miRNAs early in AD may help alleviate the mechanisms of disease and enhance therapeutic prospects.

A recently published series of papers have described the direct regulation of APP levels by miRNAs. MiR-106a was shown by Patel and colleagues to regulate APP protein levels (Patel et al. 2008); however, in a second study, the miR-20a family (consisting of miR-20a, miR-17-5p, and miR-106b, but not miR-106a) was found to be active (Hebert et al. 2009). Discrepancies such as this are relatively common among miRNA studies carried out by different groups and can largely be attributed to experimental variations in a developing research field. One important factor appears to be variations in the lengths of the $3'$ UTR cloned into the luciferase reporter vectors used to assay miRNA specificity. It is likely that $3'$ UTRs are arranged in a 3-dimensional structure that is stabilized by sequences some distance away, thus effecting the establishment of the associated RISC complex (Bartel 2009). Furthermore, additional to the seed sequence of a miRNA, the "non-seed" sequence also contributes to the specificity of miRNA regulation (Bartel 2009). Hebert et al. (2009) found that although miR-106a and miR-106b share the same seed sequence, the non-seed segment of miR-106b proved to be essential for its regulation of APP. Importantly, the tested AD patients had significantly decreased expression levels of miR-106b (Hebert et al. 2009), implying that disruption of APP regulation by miRNAs is possible during disease. In another study, miR-101 was shown to function as a negative regulator of APP expression (Vilardo et al. 2010; Long and Lahiri 2011) and Cyclooxygenase 2 (Cox-2) (Vilardo et al. 2010), a gene that promotes amyloid formation in the brain (Xiang et al. 2002). Interestingly, the overexpression of miR-101 dampens the accumulation of amyloid-β in hippocampal neurons (Vilardo et al. 2010). In this case, miR-101 has the potential to regulate both Cox-2 and APP levels, two proteins strongly associated with the development of Alzheimer's disease, suggesting a cumulative, coordinate deregulation of genes involved in the pathobiology of disease.

APP undergoes alternative splicing that produces functional isoforms of exons 7 and 8 in non-neuronal cells, while the exon 15 isoform is more abundantly expressed in neuronal cells. An increase in APP exon 7 and/or 8 isoforms in neurons leads to an increase in amyloid-β synthesis, potentially contributing to AD (Golde et al. 1990; Neve et al. 1990; Rockenstein et al. 1995). The endogenous polypyrimidine-tract-binding protein 1 (PTBP1) correlates with the presence of APP exons 7 and 8, while PTBP2 is associated with the predominant expression of exon 15. PTBP1 is partially regulated by miR-124, a miRNA that is also downregulated in AD patients, suggesting the involvement of miR-124 in affecting the alternative splicing mechanism of APP (Smith et al. 2011).

Another protein intimately involved with the progression of Alzheimer's disease is tau, which when phosphorylated functions to stabilize the microtubule network within neurons. The microtubule track spans the entire neuronal axon and is used to transport molecules from the cell body to dendrites and vice versa. In Alzheimer's disease, tau is hyperphosphorylated and begins to associate with multiple microtubules causing the tracks to intertwine, resulting in the formation of neurofibrillary tangles and eventual disruption of the molecular transport within the cells.

In primary neurons, amyloid-β peptide treatment encourages the formation of cofilin rods (Minamide et al. 2000). Similar to tau protein, the presence of cofilin rods disrupts microtubule bundles in neurites, interfering with neuritic transport and neuronal structure and activity (Maloney and Bamburg 2007; Davis et al. 2009). Recent evidence links the formation of cofilin aggregates to the formation of tau neurofibrils (Whiteman et al. 2009). In 2010, Yao and colleagues demonstrated that miR-107 and miR-103 are both able to regulate cofilin levels (Yao et al. 2010). Furthermore, they showed that both of these miRNAs were decreased in a transgenic mouse model of AD. Levels of cofilin were similarly increased, suggesting that miRNA dysregulation can also contribute to the cytoskeletal pathology that accompanies AD progression.

Another biological process that is affected during Alzheimer's disease, and common to neurodegenerative conditions, is inflammatory signaling. Accumulation of amyloid-β 1-42 fragments functions as proinflammatory mediators, activating an inflammatory response that further perpetuates the pathobiology of disease. MiR-146a, a well-studied miRNA with anti-inflammatory function (Sonkoly and Pivarcsi 2009), is increased in AD. This miRNA is able to suppress the expression of both complement factor H (CFH) (Lukiw et al. 2008) and interleukin-1 receptor-associated kinase 1 (IRAK-1) (Cui et al. 2010), and may in part modulate a potentially harmful inflammatory response in AD brain.

Ultimately, neurons are lost during AD, and apoptosis is a potential molecular mechanism by which death occurs. The majority of the Alzheimer's disease–related miRNAs reported so far are downregulated during disease, thus relieving regulatory restraints on the expression of propathogenic proteins. However, increased expression of miR-34a observed in brain samples from AD patients (Wang et al. 2009) may function to perpetuate neuronal death by decreasing the expression of the antiapoptotic protein B-cell lymphoma 2 (BCL2). Interestingly, miR-106b has also been shown to target the TGF-B type II receptor, a protein that may have a neuroprotective role in a transgenic mouse model of Alzheimer's disease (Wang et al. 2010), thereby contributing to pathogenesis. Given the previously described role of miR-106b in the processing of APP, this illustrates the propensity for dysregulated miRNAs to influence multiple pathways that are potentially functionally unrelated. A further example of this is observed with miR-29a which targets both BACE1 and NAV3 (neurone navigator 3) protein that is an important regulator of axonal guidance (Shioya et al. 2010). Interestingly, only mRNA expression levels, not the protein levels, are elevated in AD (Shioya et al. 2010). A multifactorial approach to understanding the complexities of miRNA function in neurodegeneration is therefore vital.

18.2.2 Parkinson's Disease

Parkinson's disease (PD) affects roughly 1% of the elderly population (Savitt et al. 2006) by causing severe degeneration of dopaminergic neurons in the substantia nigra (SN). Although current treatments exist, they only target symptoms and involve dopamine replacement therapy, which can mitigate some of the symptoms of disease but does not counteract progressive degeneration (Zesiewicz et al. 2010). To add further complexity, the cause of PD is primarily sporadic in nature and linked to numerous gene disruptions including α-synuclein (α-syn), leucine-rich repeat kinase 2 (LRRK2), parkin, PTEN-induced kinase 1 (PINK1), and DJ-1. Abnormal accumulation of α-synuclein protein in the brain forming Lewy bodies is a hallmark of disease. However, it is not known whether these fibrillar aggregates directly contribute to cell death perhaps by inducing oxidative stress or are by-products of the disease process (Harraz et al. 2011).

MiRNA profiling of tissues from the midbrain of PD patients revealed the significant reduction of miR-133b, a miRNA normally specifically expressed in midbrain DNs, in comparison to similar samples from a normal pool. It was also found to be severely reduced in mouse dopamine deficiency models. Overexpression of this miR-133b leads to the decrease in dopamine release from neurons, suggesting a direct link to dopamine homeostasis (Kim et al. 2007). Pituitary homeobox 3 (PITX3), a transcription factor implicated in PD (Fuchs et al. 2009), was recently identified as a miR-133b target. Interestingly, PITX3 mutant mice exhibit loss of dopaminergic neurons in the substantia nigra (Hwang et al. 2003), suggesting the existence of a negative feedback loop between PITX3 and miR-133b (Kim et al. 2007).

Computational methodologies have been used to predict miRNA binding sites within genes linked to PD, such as α-syn and fibroblast growth factor 20 (FGF20). These miRNAs have subsequently been further investigated as candidate regulators of the disease process. The brain-enriched miRNAs, miR-7 and miR-153, were predicted to target α-syn and were duly shown to be decreased in the SN of PD mouse models (Junn et al. 2009; Doxakis 2010). MiR-7 has a neuroprotective role by preventing oxidative stress, and miR-7 inhibition causes neuronal apoptosis. Depletion of these miRNAs resulting in a concomitant increase in α-syn levels in PD brain would indicate a functional role of these miRNAs in the disease process; however, this has yet to be confirmed. Fibroblast growth factor 20 (FGF20) disruptions are associated with a higher risk of PD, and single-nucleotide polymorphisms (SNP) in its 3′ UTR has been identified in some PD patients (Wang et al. 2008a). One such SNP is within a miR-433 binding site, leading to decreased miR-433 binding efficiency. In an experimental model, decreased miR-433 binding results in FGF20 overexpression and a concomitant increase in α-syn protein levels. Moreover, human brains with the miR-433 SNP have higher levels of α-syn accumulation than those without the mutation (Wang et al. 2008a). SNPs in the 3′ UTRs of genes associated with neurodegenerative disease may well contribute to the pathogenesis of neurodegenerative disease and should be a focus for future study.

18.2.3 Huntington's Disease

Huntington's disease is an autosomal dominant neurodegenerative disorder that affects cortical and striatal neurons. The genetic factor is the insertion of a CAG trinucleotide repeat expansion in the *huntington* gene that codes for the huntingtin protein (Htt) (The Huntington's Disease Collaborative Research Group 1993). This results in the introduction of at least 36 glutamate residues that expands the N terminus, altering the protein's structure and function. MiRNA expression profiling of human cortex samples from HD patients, relative to healthy controls, revealed a number of dysregulated miRNAs of which many are neuronal specific (Johnson et al. 2008; Packer et al. 2008; Marti et al. 2010). These included decreases in the amounts of miR-9, miR-9*, miR-29b, miR-124a, miR-132, miR-135b, miR-139, miR-212, and miR-218 (Johnson et al. 2008; Packer et al. 2008) and upregulation of miR-29a and miR-330 (Johnson et al. 2008). While specific functions of the majority of these miRNAs have not been determined, it is known that miR-9, miR-9*, and miR-124a are regulated by the repressor element 1-silencing transcription factor (REST), a master regulator of neuronal genes that plays a role in HD (Wu and Xie 2006).

REST is normally sequestered by functional Htt in the cytoplasm of neurons where it remains inactive. Mutant Htt is unable to bind REST which results in the translocation of Htt to the nucleus and subsequent binding to the repressor element 1/neuron-restrictive silencer element (RE1/NRSE) upstream of numerous neuronally expressed genes and miRNAs. Interestingly, miR-9, miR-9*, miR-132, and miR-124a are all posttranscriptional regulators of REST (Wu and Conaco et al. 2006; Wu and Xie 2006; Packer et al. 2008). In addition, miR-9/miR-9* directly bind to the 3′ UTRs of REST and the REST corepressor 1 (CoREST), respectively. MiR-132 targets the protein MeCP2 (Klein et al. 2007) that can interact with REST and CoREST to suppress transcription (Lunyak et al. 2002). In neural progenitors, REST inhibits miR-124a expression, allowing the persistence of non-neuronal transcripts. During differentiation into mature neurons, REST leaves the miR-124a RE1 site and the non-neuronal transcripts are selectively degraded. Mature miR-124a also targets a splicing factor, PTBP1, thus tipping the balance toward a brain-specific alternative pre-mRNA splicing pattern by PTBP2 (Makeyev et al. 2007). These data suggest that REST/CoREST related miRNAs are involved in a double feedback loop regulatory mechanism. This complex regulatory device for gene expression has previously been described in some systems where the levels of multiple, key proteins are critical for miRNA function (Tsang et al. 2007).

MiR-30a was recently shown to be increased in brain tissue samples from HD patients (Marti et al. 2010) which targets the prosurvival brain derived growth factor (BDNF) (Mellios et al. 2008) that also contains an RE1/NRS element within its promoter. Hence, REST inhibition of BDNF transcription contributes to neuronal dysfunction and death (Zuccato et al. 2007). Indeed, decreased BDNF mRNA and protein levels are also seen in HD patients (Zuccato et al. 2001). These data further emphasize the important contribution of miRNAs to gene regulation in HD.

Parallel sequencing of miRNAs in the frontal cortex and striatum regions of HD patients not only identified numerous dysregulated miRNAs (both having increased and decreased expression levels) but also distinguished numerous new IsomiRs (Marti et al. 2010). IsomiRs, highly abundant in HD tissues, are miRNAs that exhibit variation from their "reference" sequences and $5'$ trimming modifications from sequencing data. The putative targets of the seed-region IsomiRs suggest that their altered expression may contribute to aberrant gene expression in HD. Nevertheless, the role of miRNA variation in biological processes and the mechanisms by which IsomiRs are selectively generated have as yet to be determined.

Another interesting finding that appears to be a factor common to a number of neurodegenerative diseases is the induction of changes to the RNA silencing machinery itself. Htt protein actually sequesters within RNA processing bodies (P-bodies) where it colocalizes with Argonaute 2 (Ago2). Mutant Htt transgenic mice were found to have reduced numbers of P-bodies in cortical neurons (Savas et al. 2008), resulting in reduced gene silencing activity in comparison with wild-type mice. Thereby, Htt may function as a coaccessory to Ago2 and its mutation in disease may lead to aberrations in the normal activity of Ago2 in miRNA-mediated silencing.

18.2.4   Prion Disease

Transmissible spongiform encephalopathies (TSEs), or prion diseases, comprise a group of rare but fatal neurodegenerative disorders that affect both humans and animals (Aguzzi and O'Connor 2010). Unique among neurodegenerative diseases, TSEs have the ability to be transmitted by template-dependent conversion of the normal prion protein, PrP^C, to an abnormal, pathogenic isoform, PrP^{Sc}. In addition to transmission of PrP^{Sc} in contaminated food or through surgical procedures, the initial acquisition of PrP^{Sc} can also occur through spontaneous generation or genetic mutation of the prion gene, *PRNP* (Aguzzi and O'Conner 2010; Lloyd et al. 2011). In prion diseases, the replication of PrP^{Sc} is the cause of neuronal damage and death, whereas in other neurodegenerative diseases, the toxic trigger is less evident. Although there are key differences in the proximal triggers and markers of different neurodegenerative diseases, it is becoming clear that multiple convergent downstream pathways are stimulated. Animal models of prion disease have clear advantages for studying these complexities; not least, they are extremely well defined and lead to a clear endpoint – the death of the animal. Similar models for other neurodegenerative diseases rely on transgenic modifications which simulate only certain facets of degeneration. Thus, prion infection models represent an excellent system to identify overarching pathological cascades and targets for therapeutic intervention.

The first indication of miRNA involvement in prion disease pathology was the identification of 15 miRNAs deregulated at end-stage of disease in mouse models (Saba et al. 2008). The majority of these miRNAs were increased and included miR-342-3p, miR-328, miR-128, miR-146a, and miR-191, while miR-338-3p and miR-337-3p were both downregulated. Using bioinformatic software such as TargetScan, miRBase, and PicTar (reviewed in Majer and Booth 2010) along with luciferase assays, the early growth response 1 (EGR1) gene was identified as a target of miR-191 (Saba et al. 2008). EGR1 is a transcription factor that potentially regulates multiple genes involved in neuronal function (Beckmann and Wilce 1997; Harada et al. 2001; Jones et al. 2001; James et al. 2006) and was observed to be downregulated at end-stage of prion disease (Sorensen et al. 2008). A number of the dysregulated miRNAs identified in this study have also been described in investigations of other neurodegenerative diseases. For example, miR-128, miR-328, and miR-146a were all aberrantly expressed in Alzheimer's disease (Lukiw 2007; Lukiw et al. 2008; Boissonneault et al. 2009; Cui et al. 2010) and HIV-induced neurodegenerative disease (Eletto et al. 2008) as well as in prion disease (Montag et al. 2009). The identification of aberrantly expressed miRNAs found in common between different neurodegenerative diseases echo's similarities identified within gene expression studies among multiple diseases. Taken together, this suggests that it may be possible to identify a generalized pattern of transcriptional dysregulation between many neurodegenerative disease processes.

Aberrantly expressed miRNAs have also been identified in BSE-infected macaques and CJD patients (Montag et al. 2009). Two miRNAs, miR-342-3p and miR-494, were significantly upregulated in BSE-infected macaques, while miR-342-3p was also upregulated in brain samples from CJD patients. Overall, miR-342-3p was found to be upregulated in human samples (CJD) and two animal models (scrapie-infected mice and BSE-infected primates) (Saba et al. 2008; Montag et al. 2009) suggesting it to be a strong candidate for consistent involvement in prion pathogenesis across various host and prion species.

18.2.5   Amyotrophic Lateral Sclerosis

Amyotrophic lateral sclerosis (ALS) is a fatal motor system disease leading to rapid neurodegeneration of motor neurons (MNs) located in the ventral horn of the spinal cord, causing atrophy and paralysis of limbs and respiratory muscles (Mulder et al. 1986). Little is known about the causative trigger for ALS. MNs develop proteinaceous inclusions in their cell bodies and axons, although in contrast to that seen in Alzheimer's, Parkinson's, Huntington's, and prion diseases, these protein aggregates do not form amyloid. Inclusions usually contain ubiquitin and often the superoxide dismutase-1 (SOD1) which is associated with dominantly inherited familial forms of ALS (Bruijn et al. 2004). Additionally, mutations in RNA processing proteins such as the transactive response (TAR) DNA binding protein 43 kDa (TDP-43) (Arai et al. 2006) and the fused in sarcoma/translocation in liposarcoma (FUS/TLS) (Kwiatkowski et al. 2009; Vance et al. 2009) have also been localized to neuronal inclusion bodies. Interestingly, TDP-43 and FUS/TLS proteins are associated with components of the Drosha microprocessor complexes

(Gregory et al. 2004). Abnormal aggregation of these proteins inevitably disrupts the processing of pri-miRNA to pre-miRNAs in the nucleus (Ling et al. 2010). It has been shown that TDP-43 knockdown in cultured cells results in changes to the total miRNA population. Nevertheless, it remains unknown whether TDP-43 affects the processing of Drosha-free miRNA that are expressed from introns (miRtrons) (Buratti et al. 2010).

The extent of miRNA function in MN survival was assessed in Dicer ablation studies that specifically hindered miRNA activity in postmitotic somatic MNs (MNDicer^{mut} mice). Interestingly, MNDicer^{mut} showed progressive locomotion dysfunction, atrophy, and neuronal degeneration, strongly indicating an essential function of miRNAs in MN survival (Haramati et al. 2010). Significant downregulation of miR-9 and miR-9* was observed in both MNDicer^{mut} and spinal muscular atrophy animal models where miR-9 was predicted to regulate the expression of the heavy neurofilament subunit (NF-H) (Haramati et al. 2010). Defects in the intermediate filament system have previously been implicated in ALS disease, and disruption of the coordinated expression of neurofilament genes by miRNAs may contribute to the cause of this observation (Figlewicz et al. 1994; Al-Chalabi et al. 1999).

Considering that ALS affects neurons of the motor system, alterations of miRNA expression levels specifically at the neuromuscular junction have been further investigated. One of the miRNAs identified to be increased in a SOD1 ALS mouse model was miR-206. Deletion of miR-206 in G93A-*Sod1* mice did not affect the timing of disease onset but led to accelerated disease progression and shortened survival times (Williams et al. 2009). Further investigation suggested that miR-206 is important for efficient regeneration of neuromuscular synapses following injury by regulating histone deacetylase 4 (HDAC4) levels, a protein that hinders the repair of neurons (Williams et al. 2009).

18.2.6   Multiple Sclerosis

Multiple sclerosis (MS) is the result of chronic inflammation within the nervous system that causes damage to the myelin sheath, the protective covering surrounding nerve cells. It is a relatively common disease affecting approximately 1–2 people per 1000 with the incidence rate appearing to be on the rise (Hauser and Oksenberg 2006). Typically, MS patients exhibit progressive deterioration of neuronal function and have 5–10 years shorter life expectancies. Although symptom manifestation varies between relapsing and progressive forms of MS, neuronal deterioration is observed in both. The pathogenic hallmarks of MS diseases include autoimmunity, an uncontrolled activation of inflammatory cells within the brain that leads to demyelination of axons and destruction of oligodendrocytes, forming white matter lesions (Hauser and Oksenberg 2006). The trigger that activates this degenerative cascade remains unknown, but research suggests a combination of genetic and environmental factors.

Increased understanding of this inflammatory process is required prior to the development of effective therapies for MS. Otaegui and colleagues (2009) obtained blood samples from 21 individuals consisting of 9 MS patients in remission, 4 MS patients during relapse, and 8 healthy controls. Peripheral blood mononuclear cells (PBMCs) were isolated from whole blood and miRNA profiling revealed increased expression of miR-18b and miR-599 in relapse samples, while miR-96 levels were increased in remission samples (Otaegui et al. 2009). Bioinformatic analysis suggested that potentially important targets of miR-96 are involved in interleukin or wnt signaling pathways, although these have yet to be experimentally validated.

$T_H 1$ and T_H-17 cells have significant involvement in disease pathogenesis by being critically involved in $CD4^+$ T-cell-mediated autoimmunity. In a recent study, miRNAs regulating T_H-17 cell differentiation were identified in peripheral blood leukocytes (Du et al. 2009). MiR-326 was found to be increased in MS patients, and it was further determined that Est-1, an inhibitor of T_H-17 differentiation, is a target of this miRNA (Du et al. 2009). Manipulation of miR-326 expression levels in an in vivo MS model system affected T_H-17 population numbers: decreased miR-326 levels produced fewer T_H-17 cells which in turn resulted in milder experimental autoimmune encephalomyelitis (EAE) (Du et al. 2009). In contrast, by overexpressing miR-326, more T_H-17 cells were produced and a faster onset of EAE was observed (Du et al. 2009).

Profiling miRNAs in CD4+, CD8+, T cells, and B cells from peripheral blood of relapsing-remitting MS patients revealed increases in expression of miR-485-3p, miR-376a, miR-193a, miR-126, and in particular a marked upregulation of miR-17, known to function in autoimmunity (Lindberg et al. 2010). Stimulation of $CD4^+$ cells with anti-CD3/CD28 resulted in the significant upregulation of miR-17 and miR-193a. The authors suggested that miR-17 directly or indirectly regulates phosphatase and tensin homolog (PTEN), phosphatidylinositol 3-kinase regulatory subunit 1 (PI3KR1), proapoptotic member of the Bcl-2 family (Bim), and transcription factor 1 of the elongation 2 factor family (E2F1) during $CD4^+$ T-cell stimulation. In turn, miR-193a has been involved in apoptosis by influencing the activation of the caspase cascade (Ovcharenko et al. 2007). Further research to validate direct targets of miR-17 and miR-193a needs to be performed to determine the extent of their contribution to disease.

Another whole blood miRNA transcriptome study that employed a relatively large MS cohort determined the downregulation of 26 miRNAs across all samples from primary progressive, secondary progressive, and relapsing remitting MS disease. MiR-17 and miR-20a, interestingly encoded in the same cistron, were the most significantly downregulated as compared to controls (Cox et al. 2010). This is in contrast to the previously described upregulation of miR-17 in $CD4^+$ cells of MS patients; the reason for this discrepancy is unknown. MiR-17 and miR-20a were postulated to inhibit T-cell activation in this study, but the mechanism remains to be resolved.

Only one study to date has investigated miRNA dysregulation in MS lesions; three miRNAs, miR-34a, miR-155, and miR-326, were associated with active lesions (Junker et al. 2009). All three of these miRNAs targeted CD47, a gene that is normally expressed on resident host cells and is considered a "marker of self" that functions to inhibit macrophage activation (Kinchen and Ravichandran 2008). This gene was also found to be downregulated by 50% in active MS lesions as compared to control white matter samples (Junker et al. 2009). Perhaps in the MS lesion environment these 3 miRNAs remove the inhibitory effects imposed by resident cells on macrophage activation, allowing these macrophages to initiate myelin phagocytosis.

18.2.7   Fragile X Syndrome and Fragile X–Associated Tremor/Ataxia Syndrome

Fragile X syndrome (FXS) is the most common inherited cause of mental retardation that occurs due to the silencing of the fragile X mental retardation 1 (FMR1) gene (Pieretti et al. 1991; Verkerk et al. 1991). In disease, a massive expansion of CGG repeats in the 5' untranslated region (UTR) of *FMR1* becomes hypermethylated and leads to gene silencing so that the protein product, FMR protein (FMRP), is lacking. Although no neurodegeneration is seen in fragile X patients, the FMR1 gene codes for a protein that is intimately linked to the miRNA processing machinery. Interestingly, a common "premutation" form of the gene exists in some individuals in which FMRP contains a number of CGG repeats but is still expressed. These individuals suffer from an adult onset progressive neurodegenerative disorder termed fragile X tremor ataxia syndrome (FXTAS) that is characterized by intranuclear ubiquitin-positive inclusions in neurons and astrocytes of the cerebellum and cerebral cortex (Willemsen et al. 2003; Raske and Hagerman 2009).

FMRP is an RNA-binding protein abundantly expressed in neuronal dendrites and spines (Feng et al. 1997) which functions to suppress translation (Laggerbauer et al. 2001; Lin et al. 2006). This RNA-binding protein is important during neuronal development and synaptogenesis; *Fmr1* knockout mice have abnormal dendritic spines (Comery et al. 1997; Nimchinsky et al. 2001), a phenotype evident in human patients of FXS (Hinton et al. 1991). FMRP has been shown to interact with the Dicer and Argonaute 2 (Ago2) proteins of the miRNA machinery, suggesting that the functional activity of FMRP is mediated by the miRNA complex (Jin et al. 2004; Plante et al. 2006). More specifically, the absence of FMRP in *Drosophila* resulted in the lowered abundance of the Dicer/Ago complexes which partially decreases miRNA-124a expression levels (Xu et al. 2008). Hence, FMRP can modulate the processing of miR-124a during development but is not essential for miRNA biogenesis (Xu et al. 2008).

In lieu of the fact that FMRP is localized to dendritic spines and functions in synaptogenesis, Edbauer et al. (2010) studied whether miRNAs that associate with FMRP may regulate these neuronal properties in mice. The authors found that miR-125b and miR-132 both affect spine morphology and their function requires FMRP (Edbauer et al. 2010). Further investigations led to the identification of the NR2A, a subunit of the NMDA receptor, to be a target of FMRP-associated miR-125b (Edbauer et al. 2010). It is known that several functions of NMDA-receptor-dependent plasticity are impaired during FXS, which may be partially explained by the lack of FMRP-regulated miRNA function. Interestingly, both these miRNAs have been implicated in neurodegeneration which may indicate some convergence of pathways triggered during disease prior to neuronal death.

Conflicting results on the role of miRNAs in the regulation of FMR1 have been reported. Recent evidence points to FMRP regulation by miRNAs in which miR-19b, miR-302b*, and miR-323-3p repress gene expression in HEK293T cells (Yi et al. 2010). Nevertheless, FMR1 mRNA expression levels did not change in Dicer knockout cells (Cheever et al. 2010), while in vivo studies have yet to be performed.

In FXTAS, the expanded FMR1 mRNA is present within the intranuclear inclusions found in neurons and astrocytes of patients, which suggests that the mRNA itself is important for the neurodegenerative phenotype. It has been shown that the pri-miRNA processing complex (microprocessor comprising Drosha and DGCR8 proteins) can interact directly with CGG-repeat mRNA (Faller et al. 2010; Sellier et al. 2010). This results in protein sequestration and subsequently a reduced capability to process pri-miRNA to pre-miRNA. These data imply that CGG-repeat-induced pathogenesis in FXTAS may involve sequestration of proteins engaged in the miRNA processing machinery.

18.3   The Challenges of Studying MicroRNAs in CNS Tissue

Although a number of genomic investigations of miRNA expression have been performed, there is much ambiguity and seemingly contradictory reports of under- or over-expression of particular miRNAs. A crucial contributing factor to explain this is undoubtedly the vast complexity of brain tissue combined with the extended periods of time in which disease progression occurs in both patients and animal models. Brain tissue is made up of a myriad of neuronal cell types working together in intricate cell networks. Often, one cell performs a rapid response involving biochemical and genetic alterations, while adjacent cells act in an opposite manner to dampen and counter the neighbors' activity. Adding to the complexity is the multifarious array of cell types that both support and overlay their own functions upon these neural networks. These cell types include astrocytes, microglia, and oligodendrocytes that can outnumber neurons by up to 20:1. Therefore, in a tissue sample, transcriptome profiles are representative of numerous cell types which makes the determination of changes specific to a particular cell type difficult. In particular, neurons whose perturbed function likely contributes mostly to the clinical phenotype and eventual death of patients are likely to be swamped by the changes seen in multiplying support cells.

We believe that a potential way to counteract these issues is the use of Laser Capture Microdissection to carefully remove cell bodies from frozen brain sections. This methodology has numerous advantages including: (1) the ability to cut out groups of similarly functioning neurons such as those enriched in hippocampus CA regions; (2) the option of prestaining sections to identify particular cell types, or for example, prion-infected cells; and (3) the potential to apply this technology for tissues from both animal models and human samples. Of course, numerous practical challenges exist such as the requirement for rapid preservation of tissue to ensure miRNA integrity, the requirement for rapid labeling protocols for use prior to dissection, difficulty in obtaining samples completely free of contaminating tissue, and the challenge of working with small quantities of RNAs including the use of RNA amplification techniques. However, a number of reports have shown the efficacy of this type of methodology (Hoefig et al. 2010), and work in our lab has been successful in determining very early changes in miRNA profiles that are specific to hippocampal neurons (manuscript in preparation).

Other methods likely to be useful when looking at animal models of neurodegenerative diseases are new lines of transgenic mice in which specific neuronal populations are tagged and could be isolated following disruption of brain tissue (Livet et al. 2007; Beirowski et al. 2005). Increasingly sophisticated approaches for the targeted expression of Cre suggest that this unique approach can be applied across systems, such as to insert novel markers within specific neuronal populations (Nelson et al. 2006). These markers can then be used to purify tagged cells by techniques such as automated fluorescent cell sorting (FACS) for transcriptome analysis. Crossing these lines with mouse models of neurodegenerative disease should provide exciting new avenues of exploration. Nevertheless, caution should be exercised since tagging proteins may well infer a biological change to the system (Comley et al. 2011).

18.4 The Use of MicroRNAs as Potential Biomarkers for Neurological Disorders

A biomarker is defined as a measurable biological component that is able to discriminate between normal biological processes, pathological states, or pharmacological responses to therapy. An ideal biomarker would also be noninvasive, cost-effective, translatable from animal models to humans, and be detectable early in the disease course. Biomarker discovery has expanded from genomic, transcriptomic, and proteomic avenues to also include miRnomic analysis. For example, miRNA expression profiling to discriminate between cancer and normal tissue (Lodes et al. 2009; Zhao et al. 2010; Bansal et al. 2011), to identify the tissue origin of metastatic cancer (Rosenfeld et al. 2008), and to predict prognosis of disease (Liu et al. 2011) has had some success. Furthermore, different physiological or pathological conditions have been associated with unique "signatures" of deregulated miRNAs detected in bodily fluids (Gilad et al. 2008; Hanke et al. 2010; Weber et al. 2010), highlighting the potential utility of monitoring multiple miRNA biomarkers for disease conditions. Changes in miRNA expression profiles associated with neuro-degenerative diseases may well serve as biomarkers for these disorders which require improved methods of diagnosis.

Cell-free nucleic acids such as miRNAs have been readily detected in numerous body fluids and have been exploited as potentially useful biomarkers of disease (Li et al. 2007; Gilad et al. 2008; Park et al. 2009; Hanke et al. 2010; Weber et al. 2010; Zubakov et al. 2010). Interestingly, miRNAs present in the blood are stabilized by associating with ribonucleoprotein complexes such as RISC or encapsulated in exosomes or microvesicles (Valadi et al. 2007; Hunter et al. 2008; Arroyo et al. 2011). This increased stability of circulating miRNAs is conducive to their exploita-tion as biomarkers over mRNA sampling. Interestingly, tissue-specific miRNAs, such as the brain-enriched miR-124, have been detected in plasma samples following stroke in rats (Laterza et al. 2009). If miRNA markers of brain injury are circulating in blood samples, perhaps certain miRNA-specific markers of neurodegenerative diseases may also be detected in bodily fluids. To date, few studies have been reported in regard to miRNA biomarker discovery for neurodegenerative disease.

Analysis of cerebrospinal fluid (CSF) samples from 10 AD patients revealed the presence of 60 deregulated miRNAs that could be used to distinguish samples from AD versus controls (Cogswell et al. 2008). Perhaps surprisingly, many of these miRNAs have not been observed to be significantly dysregulated in affected brain regions. Some, but not all, of the miRNAs are expressed at high levels in the brain or are highly enriched in the choroid plexus at the interface between blood and CSF; however, others are expressed at very low levels, if at all in brain tissue. Many of the miRNAs detected may be derived from immune cells in the CSF; however, these may still be good indicators of disease.

Profiling miRNAs in blood mononuclear cells (BMCs) from patients with sporadic AD revealed a significant upregulation of many miRNAs including miR-34a, miR-181b, let-7f, and miR-200a (Schipper et al. 2007). Target prediction revealed that transcription/translation and synaptic activity were the functional targets of these miRNAs (Schipper et al. 2007) of which many identified target genes were previously shown to be downregulated in BMCs from AD patients (Maes et al. 2007). It is interesting to note that miR-34a was previously found to be upregulated in AD brain samples (Wang et al. 2009), while let-7f was upregulated in the CSF of AD patients (Cogswell et al. 2008). Perhaps, these miRNAs may be specific markers of Alzheimer's disease.

Due to the autoimmune nature of multiple sclerosis, many publications investigated the presence of deregulated miRNAs in either whole blood or periph-eral blood leukocytes. Specifically, increased levels of miR-326 have been documented in relapsing multiple sclerosis patients, but not in similar diseases, suggesting that miR-326 may be a potential biomarker and/or good therapeutic candidate for MS (Du et al. 2009). Nevertheless, a second study failed to recognize miR-326 as a determinant of MS but instead, identified 165 miRNAs deregulated in patients with relapsing MS as compared with healthy controls. MiR-145 was the single best candidate marker, exhibiting over 89.5% specificity, sensitivity, and accuracy (Keller et al. 2009). However, to date, miRNA-145 failed to be identified in other MS-related studies.

Although these few studies highlight an inconsistency, they nevertheless provide an impetus for further forays into the use of CSF and/or serum miRNAs as biomarkers for early and specific diagnosis of neurodegenerative diseases.

18.5   MicroRNA-Based Therapeutics Targeting Neurodegenerative Diseases

Manipulating the expression levels of disease-related miRNAs in the hope of curing or prolonging disease onset is an exciting avenue of exploration. Introduction of mature miRNA mimics, pre-miRNAs, or lentiviral-based vector systems encoding miRNAs can be used to effectively increase levels of endogenous miRNAs in vitro and in vivo. However, whereas a single cellular target is paramount in conventional drugs, miRNAs have numerous molecular targets raising the possibility of perturbation of multiple "unwanted" cellular functions. Robust promoters may drive the expression of miRNAs to above physiological levels, exacerbating these effects. Furthermore, these abundantly expressed miRNAs may outcompete other endogenous miRNAs for the RISC machinery, resulting in additional off-target physiological effects. For these reasons, adequate dose optimization must be achieved before clinical application.

Conversely, neutralizing the endogenous miRNA functions can be achieved by introducing synthetic anti-microRNAs or "antagomirs." These are single-stranded, antisense oligonucleotides that have perfect sequence complementarity to the miRNA target, thus interfering with miRNA function (Krutzfeldt et al. 2005). The exact mechanism of action remains ambiguous; however, this approach has been used successfully in vivo to prevent gene repression by the targeting miRNA. Modified LNA-antimiRs designed against miR-122 administered over long-term periods successfully reduced the amount of miR-122 and did not show toxicity in nonhuman primates or mice (Elmen et al. 2008a, b). Hence, the use of modified oligonucleotides through noninvasive routes such as intravenous injections or CSF infusion provides a potential nontoxic miRNA-based therapy. Another way to interfere with miRNA function is by scavenging away the miRNA itself by providing competitive binding targets, otherwise known as "sponging" (Ebert et al. 2007). These miRNA decoys contain multiple microRNA recognition elements (MREs) that compete with endogenous miRNA–mRNA interactions. These MRE sequences are plasmid-expressed and driven by strong promoters to effectively neutralize the targeted endogenous miRNAs. A second benefit of this methodology is that it addresses the redundancy often seen between independently expressed members of a miRNA family that all share the same seed sequence. As the interaction between the decoy and the miRNA is based on the seed sequence composition, the effective removal of the entire seed-specific miRNA family is possible (Ebert et al. 2007). Furthermore, multiple endogenous miRNAs may be removed from the system due to the presence of different MRE sequences found on the sponge. This allows for the removal of repressed genes that are targeted by multiple miRNAs. Similarly, tiny LNAs have recently been developed that consist of 8-mer-long LNA oligonucleotides that are complementary to the miRNA seed regions and effectively repress miRNA families with negligible off-target effects (Obad et al. 2011). Furthermore, tiny LNAs can be delivered systemically when combined with phosphorothioate backbone leading to long-lasting silencing of miRNA function (Obad et al. 2011).

The complexity of neurodegenerative diseases poses an immediate challenge to the design of effective miRNA-based therapeutics. The triggers responsible for disease remain poorly understood as do the molecular pathways that lead to neuron damage and death. RNAi-based therapies have previously been used to successfully decrease protein levels implicated in neurodegenerative disease progression. For example, a lentiviral-based siRNA expression vector specific for BACE1 resulted in decreased APP cleavage in vivo (Singer et al. 2005), while targeting mutant allele-specific APP decreased disease-related behavior and pathology (Rodriguez-Lebron et al. 2009). Also, inhibiting the expression of PrP^C interferes with prion disease progression (Bueler et al. 1994; Pfeifer et al. 2006; White et al. 2008) and, surprisingly, reversal of prion neuropathology when administration of RNAi-based treatment occurs early in the course of disease (Mallucci et al. 2003). Targeting α-synuclein in PD by injection of shRNA-specific lentiviral vector systems (Sapru et al. 2006) or infusion of chemically modified siRNA (Lewis et al. 2008) significantly reduced protein levels. Similar methods were employed to target the mutant Htt, reflective of HD, leading to effective allele inhibition and improved neuropathology and behavior associated with disease (Harper et al. 2005; Boudreau et al. 2009). Interestingly, introducing mismatch mutations to the siRNA, which mirror miRNA function, showed an increased selectivity and potency of inhibiting mutant Htt expression (Hu et al. 2010), strongly suggesting that miRNAs may be better candidates for RNA-based therapies. MiRNA drugs have an added advantage over siRNA- or shRNA-based therapies in that they do not appear to have the same issues with toxicity (McBride et al. 2008; Boudreau et al. 2009).

Delivery of any drug to the brain by peripheral administration is extremely challenging. The blood-brain barrier (BBB) is a complex organization of cerebral endothelial cells and their basal lamina, which are surrounded and supported by astrocytes and perivascular macrophages. The BBB effectively protects the CNS from unwanted, and typically harmful, compounds from entering the delicate CNS. For the most part, the BBB is very effective at executing this function, but it poses a considerable challenge for drug delivery to the CNS. Recent advancements include the development of nanoparticle or viral evasion strategies. Numerous delivery shuttles have been designed to successfully deliver siRNAs into the CNS but at low frequencies (Leng et al. 2009). Furthermore, additional hyperfusion chemicals may be needed to effectively open the tight junctions of the BBB, allowing for greater entry of these compounds to the brain. The temporary disruption of the BBB may pose further risks for patients with neurodegenerative disease, for example, allowing access to peripheral immune cells that may aggravate the condition of the patient. One method to surmount these challenges was recently developed by Kumar et al. (2007). This method consists of a 29-amino acid peptide derived from the rabies virus glycoprotein (RVG) that specifically binds to nicotinic acetylcholine receptors (nAchRs) (Lentz et al. 1982; Lafon 2005) and has been found to efficiently transverse the BBB. SiRNA can be bound to an RVG peptide that has been modified with a stretch of 9 synthesized arginine residues (RVG-9R). The intravenous injection of RVG-9R/siRNA complexes was shown to enter the CNS without inducing inflammatory cytokines or antipeptide antibodies (Kumar et al. 2007). However, siRNA can be degraded from these complexes in serum, reducing its efficacy. An alternative strategy would be to encapsulate siRNAs with liposomal nanoparticles shown to increase stability in serum (Leng et al. 2009). To this end, RVG linked to LSPCs (Pulford et al. 2010) or to a disulfide polyethyleneimine (SSPEI) nanomaterial showed increased stability in the blood (Hwang et al. 2011). Successful delivery of miR-124a to the brain was observed when complexes were injected into the blood, while a greater abundance transversed the BBB when mannitol was used to perfuse the brain (Hwang et al. 2011). For all reported RVG-RNA complexes, the small RNA species delivered to the brain conferred functional activity through an unknown release mechanism. It should be noted that the specificity of this delivery system is based on the RVG peptide being able to recognize nAchRs which are not exclusively found in neurons but on other cells such as macrophages (Kim et al. 2010). Perhaps, specificity of RVG complexes for targeting subsets of neurons may be obtained by incorporating additional recognition components to these complexes.

In summary, successful manipulation of miRNA expression levels has been obtained both in in vitro and in vivo systems (highlighted in Fig. 18.1). The specificity of miRNA function and the specific non-invasive delivery to the CNS are two major challenges that remain to be overcome before miRNA-based therapies can be used against neurodegenerative diseases in the clinic. Nevertheless, phase II clinical trials are already underway for the use of LNA-based antisense molecule against miR-122 in Hepatitis C virus affected patients, illustrating the tremendous rate of progress toward novel miRNA drugs since their demonstration in mammals only a decade ago. Although many challenges remain to be resolved, our increasing understanding of the involvement of miRNAs in neurodegenerative diseases presents exciting opportunities for the design of miRNA-based therapies to combat some of humankind's most common and debilitating diseases.

18.6   Future Perspectives

Although a relatively small number of studies have been reported to date in regard to miRNA function in neurodegenerative conditions, several common themes have been uncovered. One of the most intriguing is the proven importance of miRNAs in

a. Examples of miRNA deregulation during AD

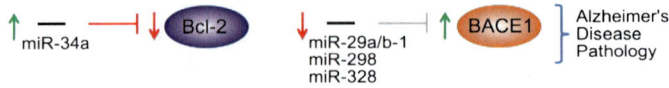

b. Potential use of miRNA therapies

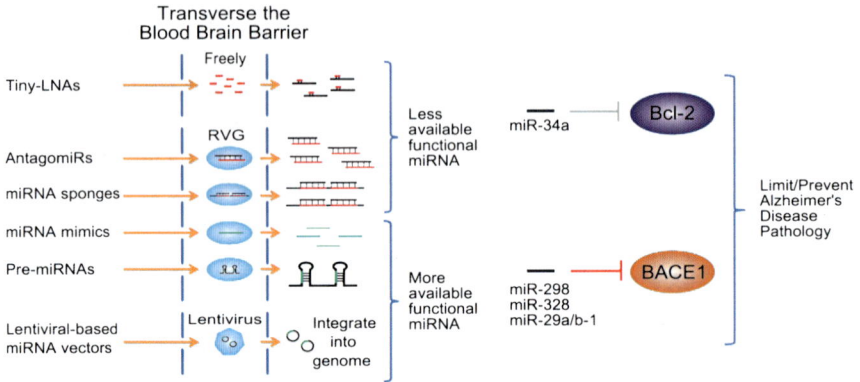

Fig. 18.1 The methods successfully employed for in vivo miRNA manipulation. Some miRNAs that are implicated in Alzheimer's disease are used as examples (**a**) for the potential employment of currently designed therapies to combat this disease (**b**). Overall, *green arrows* represent an upregulation of the indicated miRNA(s)/target protein, while *red arrows* reflect miRNA or target protein downregulation

long-term neuronal function and survival. It is interesting to note that protein misfolding, as observed with TGD-43, is able to ablate miRNA activity by sequestering Drosha-associated proteins in not only ALS disease but also in other neurodegenerative diseases such as AD (Herman et al. 2011; Wilson et al. 2011) and Parkinson's disease (Tian et al. 2011). It is also evident that the function of FMRP is also closely associated with the miRNA biogenesis pathway. FMRP is able to modulate the expression of miRNAs and may be conversely regulated by miRNAs in vivo, a potential contributing pathway to disease pathology. Perhaps miRNA biogenesis and processing machinery is affected by similar mechanisms in other protein misfolding diseases. MiRNA expression deregulation appears to be a common feature between all neurodegenerative diseases, including early stages that occur prior to the development of clinical symptoms. In our own studies on prion disease, the earliest detectable changes in affected neurons may surpass molecular buffering mechanisms evoked by homeostatic circuits, perpetuating a stressed state that ultimately leads to neuronal demise. Identification of some of these changes may provide biomarkers indicative of early stages of disease, such as dysregulation of miR-9/9* in HD and AD. This review represents the tip of the iceberg in the identification of the miRNAs and their targets involved in neurodegeneration. The future of miRNA research will undoubtedly lead to breakthroughs in understanding the diagnosis and treatment of this group of devastating diseases.

References

Aguzzi A, O'Connor T (2010) Protein aggregation diseases: pathogenicity and therapeutic perspectives. Nat Rev Drug Discov 9(3):237–248

Al-Chalabi A, Andersen PM, Nilsson P, Chioza B, Andersson JL, Russ C, Shaw CE, Powell JF, Leigh PN (1999) Deletions of the heavy neurofilament subunit tail in amyotrophic lateral sclerosis. Hum Mol Genet 8(2):157–164

Arai T, Hasegawa M, Akiyama H, Ikeda K, Nonaka T, Mori H, Mann D, Tsuchiya K, Yoshida M, Hashizume Y, Oda T (2006) TDP-43 is a component of ubiquitin-positive tau-negative inclusions in frontotemporal lobar degeneration and amyotrophic lateral sclerosis. Biochem Biophys Res Commun 351(3):602–611

Arroyo JD, Chevillet JR, Kroh EM, Ruf IK, Pritchard CC, Gibson DF, Mitchell PS, Bennett CF, Pogosova-Agadjanyan EL, Stirewalt DL, Tait JF, Tewari M (2011) Argonaute2 complexes carry a population of circulating microRNAs independent of vesicles in human plasma. Proc Natl Acad Sci USA 108(12):5003–5008

Bansal A, Lee IH, Hong X, Anand V, Mathur SC, Gaddam S, Rastogi A, Wani SB, Gupta N, Visvanathan M, Sharma P, Christenson LK (2011) Feasibility of MicroRNAs as biomarkers for Barrett's esophagus progression: a pilot cross-sectional, phase 2 biomarker study. Am J Gastroenterol 106:1055–1063

Bartel DP (2009) MicroRNAs: target recognition and regulatory functions. Cell 136(2):215–233

Beckmann AM, Wilce PA (1997) Egr transcription factors in the nervous system. Neurochem Int 31(4):477–510

Beirowski B, Adalbert R, Wagner D, Grumme DS, Addicks K, Ribchester RR, Coleman MP (2005) The progressive nature of wallerian degeneration in wild-type and slow wallerian degeneration (WldS) nerves. BMC Neurosci 6:6

Bernstein E, Kim SY, Carmell MA, Murchison EP, Alcorn H, Li MZ, Mills AA, Elledge SJ, Anderson KV, Hannon GJ (2003) Dicer is essential for mouse development. Nat Genet 35 (3):215–217

Boissonneault V, Plante I, Rivest S, Provost P (2009) MicroRNA-298 and microRNA-328 regulate expression of mouse beta-amyloid precursor protein-converting enzyme 1. J Biol Chem 284 (4):1971–1981

Boudreau RL, McBride JL, Martins I, Shen S, Xing Y, Carter BJ, Davidson BL (2009) Nonallele-specific silencing of mutant and wild-type huntingtin demonstrates therapeutic efficacy in Huntington's disease mice. Mol Ther 17(6):1053–1063

Bruijn LI, Miller TM, Cleveland DW (2004) Unraveling the mechanisms involved in motor neuron degeneration in ALS. Annu Rev Neurosci 27:723–749

Bueler H, Raeber A, Sailer A, Fischer M, Aguzzi A, Weissmann C (1994) High prion and PrPSc levels but delayed onset of disease in scrapie-inoculated mice heterozygous for a disrupted PrP gene. Mol Med (Cambridge, MA) 1(1):19–30

Buratti E, De Conti L, Stuani C, Romano M, Baralle M, Baralle F (2010) Nuclear factor TDP-43 can affect selected microRNA levels. FEBS J 277(10):2268–2281

Cheever A, Blackwell E, Ceman S (2010) Fragile X protein family member FXR1P is regulated by microRNAs. RNA (New York, NY) 16(8):1530–1539

Choi PS, Zakhary L, Choi WY, Caron S, Alvarez-Saavedra E, Miska EA, McManus M, Harfe B, Giraldez AJ, Horvitz HR, Schier AF, Dulac C (2008) Members of the miRNA-200 family regulate olfactory neurogenesis. Neuron 57(1):41–55.

Cogswell JP, Ward J, Taylor IA, Waters M, Shi Y, Cannon B, Kelnar K, Kemppainen J, Brown D, Chen C, Prinjha RK, Richardson JC, Saunders AM, Roses AD, Richards CA (2008) Identification of miRNA changes in Alzheimer's disease brain and CSF yields putative biomarkers and insights into disease pathways. J Alzheimers Dis 14(1):27–41

Comery TA, Harris JB, Willems PJ, Oostra BA, Irwin SA, Weiler IJ, Greenough WT (1997) Abnormal dendritic spines in fragile X knockout mice: maturation and pruning deficits. Proc Natl Acad Sci USA 94(10):5401–5404

Comley LH, Wishart TM, Baxter B, Murray LM, Nimmo A, Thomson D, Parson SH, Gillingwater
    TH (2011) Induction of cell stress in neurons from transgenic mice expressing yellow fluores-
    cent protein: Implications for neurodegeneration research. PLoS One 6(3):e17639.
Conaco C, Otto S, Han JJ, Mandel G (2006) Reciprocal actions of REST and a microRNA promote
    neuronal identity. Proc Natl Acad Sci USA 103(7):2422–2427
Cox MB, Cairns MJ, Gandhi KS, Carroll AP, Moscovis S, Stewart GJ, Broadley S, Scott RJ, Booth
    DR, Lechner-Scott J, Consortium ANMSG (2010) MicroRNAs miR-17 and miR-20a inhibit T
    cell activation genes and are under-expressed in MS whole blood. PloS One 5(8):e12132
Cui JG, Li YY, Zhao Y, Bhattacharjee S, Lukiw WJ (2010) Differential regulation of interleukin-1
    receptor-associated kinase-1 (IRAK-1) and IRAK-2 by microRNA-146a and NF-kappaB in
    stressed human astroglial cells and in Alzheimer disease. J Biol Chem 285(50):38951–38960
Davis RC, Maloney MT, Minamide LS, Flynn KC, Stonebraker MA, Bamburg JR (2009) Mapping
    cofilin-actin rods in stressed hippocampal slices and the role of cdc42 in amyloid-beta-induced
    rods. J Alzheimers Dis 18(1):35–50
Davis TH, Cuellar TL, Koch SM, Barker AJ, Harfe BD, McManus MT, Ullian EM (2008)
    Conditional loss of dicer disrupts cellular and tissue morphogenesis in the cortex and hippo-
    campus. J Neurosci 28(17):4322–4330
De Pietri Tonelli D, Pulvers JN, Haffner C, Murchison EP, Hannon GJ, Huttner WB (2008)
    miRNAs are essential for survival and differentiation of newborn neurons but not for expansion
    of neural progenitors during early neurogenesis in the mouse embryonic neocortex. Develop-
    ment (Cambridge, England) 135(23):3911–3921
Doxakis E (2010) Post-transcriptional regulation of alpha-synuclein expression by mir-7 and mir-
    153. J Biol Chem 285(17):12726–12734
Du C, Liu C, Kang J, Zhao G, Ye Z, Huang S, Li Z, Wu Z, Pei G (2009) MicroRNA miR-326
    regulates TH-17 differentiation and is associated with the pathogenesis of multiple sclerosis.
    Nat Immunol 10(12):1252–1259
Ebert MS, Neilson JR, Sharp PA (2007) MicroRNA sponges: competitive inhibitors of small
    RNAs in mammalian cells. Nat Methods 4(9):721–726
Edbauer D, Neilson JR, Foster KA, Wang CF, Seeburg DP, Batterton MN, Tada T, Dolan BM,
    Sharp PA, Sheng M (2010) Regulation of synaptic structure and function by FMRP-associated
    microRNAs miR-125b and miR-132. Neuron 65(3):373–384
Eletto D, Russo G, Passiatore G, Del Valle L, Giordano A, Khalili K, Gualco E, Peruzzi F (2008)
    Inhibition of SNAP25 expression by HIV-1 Tat involves the activity of mir-128a. J Cell
    Physiol 216(3):764–770
Elmen J, Lindow M, Schutz S, Lawrence M, Petri A, Obad S, Lindholm M, Hedtjarn M, Hansen
    HF, Berger U, Gullans S, Kearney P, Sarnow P, Straarup EM, Kauppinen S (2008a) LNA-
    mediated microRNA silencing in non-human primates. Nature 452(7189):896–899
Elmen J, Lindow M, Silahtaroglu A, Bak M, Christensen M, Lind-Thomsen A, Hedtjarn M, Hansen
    JB, Hansen HF, Straarup EM, McCullagh K, Kearney P, Kauppinen S (2008b) Antagonism of
    microRNA-122 in mice by systemically administered LNA-antimiR leads to up-regulation of a
    large set of predicted target mRNAs in the liver. Nucleic Acids Res 36(4):1153–1162
Faller M, Toso D, Matsunaga M, Atanasov I, Senturia R, Chen Y, Zhou ZH, Guo F (2010) DGCR8
    recognizes primary transcripts of microRNAs through highly cooperative binding and forma-
    tion of higher-order structures. RNA 16(8):1570–1583
Feng Y, Gutekunst CA, Eberhart DE, Yi H, Warren ST, Hersch SM (1997) Fragile X mental
    retardation protein: nucleocytoplasmic shuttling and association with somatodendritic
    ribosomes. J Neurosci 17(5):1539–1547
Ferri CP, Prince M, Brayne C, Brodaty H, Fratiglioni L, Ganguli M, Hall K, Hasegawa K, Hendrie
    H, Huang Y, Jorm A, Mathers C, Menezes PR, Rimmer E, Scazufca M, Alzheimer's Disease I
    (2005) Global prevalence of dementia: a Delphi consensus study. Lancet 366
    (9503):2112–2117
Figlewicz DA, Krizus A, Martinoli MG, Meininger V, Dib M, Rouleau GA, Julien JP (1994)
    Variants of the heavy neurofilament subunit are associated with the development of
    amyotrophic lateral sclerosis. Hum Mol Genet 3(10):1757–1761

Fratiglioni L, Qiu C (2009) Prevention of common neurodegenerative disorders in the elderly. Exp Gerontol 44(1–2):46–50

Fuchs J, Mueller JC, Lichtner P, Schulte C, Munz M, Berg D, Wullner U, Illig T, Sharma M, Gasser T (2009) The transcription factor PITX3 is associated with sporadic Parkinson's disease. Neurobiol Aging 30(5):731–738

Fukagawa T, Nogami M, Yoshikawa M, Ikeno M, Okazaki T, Takami Y, Nakayama T, Oshimura M (2004) Dicer is essential for formation of the heterochromatin structure in vertebrate cells. Nat Cell Biol 6(8):784–791

Gilad S, Meiri E, Yogev Y, Benjamin S, Lebanony D, Yerushalmi N, Benjamin H, Kushnir M, Cholakh H, Melamed N, Bentwich Z, Hod M, Goren Y, Chajut A (2008) Serum microRNAs are promising novel biomarkers. PloS One 3(9):e3148

Golde TE, Estus S, Usiak M, Younkin LH, Younkin SG (1990) Expression of beta amyloid protein precursor mRNAs: recognition of a novel alternatively spliced form and quantitation in Alzheimer's disease using PCR. Neuron 4(2):253–267

Gregory RI, Yan KP, Amuthan G, Chendrimada T, Doratotaj B, Cooch N, Shiekhattar R (2004) The microprocessor complex mediates the genesis of microRNAs. Nature 432(7014): 235–240

Hanke M, Hoefig K, Merz H, Feller AC, Kausch I, Jocham D, Warnecke JM, Sczakiel G (2010) A robust methodology to study urine microRNA as tumor marker: microRNA-126 and microRNA-182 are related to urinary bladder cancer. Urol Oncol 28(6):655–661

Harada T, Morooka T, Ogawa S, Nishida E (2001) ERK induces p35, a neuron-specific activator of Cdk5, through induction of Egr1. Nat Cell Biol 3(5):453–459

Haramati S, Chapnik E, Sztainberg Y, Eilam R, Zwang R, Gershoni N, McGlinn E, Heiser PW, Wills AM, Wirguin I, Rubin LL, Misawa H, Tabin CJ, Brown R Jr, Chen A, Hornstein E (2010) miRNA malfunction causes spinal motor neuron disease. Proc Natl Acad Sci USA 107(29):13111–13116

Harper SQ, Staber PD, He X, Eliason SL, Martins IH, Mao Q, Yang L, Kotin RM, Paulson HL, Davidson BL (2005) RNA interference improves motor and neuropathological abnormalities in a Huntington's disease mouse model. Proc Natl Acad Sci USA 102(16):5820–5825

Harraz MM, Dawson TM, Dawson VL (2011) MicroRNAs in Parkinson's disease. J Chem Neuroanat 42:127–130

Hauser SL, Oksenberg JR (2006) The neurobiology of multiple sclerosis: genes, inflammation, and neurodegeneration. Neuron 52(1):61–76

Hebert SS, Horre K, Nicolai L, Papadopoulou AS, Mandemakers W, Silahtaroglu AN, Kauppinen S, Delacourte A, De Strooper B (2008) Loss of microRNA cluster miR-29a/b-1 in sporadic Alzheimer's disease correlates with increased BACE1/beta-secretase expression. Proc Natl Acad Sci USA 105(17):6415–6420

Hebert SS, Horre K, Nicolai L, Bergmans B, Papadopoulou AS, Delacourte A, De Strooper B (2009) MicroRNA regulation of Alzheimer's Amyloid precursor protein expression. Neurobiol Dis 33(3):422–428

Herman AM, Khandelwal PJ, Stanczyk BB, Rebeck GW, Moussa CE (2011) beta-Amyloid triggers ALS-associated TDP-43 pathology in AD models. Brain Res 1386:191–199

Hinton VJ, Brown WT, Wisniewski K, Rudelli RD (1991) Analysis of neocortex in three males with the fragile X syndrome. Am J Med Genet 41(3):289–294

Hoefig KP, Heissmeyer V (2010) Measuring microRNA expression in size-limited FACS-sorted and microdissected samples. Methods Mol Biol 667:47–63

Hu J, Liu J, Corey DR (2010) Allele-selective inhibition of huntingtin expression by switching to an miRNA-like RNAi mechanism. Chem Biol 17(11):1183–1188

Hunter MP, Ismail N, Zhang X, Aguda BD, Lee EJ, Yu L, Xiao T, Schafer J, Lee ML, Schmittgen TD, Nana-Sinkam SP, Jarjoura D, Marsh CB (2008) Detection of microRNA expression in human peripheral blood microvesicles. PloS One 3(11):e3694

Hwang DY, Ardayfio P, Kang UJ, Semina EV, Kim KS (2003) Selective loss of dopaminergic neurons in the substantia nigra of Pitx3-deficient aphakia mice. Brain Res Mol Brain Res 114 (2):123–131

Hwang DW, Son S, Jang J, Youn H, Lee S, Lee D, Lee YS, Jeong JM, Kim WJ, Lee DS (2011) A brain-targeted rabies virus glycoprotein-disulfide linked PEI nanocarrier for delivery of neurogenic microRNA. Biomaterials 32:4968–4975

James AB, Conway AM, Morris BJ (2006) Regulation of the neuronal proteasome by Zif268 (Egr1). J Neurosci 26(5):1624–1634

Jin P, Zarnescu DC, Ceman S, Nakamoto M, Mowrey J, Jongens TA, Nelson DL, Moses K, Warren ST (2004) Biochemical and genetic interaction between the fragile X mental retardation protein and the microRNA pathway. Nat Neurosci 7(2):113–117

Johnson R, Zuccato C, Belyaev ND, Guest DJ, Cattaneo E, Buckley NJ (2008) A microRNA-based gene dysregulation pathway in Huntington's disease. Neurobiol Dis 29(3):438–445

Jones MW, Errington ML, French PJ, Fine A, Bliss TV, Garel S, Charnay P, Bozon B, Laroche S, Davis S (2001) A requirement for the immediate early gene Zif268 in the expression of late LTP and long-term memories. Nat Neurosci 4(3):289–296

Junker A, Krumbholz M, Eisele S, Mohan H, Augstein F, Bittner R, Lassmann H, Wekerle H, Hohlfeld R, Meinl E (2009) MicroRNA profiling of multiple sclerosis lesions identifies modulators of the regulatory protein CD47. Brain 132(Pt 12):3342–3352

Junn E, Lee KW, Jeong BS, Chan TW, Im JY, Mouradian MM (2009) Repression of alpha-synuclein expression and toxicity by microRNA-7. Proc Natl Acad Sci USA 106 (31):13052–13057

Kanellopoulou C, Muljo SA, Kung AL, Ganesan S, Drapkin R, Jenuwein T, Livingston DM, Rajewsky K (2005) Dicer-deficient mouse embryonic stem cells are defective in differentiation and centromeric silencing. Genes Dev 19(4):489–501

Kawase-Koga Y, Otaegi G, Sun T (2009) Different timings of Dicer deletion affect neurogenesis and gliogenesis in the developing mouse central nervous system. Dev Dyn 238(11): 2800–2812

Kawase-Koga Y, Low R, Otaegi G, Pollock A, Deng H, Eisenhaber F, Maurer-Stroh S, Sun T (2010) RNAase-III enzyme Dicer maintains signaling pathways for differentiation and survival in mouse cortical neural stem cells. J Cell Sci 123(Pt 4):586–594

Keller A, Leidinger P, Lange J, Borries A, Schroers H, Scheffler M, Lenhof HP, Ruprecht K, Meese E (2009) Multiple sclerosis: microRNA expression profiles accurately differentiate patients with relapsing-remitting disease from healthy controls. PloS One 4(10):e7440

Kim J, Inoue K, Ishii J, Vanti WB, Voronov SV, Murchison E, Hannon G, Abeliovich A (2007) A MicroRNA feedback circuit in midbrain dopamine neurons. Science (New York, NY) 317 (5842):1220–1224

Kim SS, Ye C, Kumar P, Chiu I, Subramanya S, Wu H, Shankar P, Manjunath N (2010) Targeted delivery of siRNA to macrophages for anti-inflammatory treatment. Mol Ther 18(5):993–1001

Kinchen JM, Ravichandran KS (2008) Phagocytic signaling: you can touch, but you can't eat. Curr Biol 18(12):R521–524

Klein ME, Lioy DT, Ma L, Impey S, Mandel G, Goodman RH (2007) Homeostatic regulation of MeCP2 expression by a CREB-induced microRNA. Nat Neurosci 10(12):1513–1514

Krutzfeldt J, Rajewsky N, Braich R, Rajeev KG, Tuschl T, Manoharan M, Stoffel M (2005) Silencing of microRNAs in vivo with 'antagomirs'. Nature 438(7068):685–689

Kumar P, Wu H, McBride JL, Jung KE, Kim MH, Davidson BL, Lee SK, Shankar P, Manjunath N (2007) Transvascular delivery of small interfering RNA to the central nervous system. Nature 448(7149):39–43

Kwiatkowski TJ Jr, Bosco DA, Leclerc AL, Tamrazian E, Vanderburg CR, Russ C, Davis A, Gilchrist J, Kasarskis EJ, Munsat T, Valdmanis P, Rouleau GA, Hosler BA, Cortelli P, de Jong PJ, Yoshinaga Y, Haines JL, Pericak-Vance MA, Yan J, Ticozzi N, Siddique T, McKenna-Yasek D, Sapp PC, Horvitz HR, Landers JE, Brown RH Jr (2009) Mutations in the FUS/TLS gene on chromosome 16 cause familial amyotrophic lateral sclerosis. Science (New York, NY) 323(5918):1205–1208

Lafon M (2005) Rabies virus receptors. J Neurovirol 11(1):82–87

Laggerbauer B, Ostareck D, Keidel EM, Ostareck-Lederer A, Fischer U (2001) Evidence that fragile X mental retardation protein is a negative regulator of translation. Hum Mol Genet 10 (4):329–338

Laterza OF, Lim L, Garrett-Engele PW, Vlasakova K, Muniappa N, Tanaka WK, Johnson JM, Sina JF, Fare TL, Sistare FD, Glaab WE (2009) Plasma MicroRNAs as sensitive and specific biomarkers of tissue injury. Clin Chem 55(11):1977–1983

Leng Q, Woodle MC, Lu PY, Mixson AJ (2009) Advances in systemic siRNA delivery. Drugs Future 34(9):721

Lentz TL, Burrage TG, Smith AL, Crick J, Tignor GH (1982) Is the acetylcholine receptor a rabies virus receptor? Science (New York, NY) 215(4529):182–184

Lewis J, Melrose H, Bumcrot D, Hope A, Zehr C, Lincoln S, Braithwaite A, He Z, Ogholikhan S, Hinkle K, Kent C, Toudjarska I, Charisse K, Braich R, Pandey RK, Heckman M, Maraganore DM, Crook J, Farrer MJ (2008) In vivo silencing of alpha-synuclein using naked siRNA. Mol Neurodegener 3:19

Li J, Smyth P, Flavin R, Cahill S, Denning K, Aherne S, Guenther SM, O'Leary JJ, Sheils O (2007) Comparison of miRNA expression patterns using total RNA extracted from matched samples of formalin-fixed paraffin-embedded (FFPE) cells and snap frozen cells. BMC Biotechnol 7:36

Lin SL, Chang SJ, Ying SY (2006) First in vivo evidence of microRNA-induced fragile X mental retardation syndrome. Mol Psychiatr 11(7):616–617

Lindberg RL, Hoffmann F, Mehling M, Kuhle J, Kappos L (2010) Altered expression of miR-17-5p in CD4+ lymphocytes of relapsing-remitting multiple sclerosis patients. Eur J Immunol 40(3):888–898

Ling SC, Albuquerque CP, Han JS, Lagier-Tourenne C, Tokunaga S, Zhou H, Cleveland DW (2010) ALS-associated mutations in TDP-43 increase its stability and promote TDP-43 complexes with FUS/TLS. Proc Natl Acad Sci USA 107(30):13318–13323

Liu XG, Zhu WY, Huang YY, Ma LN, Zhou SQ, Wang YK, Zeng F, Zhou JH, Zhang YK (2011) High expression of serum miR-21 and tumor miR-200c associated with poor prognosis in patients with lung cancer. Med Oncol (Northwood, London, England) (Epub ahead of print)

Livet J, Weissman TA, Kang H, Draft RW, Lu J, Bennis RA, Sanes JR, Lichtman JW (2007) Transgenic strategies for combinatorial expression of fluorescent proteins in the nervous system. Nature 450(7166):56–62

Lloyd S, Mead S, Collinge J (2011) Genetics of prion disease. Top Curr Chem 305:1–22

Lodes MJ, Caraballo M, Suciu D, Munro S, Kumar A, Anderson B (2009) Detection of cancer with serum miRNAs on an oligonucleotide microarray. PloS One 4(7):e6229

Long JM, Lahiri DK (2011) MicroRNA-101 downregulates Alzheimer's amyloid-beta precursor protein levels in human cell cultures and is differentially expressed. Biochem Biophys Res Commun 404(4):889–895

Lukiw WJ (2007) Micro-RNA speciation in fetal, adult and Alzheimer's disease hippocampus. Neuroreport 18(3):297–300

Lukiw WJ, Zhao Y, Cui JG (2008) An NF-kappaB-sensitive micro RNA-146a-mediated inflammatory circuit in Alzheimer disease and in stressed human brain cells. J Biol Chem 283 (46):31315–31322

Lunyak VV, Burgess R, Prefontaine GG, Nelson C, Sze SH, Chenoweth J, Schwartz P, Pevzner PA, Glass C, Mandel G, Rosenfeld MG (2002) Corepressor-dependent silencing of chromosomal regions encoding neuronal genes. Science 298(5599):1747–1752

Maes OC, Xu S, Yu B, Chertkow HM, Wang E, Schipper HM (2007) Transcriptional profiling of Alzheimer blood mononuclear cells by microarray. Neurobiol Aging 28(12):1795–1809

Majer A, Booth SA (2010) Computational methodologies for studying non-coding RNAs relevant to central nervous system function and dysfunction. Brain Res 1338:131–145

Makeyev EV, Zhang J, Carrasco MA, Maniatis T (2007) The MicroRNA miR-124 promotes neuronal differentiation by triggering brain-specific alternative pre-mRNA splicing. Mol Cell 27(3):435–448

Mallucci G, Dickinson A, Linehan J, Klohn PC, Brandner S, Collinge J (2003) Depleting neuronal PrP in prion infection prevents disease and reverses spongiosis. Science (New York, NY) 302 (5646):871–874

Maloney MT, Bamburg JR (2007) Cofilin-mediated neurodegeneration in Alzheimer's disease and other amyloidopathies. Mol Neurobiol 35(1):21–44

Marti E, Pantano L, Banez-Coronel M, Llorens F, Minones-Moyano E, Porta S, Sumoy L, Ferrer I, Estivill X (2010) A myriad of miRNA variants in control and Huntington's disease brain regions detected by massively parallel sequencing. Nucleic Acids Res 38(20):7219–7235

McBride JL, Boudreau RL, Harper SQ, Staber PD, Monteys AM, Martins I, Gilmore BL, Burstein H, Peluso RW, Polisky B, Carter BJ, Davidson BL (2008) Artificial miRNAs mitigate shRNA-mediated toxicity in the brain: Implications for the therapeutic development of RNAi. Proc Natl Acad Sci U S A 105(15):5868–5873

Mellios N, Huang HS, Grigorenko A, Rogaev E, Akbarian S (2008) A set of differentially expressed miRNAs, including miR-30a-5p, act as post-transcriptional inhibitors of BDNF in prefrontal cortex. Hum Mol Genet 17(19):3030–3042

Minamide LS, Striegl AM, Boyle JA, Meberg PJ, Bamburg JR (2000) Neurodegenerative stimuli induce persistent ADF/cofilin-actin rods that disrupt distal neurite function. Nat Cell Biol 2 (9):628–636

Montag J, Hitt R, Opitz L, Schulz-Schaeffer WJ, Hunsmann G, Motzkus D (2009) Upregulation of miRNA hsa-miR-342–3p in experimental and idiopathic prion disease. Mol Neurodegener 4:36

Mulder DW, Kurland LT, Offord KP, Beard CM (1986) Familial adult motor neuron disease: amyotrophic lateral sclerosis. Neurology 36(4):511–517

Nelson SB, Sugino K, Hempel CM (2006) The problem of neuronal cell types: A physiological genomics approach. Trends Neurosci 29(6):339–345

Neve RL, Rogers J, Higgins GA (1990) The Alzheimer amyloid precursor-related transcript lacking the beta/A4 sequence is specifically increased in Alzheimer's disease brain. Neuron 5(3):329–338

Nimchinsky EA, Oberlander AM, Svoboda K (2001) Abnormal development of dendritic spines in FMR1 knock-out mice. J Neurosci 21(14):5139–5146

Nunez-Iglesias J, Liu CC, Morgan TE, Finch CE, Zhou XJ (2010) Joint genome-wide profiling of miRNA and mRNA expression in Alzheimer's disease cortex reveals altered miRNA regulation. PloS One 5(2):e8898

O'Brien RJ, Wong PC (2010) Amyloid precursor protein processing and Alzheimers disease. Annu Rev Neurosci 34:185–204

Obad S, dos Santos CO, Petri A, Heidenblad M, Broom O, Ruse C, Fu C, Lindow M, Stenvang J, Straarup EM, Hansen HF, Koch T, Pappin D, Hannon GJ, Kauppinen S (2011) Silencing of microRNA families by seed-targeting tiny LNAs. Nat Genet 43(4):371–378

Otaegui D, Baranzini SE, Armananzas R, Calvo B, Munoz-Culla M, Khankhanian P, Inza I, Lozano JA, Castillo-Trivino T, Asensio A, Olaskoaga J, Lopez de Munain A (2009) Differential micro RNA expression in PBMC from multiple sclerosis patients. PloS One 4(7):e6309

Ovcharenko D, Kelnar K, Johnson C, Leng N, Brown D (2007) Genome-scale microRNA and small interfering RNA screens identify small RNA modulators of TRAIL-induced apoptosis pathway. Cancer Res 67(22):10782–10788

Packer AN, Xing Y, Harper SQ, Jones L, Davidson BL (2008) The bifunctional microRNA miR-9/miR-9* regulates REST and CoREST and is downregulated in Huntington's disease. J Neurosci 28(53):14341–14346

Park NJ, Zhou H, Elashoff D, Henson BS, Kastratovic DA, Abemayor E, Wong DT (2009) Salivary microRNA: discovery, characterization, and clinical utility for oral cancer detection. Clin Cancer Res 15(17):5473–5477

Patel N, Hoang D, Miller N, Ansaloni S, Huang Q, Rogers JT, Lee JC, Saunders AJ (2008) MicroRNAs can regulate human APP levels. Mol Neurodegener 3:10

Pfeifer A, Eigenbrod S, Al-Khadra S, Hofmann A, Mitteregger G, Moser M, Bertsch U, Kretzschmar H (2006) Lentivector-mediated RNAi efficiently suppresses prion protein and prolongs survival of scrapie-infected mice. J Clin Investig 116(12):3204–3210

Pieretti M, Zhang FP, Fu YH, Warren ST, Oostra BA, Caskey CT, Nelson DL (1991) Absence of expression of the FMR-1 gene in fragile X syndrome. Cell 66(4):817–822

Plante I, Davidovic L, Ouellet DL, Gobeil LA, Tremblay S, Khandjian EW, Provost P (2006) Dicer-derived microRNAs are utilized by the fragile X mental retardation protein for assembly on target RNAs. J Biomed Biotechnol 2006(4):64347

Pulford B, Reim N, Bell A, Veatch J, Forster G, Bender H, Meyerett C, Hafeman S, Michel B, Johnson T, Wyckoff AC, Miele G, Julius C, Kranich J, Schenkel A, Dow S, Zabel MD (2010) Liposome-siRNA-peptide complexes cross the blood-brain barrier and significantly decrease PrP on neuronal cells and PrP in infected cell cultures. PloS One 5(6):e11085

Raske C, Hagerman PJ (2009) Molecular pathogenesis of fragile X-associated tremor/ataxia syndrome. J Investig Med 57(8):825–829

Rockenstein EM, McConlogue L, Tan H, Power M, Masliah E, Mucke L (1995) Levels and alternative splicing of amyloid beta protein precursor (APP) transcripts in brains of APP transgenic mice and humans with Alzheimer's disease. J Biol Chem 270(47):28257–28267

Rodriguez-Lebron E, Gouvion CM, Moore SA, Davidson BL, Paulson HL (2009) Allele-specific RNAi mitigates phenotypic progression in a transgenic model of alzheimer's disease. Mol Ther 17(9):1563–1573

Rosenfeld N, Aharonov R, Meiri E, Rosenwald S, Spector Y, Zepeniuk M, Benjamin H, Shabes N, Tabak S, Levy A, Lebanony D, Goren Y, Silberschein E, Targan N, Ben-Ari A, Gilad S, Sion-Vardy N, Tobar A, Feinmesser M, Kharenko O, Nativ O, Nass D, Perelman M, Yosepovich A, Shalmon B, Polak-Charcon S, Fridman E, Avniel A, Bentwich I, Bentwich Z, Cohen D, Chajut A, Barshack I (2008) MicroRNAs accurately identify cancer tissue origin. Nat Biotechnol 26(4):462–469

Saba R, Goodman CD, Huzarewich RL, Robertson C, Booth SA (2008) A miRNA signature of prion induced neurodegeneration. PLoS One 3(11):e3652

Sapru MK, Yates JW, Hogan S, Jiang L, Halter J, Bohn MC (2006) Silencing of human alpha-synuclein in vitro and in rat brain using lentiviral-mediated RNAi. Exp Neurol 198(2):382–390

Savas JN, Makusky A, Ottosen S, Baillat D, Then F, Krainc D, Shiekhattar R, Markey SP, Tanese N (2008) Huntington's disease protein contributes to RNA-mediated gene silencing through association with Argonaute and P bodies. Proc Natl Acad Sci USA 105(31):10820–10825

Savitt JM, Dawson VL, Dawson TM (2006) Diagnosis and treatment of Parkinson disease: molecules to medicine. J Clin Investig 116(7):1744–1754

Schaefer A, O'Carroll D, Tan CL, Hillman D, Sugimori M, Llinas R, Greengard P (2007) Cerebellar neurodegeneration in the absence of microRNAs. J Exp Med 204(7):1553–1558

Schipper HM, Maes OC, Chertkow HM, Wang E (2007) MicroRNA expression in Alzheimer blood mononuclear cells. Gene Regul Syst Bio 1:263–274

Sellier C, Hagerman P, Willemsen R, Charlet-Berguerand N (2010) DROSHA/DGCR8 sequestration by expanded CGG repeats leads to global micro-RNA processing alteration in FXTAS patients. 12th International Fragile X Conference; July 21–25; Detroit, MI

Shin D, Shin JY, McManus MT, Ptacek LJ, Fu YH (2009) Dicer ablation in oligodendrocytes provokes neuronal impairment in mice. Ann Neurol 66(6):843–857

Shioya M, Obayashi S, Tabunoki H, Arima K, Saito Y, Ishida T, Satoh J (2010) Aberrant microRNA expression in the brains of neurodegenerative diseases: miR-29a decreased in Alzheimer disease brains targets neurone navigator 3. Neuropathol Appl Neurobiol 36(4):320–330

Singer O, Marr RA, Rockenstein E, Crews L, Coufal NG, Gage FH, Verma IM, Masliah E (2005) Targeting BACE1 with siRNAs ameliorates alzheimer disease neuropathology in a transgenic model. Nat Neurosci 8(10):1343–1349

Smith P, Al Hashimi A, Girard J, Delay C, Hebert SS (2011) In vivo regulation of amyloid precursor protein neuronal splicing by microRNAs. J Neurochem 116(2):240–247

Sonkoly E, Pivarcsi A (2009) microRNAs in inflammation. Int Rev Immunol 28(6):535–561

Sorensen G, Medina S, Parchaliuk D, Phillipson C, Robertson C, Booth SA (2008) Comprehensive transcriptional profiling of prion infection in mouse models reveals networks of responsive genes. BMC Genomics 9:114

The Huntington's Disease Collaborative Research Group (1993) A novel gene containing a trinucleotide repeat that is expanded and unstable on Huntington's disease chromosomes. Cell 72(6):971–983

Tian T, Huang C, Tong J, Yang M, Zhou H, Xia XG (2011) TDP-43 potentiates alpha-synuclein toxicity to dopaminergic neurons in transgenic mice. Int J Biol Sci 7(2):234–243

Tsang J, Zhu J, van Oudenaarden A (2007) MicroRNA-mediated feedback and feedforward loops are recurrent network motifs in mammals. Mol Cell 26(5):753–767

Valadi H, Ekstrom K, Bossios A, Sjostrand M, Lee JJ, Lotvall JO (2007) Exosome-mediated transfer of mRNAs and microRNAs is a novel mechanism of genetic exchange between cells. Nat Cell Biol 9(6):654–659

Vance C, Rogelj B, Hortobagyi T, De Vos KJ, Nishimura AL, Sreedharan J, Hu X, Smith B, Ruddy D, Wright P, Ganesalingam J, Williams KL, Tripathi V, Al-Saraj S, Al-Chalabi A, Leigh PN, Blair IP, Nicholson G, de Belleroche J, Gallo JM, Miller CC, Shaw CE (2009) Mutations in FUS, an RNA processing protein, cause familial amyotrophic lateral sclerosis type 6. Science (New York, NY) 323(5918):1208–1211

Verkerk AJ, Pieretti M, Sutcliffe JS, Fu YH, Kuhl DP, Pizzuti A, Reiner O, Richards S, Victoria MF, Zhang FP (1991) Identification of a gene (FMR-1) containing a CGG repeat coincident with a breakpoint cluster region exhibiting length variation in fragile X syndrome. Cell 65 (5):905–914

Vilardo E, Barbato C, Ciotti M, Cogoni C, Ruberti F (2010) MicroRNA-101 regulates amyloid precursor protein expression in hippocampal neurons. J Biol Chem 285(24):18344–18351

Visvanathan J, Lee S, Lee B, Lee JW, Lee SK (2007) The microRNA miR-124 antagonizes the anti-neural REST/SCP1 pathway during embryonic CNS development. Genes Dev 21 (7):744–749

Wang G, van der Walt JM, Mayhew G, Li YJ, Zuchner S, Scott WK, Martin ER, Vance JM (2008a) Variation in the miRNA-433 binding site of FGF20 confers risk for Parkinson disease by overexpression of alpha-synuclein. Am J Hum Genet 82(2):283–289

Wang WX, Rajeev BW, Stromberg AJ, Ren N, Tang G, Huang Q, Rigoutsos I, Nelson PT (2008b) The expression of microRNA miR-107 decreases early in Alzheimer's disease and may accelerate disease progression through regulation of beta-site amyloid precursor protein-cleaving enzyme 1. J Neurosci 28(5):1213–1223

Wang X, Liu P, Zhu H, Xu Y, Ma C, Dai X, Huang L, Liu Y, Zhang L, Qin C (2009) miR-34a, a microRNA up-regulated in a double transgenic mouse model of Alzheimer's disease, inhibits bcl2 translation. Brain Res Bull 80(4–5):268–273

Wang H, Liu J, Zong Y, Xu Y, Deng W, Zhu H, Liu Y, Ma C, Huang L, Zhang L, Qin C (2010) miR-106b aberrantly expressed in a double transgenic mouse model for Alzheimer's disease targets TGF-beta type II receptor. Brain Res 1357:166–174

Wang WX, Huang Q, Hu Y, Stromberg AJ, Nelson PT (2011) Patterns of microRNA expression in normal and early alzheimer's disease human temporal cortex: White matter versus gray matter. Acta Neuropathol 121(2):193–205

Weber JA, Baxter DH, Zhang S, Huang DY, Huang KH, Lee MJ, Galas DJ, Wang K (2010) The microRNA spectrum in 12 body fluids. Clin Chem 56(11):1733–1741

White MD, Farmer M, Mirabile I, Brandner S, Collinge J, Mallucci GR (2008) Single treatment with RNAi against prion protein rescues early neuronal dysfunction and prolongs survival in mice with prion disease. Proc Natl Acad Sci USA 105(29):10238–10243

Whiteman IT, Gervasio OL, Cullen KM, Guillemin GJ, Jeong EV, Witting PK, Antao ST, Minamide LS, Bamburg JR, Goldsbury C (2009) Activated actin-depolymerizing factor/cofilin sequesters phosphorylated microtubule-associated protein during the assembly of Alzheimer-like neuritic cytoskeletal striations. J Neurosci 29(41):12994–13005

Willemsen R, Hoogeveen-Westerveld M, Reis S, Holstege J, Severijnen LA, Nieuwenhuizen IM, Schrier M, van Unen L, Tassone F, Hoogeveen AT, Hagerman PJ, Mientjes EJ, Oostra BA (2003) The FMR1 CGG repeat mouse displays ubiquitin-positive intranuclear neuronal inclusions; implications for the cerebellar tremor/ataxia syndrome. Hum Mol Genet 12(9):949–959

Williams AH, Valdez G, Moresi V, Qi X, McAnally J, Elliott JL, Bassel-Duby R, Sanes JR, Olson EN (2009) MicroRNA-206 delays ALS progression and promotes regeneration of neuromuscular synapses in mice. Science (New York, NY) 326(5959):1549–1554

Wilson AC, Dugger BN, Dickson DW, Wang DS (2011) TDP-43 in aging and Alzheimer's disease - a review. Int J Clin Exp Pathol 4(2):147–155

Wu J, Xie X (2006) Comparative sequence analysis reveals an intricate network among REST, CREB and miRNA in mediating neuronal gene expression. Genome Biol 7(9):R85

Xiang Z, Ho L, Yemul S, Zhao Z, Qing W, Pompl P, Kelley K, Dang A, Teplow D, Pasinetti GM (2002) Cyclooxygenase-2 promotes amyloid plaque deposition in a mouse model of Alzheimer's disease neuropathology. Gene Expr 10(5–6):271–278

Xu XL, Li Y, Wang F, Gao FB (2008) The steady-state level of the nervous-system-specific microRNA-124a is regulated by dFMR1 in Drosophila. J Neurosci 28(46):11883–11889

Yao J, Hennessey T, Flynt A, Lai E, Beal MF, Lin MT (2010) MicroRNA-related cofilin abnormality in Alzheimer's disease. PloS One 5(12):e15546

Yi YH, Sun XS, Qin JM, Zhao QH, Liao WP, Long YS (2010) Experimental identification of microRNA targets on the 3' untranslated region of human FMR1 gene. J Neurosci Methods 190(1):34–38

Zesiewicz TA, Sullivan KL, Arnulf I, Chaudhuri KR, Morgan JC, Gronseth GS, Miyasaki J, Iverson DJ, Weiner WJ, Quality Standards Subcommittee of the American Academy of N (2010) Practice Parameter: treatment of nonmotor symptoms of Parkinson disease: report of the quality standards subcommittee of the American academy of neurology. Neurology 74 (11):924–931

Zhao H, Shen J, Medico L, Wang D, Ambrosone CB, Liu S (2010) A pilot study of circulating miRNAs as potential biomarkers of early stage breast cancer. PloS One 5(10):e13735

Zong Y, Wang H, Dong W, Quan X, Zhu H, Xu Y, Huang L, Ma C, Qin C (2011) miR-29c regulates BACE1 protein expression. Brain Res 1395:108–115

Zubakov D, Boersma AW, Choi Y, van Kuijk PF, Wiemer EA, Kayser M (2010) MicroRNA markers for forensic body fluid identification obtained from microarray screening and quantitative RT-PCR confirmation. Int J Legal Med 124(3):217–226

Zuccato C, Ciammola A, Rigamonti D, Leavitt BR, Goffredo D, Conti L, MacDonald ME, Friedlander RM, Silani V, Hayden MR, Timmusk T, Sipione S, Cattaneo E (2001) Loss of huntingtin-mediated BDNF gene transcription in Huntington's disease. Science (New York, NY) 293(5529):493–498

Zuccato C, Belyaev N, Conforti P, Ooi L, Tartari M, Papadimou E, MacDonald M, Fossale E, Zeitlin S, Buckley N, Cattaneo E (2007) Widespread disruption of repressor element-1 silencing transcription factor/neuron-restrictive silencer factor occupancy at its target genes in Huntington's disease. J Neurosci 27(26):6972–6983

Chapter 19
siRNA Therapeutic Design: Tools and Challenges

Amanda P. Malefyt, Phillip A. Angart, Christina Chan, and S. Patrick Walton

Abstract Current RNA-based therapeutics are principally focused toward activating the RNA interference (RNAi) pathway through exogenous administration of short interfering RNAs (siRNAs) and sometimes short hairpin RNAs (shRNAs). The promise of RNAi-based therapeutics arises from their broad applicability and excellent specificity. This chapter reviews siRNA design strategies for improving intracellular interactions with the RNAi pathway proteins as well as key characteristics required for the design of optimal delivery vehicles to maximize specific silencing in only the cells of interest. The status of previous and ongoing clinical trials will be described as these provide insight for overcoming future challenges for long-term use of RNAi as a therapeutic modality.

Keywords Asymmetry • clinical trials • immune response • nanoparticles • nanotechnology • nucleic acid delivery • off-target effects • RNAi • siRNA • therapeutics

19.1 Introduction

While the list of regulatory RNAs in human cells continues to grow, the principal focus for RNA-based therapeutics remains on leveraging the RNA interference (RNAi) pathway through the exogenous delivery of short interfering RNAs (siRNAs) and, to a lesser degree, short hairpin RNAs (shRNAs). The versatility of RNAi to target essentially any protein and treat a diversity of diseases makes it an ideal addition to the current repertoire of therapeutic modalities that principally includes small molecules and recombinant proteins.

S. Patrick Walton (✉)
Department of Chemical Engineering and Materiels Science, Michigan State University, East Lansing, MI, USA
e-mail: spwalton@egr.msu.edu

B. Mallick (eds.), *Regulatory RNAs*, DOI 10.1007/978-3-642-22517-8_19,
© Springer-Verlag Berlin Heidelberg (outside the USA) 2012

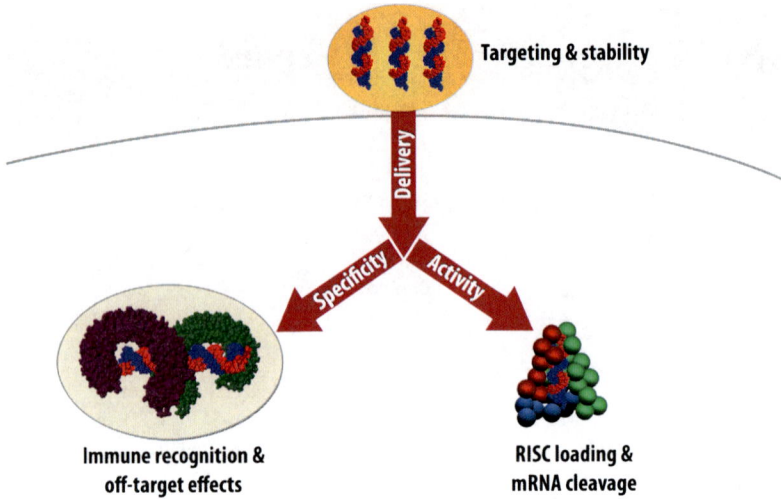

Fig. 19.1 siRNA therapeutic development. The development of optimized siRNA therapeutics must focus on three related but distinct functions: delivery of active siRNAs to the tissues and cells of interest, strong silencing activity against the target of interest, and avoiding nonspecific effects.

In this chapter, we will focus on therapeutic applications of RNAi and the types of choices that must be made in the development of this new class of therapeutics (Fig. 19.1). Foci include the features of the siRNA molecule itself that are important for maximizing silencing activity and how to design delivery vehicles to transport siRNAs to their intended location. We will also describe the status of RNAi-based therapeutics currently in clinical trials, as well as the challenges that need to be overcome for the long-term success of future clinical trials.

19.2 siRNA Design

In RNAi, the siRNA is both the initiator of the pathway as well as the component that provides target gene specificity. siRNAs, which are double-stranded ribonucleic acids (dsRNAs), must interact with a series of proteins that select one of the two strands as the active strand. Throughout the pathway, siRNAs, in both double-stranded and single-stranded forms, interact with a variety of cellular proteins and the target mRNA. Design rules for siRNAs then should maximize the useful interactions with the proteins and target while minimizing those that lead to reduced specific activity or increased nonspecific activity. A firm grasp of the RNAi mechanism is necessary for the construction of a complete set of design rules.

19.2.1 Details of the RNAi Mechanism

In human cells, the minimal components of RNAi required to silence the expression of a gene are a single-stranded siRNA and the protein argonaute 2 (Ago2) (Rivas et al. 2005; Liu et al. 2004). This RNA–protein complex is called the active RNA-induced silencing complex (RISC) and cleaves the target mRNA at the center of the region of the mRNA that is complementary to the siRNA strand (Hammond et al. 2000; Rivas et al. 2005; Elbashir et al. 2001b; Leuschner et al. 2006). When initiated with double-stranded siRNAs, formation of an active RISC involves the loading of one siRNA strand, called the guide strand, into Ago2, while the other strand, the passenger strand, is nicked by Ago2 to yield the active RISC (Tomari et al. 2004b; Miyoshi et al. 2005; Matranga et al. 2005; Rand et al. 2005; Leuschner et al. 2006). As either siRNA strand can be loaded into the active RISC, to ensure correct targeting, the strand selected as the guide strand must be the strand that is complementary to the intended target mRNA and, moreover, the targeted sequence must be unique within the transcriptome. If the incorrect strand is loaded into RISC (or if the correct strand is complementary to multiple target mRNAs), it can result in deleterious "off-target" effects (Jackson et al. 2003; Jackson and Linsley 2010). While Ago2 is minimally required for formation of an active RISC, other proteins and protein complexes, including Dicer, TRBP, PACT, and C3PO, are known to be central to the pathway or closely associated with proteins central to the pathway, with their functional roles still incompletely defined (Liu et al. 2004; Hammond et al. 2000; Kok et al. 2007; Lee et al. 2006; Ye et al. 2011; Liu et al. 2009). However, in some cases, it has been found that these proteins interact more favorably with siRNAs of particular sequences or structures. The remainder of this section focuses on the characteristics of the siRNA that are known to be recognized by RNAi pathway proteins and, hence, can be applied to designing siRNAs for optimal activity and selectivity in humans.

19.2.2 General Structural Features of siRNAs

The unique structure of siRNAs, ~19 bp duplexes with 2 nt overhangs on each $3'$ end, is a result of the endonucleolytic processing of longer dsRNAs by Dicer (Fig. 19.2) (Bernstein et al. 2001; Zamore et al. 2000; Elbashir et al. 2001a; Zhang et al. 2002a, 2004; Lima et al. 2009; Sakurai et al. 2011). To be recognized by the pathway proteins, siRNAs must also have a $5'$ phosphate (Nykänen et al. 2001). Fortuitously, synthetic siRNAs are rapidly phosphorylated upon entry to cells by the protein Clp1 (Weitzer and Martinez 2007). Thus, exogenous siRNAs synthesized with the canonical siRNA structure are active as silencers (Lima et al. 2009; Sakurai et al. 2011). For additional information, please see the chapter 5 (this volume).

Passenger Strand (Sense)

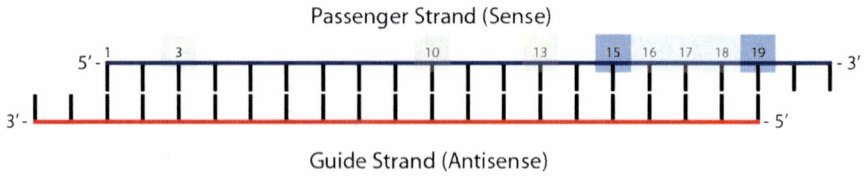

Guide Strand (Antisense)

Fig. 19.2 Structure of an siRNA. The canonical siRNA structure consists of two complementary RNA strands with 19 bp and two nucleotide overhangs on each 3′ end. Current design criteria involve nucleotide preferences at specific positions (indicated by *shading*) as well as general recommendations for overall and terminal base content (see Table 19.1).

19.2.3   Factors Influencing Silencing Activity

Understanding the silencing activity of an siRNA requires knowledge of how siRNAs activate the pathway and how siRNAs and the pathway proteins subsequently mediate silencing. Investigation of these molecular scale interactions is guided in part by large dataset analyses, which can be used to find features characteristic of active and inactive siRNAs (Fig. 19.2 and Table 19.1) (Reynolds et al. 2004; Ui-Tei et al. 2004; Huesken et al. 2005; Shabalina et al. 2006). Analysis of these datasets has helped discern the impact of positional base preferences, overall hybridization stability, and local hybridization stability. Design of an siRNA with knowledge of these interactions can be used to minimize the recognition of siRNAs by cellular immune responses, ensure proper strand selection, and maximize RISC turnover and stability.

19.2.3.1   Asymmetry

One of the major challenges in siRNA design is ensuring selection of the antisense strand as the guide strand. Selective incorporation of one of the siRNA strands is referred to as asymmetry. TRBP, Dicer, and Ago2 directionally bind siRNAs based upon differences localized to the termini of the siRNA (Gredell et al. 2010; Noland et al. 2011; Frank et al. 2010; Matranga et al. 2005), although there are discrepancies as to what factors contribute to asymmetry and the roles of the various pathway proteins in the recognition of asymmetry. In *Drosophila*, R2D2 (a TRBP homolog) senses and preferentially binds to the termini with the greater hybridization stability, thereby orienting the less stably hybridized end to be bound by Dcr-2 (a Dicer homolog), followed by association of Ago2 with the ternary complex (Liu 2003; Tomari et al. 2004a, b; Liu et al. 2004). It is thought that the end with greater hybridization stability adopts a more stable A-form helix that is more strongly bound by the dsRNA binding domains present in R2D2 (Tomari et al. 2004a, b). In humans, asymmetry sensing and the formation of an active RISC are less well understood, though it is known to be different in some ways than the *Drosophila* pathway (Wang et al. 2009; Sakurai et al. 2011). Methods used to

Table 19.1 Selected siRNA sequences from active clinical trials analyzed according to criteria for siRNA design (Reynolds et al. 2004)

Reynolds criterion	Rationale	CALAA-01			Bevasiranib	ALN-RSV01	TD101	AGN211745	QPI-1002
		siR2Ap5	siR2Bp5	siR2Bp6					
A base other than "G" or "C" at 19 (sense strand)	Ago2 specificity loop (Frank et al. 2010)		✓						✓
Absence of internal repeats	Unstructured guide strand (Patzel et al. 2005)								
A base other than "G" at position 13 (sense strand)	Unknown	✓		✓	✓	✓			✓
An "A" base at position 19 (sense strand)	Ago2 specificity loop (Frank et al. 2010)		✓				✓		✓
At least 3 "A/U" bases at positions 15–19 (sense strand)	Asymmetry (Mathews et al. 1999; Lu and Mathews 2008a; Schwarz et al. 2003; Hutvagner 2005)		✓	✓		✓	✓		✓
30–52% G/C content	mRNA and siRNA unwinding (Matveeva et al. 2010; Birmingham et al. 2007)	✓	✓	✓		✓	✓	✓	✓
A "U" base at position 10 (sense strand)	Slicing site geometry (Sashital and Doudna 2010)	✓							✓
An "A" base at position 3 (sense strand)	Unknown	✓							

predict asymmetry in relation to silencing activity may depend on the nucleotides at the 5'-termini (Gredell et al. 2010; Seitz et al. 2011; Frank et al. 2010), hybridization stability of the termini (Mathews et al. 1999; Lu and Mathews 2008b; Schwarz et al. 2003; Hutvagner 2005), or both (Gredell et al. 2010; Seitz et al. 2011; Frank et al. 2010; Walton et al. 2010).

Early predictions of asymmetry depended only upon differential hybridization stability of the four-terminal hybridized nucleotides (Schwarz et al.; Hutvagner 2005; Khvorova et al. 2003; Tomari et al. 2004b). More recent findings have pointed to positional base preferences at the termini of miRNAs and more active siRNAs (Seitz et al. 2011; Gredell et al. 2010; Walton et al. 2010). At this point, it is clear that more active siRNAs and miRNAs contain A or U base pairs at the 5' position of the guide strand (Reynolds et al. 2004; Gredell et al. 2010; Seitz et al. 2011). Ago2 recognizes this position using a nucleotide specificity loop that does not exist in other Ago2 homologs (Frank et al. 2010). Based on these studies, asymmetry predictions for RNAi should take into account nucleotide preferences at the 5' termini.

19.2.3.2 mRNA Target Structures

Another variable that can effect gene silencing and RISC activity is the secondary structure of the target mRNA (Fig. 19.3) (Vickers et al. 2003; Ameres et al. 2007; Bohula et al. 2003; Brown et al. 2005; Kretschmer-Kazemi Far and Sczakiel 2003). There are a variety of different methods to predict the efficiency of silencing based upon secondary structure (Bohula et al. 2003; Overhoff et al. 2005; Schubert et al. 2005; Shao et al. 2007; Kiryu et al. 2011; Lu and Mathews 2008b). The findings of these reports indicate that regions of low secondary structure tend to be better targets for siRNA-mediated silencing. mRNAs with little secondary structure at the 3' and 5' ends of the siRNA target region tend to be silenced more efficiently as compared to other structures (Yoshinari et al. 2004; Gredell et al. 2008; Vickers et al. 2003). As the prediction and experimental characterization of target mRNA structures continues to improve, it is expected that exact specifications for the impact of target structure on siRNA design will be more clearly delineated, including an understanding of any impact due to mRNA tertiary structures.

19.2.4 siRNA-Mediated Cytotoxicity and Immune Response Activation

In addition to unintended effects resulting from targeting the wrong gene or by interfering with the endogenous RNAi functionality (Grimm et al. 2006; Sioud 2007), siRNAs are detected by the human innate immune system by a series of surface and cytoplasmic receptors (Samuel-Abraham and Leonard 2010). These

Fig. 19.3 Possible secondary structures within an mRNA. mRNA molecules adopt complex secondary and, to a lesser degree, tertiary structures that make some locations highly accessible (*green shading*), partially accessible (*yellow shading*), or highly inaccessible (*red shading*).

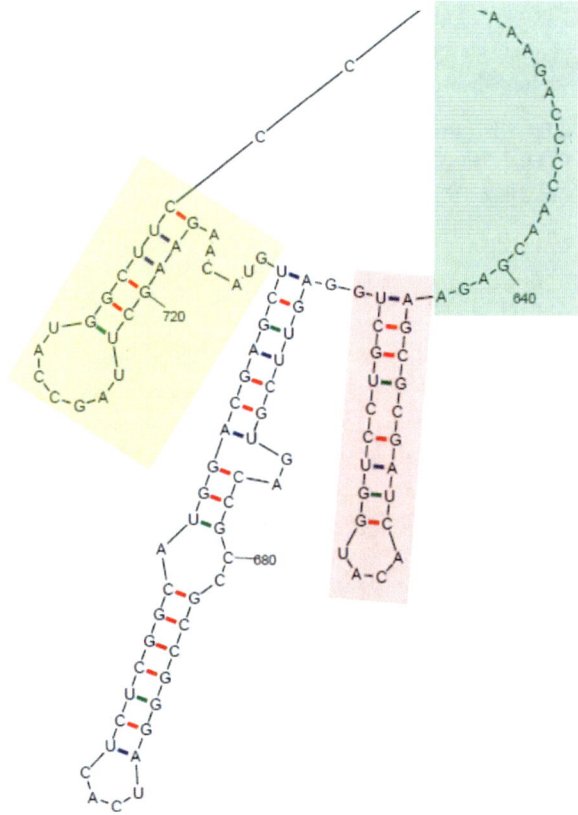

receptors can recognize either specific sequences or the A-form helix of dsRNA, and their expression levels vary across different cell types (Iwasaki and Medzhitov 2004; Judge and Maclachlan 2008). Immune recognition poses a significant concern in the development of siRNA therapeutics and presents a challenge in the design of an siRNA structure that efficiently knocks down its target gene and does so with minimal nonspecific effects. Immunogenicity can be mitigated with the use of an appropriate delivery vehicle and chemical modifications to the ribose sugar, aromatic base, and phosphate backbone, but these modifications must be made without compromising specific silencing activity.

19.2.4.1   Sequence-Specific Off-Target Effects

Off-target effects occur when RISC cleaves the wrong mRNA. These effects can be caused by targeting an mRNA sequence that is not unique, by the incorporation of the wrong strand into RISC, by the incorporation of a degraded fragment of mRNA into RISC, or by activating a miRNA pathway by designing an siRNA with miRNA seed region homology (Jackson et al. 2003; Jackson and Linsley 2010; Bartel

2009). Sequence-specific off-target effects can generally be avoided through comparison of the potential siRNA sequence against the wealth of available sequence information using utilities such as:

BLAST (http://blast.ncbi.nlm.nih.gov/Blast.cgi)
RefSeq (http://www.ncbi.nlm.nih.gov/RefSeq)
miRBase (http://www.mirbase.org/)

19.2.4.2 Surface Receptor Responses

Toll-like receptors (TLRs) are a set of transmembrane proteins that are expressed primarily within endosomes and lysosomes of immune cells (Takeda et al. 2003). TLR7 and TLR8 activate an immune cascade after binding specific ssRNA sequences called pathogen-associated molecular patterns (PAMPs) (Hornung et al. 2005; Judge et al. 2005; Sioud 2005, 2006; Diebold et al. 2004, 2006), resulting in the production of type I interferons, primarily interferon-α (IFN-α), and inflammatory cytokines, including interleukin-6 (IL-6) and tumor necrosis factor-α (TNF-α) (Liu 2005; Judge and Maclachlan 2008; Gorden et al. 2005; Hornung et al. 2005; Judge et al. 2005). siRNAs are believed to activate TLR7 and TLR8 responses following melting in the acidic endosomes (Sioud 2005; Goodchild et al. 2009). The severity of the response can also vary significantly with different sequence motifs and cell types (Hornung et al. 2005; Judge et al. 2005; Sioud 2005; Diebold et al. 2004, 2006; Iwasaki and Medzhitov 2004). siRNAs may also activate TLR3, a sequence-independent receptor for dsRNA found on the cell membrane and within endosomes (Alexopoulou et al. 2001; Karikó et al. 2004). This information has led to the development of siRNAs that can avoid a TLR response by using a delivery vehicle that does not enter the cell by endosomal integration as well as by blocking TLR binding to the siRNA by changing the 2′-OH of uridine bases to a 2′-OMe group (Robbins et al. 2009).

19.2.4.3 Cytoplasmic siRNA Recognition

OAS1, PKR, and RIG-I are all nonspecific cytoplasmic receptors for exogenous dsRNA (Gantier and Williams 2007; Samuel-Abraham and Leonard 2010) and siRNAs (Hornung et al. 2006). These cytoplasmic receptors have more uniform expression levels across all cell types compared to TLRs and cannot be easily avoided by using a cell-specific delivery vehicle. PKR and OAS1 recognize dsRNA, and the level of immune response depends on the length of the dsRNA bound to the receptor. PKR generally requires dsRNA greater than 30 bp for dimerization and subsequent activation, although it has been found to bind to dsRNA as short as 16 bp in length (Manche et al. 1992; Bevilacqua and Cech 1996). PKR recognition of siRNAs may be facilitated by its interactions with the RNAi-associated proteins PACT and TRBP, which regulate PKR phosphorylation

(Kok et al. 2007; Lee et al. 2006). OAS1 can bind dsRNA as short as 19 bp, and its activation is particularly sensitive to the presence of the motif $NNWW(N_9)WGN$ (Kodym et al. 2009). OAS1 activates RNase L, which nonspecifically degrades ssRNA (Minks et al. 1979). The degraded products of RNase L can subsequently activate immune responses in neighboring cells (Malathi et al. 2007). RIG-I is strongly activated when bound by a blunt-ended dsRNA with a $5'$-triphosphate (Schlee and Hartmann 2010; Schlee et al. 2009; Yoneyama et al. 2004; Hornung et al. 2006; Kato et al. 2008; Kim et al. 2004). Nonetheless, RIG-I can still be activated by siRNAs (Pichlmair et al. 2006; Marques et al. 2006). Taken together, the evidence shows that siRNAs do not initiate most cytoplasmic immune responses. However, moving forward, it will be important to further define, and hence design around, the molecular events that do result in immune stimulation upon siRNA administration.

19.2.5 Chemical Modifications and Other Exogenous siRNA-Like Structures

siRNA characteristics can be manipulated through changes in the length of either siRNA strand or by the addition of chemical modifications to the phosphate backbone, ribose sugar, or aromatic base. With the large number of potential modifications that can be made, it is difficult to know which to use to generate the desired effect. To further complicate design, chemical modifications used in tandem require subsequent structural optimization (Dande et al. 2006). For the interested reader, more detailed analyses of the investigation and application of chemical modifications in siRNA design have been described previously (Guo et al. 2010a; Bramsen et al. 2009).

Various structural designs have been found to enhance siRNA activity as well as mitigate unintended effects. For clarity, these structures are given different names that are indicative of their unique structures/modes of action. For instance, aiRNAs are asymmetric interfering RNAs, where asymmetric refers to the differences between the length of the guide strand and passenger strand. aiRNAs are designed to contain a passenger strand shorter than the guide strand and are shown to have potentially stronger and longer-lasting silencing than standard siRNAs (Sun et al. 2008; Chu and Rana 2008). Examples of other structures include small internally segmented interfering RNA (sisiRNA), double-guide siRNA (dgRNA), and long interfering RNA (liRNA) (Bramsen et al. 2007; Hossbach et al. 2006; Chang et al. 2011). In some cases, siRNAs have been used intentionally for their immunostimulatory effect, termed immunostimulatory RNA (isRNA), and have desirable therapeutic attributes despite their inability to knock down a target gene (Schlee et al. 2006). The future of siRNA designs will likely include at least some modifications to the canonical siRNA design; what is still not clear is how best to select optimal sets of modifications for clinical applications.

19.3 Delivery

The direct administration of siRNAs has shown some success when delivered topically to areas such as the eye, skin, or vagina or locally to the lung, brain, or isolated tumors (Whitehead et al. 2009). However, this requires direct and sometimes intrusive access to the areas of interest. When delivered systemically, naked siRNAs have minimal in vivo success due to degradation by serum nucleases and early filtration through the renal system (Soutschek et al. 2004; van de Water et al. 2006). Additionally, naked siRNAs cannot be targeted directly to the tissues/cells of interest. As a result, it is preferred to utilize some type of carrier to protect the siRNA cargo and aid in its delivery to the cells of interest. Delivery approaches are generally divided between biological (viral or bacterial) (Seow and Wood 2009) and nonviral methods. With respect to RNAi, viral vectors are primarily useful for chronic therapies requiring long-term expression of shRNAs. Bacterial vectors are less common, but nonetheless still an area of continuing investigation (Xiang et al. 2006). Nonviral delivery vehicles, which can deliver either siRNAs or shRNAs for transient therapies, are further categorized according to the chemical or physical properties into several different groups such as lipids, polymers, or solid-core particles. These carriers typically range in size from 50 to 200 nm and are sometimes termed nanoparticles (NPs). Each of these approaches has shown success in cell culture and in some in vivo studies (Shim and Kwon 2010). Nonetheless, refinement of currently available vehicles is likely necessary for development of highly effective, noncytotoxic, tissue-specific, systemic delivery vehicles.

19.3.1 Viral Vectors

Viral-based delivery can be achieved using adenoviruses (ADs), adeno-associated viruses (AAVs), or retroviruses such as lentiviruses (Ghosh et al. 2006; Coura and Nardi 2007; Cockrell and Kafri 2007). ADs are large viruses (60–90 nm in diameter) and are typically transported into the cell through clathrin-coated endocytosis (Ghosh et al. 2006). Their large size allows for the highest nucleic acid packing capacity among viral vectors. ADs naturally deliver their cargo to the nucleus where they infect their host cells with viral, double-stranded DNA. ADs can be purified to high concentration and have shown success in infecting both dividing and nondividing cells (Bain et al. 2004). AAVs are modified forms of adenoviruses in which the majority of the viral DNA has been removed and replaced with single-stranded DNA constructs (Coura and Nardi 2007). AAVs are nonpathogenic on their own, having to rely on helper viruses as well as the host cells' polymerases for replication. Due to their simpler structure as compared to ADs, AAVs have a much smaller packaging capacity (average size is ~22 nm in diameter) (Coura and Nardi 2007). AAVs can often avoid immune response and provide stable transgene expression over time, however at a level much lower than other viruses

(Gao et al. 2002). Retroviruses, such as lentiviruses, use envelope fusion to gain access into target cells and are the most common form of virus used for gene therapy (Ghosh et al. 2006). Unlike ADs which require delivering DNA to the nucleus, retroviruses act by delivering RNA into the cell and using reverse transcription to incorporate their DNA into the genome. Lentiviral vectors have reported improved success in transducing primary cells over other viral vector systems (Cockrell and Kafri 2007). RNAi-based strategies initiated by viral infection rely on transduction of DNA encoding shRNAs, which are then processed and enter the RNAi pathway (Grimm et al. 2006).

Viral delivery systems are highly efficient in delivering their therapeutic cargo but do so with the concomitant risk of immunogenicity (Kaiser 2007; Hartman et al. 2008), though this is not solely a shortcoming of viral delivery systems (Robbins et al. 2009). To this point, AAVs are advantageous due to their reduced immunogenicity relative to ADs. Even if innate immune recognition is avoided, adaptive immunity can lead to decreased viral vector efficiency over time (Ghosh et al. 2006). Additionally, all viral vectors, however modified, run the risk of uncontrolled insertional mutagenesis with active viruses or untargeted cells (Cockrell and Kafri 2007).

Generally, viral promoters result in strong expression of viral proteins by the infected cell. For shRNA expression, such high expression results in competition between endogenous miRNAs and viral shRNAs utilizing the same RNAi pathway, saturating the shared nuclear exporter, exportin-5 (Grimm et al. 2006). As a result, normal miRNA functionality is impeded, leading to cell dysregulation and death. Negative cell responses can be mitigated by using smaller doses of viral vectors or using weaker promoters for shRNA expression but must be balanced against lower efficacy. Despite their slower activation relative to other viral vectors, lentiviral vectors seem to be the most feasible choice for shRNA transduction due to their relatively broader infectivity.

19.3.2 Lipid-Based Vehicles

Among lipid-based approaches, the majority of delivery vehicles are based on cationic lipids. The use of cationic lipids allows facile complex formation with anionic nucleic acids. The resulting complexes, sometimes referred to as lipoplexes, protect siRNAs from serum degradation during transport to cells of interest (Buyens et al. 2008). The first reported cationic transfection lipid was N-[1-(2,3-dioleyloxy)propyl]-N,N,N-trimethylammonium chloride (DOTMA) (Malone et al. 1989). DOTMA was subsequently modified to create a second transfection reagent, N-[1-(2,3-Dioleoyloxy)propyl]-N,N,N-trimethylammonium methyl-sulfate (DOTAP) (Ren et al. 2000). Their common structure consists of a quaternary amine head group and a glycerol-based backbone linked to two long hydrocarbon chains. A variety of cationic lipids are now available commercially, with their principal application being siRNA delivery in cell culture.

For improved delivery specificity, targeting ligands have been conjugated to the lipids (e.g., such as transferrin to mediate uptake via the transferrin receptor) (Cardoso et al. 2007). High-throughput screening approaches have been used to identify the important characteristics of successful lipid-based delivery agents (Akinc et al. 2008). Based on this study, lipid-based reagents should include amide linkages, two or more alkyl tails of 8–12 carbon length, as well as the presence of secondary amines, though the incorporation of these characteristics does not ensure successful siRNA delivery. Interestingly, it has been reported that combinations of ineffective lipid-based reagents can yield more active reagents through synergistic effects (Whitehead et al. 2011). It remains to be seen if lipids in development can overcome the significant cytotoxicity that to date has limited their use in vivo (Zhang et al. 2007).

Endogenous liposome-like structures termed exosomes have gained attention recently as potential siRNA delivery vehicles (Thery 2011). Exosomes are small membrane vesicles, averaging 100 nm in diameter, that are released from most cell types. Originating from endocytotic vesicles, their existence has been known for over 25 years; however, their ability to carry RNA molecules, including miRNAs, between cells is a relatively recently discovered phenomenon (Valadi et al. 2007). Exosomes have been shown to cross the blood-brain barrier, another advantage for in vivo application. Moreover, exogenous exosomes can be generated containing targeting peptides for targeted delivery applications (Alvarez-Erviti et al. 2011).

19.3.3 Polymer-Based Vehicles

Polymeric vehicles have been used for the delivery of both plasmids and siRNAs. As with lipid-based vehicles, polymers used for nucleic acid delivery are typically cationic to allow for self-assembly of the polymer-nucleic acid complexes, sometimes termed polyplexes. Linear and branched polyethylenimines (LPEI, BPEI) have been routinely used, despite their significant cytotoxicity (Burke and Pun 2008). Current polymer systems lack the efficacy of lipid-based systems. Modifications of single polymer systems have been explored, such as adding poly (ethylene glycol) (PEG) (Mao et al. 2006) or ethyl acrylate (Zintchenko et al. 2008), and have been found to increase in vivo delivery efficiency by increasing circulation time and decreasing toxicity.

Other polymer systems that have been used for nucleic acid delivery include poly(β-amino esters) (PBAE) (Lynn and Langer 2000), poly(amidoamine) (PAMAM) dendrimers (Tang et al. 1996), and chitosan (Katas and Alpar 2006; Liu et al. 2007). In order to combine positive attributes from varied synthetic polymers, combinations of polymers to create diblock or triblock polymers have also been tested as successful methods of improving nucleic acid delivery in cell culture. These include polyvinyl alcohol/poly(D,L-lactide-co-glycolide) (PVA-b-PLGA) (Nguyen et al. 2008) or poly(ethylene oxide) (PEO)/poly(ε-caprolactone) (PEO-b-PCL) (Xiong et al. 2009). While cationic polymer systems rely on the electrostatic attraction of the siRNA for complex formation, covalent attachment of the siRNA to the polymer by disulfide bonds has also shown some success (Rozema et al. 2007). It is believed that the presence of disulfide bonds, which are reduced through cellular levels of glutathione and thioredoxin, can also aid in the intracellular release of the complex. The numerous and varied functional groups that are available on polymers allow for extensive modifications to tune the properties of the polymers for siRNA delivery applications. Rules for such modifications are just beginning to emerge (Siegwart et al. 2011; Portis et al. 2010).

19.3.4 Solid-Core Particles

A third type of delivery vehicle currently under development uses solid-core particles such as iron, gold, or silica (Veiseh et al. 2011; Rosi et al. 2006; Hom et al. 2010). One advantage of these NPs is tighter control of the size and structure of complexes formed. Also, solid-core particles have the potential to provide enhanced imaging and diagnostic signals for in vivo applications.

Iron NPs can be coated with other compounds to improve delivery efficiency. Commercially available iron NPs (e.g., SilenceMag, OZ Biosciences) rely on a mixture of solid-core iron particles with lipids to create a combination delivery reagent. Other iron NPs utilize cationic polymer coatings or polymer peptides to help bind siRNAs and improve delivery efficiency (Veiseh et al. 2011). Delivery of iron NPs can be controlled with a magnetic field, termed magnetofection (Lee et al. 2011). This can enhance delivery to the cells of interest and increase transfection rates.

siRNAs have been conjugated to gold NPs via disulfide bonds (Rosi et al. 2006). This improves their nuclease stability over free siRNAs (Patel et al. 2011). In addition, combinations between delivery systems have been examined, such as using a gold NP modified with PEG and/or PBAE polymers (Lee et al. 2008; Lee et al. 2009) or PEI and polyanhydrides (PAH) in a layer-by-layer approach (Guo et al. 2010b). Gold NPs modified with folate for receptor targeting and near-IR photoactivity have shown selective in vivo activity in tumors (Lu et al. 2010) indicating the ability for receptor-mediated targeting.

Increasingly, silica NPs are being studied for siRNA delivery. Similar to other solid-core NPs, silica particle size can be finely tuned, and particles can be easily modified with the addition of polymers such as PEI (Hom et al. 2010). Furthermore, the porous nature of silica particles allows for codelivery of siRNAs with other small molecule drugs such as doxorubicin for cancer treatment (Meng et al. 2010). However, solid-core particles create additional barriers to in vivo applications since their size and insolubility also play a role in their ability to traverse the circulation. It has been shown that sedimentation of solid-core gold NPs in two-dimensional cell culture has a direct effect on cellular uptake (Cho et al. 2011).

19.3.5   Mechanism of Entry

In order to activate the RNAi pathway, siRNA complexes must enter the cell and access the cytoplasm. Most often, this occurs through macropinocytosis or endocytosis (Dausend et al. 2008). Macropinocytosis results from a cell membrane reaching out and enveloping the delivery vehicle, allowing for entry of larger (>150 nm) particles. Endocytosis can be subdivided into categories such as receptor-mediated, clathrin dependent, caveolae, or lipid raft based (Dausend et al. 2008). These methods create internalized vesicles of delivery vehicles that, while intracellular, do not provide access to the cytoplasm. Although the exact means of delivery for each vehicle has not been defined, the two main hypotheses for entry into the cytoplasm from vesicles are through the proton sponge effect (Sonawane et al. 2003; Akinc et al. 2005) or a membrane fusion event (Lu et al. 2009).

According to the proton sponge hypothesis, amine-containing vehicles, especially polymers containing many amines, are endocytosed through normal means. As the endosome acidifies, the buffering capacity of the amines draws in an excess of protons and concomitantly an excess of chloride ions. Osmotic swelling then causes the endosome to burst, releasing the contents into the cytoplasm of the cell (Cho et al. 2003; Sonawane et al. 2003; Akinc et al. 2005). Other delivery vehicles utilize fusogenic peptides to create more potent and active endosomal escape (Kwon et al. 2008).

While it has been proven that all NP delivery vehicles display some type of endocytotic uptake (Medina-Kauwe et al. 2005), for some lipid vehicles, endocytosis is not the only pathway through which silencing can be achieved. In one study, it was shown that normal conditions result in 95% of lipoplexes being taken via endocytosis (Lu et al. 2009). However, when endocytosis was blocked, lipid/siRNA complexes still accumulated inside the cell and caused silencing. It is believed the access occurred through direct fusion of the lipoplex with the lipid bilayer membrane, resulting in delivery directly to the cytoplasm of the cell.

19.3.6   Two-Dimensional In Vitro, Three-Dimensional In Vitro, and In Vivo Transition Challenges

Typical laboratory experiments involve treating cells grown on a flat cell culture plate where the treatment is added directly on top of the cells. However, this does not mimic the cells' natural environment. In vivo, cells are embedded in tissues, with extracellular matrix and cell contacts in all directions. In the development of novel siRNA therapeutics, few good cell culture-based models exist for examining the differences in delivery to three-dimensional cultures versus two-dimensional cultures. The development of three-dimensional collagen (Ishihara et al. 2010), fibronectin (Zhou et al. 2008), or synthetic PEG (Raeber et al. 2005) hydrogels to study cell morphology and migration has often been used as an intermediate step between cell culture and in vivo models. Current three-dimensional studies involving siRNA treatments either resort to pretreating cells before embedding in the matrix (Ivanov et al. 2008), treatment of cells growing on top of a hydrogel (Lei et al. 2010), or nucleic acid–containing hydrogel scaffolds implanted into animal models (Andersen et al. 2010). In three-dimensional systems, siRNA delivery is limited, resulting in longer transfection times as well as reduced delivery efficiency. Many commercially available transfection reagents can be ineffective when attempting to treat cells embedded within a three-dimensional matrix (Zhang et al. 2010). Future development of reagents for in vivo delivery will depend on the availability of three-dimensional cell culture model systems.

Despite the multitude of candidates, there is no single delivery vehicle that can guarantee reliable, consistent siRNA delivery to all cell types. This is attributable both to a need to optimize the vehicles and to the lack of good model systems for evaluating in vivo vectors. The best vehicle choice may also depend on the disease and cellular targets, as generalized toxicity can be an acceptable response if confined to, for instance, cancer cells. The risk of immune complications often hinders viral vector development even though they can show highly efficient delivery. Lipid vehicles provide simple and efficient transfection and are the standard for cell culture as well as large-scale bioreactors where toxicity is not a concern. Solid-core particles can be highly modified without altering their structure. While often noncytotoxic at low concentrations, accumulation of these typically nondegradable particles is a potential concern for long-term or repeated use. Polymer vehicles provide the greatest variety of constructs but currently lack the efficiency attainable with viral vectors. However, their potential for extensive modification, in concert with their potential biodegradability, seems to make them the likeliest candidates for future siRNA delivery applications.

19.4 Clinical Challenges and Successes

There are several ongoing clinical trials utilizing siRNA therapeutics to treat macular degeneration, solid tumor cancers, acute kidney failure, and viral infections such as hepatitis C (HCV) and respiratory syncytial virus (RSV) (Davidson and McCray 2011), but, as yet, no clinical trials have progressed to FDA approval. Although the time frame for therapeutic development from discovery to FDA approval can vary markedly depending on the drug type and its target, RNAi therapeutics are not moving through the developmental pipeline as quickly as initially hoped. As a result, major pharmaceutical companies have started cutting funding toward development of RNAi therapeutics (Pollack 2011; Krieg 2011). Current therapies in clinical trials are investigating a variety of involved siRNA designs that leverage chemical modifications and tailored delivery vehicles (Table 19.2), though current siRNA designs do not incorporate many of the established criteria for maximizing siRNA activity (Table 19.1).

Table 19.2 The status of current clinical trials for RNAi therapy (from www.clinicaltrials.gov)

Class	Disease	Name	Target	Sponsor	Status	Administration	Delivery	Unique features	Reference
Skin disorders	Pachyonychia congenita	TD101	Keratin 6A N171K mutant	Pachyonychia congenita project	Completed, phase I	Intradermal injection	Not specified	Specific targeting of dominant-negative mRNA	Leachman et al. (2008) and Hickerson et al. (2008)
Ocular and retinal disorders	Nonarteritic anterior ischemic optic neuropathy	QPI-1007	Caspase 2	Quark Pharm. Inc.	Active, phase I	Intravitreal injection	Not specified		
	Age-related macular degeneration, choroidal neovascularization	AGN211745	VEGFR1	siRNA Therapeutics Inc	Completed, phase I, II	Intravitreal injection	Not specified	siRNA modifications: deoxythymidine overhangs, phosphorothioate linkage, inverted 2'-deoxy abasic nucleotides	Shen et al. (2006)
	Glaucoma	SYL040012	β2 adrenergic receptor	Sylentis	Active, phase I, II	Topical, eye drops	Not specified		Jimenez et al. (2011)
	Diabetic macular edema	Bevasiranib	VEGF	Opko Health, Inc.	Completed, phase II	Intravitreal injection	Not specified		
	Macular degeneration	Bevasiranib	VEGF	Opko Health, Inc.	Completed, phase II	Intravitreal injection	Not specified		
Cancer	Solid tumors	FANG	Furin	Gradalis, Inc.	Active, phase II	Injection of treated cells	Ex vivo electroporation	Treated cells express plasmid and shRNA	
	Solid tumors	CALAA-01	RRM2	Calando Pharm.	Active, phase I	Intravenous injection	Cyclodextrin polymer with transferrin receptor target	siRNA pool	Davis (2009)
	Chronic myeloid leukemia	SPC2996	BCL-2	Santaris Pharm.	Ongoing, phase I, II	Systemic administration	Not specified	16-mer oligonucleotide	Hansen et al. (2006)
	Solid tumors	ALN-VSP02	VEGF, kinesin spindle protein	Alnylam Pharm.	Active, phase I	Intravenous injection	Not specified		
	Pancreas adenocarcinoma	siG12D LODER	KRASG12D	Silenseed Ltd.	Active, phase I	Localized injection of siRNA-	Miniature biodegradable		

	Drug	Target	Company	Status	Route	Formulation/polymeric matrix	Description	Reference
Metastatic melanoma	NCT00672542	LMP2, LMP7, and MECL1	Duke University	Active, phase I	Intradermal injection of in vitro–treated monocytes	Ex vivo electroporation	siRNA pool targeting iP subunits LMP2, LMP7, MECL-1	Dannull et al. (2007)
Advanced, recurrent, or metastatic solid malignancies	Atu027	PKN3	Silence Therapeutics	Active, phase I	Intravenous injection	Liposomal siRNA formulation	Methylated siRNA sequences	Aleku et al. (2008)
Kidney disorders								
Delayed graft function kidney transplant	QPI-1002/I5NP	p53	Quark Pharm., Inc.	Active, phase I, II	Intravenous injection	Not specified	Methylated siRNA sequences	Feinstein (2011)
Kidney injury acute renal failure	QPI-1002/I5NP	p53	Quark Pharm., Inc.	Completed, phase I	Intravenous injection	Not specified		
Antiviral								
Hepatitis C virus	SPC3649/miravirsen	miR-122	Santaris Pharm	Active, phase II	Subcutaneous injection	Not specified	miRNA antagonist	
RSV in volunteers	ALN-RSV01	RSV nucleocapsid	Alnylam Pharm.	Completed, phase II	Inhalent	none specified		Zhang et al. (2002b), Johnson et al. (2007), Alvarez et al. (2009) and Meyers (2011)
RSV in lung transplant patients	ALN-RSV01	RSV nucleocapsid	Alnylam Pharm.	Completed, phase I	Inhalent	none specified		
RSV in lung transplant patients	ALN-RSV01	RSV nucleocapsid	Alnylam Pharm.	Active, phase II	Inhalent	none specified		

ALN-RSV01 is an siRNA targeted against the nucleocapsid mRNA of RSV, a gene critical to viral replication (Alvarez et al. 2009). RSV-caused bronchiolitis can lead to hospitalization of infants and is implicated in the development of other illnesses including asthma and reactive airway disease (Leader and Kohlhase 2002; Kalina and Gershwin 2004; Peebles 2004; Psarras et al. 2004). To date, there is no vaccine developed to prevent RSV; although other treatments do exist, their use is limited (Ventre and Randolph 2007; Stevens and Hall 2004). ALN-RSV01 has progressed to a phase IIb clinical trial and is currently the closest RNAi therapeutic to reaching FDA approval (Zamora et al. 2011). RSV infections are specifically amenable to RNAi therapeutics because the infection is superficial, localized to the airway epithelial cells, and is easily accessible by inhaled delivery of naked siRNA (Zhang et al. 2002b; Johnson et al. 2007; Alvarez et al. 2009; Meyers 2011).

The selection of the ALN-RSV01 siRNA began with the identification of a 19 bp conserved region of the virus followed by screening for siRNAs with fewer than 17 bp homology with any host genes (Alvarez et al. 2009). ALN-RSV01 was found to be particularly effective in the knockdown of the nucleocapsid protein in comparison to other candidate siRNAs (Alvarez et al. 2009). The ALN-RSV01 siRNA is highly asymmetric; however, the first nucleotide of the guide strand is a cytosine and not a uracil or adenine, which have been found to mediate more efficient siRNA interactions with Ago2. While the current clinical trial does not use a chemically modified siRNA, the patent filed by Alnylam Pharmaceuticals explores many different chemical modifications including the addition of $2'$-OMe to the ribose backbone and hydroxyproline to the $5'$ termini (Meyers 2010).

OPKO Health's Bevasiranib was the first siRNA therapeutic to reach stage III clinical trials. As an early frontrunner, the naked siRNA was injected intravitreally to target the vascular endothelial growth factor (VEGF) for use in the treatment of age-related macular degeneration (Dejneka et al. 2008). The phase III clinical trial relied on combinatorial therapy of Bevasiranib combined with monoclonal antibody therapy (Singerman 2009). However, the company withdrew the trial in 2009 due to lack of significant activity in reducing vision loss and is currently looking for alternative delivery methods such as increased dosing or the use of some type of siRNA delivery vehicle to improve efficacy (Rubin 2009).

The first targeted siRNA delivery drug to reach clinical trials is Calando Pharmaceutical's CALAA-01. The siRNA, designed to target tumor growth factors, is delivered in a protective cyclodextrin polymer NP, modified with human transferrin to target binding to transferrin receptors, which are typically upregulated in cancer cells (Davis 2009). While still in the phase I stage, the advantage to this upcoming therapeutic is that it addresses required design features previously discussed, including a polymeric vehicle with a cationic backbone to bind and protect nucleic acids, components that minimize immunogenicity, PEG to increase intravenous circulation time, complex size small enough to access the tumors but large enough to minimize kidney filtration, receptor targeting for enhanced cell specificity, and efficient endosomal release.

19.5 Conclusion

Current clinical trials use siRNAs based largely on early design rules and experimental validation of a limited number of candidates (e.g., ALN-RSV01). While siRNA sequence selection and design has proven more complex than initially anticipated, improved understanding of the RNAi mechanism has led to the development of siRNAs with improved functionality. Moving forward, the development of siRNA therapeutics will likely be informed by a better understanding of the RNAi pathway in regard to the proteins involved and how they recognize siRNAs, leading to the rational inclusion of modifications to the canonical siRNA design.

While improving the function of siRNA itself is important, the majority of ongoing siRNA clinical trials focus on accessible, localized targets (lungs, eyes, solid tumors) or targets that are natural filtering agents (liver, kidneys), indicating that the key barrier to successful siRNA therapies is delivery. Targeted delivery agents and increased knowledge into how siRNAs traverse complex tissues are important areas of continued study.

Following continued optimization of the multitude of design variables discussed, future challenges for therapeutic development include techniques for the reliable large-scale processing of both siRNAs as well as complex delivery vehicles. Nonetheless, comprehensive knowledge obtained outlining design guidelines as well as current clinical trial successes and failures provide a strong starting point for the continued development of RNAi therapeutics.

References

Akinc A, Thomas M, Klibanov AM, Langer R (2005) Exploring polyethylenimine-mediated DNA transfection and the proton sponge hypothesis. J Gene Med 7(5):657–663. doi:10.1002/jgm.696

Akinc A, Zumbuehl A, Goldberg M, Leshchiner ES, Busini V, Hossain N, Bacallado SA, Nguyen DN, Fuller J, Alvarez R, Borodovsky A, Borland T, Constien R, de Fougerolles A, Dorkin JR, Jayaprakash KN, Jayaraman M, John M, Koteliansky V, Manoharan M, Nechev L, Qin J, Racie T, Raitcheva D, Rajeev KG, Sah DWY, Soutschek J, Toudjarska I, Vornlocher HP, Zimmermann TS, Langer R, Anderson DG (2008) A combinatorial library of lipid-like materials for delivery of RNAi therapeutics. Nat Biotechnol 26(5):561–569. doi:10.1038/nbt1402

Aleku M, Schultz P, Keil O, Santel A, Schaeper U, Dieckhoff B, Janke O, Endruschat J, Durieux B, Roder N, Loffler K, Lange C, Fechtner M, Mopert K, Fisch G, Dames S, Arnold W, Jochims K, Giese K, Wiedenmann B, Scholz A, Kaufmann J (2008) Atu027, a liposomal small interfering RNA formulation targeting protein kinase N3, inhibits cancer progression. Cancer Res 68(23):9788–9798. doi:10.1158/0008-5472.CAN-08-2428

Alexopoulou L, Holt AC, Medzhitov R, Flavell RA (2001) Recognition of double-stranded RNA and activation of NF-kappa B by Toll-like receptor 3. Nature 413(6857):732–738. doi:10.1038/35099560

Alvarez R, Elbashir S, Borland T, Toudjarska I, Hadwiger P, John M, Roehl I, Morskaya SS, Martinello R, Kahn J, Van Ranst M, Tripp RA, DeVincenzo JP, Pandey R, Maier M, Nechev L, Manoharan M, Koteliansky V, Meyers R (2009) RNA interference-mediated silencing of the respiratory syncytial virus nucleocapsid defines a potent antiviral strategy. Antimicrob Agents Chemother 53(9):3952–3962. doi:10.1128/aac.00014–09

Alvarez-Erviti L, Seow YQ, Yin HF, Betts C, Lakhal S, Wood MJA (2011) Delivery of siRNA to the mouse brain by systemic injection of targeted exosomes. Nat Biotechnol 29(4):341–345. doi:10.1038/nbt.1807

Ameres SL, Martinez J, Schroeder R (2007) Molecular basis for target RNA recognition and cleavage by human RISC. Cell 130(1):101–112. doi:10.1016/j.cell.2007.04.037

Andersen MO, Nygaard JV, Burns JS, Raarup MK, Nyengaard JR, Bunger C, Besenbacher F, Howard KA, Kassem M, Kjems J (2010) siRNA nanoparticle functionalization of nanostructured scaffolds enables controlled multilineage differentiation of stem cells. Mol Ther 18(11):2018–2027. doi:10.1038/mt.2010.166

Bain JR, Schisler JC, Takeuchi K, Newgard CB, Becker TC (2004) An adenovirus vector for efficient RNA interference-mediated suppression of target genes in insulinoma cells and pancreatic islets of Langerhans. Diabetes 53(9):2190–2194. doi:10.2337/diabetes.53.9.2190

Bartel DP (2009) MicroRNAs: target recognition and regulatory functions. Cell 136(2):215–233. doi:10.1016/j.cell.2009.01.002

Bernstein E, Caudy AA, Hammond SM, Hannon GJ (2001) Role for a bidentate ribonuclease in the initiation step of RNA interference. Nature 409(6818):363–366. doi:10.1038/35053110

Bevilacqua PC, Cech TR (1996) Minor-groove recognition of double-stranded RNA by the double-stranded RNA-binding domain from the RNA-activated protein kinase PKR. Biochemistry 35(31):9983–9994. doi:10.1021/bi9607259

Birmingham A, Anderson E, Sullivan K, Reynolds A, Boese Q, Leake D, Karpilow J, Khvorova A (2007) A protocol for designing siRNAs with high functionality and specificity. Nat Protoc 2 (9):2068–2078. doi:10.1038/nprot.2007.278

Bohula EA, Salisbury AJ, Sohail M, Playford MP, Riedemann J, Southern EM, Macaulay VM (2003) The efficacy of small interfering RNAs targeted to the type 1 insulin-like growth factor receptor (IGF1R) is influenced by secondary structure in the IGF1R transcript. J Biol Chem 278(18):15991–15997. doi:10.1074/jbc.M300714200

Bramsen JB, Laursen MB, Damgaard CK, Lena SW, Babu BR, Wengel J, Kjems J (2007) Improved silencing properties using small internally segmented interfering RNAs. Nucleic Acids Res 35(17):5886–5897. doi:10.1093/nar/gkm548

Bramsen JB, Laursen MB, Nielsen AF, Hansen TB, Bus C, Langkjaer N, Babu BR, Hojland T, Abramov M, Van Aerschot A, Odadzic D, Smicius R, Haas J, Andree C, Barman J, Wenska M, Srivastava P, Zhou C, Honcharenko D, Hess S, Muller E, Bobkov GV, Mikhailov SN, Fava E, Meyer TF, Chattopadhyaya J, Zerial M, Engels JW, Herdewijn P, Wengel J, Kjems J (2009) A large-scale chemical modification screen identifies design rules to generate siRNAs with high activity, high stability and low toxicity. Nucleic Acids Res 37(9):2867–2881. doi:10.1093/nar/gkp106

Brown KM, Chu C-Y, Rana TM (2005) Target accessibility dictates the potency of human RISC. Nat Struct Mol Biol 12(5):469–470. doi:10.1038/nsmb931

Burke RS, Pun SH (2008) Extracellular barriers to in Vivo PEI and PEGylated PEI polyplex-mediated gene delivery to the liver. Bioconjug Chem 19(3):693–704. doi:10.1021/bc700388u

Buyens K, Lucas B, Raemdonck K, Braeckmans K, Vercammen J, Hendrix J, Engelborghs Y, De Smedt SC, Sanders NN (2008) A fast and sensitive method for measuring the integrity of siRNA-carrier complexes in full human serum. J Control Release 126(1):67–76. doi:10.1016/j.jconrel.2007.10.024

Cardoso ALC, Simoes S, de Almeida LP, Pelisek J, Culmsee C, Wagner E, de Lima MCP (2007) SiRNA detivery by a transferrin-associated lipid-based vector: a non-viral strategy to mediate gene silencing. J Gene Med 9(3):170–183. doi:10.1002/jgm.1006

Chang CI, Lee TY, Dua P, Kim S, Li CJ, Lee D-K (2011) Long double-stranded rna-mediated rna interference and immunostimulation: long interfering double-stranded rna as a potent anticancer therapeutics. Oligonucleotides 21(3):149–155. doi:10.1089/nat.2011.0296

Cho YW, Kim JD, Park K (2003) Polycation gene delivery systems: escape from the endosomes to the cytosol. J Pharm Pharmacol 55:721–734

Cho EC, Zhang Q, Xia YN (2011) The effect of sedimentation and diffusion on cellular uptake of gold nanoparticles. Nat Nanotechnol 6(6):385–391. doi:10.1038/nnano.2011.58

Chu C-Y, Rana TM (2008) Potent RNAi by short RNA triggers. RNA 14(9):1714–1719. doi:10.1261/rna.1161908

Cockrell AS, Kafri T (2007) Gene delivery by lentivirus vectors. Mol Biotechnol 36(3):184–204. doi:10.1007/s12033–007–0010–8

Coura RD, Nardi NB (2007) The state of the art of adeno-associated virus-based vectors in gene therapy. Virol J 4. doi:10.1186/1743–422x-4–99

Dande P, Prakash TP, Sioufi N, Gaus H, Jarres R, Berdeja A, Swayze EE, Griffey RH, Bhat B (2006) Improving RNA interference in mammalian cells by 4'-thio-modified small interfering RNA (siRNA): effect on siRNA activity and nuclease stability when used in combination with 2'-O-alkyl modifications. J Med Chem 49(5):1624–1634. doi:10.1021/jm050822c

Dannull J, Lesher DT, Holzknecht R, Qi WN, Hanna G, Seigler H, Tyler DS, Pruitt SK (2007) Immunoproteasome down-modulation enhances the ability of dendritic cells to stimulate antitumor immunity. Blood 110(13):4341–4350. doi:10.1182/blood-2007–04–083188

Dausend J, Musyanovych A, Dass M, Walther P, Schrezenmeier H, Landfester K, Mailander V (2008) Uptake mechanism of oppositely charged fluorescent nanoparticles in hela cells. Macromol Biosci 8(12):1135–1143. doi:10.1002/mabi.200800123

Davidson BL, McCray PB (2011) Current prospects for RNA interference-based therapies. Nat Rev Genet 12(5):329–340. doi:10.1038/nrg2968

Davis ME (2009) The first targeted delivery of sirna in humans via a self-assembling, cyclodextrin polymer-based nanoparticle: from concept to clinic. Mol Pharm 6(3):659–668. doi:10.1021/mp900015y

Dejneka NS, Wan SH, Bond OS, Kornbrust DJ, Reich SJ (2008) Ocular biodistribution of bevasiranib following a single intravitreal injection to rabbit eyes. Mol Vis 14 (116–19):997–1005

Diebold SS, Kaisho T, Hemmi H, Akira S, Reise Sousa C (2004) Innate antiviral responses by means of TLR7-mediated recognition of single-stranded RNA. Science 303(5663):1529–1531. doi:10.1126/science.1093616

Diebold SS, Massacrier C, Akira S, Paturel C, Morel Y, Reise Sousa C (2006) Nucleic acid agonists for toll-like receptor 7 are defined by the presence of uridine ribonucleotides. Eur J Immunol 36(12):3256–3267. doi:10.1002/eji.200636617

Elbashir SM, Harborth J, Lendeckel W, Yalcin A, Weber K, Tuschl T (2001a) Duplexes of 21-nucleotide RNAs mediate RNA interference in cultured mammalian cells. Nature 411 (6836):494–498. doi:10.1038/35078107

Elbashir SM, Lendeckel W, Tuschl T (2001b) RNA interference is mediated by 21- and 22-nucleotide RNAs. Genes Dev 15(2):188–200. doi:10.1101/gad.862301

Feinstein E (2011) Prevention and treatment of acute renal failure and other kidney diseases by inhibition of p53 by siRNA. 7910566

Frank F, Sonenberg N, Nagar B (2010) Structural basis for 5'-nucleotide base-specific recognition of guide RNA by human AGO2. Nature 465(7299):818–822. doi:10.1038/nature09039

Gantier MP, Williams BRG (2007) The response of mammalian cells to double-stranded RNA. Cytok Growth Fact Rev 18(5–6):363–371. doi:10.1016/j.cytogfr.2007.06.016

Gao GP, Alvira MR, Wang LL, Calcedo R, Johnston J, Wilson JM (2002) Novel adeno-associated viruses from rhesus monkeys as vectors for human gene therapy. Proc Natl Acad Sci USA 99 (18):11854–11859. doi:10.1073/pnas.182412299

Ghosh SS, Gopinath P, Ramesh A (2006) Adenoviral vectors—a promising tool for gene therapy. Appl Biochem Biotechnol 133(1):9–29. doi:10.1385/abab:133:1:9

Goodchild A, Nopper N, King A, Doan T, Tanudji M, Arndt GM, Poidinger M, Rivory LP, Passioura T (2009) Sequence determinants of innate immune activation by short interfering RNAs. BMC Immunol 10:40. doi:10.1186/1471-2172-10-40

Gorden KB, Gorski KS, Gibson SJ, Kedl RM, Kieper WC, Qiu XH, Tomai MA, Alkan SS, Vasilakos JP (2005) Synthetic TLR Agonists reveal functional differences between human TLR7 and TLR8. J Immunol 174(3):1259–1268

Gredell JA, Berger AK, Walton SP (2008) Impact of target mRNA structure on siRNA silencing efficiency: a large-scale study. Biotechnol Bioeng 100(4):744–755. doi:10.1002/bit.21798

Gredell JA, Dittmer MJ, Wu M, Chan C, Walton SP (2010) Recognition of siRNA asymmetry by TAR RNA binding protein. Biochemistry 49(14):3148–3155. doi:10.1021/bi902189s

Grimm D, Streetz KL, Jopling CL, Storm TA, Pandey K, Davis CR, Marion P, Salazar F, Kay MA (2006) Fatality in mice due to oversaturation of cellular microRNA/short hairpin RNA pathways. Nature 441(7092):537–541. doi:10.1038/nature04791

Guo PX, Coban O, Snead NM, Trebley J, Hoeprich S, Guo SC, Shu Y (2010a) Engineering RNA for targeted siRNA delivery and medical application. Adv Drug Deliv Rev 62(6):650–666. doi:10.1016/j.addr.2010.03.008

Guo ST, Huang YY, Jiang QA, Sun Y, Deng LD, Liang ZC, Du QA, Xing JF, Zhao YL, Wang PC, Dong AJ, Liang XJ (2010b) Enhanced gene delivery and siRNA silencing by gold nanoparticles coated with charge-reversal polyelectrolyte. ACS Nano 4(9):5505–5511. doi:10.1021/nn101638u

Hammond SM, Bernstein E, Beach D, Hannon GJ (2000) An RNA-directed nuclease mediates post-transcriptional gene silencing in Drosophila cells. Nature 404(6775):293–296. doi:10.1038/35005107

Hansen BJ, Westergaard M, Frieden M, Hansen HF, Kjaerulff LS, Thrue CA, Wissenbach MH, Dalby LW, Kearney P, Oerum O (2006) SPC2996-A Bcl-2 RNA antagonist being studied in chronic lymphocytic leukemia. J Clin Oncol 24:6610

Hartman ZC, Appledorn DM, Amalfitano A (2008) Adenovirus vector induced innate immune responses: impact upon efficacy and toxicity in gene therapy and vaccine applications. Virus Res 132(1–2):1–14. doi:10.1016/j.virusres.2007.10.005

Hickerson RP, Smith FJD, Reeves RE, Contag CH, Leake D, Leachman SA, Milstone LM, McLean WHI, Kaspar RL (2008) Single-nucleotide-specific siRNA targeting in a dominant-negative skin model. J Investig Dermatol 128(3):594–605. doi:10.1038/sj.jid.5701060

Hom C, Lu J, Liong M, Luo HZ, Li ZX, Zink JI, Tamanoi F (2010) Mesoporous silica nanoparticles facilitate delivery of siRNA to shutdown signaling pathways in mammalian cells. Small 6(11):1185–1190. doi:10.1002/smll.200901966

Hornung V, Guenthner-Biller M, Bourquin C, Ablasser A, Schlee M, Uematsu S, Noronha A, Manoharan M, Akira S, de Fougerolles A, Endres S, Hartmann G (2005) Sequence-specific potent induction of IFN-alpha by short interfering RNA in plasmacytoid dendritic cells through TLR7. Nat Med 11(3):263–270. doi:10.1038/nm1191

Hornung V, Ellegast J, Kim S, Brzózka K, Jung A, Kato H, Poeck H, Akira S, Conzelmann K-K, Schlee M, Endres S, Hartmann G (2006) 5′-Triphosphate RNA is the ligand for RIG-I. Science 314(5801):994–997. doi:10.1126/science.1132505

Hossbach M, Gruber J, Osborn M, Weber K, Tuschl T (2006) Gene silencing with siRNA duplexes composed of target-mRNA-complementary and partially palindromic or partially complementary single-stranded siRNAs. RNA Biol 3(2):82–89

Huesken D, Lange J, Mickanin C, Weiler J, Asselbergs F, Warner J, Meloon B, Engel S, Rosenberg A, Cohen D, Labow M, Reinhardt M, Natt F, Hall J (2005) Design of a genome-wide siRNA library using an artificial neural network. Nat Biotechnol 23(8):995–1001. doi:10.1038/nbt1118

Hutvagner G (2005) Small RNA asymmetry in RNAi: function in RISC assembly and gene regulation. FEBS Lett 579(26):5850–5857. doi:10.1016/j.febslet.2005.08.071

Ishihara S, Haga H, Yasuda M, Mizutani T, Kawabata K, Shirato H, Nishioka T (2010) Integrin beta 1-dependent invasive migration of irradiation-tolerant human lung adenocarcinoma cells in 3D collagen matrix. Biochem Biophys Res Commun 396(3):651–655. doi:10.1016/j.bbrc.2010.04.150

Ivanov AI, Hopkins AM, Brown GT, Gerner-Smidt K, Babbin BA, Parkos CA, Nusrat A (2008) Myosin II regulates the shape of three-dimensional intestinal epithelial cysts. J Cell Sci 121 (11):1803–1814. doi:10.1242/jcs.015842

Iwasaki A, Medzhitov R (2004) Toll-like receptor control of the adaptive immune responses. Nat Immunol 5(10):987–995. doi:10.1038/ni1112

Jackson AL, Linsley PS (2010) Recognizing and avoiding siRNA off-target effects for target identification and therapeutic application. Nat Rev Drug Discov 9(1):57–67. doi:10.1038/nrd3010

Jackson AL, Bartz SR, Schelter J, Kobayashi SV, Burchard J, Mao M, Li B, Cavet G, Linsley PS (2003) Expression profiling reveals off-target gene regulation by RNAi. Nat Biotechnol 21 (6):635–637. doi:10.1038/nbt831

Jimenez AI, Sesto A, Gascon I, Roman JP, Gonzalez de Buitrago G (2011) Methods and compositions for the treatment of eye disorders with increased intraocular pressure. 7902169, 8 Mar 2011

Johnson JE, Gonzales RA, Olson SJ, Wright PF, Graham BS (2007) The histopathology of fatal untreated human respiratory syncytial virus infection. Mod Pathol 20(1):108–119. doi:10.1038/modpathol.3800725

Judge A, Maclachlan I (2008) Overcoming the innate immune response to small interfering RNA. Hum Gene Ther 19(2):111–124. doi:10.1089/hum.2007.179

Judge AD, Sood V, Shaw JR, Fang D, McClintock K, MacLachlan I (2005) Sequence-dependent stimulation of the mammalian innate immune response by synthetic siRNA. Nat Biotechnol 23 (4):457–462. doi:10.1038/nbt1081

Kaiser J (2007) Clinical research: death prompts a review of gene therapy vector. Science 317 (5838):580. doi:10.1126/science.317.5838.580

Kalina WV, Gershwin LJ (2004) Progress in defining the role of RSV in allergy and asthma: from clinical observations to animal models. Clin Dev Immunol 11(2):113–119. doi:10.1080/10446670410001722131

Karikó K, Bhuyan P, Capodici J, Weissman D (2004) Small interfering RNAs mediate sequence-independent gene suppression and induce immune activation by signaling through toll-like receptor 3. J Immunol 172(11):6545–6549

Katas H, Alpar HO (2006) Development and characterisation of chitosan nanoparticles for siRNA delivery. J Control Release 115(2):216–225. doi:10.1016/j.jconrel.2006.07.021

Kato H, Takeuchi O, Mikamo-Satoh E, Hirai R, Kawai T, Matsushita K, Hiiragi A, Dermody TS, Fujita T, Akira S (2008) Length-dependent recognition of double-stranded ribonucleic acids by retinoic acid-inducible gene-1 and melanoma differentiation-associated gene 5. J Exp Med 205 (7):1601–1610. doi:10.1084/jem.20080091

Khvorova A, Reynolds A, Jayasena SD (2003) Functional siRNAs and miRNAs exhibit strand bias. Cell 115(2):209–216. doi:10.1016/S0092-8674(03)00801-8

Kim D-H, Longo M, Han Y, Lundberg P, Cantin E, Rossi JJ (2004) Interferon induction by siRNAs and ssRNAs synthesized by phage polymerase. Nat Biotechnol 22(3):321–325. doi:10.1038/nbt940

Kiryu H, Terai G, Imamura O, Yoneyama H, Suzuki K, Asai K (2011) A detailed investigation of accessibilities around target sites of siRNAs and miRNAs. Bioinformatics 27(13):1788–1797. doi:10.1093/bioinformatics/btr276

Kodym R, Kodym E, Story MD (2009) 2'-5'-Oligoadenylate synthetase is activated by a specific RNA sequence motif. Biochem Biophys Res Commun 388(2):317–322. doi:10.1016/j.bbrc.2009.07.167

Kok KH, Ng MHJ, Ching YP, Jin DY (2007) Human TRBP and PACT directly interact with each other and associate with dicer to facilitate the production of small interfering RNA. J Biol Chem 282(24):17649–17657. doi:10.1074/jbc.M611768200

Kretschmer-Kazemi Far R, Sczakiel G (2003) The activity of siRNA in mammalian cells is related to structural target accessibility: a comparison with antisense oligonucleotides. Nucleic Acids Res 31(15):4417–4424. doi:10.1093/nar/gkg649

Krieg AM (2011) Is RNAi dead? Mol Ther 19(6):1001–1002. doi:10.1038/mt.2011.94

Kwon EJ, Bergen JM, Pun SH (2008) Application of an HIV gp41-derived peptide for enhanced intracellular trafficking of synthetic gene and siRNA delivery vehicles. Bioconjug Chem 19 (4):920–927. doi:10.1021/bc700448h

Leachman SA, Hickerson RP, Hull PR, Smith FJD, Milstone LM, Lane EB, Bale SJ, Roop DR, McLean WHI, Kaspar RL (2008) Therapeutic siRNAs for dominant genetic skin disorders including pachyonychia congenita. J Dermatol Sci 51(3):151–157. doi:10.1016/j.jdermsci.2008.04.003

Leader S, Kohlhase K (2002) Respiratory syncytial virus-coded pediatric hospitalizations, 1997 to 1999. Pediatr Infect Dis J 21(7):629–632. doi:10.1097/00006454-200207000-00005

Lee Y, Hur I, Park SY, Kim YK, Suh MR, Kim VN (2006) The role of PACT in the RNA silencing pathway. EMBO J 25(3):522–532. doi:10.1038/sj.emboj.7600942

Lee SH, Bae KH, Kim SH, Lee KR, Park TG (2008) Amine-functionalized gold nanoparticles as non-cytotoxic and efficient intracellular siRNA delivery carriers. Int J Pharm 364(1):94–101. doi:10.1016/j.ijpharm.2008.07.027

Lee JS, Green JJ, Love KT, Sunshine J, Langer R, Anderson DG (2009) Gold, Poly(Beta-Amino Ester) nanoparticles for small interfering RNA delivery. Nano Lett 9(6):2402–2406. doi:10.1021/nl9009793

Lee S, Shim G, Kim S, Kim YB, Kim CW, Byun Y, Oh YK (2011) Enhanced transfection rates of small-interfering RNA using dioleylglutamide-based magnetic lipoplexes. Oligonucleotides 21(3):165–172. doi:10.1089/nat.2010.0274

Lei YG, Ng QKT, Segura T (2010) Two and three-dimensional gene transfer from enzymatically degradable hydrogel scaffolds. Microsc Res Tech 73(9):910–917. doi:10.1002/jemt.20840

Leuschner PJF, Ameres SL, Kueng S, Martinez J (2006) Cleavage of the siRNA passenger strand during RISC assembly in human cells. EMBO Rep 7(3):314–320. doi:10.1038/sj.embor.7400637

Lima WF, Murray H, Nichols JG, Wu H, Sun H, Prakash TP, Berdeja AR, Gaus HJ, Crooke ST (2009) Human Dicer binds short single-strand and double-strand RNA with high affinity and interacts with different regions of the nucleic acids. J Biol Chem 284(4):2535–2548. doi:10.1074/jbc.M803748200

Liu Q (2003) R2D2, a bridge between the initiation and effector steps of the drosophila RNAi pathway. Science 301(5641):1921–1925. doi:10.1126/science.1088710

Liu Y-J (2005) IPC: professional type 1 interferon-producing cells and plasmacytoid dendritic cell precursors. Annu Rev Immunol 23:275–306. doi:10.1146/annurev.immunol.23.021704.115633

Liu J, Carmell MA, Rivas FV, Marsden CG, Thomson JM, Song J-J, Hammond SM, Joshua-Tor L, Hannon GJ (2004) Argonaute2 Is the catalytic engine of mammalian RNAi. Sci Signal 305 (5689):1437. doi:10.1126/science.1102513

Liu XD, Howard KA, Dong MD, Andersen MO, Rahbek UL, Johnsen MG, Hansen OC, Besenbacher F, Kjems J (2007) The influence of polymeric properties on chitosan/siRNA nanoparticle formulation and gene silencing. Biomaterials 28(6):1280–1288. doi:10.1016/j.biomaterials.2006.11.004

Liu Y, Ye X, Jiang F, Liang C, Chen D, Peng J, Kinch LN, Grishin NV, Liu Q (2009) C3PO, an endoribonuclease that promotes RNAi by facilitating RISC activation. Science 325 (5941):750–753. doi:10.1126/science.1176325

Lu ZJ, Mathews DH (2008a) Efficient siRNA selection using hybridization thermodynamics. Nucleic Acids Res 36(2):640–647. doi:10.1093/nar/gkm920

Lu ZJ, Mathews DH (2008b) Efficient siRNA selection using hybridization thermodynamics. Nucleic Acids Res 36(2):640–647. doi:10.1093/nar/gkm920

Lu JJ, Langer R, Chen JZ (2009) A novel mechanism is involved in cationic lipid-mediated functional siRNA delivery. Mol Pharm 6(3):763–771. doi:10.1021/mp900023v

Lu W, Zhang GD, Zhang R, Flores LG, Huang Q, Gelovani JG, Li C (2010) Tumor site-specific silencing of NF-kappa B p65 by targeted hollow gold nanosphere-mediated photothermal transfection. Cancer Res 70(8):3177–3188. doi:10.1158/0008-5472.can-09-3379

Lynn DM, Langer R (2000) Degradable poly(beta-amino esters): synthesis, characterization, and self-assembly with plasmid DNA. J Am Chem Soc 122(44):10761–10768. doi:10.1021/ja0015388

Malathi K, Dong B, Gale M, Silverman RH (2007) Small self-RNA generated by RNase L amplifies antiviral innate immunity. Nature 448(7155):816–819. doi:10.1038/nature06042

Malone RW, Felgner PL, Verma IM (1989) CATIONIC liposome-mediated RNA transfection. Proc Natl Acad Sci USA 86(16):6077–6081. doi:10.1073/pnas.86.16.6077

Manche L, Green SR, Schmedt C, Mathews MB (1992) Interactions between double-stranded RNA regulators and the protein kinase DAI. Mol Cell Biol 12(11):5238–5248

Mao SR, Neu M, Germershaus O, Merkel O, Sitterberg J, Bakowsky U, Kissel T (2006) Influence of polyethylene glycol chain length on the physicochemical and biological properties of poly (ethylene imine)-graft-poly(ethylene glycol) block copolymer/SiRNA polyplexes. Bioconjug Chem 17(5):1209–1218. doi:10.1021/bc060129j

Marques JT, Devosse T, Wang D, Zamanian-Daryoush M, Serbinowski P, Hartmann R, Fujita T, Behlke MA, Williams BR (2006) A structural basis for discriminating between self and nonself double-stranded RNAs in mammalian cells. Nat Biotechnol 24(5):559–565. doi:10.1038/nbt1205

Mathews D, Sabina J, Zuker M, Turner D (1999) Expanded sequence dependence of thermodynamic parameters improves prediction of RNA secondary structure. J Mol Biol 288 (5):911–940. doi:10.1006/jmbi.1999.2700

Matranga C, Tomari Y, Shin C, Bartel DP, Zamore PD (2005) Passenger-strand cleavage facilitates assembly of siRNA into Ago2-containing RNAi enzyme complexes. Cell 123 (4):607–620. doi:10.1016/j.cell.2005.08.044

Matveeva OV, Kang Y, Spiridonov AN, Saetrom P, Nemtsov VA, Ogurtsov AY, Nechipurenko YD, Shabalina SA (2010) Optimization of duplex stability and terminal asymmetry for shRNA design. PLoS One 5(4):e10180. doi:10.1371/journal.pone.0010180

Medina-Kauwe LK, Xie J, Hamm-Alvarez S (2005) Intracellular trafficking of nonviral vectors. Gene Ther 12(24):1734–1751. doi:10.1038/sj.gt.3302592

Meng HA, Liong M, Xia TA, Li ZX, Ji ZX, Zink JI, Nel AE (2010) Engineered design of mesoporous silica nanoparticles to deliver doxorubicin and P-glycoprotein siRNA to overcome drug resistance in a cancer cell line. ACS Nano 4(8):4539–4550. doi:10.1021/nn100690m

Meyers R (2010) Methods and compostions for prevention or treatment of RSV infection using modified duplex RNA molecules. WO 2010/048590 A1, 29 Apr 2010

Meyers R (2011) RNAi modulation of RSV and therapeutic uses thereof. US Patent 7,981,869 B2, 19 July 2011

Minks MA, Benvin S, Maroney PA, Baglioni C (1979) Metabolic stability of 2'S'oligo (A) and activity of 2' S'oligo (A)-dependent endonuclease in extracts of interferon-treated and control HeLa cells. Nucleic Acids Res 6(2):767–780

Miyoshi K, Tsukumo H, Nagami T, Siomi H, Siomi MC (2005) Slicer function of Drosophila argonautes and its involvement in RISC formation. Genes Dev 19(23):2837–2848. doi:10.1101/gad.1370605

Nguyen J, Steele TWJ, Merkel O, Reul R, Kissel T (2008) Fast degrading polyesters as siRNA nano-carriers for pulmonary gene therapy. J Control Release 132(3):243–251. doi:10.1016/j.jconrel.2008.06.010

Noland CL, Ma E, Doudna JA (2011) siRNA repositioning for guide strand selection by human Dicer complexes. Mol cell 43(1):110–121. doi:10.1016/j.molcel.2011.05.028

Nykänen A, Haley B, Zamore PD (2001) ATP requirements and small interfering RNA structure in the RNA interference pathway. Cell 107(3):309–321. doi:10.1016/S0092-8674(01)00547-5

Overhoff M, Alken M, Far RK-K, Lemaitre M, Lebleu B, Sczakiel G, Robbins I (2005) Local RNA target structure influences siRNA efficacy: a systematic global analysis. J Mol Biol 348 (4):871–881. doi:10.1016/j.jmb.2005.03.012

Patel PC, Hao L, Yeung WSA, Mirkin CA (2011) Duplex end breathing determines serum stability and intracellular potency of siRNA-Au NPs. Mol Pharm 8:6

Patzel V, Rutz S, Dietrich I, Koberle C, Scheffold A, Kaufmann SHE (2005) Design of siRNAs producing unstructured guide-RNAs results in improved RNA interference efficiency. Nat Biotechnol 23(11):1440–1444. doi:10.1038/nbt1151

Peebles RS Jr (2004) Viral infections, atopy, and asthma: is there a causal relationship? J Allergy Clin Immunol 113(1 Suppl):S15–S18. doi:10.1016/j.jaci.2003.10.033

Pichlmair A, Schulz O, Tan CP, Naslund TI, Liljestrom P, Weber F, Reise Sousa C (2006) RIG-I-mediated antiviral responses to single-stranded RNA bearing 5′-phosphates. Science 314 (5801):997–1001. doi:10.1126/science.1132998

Pollack A (2011) Drugmakers' fever for the power of RNA interference has cooled. New York Times, 8 Feb 2011

Portis AM, Carballo G, Baker GL, Chan C, Walton SP (2010) Confocal microscopy for the analysis of siRNA delivery by polymeric nanoparticles. Microsc Res Tech 73(9):878–885. doi:10.1002/jemt.20861

Psarras S, Papadopoulos NG, Johnston SL (2004) Pathogenesis of respiratory syncytial virus bronchiolitis-related wheezing. Paediatr Respir Rev 5(Suppl A):S179–S184. doi:10.1016/s1526-0542(04)90034-6

Raeber GP, Lutolf MP, Hubbell JA (2005) Molecularly engineered PEG hydrogels: a novel model system for proteolytically mediated cell migration. Biophys J 89(2):1374–1388. doi:10.1529/biophysj.104.050682

Rand TA, Petersen S, Du F, Wang X (2005) Argonaute2 cleaves the anti-guide strand of siRNA during RISC activation. Cell 123(4):621–629. doi:10.1016/j.cell.2005.10.020

Ren T, Song YK, Zhang G, Liu D (2000) Structural basis of DOTMA for its high intravenous transfection activity in mouse. Gene Ther 7(9):764–768. doi:10.1038/sj.gt.3301153

Reynolds A, Leake D, Boese Q, Scaringe S, Marshall WS, Khvorova A (2004) Rational siRNA design for RNA interference. Nat Biotechnol 22(3):326–330. doi:10.1038/nbt936

Rivas FV, Tolia NH, Song JJ, Aragon JP, Liu JD, Hannon GJ, Joshua-Tor L (2005) Purified Argonaute2 and an siRNA form recombinant human RISC. Nat Struct Mol Biol 12 (4):340–349. doi:10.1038/nsmb918

Robbins M, Judge A, MacLachlan I (2009) siRNA and innate immunity. Oligonucleotides 19 (2):89–101. doi:10.1089/oli.2009.0180

Rosi NL, Giljohann DA, Thaxton CS, Lytton-Jean AKR, Han MS, Mirkin CA (2006) Oligonucleotide-modified gold nanoparticles for intracellular gene regulation. Science 312 (5776):1027–1030. doi:10.1126/science.1125559

Rozema DB, Lewis DL, Wakefield DH, Wong SC, Klein JJ, Roesch PL, Bertin SL, Reppen TW, Chu Q, Blokhin AV, Hagstrom JE, Wolff JA (2007) Dynamic polyconjugates for targeted in vivo delivery of siRNA to hepatocytes. Proc Natl Acad Sci USA 104(32):12982–12987. doi:10.1073/pnas.0703778104

Rubin SD (2009) OPKO health announces update on phase III clinical trial of Bevasiranib

Sakurai K, Amarzguioui M, Kim D-H, Alluin J, Heale B, M-s S, Gatignol A, Behlke MA, Rossi JJ (2011) A role for human Dicer in pre-RISC loading of siRNAs. Nucleic Acids Res 39 (4):1510–1525. doi:10.1093/nar/gkq846

Samuel-Abraham S, Leonard JN (2010) Staying on message: design principles for controlling nonspecific responses to siRNA. FEBS J 277(23):4828–4836. doi:10.1111/j.1742-4658.2010.07905.x

Sashital DG, Doudna JA (2010) Structural insights into RNA interference. Curr Opin Struct Biol 20(1):90–97. doi:10.1016/j.sbi.2009.12.001

Schlee M, Hartmann G (2010) The chase for the RIG-I ligand—recent advances. Mol Ther 18 (7):1254–1262. doi:10.1038/mt.2010.90

Schlee M, Hornung V, Hartmann G (2006) siRNA and isRNA: two edges of one sword. Mol Ther 14(4):463–470. doi:10.1016/j.ymthe.2006.06.001

Schlee M, Roth A, Hornung V, Hagmann CA, Wimmenauer V, Barchet W, Coch C, Janke M, Mihailovic A, Wardle G, Juranek S, Kato H, Kawai T, Poeck H, Fitzgerald KA, Takeuchi O, Akira S, Tuschl T, Latz E, Ludwig J, Hartmann G (2009) Recognition of 5′-triphosphate by RIG-I helicase requires short blunt double-stranded RNA as contained in panhandle of negative-strand virus. Immunity 31(1):25–34. doi:10.1016/j.immuni.2009.05.008

Schubert S, Grünweller A, Erdmann VA, Kurreck J (2005) Local RNA target structure influences siRNA efficacy: systematic analysis of intentionally designed binding regions. J Mol Biol 348 (4):883–893. doi:10.1016/j.jmb.2005.03.011

Schwarz D, Hutvagner G, Du T, Xu Z, Aronin N, Zamore P (2003) Asymmetry in the assembly of the RNAi enzyme complex. Cell 115(2):199–208

Seitz H, Tushir J, Zamore P (2011) A 5′-uridine amplifies miRNA/miRNA* asymmetry in Drosophila by promoting RNA-induced silencing complex formation. Silence 2:1–10

Seow Y, Wood MJ (2009) Biological gene delivery vehicles: beyond viral vectors. Mol Ther 17 (5):767–777. doi:10.1038/mt.2009.41

Shabalina SA, Spiridonov AN, Ogurtsov AY (2006) Computational models with thermodynamic and composition features improve siRNA design. BMC Bioinforma 7:65. doi:10.1186/1471–2105-7–65

Shao Y, Chan CY, Maliyekkel A, Lawrence CE, Roninson IB, Ding Y (2007) Effect of target secondary structure on RNAi efficiency. RNA 13(10):1631–1640. doi:10.1261/rna.546207

Shen J, Samul R, Silva RL, Akiyama H, Liu H, Saishin Y, Hackett SF, Zinnen S, Kossen K, Fosnaugh K, Vargeese C, Gomez A, Bouhana K, Aitchison R, Pavco P, Campochiaro PA (2006) Suppression of ocular neovascularization with siRNA targeting of VEGF receptor 1. Gene Ther 13(3):225–234. doi:10.1038/sj.gt.3302641

Shim MS, Kwon YJ (2010) Efficient and targeted delivery of siRNA in vivo. FEBS J 277 (23):4814–4827. doi:10.1111/j.1742–4658.2010.07904.x

Siegwart DJ, Whitehead KA, Nuhn L, Sahay G, Cheng H, Jiang S, Ma M, Lytton-Jean A, Vegas A, Fenton P, Levins CG, Love KT, Lee H, Cortez C, Collins SP, Li YF, Jang J, Querbes W, Zurenko C, Novobrantseva T, Langer R, Anderson DG (2011) Combinatorial synthesis of chemically diverse core-shell nanoparticles for intracellular delivery. Proc Natl Acad Sci USA. doi:10.1073/pnas.1106379108

Singerman L (2009) Combination therapy using the small interfering RNA bevasiranib. Retina-J Retin Vitr Dis 29(6):S49–S50

Sioud M (2005) Induction of inflammatory cytokines and interferon responses by double-stranded and single-stranded siRNAs is sequence-dependent and requires endosomal localization. J Mol Biol 348(5):1079–1090. doi:10.1016/j.jmb.2005.03.013

Sioud M (2006) Innate sensing of self and non-self RNAs by toll-like receptors. Trends Mol Med 12(4):167–176. doi:10.1016/j.molmed.2006.02.004

Sioud M (2007) RNA interference and innate immunity. Adv Drug Deliver Rev 59(2–3):153–163. doi:10.1016/j.addr.2007.03.006

Sonawane ND, Szoka FC, Verkman AS (2003) Chloride accumulation and swelling in endosomes enhances DNA transfer by polyamine-DNA polyplexes. J Biol Chem 278(45):44826–44831. doi:10.1074/jbc.M308643200

Soutschek J, Akinc A, Bramlage B, Charisse K, Constien R, Donoghue M, Elbashir S, Geick A, Hadwiger P, Harborth J, John M, Kesavan V, Lavine G, Pandey RK, Racie T, Rajeev KG, Rohl I, Toudjarska I, Wang G, Wuschko S, Bumcrot D, Koteliansky V, Limmer S, Manoharan M, Vornlocher HP (2004) Therapeutic silencing of an endogenous gene by systemic administration of modified siRNAs. Nature 432(7014):173–178. doi:10.1038/nature03121

Stevens TP, Hall CB (2004) Controversies in palivizumab use. Pediatr Infect Dis J 23 (11):1051–1052. doi:10.1097/01.inf.0000145759.71531.d8

Sun X, Rogoff HA, Li CJ (2008) Asymmetric RNA duplexes mediate RNA interference in mammalian cells. Nat Biotechnol 26(12):1379–1382. doi:10.1038/nbt.1512

Takeda K, Kaisho T, Akira S (2003) Toll-like receptors. Annu Rev Immunol 21(1):335–376. doi:10.1146/annurev.immunol.21.120601.141126

Tang MX, Redemann CT, Szoka FC (1996) In vitro gene delivery by degraded polyamidoamine dendrimers. Bioconjug Chem 7(6):703–714. doi:10.1021/bc9600630

Thery C (2011) Exosomes: secreted vesicles and intercellular communications. F1000 Biology Reports 3

Tomari Y, Du T, Haley B, Schwarz D, Bennett R, Cook H, Koppetsch B, Theurkauf W, Zamore P (2004a) RISC assembly defects in the Drosophila RNAi mutant armitage. Cell 116(6):831–841

Tomari Y, Matranga C, Haley B, Martinez N, Zamore PD (2004b) A protein sensor for siRNA asymmetry. Science 306(5700):1377–1380. doi:10.1126/science.1102755

Ui-Tei K, Naito Y, Takahashi F, Haraguchi T, Ohki-Hamazaki H, Juni A, Ueda R, Saigo K (2004) Guidelines for the selection of highly effective siRNA sequences for mammalian and chick RNA interference. Nucleic Acids Res 32(3):936–948. doi:10.1093/nar/gkh247

Valadi H, Ekstrom K, Bossios A, Sjostrand M, Lee JJ, Lotvall JO (2007) Exosome-mediated transfer of mRNAs and microRNAs is a novel mechanism of genetic exchange between cells. Nat Cell Biol 9(6):U654–U672. doi:10.1038/ncb1596

van de Water FM, Boerman OC, Wouterse AC, Peters JGP, Russel FGM, Masereeuw R (2006) Intravenously administered short interfering RNA accumulates in the kidney and selectively suppresses gene function in renal proximal tubules. Drug Metab Dispos 34(8):1393–1397. doi:10.1124/dmd.106.009555

Veiseh O, Kievit FM, Mok H, Ayesh J, Clark C, Fang C, Leung M, Arami H, Park JO, Zhang MQ (2011) Cell transcytosing poly-arginine coated magnetic nanovector for safe and effective siRNA delivery. Biomaterials 32(24):5717–5725. doi:10.1016/j.biomaterials.2011.04.039

Ventre K, Randolph AG (2007) Ribavirin for respiratory syncytial virus infection of the lower respiratory tract in infants and young children. Cochrane Database of Syst Rev (1):Cd000181. doi:10.1002/14651858.CD000181.pub3

Vickers TA, Koo S, Bennett CF, Crooke ST, Dean NM, Baker BF (2003) Efficient reduction of target RNAs by small interfering RNA and RNase H-dependent antisense agents. A comparative analysis. J Biol Chem 278(9):7108–7118. doi:10.1074/jbc.M210326200

Walton SP, Wu M, Gredell JA, Chan C (2010) Designing highly active siRNAs for therapeutic applications. FEBS J 277(23):4806–4813. doi:10.1111/j.1742-4658.2010.07903.x

Wang HW, Noland C, Siridechadilok B, Taylor DW, Ma EB, Felderer K, Doudna JA, Nogales E (2009) Structural insights into RNA processing by the human RISC-loading complex. Nat Struct Mol Biol 16(11):1148–1153. doi:10.1038/nsmb.1673

Weitzer S, Martinez J (2007) The human RNA kinase hClp1 is active on 3′ transfer RNA exons and short interfering RNAs. Nature 447(7141):222–227. doi:10.1038/nature05777

Whitehead KA, Langer R, Anderson DG (2009) Knocking down barriers: advances in siRNA delivery. Nat Rev Drug Discov 8(2):129–138. doi:10.1038/nrd2742

Whitehead KA, Sahay G, Li GZ, Love KT, Alabi CA, Ma M, Zurenko C, Querbes W, Langer R, Anderson DG (2011) Synergistic silencing: combinations of lipid-like materials for efficacious siRNA delivery. Mol Ther:7. doi:10.1038/mt.2011.141

Xiang SL, Fruehauf J, Li CJ (2006) Short hairpin RNA-expressing bacteria elicit RNA interference in mammals. Nat Biotechnol 24(6):697–702. doi:10.1038/nbt1211

Xiong XB, Uludag H, Lavasanifar A (2009) Biodegradable amphiphilic poly(ethylene oxide)-block-polyesters with grafted polyamines as supramolecular nanocarriers for efficient siRNA delivery. Biomaterials 30(2):242–253. doi:10.1016/j.biomaterials.2008.09.025

Ye X, Huang N, Liu Y, Paroo Z, Huerta C, Li P, Chen S, Liu Q, Zhang H (2011) Structure of C3PO and mechanism of human RISC activation. Nat Struct Mol Biol 18(6):650–657. doi:10.1038/nsmb.2032

Yoneyama M, Kikuchi M, Natsukawa T, Shinobu N, Imaizumi T, Miyagishi M, Taira K, Akira S, Fujita T (2004) The RNA helicase RIG-I has an essential function in double-stranded RNA-induced innate antiviral responses. Nat Immunol 5(7):730–737. doi:10.1038/ni1087

Yoshinari K, Miyagishi M, Taira K (2004) Effects on RNAi of the tight structure, sequence and position of the targeted region. Nucleic Acids Res 32(2):691–699. doi:10.1093/nar/gkh221

Zamora MR, Budev M, Rolfe M, Gottlieb J, Humar A, DeVincenzo J, Vaishnaw A, Cehelsky J, Albert G, Nochur S, Gollob JA, Glanville AR (2011) RNA interference therapy in lung transplant patients infected with respiratory syncytial virus. Am J Respir Crit Care Med 183 (4):531–538. doi:10.1164/rccm.201003–0422OC

Zamore PD, Tuschl T, Sharp PA, Bartel DP (2000) RNAi: double-stranded RNA directs the ATP-dependent cleavage of mRNA at 21 to 23 nucleotide intervals. Cell 101(1):25–33. doi:10.1016/S0092–8674(00)80620–0

Zhang H, Kolb FA, Brondani V, Billy E, Filipowicz W (2002a) Human Dicer preferentially cleaves dsRNAs at their termini without a requirement for ATP. EMBO J 21(21):5875–5885

Zhang LQ, Peeples ME, Boucher RC, Collins PL, Pickles RJ (2002b) Respiratory syncytial virus infection of human airway epithelial cells is polarized, specific to ciliated cells, and without obvious cytopathology. J Virol 76(11):5654–5666. doi:10.1128/jvi.76.11.5654–5666.2002

Zhang H, Kolb FA, Jaskiewicz L, Westhof E, Filipowicz W (2004) Single processing center models for human Dicer and bacterial RNase III. Cell 118(1):57–68. doi:10.1016/j.cell.2004.06.017

Zhang SB, Zhao B, Jiang HM, Wang B, Ma BC (2007) Cationic lipids and polymers mediated vectors for delivery of siRNA. J Control Release 123(1):1–10. doi:10.1016/j.jconrel.2007.07.016

Zhang HY, Lee MY, Hogg MG, Dordick JS, Sharfstein ST (2010) Gene delivery in three-dimensional cell cultures by superparamagnetic nanoparticles. ACS Nano 4(8):4733–4743. doi:10.1021/nn9018812

Zhou XM, Rowe RG, Hiraoka N, George JP, Wirtz D, Mosher DF, Virtanen I, Chernousov MA, Weiss SJ (2008) Fibronectin fibrillogenesis regulates three-dimensional neovessel formation. Genes Dev 22(9):1231–1243. doi:10.1101/gad.1643308

Zintchenko A, Philipp A, Dehshahri A, Wagner E (2008) Simple modifications of branched PEI lead to highly efficient siRNA carriers with low toxicity. Bioconjug Chem 19(7):1448–1455. doi:10.1021/bc800065f

Chapter 20
Artificial MicroRNA and Its Applications

Pranjal Yadava and Sunil Kumar Mukherjee

Abstract Enhanced understanding of cellular microRNA (miRNA) biogenesis machinery has allowed researchers to engineer synthetic or artificial miRNAs (amiRNAs) that can be designed to direct efficient silencing of any transcript. The amiRNA technology has not only widened the existing gene silencing tool kit but also offers several distinct improvements over existing RNAi approaches, primarily based on siRNA generating hairpin RNA precursors. amiRNAs have already been applied to a wide range of agricultural and medical applications. This chapter discusses various aspects of miRNA processing, design principles of amiRNA expression vectors and their application.

Keywords Artificial microRNA • Gene silencing • microRNA

20.1 Introduction to MicroRNAs

MicroRNAs (miRNAs) are a class of 19–24 nucleotide (nt) nonprotein-coding RNAs, which constitute an important component of posttranscriptional as well as transcriptional regulation of gene expression (Ambros et al. 2003; Bartel 2004; Kim et al. 2008; Younger and Corey 2011). Pioneering research during last decade was successful in not only deciphering important players of cellular miRNA biogenesis machinery but also unraveling their important roles in development and disease. Mature miRNAs are processed from imperfect duplex regions of long primary RNA polymerase II (Pol II)-generated transcripts, termed pri-miRNAs (Cai et al. 2004). These precursors are sequentially processed into a mature miRNAs by proteins involved in cellular miRNA biogenesis machinery (Table 20.1).

S.K. Mukherjee (✉)
International Centre for Genetic Engineering and Biotechnology, Aruna Asaf Ali Marg,
New Delhi, India
e-mail: sunilm@icgeb.res.in

B. Mallick (eds.), *Regulatory RNAs*, DOI 10.1007/978-3-642-22517-8_20,
© Springer-Verlag Berlin Heidelberg (outside the USA) 2012

Table 20.1 Important proteins involved in miRNA biogenesis machinery

Organism	Protein	Location	Function
Plants	DCL1	Nuclear	dsRNA-specific RNAse III involved in dicing pri-miRNAs
	HYL1	Nuclear	Interacts with DCL1 and confers stability to miRNA precursors
	HEN1	Nuclear	Protects miRNA/miRNA* duplex 3′ end methylation
	Serrate	Nuclear	Binds to pre-miRNAs in association with DCL1 and HYL1 and helps in processing
	HASTY	Nuclear membrane	Export of miRNA/miRNA* duplex to cytoplasm
	SDE3/RDR	Nucleocytoplasmic	Performs catalysis of ds long RNA generation that can initiate different RNAi pathways
	AGO1	Nucleocytoplasmic	miRNA-catalyzed target cleavage
Animals	Drosha	Nuclear	RNase III-type enzyme that binds dsRNA with characteristic structures and generates pre-miRNA forms by cleaving pri-miRNAs
	DGCR8/Pasha	Nuclear	dsRNA-binding protein assists Drosha function
	Exportin5/ RanGTPase	Nucleocytoplasmic	Transports pre-miRNA from nucleus to cytoplasm
	Dicer	Cytoplasmic	dsRNA-specific RNAse III involved in processing pre-miRNA in cytoplasm to form miRNA/miRNA* duplex

In animals, the pri-miRNA is processed in the nucleus into hairpin pre-miRNA by the double-strand specific ribonuclease III (RNase III) enzyme, Drosha, and its cognate RNA-binding proteins. Subsequent to nuclear processing, the pre-miRNA is transported to the cytoplasm via exportin-5 (Exp-5) (Yi et al. 2005; Lund et al. 2004), where it is digested by a second double-stranded RNA (dsRNA)-specific RNAse III enzyme, called Dicer and the associated proteins. The resulting 19–24 nt miRNA duplex is then unwound, and one of the strand preferentially associates with a complex known as the RNA-induced silencing complex (RISC) that participates in repressing the target transcript, resulting in its reduced accumulation or translation.

In contrast to animals, in plants both the pri-and pre-miRNA are processed in the nucleus by the dsRNA-specific RNAse III enzyme, DCL1 in combination with HYPONASTIC LEAVES1 (HYL1), and Serrate (SE) to release the miRNA duplex (Tang et al. 2003). Another protein specific to plants called Hua-Enhancer1 (HEN1) methylates the terminal sugar residues of miRNA to prevent 3′ end uridylation and increases the stability of miRNA. DCL1, HYL1, and SE are known to interact with each other and are colocalized in the nuclear bodies, called dicing bodies or D-bodies (Fang and Spector 2007). The miRNA:miRNA* duplex formed in the nucleus is transported to the cytoplasm by Exp-5 homolog, HASTY (Bollman et al. 2003), where it is unwound, and usually one of the strand (miRNA) is incorporated into RISC to guide it to repress target mRNA(s). Argonaute (AGO) protein of the RISC carries out the actual slicing function with the help of its PAZ and PIWI domains. The PAZ domain allows for interaction with DCL1 and other proteins and is also believed to help align the miRNA with its target mRNA. The structure of the PIWI domain is related to RNase H enzyme, and it is believed to be involved in miRNA-catalyzed target cleavage. Earlier, it was believed that plant miRNAs lead to cleavage of the target, while in animals, translational repression is the predominant mode of miRNA-directed repression of gene expression; however, it has now been demonstrated that translational repression also occurs in plants (Brodersen et al. 2008). miRNA biogenesis operates under a feedback regulation, as the DCL1 and AGO1 genes are themselves regulated by miRNAs. DCL1 is targeted by miR162 (Xie et al. 2003), while AGO1 is targeted by miRNA168.

Over the years, few noncanonical miRNA biogenesis pathways have been discovered especially in animals, where miRNAs were found to be processed from introns (termed as mirtrons) (Ruby et al. 2007), viral tRNA-like sequences (Bogerd et al. 2010), or snoRNA precursors (Ono et al. 2011). Dicer-independent but AGO2-dependent miRNA maturation, viz., formation of miRNA451 which is highly conserved in animal kingdom, is also known (Yang et al. 2010). In a similar manner, upregulation of few miRNAs are also observed in transformed cell lines deficient in Dicer activity. In another phenomenon sometimes referred to as RNA activation (RNAa), animal miRNAs are shown to upregulate translation as well, adding to further complexity (Vasudevan et al. 2007; Huang et al. 2010). miRNAs have been found to be involved in modulation of almost every significant biological process, like development, signaling, and stress response.

20.2   Determinants of miRNA Processing

A pri-miRNA contains the following structural features: a terminal loop region, a mostly dsRNA stem encompassing the miRNA:miRNA* duplex and a ~1 helical turn extension, and flanking ssRNA (Fig. 20.1). Emerging evidences indicate that architecture of stem and loop region of miRNA precursor has profound influence on its correct recognition and accurate processing. The miRNA precursors exhibit terminal loop of varying sizes and possess diverse structure destabilizing motifs (bulges) in the stem region. The bulges, loops, and asymmetry of miRNA:miRNA* duplex are a function of free energy values of RNA secondary structure, which in turn is further modulated by non-Watson–Crick base pairing and other parameters like temperature and strand concentration. Most of the miRNA precursors are processed in a "base-to-loop" mode in which the initial Drosha-like cut is located close to the base of the stem (Werner et al. 2010, Mateos et al. 2009, Kim 2005).

Most animal pri-miRNAs are processed by two cleavages, the first at a loop-distal site 11 nt from the end of the hairpin and the second 22 nt beyond the first. In plants, the first cleavage is often at 15 nt from an unpaired region and lower stem is most critical for miRNA biogenesis. Closing bulges immediately below the

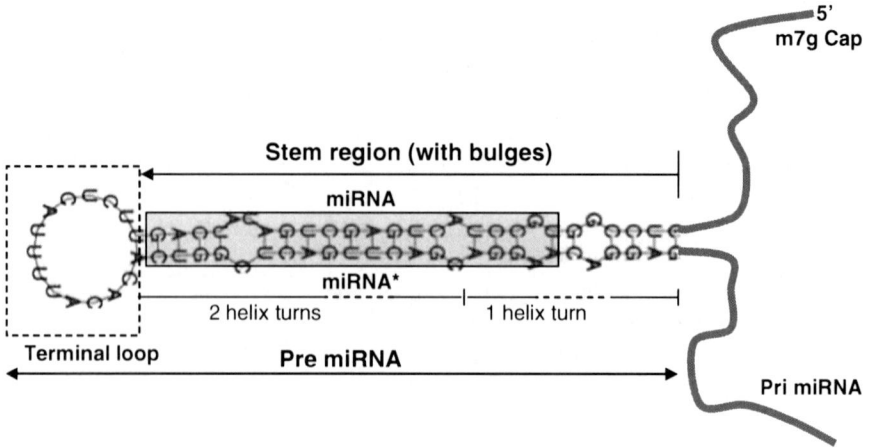

Fig. 20.1 Structural features of an miRNA precursor. The pri-miRNA consists of one or more pre-miRNA(s), which harbor mature miRNA. Pre-miRNA consists of a central stem region of ~3 helical turns (not shown on scale here) and a terminal loop region. The mature miRNA/miRNA* reside in the stem region (*shaded region*). The pre-miRNA is sequentially processed by Dicer (viz., DCL1) to release miRNA:miRNA* duplex. The structural parameters of plant pre-miRNAs are different from those of the animal pre-miRNAs

loop-distal cleavage sites increases the accumulation of accurately cleaved precursor miRNAs but decreases the abundance of the mature miRNAs. Pri-miRNAs with an unpaired lower stem would not process, while variants with a perfectly paired middle or upper stem were processed normally (Song et al. 2010).

Other studies have also demonstrated that in plants, secondary structure of the lower stem region is most important and a shift or introduction of a major bulge in this region could move or disrupt the initial cleavage site. If the bulge in the lower stem is altogether removed, processing of the precursor is highly inaccurate or inefficient. Thus, this bulge or loop above the unpaired region in the lower stem defines a required structure that ensures accurate and efficient processing of plant miRNA precursors. On the other hand, most point mutations in the terminal loop had minimal impact on miRNA maturation. In case of partial deletion of the upper loop, mature miR172 was still detected, but a complete deletion of the terminal loop completely abolished mature miR172 accumulation (Mateos et al. 2009). However, the structure at the upper junction of the miRNA:miRNA* duplex and terminal loop is important for processing because the second DCL1 cut is located in this region. It is extremely important to note that, unlike animal miRNAs, miRNA:miRNA* duplex region of plant miRNAs are tolerant to point mutations. This consideration has important bearing on designing the artificial microRNAs (amiRNAs) which have been illustrated in later sections.

The rules mentioned above hold true for majority of the miRNA precursors, but there are exceptions also. One notable exception is in pri-miR319a, which has been most extensively used in designing many plant amiRNAs. Unlike other precursors, pri-miR319a exhibits a loop-to-base processing (Bologna et al. 2009).

Conventional miRNA biogenesis pathway presumes that only one of the strands of the miRNA:miRNA* duplex is incorporated into the functional RISC complex, while the other strand is degraded. However, with the emerging evidences from the next generation sequencing data, it is clear that the miRNA* strand also accumulates in various tissues and under varying physiological conditions. Generally, the animal-miRNA strand associates with AGO1, while the animal-miRNA* strand has affinity toward AGO2 (Czech et al. 2009). This discrimination in the cellular abundance of miRNA and miRNA* is largely governed by mismatches in the miRNA:miRNA* duplex at positions 9 and/or 10 relative to the miRNA strand, with thermodynamic asymmetry of the 5′ end playing a subsidiary role. In plants, 5′ nucleotides play a role in specific AGO interactions. The functionality of the miRNA*s has been demonstrated using the reporter-based RNAi sensor constructs and the targets of a small number of miRNA* sequences have been identified (Yang et al. 2011). Thus, the ability to distinguish miRNA from miRNA* is crucial for not only correct functional roles of endogenous miRNAs, but also it has important significance for amiRNA design. Apart from structure, specific pri-miRNA binding proteins are also found to modulate pri-miRNA processing. A group of proteins, known as SMADs directly binds to stem regions of a group of animal pri-miRNA and promote their processing by Drosha (Davis et al. 2010).

20.3   Artificial miRNAs (amiRNAs)

Many excellent reviews on biogenesis and silencing functions of natural miRNAs that are encoded by genomes of several eukaryotic species are available in the literature (Djuranovic et al. 2011). These functions are entirely dependent on the abundance of the cellular factors as mentioned in Table 20.1. However, presence of such factors could be exploited to broaden or engineer the silencing purview carried out by the eukaryotic cells. The described details of miRNA processing determinants amply suggest that pre-miRNA processing is largely dependent on precursor structure and not on the miRNA/miRNA* sequence. Accordingly, it has been shown that altering several nucleotides within the miRNA/miRNA* strands of the miRNA precursor transcript has no bearing on mature miRNA biogenesis and maturation, as long as the overall secondary structure of the modified precursor remains unchanged. This observation has been utilized to alter endogenous miRNA precursors to produce synthetic or artificial mature miRNAs (amiRNAs) designed to target any gene of interest. amiRNA technology was first used to knock down gene expression in human cell lines (Zeng et al. 2002). A nice example could be drawn from usages of the backbone of pre-miRNA451. As mentioned earlier, the formation of miRNA451 is dicer independent. Hence using the backbone of pre-miRNA451, many different miRNAs could be expressed in cells lacking the Dicer activity (Yang et al. 2010; Cifuentes et al. 2010). This technology, i.e., expression of any designed mature miRNA sequence using the backbone of known pre-miRNA, was subsequently employed to specifically downregulate gene expression in transgenic plants (Schwab et al. 2006; Alvarez et al. 2006). While sense transgene–mediated silencing and intron-spliced hairpin RNA constituted the first and second generation of gene silencing technologies, respectively, amiRNA technology could be regarded as a third generation of gene silencing technologies.

20.4 Design Parameters of amiRNAs

amiRNAs are essentially designed to mimic the natural miRNAs. Their sequences are designed according to the determinants of plant miRNA target selection, such that the 21-nt RNA specifically silences its target gene(s). amiRNAs are intended to be similar to natural miRNAs by three major criteria (1) express a 5′ terminal uracil (found in most of the natural plant and animal miRNAs), (2) display 5′ instability relative to the amiRNA* duplex strand, and (3) possess adenosine at position 10 of the mature amiRNA strand (Reynolds et al. 2004; Mallory et al. 2004). In the stem region of miRNA precursor containing the miRNA/miRNA* duplex, the end of one duplex strand is normally thermodynamically less stable. The strand with lower thermodynamic stability at its 5′ end (5′ instability) is preferentially incorporated into activated RISC (Khvorova et al. 2003; Schwarz et al. 2003). This principle of strand asymmetry of natural miRNAs is unfailingly followed in amiRNA design. The amiRNA and amiRNA* of the artificial precursor should maintain the same structural relationship (in terms of mismatches and bulges) with each other as found in the corresponding miRNA and miRNA* of its natural precursor.

For the amiRNA to be effective in repressing its intended target, pairing of the target to the 5′ portion of the amiRNA (positions 2–12) is most important, and this region should not have any mismatch and rarely any more than one. Presence of up to two mismatches in the 5′ part can only be compensated by perfect pairing in the 3′ portion. Mismatches at the presumptive cleavage site (between positions 10 and 11) and more than two consecutive mismatches in the 3′ part should be avoided. Usually, 1–3 mismatches are deliberately kept in the 3′ part of the amiRNA to reduce the likelihood that an amiRNA would act as a primer for RNA-dependent RNA polymerases (RdRP), and thereby trigger secondary RNAi. The likelihood of RdRP-mediated transitivity can be checked by determining potential 21-mer secondary siRNAs for the amiRNA target gene from both the strands of a 200–300-bp region, surrounding the initial binding site of the amiRNA. Probable targets of these secondary siRNAs can be recognized using miRNA/siRNA target prediction algorithms. As a thumb rule, the overall free energy of amiRNA–target pairing should not exceed −30 kcal/mole.

The amiRNA should be designed to target only the gene of interest to limit and/or remove the chances of off-target silencing. The specificity of the amiRNAs can be tested by various miRNA target prediction algorithms. While doing this investigation, special attention should be given to the matches between 2 and 8 nucleotide positions of amiRNAs, as base pairing to this so-called seed region between positions 2 and 8 is often enough for target recognition in animals.

In contrast to miRNAs in animals, natural plant miRNAs have a very narrow action spectrum and target only mRNAs with few mismatches.

An automated Web tool that incorporates many of the above described parameters has been developed by Dr Detlef Weigel's laboratory at Max Planck Institute of Developmental Biology, Tubingen, and is publicly available at wmd3. weigelworld.org. The Web site also contains detailed protocols for pre-amiRNA synthesis.

20.5   Pre-amiRNA Construction and New amiRNA Vectors

Following its design, the amiRNA can be incorporated into a suitable precursor which would help deliver the amiRNA into a cell. In the cell, the precursor would be recognized by cellular miRNA biogenesis machinery which would dice it like any other natural pre-miRNA and release the amiRNA to carry out its intended silencing. For this purpose, the amiRNA is brought in a backbone of a natural miRNA precursor by replacing the resident miRNA and miRNA* of natural miRNA precursor by designer amiRNA and amiRNA* using site directed mutagenesis. This mutagenesis can be achieved by a series of overlapping PCRs using appropriately designed oligonucleotides (Schwab et al. 2006), in which the pre-amiRNA is synthesized in parts and later fused together using two terminal oligonucleitides to get the full-length amiRNA precursor. As a simpler alternate strategy, the full-length pre-amiRNA gene itself can be commercially synthesized.

In most of the amiRNA studies in plants, the natural precursor structures of ath-miRNA159a, ath-miRNA164b, ath-miRNA172a, ath-miRNA319a, and osa-miRNA528 have been successfully used. Recently, a simple amiRNA vector (pAmiR169d), which is based on the structure of *Arabidopsis* miRNA169d precursor (pre-miRNA169d), was designed (Liu et al. 2010). As the processing efficiency of the various vectors varies widely, the abundance of the mature amiRNA accumulation also varies accordingly. Thus, the choice of the vector backbone is crucial to generating the appropriate amount of intracellular amiRNA.

In mammals, precursors of miRNA-16, miRNA-206, miRNA-331, and many other natural precursors have been used to engineer amiRNAs. New amiRNA vectors for concurrent targeting of multiple genes have also been developed in mammalian systems (Hu et al. 2009). The optimum number of concatenated amiRNA precursors in a multi-amiRNA expression vector should not be more than four. The relative position of an amiRNA in the multi-amiRNA expression vector has no apparent influence on its gene silencing activity (Hu et al. 2010). Inducible Pol II promoters can be used to overexpress authentic miRNAs in cell culture (Zeng et al. 2005). A modified *Cabbage leaf curl virus* vector has been designed to express artificial as well as endogenous miRNAs in plants. Using this viral miRNA expression system, it was demonstrated that amiRNA-based virus-induced gene silencing (VIGS) or "miRNA VIGS" offered efficient silencing of the expression of the endogenous genes PDS, Su, CLA1, and SGT1 in *Nicotiana benthamiana* infiltrated leaves (Tang et al. 2010b). Furthermore, the "Highly Efficient gene Silencing Compatible vector" (HESC vector) is a new amiRNA plant expression vector for rice and is deemed suitable for use in a systems biology approach for functional genomic research (Wang et al. 2010b). Similarly, a novel approach to construct plant amiRNA expression vectors with seamless enzyme-free cloning (SEFC) and mating-assisted genetically integrated cloning (MAGIC) has been tested to generate more than 200 amiRNA vectors in a high-throughput fashion (Yan et al. 2011).

20.6 Applications of amiRNA Technology

amiRNAs are generally seen as a third-generation RNAi technology and have found a wide range of applications in basic research, therapeutic medicine, and plant biotechnology.

20.6.1 amiRNA in Basic Research

amiRNA screens are becoming the method of choice for large-scale functional genomics studies in both animal and plant systems. The Arabidopsis 2010 project, which aimed to identify function of all the *Arabidopsis* genes by the year 2010, has used the amiRNA approach to help achieve this goal (Small 2007). Each of the estimated 22,000 *Arabidopsis* genes is targeted by three unique amiRNAs, making this the first amiRNA-based genome-wide resource for plant RNAi. These constructs use the ath-miRNA319a backbone and CaMV 35S promoter. Recently, amiRNA-based targeting of *AGAMOUS-LIKE 6* (*AGL6*) gene in *Arabidopsis* revealed its novel role in regulation of circadian clock (Yoo et al. 2011). In a study, using an amiRNA approach to simultaneously silence the three SHINE (SHN) clade members, it was revealed that these transcription factors act redundantly to shape the surface and morphology of *Arabidopsis* flowers (Shi et al. 2011). Similarly, amiRNA-mediated targeting led to identification of an *Arabidopsis* plasma membrane–located ATP transporter (PM-ANT1) important for anther development and autogamy (Rieder and Neuhaus 2011). In another study, two different amiRNA constructs were designed to specifically downregulate two different subsets of phenylalanine ammonia-lyase (*PAL)* genes in *Populus* tree, revealing differential regulation within the gene family (Shi et al. 2010).

For use in animal cell culture–based systems, amiRNA libraries have been produced to target mammalian genes like mouse p53 ORF (Xue et al. 2009). Such an enzymatically prepared library has the potential to target the whole transcriptome for genome-wide RNAi screening, or a randomized amiRNA library to search for functional amiRNAs. amiRNA technology has been used to develop new mouse models of human diseases. Autosomal dominant polycystic kidney disease (ADPKD) is one of the most common life-threatening inherited diseases, and the *PKD1* gene is responsible for most cases of this disease. Previous efforts to develop a mouse model recapitulating this disease have been unsuccessful owing to complexities posed by "haploinsufficiency" phenomenon exhibited by *PKD1*. Recently, ubiquitin B driven co-cistronic expression of two amiRNAs targeting *PKD1* and an Emerald GFP reporter in transgenic mice helped in creating an ideal mouse model for studying ADPKD (Wang et al. 2010a).

20.6.2 amiRNA in Medicine

In medicine, amiRNAs are finding application toward developing therapies against dreaded human diseases—infectious as well as metabolic or neurological disorders.

amiRNA against a GPCR family chemokine receptor CXCR4 can serve as an alternative means of therapy to lower CXCR4 expression and to block the invasion and metastasis of breast cancer cells (Liang et al. 2007). AmiRNA expression vector successfully targeting prostate apoptosis response-4 (PAR4) gene in SW620 cells has been developed which may lead to future therapy against human colorectal cancer (Han et al. 2010).

AmiRNAs can be used to enhance immunogenicity of DNA vaccines by silencing cellular apoptotic and antiviral pathway that limit maximal antigen expression. In a recent study, DNA vaccine vectors co-expressing amiRNA with HIV-1 envelope (Env) antigens were found to influence the magnitude or quality of the immune responses to Env in mice. Vaccinating BALB/c mice with a DNA vaccine vector delivering amiRNA targeting cellular antiviral protein PERK was able to augment the generation of Env-specific T cell immunity (Wheatley et al. 2011). Further trials will be needed to ascertain if such novel approach can lead to an effective AIDS vaccine for eventual human use.

amiRNA-based antiviral therapy has shown promise for circumventing viral mutations and targeted delivery, two major constrains faced by current RNAi-based therapies. Ye et al. (2011) designed two amiRNAs targeting 3'UTR of myocarditis causing coxsackievirus B3 (CVB3) genome with mismatches to the central region of their targeting sites. To achieve specific delivery, amiRNAs were linked to the folate-conjugated bacterial phage packaging RNA (pRNA) which delivered the complexes into HeLa cells, a group of folate receptor positive cancer cells widely used as an in vitro model for CVB3 infection, via folate-mediated specific internalization. It was found that the designed pRNA–amiRNA conjugates were tolerable to target mutations with little effect on triggering interferon induction. amiRNAs have also been used to target other important mammalian viruses like hepatitis B (Gao et al. 2008) and rabies (Israsena et al. 2009).

In a study demonstrating the potential of amiRNA-based therapy for veterinary use, amiRNA targeting foot-and-mouth disease virus (FMDV) 3D gene, were found to efficiently inhibit FMDV replication in vitro (Du et al. 2011).

An interesting study supports the potential use of amiRNA as a therapeutic agent for the treatment of alcohol dependence. In the brain, the stress system plays an important role in motivating continued alcohol use and relapse. The neuropeptide substance P and the neurokinin-1 receptor (NK1R) are involved in the stress response and drug reward systems. Lentivirus vector–based delivery of amiRNA targeting NK1R into the mouse brain leads to decreased voluntary alcohol consumption by these mice (Baeka et al. 2010). amiRNA-based therapy of neurological disorders is more desirable over small hairpin RNAs (shRNAs) as amiRNAs can mitigate shRNA-mediated toxicity in the brain (McBride et al. 2008).

In another study pointing toward potential of amiRNA-based therapy for augmenting medical transplantation procedures, knockdown of a transmembrane protein neuropilin-2 (NP2) by amiRNA improved corneal graft survival by selectively inhibiting lymphangiogenesis promoted immune rejection (Tang et al. 2010a).

20.6.3 amiRNA in Plant Biotechnology

In addition to Arabidopsis, amiRNA technology has been validated in rice (Warthmann et al. 2008), moss (Khraiwesh et al. 2008), Chlamydomonas, tobacco, and tomato (Alvarez et al. 2006).

amiRNA technology was used to develop a method for customized expression of flowering—a trait of significant biotechnological interest. Ethanol-inducible expression of a heterologous FT gene from the model legume Medicago truncatula (Medicago) was able to rescue the late-flowering phenotype of Arabidopsis transgenics overexpressing an amiRNA targeting endogenous FT gene (Yeoh et al. 2011).

Application of this technology to achieve plant virus resistance has been proved successful for several viruses and in model plants (Niu et al. 2006; Qu et al. 2007; Ai et al. 2011). However, the potential of this approach has not been tested on geminiviruses or in crop plants (Shepherd et al. 2009). Now, amiRNAs have been shown to be effective in engineering geminivirus resistance in tomato (Yadava and Mukherjee 2010). Two pre-amiRNAs targeting the conserved regions of Tomato leaf curl virus (ToLCV)—Replicase (Rep) along with AC4 and AC2 RNAi suppressors separately—were designed using the ath-miRNA319a backbone in our laboratory. In computational analysis, the designed amiRNAs were found to bind to the viral targets, namely, Rep/AC4 and Rep/AC2, effectively and specifically without any off-target effects. These amiRNAs were processed in vitro, as previously demonstrated for endogenous miRNA precursors, using either Arabidopsis inflorescence or wheat germ extracts; and the mature amiRNAs were also detected in vivo in stably transformed tobacco and tomato plants. The transgenic plants overexpressing the amiRNAs were resistant to Tomato leaf curl New Delhi virus (ToLCNDV) bipartite agroinfectious clones. A small number of the transgenic lines showed complete immunity to viral infection. These transgenic lines also inhibited the mini-viral DNA replication (Fig. 20.2). In conclusion, the

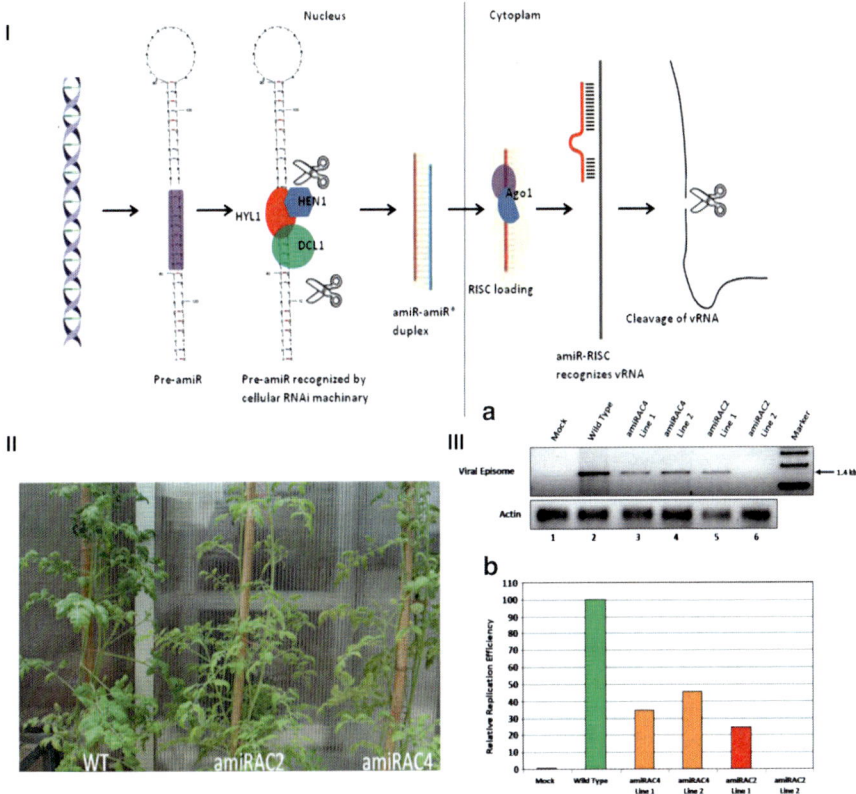

Fig. 20.2 amiRNA-mediated virus resistance. (I) Working hypothesis. The amiRNA gene is introduced in the nuclear genome of tomato plants using stable transgenesis. The transgene is designed to mimic natural miRNA biogenesis process. As soon as the transgene is transcribed, it assumes a characteristic "hairpin"-type secondary structure, termed pri-amiRNA. This secondary structure is a substrate for cellular DCL1, which together with other accessory proteins, like HEN1, HYL1, etc., sequentially cleaves pri-amiRNA into 20-mer duplex RNA (amiRNA: amiRNA* duplex). The duplex is exported from the nucleus to the cytoplasm with the help of HASTY. Because the principle of strand asymmetry is inbuilt during amiRNA transgene design, one of the strands of amiRNA:amiRNA* duplex is preferentially incorporated into the RISC (RISC loading) following unwinding. The amiRNA-activated RISC is ready to cleave/repress cognate RNA targets in the cytoplasm, which is dependent on the loaded amiRNA sequence. The amiRNA has been specifically designed to bind to and suppress ToLCV Rep/AC2/AC4 mRNA. Thus, amiRNA strategy uses cell's own miRNA biogenesis pathway to produce anti-ToLCV miRNAs leading to containment of the virus. (II) amiRNA transgenic tomato resists viral challenge. Response of wild-type and amiRNA transgenic tomato lines to *Agrobacterium*-delivered ToLCNDV infectious clones at 60 dpi is shown. Both amiRNA–AC4 and amiRNA–AC2 transgenic lines were challenged with an agroinfectious clone of a bipartite virus ToLCNDV, harboring both DNA A and DNA B (A + B). At 60 days post infiltration (dpi) A + B inoculated wild-type plant (WT) developed severe curling. However, most of the transgenic lines remained largely healthy, without much curling. (III) Amplicon assay for measuring viral replication efficiency. (**a**) Detection of viral episome in wild-type and transgenic tomato plants. At 15 dpi, the viral episome was detected, using a PCR-based strategy, in wild-type plant. However, all the four transgenic

results of this study demonstrate that amiRNAs can be effectively deployed to engineer ToLCV resistance in tomato. This study opens up the possibility of employing the amiRNA strategy to target other economically important viruses in other important crop species.

20.7   Advantages of amiRNA Technology

amiRNAs have several unique advantages over other RNAi technologies for functional genomic applications. amiRNAs are likely to be particularly useful for targeting groups of closely related genes, including tandemly arrayed genes. Approximately 4,000 genes in *Arabidopsis* are found in tandem arrays, and no convenient tool has been reported to generate knockout lines. Because of their exquisite specificity, amiRNAs could possibly be adapted for allele-specific knockouts. There is a substantial level of alternative splicing, and amiRNAs have the potential to target only specific splice variants. Unlike conventional hairpin mediated RNAi, in which small RNAs are generated from both the strands, amiRNAs have the advantage of being strand specific. Moreover, amiRNA sequences can be optimized for high efficiency since they are always produced from the same locus in their precursors. Most importantly, amiRNA-induced mutants can be complemented by amiRNA resistant targets, where silent mutations can be introduced in the amiRNA target site, disrupting amiRNA-mediated degradation (Schwab et al. 2006).

In the hairpin RNAi approach, multiple siRNAs are formed from one precursor. As the exact positions of Dicer/DCL2/DCL3 cleavage are not known, the $5'$ ends of siRNAs cannot be accurately determined. Also, the parameters determining targets of siRNAs are yet not fully known. Prediction of small RNA targets other than the perfectly complementary intended targets is difficult. In contrast, amiRNAs are produced from pre-amiRNAs which preferentially generate only one single stable mature amiRNA. Since, the determinants of amiRNA target selection have been determined, the complete target spectrum of amiRNA is readily identifiable (Schwab et al. 2006). siRNAs are perfectly complementary to their targets, and thus their binding to the target can lead to transitivity, i.e., RdRP-dependent amplification and generation of secondary siRNAs. The promiscuity of secondary siRNAs may lead to off-target effects in RNAi-based transgenic plants. On the other hand, amiRNAs can be deliberately designed to include few mismatches with

---

Fig. 20.2 (continued) lines showed much reduced accumulation of the viral episome. A few lines, for example, the one represented in lane #6, did not show any episome activity at all, reflecting their robustness to resist the virus. Expectedly, no episome was detected in the mock inoculated plants. (**b**) The intensity of episomal bands was measured using ImageQuant software and normalized with respect to actin control. Relative replication efficiency as percentage of that of wild type is shown as bar graph. (Yadava et al. 2010; Yadava 2010)

respect to their target, thereby altogether avoiding complications of transitivity. These mismatches would not have any effect on silencing as even imperfectly matched miRNAs are known to efficiently repress their targets. One of the major concerns of using RNAi in developing transgenic crops is the potential pleiotropic effects caused by off-targeting. This concern is most satisfactorily addressed by amiRNAs, which paves the way for their utilization in developing novel traits in GM crops (Yadava 2011).

Earlier, it was believed that unlike conventional siRNA-based RNAi, amiRNAs can function in tissue-specific and inducible manner, as they have only limited cell-autonomous effect (Schwab et al. 2006). However, question of systemicity of amiRNA has been revisited recently and emerging evidence points out noncell-autonomous nature of amiRNAs as well as trans-acting siRNAs. Nonautonomous effects of the miRNA were seen to be triggered by several different miRNA precursors deployed as backbones (Felippes et al. 2011).

siRNA-based silencing is known to be compromised in extremes of temperature. On the other hand, miRNA biogenesis machinery is more robust as many natural miRNAs are produced by the organism in varying conditions, including extremes of temperatures. Thus, amiRNA-based silencing is expected to be more resilient, and amiRNA transgenics using precursors that express optimally in such conditions of abiotic stresses can possibly be widely adapted to temperate and tropical agroecosystems.

20.8   Perspectives

The advent of amiRNA technology has widened the available options to direct efficient gene silencing. Although, determinants of pre-miRNA processing are not yet fully characterized, there is already a rush to use amiRNA technology for various biotechnological goals using and maintaining the endogenous backbones of tested endogenous miRNAs such as ath-miRNA159a, ath-miRNA164b, ath-miRNA172a, ath-miRNA319a, and osa-miRNA528. Plant amiRNA expression vectors are robust to alterations in the stem region and different amiRNA precursors can be combined with various promoters to achieve desired abundance of mature amiRNA offering optimum silencing. Nevertheless, with a more detailed understanding of some of the more important structural features that control the efficiency of plant pri-miRNA processing, it may be possible to engineer miRNA precursors with different processing abilities. One of the important aspects, often ignored with respect to amiRNA design is that of RNA editing of amiRNA precursors. RNA editing of primary transcripts by ADARs (adenosine deaminases acting on RNA) modifies adenosine (A) into inosine (I). Because the base pairing properties of inosine are similar to those of guanosine (G), A-to-I editing of miRNA precursors may change their sequence, base pairing, and structural properties and can influence their further processing as well as their target recognition abilities. Several examples of editing-mediated regulation of miRNA processing have been described. Very often, it is required to express multiple amiRNAs to generate a desired phenotype or agricultural trait. Duan et al. (2008) expressed three amiRNAs by ligating three pre-amiRNAs and keeping the ligated product under the 35S transcription unit. These three small RNAs targeted three different accessible regions of the conserved 3'UTR of the RNA genomes of various *Cucumber mosaic virus* (CMV) isolates. The transgenic tobacco or *Arabidopsis* expressing the three amiRNAs showed extreme resistance to the challenge CMV. However, this technology has limited applications so far because of the unavailability of a suitable plant pri-miRNA vector that can give rise to multiple pre-miRNAs in plants. Though such vectors are well known in animals, the corresponding plant vectors remain to be discovered. Such plant vectors will usher in a new phase of plant biotechnology. Another important area of amiRNA research would be to explore their usage in transcriptional gene silencing. So far, the roles of amiRNAs have remained limited in posttranscriptional gene silencing. Revealing their usages in formation of heterochromatin would definitely show the novel means in control of human disease.

Acknowledgments PY is thankful to Council of Scientific and Industrial Research, India, for the Shyama Prasad Mukherjee Fellowship, the award of which enabled PY to develop the amiRNA transgenics, and also to the Management of Ankur Seeds Pvt Ltd. Some part of the research reported here was possible due to a grant awarded by DBT, Govt. of India, to SM.

References

Ai T, Zhang L, Gao Z et al (2011) Highly efficient virus resistance mediated by artificial microRNAs that target the suppressor of PVX and PVY in plants. Plant Biol (Stuttg) 13:304–316. doi:10.1111/j.1438-8677.2010.00374.x

Alvarez JP, Pekker I, Goldshmidt A, Blum E, Amsellem Z, Eshed Y (2006) Endogenous and synthetic microRNAs stimulate simultaneous, efficient, and localized regulation of multiple targets in diverse species. Plant Cell 18:1134–1151

Ambros V, Lee RC, Lavanway A, Williams PT, Jewell D (2003) MicroRNAs and other tiny endogenous RNAs in *C. elegans*. Curr Biol 13:807–818

Baeka MN, Junga KH, Haldera D, Choia MR, Leea B, Leeb B, Jungb MH, Choib I, Chungc M, Ohd D, Chaia YG (2010) Artificial microRNA-based neurokinin-1 receptor gene silencing reduces alcohol consumption in mice. Neurosci 475:124–128. doi:10.1016/j.neulet.2010.03.051

Bartel DP (2004) MicroRNAs: genomics, biogenesis, mechanism, and function. Cell 116:281–297

Bogerd HP, Karnowski HW, Cai X, Shin J, Pohlers M, Cullen BR (2010) A mammalian herpesvirus uses noncanonical expression and processing mechanisms to generate viral MicroRNAs. Mol Cell 37:135–142

Bollman KM, Aukerman MJ, Park M, Hunter C, Berardini TZ, Poethig RS (2003) HASTY, the Arabidopsis ortholog of exportin 5/MSN5, regulates phase change and morphogenesis. Development 130:1493–1504

Bologna NG, Mateos JL, Bresso EG, Palatnik JF (2009) A loop-to-base processing mechanism underlies the biogenesis of plant microRNAs miR319 and miR159. EMBO J 28 (23):3646–3656. doi:10.1038/emboj.2009.292

Brodersen P, Sakvarelidze-Achard L, Bruun-Rasmussen M, Dunoyer P, Yamamoto YY, Sieburth L, Voinnet O (2008) Widespread translational inhibition by plant miRNAs and siRNAs. Science 320:1185–1190

Cai X, Hagedorn CH, Cullen BR (2004) Human microRNAs are processed from capped, polyadenylated transcripts that can also function as mRNAs. RNA 10:1957–1966

Cifuentes D, Xue H, Taylor DW, Patnode H, Mishima Y, Cheloufi S, Ma E, Mane S, Hannon GJ, Lawson ND, Wolfe SA, Giralde AJ (2010) A novel microRNA processing pathway independent of Dicer requires AGO2 catalytic activity. Science 328:1694–1698

Czech B, Zhou R, Erlich Y, Brennecke J, Binari R, Villatta C, Gordon A, Perrimon N, Hannon GJ (2009) Hierarchical rules for Argonate loading in Drosophila. Mol Cell 36:445–456

Davis BN, Hilyard AC, Nguyen PH, Lagna G, Hata A (2010) Smad proteins bind a conserved RNA sequence to promote microRNA maturation by Drosha. Mol Cell 13:373–84

Djuranovic S, Nahvi A, Green R (2011) A parsimonious model for gene regulation by miRNAs. Science 331:550–553. doi:10.1126/science.1191138

Du J, Gao S, Luo J et al (2011) Effective inhibition of foot-and-mouth disease virus (FMDV) replication in vitro by vector-delivered microRNAs targeting the 3D gene. Virol J 8:292. doi:10.1186/1743-422X-8-292

Duan C-G, Wang C-H, Fang R-X, Guo H-S (2008) Artificial microRNAs highly accessible to targets confer efficient virus resistance in plants. J Virol 82:11084–11085

Fang Y, Spector DL (2007) Identification of nuclear dicing bodies containing proteins for MicroRNA biogenesis in living Arabidopsis plants. Curr Biol 17:818–823

Felippes FFD, Ott F, Weigel D (2011) Comparative analysis of non-autonomous effects of tasiRNAs and miRNAs in Arabidopsis thaliana. Nucleic Acids Res 39(7):2880–2889. doi:10.1093/nar/gkq1240

Gao Y, Yu L, Wei W, Li J, Luo Q, Shen J (2008) Inhibition of hepatitis B virus gene expression and replication by artificial microRNA. World J Gastroenterol 14:4684–4689

Han N, Chu LS, Cao J et al. (2010) [Construction and application of an artificial microRNA expression vector for inhibiting PAR4] 26(11):1105–7

Hu T, Fu Q, Chen P, Ma L, Sin O, Guo D (2009) Construction of an artificial MicroRNA expression vector for simultaneous inhibition of multiple genes in mammalian cells. Int J Mol Sci 10(5):2158–68

Hu T, Chen P, Fu Q, Liu Y, Ishaq M, Li J, Ma L, Guo D (2010) Comparative studies of various artificial microRNA expression vectors for RNAi in mammalian cells. Mol Biotechnol 46:34–40

Huang V, Qin Y, Wang J, Wang X, Place RF, Lin G, Lue TF, Li L (2010) RNAa is conserved in mammalian cells. PLoS One 5:e8848

Israsena N, Supavonwong P, Ratanasetyuth N, Khawplod P, Hemachudha T (2009) Inhibition of rabies virus replication by multiple artificial microRNAs. Antiviral Res 84:76–8

Khraiwesh B, Ossowski S, Weigel D, Reski R, Frank W (2008) Specific gene silencing by artificial microRNAs in Physcomitrella patens: an alternative to targeted gene knockouts. Plant Physiol 148:684

Khvorova A, Reynolds A, Jayasena SD (2003) Functional siRNAs and miRNAs exhibit strand bias. Cell 115:209–216

Kim VN (2005) MicroRNA biogenesis: coordinated cropping and dicing. Nat Rev Mol Cell Biol 6:376–385

Kim DH, Sætrom P, Snøve O, Rossi JJ (2008) MicroRNA-directed transcriptional gene silencing in mammalian cells. Proc Natl Acad Sci USA 105:16230–16235. doi:10.1073/pnas.0808830105

Liang Z, Wu H, Reddy S, Zhu A, Wang S, Blevins D, Yoon Y, Zhang Y, Shim H (2007) Blockade of invasion and metastasis of breast cancer cells via targeting CXCR4 with an artificial microRNA. Biochem Biophys Res Commun 363:542–546

Liu C, Zhang L, Sun J, Luo Y, Wang M, Fan Y, Wang L (2010) A simple artificial microRNA vector based on ath-miR169d precursor from Arabidopsis. Mol Biol Rep 37:903–909

Lund E, Güttinger S, Calado A, Dahlberg JE, Kutay U (2004) Nuclear export of microRNA precursors. Science 303:95–98

Mallory AC, Reinhart BJ, Jones-Rhoades MW, Tang G, Zamore PD, Barton MK, Bartel DP (2004) MicroRNA control of PHABULOSA in leaf development: importance of pairing to the microRNA 5′ region. EMBO J 23:3356–3364

Mateos JL, Bologna NG, Chorostecki U, Palatnik JF (2009) Identification of MicroRNA processing determinants by random mutagenesis of Arabidopsis MIR172a precursor. Curr Biol 20:49–54. doi:10.1016/j.cub.2009.10.072

McBride JL, Boudreau RL, Harper SQ, Staber PD, Monteys AM, Martins I, Gilmore BL, Burstein H, Peluso RW, Polisky B, Carter BJ, Davidson BL (2008) Artificial miRNAs mitigate shRNA-mediated toxicity in the brain: implications for the therapeutic development of RNAi. Proc Natl Acad Sci USA 105:5868–5873

Niu QW, Lin SS, Reyes JL, Chen KC, Wu HW, Yeh SD, Chua NH (2006) Expression of artificial microRNAs in transgenic Arabidopsis thaliana confers virus resistance. Nat Biotechnol 24:1420–1428

Ono M, Scot MS, Yamada K, Avolio F, Barton GJ, Lamond AI (2011) Identification of human miRNA precursors that resemble box C/D snoRNA. Nucleic Acids Res 39:3879–3891

Qu J, Ye J, Fang R (2007) Artificial microRNA-mediated virus resistance in plants. J Virol 81:6690

Reynolds A, Leake D, Boese Q, Scaringe S, Marshall WS, Khvorova A (2004) Rational siRNA design for RNA interference. Nat Biotechnol 22:326–330

Rieder B, Neuhaus HE (2011) Identification of an *Arabidopsis* Plasma membrane—located ATP transporter important for anther development. Plant Cell Online. doi:10.1105/tpc.111.084574

Ruby JG, Jan CH, Bartel DP (2007) Intronic microRNA precursors that bypass Drosha processing. Nature 448:83–86

Schwab R, Ossowski S, Riester M, Warthmann N, Weigel D (2006) Highly specific gene silencing by artificial microRNAs in Arabidopsis. Plant Cell 18:1121

Schwarz DS, Hutvágner G, Du T, Xu Z, Aronin N, Zamore PD (2003) Asymmetry in the assembly of the RNAi enzyme complex. Cell 115:199–208

Shepherd DN, Martin DP, Thomson JA (2009) Transgenic strategies for developing crops resistant to geminiviruses. Plant Sci 176:1–11

Shi R, Yang C, Lu S, Sederoff R, Chiang VL (2010) Specific down-regulation of *PAL* genes by artificial microRNAs in *Populus trichocarpa*. Planta 232:1281–1288. doi:10.1007/s00425-010-1253-3

Shi JX, Malitsky S, De Oliveira S, Branigan C, Franke RB et al (2011) SHINE transcription factors act redundantly to pattern the archetypal surface of Arabidopsis flower organs. PLoS Genet 7 (5):e1001388. doi:10.1371/journal.pgen.1001388

Song L, Michael J, Axtell FNV (2010) RNA secondary structural determinants of miRNA precursor processing in Arabidopsis. Curr Biol 20:37–41. doi:10.1016/j.cub.2009.10.076

Tang G, Reinhart BJ, Bartel DP, Zamore PD (2003) A biochemical framework for RNA silencing in plants. Genes Dev 17:49–63

Tang X, Sun J, Wang X, Du L, Liu P (2010a) Blocking neuropilin-2 enhances corneal allograft survival by selectively inhibiting lymphangiogenesis on vascularized beds. Mol Vis 16:2354–2361

Tang Y, Wang F, Zhao J, Xie K, Hong Y, Liu Y (2010b) Virus-Based MicroRNA expression for gene functional analysis in plants. Plant Physiol 153:632–641

Vasudevan S, Tong Y, Steitz JA (2007) Switching from repression to activation: MicroRNAs can up-regulate translation. Science 318:1931–1934

Wang E, HsiehLi H, Chiou Y et al (2010a) Progressive renal distortion by multiple cysts in transgenic mice expressing artificial microRNAs against Pkd1. J Pathol 222:238–248. doi:10.1002/path.2765

Wang X, Yang Y, Yu C, Zhou J, Cheng Y, Yan C, Chen J (2010b) A highly efficient method for construction of rice artificial MicroRNA vectors. Mol Biotechnol 3:211–218. doi:10.1007/s12033-010-9291-4

Warthmann N, Chen H, Ossowski S, Weigel D, Hervé P (2008) Highly specific gene silencing by artificial miRNAs in rice. PLoS One 3:e1829

Werner S, Wollmann H, Schneeberger K, Weigel D (2010) Structure determinants for accurate processing of miR172a in Arabidopsis thaliana. Curr Biol 20:42–48. doi:10.1016/j.cub.2009.10.073

Wheatley AK, Kramski M, Alexander MR, Toe JG, Center RJ, Purcell DFJ (2011) Co-expression of miRNA targeting the expression of PERK, but not PKR, enhances cellular immunity from an HIV-1 Env DNA vaccine. PLoS One 6(3):e18225

Xie Z, Kasschau KD, Carrington JC (2003) Negative feedback regulation of Dicer-Like1 in Arabidopsis by microRNA-guided mRNA degradation. Curr Biol 13:784–789

Xue L, Yuan Q, Yang Y, Wu J (2009) Enzymatic preparation of an artificial microRNA library. Biochem Biophys Res Commun 390:791–796

Yadava P (2010) Designing artificial microRNAs as a combat strategy against a plant geminivirus. PhD thesis. ICGEB-Jawaharlal Nehru University

Yadava P (2011) Artificial microRNA: a third generation RNAi technology. In: Gaur RK, Gafni Y, Sharma P, Gupta PK (eds) iRNA technology. Science Publishers, New Hampshire

Yadava P, Mukherjee SK (2010) Engineering geminivirus resistance in tomatoes using artificial microRNAs. Keystone Symposium on RNA Silencing Mechanisms in Plants, Santa Fe, NM, USA, 21–26 Feb 2010

Yadava P, Suyal G, Mukherjee SK (2010) Begomovirus DNA replication and pathogenecity. Curr Sci 98:360–369

Yan H, Deng X, Cao Y et al (2011) A novel approach for the construction of plant amiRNA expression vectors. J Biotechnol 151:9–14. doi:10.1016/j.jbiotec.2010.10.078

Yang JS, Maurine T, Robine N, Rasmussen KD, Jeffrey KL, Chanwani R, Papapetroud EP, Sadelain M, O'Carrol D, Lai EC (2010) Conserved vertebrate miR-451 provide a platform for Dicer-independent, AGO2 mediated microRNA biogenesis. Proc Natl Acad Sci USA 107:15163–15168

Yang JS, Phillips MD, Betel D, Mu P, Ventura A, Siepel AC, Chen KC, Lai EC (2011) Widespread regulatory activity of vertebrate microRNA* species. RNA 17:312–326

Ye X, Liu Z, Hemida MG, Yang D (2011) Targeted delivery of mutant tolerant anti-coxsackievirus artificial microRNAs using folate conjugated bacteriophage Phi29 pRNA. PLoS One 6(6):e21215. doi:10.1371/journal.pone.0021215

Yeoh CC, Balcerowicz M, Laurie R, Macknight R, Putterill J (2011) Developing a method for customized induction of flowering. BMC Biotechnol 11:36. doi:10.1186/1472-6750-11-36

Yi R, Doehle BP, Qin Y, Macara IG, Cullen BR (2005) Overexpression of Exportin 5 enhances RNA interference mediated by short hairpin RNAs and microRNAs. RNA 11:220–226

Yoo SK, Hong SM, Lee JS, Ahn JH (2011) A genetic screen for leaf movement mutants identifies a potential role for *AGAMOUS-LIKE* 6 (*AGL6*) in circadian-clock control. Mol Cells 31:281–287. doi:10.1007/s10059-011-0035-5

Younger ST, Corey DR (2011) Transcriptional gene silencing in mammalian cells by miRNA mimics that target gene promoters. Nucleic Acids Res. doi:10.1093/nar/gkr155

Zeng Y, Wagner EJ, Cullen BR (2002) Both natural and designed micro RNAs can inhibit the expression of cognate mRNAs when expressed in human cells. Mol Cell 9:1327–1333

Zeng Y, Cai X, Cullen BR (2005) Use of RNA polymerase II to transcribe artificial microRNAs. Methods Enzymol 392:371

Chapter 21
Deep Sequencing of MicroRNAs in Cancer: Expression Profiling and Its Applications

Ândrea Ribeiro-dos-Santos, Aline Maria Pereira Cruz, and Sylvain Darnet

Abstract MicroRNAs are small, non-coding RNA molecules that regulate genes post-transcriptionally through the degradation of target messenger RNA or the suppression of protein synthesis. According to recent findings, alterations in the patterns of microRNA expression can lead to the loss or gain of function of particular gene targets, which may then act in carcinogenesis as tumour suppressors or oncogenes, respectively. To study the correlation of cancer progression with microRNA expression, next-generation sequencing technology is a powerful tool to obtain genome-wide expression levels of small RNAs. With this technology, sequencing coverage of the microRNA transcriptome is high and allows the identification of novel microRNAs, the detection of microRNA polymorphisms and the quantification of individual microRNAs by digital gene expression analysis. Thus, the expression profiles of microRNAs can provide accurate diagnoses, therapy effect predictions and prognoses, and they can act as biomarkers for various types of cancer when characterised by next-generation sequencing technologies.

Keywords Applications • cancer • deep sequencing • expression profiling • miRNAs • NGS

Abbreviations

DGE	Digital gene expression
isoMir	miRNA isoform
miRNA	MicroRNA
miRnome	Full set of microRNAs

Â. Ribeiro-dos-Santos (✉)
Laboratório de Genética Humana e Médica/Laboratório de Biologia Computacional, Instituto de Ciências Biológicas, Universidade Federal do Pará, Belém, Pará, Brazil
e-mail: akely@ufpa.br; andrea.santos@pq.cnpq.br

B. Mallick (eds.), *Regulatory RNAs*, DOI 10.1007/978-3-642-22517-8_21,

NGS	Next-generation sequencer
onco-miR	MicroRNA that acts as oncogene
pre-miRNA	miRNA precursor
pri-miRNA	Primary miRNA
qRT-PCR	Real-time quantitative reverse transcription PCR
QV	Quality value
siRNA	Small interference RNA
TS-miR	Tumour suppressor microRNA
WG-smRNA-Seq	Whole-genome small RNA sequencing

21.1   Introduction

Recent advances in the study of microRNAs, small regulatory RNAs, have opened new possibilities to understand the mechanisms regulating carcinogenesis (Zimmerman and Wu 2011). The microRNA regulatory network is complex and extensive: in humans, the prediction is that 1,400 mature miRNAs, encoded in 3% of the genome, regulate over 60% of protein-coding genes (Friedman et al. 2009). The major advance is the evidence that the expression of microRNAs is specific to each type of cancer and its developmental stage (Lu et al. 2005). The new challenge in cancer biology, then, is to characterise the miRnomes, the sets of microRNAs expressed in a specific tissue (or cell) before and during cancer development (Garzon et al. 2006). Comparing the miRnomes, it may be possible to predict the function of each microRNA and verify its association with one type or stage of cancer to define biomarkers (Perrotti and Eiring 2010; Weidhaas 2010).

   To respond to this challenge, the characterisation of miRnomes could be based on approaches using hybridisation and sequencing (Sharma and Vogel 2009). Hybridisation technologies, such as microarrays, allow the quantification of the relative expression levels of RNA species across different conditions (Sharma and Vogel 2009; Wang et al. 2009a). Hybridisation methods have technical limitations, however, including high levels of background signal due to cross-hybridisation, which limits the dynamic range of detection. The design of microarrays also depends on pre-existing knowledge, which means that novel microRNA molecules and transcript variants cannot be detected (Wang et al. 2009a).

   Sequencing-based technologies are based on the sequencing of a small RNA cDNA library, or microRNA cDNA library obtained by enrichment for microRNA from a small RNA cDNA library. The library-sequencing approach has the great advantage of allowing the identification of each molecule present in the sample (Nagalakshmi et al. 2010; Wang et al. 2009c). Some methods, however, such as Sanger sequencing of cDNA, have technical limitations: the library-sequencing coverage is so low that the miRnome is only partially characterised, and gene expression data are reliable only for the most highly expressed genes (Wang et al. 2009a). Next-generation sequencer (NGS) technologies can overcome these limitations by generating high numbers of reads, increasing the microRNA cDNA library sampling and providing a deeper and more complete view of the miRnome.

These new technologies, known as RNA-Seq, or deep sequencing, have been successfully applied to the human miRnome (Hawkins et al. 2010; Lister et al. 2009; Wang et al. 2009a; Morozova et al. 2009). Human small RNA transcriptome, which comprises functional non-coding RNAs shorter than 50 nt, has been estimated to account for approximately 1–2% of the human genome (Borel et al. 2008). Considering that the size of the human genome is approximately 3.1 GB, the small RNA transcriptome size is approximately 31–62 MB, and this amount of sequence can be rapidly analysed in one NGS experiment. Such high-throughput datasets ensure the accurate and complete, or quasi-complete, identification of the microRNA repertoire of the sequenced library. Sequencing the entire population of microRNAs in a sample provides a direct means to identify most, if not all, small RNA species that are present in the sample (Mishra 2009). Considering that a NGS run could reach 20 Gb, the sequencing coverage is very high, more than $300\times$, and allows for highly sensitive microRNA quantification. The range of microRNA detection is large, from one to several thousand copies. This high coverage permits reliable nucleotide polymorphism detection because it is generally accepted that transcriptome coverage of greater than $20\times$ is required for good reliability; however, it has been shown that coverage of greater than $40\times$ is needed for the reliable detection of polymorphisms related to cancer (Meyerson et al. 2010; Goya et al. 2010). The use of deep sequencing to characterise the miRnome has been revolutionary for human genetics and is a powerful tool for studying and understanding miRnome variations in cellular processes such as carcinogenesis (Fabbri and Calin 2010; Negrini et al. 2009).

21.2 Methods

This section describes the NGS technologies used to perform transcriptome deep sequencing and, based on recent cancer genetic studies, describes the strategies that could be used to profile the human miRnome, including details about sampling, small RNA cDNA library preparation, template preparation and sequencing, bioinformatics analysis and statistical post-analysis interpretation.

21.2.1 Next-Generation Sequencing Technologies

Sequencing technologies have significantly evolved over the last 5 years. Using new sequencing chemistries, novel sequencing platforms have been developed and optimised to generate high numbers of reads from a unique library. In comparison to the Sanger sequencing protocol, which was the first and remains the most commonly used sequencing protocol in molecular biology, NGS platforms produce short or ultra-short reads and have a slightly inferior accuracy. However, during one full run on an NGS platform, several million reads are generated in a few days at a relatively low cost, allowing the rapid sequencing of genomes or transcriptomes with high coverage, which is typically not feasible using Sanger sequencing (Morozova et al. 2009).

21.2.1.1 454-Roche Pyrosequencing

The 454-Roche method is based on massive parallel sequencing using pyrosequencing technology (Margulies et al. 2005; Thomas and Harkins 2008). Using emulsion PCR, each single DNA template is attached by amplification to an individual bead, forming a clonal colony. Each single bead is deposited into microwells that are inside a 454 sequencer, wherein the nucleotide-incorporation event results in pyrophosphate release and well-localised luminescence during the sequencing cycle. The luminescence signal for each well and cycle is recorded, and data integration generates sequence reads. The last version of the 454 FLX sequencer generates approximately one million sequence reads per run, with read lengths of 500 bases, for a total sum of 500 MB of data (Thomas and Harkins 2008).

21.2.1.2 Illumina Sequencing

The NGS Solexa platform is based on sequence by synthesis using reversible dye-terminator sequencing (Illumina 2010). A bridge amplification is performed in order to fix DNA molecules on slides and form sequence clusters, wherein each cluster consists of one single DNA molecule. Each ddNTP is fluorescently labelled, and only one nucleotide can be added in each cycle due to a terminal $3'$ blocker, which is chemically removed before each cycle. After each cycle, a camera takes images of the slide, and the emitted signal permits the nucleotide that is added to each sequence cluster to be identified. Integrating the information by cluster and cycle, a sequence read dataset is obtained. This NGS platform, Genome Analyser II, generates short reads, which are approximately 35–100 bp long for the latest version, and, when using a 30-bp-long read, is able to generate 10 GB of data per run (Illumina 2010).

21.2.1.3 SOLiD System

SOLiD technology is based on other sequencing chemistry methods. Each sequencing step is based on the addition of labelled oligonucleotides by hybridisation and ligation, which is a methodology that differs from the two other NGS systems, which are based on an extension step that uses DNA polymerase (Applied Biosystems 2011). Each single molecule of DNA is attached to a bead by emulsion PCR, and all of the beads are deposited onto a glass slide. The sequencing reaction is based on the use of a labelled oligonucleotide mixture. Each labelled oligonucleotide (8-mers) has a degenerated sequence at only the $5'$ extremity, wherein the first two nucleotides are known. The oligonucleotides are used as probes that are annealed with the template and allow for the detection of the dimer that is present at the 3′ end of the template. After annealing and ligation of the oligonucleotides, the detection is performed with only four dyes and a dibase encoding system that allows for the identification of all 16 dinucleotides (Applied Biosystems 2011). In order to obtain the complete sequence, the sequencing reaction is performed in different phases (N-1, N) that allow for an overlap of information on the sequence template. Using this methodology, each base is sequenced twofold with the dimer system, which increases the accuracy of the read dataset and enables the distinguishing of sequencing error and polymorphism variation (Applied Biosystems 2011). The latest version of the SOLiD system, 5500xl, can generate up to 30 Gb of data with read lengths of 75 bp (Applied Biosystems 2011).

The three aforementioned NGS technologies are currently ubiquitously used for the sequencing of miRnomes or small RNA cDNA libraries (Table 21.1); however, many other NGS platforms are in use on a laboratory scale, such as the Helicos and Pacific sequencers (Hayden 2009). The third generation of sequencer, which offers more efficient sequencing technology, has already been developed and will be employed to characterise the human transcriptome and genome (Hayden 2009).

Table 21.1 Deep-sequencing experiments performed to explore human miRnome in cancer disease

Cancer type	Sample	NGS platform	Raw dataset for library (10^3 reads)	Bioinformatics tools	Reference
Breast	2 cell lines	454	180	Vmatch	Nygaard et al. (2009)
	5 tissues	454	180	Vmatch	Nygaard et al. (2009)
Embryonal	10 tissues	SOLiD	19,000	MAQ	Schulte et al. (2010)
Gastric	1 tissue	SOLiD	5,000	RNA2MAP	Ribeiro-dos-Santos et al. (2010)
Hepatocellular	44 tissues	454	500	BLAST/BLAT	Mizuguchi et al. (2011)
	3 cell lines	454	500	BLAST/BLAT	Mizuguchi et al. (2011)
Kidney	6 tissues	Illumina	10,000	Novalign	Weng et al. (2010)
Leukaemia/	3 cell lines	Illumina	5,000	BLAST	Vaz et al. (2010)
lymphomas	31 cell lines	Illumina	11,000	miRDeep	Jima et al. (2010)
Lung	2 cell lines	454	100	Not. infor.	Tarasov et al. (2007)
Nasopharyngeal	2 cell lines	Illumina	5,000	Bowtie	Liao et al. (2010)
	4 tissues	454	20	BLAST	Zhu et al. (2009)
Prostate	4 cell lines	454	45	BLAST	Mitchell et al. (2008)
	2 cell lines	454	20	BLAST	Sun et al. (2011)
	4 pool tissues	454	–	Not. infor.	Szczyrba et al. (2010)
Ovarian	2 cell lines	454	200	BLAST	Wyman et al. (2009)
	3 tissues	454	200	BLAST	Wyman et al. (2009)

21.2.2   NGS Strategies for Profiling miRnome Variations in Cancer

21.2.2.1   The Sampling of a Small RNA cDNA Library Preparation in Cancer Studies

Sample preparation is of major importance for NGS and for assessing the quality of a library preparation. Table 21.1 lists recent published studies regarding the deep sequencing of the miRnome in the context of cancer studies. NGS experiments have been performed using human cell lines or tissues that were directly collected from patients. In Table 21.1, it can be seen that there have been nine miRnome sequencing experiments using carcinoma and teratoma cell lines. Working with stem cells and carcinoma cell lines is easy due to the capability of experimental reproduction, the ability to confirm subsequent results with other approaches and the ability to obtain a high quantity of material, which is a great advantage to the development of a miRnome library (MacLeod et al. 2008). In order to study miRnome variation in cell lines, miRNA expression should be compared to those of other cell lines or to normal tissues from which the cell lines were isolated. Unlike other methods of detection, which may require large amounts of input RNA, the NGS system allows for the sequencing of small RNA transcriptomes from samples that are as small as 10–500 ng of total RNA, which enables the use of tissues or tissue fragments (Wang et al. 2009a).

All of the experimental designs depicted in Table 21.1 are based on the binary comparison(s) of tumour(s) versus controls (or normal). Experiments that use tissues are more complex due to the difficulties that are associated with collecting tumour tissue and defining control tissue (extra-tumour, adjacent tissue, total blood or serum), which must be from the same patient. One other difficulty is the presence of cell layers and heterogeneity in a tumour (Wang et al. 2009b). RNA extraction can be performed with a complete tumour or with only a fragment or piece that is obtained by microdissection. For RNA extraction, tissue samples can be fresh, frozen, formalin- or paraformalin-fixed or paraffin-embedded (Ambion 2010).

Sample preparation is of major importance for NGS experiments in order to define what biological information can be extracted and ensure the preparation of a high-quality small RNA cDNA library.

21.2.2.2   Small RNA cDNA Library Preparation for NGS Platforms

Figure 21.1 depicts the different steps that are required to obtain a small RNA cDNA library from samples such as cell line cultures or tissues, based on an amplification-based protocol. Two kinds of protocols are commonly used, and the greatest difference between the protocols is the step that converts the RNA to double-stranded DNA (Tian et al. 2010). The protocol, which is optimised for the 454 and Illumina platforms, is based on the successive ligation of 5' and 3' adapters and then reverse transcription followed by PCR amplification (Fig. 21.1a) (Lu et al. 2007;

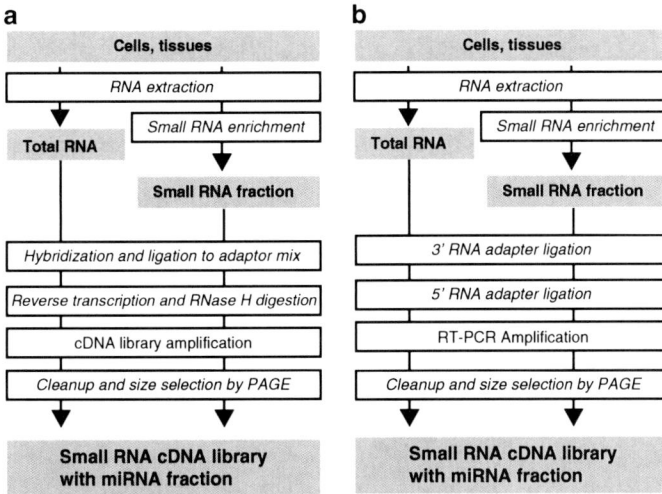

Fig. 21.1 Small RNA cDNA library preparation methods for NGS platform. (**a**) Optimised protocol for SOLiD platform. The library preparation could be performed with total RNA or enriched small RNA fraction. First step is a hybridisation and ligation of two adaptors. Following step is the reverse transcription and cDNA library amplification. Finally, a cleanup and size selection of library, about 100 bp (small RNA + 2 adaptors length), is performed by poly-acrylamide gel electrophoresis. (**b**) Optimised protocol for Illumina and 454 platform. This protocol is without hybridisation step and includes a direct ligation followed by RT-PCR amplification

Thomas and Ansel 2010). The other protocol, which is more specific to the SOLiD system, is based on the hybridisation of adapters with degenerated extremities, followed by reverse transcription and library amplification by PCR (Fig. 21.1b) (Applied Biosystems 2011). For both protocols, the final step involves small RNA cDNA gel separation and size selection by PAGE. Optionally, a preliminary first step of small RNA enrichment can be performed if the estimated concentration of miRNA is less than 0.5% of the total RNA so as to optimise the protocol (Applied Biosystems 2011). This preliminary small RNA enrichment step can be performed using gel separation or a commercial kit, such as the miRvana miRNA isolation kit (Ambion – Life Technologies, USA). Unlike larger RNA molecules, small RNAs can be directly sequenced after adaptor ligation, without fragmentation into smaller pieces (200–500 bp), as their size is compatible with most deep-sequencing technologies. The quantification of the small RNA, or miRNA, fraction can be performed using the NanoDrop Spectrophotometer and the Agilent 2100 Bioanalyser with the RNA 6000 Nano kit and Small RNA Chip kit.

For miRnome studies, a full NGS run provides ultra-high coverage, and one good option is to sequence many libraries in the same run. For example, the SOLiD system permits a slide to be divided into 4, 8 or 16 quadrants in order to perform parallel sequencing. Another, more cost-effective option is to mix several libraries in one run using multiplex identifiers (Parameswaran et al. 2007). This method is based on the use of indexing nucleotides, which are termed 'barcodes', and these can be added to the individual cDNA libraries such that the origins of the sequences can be traced. Sequencing several libraries in one run has significant advantages, including increasing the reproducibility and avoiding sequencing variations either between different runs or, in a unique run, between positions on the slide (Applied Biosystems 2011). In order to prepare a multiplex sample for NGS platforms, it is better to mix the libraries at the same concentration, but it is necessary to determine the concentrations of the libraries in order to do this. Depending on the barcode length, it is possible to codify and sequence up to 96 samples in a unique run (Parameswaran et al. 2007). It is possible to add a synthetic miRNA-like sequence with a known sequence and concentration, such as the miSPIKE™ RNA control oligonucleotide from Integrated DNA Technologies (Integrated DNA Technologies 2010) to each library, which can then be used as an internal RNA control in order to normalise the gene expression data.

Small RNA deep sequencing involves several manipulation stages during the production of cDNA libraries, which can complicate its use in profiling different types of transcripts (Tian et al. 2010). For example, a study comparing the sequencing of cloning libraries has shown that SOLiD data differ from Illumina data and that this may be related to distinct methods of adapter ligation. In addition, the high correlation between qRT-PCR results and SOLiD data might be due to the similarities of the hybridisation-based methods (Tian et al. 2010; Wang et al. 2009a).

miRnome library preparation can be considered to be a limiting step, specifically due to the associated amplification steps, and further optimisation can be performed in order to increase NGS reliability and improve the estimation of miRNA abundance (Kircher and Kelso 2010). New NGS library preparation protocols, which are based on direct and single-molecule sequencing, have been published and have demonstrated that the determination of RNA abundance is more reliable without the library amplification step, thereby avoiding the plateau problem of highly represented RNA and increasing the detection of molecules with low abundances (Blow 2009; Bailo and Deckert 2008; Ozsolak et al. 2009; Sam et al. 2011). For more information on targeted library construction techniques, please see chapter 10 of this volume.

21.2.2.3 Template Preparation and Sequencing

With the Sanger method, the sequencing reaction occurs in a PCR tube, and the sequencing products are sorted and identified by capillary electrophoresis. The requirement for electrophoresis by each sequencing reaction limits the throughput of this methodology. For NGS platforms, the DNA molecules are fixed on a solid matrix, such as microbeads that have been distributed on a glass slide or microwells (Kircher and Kelso 2010). This system allows millions of DNA molecules to be individually deposited, such that a high number of sequencing reactions can be performed in situ and in parallel. One sequencing cycle consists of the incorporation of labelled nucleotides by polymerase extension or oligonucleotide ligation, followed by the detection and identification of the incorporated labelled nucleotides. The detection is based on the use of a camera system that records pictures of the complete slide for each cycle. The emitted signal for each spot, which corresponds to an individual DNA molecule, provides one piece of sequence information at a time, enabling parallel sequencing of all of the deposited molecules (Kircher and Kelso 2010).

For the 454-Roche and SOLiD NGS platforms, the DNA molecules of a library are attached individually and clonally to microbeads using emulsion PCR. For the 454-Roche platform, each bead is deposited into a microwell, where all of the sequencing reactions are performed (Thomas and Harkins 2008). For the SOLiD platform, all of the beads are deposited onto a glass slide (Applied Biosystems 2011). For the Illumina platform, the DNA molecules are fixed onto a slide by bridge amplification, such that each DNA molecule forms a clonal cluster, and all of the clusters are distributed across a complete slide (Illumina 2010).

For the three NGS systems that are commonly used to sequence the miRnome, template preparation is important and can influence the sequencing quality and the sequence read depth. For each platform, one bead (or cluster) corresponds to one read, and the number of deposited beads affects the read number and the 'deepness' of the sequencing (Kircher and Kelso 2010). For the SOLiD system, the quantity of deposited beads is approximately 150–300 million per slide, with the major limit being the camera resolution (Applied Biosystems 2011). For each bead or cluster, the camera records the emitted signal, and the camera resolution determines the minimum distance between two beads that is necessary to obtain a good signal (Metzker 2010).

There are differences between the three major NGS platforms (Kircher and Kelso 2010). Due to sequencing chemistry differences, the raw data generated by the three platforms differ in terms of read length and read number. The 454-Roche system generates read lengths of approximately 400–500 bp with up to 600 MB of total data, the Illumina system produces 100-bp reads and 200 GB of total data, and the SOLiD system gives 75-bp reads and 100 GB of total data. For miRnome sequencing, the read length is not limiting because all mature miRNAs are shorter than the read length, ranging from 18 to 25 nt. Only when isolating pri- or pre-miRNAs would a larger read length be of interest. The output dataset size is important because the higher the read number, the higher the overall sequencing coverage and the higher the number of miRnome libraries that can be sequenced in one run. Table 21.1 lists publications that have described miRnome sequencing using NGS platforms. The raw sequence dataset size is variable and depends on the platform. For example, when using the 454-Roche systems, 20,000–500,000 reads were obtained, whereas the Illumina and SOLiD systems produced 5–10 million and 5–19 million reads, respectively. One experiment was performed with a pool of tissues, and one made use of multiplexed libraries with barcode identifiers (Szczyrba et al. 2010; Ribeiro-dos-Santos et al. 2010, respectively).

After sequencing and the generation of the raw sequence dataset, bioinformatics tools are used to extract the biological information from the raw sequence, such as miRNA sequences and variations in miRNA expression.

21.2.3   Bioinformatics Analysis of Small RNA Deep-Sequencing Dataset

NGS platforms generate large datasets with high numbers of sequence reads. Retrieving the biological information from these large datasets is a challenging bioinformatics task (Horner et al. 2010). The objective is to process the large datasets and to certify the reliability of detection and the unambiguous quantification of small RNAs and microRNAs obtained from the raw datasets. This task is difficult because not all RNA species of the human small RNA transcriptome have been characterised, and the detection of novel sequences or isoforms in NGS datasets requires specialised algorithms (Horner et al. 2010). Statistical methods and bioinformatics pipelines have been developed to optimise the processing of such datasets.

21.2.3.1   Pipelines for Small RNA Deep-Sequencing Dataset Analysis

In NGS, read datasets that are obtained from small RNA cDNA library–sequencing experiments are expected to identify sequences of interest, i.e., microRNAs (mature miRNA, miRNA* and pri-miRNA fragments); however, many other sequences are also detected, such as other small RNAs (snoRNA, piwiRNA and other kinds), biological contaminants (rRNA and tRNA), sequencing reagent contaminants (adapters) and aberrant sequences (which are randomly generated by the sequencing system) (Hackenberg et al. 2009; Huang et al. 2010; Wang et al. 2009b). To annotate the reads, two strategies are commonly used to analyse miRnomes, as listed in Table 21.1 and illustrated in Fig. 21.2.

In the first method, the pipeline (Fig. 21.2a) is based on successive rounds of detection of known molecules. The identified sequences are systematically subtracted from the dataset. The remaining sequence set is mapped to the complete human genome and analysed with algorithms to predict novel small RNAs and microRNAs. To optimise the read alignment against a genome reference sequence, the first step of the pipeline includes an optional filtering step that excludes reads with low sequence quality or short lengths. The filtering step can also involve trimming reads using QV (quality value) or fixed parameters (Friedlander et al. 2008; Wang et al. 2009b). As with Sanger methods, all nucleotides that are sequenced with an NGS system have a QV, which indicates the probability that the correct nucleotide is at a particular position. Filtering the raw data and excluding all of the reads with low reliabilities has the significant advantage of decreasing

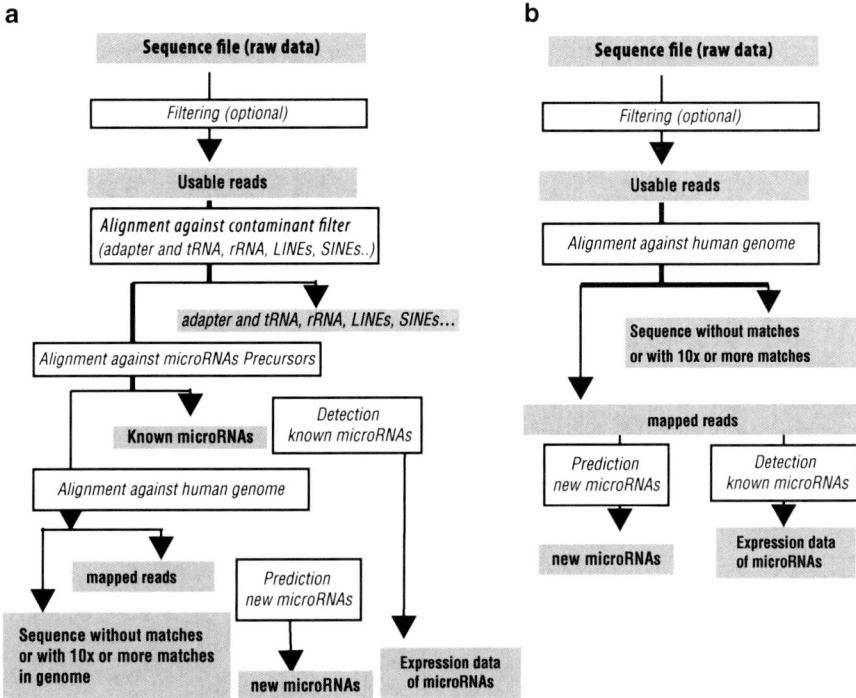

Fig. 21.2 Bioinformatics pipeline for miRnome NGS dataset analysis. (**a**) Pipeline workflow based on step by step analysis. After an optional filtering, each kind of small RNAs are detected in the sequence dataset. Finally the no-identified fraction is mapped against human genome to annotated novel small RNA sequences, as novel microRNAs. (**b**) Pipeline workflow based on annotation strategy. After a previous filtering step, all reads are mapped on human genome and annotated, aiming to identify microRNAs

the size of the dataset to be analysed and reducing the computation time required for further sequence comparisons. To optimise miRNA detection, other pre-processing methods can be used, such as trimming of the 3' extremities of reads to obtain reads that are comparable in size to mature miRNAs, and/or the selection of reads with high QVs in the first ten nucleotides, which are thought to contain the 'seeds' for miRNA binding (Ribeiro-dos-Santos et al. 2010).

Table 21.1 indicates the alignment algorithm used to compare the read datasets to the corresponding reference sequences. Depending on the size of the dataset and read, the type of algorithm employed in this context differs. A BLAST algorithm has been employed for datasets generated using the 454-Roche platform (Altschul et al. 1990). For datasets with a high number of short reads, such as those that originate from the Illumina and SOLiD systems, algorithms with greater efficiencies in aligning short reads, such as MAQ, bowtie or RNA2MAP, are used (Langmead et al. 2009; Li et al. 2008; Applied Biosystems 2011). As shown in Fig. 21.2, the first step in the analysis of sequence reads is to align the sequences to a set of potential contaminant sequences, which includes adapter sequences, primers and sequences from previously identified small RNA molecules. The second step is the detection of miRNAs, which are aligned against the sequences of pre-miRNAs. It is more informative to align against pre-miRNAs than against mature microRNAs because alignment against pre-miRNAs allows for the detection of miRNAs and other fragments, such as loop sequences. The miRBase sequence database has been used in all of the studies that have employed this type of pipeline. The latest version of miRBase (17.0) has 1,426 human miRNA entries (Griffiths-Jones et al. 2006).

Alignment parameters must be adjusted to optimise miRNA detection without a loss of reliability. For example, the number of mismatches is an important factor because without mismatches, the reliability of detection is high and the detection of distinct isoMir, multiple sequence variants of a mature miRNA, is possible; however, this may involve discarding reads that have even one sequencing error or sequences that are polymorphic at only one position (Guo et al. 2011; Horner et al. 2010). Many pipelines have an option to align and select all of the reads that match one miRNA sequence reference or to select only the reads that align against a particular reference sequence without ambiguity, such as with other isomiRs. The pipeline generates several files that list all reads with their matches against the pre-miRNA reference. One file has the read counts, and another file lists the homologous positions between the reads and the reference, which can be used to visualise the alignment in a genome browser such as UCSC genome browser. For more discussion on handling miRNA mapping, please see the discussion on miRNA cross-mapping in the chapter 10 of this volume.

In a subsequent step, the remaining reads are aligned against the human genome to discover new small transcripts and novel molecules (sequences that were not included in previous comparisons). The read position in the genome can be highly informative and permits the annotation of molecular function (Borel et al. 2008). For example, the use of other genomic information, such as sequence conservation in other species or known transcript annotations, could potentially lead to the discovery of new molecules, including novel miRNAs. Other ways to predict new miRNAs include the use of algorithms that identify possible pre-miRNA structures based on thermodynamic properties (Friedlander et al. 2008; Agarwal et al. 2010). In many studies, it has been observed that a relatively high proportion of reads did not map to the reference genome. The set of unmapped reads comprises reads that match more than $10\times$ of the reference genome and those that do not match the genome at all. Reads without matches could be aberrant sequences randomly generated by sequencing errors, RNA sequences that originated from RNA degradation or fragments generated during RNA processing by enzymes, including the Dicer or splicing complex enzymes (Guo et al. 2011). In the final analysis, the pipeline will have identified the miRNA sequences, known or novel, with a read count number for each that is used to perform digital gene expression (DGE) analysis (Linsen et al. 2009; Audic and Claverie 1997).

The second method, shown in Fig. 21.2b, is primarily based on mapping all of the reads against the human reference genome and, secondarily, performs a functional annotation of the mapped reads. This functional annotation is based on all of the information that is available for the genomic positions of the mapped reads. All of the reads that are mapped to characterised regions can be annotated, whereas reads that are mapped to regions that lack such information require further investigation (Borel et al. 2008). One uncharacterised region with many read matches could be investigated to reveal novel microRNAs using algorithms, such as miRDeep (Friedlander et al. 2008). As is the case for the other pipeline strategy, the final step involves counting the numbers of matching reads for all of the miRNAs that were identified from the sequence dataset.

21.2.3.2 Digital Gene Expression Analysis and Profiling

Processing the data via the bioinformatics pipeline results in a list of all of the identified miRNAs and their relative abundances based on read counts. DGE analysis is based on the read count number of each RNA molecule.

There are many considerations for the comparison of different conditions, such as normalisation of the read count number. The read number can be normalised by the total number of reads or the number of mapped reads to correct for the deepness of each run for which the results are typically expressed in rpm (reads per million) units. Alternatively, normalisation can be performed by using an endogenous control transcript, such as snoU6, which is a small RNA with a relatively constant concentration. Synthetic internal controls can also be used, such as 20- or 30-mer oligonucleotides of known concentrations that are added during the library preparation. Having an endogenous or internal control greatly increases the reliability of further DGE results and allows for comparisons among libraries that have been prepared from different tissues or tumour stages.

It is possible to profile miRNA expression patterns in tissues or cell lines based on the relative expression levels of miRNAs. The aim of profiling is to define miRNAs with specific expression patterns, such as specificity to one tissue or stage of development, that could be used as biomarkers for further investigation ('t Hoen et al. 2008). With the proper experimental design, data comparisons can be based on different statistical methods. All of the studies listed in Table 21.1 have descriptive and graphical data representations, which depict the miRNA read numbers (or their log transformations) or the normalised values for different conditions. These values are used to visualise the global variations between conditions and samples. In many studies, the ratios between conditions, fold changes or Z-factors have been calculated to quantify variation. To evaluate the significance of the variation, t-tests or Vencio tests based on Bayesian methods can be performed (Vencio et al. 2006).

To investigate the specificity of miRNA expression, approaches based on clustering (hierarchical and unsupervised) or Venn diagrams have been used in many of the studies that are listed in Table 21.1. Venn diagrams, clustering dendrograms and heat maps are helpful to represent comparative miRNA expression levels and define groups or clusters of miRNA expression relative to particular conditions or samples. The group and cluster definitions enable the establishment of miRNA expression profiles for tissues and tumours at different stages of development.

The defined profiles, which are used to define miRNA expression models in humans, need to be validated by other methods. In the studies listed in Table 21.1, the correlation between the NGS results and qRT-PCR has been approximately 80–95%, demonstrating that the NGS results are reliable estimates of microRNA expression.

21.3 Deep-Sequencing Applications in Cancer Research

21.3.1 The Roles of miRNA in Carcinogenesis

In several types of cancer, genomic stability is lost through the disturbance of minute mechanisms that regulate the balance between cell proliferation and apoptosis. These genetic changes include a large range of events, such as chromosomal rearrangements, amplification, viral integration, microsatellite changes, insertions/ deletions, SNPs and interactions with pathogens. Studies of these events, which are associated with genomic instability, have lead to the discovery of novel genes that are linked to critical regulatory pathways, including microRNAs (Yasui 2005).

However, single genetic alterations are often insufficient to explain the complexity of the aberrations observed in cancer cells. Therefore, epigenetic phenomena, which are defined as heritable changes in gene activity without changes (damage) to the DNA sequence, are thought to contribute to the initiation and progression of cancer. Examples of such changes include the hypermethylation of tumour suppressor genes, the global DNA hypomethylation and post-translational histone modifications. These phenomena explain the aberrant expression of miRNAs in cancer and show that microRNAs can regulate the enzymes that are involved in the methylation of CpG islands in tumour suppressor genes, conferring the role of pathway component in malignant phenotype or executor of specific epigenetic events on microRNAs. Thus, genetic changes that are complemented by epigenetic changes can shed new light on a mechanism that partially explains the misregulation of miRNA expression in cancer (Di Leva and Croce 2010; Zhang 2010).

In general, spontaneous carcinogenesis originates from multiple genetic and epigenetic events. However, in recent years, genome-wide studies have shown that miRNA genes are frequently located within regions of loss of heterozygosity, amplification regions, fragile sites and other cancer-associated genomic regions (Calin et al. 2004). It has been demonstrated that miRNAs are involved in human tumourigenesis, thus revealing a new paradigm in the molecular architecture of human cancer (Negrini et al. 2009).

Calin et al. (2002) provided the first evidence linking miRNAs to human cancers by showing that frequent deletions and downregulation of the microRNA genes miR-15 and miR-16 at 13q14 occurred in chronic lymphocytic leukaemia (CLL).

The target of miR-15/16 is the mRNA of the anti-apoptotic oncogene Bcl-2. Deletion thus results in overexpression of Bcl-2, leading to the initiation of stepwise leukaemogenesis (Cimmino et al. 2005). Therefore, miR-15 and miR-16 are natural antisense Bcl-2 interactors that could be used for the treatment of Bcl-2-overexpressing tumours (Guo et al. 2009; Varol et al. 2011).

Numerous profiling studies have found significantly dysregulated miRNA expression in various cancer, showing patterns of both downregulation and upregulation. These results suggest that microRNAs have a dual role in tumourigenesis, such that alterations in a microRNA expression pattern may lead to the loss or gain of gene function. Thus, microRNAs may act as either tumour oncogenes (onco-miR) and tumour suppressors (TS-miR) during tumour development and progression (Varol et al. 2011; Wang et al. 2010). Another well-known example of a tumour suppressor is let-7 that targets Ras mRNA. When expressed at low levels, it is associated with non-small cell lung cancers, and a poor prognosis is associated with overexpression of the Ras oncoprotein, (Belinsky et al. 2008).

In contrast, overexpression of oncogenic miRNAs induces tumour development. The miR-17-92 gene cluster, the first onco-miR ever identified, accelerates the development of human B-cell lymphoma. Overexpression of the miR-17-92 cluster has been shown to interact with c-myc expression to accelerate tumour development (Barbarotto et al. 2008; Varol et al. 2011). MicroRNAs act as both a tumour suppressor and an oncogene.

It is believed that this miRNA is a component of distinct pathways with different effects on cell growth, survival and proliferation. These effects must correspond to particular cell types and gene expression patterns (Calin and Croce 2006). A study by Zhang et al. (2008) demonstrated that miR-21 overexpression resulted in cell proliferation. Changes in miR-21 expression caused different effects in the gastric tumour cell line AGS, triggering tumour development when stimulated and reducing the proliferation and invasion that leads to apoptosis when repressed (Zhang et al. 2008). In contrast, another study has identified a mutation in the p53 protein (responsible for inducing miR-34 transcription) that promotes cell proliferation as a consequence of its low expression (Zimmerman and Wu 2011). Yao et al. (2009) detailed the location, mechanisms (tumour suppressor and oncogene) and targets of the microRNAs that are involved in gastric carcinogenesis.

A publication by Ribeiro-dos-Santos et al. (2010) described the expression profile of the miRnome of healthy gastric tissue. The authors, via ultra-deep sequencing on the SOLiD platform using barcodes (multiplexing), identified a group of 15 miRNAs that are highly expressed in gastric tissue (Fig. 21.3). Subsequently, the expression of these miRNAs was validated in 10 healthy individuals by qRT-PCR, with a significant correlation of 83.97% (P = 0.05). This study aimed to validate and characterise the normal miRNA profiles of human gastric tissue to establish a reference profile for healthy individuals.

The establishment of normal miRNA profiles in human tissue is aimed at establishing a reference profile of healthy individuals that can be compared to profiles from various tissues and organs that are affected by cancer. Therefore, these results, in addition to the tumour cell line studies, have provided new information about the pathogenesis of gastric cancer as mediated by complex

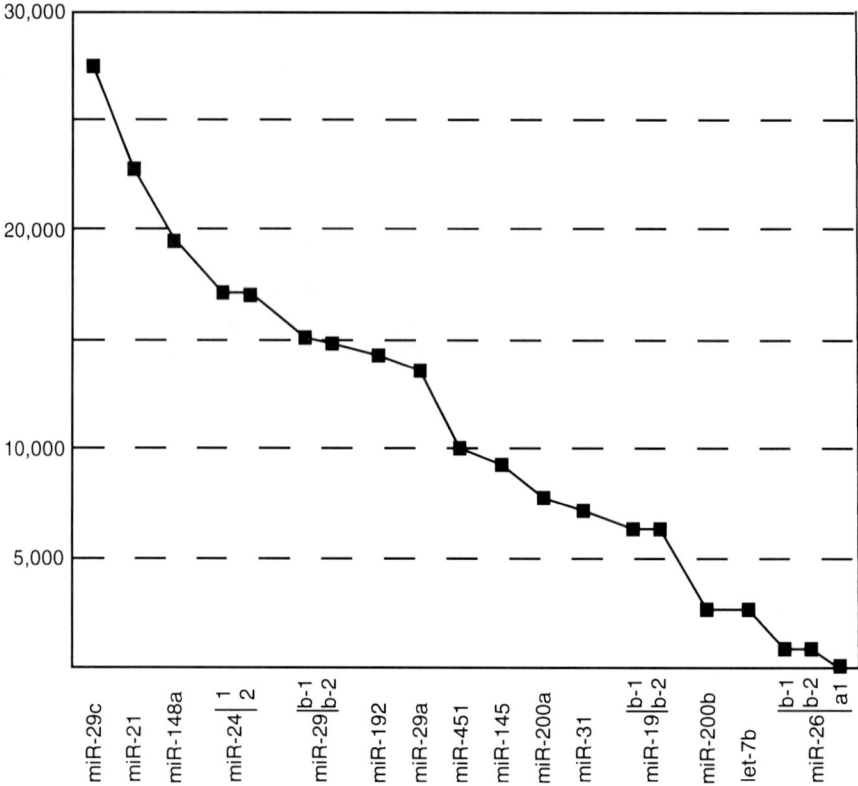

Fig. 21.3 miRNAs highly expressed in human gastric tissue. Read counts obtained from ultra-deep SOLiD sequencing of a small RNA library prepared from gastric tissue

miRNA regulatory pathways. Valuable results can be obtained from the analyses of miRNA expression profiles that elucidate the functions of miRNA regulatory pathways in malignant tumours and reveal a common method for directing the processes of carcinogenesis. All of this information leads to a better understanding of advanced gastric cancer and possible improvements in the diagnosis and treatment of the disease (Tie et al. 2010).

21.3.2 MicroRNA Signatures

Since the discoveries of microRNAs, various profiling studies have found significantly dysregulated microRNA profiles in various cancers, which suggests that aberrant miRNA expression profiles could be used as biomarkers. MicroRNAs can serve as biomarkers because they are an abundant class of molecules that are expressed with high specificity in a given tissue and are involved in regulating a large number of human genes. Using microRNAs as biomarkers could improve the identification of patients who would benefit from more aggressive and specific therapies (Zhang et al. 2008).

This tool adds to the traditional diagnostic measures that are available for samples that have been obtained from various biological sources, such as tissue (biopsy or surgical resection), plasma, blood, faeces, urine, cerebrospinal fluid, peritoneal fluid and saliva (Wang et al. 2010; Zhou et al. 2010; Mitchell et al. 2008). The development of sensitive and specific biomarkers for cancer will improve cancer management and early detection (Varol et al. 2011; Wang et al. 2010), helping to provide an adequate time window to prevent metastatic disease and death (Yachida et al. 2010). However, miRNA-related biomarkers should not be limited to miRNA expression but should also include miRNA-related SNPs, methylation changes and mutations (Xie et al. 2010).

The literature indicates that global expression of miRNAs is higher in normal tissues than in the homologous tumour tissue, which suggests that the normal miRNA expression pattern is important for maintaining the integrity of cellular differentiation processes (Lu et al. 2005). However, as there are aberrant expression of microRNA in tumour tissue, these specific molecular signature may be useful in which the first stage involves screening, early detection, classic diagnosis and prognosis (Volinia et al. 2006) (Table 21.2). Table 21.1 shows the miRnomes that have been sequenced using NGS, all of which include profiling of different kinds of cancer, such as breast, embryonic, gastric, hepatocellular, kidney, leukaemia, lymphomas, lung, nasopharyngeal, prostate and ovarian cancers. One common observation relevant to this table is that the miRnome profile varies during carcinogenesis, with many microRNAs becoming upregulated or downregulated during the process.

Table 21.2 Analysis of miRNA expression profiles in different stages of carcinogenesis

Stages of carcinogenesis	Features	Conduct	MicroRNA expression profile
Initiation	Founder mutation	Screening	Identifies early molecular changes (biomarkers)
Promotion	Presence of parental clones	Early diagnosis (asymptomatic disease)	Determines the precise tumour staging, group and hierarchical tissue
Progression	Advantages of survival, motility and invasion are acquired according to the clonal evolution and observation of subclones	Classic diagnosis with symptoms installed	Determines embryonic origin in undifferentiated tumours
Metastasis	Angiogenic factors, anti-apoptotic proteins, among other factors are active	Prognosis	Detects recurrence or other metastatic organs related to microRNA involved

For example, Mizuguchi et al. demonstrated that sequencing-based miRNA clustering could be used to identify patients with a high potential for early tumour recurrence after liver surgery (Mizuguchi et al. 2011). Employing Illumina sequencing, Chen et al. (2008) sequenced all of the serum miRNAs of healthy Chinese subjects and identified over 100 and 91 serum miRNAs in male and female subjects, respectively. The authors also identified expression patterns of serum miRNAs specific to lung cancer, colorectal cancer and diabetes, providing evidence that serum miRNAs contain fingerprints for various diseases. Using these analyses, the authors concluded that serum miRNAs could potentially serve as biomarkers for the detection of various cancer types and other diseases. Liu et al. (2010) demonstrated that plasma miR-31 levels were significantly elevated in oral squamous cell carcinoma patients relative to age- and sex-matched controls. In addition, plasma miR-31 levels in patients are remarkably reduced after tumour resection, suggesting that this marker is tumour associated. The authors also demonstrated the feasibility of detecting increases in miR-31 in the saliva of patients. Analysing miRNA profiles facilitates the detection of coregulation patterns and the identification of miRNAs that seem to be specific to the tumour or a tissue development stage.

21.3.3 Perspectives on miRNA Use in Cancer Therapy

The association of a significantly dysregulated miRNA expression profile with the pathogenesis and progression of cancer illustrates the large potential for utilising microRNAs as targets for therapeutic intervention. The basic strategy of current microRNA-based treatment methods is either to antagonise the expression of target miRNAs with antisense technology or to restore or strengthen the function of particular microRNAs to inhibit the expression of certain protein-coding genes (Wang et al. 2010).

The therapeutic application of microRNAs or any microRNA-related molecules will rely on the development of efficient delivery strategies based on viral vectors or nonviral nanoparticles (Aigner 2011). Despite many encouraging advances, the application of these therapeutic methods against cancer is just beginning. There is a convergence of screening results regarding the functionality of regulatory networks of miRNAs (in vitro or in vivo), which is needed to reduce the gap between scientific research and clinical applicability. Our understanding of the efficiency of the methods and techniques in terms of reducing adverse reactions is inadequate, as is a full understanding of their applicability to non-invasive early diagnosis. These processes are important for the development of therapy and individualised medicine, and they are relevant to the timely prognosis of many cancer types that are typically only discovered at advanced stages, with extensive invasion and metastasis to other tissues (Oue et al. 2005; Tie et al. 2010).

Acknowledgements The work was supported by the Genoma Paraense de Genomica e Proteomica Project (Governo do Para/SEDECT/FAPESPA), PROPESP/UFPA, FADESP and CAPES (Coordenacao de Aperfeicoamento de Pessoal de Nivel Superior). Amanda Barros is warmly acknowledged for assistance in editing the manuscript.

References

Agarwal S, Vaz C, Bhattacharya A, Srinivasan A (2010) Prediction of novel precursor miRNAs using a context-sensitive hidden Markov model (CSHMM). BMC Bioinformatics 11:S1–S29. doi:10.1186/1471-2105-11-S1-S29

Aigner A (2011) MicroRNAs (miRNAs) in cancer invasion and metastasis: therapeutic approaches based on metastasis-related miRNAs. J Mol Med 89(5):445–457. doi:10.1007/s00109-010-0716-0

Altschul SF, Gish W, Miller W, Myers EW, Lipman DJ (1990) Basic local alignment search tool. J Mol Biol 215(3):403–410. doi:10.1006/jmbi.1990.9999 S0022-2836(05), 80360-2 [pii]

Ambion (2010) Accurate, sensitive quantification of microRNAs from FFPE tissues. Application Notes

Applied Biosystems (2011) SOLiD system barcoding. Applied Biosystems Technical Notes. http://www3.appliedbiosystems.com/cms/groups/mcb_marketing/documents/generaldocuments/cms_057554.pdf. Accessed 11 April 2011

Audic S, Claverie J (1997) The significance of digital gene expression profiles. Genome Res 7 (10):986–995. doi:10.1101/gr.7179508

Bailo E, Deckert V (2008) Tip-enhanced Raman spectroscopy of single RNA strands: towards a novel direct-sequencing method. Angew Chem Int Ed Engl 47(9):1658–1661. doi:10.1002/anie.200704054

Barbarotto E, Schmittgen TD, Calin GA (2008) MicroRNAs and cancer: profile, profile, profile. Int J Cancer 122(5):969–977. doi:10.1002/ijc.23343

Belinsky MG, Rink L, Cai KQ, Ochs MF, Eisenberg B, Huang M, von Mehren M, Godwin AK (2008) The insulin-like growth factor system as a potential therapeutic target in gastrointestinal stromal tumors. Cell Cycle 7(19):2949–2955. doi:6760 [pii]

Blow N (2009) Transcriptomics: the digital generation. Nature 458(7235):239–242. doi:458239a [pii]10.1038/458239a

Borel C, Gagnebin M, Gehrig C, Kriventseva EV, Zdobnov EM, Antonarakis SE (2008) Mapping of small RNAs in the human ENCODE regions. Am J Hum Genet 82(4):971–981. doi:S0002-9297(08), 00208-5 [pii] 10.1016/j.ajhg.2008.02.016

Calin GA, Dumitru CD, Shimizu M, Bichi R, Zupo S, Noch E, Aldler H, Rattan S, Keating M, Rai K, Rassenti L, Kipps T, Negrini M, Bullrich F, Croce CM (2002) Frequent deletions and down-regulation of micro-RNA genes miR15 and miR16 at 13q14 in chronic lymphocytic leukemia. Proc Natl Acad Sci USA 99(24):15524–15529. doi:10.1073/pnas.242606799242606799 [pii]

Calin GA, Sevignani C, Dumitru CD, Hyslop T, Noch E, Yendamuri S, Shimizu M, Rattan S, Bullrich F, Negrini M, Croce CM (2004) Human microRNA genes are frequently located at fragile sites and genomic regions involved in cancers. Proc Natl Acad Sci USA 101 (9):2999–3004. doi:10.1073/pnas.0307323101030732310 [pii]

Calin G, Croce C (2006) MicroRNA signatures in human cancers. Nat Rev Cancer 6(11):857–866. doi:10.1038/nrc199710.1038/nrc1997

Chen A, Luo M, Yuan G, Yu J, Deng T, Zhang L, Zhou Y, Mitchelson K, Cheng J (2008) Complementary analysis of microrna and mRNA expression during phorbol 12-myristate 13-acetate (tpa)-induced differentiation of hl-60 cells. Biotechnol Lett 30(12):2045–2052. doi:10.1007/s10529-008-9800-8

Cimmino A, Calin GA, Fabbri M, Iorio MV, Ferracin M, Shimizu M, Wojcik SE, Aqeilan RI, Zupo S, Dono M, Rassenti L, Alder H, Volinia S, Liu CG, Kipps TJ, Negrini M, Croce CM (2005) miR-15 and miR-16 induce apoptosis by targeting BCL2. Proc Natl Acad Sci USA 102 (39):13944–13949. doi:0506654102 [pii]10.1073/pnas.0506654102

Di Leva G, Croce CM (2010) Roles of small RNAs in tumor formation. Trends Mol Med 16 (6):257–267. doi:S1471-4914(10), 00052-3 [pii]10.1016/j.molmed.2010.04.001

Fabbri M, Calin GA (2010) Epigenetics and miRNAs in human cancer. Adv Genet 70:87–99. doi: B978-0-12-380866-0.60004-6 [pii] 10.1016/B978-0-12-380866-0.60004-6

Friedlander MR, Chen W, Adamidi C, Maaskola J, Einspanier R, Knespel S, Rajewsky N (2008) Discovering microRNAs from deep sequencing data using miRDeep. Nat Biotechnol 26 (4):407–415. doi:nbt1394 [pii] 10.1038/nbt1394

Friedman RC, Farh KKH, Burge CB, Bartel DP (2009) Most mammalian mRNAs are conserved targets of microRNAs. Genome Res 19(1):92–105. doi:10.1101/Gr.082701.108

Garzon R, Fabbri M, Cimmino A, Calin G, Croce C (2006) MicroRNA expression and function in cancer. Trends Mol Med 12(12):580–587. doi:10.1016/j.molmed.2006.10.006

Garzon R, Marcucci G, Croce CM (2010) Targeting microRNAs in cancer: rationale, strategies and challenges. Nat Rev Drug Discov 9(10):775–789. doi:nrd3179 [pii] 10.1038/nrd3179

Goya R, Sun MG, Morin RD, Leung G, Ha G, Wiegand KC, Senz J, Crisan A, Marra MA, Hirst M, Huntsman D, Murphy KP, Aparicio S, Shah SP (2010) SNVMix: predicting single nucleotide variants from next-generation sequencing of tumors. Bioinformatics 26(6):730–736. doi: btq040 [pii] 10.1093/bioinformatics/btq040

Griffiths-Jones S, Grocock R, van Dongen S, Bateman A, Enright A (2006) miRBase: microRNA sequences, targets and gene nomenclature. Nucleic Acids Res 34(Database issue):D140–144. doi:10.1093/nar/gkj112

Guo CJ, Pan Q, Li DG, Sun F, Liu BW (2009) miR-15b and miR-16 are implicated in activation of the rat hepatic stellate cell: an essential role for apoptosis. J Hepatol 50(4):766–778. doi: S0168-8278(09), 00008-7 [pii] 10.1016/j.jhep. 2008.11.025

Guo L, Liang T, Lu Z (2011) A comprehensive study of multiple mapping and feature selection for correction strategy in the analysis of small RNAs from SOLiD sequencing. Biosystems. doi: S0303-2647(11), 00008-6 [pii] 10.1016/j.biosystems.2011.01.004

Hackenberg M, Sturm M, Langenberger D, Falcon-Perez JM, Aransay AM (2009) miRanalyzer: a microRNA detection and analysis tool for next-generation sequencing experiments. Nucleic Acids Res 37(Web Server issue):W68–76. doi:gkp347 [pii] 10.1093/nar/gkp347

Hawkins RD, Hon GC, Ren B (2010) Next-generation genomics: an integrative approach. Nat Rev Genet 11(7):476–486. doi:nrg2795 [pii] 10.1038/nrg2795

Hayden EC (2009) Genome sequencing: the third generation. Nature 457(7231):768–769. doi:10.1038/News.2009.86

Horner DS, Pavesi G, Castrignano T, De Meo PD, Liuni S, Sammeth M, Picardi E, Pesole G (2010) Bioinformatics approaches for genomics and post genomics applications of next-generation sequencing. Brief Bioinform 11(2):181–197. doi:bbp046 [pii] 10.1093/bib/bbp046

Huang PJ, Liu YC, Lee CC, Lin WC, Gan RR, Lyu PC, Tang P (2010) DSAP: deep-sequencing small RNA analysis pipeline. Nucleic Acids Res 38(Web Server issue):W385–391. doi:gkq392 [pii] 10.1093/nar/gkq392

Illumina (2010) Illumina sequencing technology. Illumina. http://www.illumina.com/documents/ products/techspotlights/techspotlight_sequencing.pdf. Accessed 11 April 2011

Integrated_DNA_Technologies (2010) miRCat, miRCat-33 microRNA cloning kit technical manual. Integrated DNA technologies. http://cdn.idtdna.com/Support/Technical/TechnicalBulletinPDF/miRCat_User_Guide.pdf. Accessed 11 April 2011

Jima DD, Zhang J, Jacobs C, Richards KL, Dunphy CH, Choi WW, Yan Au W, Srivastava G, Czader MB, Rizzieri DA, Lagoo AS, Lugar PL, Mann KP, Flowers CR, Bernal-Mizrachi L, Naresh KN, Evens AM, Gordon LI, Luftig M, Friedman DR, Weinberg JB, Thompson MA, Gill JI, Liu Q, How T, Grubor V, Gao Y, Patel A, Wu H, Zhu J, Blobe GC, Lipsky PE, Chadburn A, Dave SS (2010) Deep sequencing of the small RNA transcriptome of normal and malignant human B cells identifies hundreds of novel microRNAs. Blood 116(23):e118–127. doi:blood-2010-05-285403 [pii] 10.1182/blood-2010-05-285403

Kircher M, Kelso J (2010) High-throughput DNA sequencing—concepts and limitations. Bioessays 32(6):524–536. doi:10.1002/bies.200900181

Langmead B, Trapnell C, Pop M, Salzberg SL (2009) Ultrafast and memory-efficient alignment of short DNA sequences to the human genome. Genome Biol 10(3):-. doi:Artn R25 Doi 10.1186/Gb-2009-10-3-R25

Li H, Ruan J, Durbin R (2008) Mapping short DNA sequencing reads and calling variants using mapping quality scores. Genome Res 18(11):1851–1858. doi:10.1101/Gr.078212.108

Liao JY, Ma LM, Guo YH, Zhang YC, Zhou H, Shao P, Chen YQ, Qu LH (2010) Deep sequencing of human nuclear and cytoplasmic small RNAs reveals an unexpectedly complex subcellular distribution of miRNAs and tRNA 3' trailers. PLoS One 5(5):e10563. doi:10.1371/journal.pone.0010563

Linsen S, de Wit E, Janssens G, Heater S, Chapman L, Parkin R, Fritz B, Wyman S, de Bruijn E, Voest E, Kuersten S, Tewari M, Cuppen E (2009) Limitations and possibilities of small RNA digital gene expression profiling. Nat Methods 6(7):474–476. doi:10.1038/nmeth0709-474

Lister R, Gregory BD, Ecker JR (2009) Next is now: new technologies for sequencing of genomes, transcriptomes, and beyond. Curr Opin Plant Biol 12(2):107–118. doi:10.1016/J.Pbi.2008.11.004

Liu R, Zhang C, Hu Z, Li G, Wang C, Yang C, Huang D, Chen X, Zhang H, Zhuang R, Deng T, Liu H, Yin J, Wang S, Zen K, Ba Y, Zhang CY (2010) A five-microRNA signature identified from genome-wide serum microRNA expression profiling serves as a fingerprint for gastric cancer diagnosis. Eur J Cancer. doi:S0959-8049(10), 01064-6 [pii]10.1016/j.ejca.2010.10.025

Lu J, Getz G, Miska E, Alvarez-Saavedra E, Lamb J, Peck D, Sweet-Cordero A, Ebert B, Mak R, Ferrando A, Downing J, Jacks T, Horvitz H, Golub T (2005) MicroRNA expression profiles classify human cancers. Nature 435(7043):834–838. doi:10.1038/nature03702

Lu C, Meyers B, Green P (2007) Construction of small RNA cDNA libraries for deep sequencing. Methods 43(2):110–117. doi:10.1016/j.ymeth.2007.05.002

MacLeod RAF, Nagel S, Scherr M, Schneider B, Dirks WG, Uphoff CC, Quentmeier H, Drexler HG (2008) Human leukemia and lymphoma cell lines as models and resources. Curr Med Chem 15(4):339–359

Margulies M, Egholm M, Altman WE, Attiya S, Bader JS, Bemben LA, Berka J, Braverman MS, Chen YJ, Chen Z, Dewell SB, Du L, Fierro JM, Gomes XV, Godwin BC, He W, Helgesen S, Ho CH, Irzyk GP, Jando SC, Alenquer ML, Jarvie TP, Jirage KB, Kim JB, Knight JR, Lanza JR, Leamon JH, Lefkowitz SM, Lei M, Li J, Lohman KL, Lu H, Makhijani VB, McDade KE, McKenna MP, Myers EW, Nickerson E, Nobile JR, Plant R, Puc BP, Ronan MT, Roth GT, Sarkis GJ, Simons JF, Simpson JW, Srinivasan M, Tartaro KR, Tomasz A, Vogt KA, Volkmer GA, Wang SH, Wang Y, Weiner MP, Yu P, Begley RF, Rothberg JM (2005) Genome sequencing in microfabricated high-density picolitre reactors. Nature 437(7057):376–380. doi:nature03959 [pii] 10.1038/nature03959

Metzker ML (2010) Sequencing technologies—the next generation. Nat Rev Genet 11(1):31–46. doi:nrg2626 [pii] 10.1038/nrg2626

Meyerson M, Gabriel S, Getz G (2010) Advances in understanding cancer genomes through second-generation sequencing. Nat Rev Genet 11(10):685–696. doi:nrg2841 [pii] 10.1038/nrg2841

Mishra PJ (2009) MicroRNA polymorphisms: a giant leap towards personalized medicine. Per Med 6(2):119–125. doi:10.2217/17410541.6.2.119

Mitchell PS, Parkin RK, Kroh EM, Fritz BR, Wyman SK, Pogosova-Agadjanyan EL, Peterson A, Noteboom J, O'Briant KC, Allen A, Lin DW, Urban N, Drescher CW, Knudsen BS, Stirewalt DL, Gentleman R, Vessella RL, Nelson PS, Martin DB, Tewari M (2008) Circulating microRNAs as stable blood-based markers for cancer detection. Proc Natl Acad Sci USA 105(30):10513–10518. doi:10.1073/Pnas.0804549105

Mizuguchi Y, Mishima T, Yokomuro S, Arima Y, Kawahigashi Y, Shigehara K, Kanda T, Yoshida H, Uchida E, Tajiri T, Takizawa T (2011) Sequencing and bioinformatics-based analyses of the microRNA transcriptome in hepatitis B-related hepatocellular carcinoma. PLoS One 6(1):e15304. doi:10.1371/journal.pone.0015304

Morozova O, Hirst M, Marra MA (2009) Applications of new sequencing technologies for transcriptome analysis. Annu Rev Genomics Hum Genet 10:135–151. doi:10.1146/annurev-genom-082908-145957

Nagalakshmi U, Waern K, Snyder M (2010) RNA-Seq: a method for comprehensive transcriptome analysis. Curr Protoc Mol Biol 4:4.11.1–13. doi:10.1002/0471142727.mb0411s89

Negrini M, Nicoloso MS, Calin GA (2009) MicroRNAs and cancer—new paradigms in molecular oncology. Curr Opin Cell Biol 21(3):470–479. doi:S0955-0674(09), 00064-7 [pii] 10.1016/j.ceb.2009.03.002

Nygaard S, Jacobsen A, Lindow M, Eriksen J, Balslev E, Flyger H, Tolstrup N, Moller S, Krogh A, Litman T (2009) Identification and analysis of miRNAs in human breast cancer and teratoma samples using deep sequencing. BMC Med Genomics 2:35. doi:1755-8794-2-35 [pii]10.1186/1755-8794-2-35

Oue N, Aung PP, Mitani Y, Kuniyasu H, Nakayama H, Yasui W (2005) Genes involved in invasion and metastasis of gastric cancer identified by array-based hybridization and serial analysis of gene expression. Oncology 69(suppl 1):17–22. doi:86627 [pii]10.1159/000086627

Ozsolak F, Platt AR, Jones DR, Reifenberger JG, Sass LE, McInerney P, Thompson JF, Bowers J, Jarosz M, Milos PM (2009) Direct RNA sequencing. Nature 461(7265):814–818. doi:nature08390 [pii]10.1038/nature08390

Parameswaran P, Jalili R, Tao L, Shokralla S, Gharizadeh B, Ronaghi M, Fire A (2007) A pyrosequencing-tailored nucleotide barcode design unveils opportunities for large-scale sample multiplexing. Nucleic Acids Res 35(19):e130. doi:10.1093/nar/gkm760

Perrotti D, Eiring AM (2010) The new role of microRNAs in cancer. Future Oncol 6(8):1203–1206. doi:10.2217/fon.10.78

Ribeiro-dos-Santos A, Khayat AS, Silva A, Alencar DO, Lobato J, Luz L, Pinheiro DG, Varuzza L, Assumpcao M, Assumpcao P, Santos S, Zanette DL, Silva WA Jr, Burbano R, Darnet S (2010) Ultra-deep sequencing reveals the microRNA expression pattern of the human stomach. PLoS One 5(10):e13205. doi:10.1371/journal.pone.0013205

Sam LT, Lipson D, Raz T, Cao X, Thompson J, Milos PM, Robinson D, Chinnaiyan AM, Kumar-Sinha C, Maher CA (2011) A comparison of single molecule and amplification based sequencing of cancer transcriptomes. PLoS One 6(3):e17305. doi:10.1371/journal.pone.0017305

Schulte JH, Marschall T, Martin M, Rosenstiel P, Mestdagh P, Schlierf S, Thor T, Vandesompele J, Eggert A, Schreiber S, Rahmann S, Schramm A (2010) Deep sequencing reveals differential expression of microRNAs in favorable versus unfavorable neuroblastoma. Nucleic Acids Res 38(17):5919–5928. doi:gkq342 [pii]10.1093/nar/gkq342

Sharma CM, Vogel J (2009) Experimental approaches for the discovery and characterization of regulatory small RNA. Curr Opin Microbiol 12(5):536–546. doi:S1369-5274(09), 00111-8 [pii]10.1016/j.mib.2009.07.006

Sun D, Lee YS, Malhotra A, Kim HK, Matecic M, Evans C, Jensen RV, Moskaluk CA, Dutta A (2011) miR-99 family of MicroRNAs suppresses the expression of prostate-specific antigen and prostate cancer cell proliferation. Cancer Res 71(4):1313–1324. doi:0008-5472.CAN-10-1031 [pii]10.1158/0008-5472.CAN-10-1031

Szczyrba J, Loprich E, Wach S, Jung V, Unteregger G, Barth S, Grobholz R, Wieland W, Stohr R, Hartmann A, Wullich B, Grasser F (2010) The microRNA profile of prostate carcinoma obtained by deep sequencing. Mol Cancer Res 8(4):529–538. doi:1541-7786.MCR-09-0443 [pii]10.1158/1541-7786.MCR-09-0443

't Hoen P, Ariyurek Y, Thygesen H, Vreugdenhil E, Vossen R, de Menezes R, Boer J, van Ommen G, den Dunnen J (2008) Deep sequencing-based expression analysis shows major advances in robustness, resolution and inter-lab portability over five microarray platforms. Nucleic Acids Res 36(21):e141. doi:10.1093/nar/gkn705

Tarasov V, Jung P, Verdoodt B, Lodygin D, Epanchintsev A, Menssen A, Meister G, Hermeking H (2007) Differential regulation of microRNAs by p53 revealed by massively parallel Sequencing—miR-34a is a p53 target that induces apoptosis and G(1)-arrest. Cell Cycle 6 (13):1586–1593

Thomas MF, Ansel KM (2010) Construction of small RNA cDNA libraries for deep sequencing. Methods Mol Biol 667:93–111. doi:10.1007/978-1-60761-811-9_7

Thomas Jarvie, Harkins T (2008) Transcriptome sequencing with the Genome sequencer FLX system. Roche. http://www.nature.com/nmeth/journal/v5/n9/full/nmeth.f.220.html. Accessed 11 April 2011

Tian G, Yin X, Luo H, Xu X, Bolund L, Zhang X (2010) Sequencing bias: comparison of different protocols of microRNA library construction. BMC Biotechnol 10:64. doi:1472-6750-10-64 [pii]10.1186/1472-6750-10-64

Tie J, Pan Y, Zhao L, Wu K, Liu J, Sun S, Guo X, Wang B, Gang Y, Zhang Y, Li Q, Qiao T, Zhao Q, Nie Y, Fan D (2010) MiR-218 inhibits invasion and metastasis of gastric cancer by targeting the Robo1 receptor. PLoS Genet 6(3):e1000879. doi:10.1371/journal.pgen.1000879

Varol N, Konac E, Gurocak OS, Sozen S (2011) The realm of microRNAs in cancers. Mol Biol Rep 38(2):1079–1089. doi:10.1007/s11033-010-0205-0

Vaz C, Ahmad HM, Sharma P, Gupta R, Kumar L, Kulshreshtha R, Bhattacharya A (2010) Analysis of microRNA transcriptome by deep sequencing of small RNA libraries of peripheral blood. BMC Genomics 11:288. doi:1471-2164-11-288 [pii]10.1186/1471-2164-11-288

Vencio RZ, Patrao DF, Baptista CS, Pereira CA, Zingales B (2006) BayBoots: a model-free Bayesian tool to identify class markers from gene expression data. Genet Mol Res 5 (1):138–142. doi:XM6 [pii]

Volinia S, Calin GA, Liu CG, Ambs S, Cimmino A, Petrocca F, Visone R, Iorio M, Roldo C, Ferracin M, Prueitt RL, Yanaihara N, Lanza G, Scarpa A, Vecchione A, Negrini M, Harris CC, Croce CM (2006) A microRNA expression signature of human solid tumors defines cancer gene targets. Proc Natl Acad Sci USA 103(7):2257–2261. doi:0510565103 [pii]10.1073/pnas.0510565103

Wang K, Li J, Li ST, Bolund L, Wiuf C (2009a) Estimation of tumor heterogeneity using CGH array data. BMC Bioinformatics 10:12. doi:10.1186/1471–2105–10–12

Wang WC, Lin FM, Chang WC, Lin KY, Huang HD, Lin NS (2009b) miRExpress: analyzing high-throughput sequencing data for profiling microRNA expression. BMC Bioinformatics 10:328. doi:1471-2105-10-328 [pii]10.1186/1471-2105-10-328

Wang Z, Gerstein M, Snyder M (2009c) RNA-Seq: a revolutionary tool for transcriptomics. Nat Rev Genet 10(1):57–63. doi:nrg2484 [pii]10.1038/nrg2484

Wang J, Wang Q, Liu H, Hu B, Zhou W, Cheng Y (2010) MicroRNA expression and its implication in the diagnosis and therapeutic strategies of gastric cancer. Cancer Lett 297 (2):137–143. doi:S0304-3835(10), 00377-0 [pii]10.1016/j.canlet.2010.07.018

Weng L, Wu X, Gao H, Mu B, Li X, Wang JH, Guo C, Jin JM, Chen Z, Covarrubias M, Yuan YC, Weiss LM, Wu H (2010) MicroRNA profiling of clear cell renal cell carcinoma by whole-genome small RNA deep sequencing of paired frozen and formalin-fixed, paraffin-embedded tissue specimens. J Pathol 222(1):41–51. doi:10.1002/path.2736

Weidhaas J (2010) Using micrornas to understand cancer biology. Lancet Oncol 11(2):106–107. doi:10.1016/S1470-2045(09)70386-9

Wyman SK, Parkin RK, Mitchell PS, Fritz BR, O'Briant K, Godwin AK, Urban N, Drescher CW, Knudsen BS, Tewari M (2009) Repertoire of microRNAs in epithelial ovarian cancer as determined by next generation sequencing of small RNA cDNA libraries. PLoS One 4(4): e5311. doi:10.1371/journal.pone.0005311

Xie L, Qian X, Liu B (2010) MicroRNAs: novel biomarkers for gastrointestinal carcinomas. Mol Cell Biochem 341(1–2):291–299. doi:10.1007/s11010-010-0463-0

Yachida S, Jones S, Bozic I, Antal T, Leary R, Fu B, Kamiyama M, Hruban RH, Eshleman JR, Nowak MA, Velculescu VE, Kinzler KW, Vogelstein B, Iacobuzio-Donahue CA (2010)

Distant metastasis occurs late during the genetic evolution of pancreatic cancer. Nature 467(7319):1114–1117. doi:nature09515 [pii] 10.1038/nature09515

Yao Y, Suo AL, Li ZF, Liu LY, Tian T, Ni L, Zhang WG, Nan KJ, Song TS, Huang C (2009) MicroRNA profiling of human gastric cancer. Mol Med Rep 2(6):963–970. doi:10.3892/mmr_00000199

Yasui W, Oue N, Aung PP, Matsumura S, Shutoh M, Nakayama H (2005) Molecular-pathological prognostic factors of gastric cancer: a review. Gastric Cancer 8(2):86–94. doi:10.1007/s10120-005-0320-0

Zhang Z, Li Z, Gao C, Chen P, Chen J, Liu W, Xiao S, Lu H (2008) miR-21 plays a pivotal role in gastric cancer pathogenesis and progression. Lab Invest 88(12):1358–1366. doi:labinvest200894 [pii]10.1038/labinvest.2008.94

Zhang S, Chen L, Jung EJ, Calin GA (2010) Targeting microRNAs with small molecules: from dream to reality. Clin Pharmacol Ther 87(6):754–758. doi:clpt201046 [pii]10.1038/clpt.2010.46

Zhou H, Guo JM, Lou YR, Zhang XJ, Zhong FD, Jiang Z, Cheng J, Xiao BX (2010) Detection of circulating tumor cells in peripheral blood from patients with gastric cancer using microRNA as a marker. J Mol Med 88(7):709–717. doi:10.1007/s00109-010-0617-2

Zhu JY, Pfuhl T, Motsch N, Barth S, Nicholls J, Grasser F, Meister G (2009) Identification of novel Epstein-Barr virus microRNA genes from nasopharyngeal carcinomas. J Virol 83 (7):3333–3341. doi:JVI.01689-08 [pii]10.1128/JVI.01689-08

Zimmerman AL, Wu S (2011) MicroRNAs, cancer and cancer stem cells. Cancer Lett 300 (1):10–19. doi:S0304-3835(10), 00464-7 [pii]10.1016/j.canlet.2010.09.019

Index

B. Mallick (eds.), *Regulatory RNAs*, DOI 10.1007/978-3-642-22517-8,
© Springer-Verlag Berlin Heidelberg (outside the USA) 2012